13th Conference on Recent Developments in Gravity 2008

(NEB XIII)

Journal of Physics: Conference Series Volume 189

Thessaloniki, Greece
4 – 6 June 2008

ISBN: 978-1-61738-287-1

Printed from e-media with permission by:

Curran Associates, Inc.
57 Morehouse Lane
Red Hook, NY 12571

Some format issues inherent in the e-media version may also appear in this print version.

Printed by Curran Associates, Inc. (2010)

For permission requests, please contact the Institute of Physics
at the address below.

Institute of Physics
Dirac House, Temple Back
Bristol BS1 6BE UK

Phone: 44 1 17 929 7481
Fax: 44 1 17 920 0979

techtracking@iop.org

Additional copies of this publication are available from:

Curran Associates, Inc.
57 Morehouse Lane
Red Hook, NY 12571 USA
Phone: 845-758-0400
Fax: 845-758-2634
Email: curran@proceedings.com
Web: www.proceedings.com

13th Conference on Recent Developments in Gravity 2008

(NEB XIII)

Journal of Physics: Conference Series Volume 189

Thessaloniki, Greece
4 - 6 June 2008

ISBN: 978-1-61738-287-1

IOP Institute of Physics

TABLE OF CONTENTS

Sharp Bounds on the compactness of Relativistic Charged Spheres 1
H. Andréasson
An Observational Criterion to Look for an Inspiral in a Non-Kerr Spacetime 4
T. A. Apostolatos, G. Lukes-Gerakopoulos, J. Deligiannis, G. Contopoulos
Singularity Resolution in Loop Quantum Cosmology: A Brief Overview 9
A. Ashtekar
Cosmic Acceleration without Dark Energy 20
S. Basilakos, M. Plionis
Deformed Spaces and Loop Cosmology 29
M. V. Battisti
Modelling Non-Gaussianity from Foreground Contaminants 34
C. S. Carvalho
Properties of Codimension-2 Braneworlds in Six-Dimensional Lovelock Theory 47
C. Charmousis, A. Papazoglou
Canonical Quantization of Some Midi-Superspace Models in 2+1 and 3+1 Dimensions 53
T. Christodoulakis, G. Doulis, P. A. Terzis, E. Melas, T. Grammenos, G. O. Papadopoulos, A. Spanou
BTZ Solutions on Codimension-2 Braneworlds 79
B. Cuadros-Melgar, E. Papantonopoulos, M. Tsoukalas, V. Zamarias
Self-Dual Maxwell Field in 3D Gravity with Torsion and Dynamical Role of Central Charges 85
M. Blagojevic, B. Cvetkovic
Dark Energy and Inhomogeneity 91
G. F. R. Ellis
Geometric Properties of Magnetized Black Hole Event Horizons and Ergosurfaces 106
E. P. Esteban
Evolution of Supermassive Black Hole Spins in the ΛCDM Cosmology 122
N. Fanidakis, C. M. Baugh, S. Cole, C. S. Frenk
On the Tolman and Møller Mass-Energy Formulae in General Relativity 135
P. S. Florides
Hidden Symmetries and Black Holes 140
V. P. Frolov
Oscillations and Instabilities of Fast Rotating Neutron Stars 153
E. Gaertig, Kostas D. Kokkotas
Scalar-Field Cosmological and Collapse Models with General Self-Interaction Potentials 163
R. Giambò, F. Giannoni, G. Magli
Prompt Gamma-Ray Burst Emission From Gradual Energy Dissipation 175
D. Giannios
Recent Results on Short-Range Gravity Experiment 189
M. Hata, T. Akiyama, Y. Ikeda, H. Kawamura, K. Narita, K. Ninomiya, N. Ogawa, T. Sato, E. Seitaibashi, Y. Sekiguchi, R. Tsutsui, K. Yazawa, J. Murata
Brane-World Black Holes 199
P. Kanti
Gravitational Waves Versus Cosmic Strings 216
K. Kleidis
Cosmological Magnetic Fields 234
K. E. Kunze
Gravitational Waves from Gauge-Invariant Perturbations of Spherically Symmetric Spacetimes 240
P. Lasky
Neutron Star Equation of State via Gravitational Wave Observations 247
C. Markakis, J. S. Read, M. Shibata, K. Uryu, J. D. E. Creighton, J. L. Friedman, B. D. Lackey
Homogeneous and Isotropic Cosmologies in Higher Order Gravity 264
J. Miritzis
New Experimental Technique for Short-Range Gravity Measurement 274
K. Ninomiya, T. Akiyama, M. Hata, Y. Ikeda, H. Kawamura, K. Narita, N. Ogawa, T. Sato, E. Seitaibashi, Y. Sekiguchi, R. Tsutsui, K. Yazawa, J. Murata
A New Anisotropic Solution for Ellipsoidal Spaces 281
T. Papakostas

Matching of Analytical and Numerical Solutions For Neutron Stars of Arbitrary Rotation 286
G. Pappas
Asymmetrically Warped Brane Models, Bulk Photons and Lorentz Invariance 296
K. Farakos, N. E. Mavromatos, P. Pasipoularides
Time Evolution of Rapidly Rotating Stratified Neutron Stars 304
A. Passamonti
Accelerating Universe: Recent Observations and Implications for Extended Gravity Theories 310
L. Perivolaropoulos
Alternative High-z Cosmic Tracers and the Dark Energy Equation of State 341
M. Plionis, R. Terlevich, S. Basilakos, F. Bresolin, E. Terlevich, J. Melnick, I. Georgantopoulos
Two Approaches to Testing General Relativity in the Strong-Field Regime 360
D. Psaltis
Black Holes in Loop Quantum Gravity 371
R. Gambini, J. Pullin
Lattice Refinement in Loop Quantum Cosmology 376
M. Sakellariadou
Black Holes and the Quark-Gluon Plasma 393
G. Siopsis
Birefringence in Pseudo–Finsler Spacetimes 409
J. Skákala, M. Visser
Alfvén QPOs in Magnetars 417
H. Sotani, K. D. Kokkotas, N. Stergioulas
6+1 Lessons from f(R) Gravity 423
T. P. Sotiriou
Negative Mass and Repulsive Gravity in Newtonian Theory, and Consequences 434
N. K. Spyrou
The Generic Model of General Relativity 439
M. Tsamparlis
Lie Symmetries of the Geodesic Equations and Projective Collineations 473
M. Tsamparlis, A. Paliathanasis
Codimension-2 Black Hole Solutions on a Thin 3-Brane and Their Extension into the Bulk 484
M. Tsoukalas
Non-Standard Symmetries and Killing Tensors 493
M. Visinescu
Emergent Continuum Spacetime from a Random, Discrete, Partial Order 507
D. Rideout, P. Wallden
Signature-Change Events in Emergent Spacetimes with Anisotropic Scaling 515
S. Weinfurtner, A. White, M. Visser
Conical Singularities Regularized in Warped Six-Dimensional Flux Compactification and Their Induced Brane Cosmology 524
V. Zamarias

Author Index

Gravitational physics has been at the forefront of research in both Astrophysics and Cosmology. Since 1984, relativists in Greece have organized (every two years) a conference that follows progress in this field. The *13th Conference on Recent Developments in Gravity* was organized by the relativity group at the Department of Physics of the University of Thessaloniki, in June 2008. This volume contains the written versions as well as extensions of most of the talks delivered at that meeting.

The conference was attended by more than a hundred participants, half of which came from outside Greece, while it brought together nearly all Greek researchers in the areas of relativity, gravity and cosmology. The large number of participants from abroad underlines the increasingly international character of these meetings. In Thessaloniki, the plenary talks and parallel sessions covered topics on gravitational waves, relativistic astrophysics, alternative theories of gravity, mathematical relativity, cosmology and quantum gravity. The plenary speakers were A Ashtekar (Penn-State), G Ellis (Cape Town), V Frolov (Alberta), J Hough (Glasgow), P Laguna (Penn-State), R Loll (Utrecht) and D Psaltis (Arizona). In addition, invited keynote talks were presented by N Batakis (Ioannina), T Christodoulakis (Athens), S Cotsakis (Aegean), P Kanti (Ioannina), L Perivolaropoulos (Ioannina), M Plionis (NOA-Athens), J Pullin (Louisiana), G Siopsis (Tennesee) and M Tsamparlis (Athens).

The Organizing Committee of the conference comprised N Stergioulas (chair), K D Kokkotas, D Papadopoulos, C Tsagas, H Dimmelmeier, K Kleidis, K Zagouris and K Dionysopoulou while the Scientific Committee included N Spyrou (chair – Thessaloniki), P Florides (Dublin) and G Schaefer (Jena). The conference was sponsored by the Aristotle University of Thessaloniki (AUTh), the Department of Physics, the Section of Astrophysics, Astronomy and Mechanics, the Research Committee of AUTh, the Ministry of Macedonia and Thrace, the Ministry of Education and Religious Affairs, the John S Latsis Public Benefit Foundation (http://www.latsis-foundation.org), the ILIAS Network, the Municipality of Thessaloniki, the Prefecture of Thessaloniki and the British Council Greece, to all of which the organizers would like to express their gratitude for the financial and moral support.

The next (14th) conference in this series will take place in the city of Ioannina, organized by the relativity group of the University of Ioannina, in the summer of 2010.

Thessaloniki, 29 August 2009

N Stergioulas and C Tsagas

All papers published in this volume of *Journal of Physics: Conference Series* have been peer reviewed through processes administered by the proceedings Editors. Reviews were conducted by expert referees to the professional and scientific standards expected of a proceedings journal published by IOP Publishing.

Journal of Physics: Conference Series **189** (2009) 012001 doi:10.1088/1742-6596/189/1/012001

Sharp bounds on the compactness of relativistic charged spheres

Håkan Andréasson

Mathematical Sciences, University of Gothenburg
Mathematical Sciences, Chalmers University of Technology, S-41296 Göteborg, Sweden

Abstract. The problem of finding an upper bound on the mass M of a charged spherically symmetric static object with area radius R and charge $Q < M$ is addressed. This problem has resulted in a number of papers in recent years but neither a transparent nor a general inequality similar to the case without charge, i.e., $M \leq 4R/9$, has been found. Here, we discuss the surprisingly transparent inequality

$$\sqrt{M} \leq \frac{\sqrt{R}}{3} + \sqrt{\frac{R}{9} + \frac{Q^2}{3R}}.$$

The inequality holds for any solution which satisfies $p + 2p_T \leq \rho$, where $p \geq 0$ and p_T are the radial- and tangential pressures respectively and $\rho \geq 0$ is the energy density. Moreover, the inequality is sharp and in particular, sharpness holds for infinitely thin shell solutions.

1. Introduction

Black holes for which the charge or angular momentum parameter equals the mass are called extremal black holes. They are very central in black hole thermodynamics due to their vanishing surface gravity and they represent the absolute zero state of black hole physics. It is quite generally believed that extremal black holes are disallowed by nature but a proof is missing. One possibility to obtain an extremal black hole is to produce one from the collapse of an already extremal object. Previous mainly numerical studies [5, 8] have concluded that when $Q < M$ collapse always takes place at a critical stability radius R_c outside the outer horizon, and as Q approaches M, this value approaches the horizon. This is similar to the non-charged case where the Buchdahl inequality implies that collapse will take place when $R < 9M/4$, i.e., $R_c = 9M/4$, cf. [7]. In the charged case the critical value is expected to be smaller due to the Coulomb repulsion, and in particular as $Q \to M$ the stability radius does approach the outer horizon. For more information on the relation of this topic to black hole thermodynamics and extremal black holes we refer to [5], [9] and [12] and the references therein.

The problem of finding a similar bound as the classical Buchdahl bound for charged objects have resulted in several papers; some of these are analytical, cf. [12], [15], [9], [10], [13] and [16], whereas others are numerical or use a mix of numerical and analytical arguments, cf. [5], [8], and [11] to mention some of them. We refer the reader to the sources for the details of these studies but in none of them a transparent bound has been obtained (except in very special cases), on the contrary they have been quite involved and implicit. Moreover, most of these studies rely on the assumptions made by Buchdal, i.e., the energy density is assumed to be non-increasing and the pressure to be isotropic. In [1] the inequality

$$\sqrt{M} \leq \frac{\sqrt{R}}{3} + \sqrt{\frac{R}{9} + \frac{Q^2}{3R}}, \tag{1}$$

is shown to hold whenever $Q < R$ and $p + 2p_T \leq \rho$, where $p \geq 0$ and p_T are the radial- and tangential pressures respectively and $\rho \geq 0$ is the energy density. Moreover, the inequality is shown to be sharp.

In the non-charged case a general proof of the Buchdahl inequality $2m/r \leq 8/9$, in the case when $p + 2p_T \leq \rho$, was first given in [2]. A completely different proof was then given by Stalker and Karageorgis [14] where also several other situations were considered, e.g. the isotropic case where $p = p_T$. The advantage of the method in [14] (which is related to the method by Bondi [6] which however is non-rigorous) compared to the method in [2] is that it is shorter and that it is more flexible in the sense that other assumptions than $p + 2p_T \leq \rho$ can be treated. On the other hand the result in [14] is weaker than the result in [2] in the sense that the latter method implies that the steady state that saturates the inequality is unique, it is an infinitely thin shell.

The method in [14] also shows sharpness but only in the sense that there are steady states with $2m/r$ arbitrary close to 8/9, leaving open the possibility that different kinds of steady states might share this feature. Moreover, since the assumption $p + 2p_T \leq \rho$ is satisfied by solutions of the Einstein-Vlasov system it is natural to ask if there exist regular static solutions to the *coupled* system which can have $2m/r$ arbitrary close to 8/9. This question is given an affirmative answer in the non-charged case, cf. [3], where in particular it is shown that arbitrary thin shells which are regular solutions of the spherically symmetric Einstein-Vlasov system do exist. On the contrary, the matter quantities and the corresponding spacetimes constructed in [14] for showing sharpness cannot be realized by regular solutions of the Einstein-Vlasov system. The construction in [14] gives that a solution which nearly saturates the inequality $2m/r \leq 8/9$ satisfies $p+2p_T = \rho$, and in addition p_T and ρ are discontinuous. Neither of these two properties can be realized by regular solutions of the (massive) Einstein-Vlasov system. In the charged case there are no rigorous studies of solutions of the Einstein-Maxwell-Vlasov system but in [4] numerical solutions are constructed to this system which have the property that they satisfy the inequality (1) and that the difference of the left- and right hand sides in the inequality can be as small as one wishes.

The proof of the inequality (1) given in [1] is an adaption of the method in [14] to the charged case and we refer to [1] for the mathematical formulation and for the details of the proof. It should be stressed that the inequality (1) also holds anywhere inside the static object, cf. [1]. In [1] a proof is also given that infinitely thin shell solutions saturate the inequality.

References

[1] Andréasson H 2008 Sharp bounds on the critical stability radius for relativistic charged spheres *Preprint* arXiv:0804.1882
[2] Andréasson H 2008 *J. Diff. Equations* **245** 2243
[3] Andréasson H 2007 *Commun. Math. Phys.* **274** 409
[4] Andréasson H and Eklund M 2009 A numerica investigation of the steady states of the spherically symmetric Einstein-Vlasov-Maxwell (in preparation)
[5] Anninos P and Rothman T 2001 *Phys. Rev. D,* **62** 024003
[6] Bondi H 1964 *Proc. R. Soc. A* **282** 303
[7] Buchdahl H A 1959 *Phys. Rev.* **116**, 1027
[8] de Felice F, Siming L and Yunqiang Y 1999 *Class. Quantum Grav.* **16**, 2669
[9] Farrugia C J and Hajicek P 1979 *Commun. Math. Phys.* **68**, 291
[10] Fayos F, Senovilla J M M, and Torres R 2003 *Class. Quantum Grav.* **20** 2579
[11] Ghezzi C R 2005 *Phys. Rev. D* **72**, 104017
[12] Giuliani A and Rothman T 2007 *Gen. Rel. Gravitation* DOI 10.1007/s10714-007-0539-7
[13] Harko T and Mak M K 2000 *J. Math. Phys.* **41** 4752

Journal of Physics: Conference Series **189** (2009) 012001 doi:10.1088/1742-6596/189/1/012001

[14] Karageorgis P and Stalker J 2008 *Class. Quantum Grav.* **25** 195021
[15] Mak M K, Dobson P N and Harko T 2001 *Europhys. Lett.* **55**, 310
[16] Yunqiang Y and Siming L 2000 *Commun. Theor. Phys.* **33**, 571

An observational criterion to look for an inspiral in a non-Kerr spacetime

Theocharis A. Apostolatos[1], Georgios Lukes-Gerakopoulos[1,2], John Deligiannis[1], and George Contopoulos[2]

[1]Section of Astrophysics, Astronomy, and Mechanics,
Department of Physics, University of Athens,
Panepistimiopolis Zografos GR-15783, Athens, Greece
[2] Academy of Athens, Research Center for Astronomy,
Soranou Efesiou 4, GR-11527, Athens, GREECE

Abstract. In this short article we present a useful observational tool for gravitational wave detectors. More specifically, if we are looking for extreme-mass-ratio inspiraling objects in a non-Kerr spacetime, we could exploit the consequences of the KAM and the Poincaré-Birkhoff theorem which predicts plateaus in the ratio of frequencies f_ρ/f_z, that are related to a generic geodesic orbit in such a spacetime, as a function of the initial conditions of the orbit itself. While both these frequencies are changing under radiation reaction, their ratio is expected to stay stationary if it passes through such a plateau. Therefore, if detectors are able to discern the fundamental frequencies due to ρ and z oscillations of the orbit, they could in principle detect the non-Kerr-ness of the spacetime involved, just by monitoring the ratio of these two frequencies.

1. Introduction

Gair et al. [1] explored the geodesic bound orbits in a specific member of the family of spacetimes, constructed by Manko and Novikov [2] in 1990, and have shown that they lie either in a region which is not very near the center of spacetime and the corresponding trajectories in phase space look very much like the ones in a Kerr spacetime (with respect to their frequency spectra), or in a region near the center, where the orbits there exhibit chaotic behavior.

We have thoroughly analyzed the same kind of orbits in the same background and have shown that although the general picture is the one described by Gair et al, there are some details about the orbital characteristics that could be suitably exploited in a data analysis scheme to distinguish this spacetime from a Kerr spacetime. More specifically, we know from KAM (Kolmogorov-Arnold-Moser) theorem about dynamical systems, that a perturbed integrable dynamical system has most of its phase-space tori deformed but not destroyed. As a consequence the corresponding orbits are quasi-periodic and on a surface of section they define simple closed KAM curves. On the other hand, the resonant tori, that is the ones that are characterized by commensurate ratio of frequencies, disintegrate and the corresponding sections on a surface of section form a chain of islands as predicted by the Poincaré-Birkhoff theorem.

The geodesic orbits in a stationary axisymmetric metric are characterized by three fundamental frequencies f_ϕ, f_ρ, and f_z; the azimuthal frequency of the orbit and the frequencies at which the orbit oscillates with respect to ρ and z coordinates. These frequencies will, in principle, show up in the corresponding gravitational wave signal emitted when a low mass body orbits around a massive central object with a gravitational field described by such a metric. The Manko-Novikov spacetime, which can be considered as a perturbed Kerr axisymmetric spacetime (actually that was the idea behind the construction of such a spacetime) since by suitable tuning of its parameters it can be turned into a Kerr metric, has most of its 4-dimensional tori deformed with respect to the corresponding Kerr ones leading to simple closed invariant curves (KAM curves), while the resonant ones form Birkhoff chains of islands. The sequence of KAM curves correspond to *continuously different* irrational ratios of frequencies f_ρ and f_z, while the whole region of initial conditions inside a specific Birkhoff chain of islands corresponds to a specific commensurate ratio of frequencies.

In a real inspiraling binary system with extreme ratio of masses, that happens to be described by a Kerr-like, but not exactly Kerr, spacetime background (the Manko-Novikov spacetime is merely an example of such a generic case), the orbits will evolve adiabatically under radiation reaction following non-geodesic trajectories and therefore both fundamental frequencies of ρ and z (namely f_ρ and f_z) will evolve continuously. As an orbit evolves, the corresponding trajectory in phase space may pass through a Birkhoff island. Then the ratio between f_ρ and f_z remains constant, while the frequencies continue to change adiabatically. This will be a clear "smoking gun" for a non-Kerr metric that could be then correlated with other orbital characteristics related to a non-Kerr spacetime, like the ones suggested in [1], to make sure that the spacetime we are monitoring through gravitational waves can not be described by a Kerr black-hole metric. Furthermore the quantitative characteristics from the suggested non-Kerr criterion will offer complementary information for more accurate determination of the specific metric itself.

In the next section we present the case that we have analyzed in order to exhibit our idea, and we give further quantitative details of the observational non-Kerr criterion. In a forthcoming paper [3] we plan to go to a deeper investigation, by exploring the quantitative details of the plateaus of the frequency ratios in a realistic orbital evolution.

2. The Manko-Novikov metric: an example of a Non-Kerr case

We have chosen to scrutinize the same non-Kerr 'bumpy' spacetime that Gair et al [1] had used to explore the geodesic orbital characteristics when they tried to find observational signatures that distinguish such a spacetime from a Kerr spacetime. Namely, we have used the Manko-Novikov spacetime [2] which has been built as a non-linear superposition of Weyl solutions, with the Kerr solution. More specifically, we have used a member of the family of Manko-Novikov solutions, which is described by one more parameter than the corresponding Kerr metric, the q parameter which describes the deviation of the quadrupole moment of the new solution from the Kerr metric. Thus, this metric turns into an exact Kerr metric when the q parameter becomes zero. It should be noted though, that apart from the quadrupole moment, the specific Manko-Novikov solution has all higher order moments different from the corresponding Kerr moments as well, when $q \neq 0$. A detailed description of the metric can be found in [1].

We have integrated a wide range of geodesic orbits in such a spacetime numerically, by following a Runge-Kutta 7th-8th order numerical scheme. In our thorough analysis we have observed that within the outer allowed region of motion (see [1]) besides the KAM curves on the surface of section, there are chain-like regions of narrow islands between specific KAM curves. These islands are characterized by a constant ratio of fundamental frequencies that are related to the oscillations along the ρ and z axes, while this ratio is continuously and monotonically changing as we move from one KAM curve to the next one (see Fig. 2) away from the main island's center (see Fig. 1), which is the central point of the KAM curves on the surface of sections and corresponds to the degenerate curve where ρ =const. The existence of such chains of islands is expected from the Poincaré-Birkhoff theorem in perturbed integrable dynamical systems. They rise from the commensurate tori which are destroyed when one departs from an integral system (in our case the Kerr metric).

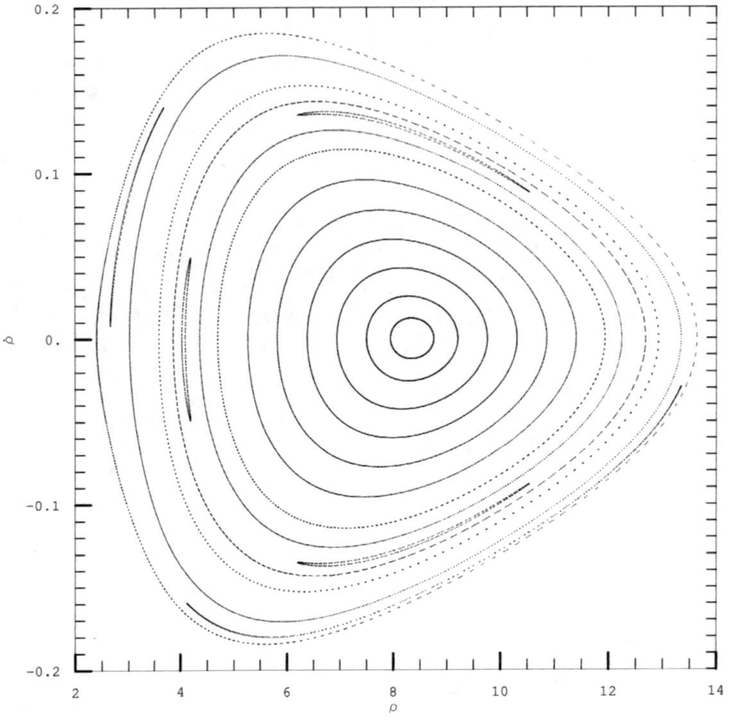

Figure 1. A sequence of invariant curves on the surface of section determined by intersections of the plane $z = 0$ by phase-space trajectories of geodesic orbits in a Manko-Novikov metric with $q = 0.95$, $a_n = 0$ for all $n > 2$, and $\chi = 0.9$; these are actually the parameters used by [1] in their plots. All curves correspond to orbits with the same physical parameters $E = 0.95$ and $L_z = 3M$, but they differ with respect to their initial conditions. Most of the curves formed are closed KAM curves, similar to the corresponding ones in a Kerr metric. On the other hand there is a number of islands formed between closed KAM curves, that correspond to resonant tori; for example, the inner triplet of islands corresponds to a resonance 1:3, while the outer doublet of islands corresponds to a resonance 1:2. The ρ coordinate is measured in units of M, while the $\dot{\rho}$ coordinate is dimensionless.

If a geodesic orbit happens to lie in such a chain of islands, it will incorporate in its spectrum two fundamental frequencies which form a ratio of integers. Moreover this will be true for a finite range of initial conditions, even though the fundamental frequencies themselves vary continuously from one set of initial conditions to another. The locking of the ratio of frequencies does not hold for geodesics that belong on KAM curves. Thus, by monitoring the gravitational waves emitted by a binary with an extreme ratio of masses we could in principle discern the fundamental frequencies f_ρ, f_z, in the spectrum of its gravitational wave signal and by following their adiabatic evolution through gravitational radiation, look for plateaus in their ratio. These plateaus would then be a true "smoking gun" for observing a non-Kerr metric.

Practically, in order to locate a chain of islands on a surface of section we have used the so-called rotation number (for a review see [4]) which is defined by the following relation

$$\nu = \lim_{N\to\infty} \sum_{i=1}^{N} \frac{\Delta\theta_i}{2\pi}, \tag{1}$$

where $\Delta\theta_i$ is the angle formed by the i-th and $(i+1)$-th successive cross-sections of the trajectory in phase space on the surface of section, with respect to the center of the main island. The rotation number is nothing but the ratio of fundamental frequencies $(f_\rho/f_z)(\text{mod}1)$. Thus, by acquiring a large number of sections we may be able to measure sufficiently accurately the ratio of frequencies and determine if we are close to commensurate ratio by approaching a fraction of small integers like the 1:3. This gives us a very useful numerical tool to search for a chain of islands. The rotation number remains stationary as long as our initial conditions cross an island and it varies when we cross its border and enter into the region of KAM curves.

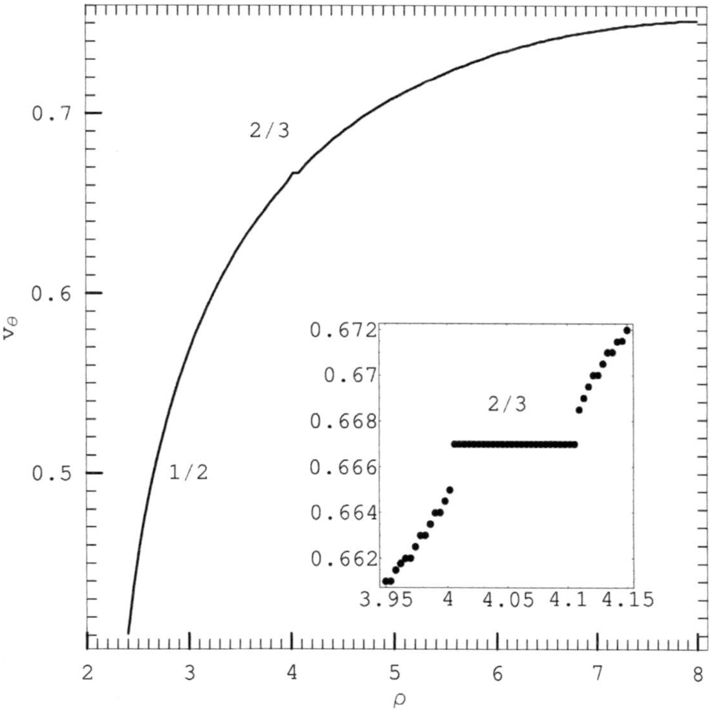

Figure 2. This is the rotation number $\nu_\theta = 1 - \nu$, where ν was defined in (1), as a function of $\rho(0)$ (the rest of initial conditions are $\dot{\rho}(0) = 0$, $z(0) = 0$). The plateau showing up at $\nu_\theta = 2/3$ corresponds to the chain of islands with ratio of frequencies $(f_\rho/f_z) = 1/3$ (see Figure 1). A smaller one at ratio $(f_\rho/f_z) = 1/2$ is not visible. The inlet shows a magnification of the plateau.

In order to exploit quantitatively the time interval that the ratio of frequencies remains constant as the orbit spirals inwards due to radiation reaction to gain information about the spacetime on which the orbit evolves, we need to evolve the orbit adiabatically and measure the effective time interval during which the orbit stays within an island of a disintegrated commensurate torus. This will be analyzed in a forthcoming paper [3].

3. General remarks and future analysis

There is a number of issues related with the potentiality of applying our proposed criterion in real physical extreme-mass-ratio binary systems. Firstly, not all orbits are expected to evolve in a way that passing through an island of constant ratio of frequencies will take place. Then, even if it passes through such a chain of islands it may happen that it misses the plateau of constant frequencies due to fast evolution of the orbit. Finally, there is no way to evolve accurately the orbit by means of radiation reaction, and the only possibility of analyzing the situation is to appeal on approximate calculations. Such an approach is expected to give us only a crude estimate of the duration of an effective plateau. All these issues will be further explored in a forthcoming paper [3].

Although our analysis is based on a specific analytic axisymmetric solution of the vacuum Einstein equations, the existence of a plateau in the evolution of the ratio of frequencies should be a generic property of all spacetimes that deviate slightly from a Kerr metric. Therefore the observational criterion of a non-Kerr metric that we suggest could be widely used.

In order to be helpful observationally, the analysis of the subject discussed here should explore as well the accuracy of discerning the existence of a plateau in the ratio of frequencies, while all frequencies are drifting slowly. This is a technical issue which we plan to investigate in the future. In [3], we present a crude quantitative analysis of the time interval of the effective plateau as a function of the masses of the binary, for a few interesting and indicative cases of orbits.

Acknowledgments

Acknowledgments T. A. Apostolatos would like to thank Kip Thorne for hospitality in Caltech during the spring of 2006 where he had the first contact with the strange Kerr-like behavior of orbits in a non-Kerr metric, as well as Chao Li and Ilya Mandel for useful discussions about the subject. This work was supported by the research funding program "Kapodistrias" with Grant No 70/4/7672. G. Lukes-Gerakopoulos was supported by the Greek Foundation of State Scholarships (IKY).

References

[1] Gair J R, Li C, Mandel I 2008 *Phys. Rev. D* **77** 024035
[2] Manko V S, Novikov I D 1992 *Class. Quant. Grav.* **9** 2477
[3] Apostolatos T A, Loukes-Gerakopoulos G, Deligiannis J, Contopoulos G *in preparation*
[4] Contopoulos G 2002 'Order and chaos in dynamical Astronomy' *Springer*

Recent Developments in Gravity (NEB XIII) IOP Publishing
Journal of Physics: Conference Series **189** (2009) 012003 doi:10.1088/1742-6596/189/1/012003

Singularity resolution in loop quantum cosmology: a brief overview

Abhay Ashtekar
Institute for Gravitation and the Cosmos, Penn State, University Park, PA 16802, U.S.A.

Abstract. A brief summary of loop quantum cosmology of homogeneous isotropic models is presented for non-experts with emphasis on the origin of and subtleties associated with the resolution of big bang and big crunch singularities. For further details, and a discussion of some frequently asked questions, the reader is directed to more comprehensive recent reviews.[1]

1. Issue of the Beginning and the End

Over the history of mankind, cosmological paradigms have evolved in interesting ways. It is illuminating to begin with a long range historical perspective by recalling paradigms that seemed obvious and most natural for centuries only to be superseded by radical shifts.

Treatise on Time, the Beginning and the End date back at least twenty five centuries. Does the flow of time have an objective, universal meaning beyond human perception? Or, is it fundamentally only a convenient, and perhaps merely psychological, notion? Did the physical universe have a finite beginning or has it been evolving eternally? Leading thinkers across cultures meditated on these issues and arrived at definite but strikingly different answers. For example, in the sixth century BCE, Gautama Buddha taught that 'a period of time' is a purely conventional notion, time and space exist only in relation to our experience, and the universe is eternal. In the Christian thought, by contrast, the universe had a finite beginning and there was debate whether time represents 'movement' of bodies or if it flows only in the soul. In the fourth century CE, St. Augustine held that time itself started with the world.

Founding fathers of modern Science from Galileo to Newton continued to accept that God created the universe. Nonetheless, their work led to a radical change of paradigm. Before Newton, boundaries between the absolute and the relative, the true and the apparent and the mathematical and the common were blurry. Newton rescued time from the psychological *and* the material world and made it objective and absolute. It now ran uniformly from the infinite past to the infinite future. This paradigm became a dogma over centuries. Philosophers often used it to argue that the universe itself *had* to be eternal. For, as Immanuel Kant emphasized, otherwise one could ask "what was there before?"

General relativity toppled this Newtonian paradigm in one fell swoop. Now the gravitational field is encoded in space-time geometry. Since geometry is a dynamical, physical entity, it is now perfectly feasible for the universe to have had a finite beginning —the big-bang— at which not only matter but *space-time itself* is born. If space is compact, matter *as well as space-time* end

[1] Dedicated to the memory of Basilis Xanthopoulos, who had an admirably pure heart, who was a passionate researcher and who founded of this series of conferences.

in the big-crunch singularity. In this respect, general relativity took us back to St. Augustine's paradigm but in a detailed, specific and mathematically precise form. In semi-popular articles and radio shows, relativists now like to emphasize that the question "what was there before?" is rendered meaningless because the notions of 'before' requires a pre-existing space-time geometry. We now have a new paradigm, a new dogma: In the Beginning there was the Big Bang.

However, the very fusion of gravity with geometry now gives rise to a new tension. In Newtonian (or Minkowskian) physics, a given physical field could become singular at a space-time point. This generally implied that the field could not be unambiguously evolved to the future of that point. However, this singularity had no effect on the global arena. Since the space-time geometry is unaffected by matter, it remains intact. Other fields could be evolved indefinitely. Trouble was limited to the one field which became ill behaved. However, because gravity is geometry in general relativity, when the gravitational field becomes singular, the continuum tares and the space-time itself ends. There is no more an arena for other fields to live in. All of physics, as we know it, comes to an abrupt halt. Physical observables associated with both matter and geometry simply diverge signalling a fundamental flaw in our description of Nature. *This is the new quandary.*

When faced with deep quandaries, one has to carefully analyze the reasoning that led to the impasse. Typically the reasoning is flawed, possibly for subtle reasons. In the present case the culprit is the premise that general relativity —with its representation of space-time as a smooth continuum— provides an accurate description of Nature arbitrarily close to the singularity. For, general relativity completely ignores quantum physics and over the last century we have learned that quantum effects become important in the physics of the small. They should in fact be dominant in parts of the universe where matter densities become enormous. Thus *the occurrence of the big bang and other singularities are predictions of general relativity, precisely in a regime where it is inapplicable!* Classical physics of general relativity does come to a halt at the big-bang and the big crunch. But this is not an indication of what really happens because use of general relativity near singularities is an extrapolation which has no physical justification whatsoever. We need a theory that incorporates not only the dynamical nature of geometry but also the ramifications of quantum physics. We need a quantum theory of gravity, a new paradigm.

These considerations suggest that *singularities of general relativity are perhaps the most promising gates to physics beyond Einstein.* They provide a fertile conceptual and technical ground in our search of the new paradigm. Consider some of the deepest conceptual questions we face today: the issue of the Beginning and the end End, the arrow of time, and the puzzle of black hole information loss. Their resolutions hinge on the true nature of singularities. In my view, considerable amount of contemporary confusion about such questions arises from our explicit or implicit insistence that singularities of general relativity are true boundaries of space-time; that we can trust causal structure all the way to these singularities; that notions such as event horizons are absolute even though changes in the metric in a Planck scale neighborhood of the singularity can move event horizons dramatically or even make them disappear altogether [1].

Over the last 2-3 years several classically singular space-times have been investigated in detail through the lens of loop quantum gravity (LQG) [2, 3, 4]. This is a non-perturbative approach to the unification of general relativity and quantum physics in which one takes Einstein's encoding of gravity into geometry seriously and elevates it to the quantum level. One is thus led to build quantum gravity using *quantum* Riemannian geometry [5, 6, 7, 8]. Both geometry and matter are *dynamical* and described *quantum mechanically* from the start. In particular, then, there is no background space-time.

The kinematical structure of the theory has been firmly established for some years now. There are also several interesting and concrete proposals for dynamics (see, in particular

[2, 3, 4, 9]). However, in my view there is still considerable ambiguity and none of the proposals is fully satisfactory. Nonetheless, over the last 2-3 years, considerable progress could be made by restricting oneself to subcases where detailed and explicit analysis is possible [10, 11, 12, 13, 14, 15, 16]. These 'mini' and 'midi' superspaces are well adapted to analyze the deep conceptual tensions discussed above. For, they encompass the most interesting of classically singular space-times —Friedman-Lemaître-Robertson-Walker (FLRW) universes with the big bang singularity and black holes with the Schwarzschild-type singularity— and analyze them in detail using symmetry reduced versions of loop quantum gravity. In all cases studied so far, classical singularities are naturally resolved and *the quantum space-time is vastly larger than what general relativity had us believe.* As a result, there is a new paradigm to analyze the old questions.

In my talk, I focused on cosmological singularities. The material I covered is discussed in greater detail in [12]. Discussion of black hole singularities and the issue of information loss can be found either in the original papers [13, 17, 15] or in a comprehensive, recent review addressed to non-experts [16]. Here, I will confine myself to a sketch of the singularity resolution in loop quantum cosmology (LQC). Section 2, will provide a conceptual setting, section 3 will summarize the main results and section 4 will present the outlook.

2. Conceptual Setting

I will restrict myself to the simplest cosmological models: FLRW space-times with a massless scalar field. I will consider both the k=1 model and the k=0 model with cosmological constant and comment on the status of more general models. These models are instructive because in classical general relativity all their solutions have a big-bang (and/or big-crunch) singularity. Therefore, a quantum resolution of these singularities is non-trivial. It is not difficult to incorporate additional matter fields and anisotropies.

Figure 1 illustrates classical dynamics for k=0 and k=1 models without a cosmological constant.[2] In the k=0 case, there are two classes of trajectories. In one the universe begins with a big-bang and expands continuously while in the other it contracts continuously into a big crunch. In the k=1 case, the universe begins with a big bang, expands to a maximum volume and then undergoes a recollapse to a big crunch singularity. Now, in quantum gravity, one does not have a single space-time in the background but rather a probability amplitude for various space-time geometries. Therefore, unlike in the classical theory, one cannot readily use, e.g., the proper time along a family of preferred observers as a clock. However, along each dynamical trajectory, the massless scalar field ϕ is monotonic. Therefore, it serves as a good 'clock variable' with respect to which the physical degrees of freedom (the density or the volume, anisotropies and other matter fields, if any) evolve. Note incidentally that, in the k=1 case, volume is double valued and cannot therefore serve as a global clock variable, while the scalar field does fulfill this role. The presence of a clock variable is *not* essential in quantum theory but its availability makes the relational dynamics easier to grasp in familiar terms. It turns out that the massless scalar field is well-suited for this purpose not only in the classical but also in the quantum theory.

If one has a fully developed quantum theory, one can proceed as follows. Choose a classical dynamical trajectory. constant of motion, this fixes the value of $p_{(\phi)}$, say $p_{(\phi)} = p^\star_{(\phi)}$. Next, choose a point $(v^\star, \phi^\star)$ on any one of these dynamical trajectories where the matter density and space-time curvature is low. This point describes the state of the FLRW universe at a late time where general relativity is expected to be valid. At the 'internal time' $\phi = \phi^\star$ construct a wave

[2] In the k=0 case, because the universe is infinite and homogeneous, to obtain a well-defined Lagrangian or Hamiltonian formulation, one has to introduce a fiducial box and restrict all integrations to it. In the k=0 figure, the volume v refers to this box. Since Lagrangian and Hamiltonian formulations are stepping stones to the path integral and canonical quantization, every quantum theory requires a fiducial box. Physical results, of course, cannot depend on the choice of this box.

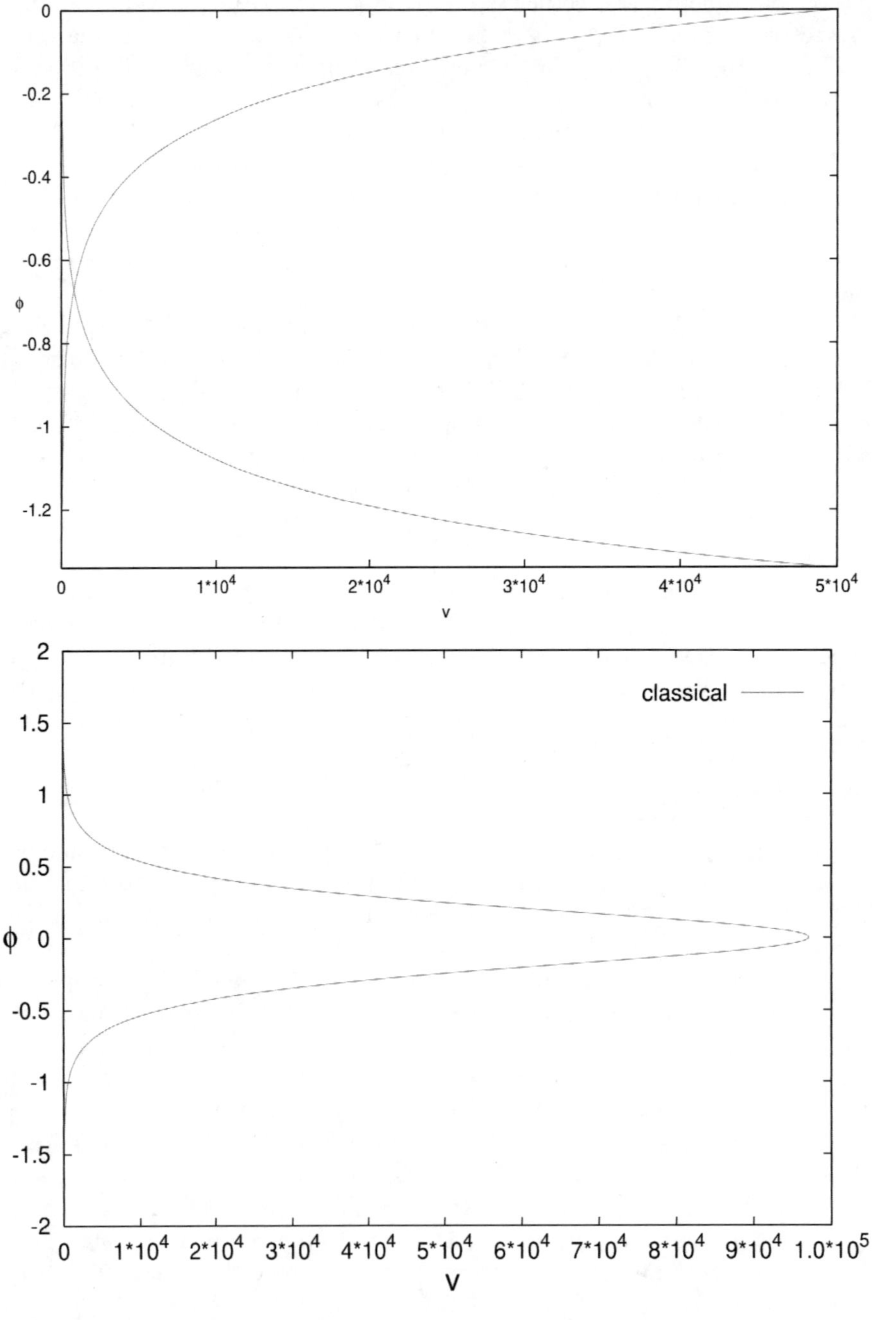

Figure 1. Dynamics of FLRW universes with zero cosmological constant and a massless scalar field. Classical trajectories are plotted in the $v - \phi$ plane, where v denotes the volume and ϕ the scalar field. *Upper panel:* k=0 trajectories. *Lower panel:* A k=1 trajectory.

packet which is sharply peaked at $v = v^\star$ and $p_{(\phi)} = p^\star_{(\phi)}$ and evolve it backward and forward in (the scalar field) time. We are then led to two questions:

i) *The infrared issue:* Does the wave packet remain peaked on the classical trajectory in the low curvature regime? Or, do quantum geometry effects accumulate over the cosmological time scales, causing noticeable deviations from classical general relativity? In particular, in the k=1 case, is there a recollapse and if so for large universes does the value $V_{\max}$ of maximum volume agree with that predicted by general relativity [18]?

ii) *The ultraviolet issue:* What is the behavior of the quantum state when the curvature grows and enters the Planck regime? Is the big-bang singularity resolved? Or, do we need to supplement dynamics with a new principle, such as the Hartle-Hawking 'no boundary proposal' [19]? What about the big-crunch? If they are both resolved, what is on the 'other side'? Does the wave packet simply disperse making space-time 'foamy' or is there a large classical universe also on the other side?

By their very construction, perturbative and effective descriptions have no problem with the first requirement. However, physically their implications can not be trusted at the Planck scale and mathematically they generally fail to provide a deterministic evolution across the putative singularity. Since the non-perturbative approaches often start from deeper ideas, it is conceivable that they could lead to new structures at the Planck scale which modify the classical dynamics and resolve the big-bang singularity. But once unleashed, do these new quantum effects naturally 'turn-off' sufficiently fast, away from the Planck regime? The universe has had some *14 billion years* to evolve since the putative big bang and even minutest quantum corrections could accumulate over this huge time period leading to observable departures from dynamics predicted by general relativity. Thus, the challenge to quantum gravity theories is to first create huge quantum effects that are capable of overwhelming the extreme gravitational attraction produced by matter densities of some 10^{94} gms/cc near the big bang, and then switching them off with extreme rapidity as the matter density falls below this Planck scale. This is a huge burden!

The question then is: How do various approaches fare with respect to these questions? The older quantum cosmology —the Wheeler-DeWitt (WDW) theory— passes the infra-red test with flying colors. But unfortunately the state follows the classical trajectory into the big bang (and in the k=1 case also the big crunch) singularity. The singularity is not resolved because *expectation values of density and curvature continue to diverge* in epochs when their classical counterparts do.

For a number of years, the failure of the WDW theory to naturally resolve the big bang singularity was taken to mean that quantum cosmology cannot, by itself, shed significant light on the quantum nature of the big bang. Indeed, for systems with a finite number of degrees of freedom we have the von Neumann uniqueness theorem which guarantees that quantum kinematics is unique. The only freedom we have is in factor ordering and this was deemed insufficient to alter the status-quo provided by the WDW theory.

The situation changed dramatically in LQG. In contrast to the WDW theory, a well established, rigorous kinematical framework *is* available in full LQG [5, 2, 3, 4]. Furthermore, this framework is uniquely singled out by the requirement of diffeomorphism invariance (or background independence) [20, 21]. If one mimics it in symmetry reduced models, one finds that a key assumption of the von-Neumann theorem is violated: *One is led to new quantum mechanics [22]!* This quantum theory is inequivalent to the WDW theory already at the kinematic level. Quantum dynamics built in this new arena agrees with the WDW theory in 'tame' situations but differs dramatically in the Planck regime, leading to a natural resolution of the big bang singularity.

3. LQC: Main Results

The main LQC results can be summarized as follows [23, 24, 25, 26, 28, 29].

Let us begin with the k=0 model without a cosmological constant. Following the strategy outlined in section 2, let us fix a point at a late time on the trajectory corresponding to an expanding classical universe, construct a quantum state which is sharply peaked at that point, and evolve it using the LQC Hamiltonian constraint. One then finds that:

- The wave packet remains sharply peaked on the classical trajectory till the matter density ρ reaches about 1% of the Planck density $\rho_{\rm Pl}$. Thus, as in the WDW theory, the LQC evolution meets the infra-red challenge successfully.
- Let us evolve the quantum state back in time, toward the singularity. In the classical solution scalar curvature and the matter energy density keep increasing and eventually diverge at the big bang. The situation is very different with quantum evolution. As the density and curvature enter the Planck scale *quantum geometry effects become dominant creating an effective repulsive force which rises very quickly, overwhelms the classical gravitational attraction, and causes a bounce thereby resolving the the big bang singularity.* See Fig 2.
- At the bounce point, the density acquires its maximum value $\rho_{\rm max}$. For the class of quantum states under discussion, numerical simulations have shown that $\rho_{\rm max}$ is *universal*, $\rho_{\rm max} \approx 0.41\rho_{\rm Pl}$ up to terms $O(\ell_{\rm Pl}^2/a_{\rm min}^2)$, independently of the details of the state and value of $p_{(\phi)}$ (provided $p_{(\phi)} \gg \hbar$ in the classical units c=G=1).
- Although in the Planck regime the peak of the wave function deviates very substantially from the general relativistic trajectory of figure 1, it follows an effective trajectory with very small fluctuations. This effective trajectory was derived [30, 31] using techniques from geometric quantum mechanics. The effective equation incorporates the leading corrections from quantum geometry. They modify the left hand side of Einstein's equations. However, to facilitate comparison with the standard form of Einstein's equations, one moves this correction to the right side through an algebraic manipulation. Then, one finds that the Friedmann equation $(\dot{a}/a)^2 = (8\pi G\,\rho/3)$ is replaced by

$$\left(\frac{\dot{a}}{a}\right)^2 = (8\pi G\,\rho/3)\left(1 - \frac{\rho}{\rho_{\rm crit}}\right). \tag{1}$$

Here, $\rho_{\rm crit} = \sqrt{3}/32\pi^2\gamma^3G^2\hbar$, where γ is the Barbero-Immirzi parameter of LQG (whose value $\gamma \sim 0.24$ is determined by the black hole entropy calculations in LQG). By plugging in numbers one finds $\rho_{\rm crit} \approx 0.41\rho_{\rm Pl}$. Thus, $\rho_{\rm crit} \approx \rho_{\rm max}$, found in numerical simulations. Furthermore, one can show analytically [28] that the spectrum of the density operator on the physical Hilbert space admits a finite upper bound $\rho_{\rm sup}$, also given by $\rho_{\rm sup} = \sqrt{3}/32\pi^2\gamma^3G^2\hbar$. This result refers to the density operator on the physical Hilbert space. Thus, *there is an excellent match between quantum theory which provides $\rho_{\rm sup}$, the effective equations which provide $\rho_{\rm crit}$ and numerical simulations which provide $\rho_{\rm max}$.*

- In classical general relativity the right side, $8\pi G\rho/3$, of the Friedmann equation is positive, whence $\dot{a}$ cannot vanish; the universe either expands forever from the big bang or contracts into the big crunch. In the LQC effective equation, on the other hand, $\dot{a}$ vanishes when $\rho = \rho_{\rm crit}$ at which a quantum bounce occurs: To the past of this event, the universe contracts while to the future, it expands. This is possible because the LQC correction $\rho/\rho_{\rm crit}$ *naturally* comes with a negative sign. This is non-trivial. In the standard brane world scenario, for example, Friedmann equation is also receives a $\rho/\rho_{\rm crit}$ correction but it it comes with a positive sign (unless one artificially makes the brane tension negative) whence the singularity is not resolved.
- Even at the onset of the standard inflationary era, the quantum correction $\rho/\rho_{\rm crit}$ is of the order 10^{-11} and hence completely negligible. Thus, LQC calculations provide *an a priori justification* for using classical general relativity during inflation.

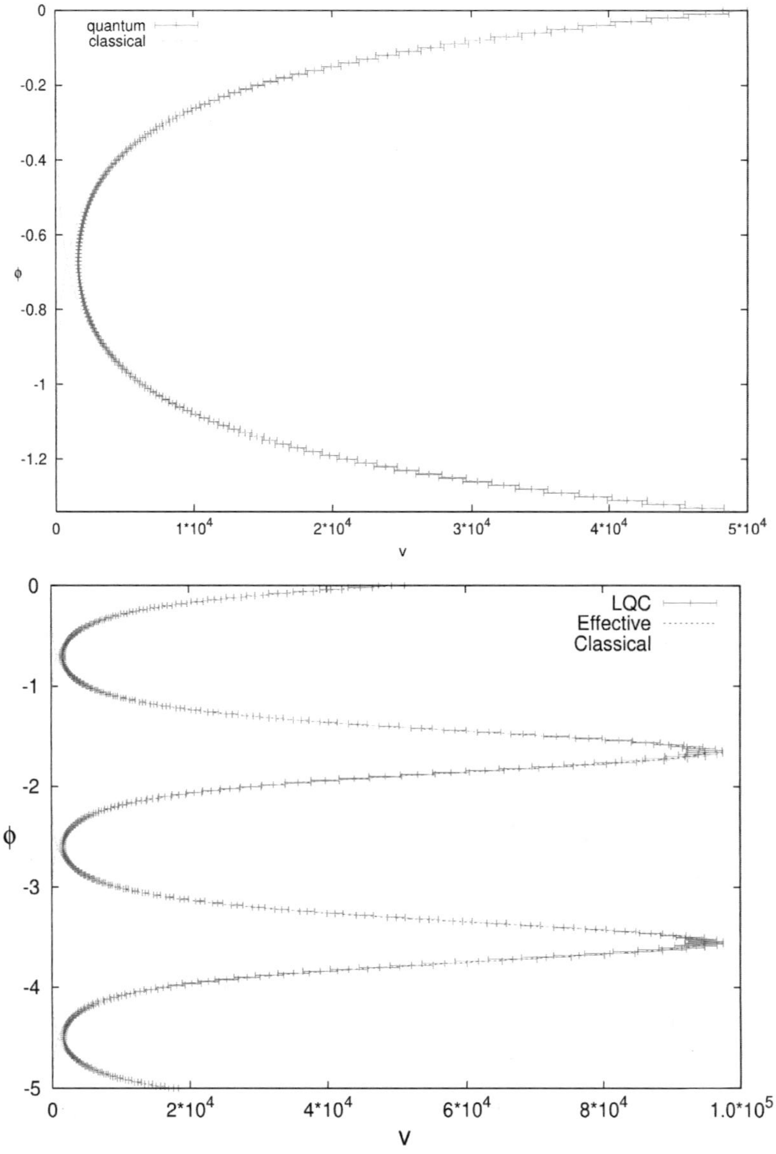

Figure 2. In the LQC evolution of models under consideration, the big bang and big crunch singularities are replaced by quantum bounces. Expectation values and dispersion of the volume operator $\hat{V}_\phi$, are compared with the classical trajectory and the trajectory from effective Friedmann dynamics. The classical trajectory deviates significantly from the quantum evolution at the Planck scale and evolves into singularities. The effective trajectory provides an excellent approximation to quantum evolution at all scales. *Upper panel:* The k=0 case. In the backward evolution, the quantum evolution follows our post big-bang branch at low densities and curvatures but undergoes a quantum bounce at matter density $\rho \sim 0.41\rho_{\rm PL}$ and joins on to the classical trajectory that was contracting to the future. *Lower panel:* The k=1 case. The quantum bounce occurs again at $\rho \sim 0.41\rho_{\rm Pl}$. Since the big bang and the big crunch singularities are resolved the evolution undergoes cycles. In this simulation $p^\star_{(\phi)} = 5\times10^3$, $\Delta p_{(\phi)}/p^\star_{(\phi)} = 0.018$, and $v^\star = 5\times10^4$.

• The analysis has been extended to include a cosmological constant. In the case when it is negative, the classical universe starts out with a big bang, expands to a maximum volume and then undergoes a recollapse to a big crunch singularity. The recollapse is faithfully reproduced by the LQC evolution. However, in contrast to general relativity and the WDW theory, both the big-bang and the big-crunch singularities are resolved [32]. Thus, in this case, the LQC evolution leads to a 'cyclic' universe (as in the k=1 model discussed below).

• The case of a positive cosmological constant is more subtle [33]. Now, as in the $\Lambda = 0$ case, the classical theory admits two types of trajectories. One starts with a big bang and expands to infinity while the other starts out with infinite volume and contracts into a big crunch. But, in contrast to the $\Lambda = 0$ case, they attain an infinite volume at a *finite* value $\phi_{\rm max}$ of ϕ. The energy density $\rho|_\phi$ at the 'internal time' ϕ goes to zero at $\phi_{\rm max}$. Because the ϕ 'evolution' is unitary in LQC, it yields a natural extension of the classical solution beyond $\phi_{\rm max}$. States which are semi-classical in the low $\rho|_\phi$ regime again follow an effective trajectory. Since $\rho|_\phi$ remains bounded, it is convenient to draw these trajectories in the ρ_ϕ-ϕ plane (rather than v-ϕ plane). They agree with the classical trajectories in the low ρ_ϕ regime and analytically continue the classical trajectories beyond $\rho_\phi = 0$.

• The LQC framework has also been extended to incorporate standard inflationary potential with phenomenologically viable parameters [27]. Again, the singularity is resolved. Thus, in all these cases, the principal features of the LQC evolution are robust, including the value of $\rho_{\rm crit}$.

In the closed, k=1 model, the situation is similar but there are two additional noteworthy features which reveal surprising properties of the the domain of applicability of classical general relativity.

• To start with, classical general relativity is again an excellent approximation to the LQC evolution till matter density ρ becomes about 1% of the Planck density $\rho_{\rm Pl}$ and, as the density increases further, the LQC evolution starts departing significantly. Again, quantum geometry effects become dominant creating an effective repulsive force which rises very quickly, overwhelms the classical gravitational attraction, and causes a bounce thereby resolving both the big bang and the big crunch singularities. Surprisingly these considerations apply *even to universes whose maximum radius $a_{\rm max}$ is only $23\ell_{\rm Pl}$*. For these universes, general relativity is a very good approximation in the range $8\ell_{\rm Pl} < a < 23\ell_{\rm Pl}$. The matter density acquires its minimum value $\rho_{\rm min}$ at the recollapse. The classical prediction $\rho_{\rm min} = 3/8\pi G a_{\rm max}^2$ is correct to one part in 10^5.

• The volume of the universe acquires its minimum value $V_{\rm min}$ at the quantum bounce. $V_{\rm min}$ scales linearly with $p_{(\phi)}$. Consequently, $V_{\rm min}$ can be *much* larger than the Planck size. Consider for example a quantum state describing a universe which attains a maximum radius of a megaparsec. Then the quantum bounce occurs when the volume reaches the value $V_{\rm min} \approx 5.7 \times 10^{16}\,{\rm cm}^3$, *some 10^{115} times the Planck volume.* Deviations from the classical behavior are triggered when the density or curvature reaches the Planck scale. The volume can be very large and is not the relevant scale for quantum gravity effects.

This, in all these cases, classical singularities are replaced by quantum bounces and LQC provides a rather detailed picture of the physics in the Planck regime. Furthermore, the singularity resolution does not cause infra-red problems: There is close agreement with classical general relativity away from the Planck scale. The ultraviolet-infrared tension is avoided because, although quantum geometry effects are truly enormous in the Planck regime, they die astonishingly quickly.

4. Outlook

Let us summarize the overall situation. In simple cosmological models, many of the outstanding questions have been answered in LQC in remarkable detail. The scalar field plays the role of an internal or emergent time and enables us to interpret the Hamiltonian constraint as an evolution equation. Singularity is resolved in a precise sense: While physical observables such as matter density diverge in classical solutions, giving rise to a singularity, *they are represented by bounded operators on the physical Hilbert space.* The big bang and the big crunch are naturally replaced by quantum bounces. On the 'other side' of the bounce there is again a large universe. General relativity is an excellent approximation to quantum dynamics once the matter density falls below one percent of the Planck density. Thus, LQC successfully meets both the 'ultra-violet' and 'infra-red' challenges. Furthermore results obtained in a number of models using distinct methods re-enforce one another. One is therefore led to take at least the qualitative findings seriously: *Big bang is not the Beginning nor the big crunch the End.* Quantum space-times are vastly larger than what general relativity had us believe!

How can the quantum space-times of LQC manage to be significantly larger than those in general relativity when those in the WDW theory are not? Main departures from the WDW theory occur due to *quantum geometry effects* of LQG. There is no fine tuning of initial conditions, nor a boundary condition at the singularity, postulated from outside. Furthermore, matter can satisfy all the standard energy conditions. Why then does the LQC singularity resolution not contradict the standard singularity theorems of Penrose, Hawking and others? These theorems are inapplicable because *the left hand side* of the classical Einstein's equations is modified by the quantum geometry corrections of LQC. What about the more recent singularity theorems that Borde, Guth and Vilenkin [34] proved in the context of inflation? They do not refer to Einstein's equations. But, motivated by the eternal inflationary scenario, they assume that the expansion is positive along any past geodesic. Because of the pre-big-bang contracting phase, this assumption is violated in the LQC effective theory.

While the detailed results presented in section 3 are valid only for these simplest models, partial results have been obtained also in more complicated models indicating that the singularity resolution may be robust [35, 36, 37]. In this respect there is a curious similarity with the very discovery of singularities in general relativity. They were first encountered in special examples. Although the examples were the physically most interesting ones —e.g., the big-bang and the Schwarzschild curvature singularities— at first it was thought that these space-times are singular because they are highly symmetric. It was believed that generic solutions of Einstein'e equations should be non-singular. As is well-known, this belief was shattered by the singularity theorems. Some 40 years later we have come to see that the big bang and the big crunch singularities are in fact resolved by quantum geometry effects. Is this an artifact of high symmetry? Or, are there robust *singularity resolution theorems* lurking just around the corner?

A qualitative picture that emerges is that the non-perturbative quantum geometry corrections are *'repulsive'*. While they are negligible under normal conditions, they dominate when curvature approaches the Planck scale and can halt the collapse that would classically have led to a singularity. In this respect, there is a curious similarity with the situation in the stellar collapse where a new repulsive force comes into play when the core approaches a critical density, halting further collapse and leading to stable white dwarfs and neutron stars. This force, with its origin in the Fermi-Dirac statistics, is *associated with the quantum nature of matter.* However, if the total mass of the star is larger than, say, 5 solar masses, classical gravity overwhelms this force. The suggestion from LQC is that a new repulsive force *associated with the quantum nature of geometry* comes into play and is strong enough to counter the classical, gravitational attraction, irrespective of how large the mass is. It is this force that prevents the formation of singularities. Since it is negligible until one enters the Planck regime, predictions of classical relativity on the formation of trapped surfaces, dynamical and isolated horizons would still hold. But one cannot conclude that there must be a singularity because the assumptions of the standard singularity theorems would be violated. There may be no singularities, no abrupt end to space-time where physics stops. Non-perturbative, background independent quantum physics would continue.

At first one might think that, since quantum gravity effects concern only a tiny region, whatever they may be, their influence on the global properties of space-time should be negligible whence they would have almost no bearing on the issue of the Beginning and the End. However, as we saw, once the singularity is resolved, vast new regions appear on the 'other side' ushering-in new possibilities that were totally unforeseen in the realm of Minkowski and Einstein. Which of them are realized generically? Is there a manageable classification? In the case of black holes, the singularity is again resolved but there are domains in which geometry is truly quantum: the quantum fluctuations of the metric operator are so huge near the putative singularity that the classical field obtained by taking its expectation value is a poor representation of the actual physics in these regions [15]. Presence of such regions would render classical notions of causality inadequate to understand the global structure of space-time. In particular, while one can still speak of marginally trapped surfaces and dynamical and isolated horizons in the 'tame' regions of the full quantum space-time, the notion of an event horizon turns out to be 'too global' to be meaningful. Is there perhaps a well-defined but genuinely quantum notion of causality which reduces to the familiar one on quantum states which are sharply peaked on a classical geometry? Or, do we just abandon the idea that space-time geometry dictates causality and formulate physics primarily in relational terms? There is a plethora of such exciting challenges. Their scope is vast, they force us to introduce novel concepts and they lead us to unforeseen territories. These are just the type of omens that foretell the arrival of a major paradigm shift to take us beyond Einstein's space-time continuum.

Acknowledgments: Much of this chapter is based on the work with or by Alex Corichi, Tomasz Pawlowski, Param Singh, Victor Taveras, and Kevin Vandersloot. I have also benefited from comments, suggestions and probing questions of many colleagues, especially Martin Bojowald, James Hartle, Jerzy Lewandowski, Donald Marolf, Roger Penrose, Carlo Rovelli, Bill Unruh and Madhavan Varadarajan. This work was supported in part by the NSF grant PHY0456913, the Alexander von Humboldt Foundation, the The George A. and Margaret M. Downsbrough Endowment and the Eberly research funds of Penn State.

References

[1] Hajicek P 1987 *Phys. Rev.* D **36** 1065
[2] Ashtekar A and Lewandowski J 2004 *Class. Quant. Grav.* **21** R53
[3] Rovelli C 2004 *Quantum Gravity* (Cambridge, Cambridge University Press)
[4] Thiemann T 2007 *Introduction to Modern Canonical Quantum General Relativity* (Cambridge, Cambridge University Press)
[5] Ashtekar A, Lewandowski J, Marolf D, Mourão J and Thiemann T 1995 *Jour. Math. Phys.* **36** 6456
[6] Rovelli C and Smolin L 1996 *Nucl. Phys.* B **442** 593
[7] Ashtekar A and Lewandowski J 1997 *Class. Quantum Grav.* **14** A55
[8] Ashtekar A and Lewandowski J 1997 *Adv. Theo. & Math. Phys.* **1** 388
[9] Perez A 2003 *Class. Quant. Grav.* **20** R43
[10] Bojowald M 2001 *Phys. Rev. Lett.* **86** 5227
[11] Bojowald M 2005 *Liv. Rev. Rel.* **8** 11
[12] Ashtekar A 2008 Loop Quantum Cosmology: An Overview *Preprint* arXiv:0812.0177
[13] Ashtekar A and Bojowald M 2006 *Class. Quantum Grav.* **23** 391
[14] Boehmer C G and Vandersloot K 2006 *Phys. Rev.* **76** 104030
[15] Ashtekar A, Taveras V and Varadarajan M 1800 *Phys. Rev. Lett.* **100** 211302
[16] Ashtekar A 2008 in: *Minkowski Space-time: A Hundred Years Later* Ed V Petkov (Berlin, Springer-Verlag)
[17] Ashtekar A and Bojowald M 2005 *Class. Quant. Grav.* **22** 3349
[18] Green D and Unruh W 2004 *Phys. Rev.* **70** 103502
[19] Hartle J B and Hawking S W 1983 *Phys. Rev.* D **28** 2960
[20] Lewandowski J, Okolow A, Sahlmann H and Thiemann T 2006 *Comm. Math. Phys.* **267** 703
[21] Fleishchack C 2009 *Commun. Math. Phys.* **285** 67
[22] Ashtekar A, Bojowald M and Lewandowski J 2003*Adv. Theo. Math. Phys.* **7** 233
[23] Ashtekar A, Pawlowski T and Singh P 2006 *Phys. Rev. Lett.* **96** 141301
[24] Ashtekar A, Pawlowski T and Singh P 2006 *Phys. Rev.* **73** 124038

[25] Ashtekar A, Pawlowski T and Singh P 2006 *Phys. Rev.* D **74** 084003
[26] Ashtekar A, Pawlowski T, Singh P and Vandersloot K 2006 *Phys. Rev.* D **75** 0240035; Szulc L, Kaminski W and Lewandowski J 2006 *Class. Quant. Grav.* **24** 2621
[27] Ashtekar A, Pawlowski T and Singh P 2009 Loop quantum cosmology in the pre-inflationary epoch (in preparation)
[28] Ashtekar A, Corichi A and Singh P 2008 *Phys. Rev.* D **77** 024046
[29] Corichi A and Singh P 2008 *Phys. Rev.* D **78** 024034
[30] Willis J 2004 *On the low energy ramifications and a mathematical extension of loop quantum gravity* Ph.D. Dissertation (The Pennsylvaina State University)
[31] Taveras V 2008 *Phys. Rev.* D **78** 064072
[32] Bentivegna E and Pawlowski T 2008 *Phys. Rev.* D **77** 124025
[33] Ashtekar A and Pawlowski T 2009 Loop quantum cosmology with a positive cosmological constant (in preparation)
[34] Borde A, Guth A and Vilenkin A 2003 *Phys. Rev. Lett.* **90** 151301
[35] Ashtekar A and Edward-Wilson E 2009 Loop quantum cosmology of Binachi I models (in preparation)
[36] Martin-Benito M, Garay L J and Mena Marugan G A 2008 *Phys. Rev.* D **78** 083516
[37] P. Singh 2008 Loop quantum cosmos are never singular *Preprint*

Cosmic Acceleration without dark energy

Spyros Basilakos[1], **Manolis Plionis**[2,3]

[1]Academy of Athens, Research Center for Astronomy and Applied Mathematics, Soranou Efesiou 4, 11527, Athens, Greece
[2] Institute of Astronomy & Astrophysics, National Observatory of Athens, Palaia Penteli 152 36, Athens, Greece.
[3] Instituto Nacional de Astrofísica Optica y Electrónica, AP 51 y 216, 72000, Puebla, México.

Abstract. In this work, we investigate the global dynamics of the universe within the framework of the Interacting Dark Matter (IDM) scenario. Considering that the dark matter obeys the collisional Boltzmann equation, we can obtain analytical solutions of the global density evolution, which can accommodate an accelerated expansion, equivalent to either the *quintessence* or the standard Λ models. This is possible if there is a disequilibrium between the DM particle creation and annihilation processes with the former process dominating, which creates an effective source term with negative pressure. We also find realistic solutions in which the present time is located after the inflection point.

1. Introduction

The detailed analysis of the available high quality cosmological observations ([1, 2, 3, 4] and references therein) have converged during the last decade towards a cosmic expansion history that involves a spatial flat geometry and a recent accelerating expansion of the universe. This expansion has been attributed to an energy component (the so called dark energy) with negative pressure which dominates the universe at late times and causes the observed accelerating expansion. The nature of the dark energy is still a mystery and it is one of the most fundamental current problems in physics and cosmology. Indeed, due to the absence of a physically well-motivated fundamental theory, there have been many theoretical speculations regarding the nature of the above exotic dark energy (DE) among which a cosmological constant, scalar or vector fields (see [5, 6, 7, 8, 9, 10, 11] and references therein).

Most of the recent papers in this kind of studies are based on the assumption that the DE evolves independently of the dark matter (DM). The unknown nature of both DM and DE implies that we can not preclude future surprises regarding the interactions in the dark sector. This is very important because interactions between the DM and *quintessence* could provide possible solutions to the cosmological coincidence problem. Recently, several papers have been published in this area [12, 13] proposing that the DE and DM could be coupled, assuming also that there is only one type of non-interacting DM.

However, there are other possibilities. (a) It is plausible that the dark matter is self-interacting (IDM) [14], a possibility that has been proposed to solve discrepancies between theoretical predictions and astrophysical observations, among which the gamma-ray and microwave emission from the center of our galaxy (e.g. [15, 16], [17, 18] and references therein). It has also been shown that some dark matter interactions could provide an accelerated expansion phase of the Universe [19, 20, 21]. (b) The DM could potentially contain more than one particle species, for example a mixture of cold and warm or hot dark matter [22, 23], with or without inter-component interactions.

In this work we are not concerned with the viability of the different such possibilities, nor with the properties of interacting DM models. The aim of this work is to investigate only whether there are repercussions of DM self-interactions or interactions between different species of a two-component DM, for the global dynamics of the universe and specifically whether such models can yield an accelerated phase of the cosmic expansion, without the need of the *dark energy*. Note that we do not "design" the fluid interactions to produce the desired accelerated cosmic evolution, as in some previous works (e.g. [20]), but rather we investigate whether the analytical solution space of the collisional Boltzmann equation, in the expanding Universe, allows for a late accelerated phase of the Universe.

2. Collisional Boltzmann Equation in the Expanding Universe

As we described in the introduction the DM is usually considered to contain only one type of particles that are stable and neutral. In this work, we would like to investigate the cosmological potential of a different scenario, in which the dominant "cosmic" fluid is not in equilibrium and thus the time evolution of its overall density is described by the collisional Boltzmann equation (e.g. [24]):

$$\frac{\mathrm{d}\rho}{\mathrm{d}t} + 3H(t)\rho + \kappa\rho^2 - \Psi = 0 \,, \tag{1}$$

where $H(t) \equiv \dot{\alpha}/\alpha$ is the Hubble function, κ is a constant related to the cross-section for annihilation and the mean particle velocity and Ψ is the rate of creation of the corresponding DM particle pairs. It is obvious that if we impose in the current analysis an effective pressure term such as

$$P_{\rm eff} = (\kappa\rho^2 - \Psi)/3H \,, \tag{2}$$

then the collisional Boltzmann equation reduces to the usual fluid equation: $\dot{\rho}+3H(\rho+P_{\rm eff}) = 0$. Now if the DM is collisionless or the collisional annihilation and pair creation processes are in equilibrium (i.e., $P_{\rm eff} = 0$), the corresponding solution of the above differential equation is $\rho \propto \alpha^{-3}$, as it should (where α is the scale factor of the universe). In contrast, for the case of a non-perfect fluid (i.., having a disequilibrium between the annihilation and particle pair creation processes) it becomes clear that we can either have a positive or a negative effective pressure term.

The possibility of a negative such effective pressure term, i.e., the case for which the DM particle creation term is larger than the annihilation term ($\kappa\rho^2 - \Psi < 0$), is of particular interest for its repercussions on the global dynamics of the Universe (see appendix A). Note, that in the literature, there is a wealth of papers discussing the DM particle creation processes in the context of thermodynamics (for example, e.g. [25]):

We now proceed in an attempt to analytically solve Eq. (1). To this end we change variables from t to α and thus Eq. (1) can be written:

$$\frac{\mathrm{d}\rho}{\mathrm{d}\alpha} = f(\alpha)\rho^2 + g(\alpha)\rho + R(\alpha) \,, \tag{3}$$

where

$$f(\alpha) = -\frac{\kappa}{\alpha H(\alpha)} \quad g(\alpha) = -\frac{3}{\alpha} \quad R(\alpha) = \frac{\Psi(\alpha)}{\alpha H(\alpha)} \,. \tag{4}$$

In general it is not an easy task to solve analytically Eq. (3), which is a Riccati equation, due to the fact that it is a non-linear differential equation. However, Eq. (3) could be fully solvable if (and only if) a particular solution is known. Indeed, we find that for some special cases regarding the functional form of the interactive term, such as $\Psi = \Psi(\alpha, H)$, we can derive analytical solutions (see appendix B). We have phenomenologically identified two functional forms for which we can solve the previous differential equation analytically, only one of which is of interest since it provides $\propto a^{-3}$ dependence of the scale factor. This is:

$$\Psi(\alpha) = \alpha H(\alpha)R(\alpha) = \mathcal{C}_1(n+3)\alpha^n H(\alpha) + \kappa\mathcal{C}_1^2\alpha^{2n} \,, \tag{5}$$

The general solution of equation (3) using Eq. (5) is:

$$\rho(\alpha) = \mathcal{C}_1\alpha^n + \frac{\Phi(\alpha)}{[\mathcal{C}_2 - \int_1^\alpha f(x)\Phi(x)dx]} \,, \tag{6}$$

where the kernel function $\Phi(\alpha)$ has the form:

$$\Phi(\alpha) = \alpha^{-3}\mathrm{exp}\left[-2\kappa\mathcal{C}_1\int_1^\alpha \frac{x^{n-1}}{H(x)}dx\right] \,. \tag{7}$$

Note that n, $\mathcal{C}_1$ and $\mathcal{C}_2$ are the corresponding constants of the problem.

3. Mimicking the Dark Energy

In this section we investigate the solution of our phenomenological model for $n = -3$ (see Eq. 6). The basic kernel Eq. (7) of the problem is given by: $\Phi(\alpha) = \alpha^{-3}F(\alpha)$ where

$$F(\alpha) = \exp\left[-2\kappa\mathcal{C}_1\int_1^\alpha \frac{1}{x^4 H(x)}dx\right] \,. \tag{8}$$

Note that at the present time we have $F(1) = 1$. Therefore, using the above relation and the functional form of $f(\alpha)$ (see Eq. 4), the corresponding integral in equation (6) takes the form

$$\int_1^\alpha f(x)\Phi(x)dx = -\int_1^\alpha \frac{\kappa}{x^4 H(x)}F(x)dx \,. \tag{9}$$

Considering now that the function $F(\alpha)$ is exponential, we have the following useful formula:

$$\frac{\mathrm{d}F(\alpha)}{\mathrm{d}\alpha} = -\frac{2\kappa\mathcal{C}_1 F(\alpha)}{\alpha^4 H(\alpha)} \tag{10}$$

and indeed, it is routine to perform the integration in the exponential part of Eq. (9) to obtain:

$$\int_1^\alpha f(x)\Phi(x)dx = \frac{F(\alpha)-1}{2\mathcal{C}_1} \,. \tag{11}$$

Based on this analysis it is straightforward to derive the density evolution from Eq. (6):

$$\rho(\alpha) = \alpha^{-3}\left[\mathcal{C}_1 + \frac{2\mathcal{C}_1 F(\alpha)}{\mathcal{D} - F(\alpha}\right] \,, \tag{12}$$

with $\mathcal{D} = 2\mathcal{C}_1\mathcal{C}_2 + 1$. It is obvious from Eq. (12) that when $\kappa \to 0$, our cosmological model tends to the usual $\rho \propto \alpha^{-3}$ case, as it should. Evaluating Eq. (12) at the present time ($\alpha = 1$) we obtain the present-time total *effective* cosmic density, which is: $\rho_0 = \mathcal{C}_1 + 1/\mathcal{C}_2$. It is interesting to note that although Einstein's theory requires local energy conservation, which appears to be incompatible with creation and annihilation processes, the reinterpretation of the creation/annihilation terms as an effective pressure of the medium (Eq. 2) lifts the apparent inconsistency. Below, we investigate the conditions under which Eq. 12 could provide accelerating solutions, similar to the usual *dark energy* case. Notice, that the current analysis is based on the matter era and thus, the radiation component does not play an important role in the global dynamics, since it evolves as $\propto \alpha^{-4}$.

3.1. Conditions to have an inflection point

Using the Hubble function: $H^2(\alpha) = (\dot{\alpha}/\alpha)^2 = 8\pi G\rho(\alpha)/3$, we derive the second derivative of the scale factor:

$$\ddot{\alpha} = \frac{4\pi G}{3}\left(2\alpha\rho + \alpha^2\frac{\mathrm{d}\rho}{\mathrm{d}\alpha}\right) . \tag{13}$$

Differentiating Eq. (12) with respect to the scale factor and using Eq. (10) we obtain:

$$\frac{\mathrm{d}\rho}{\mathrm{d}\alpha} = -\frac{3\mathcal{C}_1}{\alpha^4}\left[\frac{\mathcal{D}+F(\alpha)}{\mathcal{D}-F(\alpha)}\right] - \frac{4\mathcal{D}\mathcal{C}_1^2\kappa F(\alpha)}{\alpha^7 H(\alpha)[\mathcal{D}-F(\alpha)]^2} . \tag{14}$$

From Eqs. (13) and (14) we have:

$$\frac{\ddot{\alpha}}{\alpha} = -\frac{4\pi G\mathcal{C}_1 Z(\alpha)}{3H(\alpha)\alpha^6[\mathcal{D}-F(\alpha)]^2} , \tag{15}$$

where $Z(\alpha) = \alpha^3 H(\alpha)(\mathcal{D}+F)(\mathcal{D}-F) + 4\mathcal{C}_1\mathcal{D}\kappa F$. In order to have an inflection point ($\ddot{\alpha} = 0$), the function $Z(\alpha)$ should contain roots which are real and such that $\alpha \in (0,1)$. For this to be the case we must have:

$$Z(0)Z(1) < 0 \Rightarrow \lim_{\alpha\to 0} Z(\alpha) \lim_{\alpha\to 1} Z(\alpha) < 0 , \tag{16}$$

from which we obtain, for $\mathcal{C}_1 > 0$: $\kappa\mathcal{D}(\mathcal{D}-\lambda_1)(\mathcal{D}-\lambda_2) < 0$ with roots: $\lambda_{2,1} = -\tau \pm \sqrt{\tau^2+1}$ with $\tau = 2\kappa\mathcal{C}_1/H_0$. Evidently, for the $\kappa > 0$ case we have that $\lambda_1 < -1$ and $0 < \lambda_2 \leq 1$, and the range of values for which an inflection point is present in the evolution of the scale factor, α, is: $\mathcal{D} \in (-\infty, \lambda_1) \cup (0, \lambda_2)$. Furthermore, we have an additional constrain arising from the fact that H_0^2 should be positive: $H_0^2 = 8\pi G\mathcal{C}_1(\mathcal{D}+1)(\mathcal{D}-1)^{-1}/3 > 0$, implying $\mathcal{D} \in (-\infty,-1) \cup (1,+\infty)$. Joining the above restrictions we finally obtain that $\mathcal{D} \in (-\infty, \lambda_1)$. Considering the possibility of $\kappa < 0$ we have that $-1 < \lambda_1 < 0$ and $\lambda_2 \geq 1$, and a similar analysis, as for the $\kappa > 0$ case, also provides a range for which an inflection point exists: $\mathcal{D} \in (\lambda_2, +\infty)$. Although $\kappa < 0$ appears artificial, one may envision the case of a two particle type gas, and specifically a Lorentz gas with elastic scattering, in which the scattering velocity of the low-mass particles in the rest-frame of the large-mass particles could be negative, i.e., bunching back with $\delta\theta = 180°$. However, we do not investigate further such a possibility.

Since the avenue by which the IDM model provides a cosmic acceleration may appear slightly involved, we present in appendix C the correspondence between our model and the usual Λ or *dark energy* models.

Table 1. General solutions for $\mathcal{C}_1 > 0$, $\kappa > 0$ and $w_0 < 0$.

Solution	$(\mathcal{C}_1 - \rho_{m,0})$	$\mathcal{D}$	$\mathcal{C}_2$	Accel.
I	< 0	$\mathcal{D} > 1 - \frac{2\mathcal{C}_1}{\mathcal{C}_1-\rho_{m,0}} > 1$	$\mathcal{C}_2 > -\frac{1}{\mathcal{C}_1-\rho_{m,0}}$	NO
II	> 0	$\mathcal{D} < 1 - \frac{2\mathcal{C}_1}{\mathcal{C}_1-\rho_{m,0}} < -1$	$\mathcal{C}_2 < -\frac{1}{\mathcal{C}_1-\rho_{m,0}}$	YES

3.2. Relation to Quintessence *Dark Energy*

We know that for homogeneous and isotropic flat cosmologies ($\Omega_m + \Omega_Q = 1$), driven by non relativistic DM and a DE with a time varying equation of state, the density evolution of this cosmic fluid can be written as:

$$\rho(\alpha) = \alpha^{-3}\left[\rho_{m,0} + \rho_{Q,0}\mathrm{exp}\left(-3\int_1^{\alpha}\frac{w(x)}{x}dx\right)\right] , \tag{17}$$

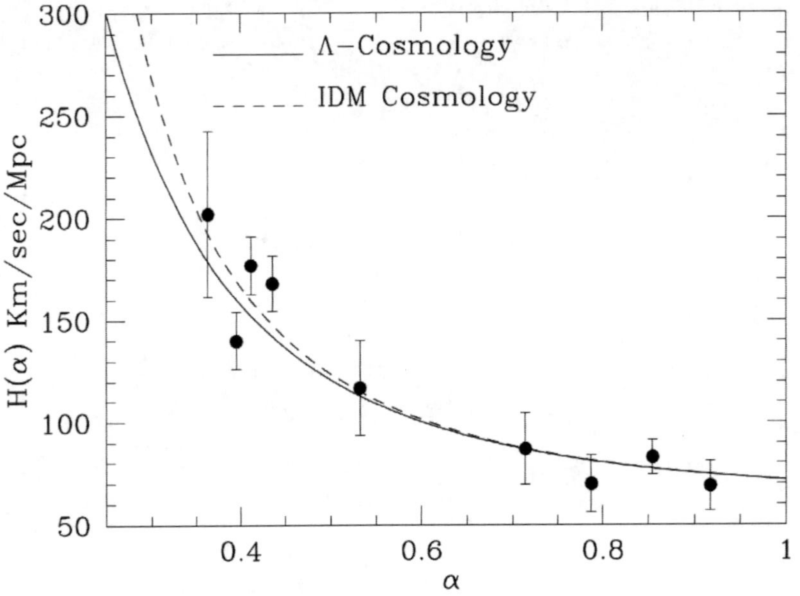

Figure 1. Comparison of the Hubble function provided by our $n = 0$ model (after fitting the constants using the early-type galaxy Hubble relation) with the traditional $\Omega_\Lambda = 0.74$, $\Omega_m = 0.26$ model. The points correspond to the galaxy data ([26]).

where $\rho_{m,0}$ and $\rho_{Q,0}$ are the present-day DM and DE densities, respectively, while $w(\alpha)$ is the equation of state parameter. Now, equating Eq. (12) and Eq. (17) we can investigate the circumstances under which DM self-interactions can act as the usual *quintessence*. At the present time ($\alpha = 1$) we see that our *effective* global density, ρ_0, can be seen as the sum: $\rho_0 = \rho_{m,0} + \rho_{Q,o}$. In the general case we have, after some algebra and a differentiation, that:

$$w(\alpha) = \frac{4\kappa\mathcal{C}_1^2 F}{3\alpha^3 H(\alpha)} \frac{\mathcal{D}}{[\mathcal{D} - F]} \frac{1}{[(\mathcal{C}_1 - \rho_{m,0})(\mathcal{D} - F) + 2F\mathcal{C}_1]}, \tag{18}$$

which evaluated at the present time we obtain:

$$w_0 = \frac{4\kappa\mathcal{C}_1^2}{3H_0} \frac{\mathcal{D}}{(\mathcal{D} - 1)} \frac{1}{[(\mathcal{C}_1 - \rho_{m,0})(\mathcal{D} - 1) + 2\mathcal{C}_1]}. \tag{19}$$

We present in Table 1 the combination of values of the constants for which $w_0 < 0$. Now we join the analysis of section 3.1 to check which of the solutions shown in Table 1 are viable solutions for our Universe, i.e., for which of these solution Eq. (16) is satisfied (i.e., the inflection point is located before the present time). This constrain excludes solution I, leaving however the solution II as a viable *quintessence* dark-energy look-alike, as far as the global dynamics is concerned.

4. Mimicking the Λ Cosmology

We will show that for $n = 0$ the global dynamics, provided by our model (see Eq. 6), is equivalent to that of the traditional Λ cosmology. To this end we use $\mathrm{d}t = \mathrm{d}\alpha/(\alpha H)$ and the basic kernel (Eq. 7) for this case becomes:

$$\Phi(\alpha) = \alpha^{-3}\exp\left[-2\kappa\mathcal{C}_1 \int_1^\alpha \frac{1}{xH(x)}dx\right] = \alpha^{-3}e^{-2\kappa\mathcal{C}_1(t-t_0)}, \tag{20}$$

where t_0 is the present age of the universe. In addition, the integral in equation (6) takes now the following form: $\int_1^\alpha f(x)\Phi(x)dx = -\kappa F(t)$ with $F(t) = \int_{t_0}^t \Phi(u)du$. Note that at the present time we have $F(t_0) = 0$. Therefore, using the above formula, the global density evolution (Eq. 6) can be written:

$$\rho(\alpha) = \mathcal{C}_1 + \alpha^{-3}\frac{e^{-2\kappa\mathcal{C}_1(t-t_0)}}{[\mathcal{C}_2 + \kappa F(t)]} \ . \tag{21}$$

As expected, at early enough times ($t \to 0$) the overall density scales according to: $\rho(\alpha) \propto a^{-3}$, while close to the present epoch the density evolves according to:

$$\rho(\alpha) \simeq \mathcal{C}_1 + \frac{\alpha^{-3}}{\mathcal{C}_2} \ , \tag{22}$$

which is approximately the corresponding evolution in the Λ cosmology, in which the term $\mathcal{C}_1$ acts as the constant-vacuum term (ρ_Λ) and the $1/\mathcal{C}_2$ term acts like matter (ρ_m).

Note that the effective pressure term (Eq. 2), for $\kappa \to 0$, becomes: $\Psi \sim 3\mathcal{C}_1 H$, which implies that: $P \sim -\Psi/3H = -\mathcal{C}_1$. Therefore, this case relates to the traditional Λ cosmology, since $\mathcal{C}_1$ corresponds to ρ_Λ (see Eq. 22). We now investigate in detail the dynamics of the $n = 0$ model.

4.1. Conditions to have an inflection point

Using the Hubble function we derive the second derivative of the scale factor, which is:

$$\frac{\ddot{\alpha}}{\alpha} = -\frac{4\pi G Z(\alpha)}{3\alpha^6 H(\alpha)[\mathcal{C}_2 + \kappa F]^2} \ , \tag{23}$$

where

$$Z(\alpha) = -2\mathcal{C}_1\alpha^6 H(\alpha)[\mathcal{C}_2 + \kappa F]^2 + \kappa e^{-4\kappa\mathcal{C}_1(t-t_0)} + \alpha^3(\mathcal{C}_2 + \kappa F)(2\kappa\mathcal{C}_1 + H)e^{-2\kappa\mathcal{C}_1(t-t_0)} \ . \tag{24}$$

As already discussed for the $n = -3$ case, in order to have an inflection point ($\ddot{\alpha} = 0$), the function $Z(\alpha)$ should contain roots which are real and such that Eq. (16) holds. For this to be the case we must have: $-\mathcal{C}_1\kappa(\mathcal{C}_2 - \lambda_1)(\mathcal{C}_2 - \lambda_2) < 0$ with roots: $\lambda_{2,1} = (\tau \pm \sqrt{\tau^2 + 2\kappa c})/c$ and $\tau = 2\kappa\mathcal{C}_1 + H_0$, $c = 4H_0\mathcal{C}_1$. For $\kappa\mathcal{C}_1 > 0$ we have $\mathcal{C}_2 \in (-\infty, \lambda_1) \cup (\lambda_2, +\infty)$, while for $\kappa\mathcal{C}_1 < 0$ we have $\mathcal{C}_2 \in (\lambda_1, \lambda_2)$. Therefore independent of the value of $\mathcal{C}_1$ there is always a range of solutions which accommodate an inflection point and therefore an accelerated phase of the scale factor.

4.2. Relation to the Standard Λ Cosmology

From Eq. (21), using the usual unit-less Ω-like parameterization, we have after some algebra that:

$$\left(\frac{H}{H_0}\right)^2 = \Omega_{1,0} + \frac{\Omega_{1,0}\Omega_{2,0}\alpha^{-3}e^{-2\kappa\mathcal{C}_1(t-t_0)}}{\Omega_{1,0} + \kappa\mathcal{C}_1\Omega_{2,0}F(t)} \ , \tag{25}$$

with $\Omega_{1,0} = 8\pi G\mathcal{C}_1/3H_0^2$ and $\Omega_{2,0} = 8\pi G/3H_0^2\mathcal{C}_2$, which in the usual Λ cosmology they correspond to Ω_Λ and Ω_m, respectively. We can now attempt to compare the Hubble function of Eq. (25) to that corresponding to the usual Λ model. To this end, we use a χ^2 minimization between our model (Eq. 25) and the Hubble relation derived directly from early type galaxies at high redshifts (Simon et al. 2005). This procedure is performed by using $\Omega_{2,0} = 0.26$, $H_0 = 72$ km/sec/Mpc and $t_0 = 13.6$ Gyrs (the age of the universe of the corresponding Λ cosmology). The resulting minimization provides $\kappa\mathcal{C}_1 = 0.0017$, $\chi^2/\text{d.f.} = 1.18$ and a Hubble relation, seen in figure 1 as the dashed line, which closely resembles the corresponding one of the traditional Λ model.

5. Conclusions

In this work we investigate analytically the evolution of the global density of the universe in the framework of an interacting DM scenario by solving analytically the collisional Boltzmann equation in an expanding Universe. The possible disequilibrium between the DM particle creation and annihilation processes creates an effective source term with negative pressure which, acting as *dark energy*, provides an accelerated expansion phase of the scale factor. Furthermore, we also find a realistic solution for which the present time is after the inflection point.

Appendix A. The main assumptions of our model

The reason why a cosmological constant or a component mimicking it leads to a cosmic acceleration is because it introduces in Friedmann's equation, which governs the global dynamics, a component which has an equation of state with negative pressure. Our model creates exactly the equivalent to the above, i.e., an effective source term which has negative pressure. The avenue throw which this is accomplished is via the collisional Boltzmann equation in an expanding universe (e.g. [24]) in which the disequilibrium between the annihilation ($\kappa\rho^2$) and particle creation (Ψ) processes provides the effective pressure: $P_{\rm eff} = (\kappa\rho^2 - \Psi)/3H$, which when negative acts exactly as a repulsive force, and therefore provides a cosmic acceleration (depending on the combination of parameters, as indicated in our Table 1). The basic assumption of this work is that the DM fluid is non-perfect, i.e., there is a disequilibrium between the particle pair annihilation and creation processes. Note that in our general solution (Eq. 6) we do not specify the direction of this disequilibrium. Only in the detailed analysis we seek solutions which provide an accelerated phase of the universe (for which of course one condition is $P < 0$), which we indeed find.

Appendix B. Solutions of the Riccati equation

With the aid of the differential equation theory we present solutions that are relevant to our Eq. (3). In general a Riccati differential equation is given by

$$y' = f(x)y^2 + g(x)y + R(x) \tag{B.1}$$

and it is fully solvable only when a particular solution is known. Below we present two case in which analytical solutions are possible.

- *Case 1:* For the case where:

$$R(x) = \mathcal{C}_1 n x^{n-1} - \mathcal{C}_1^2 x^{2n} f(x) - \mathcal{C}_1 x^n g(x) \tag{B.2}$$

  the particular solution is x^n and thus the corresponding general solution can be written as:

$$y(x) = \mathcal{C}_1 x^n + \Phi(x)\left[\mathcal{C}_2 - \int_1^x f(u)\Phi(u)du\right]^{-1}, \tag{B.3}$$

  where

$$\Phi(x) = \exp\left[\int_1^x \left(2\mathcal{C}_1 u^n f(u) + g(u)\right) du\right] \tag{B.4}$$

  and $\mathcal{C}_1, \mathcal{C}_2$ are the integration constants. Using now Eq. (4) we get $\Psi(x) = xH(x)R(x) = \mathcal{C}_1(n+3)x^n H(x) + \kappa\mathcal{C}_1^2 x^{2n}$.

- *Case 2:* For the case where:

$$R(x) = h'(x) \quad \text{with} \quad g(x) = -f(x)h(x) \tag{B.5}$$

the particular solution is $h(x)$ [in our case we have $h(x) = -3\kappa^{-1}H(x)$]. The general solution now becomes:

$$y(x) = h(x) + \Phi(x)\left[\mathcal{C}_2 - \int_1^x f(u)\Phi(u)du\right]^{-1} , \tag{B.6}$$

where

$$\Phi(x) = \exp\left[\int_1^x f(u)h(u)\right] . \tag{B.7}$$

In this framework, using Eq. (4) we finally get $\Psi(x) = xH(x)R(x) = -3\kappa^{-1}xH(x)H'(x)$.

Note that the solution of *Case 1* (for $n = -3$ and $n = 0$) is the only one providing a $\propto \alpha^{-3}$ dependence of the scale factor (see Eqs. 12, 21 and 22).

Appendix C. Correspondence between our model and the usual DE models

The necessary criteria to have cosmic acceleration and an inflection point in our past ($t_i < t_0$), are: (a) $P < 0$ and (b) $\ddot{\alpha} = 0$, which implies, from Eq. (13), that $(2\rho + \alpha \mathrm{d}\rho/\mathrm{d}\alpha) = \alpha Z(\alpha) = 0$ should have a real root within $\alpha \in (0, 1)$. Simple mathematics show that these conditions are respected both in the DE models and in our case, when:

- Dark Energy models: $P = w\rho_Q < 0$ with $w < -1/3$.
- IDM models: $P = \kappa\rho^2 - \Psi < 0$ and $Z(0)Z(1) < 0$.

For the benefit of the reader and in order to see the equivalence of the corresponding criteria ($w < -1/3$ and $Z(0)Z(1) < 0$) we briefly apply our mathematical methodology (for the inflection point) in the case of the usual *dark energy* model ($w =$ constant). Using the Friedmann equation, the second derivative of the scale factor becomes:

$$\frac{\ddot{\alpha}}{\alpha} = \frac{4\pi G}{3}\alpha Z(\alpha) \quad \text{where} \quad Z(\alpha) = \left(2\alpha^{-1}\rho + \frac{\mathrm{d}\rho}{\mathrm{d}\alpha}\right) . \tag{C.1}$$

The general condition to have an inflection point is $\ddot{\alpha} = 0$ or equivalently the equation $Z(\alpha) = 0$ should contain a real root before the present time, i.e., $\alpha \in (0, 1)$. This reads mathematically, as follows:

$$Z(0)Z(1) < 0 \Rightarrow \lim_{\alpha\to 0} Z(\alpha) \lim_{\alpha\to 1} Z(\alpha) < 0 . \tag{C.2}$$

For the simple case of the *dark energy* models with constant $w(< 0)$, the above restriction simply leads to $w < -1/3$. Indeed in this case $Z(\alpha) = (3w + 1)\alpha^{-3w}\rho_{Q,0} + \rho_{m,0}$ and from $\lim_{\alpha\to 0} Z(\alpha) \lim_{\alpha\to 1} Z(\alpha) < 0$, it can be proved easily that in order to reach an inflection point we need $w < -1/3$. In other words from a mathematical point of view the condition $\lim_{\alpha\to 0} Z(\alpha) \lim_{\alpha\to 1} Z(\alpha) < 0$ is equivalent to $w < -1/3$. For a time-varying equation of state parameter the situation becomes more complicated.

References

[1] D.N. Spergel, et al., Astrophys. J. Suplem., **170**, 377, (2007)
[2] T.M. Davis *et al.*, Astrophys. J., **666**, 716, (2007)
[3] M. Kowalski, et al., Astrophys. J., **686**, 749, (2008)
[4] E. Komatsu, et al., Astrophys. J. Suplem., **in press**, (2009), [arXiv:0803.0547]
[5] S. Weinberg, in *The cosmological constant problem, Reviews of modern physics* , **61**, 1, (1989)
[6] C. Wetterich, Astron. Astrophys. **301**, 321 (1995)
[7] R. R. Caldwell, R. Dave, and P. J. Steinhardt P. J., Phys. Rev. Lett., **80**, pp 1582, (1998)
[8] P. J. Peebles and B. Ratra, in *The cosmological constant and dark energy, Reviews of modern physics* , **75**, p. 559–606, (2003)
[9] P. Brax and J. Martin, Phys. Lett. **B468**, 40 (1999)

Journal of Physics: Conference Series **189** (2009) 012004 doi:10.1088/1742-6596/189/1/012004

[10] A. W. Brookfield, C. van de Bruck, D. F. Mota, and D. Tocchini-Valentini, Phys. Rev. Lett. **96**, 061301 (2006)

[11] C. G. Boehmer and T. Harko, Eur. Phys. J. **C50**, 423 (2007)

[12] L. Amendola and D. Tocchini-Valentini, Phys. Rev. D **64**, 043509 (2001); W. Zimdahl, D. Pavón, L. P. Chimento, Phys.Lett.B, **521**, 133, (2001)
L. Amendola, C. Quercellini, D. Tocchini-Valentini and A. Pasqui, Astrophys. J. **583**, L53 (2003); G. Mangano, G. Miele and V. Pettorino, Mod. Phys. Lett. A **18**, 831 (2003); L. P. Chimento, A. S. Jakubi, D. Pavón, W. Zimdahl, Phys.Rev.D, **67**, 083513, (2003)

[13] R. G. Cai and A. Wang, JCAP **0503**, 002 (2005); J. B. Binder and G. M. Kremer, Gen. Rel. Grav. **38**, 857 (2006); G. Huey and B. D. Wandelt, Phys. Rev. D **74**, 023519 (2006); B. Wang, C-Y, Lin, Abdalla, Phys. Lett. B., **637**, 357, (2006); R. Mainini, S. Bonometto, Phys.Rev.D, **74**, 043504, (2006); S. Das, P. S. Corasaniti, J. Khoury, Phys. Rev. D. **73**, 083509, (2006); G. M. Kremer, Gen.Relativ.Gravit., **39**, 965, (2007); A. W. Brookfield, C. van de Bruck, and L. M. H., Hall,[arXiv:0709.2297], (2007); J.-H He, B. Wang, B., JCAP, **06**, 010, (2008)

[14] D.N. Spergel and P.J. Steinhardt, Phys.Rev.Lett. **84**, 3760 (2000)

[15] R.A. Flores, J.R. Primack, Astrophys. Journal, **427**, L1 (1994)

[16] B. Moore et al. Astrophys. Journal, **524**, L19 (1999)

[17] D.. Hooper, D. P., Finkbeiner, G. Dobler, [arXiv:0705.3655], (2007)

[18] M. Regis, P. Ullio,[arXiv:0802.0234], (2008)

[19] W. Zimdahl, D.J. Schwarz, A.B. Balakin, D. Pavon, Phys. Rev. D **64**, 3501 (2001);

[20] A.B. Balakin, D. Pavon, D.J. Schwarz, W. Zimdahl, N.J.Phys, **5**, 85 (2003).

[21] J. A. S. Lima, F. E. Silva, R. C. Santos, Class. and Quantum Gravity, **25**, 205006, (2008)

[22] G. R. Farrar and P. J. E. Peebles, Astrophys. J. **604**, 1 (2004)

[23] S. S. Gubser and P. J. E. Peebles, Phys. Rev. **D70**, 123510 (2004)

[24] E. W, Kolb & M. S. Turner, *The Early Universe*, Addison-Wesley Publishing, (1990)

[25] I. Prigogine, J. Geheniau, E. Gunzig, P. Nardone, Gen. Rel.Grav. **21**, 767, (1989); M. O. Calvao, J. A. S. Lima, I. Waga, Phys.Lett. A, **162**, 223, (1992); J. A. S., Lima, A. S. M. Germano, L. R. W. Abramo, Phys.Rev. D., **53**, 4287, (1996)

[26] J., Simon, L. Verde, R. Jimenez, Phys. Rev. D., 71, 123001, (2005)

Recent Developments in Gravity (NEB XIII) IOP Publishing
Journal of Physics: Conference Series **189** (2009) 012005 doi:10.1088/1742-6596/189/1/012005

Deformed spaces and loop cosmology

Marco Valerio Battisti

ICRA and Phys. Dept. University of Rome "Sapienza", P.le A. Moro 5 00185 Rome, Italy

E-mail: battisti@icra.it

Abstract. The non-singular bouncing solution of loop quantum cosmology is reproduced by a deformed minisuperspace Heisenberg algebra. This algebra is a realization of the Snyder space, is almost unique and is related to the κ-Poincaré one. Since the sign of the deformation parameter it is not fixed, the Friedmann equation of braneworlds theory can also be obtained. Moreover, the sign is the only freedom in the picture and these frameworks are the only ones which can be reproduced by our deformed scheme. A generalized uncertainty principle for loop quantum cosmology is also proposed.

Introduction

In the last years a wide interest in the analysis of the non-commutative framework is increased (for a review see [1]). In particular, this approach can be considered a plausible candidate for describing physics at the Planck scale [2] and can be related to the intuitions of doubly special relativity (DSR) [3] which arises as a semi-classical limit of quantum gravity (see [4] and the references therein). On the other hand, a prerequisite of any quantum theory of gravity is to solve the space-time singularities predicted by the Einstein theory of general relativity. One of the most important example is the big-bang singularity appearing in the standard model of cosmology, which Friedmann dynamics is expected to be modified by quantum effects in the regime of small scale factor. Loop quantum cosmology (LQC) [5] leads to a resolution of the singularity replacing the big-bang by a big-bounce as soon as the matter energy density reaches the Planck scale, i.e. in this extreme region (quantum) gravity behaves repulsively [6].

¿From this perspective it is natural to investigate the fate of the cosmological singularity in a deformed (minisuper)-space framework [7]. Here, we consider generalized commutation relations which leave undeformed the translation group and preserve the rotational invariance, and apply this framework to the Friedmann-Robertson-Walker (FRW) Universes (for more details see [8]). As a result the effective Friedmann equation of LQC [9] naturally arises from the deformed scheme. Interesting, since the deformed Heisenberg algebra is fixed up to a sign, also the braneworlds Friedmann equation [10] is predicted by our model. In other words, this deformed phase space can be regarded, from a phenomenological point of view, as an effective framework which is able to describe the results obtained in both LQC and Randall-Sundrum braneworlds scenario. In this deformed scheme, the different predictions of such theories can be easily understood considering the opposite sign in the deformation term. It is worth noting that, a generalized uncertainty principle naturally arises in our framework and, in the braneworlds-like case, it resembles the one predicted by string theory. At the same level, a generalized uncertainty relation can also be proposed for LQC.

The paper is organized as follows. Section 1 is devoted to discuss a non-commutative space and its commutations relations. In Section 2 this framework is applied to the isotropic cosmological models analyzing the modifications induced on the Friedmann equation. Finally, in Section 3, implications of the picture are analyzed. Concluding remarks follow.

1. Deformed Heisenberg algebra

Let us start by considering a n-dimensional deformed (non-commutative) Euclidean space such that commutator between the coordinates has the non-trivial structure

$$[\tilde{x}_i, \tilde{x}_j] = \mp i\alpha M_{ij}, \tag{1}$$

where with $\tilde{x}$ we refer to the non-commutative coordinates and $\alpha > 0$ is the deformation parameter with dimension of a length square. We then demand that the rotation generators $M_{ij} = x_i p_j - x_j p_i$ satisfy the ordinary $SO(n)$ algebra and that the translation group is not deformed, i.e. $[p_i, p_j] = 0$. Is then natural to assume the commutators between M_{ij} and $\tilde{x}_k$, as well as between M_{ij} and p_k, as undeformed. This way we deal with the (Euclidean) Snyder space [11].

We now consider a general rescaling of the non-commutative coordinates $\tilde{x}_i$ in terms of a momentum-dependent function $\varphi\left(\alpha p^2\right)$ as

$$\tilde{x}_i = x_i \varphi\left(\alpha p^2\right), \tag{2}$$

i.e. we consider a realization of the algebra (1) in terms of ordinary phase space variables [12]. It is not difficult to show that the function φ is uniquely fixed (up to a sign) by our natural assumptions and reads $\varphi\left(\alpha p^2\right) = \sqrt{1 \pm \alpha p^2}$. This way the commutator between non-commutative coordinates and momenta is given by

$$[\tilde{x}_i, p_j] = i\delta_{ij}\sqrt{1 \pm \alpha p^2} \tag{3}$$

and the ordinary Heisenberg algebra is recovered as soon as $\alpha \to 0$. This deformed algebra closes in sense that all the Jacobi identities are satisfied and can be related to the κ-Poincaré one [13]. Interest in non-commutative (or deformed) phase space arises in order to mathematically describe DSR and, in particular, some of DSR models can be formulated as generalizations of Snyder model [14]. As last point, it is worth noting that in the case of minus sign, the momentum is limited from above since $p \in I \equiv (-1/\sqrt{\alpha}, 1/\sqrt{\alpha})$. Therefore, in this case, we have a truncation of the phase space at $p^2 \geq \alpha$.

2. Friedmann dynamics in the deformed framework

The FRW cosmological models are characterized by imposing the isotropy on the Cauchy surfaces which fill the space-time manifold. Isotropy reduces the phase space of GR to be two dimensional with coordinates (a, p_a), where the scale factor a is the only degree of freedom of the system. It describe the expansion of the Universe and the standard model of cosmology is based on this models [15]. The dynamics of these Universes can be obtained from the extended Hamiltonian

$$\mathcal{H}_E = \frac{2\pi G}{3} N \frac{p_a^2}{a} + \frac{3}{8\pi G} N a k - N a^3 \rho + \lambda \pi, \tag{4}$$

where λ is a Lagrange multiplier and the parameter k can be zero or ± 1 leading to the flat, open or closed Universe respectively. The term $\lambda\pi$ is introduced since π, the momentum conjugate to the lapse function N, vanishes. In the expression above $\rho = \rho(a)$ denotes a generic energy density we have introduced into the dynamics.

Let us now consider the modifications induced on the dynamics of the FRW models by the deformed Heisenberg algebra discussed above [8]. In particular, we assume the symplectic structure of the minisuperspace as deformed and thus the Poisson bracket (for any two-dimensional phase space function) appears to be

$$\{F,G\}_\alpha = \left(\frac{\partial F}{\partial a}\frac{\partial G}{\partial p_a} - \frac{\partial F}{\partial p_a}\frac{\partial G}{\partial a}\right)\sqrt{1\pm\alpha p_a^2}. \tag{5}$$

We deal with a one-dimensional mechanical system and thus the only non-trivial commutators is given by

$$\{a,p_a\} = \sqrt{1\pm\alpha p_a^2}. \tag{6}$$

Since the Poisson bracket $\{N,\pi\} = 1$ is not affected by the deformations induced by the α parameter on the system, the equations of motion $\dot{N} = \{N,\mathcal{H}_E\} = \lambda$ and $\dot{\pi} = \{\pi,\mathcal{H}_E\} = \mathcal{H} = 0$ remain unchanged[1]. On the other hand, the equations of motion of the scale factor a and its conjugate momentum p_a become modified in such an approach via the deformed symplectic geometry (5) and read

$$\begin{aligned} \dot{a} &= \{a,\mathcal{H}_E\}_\alpha = \frac{4\pi G}{3}N\frac{p_a}{a}\sqrt{1\pm\alpha p_a^2}, \\ \dot{p}_a &= \{p_a,\mathcal{H}_E\}_\alpha = N\left(\frac{2\pi G}{3}\frac{p_a^2}{a^2} - \frac{3}{8\pi G}k + 3a^2\rho + a^3\frac{d\rho}{da}\right)\sqrt{1\pm\alpha p_a^2}. \end{aligned} \tag{7}$$

The equation of motion for the Hubble rate $(\dot{a}/a)$ can be obtained solving the scalar constraint $\mathcal{H} = 0$ with respect to p_a and then considering the first equation of (7). Explicitly it becomes

$$\left(\frac{\dot{a}}{a}\right)^2 = \left(\frac{8\pi G}{3}\rho - \frac{k}{a^2}\right)\left[1 \pm \frac{3\alpha}{2\pi G}a^2\left(a^2\rho - \frac{3}{8\pi G}k\right)\right]. \tag{8}$$

We refer to this equation as the *deformed Friedmann equation* as it entails the modification arising from the deformed Heisenberg algebra previously analyzed. Of course, for $\alpha \to 0$ the ordinary Friedmann equation is recovered. Let us consider the flat FRW Universe, i.e. the $k = 0$ model. In this case the deformed equation (8) appears to be

$$\left(\frac{\dot{a}}{a}\right)^2 = \frac{8\pi G}{3}\rho\left(1 \pm \frac{\rho}{\rho_c}\right), \tag{9}$$

where $\rho_c = (2\pi G/3\alpha)\rho_P$ is the critical energy density, and ρ_P denotes the Planck one. When the limit $\alpha \to 0$ is taken into account, the critical energy density diverges leading to the ordinary dynamics. It is worth stressing that we have assumed the existence of a fundamental minimal length. In fact, as widely accepted, one of the most peculiar consequences of all promising quantum gravity theories is the existence of a fundamental cut-off length, which should be related to the Planck one [16]. Therefore, although this minimal length appears differently in distinct contexts, it is reasonable that the scale factor (the energy density) has a minimum (maximum) at the Planck scale.

The modified Friedmann equations (9) are known in literature. The one with the $(-)$ deformation term is the effective equation incorporating the LQC effects on the FRW dynamics [9]. It can be obtained using the geometric formulation of quantum mechanics [17]. On the other hand, the string inspired Randall-Sundrum braneworlds scenario leads to a modified Friedmann

[1] As usually, $\mathcal{H} = 0$ is the scalar constraint and is obtained by requiring the primary constraint $\pi = 0$ will be satisfied at all times.

equation as in (9) with the (+) sign [10]. The opposite sign of the ρ^2-term in such an equation, is the well-known key difference between the effective LQC and the Randall-Sundrum framework. In fact, the former approach leads to a non-singular bouncing cosmology while in the latter, because of the positive sign, the Hubble rate can not vanish and the Universe can not experiences a bounce (or more generally a turn-around) in the scale factor.

3. Quantum mechanical implications of the deformed picture

As we have seen, the deformed algebra (3) leads to effective dynamics of loop and braneworlds cosmologies in the minus or plus case respectively. Let us now investigate some implications of the deformed framework in the quantum theory. Firstly, we have to stress that the (±)-frameworks are, of course, physically different. More precisely, the deformed Hilbert spaces $\mathcal{F}_\pm$ underling the algebras (6) can be written as [8]

$$\mathcal{F}_\pm = L^2\left(\mathbb{R}(I), dp_a/\sqrt{1 \pm \alpha p_a^2}\right), \tag{10}$$

for the $\pm\alpha$ deformation of the ordinary Heisenberg theory, respectively. It is worth noting that these Hilbert spaces are unitarily inequivalent each other and also with respect to the ordinary one $L^2(\mathbb{R}, dp_a)$. This is not surprising since the deformation of the canonical commutation relations can be viewed, from the realization (2), as an algebra homomorphism which is a non-canonical transformation and thus it can not be implemented as an unitary transformation. New features are then introduced at both classical and quantum level. (For an application of the formalism to the harmonic oscillator problem see [8].)

A peculiar feature which deserves to be analyzed is the uncertainty principle underlying the deformed symplectic structure (5). The generalized uncertainty relation can be immediately obtained from the commutator (6) and reads

$$\Delta a \Delta p_a \geq \frac{1}{2} |\langle (1 \pm \alpha p_a^2)^{1/2} \rangle|, \tag{11}$$

by which, using the basic proprieties $\langle p_a^{2n} \rangle \geq \langle p_a^2 \rangle$ and $\langle p_a^2 \rangle = (\Delta p_a)^2 + \langle p_a \rangle^2$, the boundary of the allowed region begin

$$\Delta a = \frac{1}{2}\left|\left(\frac{1 \pm \alpha \langle p_a \rangle^2}{(\Delta p_a)^2} \pm \alpha\right)^{1/2}\right|. \tag{12}$$

As we can easily see, for an infinite uncertainty in momentum (or better when the relation $\Delta p_a \gg (\Delta p_a)^\star \equiv \sqrt{(1 \pm \alpha \langle p_a \rangle)/\alpha}$ holds), the uncertainty in the scale factor a no longer vanishes but approaches the minimal value of $\Delta a_{\text{min}} = \sqrt{\alpha}/2$. It is interesting to note that, Δa_{min} is the global minimum in the (+)-sector, while $\Delta a_{\text{min}} = 0$ is allowed in the (−)-one. This appears as soon as the dispersion on its conjugate momentum Δp_a reaches the critical value of $(\Delta p_a)^\star$.

Summarizing, in the (+)-sector a nonzero minimal uncertainty in the particle (Universe) position (the scale factor) appears. The resulting implications are quite profound. In fact, it is no longer possible to spatially localize a wave function with arbitrary precision and then no physical states which are position eigenstates exist at all since they were only formal ones [18]. As we have seen, this framework leads to the same Friedmann equation of the Randall-Sundrum braneworlds scenario and thus, it is not unexpected that the plus deformed uncertainty relation (11) contains, at the leading order in α, the string theory result $\Delta x \sim (1/\Delta p + l_s^2 \Delta p)$ [19], in which the string length l_s can be identify with $\sqrt{\alpha/2}$.

On the other hand, to obtain the Friedmann equation found in the effective loop cosmology, the (−)-deformed Heisenberg algebra is required. In this case a vanishing uncertainty in position is allowed and therefore the position eigenstates are true physical states (in the same sense of the ordinary quantum mechanics). However, differently from the Heisenberg framework, an infinite uncertainty in momentum is no longer required and $\Delta x_{\min} = 0$ appears as soon as the finite value $(\Delta p)^\star \propto 1/\sqrt{\alpha}$ is considered (we also remember that in this scheme there is a cut-off on the momentum, i.e. $|p| \leq 1/\sqrt{\alpha}$). This way, considering the minus relation (11) at the first order in α, a generalized uncertainty principle for LQC can be proposed to be $\Delta x \sim |1/\Delta p - l_L^2 \Delta p|$, where $l_L = \sqrt{\alpha/2}$ can be regarded as the loop cut-off length scale.

Conclusions

The equations of motion of the FRW models obtained in LQC and in the braneworlds scenario can be reproduced by a deformed Heisenberg algebra. This algebra in the unique one which is consistent, in the sense of the Jacobi identities, with the assumptions that both the translation and rotation groups are undeformed and that the commutator between the non-commutative coordinates is as in (1). Notably, it is also related to the κ-Poincaré algebra and the only freedom in (2) lies in the $\pm$ sign. The $(+)$-framework leads to the effective Friedmann dynamics of Randall-Sundrum braneworlds scenario, while the opposite one to that of LQC. From this perspective, the former framework is such that a vanishing uncertainty in position (the scale factor of the Universe) is not longer allowed. On the other hand, the $(-)$-scheme implies that the zero uncertainty in position appears for a finite uncertainty in the momentum, proportional to the natural cut-off of the framework. Summarizing, a non-commutative (deformed) picture which leads, at phenomenological level, to the prediction of more general theories can be formulated. The validity and applicability of this model to more complicated (and physically interesting) arenas will deserve future investigations.

References

[1] M.R.Douglas and N.A.Nekrasov, Rev.Mod.Phys. **73** (2001) 977.
[2] S.Doplicher, K.Fredenhagen and J.E.Roberts, Phys.Lett.B **331** (1994) 39.
[3] G.Amelino-Camelia, Int.J.Mod.Phys.D **11** (2002) 35; Phys.Lett.B **510** (2001) 255; J.Magueijo and L.Smolin, Phys.Rev.Lett. **88** (2002) 190403.
[4] C.Rovelli, arXiv:0808.3505; L.Smolin, arXiv:0808.3765.
[5] M.Bojowald, Living Rev.Rel. **8** (2005) 11.
[6] A.Ashtekar, T.Pawlowski and P.Singh, Phys.Rev.Lett. **96** (2006) 141301; Phys.Rev.D **73** (2006) 124038.
[7] M.V.Battisti and G.Montani, Phys.Lett.B **656** (2007) 96; Phys.Rev.D **77** (2008) 023518; B.Vakili and H.R.Sepangi, Phys.Lett.B **651** (2007) 79; A.Bina, K.Atazadeh and S.Jalalzadeh, Int.J.Theor.Phys. **47** (2008) 1354; H.Garcia-Compean, O.Obregon and C.Ramirez, Phys.Rev.Lett. **88** (2002) 161301.
[8] M.V.Battisti, arXiv:0805.1178.
[9] P.Singh, Phys.Rev.D **73** (2006) 063508; P.Singh, K.Vandersloot and G.V.Vereshchagin Phys.Rev.D **74** (2006) 043510.
[10] R.Maartens, Living Rev.Rel. **7** (2004) 7.
[11] H.S.Snyder, Phys.Rev. **71** (1947) 38.
[12] S.Meljanac, M.Milekovic and S.Pallua, Phys.Lett.B **328** (1994) 55; L.Jonke and S.Meljanac, Phys.Lett.B **526** (2002) 149.
[13] M.Maggiore, Phys.Lett.B **304** (1993) 65; Phys.Rev.D **49** (1994) 5182.
[14] J.Kowalski-Glikman, Phys.Lett.B **547** (2002) 291; J.Kowalski-Glikman and S.Nowak, Class.Quant.Grav. **20** (2003) 4799.
[15] E.W.Kolb and M.S.Turner, *The Early Universe* (Adison-Wesley Reading, 1990).
[16] L.J.Garay, Int.J.Mod.Phys.A **10** (1995) 145.
[17] V.Taveras, arXiv:0807.3325.
[18] A.Kempf, G.Mangano and R.B.Mann, Phys.Rev.D **52** (1995) 1108,; A.Kempf, J.Math.Phys. **38** (1997) 1347.
[19] D.J.Gross and P.F.Mendle, Nucl.Phys.B **303** (1988) 407; K.Konishi, G.Paffuti and P.Provero, Phys.Lett.B **234** (1990) 276.

Modelling non-Gaussianity from foreground contaminants

C.S. Carvalho[1,2,3,]

[1]Astrophysics and Cosmology Research Unit, School of Mathematical Sciences, University of KwaZulu-Natal, Durban, 4041, South Africa

[2]Laboratoire de Physique Théorique, Bâtiment 210, Université Paris-Sud, 91405 Orsay Cedex, France

[3]Instituto Superior Técnico, Departamento de Física, Avenida Rovisco Pais, 1049-001 Lisbon, Portugal

E-mail: Carla-Sofia.Carvalho@th.u-psud.fr

Abstract. We introduce a general class of models for characterizing the non-Gaussian properties of foreground contaminants in the cosmic microwave background with view towards the removal of the non-primordial non-Gaussian signal from the primordial one. This is important not only for treating temperature maps but also for characterizing the nature and origin of the primordial cosmological perturbations and thus establishing a theory of the early universe.

1. Introduction

Cosmology is presently a very active field because of the large number of observations that are becoming available and that will allow us to characterize with great precision the nature and physical origin of the primordial cosmological perturbations. A key question is whether these primordial perturbations were Gaussian.

The currently favoured best-fit cosmological models, as supported by the recent Wilkinson Microwave Anisotropy Probe (WMAP) data, are in agreement with the primordial cosmological perturbations being Gaussian to great accuracy [1]. With the WMAP data there has been a claimed detection of primordial non-Gaussianity that is currently disputed [2]. However, even the conservative official analysis hints at something, the value of f_{NL} lying within 1.9σ from the null value.

Either a detection or a more stringent constraint of f_{NL} would have profound implications for our understanding of the physics of the early Universe. In particular, it would be an observational discriminant among competing models for the generation of primordial fluctuations. An $f_{NL} \neq 0$ would be a very interesting challenge for cosmology since it would require an extensive revision of the standard cosmological model. Present observations have ruled out a great number of cosmological models already. All currently favoured models would be ruled out by such a detection. A non-detection of f_{NL} with an improved constraint on the allowed values would strongly restrict the fluctuation generation mechanisms.

[1] Present address

Due to the great improvement in sensitivity of the European Space Agency's *Planck* satellite over WMAP, understanding the nature and in particular non-Gaussian properties of foregrounds will be indispensable for the *Planck* f_{NL} analysis. For WMAP, foregrounds were not an issue because no signal was detected. Currently estimators exist [3, 4] that are proven optimal in the absence of foreground contamination and much work has been devoted to their efficient implementation [5]. However, to date there has been no comprehensive study of the impact of foregrounds. Foreground sources distributed in an anisotropic way are susceptible of being misidentified as primordial non-Gaussianity. In order to extract from the data an accurate and significant measure of f_{NL}, new estimators are needed that divide the data and are designed to mask possible contaminants. First, however, it is required to characterize the statistical properties of these spurious sources of non-Gaussianity, which requires understanding the non-linear properties of foregrounds.

This contribution reports on work in progress where we formulate a general class of models for characterizing the non-Gaussian properties of foreground contaminants in the cosmic microwave background (CMB). [6].

2. Measuring non-Gaussianity

In linear perturbation theory, the temperature anisotropy of the CMB is derived from metric perturbations by the relation $\Delta T/T \propto g_T \Delta\Phi$, where the radiation transfer function g_T depends on the scale considered.[2] Linear perturbation theory is a valid approximation because the physics of the perturbations generated during inflation is extremely weakly coupled.

In the simplest models of inflation, metric perturbations are produced by quantum fluctuations of the inflaton field. If the physics that governs the evolution of the metric perturbations is linear, since quantum fluctuations are Gaussian then so will temperature fluctuations be Gaussian. However, non-linear coupling between long and short wavelength fluctuations of the inflaton can generate weak non-Gaussianities. These will propagate into the metric perturbations and consequently into the temperature anisotropy. The primordial metric perturbations Φ can be characterized approximately by a non-linear coupling parameter, f_{NL}, such that [7]

$$\Phi(\mathbf{x}) = \Phi_L(\mathbf{x}) + f_{NL}\left[\Phi_L^2(\mathbf{x}) - \langle\Phi_L^2(\mathbf{x})\rangle\right], \tag{1}$$

where $\Phi_L(\mathbf{x})$ are Gaussian linear perturbations with $\langle\Phi_L(\mathbf{x})\rangle = 0$. In many models (e.g. single-field slow roll inflation [8, 9]), the non-linear corrections are too small to be detected. Other models (e.g. multiple field inflation [10, 11]) can generate stronger and potentially detectable non-linearity. A Gaussian distribution is characterized by vanishing odd-order moments and even-order moments defined only by the variance. Non-Gaussianity can thus be measured by correlations among an odd number of points. Here we shall use the angular three-point correlation function or bispectrum as our statistical tool since it is sensitive to weakly non-Gaussian fluctuations.

Expanding the temperature anisotropy in momentum space

$$\frac{\Delta T(\Omega)}{T} = \sum_{l,m} a_{lm} Y_{lm}(\Omega), \tag{2}$$

where the harmonic coefficients are

$$a_{lm} = 4\pi(-i)^l \int \frac{d^3k}{(2\pi)^2}\Phi(\mathbf{k}) g_{T_l}(\mathbf{k}) Y^*_{lm}(\Omega), \tag{3}$$

[2] For temperature fluctuations on super-horizon scales the Sachs–Wolfe effect dominates and g_T is a function of the parameter ω of the equation of state. At the decoupling epoch it reduces to $g_T = -1/3$ in the case of adiabatic fluctuations. On sub-horizon scales, however, g_T oscillates and we need to solve the Boltzmann photon transport equations coupled to the Einstein equations for g_T, which is found to have as parameter the decoupling time.

the bispectrum is defined as

$$B^{m_1m_2m_3}_{l_1l_2l_3} \equiv \langle a_{l_1m_1}a_{l_2m_2}a_{l_3m_3}\rangle = G^{m_1m_2m_3}_{l_1l_2l_3} b_{l_1l_2l_3}. \tag{4}$$

Here

$$G^{m_1m_2m_3}_{l_1l_2l_3} = \int d\Omega Y_{l_1m_1}(\Omega)Y_{l_2m_2}(\Omega)Y_{l_3m_3}(\Omega) \tag{5}$$

is the Gaunt integral and

$$\begin{aligned} b_{l_1l_2l_3} &= \left(\frac{2}{\pi}\right)^3 \int dx\; dk_1\; dk_2\; dk_3\; (xk_1k_2k_3)^2 B_\Phi(k_1,k_2,k_3) \\ &\times j_{l_1}(k_1x)j_{l_2}(k_2x)j_{l_3}(k_3x)\; g_{T_{l_1}}(k_1)g_{T_{l_2}}(k_2)g_{T_{l_3}}(k_3) \end{aligned} \tag{6}$$

is the reduced bispectrum defined in terms of the radiation transfer functions g_{T_l}, the Bessel functions of fractional order j_l and the primordial bispectrum $B_\Phi(k_1,k_2,k_3)$. In the flat-sky approximation, which amounts to $(l,m)\to \boldsymbol{\ell}$, where $\boldsymbol{\ell}$ is a two-dimensional wave vector in the sky, the bispectrum reduces to

$$\langle a(\boldsymbol{\ell}_1)a(\boldsymbol{\ell}_2)a(\boldsymbol{\ell}_3)\rangle = (2\pi)^2\delta^2(\boldsymbol{\ell}_1+\boldsymbol{\ell}_2+\boldsymbol{\ell}_3)B(\boldsymbol{\ell}_1,\boldsymbol{\ell}_2,\boldsymbol{\ell}_3) \tag{7}$$

which is non-vanishing only for closed triangle configurations in momentum space. Since in this approximation $G^{m_1m_2m_3}_{l_1l_2l_3} \to (2\pi)^2\delta^2(\boldsymbol{\ell}_1+\boldsymbol{\ell}_2+\boldsymbol{\ell}_3)$, then $b_{l_1l_2l_3}\to B(\boldsymbol{\ell}_1,\boldsymbol{\ell}_2,\boldsymbol{\ell}_3)$ [12].

The signal-to-noise squared for the bispectrum gives us an indication of the dominant triangle configuration to the non-Gaussian signal. We assume a scale invariant power spectrum $P(\ell) \propto \ell^{-2}$ in the flat sky approximation and $B(\ell_1,\ell_2,\ell_3) = f_{NL}\left[P(\ell_1)P(\ell_2)+P(\ell_2)P(\ell_3)+P(\ell_3)P(\ell_1)\right]$ as a result of the local non-Gaussianity ansatz. Then

$$\begin{aligned} \left(\frac{S}{N}\right)^2 &\propto \int d^2\ell_1 \int d^2\ell_2 \int d^2\ell_3 \delta^2(\boldsymbol{\ell}_1+\boldsymbol{\ell}_2+\boldsymbol{\ell}_3)\frac{B^2(\ell_1,\ell_2,\ell_3)}{P(\ell_1)P(\ell_2)P(\ell_3)} \\ &\propto f_{NL}^2 \int d^2\ell_1 \int d^2\ell_2 \int d^2\ell_3 \delta^2(\boldsymbol{\ell}_1+\boldsymbol{\ell}_2+\boldsymbol{\ell}_3)\frac{\left(\frac{1}{\ell_1^2\ell_2^2}+\frac{1}{\ell_2^2\ell_3^2}+\frac{1}{\ell_3^1\ell_1^2}\right)^2}{\frac{1}{\ell_1^2\ell_2^2\ell_3^2}} \\ &\propto f_{NL}^2 \int \frac{d\ell_<}{\ell_<}\int d^2\ell_>. \end{aligned} \tag{8}$$

For squeezed triangles, where the length of the two large sides is of order ℓ_{max} and that of the smaller side ℓ_{min}, we find that

$$\left(\frac{S}{N}\right)^2 \propto f_{NL}^2 \int \frac{d\ell_<}{\ell_<}\int d^2\ell_> \propto f_{NL}^2 \ln\left[\frac{\ell_{max}}{\ell_{min}}\right]\ell_{max}^2. \tag{9}$$

For equilateral triangles with the length of the sides of order ℓ_{max}

$$\left(\frac{S}{N}\right)^2 \propto f_{NL}^2\ell_{max}^2. \tag{10}$$

Comparing the two results we observe that the dominant contribution to the signal comes from the squeezed triangles, which introduce the logarithmic boost factor and thus couple the small-scale with the large-scale anisotropies. Should only equilateral triangles be considered, then the description of the signal would be equivalent to the separate description from different patches in the sky, which would mean losing most of the signal. The squeezed triangle configurations correspond to the local case where the non-Gaussianity was created primarily on super-horizon scales, as opposed to the equilateral case where the non-Gaussianity is primarily created at horizon-crossing. Qualitatively an $f_{NL} \neq 0$ can be understood as a modulation of the small-scale power (on scales near the resolution of the map) by the large-scale signal. Exploring further this idea, we can construct a cubic estimator of f_{NL} by filtering the signal into two contributions, namely $T_{\mathrm{low}\ \ell}$ and $T_{\mathrm{high}\ \ell}$, and considering the integral

$$E = \int d\Omega\ T_{\mathrm{low}\ \ell} T^2_{\mathrm{high}\ \ell}. \tag{11}$$

This is but a caricature of the optimal estimator given in Ref. [13] with the ideal weighting, which here serves illustration purposes. The large-scale variation in $T^2_{\mathrm{high}\ \ell}$ reflects spatial dependence in the power spectrum $P(\ell, \Omega)$ where ℓ is large and the variation in Ω is slow. In a Gaussian model, such perceived spatial dependence is simply noise from the cosmic variance and thus fortuitous. In a model with $f_{NL} \neq 0$, however, it is correlated with the large-scale temperature anisotropy.

Astronomical microwave sources as well as instrumental effects can produce spurious non-Gaussian signals which would be measured together with the primordial signal by the bispectrum and thus susceptible of being confused for primordial non-Gaussianity. Examples of such sources are [14]: the synchrotron emission, which results from the acceleration of cosmic ray electrons in magnetic fields; the free-free emission, which is produced by electron-ion scattering and is thus correlated with Hydrogen emission lines; and thermal galactic dust emission.

The separation of the CMB and foreground signal components has relied on their differing spectral and spacial distribution. However, correlations at different scales of the power of non-primordial foreground signals could not be separated from the CMB by the canonical methods. Here we aim to show that a detectable non-Gaussian signal could be mimicked by such correlations.

3. Faking non-Gaussianity

Since the a_{lm}'s are random quantities for the same sky realization, we must devise a way of generating maps of the sky with built-in non-linear foregrounds. We use motifs of a fixed shape and distributed in a probabilistic manner in the sky to model foreground sources. This resembles the galaxy counting method for the study of the large-scale structure formation [15].

Foreground realizations are generated in two steps. First a point $\boldsymbol{\xi}$ for the center of a motif is randomly selected according to a probability density function $Q(\boldsymbol{\xi})$ in the simplest case in an uncorrelated Poissonian manner. Then to each point $\boldsymbol{\theta}$ we assign a conditional probability $P(\boldsymbol{\theta}, \boldsymbol{\xi})$ of finding the point $\boldsymbol{\theta}$ for a motif centered at $\boldsymbol{\xi}$. This probability is independent of how many points are already inside the motif. We assign the value one to the motif interiors and zero to the regions between motifs. Assuming that the foregrounds are optically thin, i.e. that they do not mask the effects of each other, we take overlapping motifs to add. We also need to properly normalize these quantities so as to relate to the foreground maps. Measuring the mean number density of motifs, the probability density $Q(\boldsymbol{\xi})$ is such that when integrated over the entire sky it yields the average number of motifs $\langle N \rangle$

$$\int d^2\xi\ Q(\boldsymbol{\xi}) = \langle \text{number of motifs} \rangle\,. \tag{12}$$

The conditional probability $P(\boldsymbol{\theta}, \boldsymbol{\xi})$ yields the temperature at a point $\boldsymbol{\theta}$ given that a motif is centered at $\boldsymbol{\xi}$. Thus correlations among the ξ's as encoded in Q will model the large-scale structure, while correlations among the θ's in the same motif as encoded in P will model the small-scale structure.

We calculate the two and three-point correlation functions of the statistical map in a perturbative manner using Feynman diagrams. These correlation functions are interpreted respectively as the power spectrum and the bispectrum by assigning a temperature to the motifs. This temperature is constrained: from below by the minimum temperature that a foreground distribution must have to account for the remaining signal after removal of the astrophysical sources described previously; and from above by the maximum value of the power spectrum that a spurious source could have so as not to overshadow the background radiation.

We define the two-point correlation function of the final map $\mathcal{P}(\boldsymbol{\theta}_1, \boldsymbol{\theta}_2)$ by the sum over the two contributions to the probability that $\boldsymbol{\theta}_1$ and $\boldsymbol{\theta}_2$ are within a motif

$$\mathcal{P}(\boldsymbol{\theta}_1, \boldsymbol{\theta}_2) = \mathfrak{p}_{1,1}(\boldsymbol{\theta}_1, \boldsymbol{\theta}_2) + \mathfrak{p}_2(\boldsymbol{\theta}_1, \boldsymbol{\theta}_2), \tag{13}$$

where

$$\begin{aligned} \mathfrak{p}_{1,1} &= \frac{1}{2!}\int d^2\xi_1 \int d^2\xi_2 \; P(\boldsymbol{\theta}_1, \boldsymbol{\xi}_1)Q(\boldsymbol{\xi}_1)P(\boldsymbol{\theta}_2, \boldsymbol{\xi}_2)Q(\boldsymbol{\xi}_2) \\ \mathfrak{p}_2 &= \frac{1}{2!}\int d^2\xi_1 P(\boldsymbol{\theta}_1, \boldsymbol{\xi}_1)P(\boldsymbol{\theta}_2, \boldsymbol{\xi}_1)Q(\boldsymbol{\xi}_1). \end{aligned} \tag{14}$$

Here $\mathfrak{p}_{1,1}(\boldsymbol{\theta}_1, \boldsymbol{\theta}_2)$ is the probability that the points $\boldsymbol{\theta}_1$ and $\boldsymbol{\theta}_2$ are within different motifs centered at $\boldsymbol{\xi}_1$ and $\boldsymbol{\xi}_2$, and $\mathfrak{p}_2(\boldsymbol{\theta}_1, \boldsymbol{\theta}_2)$ is the probability that the points are within the same motif. Analogously we define the three-point correlation function $\mathcal{B}(\boldsymbol{\theta}_1, \boldsymbol{\theta}_2, \boldsymbol{\theta}_3)$ by the sum of the three contributions

$$\mathcal{B}(\boldsymbol{\theta}_1, \boldsymbol{\theta}_2, \boldsymbol{\theta}_3) = \mathfrak{p}_{1,1,1}(\boldsymbol{\theta}_1, \boldsymbol{\theta}_2, \boldsymbol{\theta}_3) + \mathfrak{p}_{2,1}(\boldsymbol{\theta}_1, \boldsymbol{\theta}_2, \boldsymbol{\theta}_3) + \mathfrak{p}_3(\boldsymbol{\theta}_1, \boldsymbol{\theta}_2, \boldsymbol{\theta}_3) \tag{15}$$

where

$$\begin{aligned} \mathfrak{p}_{1,1,1} &= \frac{1}{3!}\int d^2\xi_1 \int d^2\xi_2 \int d^2\xi_3 \; P(\boldsymbol{\theta}_1, \boldsymbol{\xi}_1)Q(\boldsymbol{\xi}_1)P(\boldsymbol{\theta}_2, \boldsymbol{\xi}_2)Q(\boldsymbol{\xi}_2)P(\boldsymbol{\theta}_3, \boldsymbol{\xi}_3)Q(\boldsymbol{\xi}_3) \\ \mathfrak{p}_{2,1} &= \frac{1}{2!}\frac{1}{2!}\int d^2\xi_1 \int d^2\xi_2 P(\boldsymbol{\theta}_1, \boldsymbol{\xi}_1)P(\boldsymbol{\theta}_2, \boldsymbol{\xi}_1)Q(\boldsymbol{\xi}_1)P(\boldsymbol{\theta}_3, \boldsymbol{\xi}_2)Q(\boldsymbol{\xi}_2) + (\text{two permutations}) \\ \mathfrak{p}_3 &= \frac{1}{3!}\int d^2\xi_1 P(\boldsymbol{\theta}_1, \boldsymbol{\xi}_1)P(\boldsymbol{\theta}_2, \boldsymbol{\xi}_1)P(\boldsymbol{\theta}_3, \boldsymbol{\xi}_1)Q(\boldsymbol{\xi}_1) \end{aligned} \tag{16}$$

are defined analogously. The combinatorial factors preceding the integrals are the number of equivalent motif configuration for indistinguishable patches in the $\boldsymbol{\xi}$–space which must be introduced to avoid multiple counting. We introduce a factor $1/n_Q!$ where n_Q is the number of Q's, and a factor $1/n_P!$ for each $\boldsymbol{\xi}$ where n_P is the number of P's shared by the same $\boldsymbol{\xi}$. For convenience we also define the one-point function

$$\mathfrak{p}_1(\boldsymbol{\theta}_1) \equiv \int d^2\xi_1 \; P(\boldsymbol{\theta}_1, \boldsymbol{\xi}_1)Q(\boldsymbol{\xi}_1). \tag{17}$$

4. A Simple Model: Uniform Distribution

For concreteness we assume the motifs to be circles of radius θ_{circ}, with sharp edges $P(\boldsymbol{\theta}, \boldsymbol{\xi}) = N_P \Theta[\theta_{circ} - |\boldsymbol{\xi} - \boldsymbol{\theta}|]$ and uniformly distributed in the sky $Q(\boldsymbol{\xi}) = const = N_Q$. Thus $N_Q = \langle N \rangle / A_{sky}$ and $N_P = T_{motif}$, which in this case are constant inside the motif and independent of the position of the motifs in the sky. [To be dimensionally correct, we must have T_{motif} normalized to the average temperature of the sky T_{sky}.] We will also discuss a non-uniform distribution of the motifs in the sky, motivated by the variation of the distribution of the large-scale structure along the latitude. This will be the working case for introducing dependence of the temperature on the position of the motifs as encoded in the distribution probability. [For a refinement of the functional form see Ref. [6].]

We compute the two and three-point correlator functions in real space for the statistical ensemble just described. The one-point function becomes

$$\mathfrak{p}_1(\boldsymbol{\theta}_1) = N_P N_Q \int d^2x_1 \, \Theta[\theta_{circ} - |\boldsymbol{x}_1|] = N_P N_Q A_{motif} \tag{18}$$

where $A_{motif} = \pi\theta^2_{circ}$. For the power spectrum we must calculate $\mathfrak{p}_{1,1}$ and $\mathfrak{p}_2$, which can be written as

$$\mathfrak{p}_{1,1}(\boldsymbol{\theta}_1, \boldsymbol{\theta}_2) = \frac{1}{2!} N_P^2 N_Q^2 \int d^2x_1 \, \Theta[\theta_{circ} - |\boldsymbol{x}_1|] \int d^2x_2 \, \Theta[\theta_{circ} - |\boldsymbol{x}_2|] = \frac{1}{2!}\mathfrak{p}_1^2 \tag{19}$$

$$\mathfrak{p}_2(\boldsymbol{\theta}_1, \boldsymbol{\theta}_2) = \frac{1}{2!} N_P^2 N_Q \int d^2x_1 \, \Theta[\theta_{circ} - |\boldsymbol{x}_1|]\Theta[\theta_{circ} - |\boldsymbol{x}_1 + \boldsymbol{\theta}_{12}|]. \tag{20}$$

Here $\boldsymbol{x}_i = \boldsymbol{\xi}_i - \boldsymbol{\theta}_i$ and $\boldsymbol{\theta}_{ij} = \boldsymbol{\theta}_i - \boldsymbol{\theta}_j$. For the bispectrum we observe that both $\mathfrak{p}_{1,1,1}$ and $\mathfrak{p}_{2,1}$ can be expressed in terms of $\mathfrak{p}_1$ and $\mathfrak{p}_2$ as follows

$$\begin{aligned}
\mathfrak{p}_{1,1,1}(\boldsymbol{\theta}_1, \boldsymbol{\theta}_2, \boldsymbol{\theta}_3) &= \frac{1}{3!} N_P^3 N_Q^3 \int d^2x_1 \, \Theta[\theta_{circ} - |\boldsymbol{x}_1|] \int d^2x_2 \, \Theta[\theta_{circ} - |\boldsymbol{x}_2|] \\
&\quad \times \int d^2x_3 \, \Theta[\theta_{circ} - |\boldsymbol{x}_3|] \\
&= \frac{1}{3!}\mathfrak{p}_1^3 \qquad (21) \\
\mathfrak{p}_{2,1}(\boldsymbol{\theta}_1, \boldsymbol{\theta}_2, \boldsymbol{\theta}_3) &= \frac{1}{2!}\frac{1}{2!} N_P^3 N_Q^2 \int d^2x_1 \, \Theta[\theta_{circ} - |\boldsymbol{x}_1|] \, \Theta[\theta_{circ} - |\boldsymbol{x}_1 + \boldsymbol{\theta}_{12}|] \\
&\quad \times \int d^2x_3 \, \Theta[\theta_{circ} - |\boldsymbol{x}_3|] + \text{(two permutations)} \\
&= \frac{1}{2!}\mathfrak{p}_1 \left[\mathfrak{p}_2(\boldsymbol{\theta}_1, \boldsymbol{\theta}_2) + \mathfrak{p}_2(\boldsymbol{\theta}_2, \boldsymbol{\theta}_3) + \mathfrak{p}_2(\boldsymbol{\theta}_3, \boldsymbol{\theta}_1)\right], \qquad (22)
\end{aligned}$$

so we have in addition to compute $\mathfrak{p}_3$ only

$$\mathfrak{p}_3(\boldsymbol{\theta}_1, \boldsymbol{\theta}_2, \boldsymbol{\theta}_3) = \frac{1}{3!} N_P^3 N_Q \int d^2x_1 \Theta[\theta_{circ} - |\boldsymbol{x}_1|] \, \Theta[\theta_{circ} - |\boldsymbol{x}_1 + \boldsymbol{\theta}_{12}|] \, \Theta[\theta_{circ} - |\boldsymbol{x}_1 + \boldsymbol{\theta}_{13}|]. \tag{23}$$

These quantities describe the probability that one, two or three points, denoted by $\boldsymbol{\theta}_1$, $\boldsymbol{\theta}_2$ and $\boldsymbol{\theta}_3$, are within the same motif centered at points $\boldsymbol{\xi}$'s which are distributed in the sky according to Q. These probabilities can also be interpreted as the measure of the overlap of motifs centered at each of these points $\boldsymbol{\theta}$'s. The integrals of the step functions about the distance from the centers at $\boldsymbol{\xi}$'s can thus be visualized as the area of the intersection of motifs centered at the $\boldsymbol{\theta}$'s. [For more details see Ref. [6].] We find that

$$\mathfrak{p}_{1,1}(\boldsymbol{\theta}_1, \boldsymbol{\theta}_2) = \frac{1}{2!}\mathfrak{p}_1^2 \tag{24}$$

$$\mathfrak{p}_2(\boldsymbol{\theta}_1, \boldsymbol{\theta}_2) = \frac{1}{2!} N_P^2 N_Q A_{1\cap 2} \tag{25}$$

where $A_{1\cap 2}$ is the area of intersection between the motifs centered at $\boldsymbol{\theta}_1$ and $\boldsymbol{\theta}_2$

$$A_{1\cap 2} = \left(2\theta^2_{circ} \arctan\left[\sqrt{\frac{4\theta^2_{circ}}{|\boldsymbol{\theta}_{12}|^2} - 1}\right] - \frac{|\boldsymbol{\theta}_{12}|^2}{2}\sqrt{\frac{4\theta^2_{circ}}{|\boldsymbol{\theta}_{12}|^2} - 1} \right) \Theta\left[2\theta_{circ} - |\boldsymbol{\theta}_{12}|\right]. \tag{26}$$

These contribute to the power spectrum. Moreover, we find that

$$\mathfrak{p}_{1,1,1}(\boldsymbol{\theta}_1,\boldsymbol{\theta}_2,\boldsymbol{\theta}_3) = \frac{1}{3!}\mathfrak{p}_1^3 \tag{27}$$

$$\mathfrak{p}_{2,1}(\boldsymbol{\theta}_1,\boldsymbol{\theta}_2,\boldsymbol{\theta}_3) = \frac{1}{2!}\mathfrak{p}_1\left[\mathfrak{p}_2(\boldsymbol{\theta}_1,\boldsymbol{\theta}_2)+\mathfrak{p}_2(\boldsymbol{\theta}_2,\boldsymbol{\theta}_3)+\mathfrak{p}_2(\boldsymbol{\theta}_3,\boldsymbol{\theta}_1)\right] \tag{28}$$

$$\mathfrak{p}_3(\boldsymbol{\theta}_1,\boldsymbol{\theta}_2,\boldsymbol{\theta}_3) = \frac{1}{3!}N_P^3 N_Q A_{1\cap 2\cap 3} \tag{29}$$

where $A_{1\cap 2\cap 3}$ is the area of intersection among the three motifs

$$\begin{aligned} A_{1\cap 2\cap 3} = & \left[\frac{1}{4}\sqrt{\left(|\boldsymbol{\theta}_{12}|^2+|\boldsymbol{\theta}_{23}|^2+|\boldsymbol{\theta}_{13}|^2\right)^2-2\left(|\boldsymbol{\theta}_{12}|^4+|\boldsymbol{\theta}_{23}|^4+|\boldsymbol{\theta}_{13}|^4\right)}-\frac{1}{2}A_{motif}\right] \\ & \times\Theta\left[2\theta_{circ}-|\boldsymbol{\theta}_{12}|\right]\Theta\left[2\theta_{circ}-|\boldsymbol{\theta}_{23}|\right]\Theta\left[2\theta_{circ}-|\boldsymbol{\theta}_{13}|\right] \\ + & \frac{1}{2}A_{1\cap 2}\Theta\left[2\theta_{circ}-|\boldsymbol{\theta}_{23}|\right]\Theta\left[2\theta_{circ}-|\boldsymbol{\theta}_{13}|\right] \\ + & \frac{1}{2}A_{2\cap 3}\Theta\left[2\theta_{circ}-|\boldsymbol{\theta}_{12}|\right]\Theta\left[2\theta_{circ}-|\boldsymbol{\theta}_{13}|\right] \\ + & \frac{1}{2}A_{1\cap 3}\Theta\left[2\theta_{circ}-|\boldsymbol{\theta}_{12}|\right]\Theta\left[2\theta_{circ}-|\boldsymbol{\theta}_{23}|\right]. \end{aligned} \tag{30}$$

These contribute to the bispectrum. It follows that

$$\mathcal{P}(\boldsymbol{\theta}_1,\boldsymbol{\theta}_2) = \frac{1}{2!}\mathfrak{p}_1^2+\mathfrak{p}_2(\boldsymbol{\theta}_1,\boldsymbol{\theta}_2) \tag{31}$$

$$\mathcal{B}(\boldsymbol{\theta}_1,\boldsymbol{\theta}_2,\boldsymbol{\theta}_3) = \frac{1}{3!}\mathfrak{p}_1^3+\frac{1}{2!}\mathfrak{p}_1\left[\mathfrak{p}_2(\boldsymbol{\theta}_1,\boldsymbol{\theta}_2)+\mathfrak{p}_2(\boldsymbol{\theta}_2,\boldsymbol{\theta}_3)+\mathfrak{p}_2(\boldsymbol{\theta}_3,\boldsymbol{\theta}_1)\right]+\mathfrak{p}_3(\boldsymbol{\theta}_1,\boldsymbol{\theta}_2,\boldsymbol{\theta}_3). \tag{32}$$

We now proceed to compute the value of f_{NL} produced. The parameter f_{NL} characterizes the amplitude of the temperature non-Gaussianity since it couples to the quadratic term of the expansion of the temperature fluctuation $\mathcal{T}\equiv\Delta T/T$ about a Gaussian distribution $\mathcal{T}_L$. Assuming the Sachs–Wolfe approximation $(\Delta T)/T=-(1/3)\Delta\Phi/c^2$ on all scales and for an infinitely thin surface of last scattering, we have in real space that

$$\mathcal{T}(\boldsymbol{\theta})=\mathcal{T}_L(\boldsymbol{\theta})+3f_{NL}\left[\mathcal{T}_L^2(\boldsymbol{\theta})-\langle\mathcal{T}_L(\boldsymbol{\theta})\rangle^2\right]. \tag{33}$$

Here the Gaussian distribution has zero mean, $\langle\mathcal{T}_L(\boldsymbol{\theta})\rangle=0$, from which it follows to leading order that

$$\mathcal{P}(\boldsymbol{\theta}_1,\boldsymbol{\theta}_2)\equiv\langle\mathcal{T}(\boldsymbol{\theta}_1)\mathcal{T}(\boldsymbol{\theta}_2)\rangle=\langle\mathcal{T}_L(\boldsymbol{\theta}_1)\mathcal{T}_L(\boldsymbol{\theta}_2)\rangle. \tag{34}$$

Since a Gaussian distribution has vanishing odd-order momenta, we find for the three-point correlation function that

$$\begin{aligned} & \mathcal{B}(\boldsymbol{\theta}_1,\boldsymbol{\theta}_2,\boldsymbol{\theta}_3)\equiv\langle\mathcal{T}(\boldsymbol{\theta}_1)\mathcal{T}(\boldsymbol{\theta}_2)\mathcal{T}(\boldsymbol{\theta}_3)\rangle \\ = & \ 6f_{NL}\left[\mathcal{P}(\boldsymbol{\theta}_1,\boldsymbol{\theta}_3)\mathcal{P}(\boldsymbol{\theta}_3,\boldsymbol{\theta}_2)+\mathcal{P}(\boldsymbol{\theta}_2,\boldsymbol{\theta}_1)\mathcal{P}(\boldsymbol{\theta}_1,\boldsymbol{\theta}_3)+\mathcal{P}(\boldsymbol{\theta}_3,\boldsymbol{\theta}_2)\mathcal{P}(\boldsymbol{\theta}_2,\boldsymbol{\theta}_1)\right]. \end{aligned} \tag{35}$$

We distinguish three cases for the three possible relations among the distances between the points $\theta's$, namely 1) $\theta_{12},\theta_{13},\theta_{23}>2\theta_{circ}$; 2) $\theta_{12}<2\theta_{circ}$ and $\theta_{13},\theta_{23}>2\theta_{circ}$; 3) $\theta_{12},\theta_{13},\theta_{23}<2\theta_{circ}$. Let us fix the size of the motif and for concreteness take the radius to be of the size of the resolution of the map, i.e. $\theta_{circ}\sim 10^{-6}$.

4.1. Case $\theta_{12}, \theta_{13}, \theta_{23} > 2\theta_{circ}$

This is the case where each point $\boldsymbol{\theta}_i$ is within a different motif and consequently only the one-point function $\mathfrak{p}_1$ and the contributions to the two and three-point functions which can be expressed in terms of $\mathfrak{p}_1$, i.e. $\mathfrak{p}_{1,1}$ and $\mathfrak{p}_{1,1,1}$, do not vanish. We find that

$$\begin{aligned} \mathcal{P}(\boldsymbol{\theta}_1, \boldsymbol{\theta}_2) &= \mathcal{P}(\boldsymbol{\theta}_1, \boldsymbol{\theta}_3) = \mathcal{P}(\boldsymbol{\theta}_2, \boldsymbol{\theta}_3) = \mathfrak{p}_{1,1} = \frac{1}{2}\mathfrak{p}_1^2, \\ \mathcal{B}(\boldsymbol{\theta}_1, \boldsymbol{\theta}_2, \boldsymbol{\theta}_3) &= \mathfrak{p}_{1,1,1} = \frac{1}{6}\mathfrak{p}_1^3, \end{aligned} \tag{36}$$

from which it follows that

$$f_{NL} = \frac{1}{6}\frac{\mathcal{B}(\boldsymbol{\theta}_1, \boldsymbol{\theta}_2, \boldsymbol{\theta}_3)}{3\mathcal{P}(\boldsymbol{\theta}_1, \boldsymbol{\theta}_2)\mathcal{P}(\boldsymbol{\theta}_2, \boldsymbol{\theta}_3)} = \frac{1}{27}\frac{1}{\mathfrak{p}_1}, \quad \mathfrak{p}_1 > 1. \tag{37}$$

4.2. Case $\theta_{12} < 2\theta_{circ}$ and $\theta_{13}, \theta_{23} > 2\theta_{circ}$

This is the case where two points, here for concreteness $\boldsymbol{\theta}_1$ and $\boldsymbol{\theta}_2$, are within the same motif. In addition to the one-point function and the contributions derived from it, we have $\mathfrak{p}_2(\boldsymbol{\theta}_1, \boldsymbol{\theta}_2) \neq 0$ and $\mathfrak{p}_{2,1}(\boldsymbol{\theta}_1, \boldsymbol{\theta}_2, \boldsymbol{\theta}_3) \neq 0$ from the contribution to the three-point function of the two-point function between $\boldsymbol{\theta}_1$ and $\boldsymbol{\theta}_2$ only. Hence

$$\begin{aligned} \mathcal{P}(\boldsymbol{\theta}_1, \boldsymbol{\theta}_2) &= \mathfrak{p}_{1,1} + \mathfrak{p}_2(\boldsymbol{\theta}_1, \boldsymbol{\theta}_2) = \frac{1}{2}\mathfrak{p}_1^2 + \mathfrak{p}_2, \quad \mathcal{P}(\boldsymbol{\theta}_1, \boldsymbol{\theta}_3) = \mathcal{P}(\boldsymbol{\theta}_2, \boldsymbol{\theta}_3) = \mathfrak{p}_{1,1} = \frac{1}{2}\mathfrak{p}_1^2, \\ \mathcal{B}(\boldsymbol{\theta}_1, \boldsymbol{\theta}_2, \boldsymbol{\theta}_3) &= \mathfrak{p}_{1,1,1} + \mathfrak{p}_{2,1}(\boldsymbol{\theta}_1, \boldsymbol{\theta}_2, \boldsymbol{\theta}_3) = \frac{1}{6}\mathfrak{p}_1^3 + \frac{1}{2}\mathfrak{p}_1\mathfrak{p}_2(\boldsymbol{\theta}_1, \boldsymbol{\theta}_2), \end{aligned} \tag{38}$$

and

$$f_{NL} = \frac{1}{6}\frac{\mathcal{B}(\boldsymbol{\theta}_1, \boldsymbol{\theta}_2, \boldsymbol{\theta}_3)}{2\mathcal{P}(\boldsymbol{\theta}_1, \boldsymbol{\theta}_2)\mathcal{P}(\boldsymbol{\theta}_2, \boldsymbol{\theta}_3) + \mathcal{P}(\boldsymbol{\theta}_1, \boldsymbol{\theta}_3)\mathcal{P}(\boldsymbol{\theta}_2, \boldsymbol{\theta}_3)} = \frac{1}{6}\frac{\frac{1}{6}\mathfrak{p}_1^3 + \frac{1}{2}\mathfrak{p}_1\mathfrak{p}_2(\boldsymbol{\theta}_1, \boldsymbol{\theta}_2)}{2\left(\frac{1}{2}\mathfrak{p}_1^2 + \mathfrak{p}_2(\boldsymbol{\theta}_1, \boldsymbol{\theta}_2)\right)\frac{1}{2}\mathfrak{p}_1^2 + \frac{1}{4}\mathfrak{p}_1^4}. \tag{39}$$

We note that $\mathfrak{p}_1^2 \propto N_Q^2 A_{motif}^2$ and $\mathfrak{p}_2 \propto N_Q A_{1\cap 2} < N_Q A_{motif}$, where N_Q is the mean number density of motifs in the sky. The relative magnitude of $\mathfrak{p}_1^2$ and $\mathfrak{p}_2$ depends on the relation between the mean number of motifs $\langle N \rangle$ and the area of the motif A_{motif}. For convenience we define $\alpha \equiv N_Q A_{motif} = \langle N \rangle A_{motif}/A_{sky}$. We must further distinguish between the following two cases. If $\alpha < \alpha^2$, i.e. $\langle N \rangle > A_{sky}/A_{motif} \sim 10^{13}$, then $\mathfrak{p}_2 < \alpha < \mathfrak{p}_1^2$. This suggests that a minimum density of motifs is required in order for the one-point function to dominate, i.e. in order for two points which are close enough to be within the same motif to be also within a second, and thus necessarily overlapping, motif. On the other hand if $\alpha > \alpha^2$, i.e. $\langle N \rangle < A_{sky}/A_{motif} \sim 10^{13}$, then both $\mathfrak{p}_1^2, \mathfrak{p}_2 < \alpha$. Here in order to discriminate the relative magnitude, we need in addition to precise the relation between $A_{1\cap 2}$ and A_{motif}, which depends on θ_{12}. If θ_{12} is sufficiently small so that $A_{1\cap 2} \sim A_{motif}$, then $\mathfrak{p}_2 \sim \alpha$ and consequently $\mathfrak{p}_2 > \mathfrak{p}_1^2$. However, if θ_{12} is very close to $2\theta_{circ}$, then $A_{1\cap 2} \ll A_{motif}$ and consequently $\mathfrak{p}_2 \ll \alpha$, so that if $\mathfrak{p}_2 < \alpha^2$ then $\mathfrak{p}_2 < \mathfrak{p}_1^2$. This is equivalent to the case $\alpha < \alpha^2$.

Hence, if $\langle N \rangle > 10^{13}$, or $\langle N \rangle < 10^{13}$ and $\theta_{12} \sim 2\theta_{circ}$, then

$$f_{NL} \sim \frac{1}{6}\frac{\frac{1}{6}\mathfrak{p}_1^3}{\frac{3}{4}\mathfrak{p}_1^4} = \frac{1}{27}\frac{1}{\mathfrak{p}_1}, \quad \mathfrak{p}_1 > 1, \tag{40}$$

whereas if $\langle N \rangle < 10^{13}$ and $\theta_{12} \ll 2\theta_{circ}$, then

$$f_{NL} \sim \frac{1}{6}\frac{\frac{1}{2}\mathfrak{p}_1\mathfrak{p}_2(\boldsymbol{\theta}_1, \boldsymbol{\theta}_2)}{\mathfrak{p}_2(\boldsymbol{\theta}_1, \boldsymbol{\theta}_2)\mathfrak{p}_1^2} = \frac{1}{12}\frac{1}{\mathfrak{p}_1}, \quad \mathfrak{p}_1 < 1. \tag{41}$$

4.3. Case $\theta_{12}, \theta_{13}, \theta_{23} < 2\theta_{circ}$

This is the case where the three points are within the same motif. All the terms contribute to the two and three-point functions. In order to determine which contribution dominates for a mean number of motifs $\langle N \rangle$ distributed in the sky according to Q, we do an analysis similar to that above. Thus if $\langle N \rangle > 10^{13}$

$$f_{NL} \sim \frac{1}{6}\frac{\frac{1}{6}\mathfrak{p}_1^3}{\frac{3}{4}\mathfrak{p}_1^4} = \frac{1}{27}\frac{1}{\mathfrak{p}_1}, \quad \mathfrak{p}_1 > 1. \tag{42}$$

However, if $\langle N \rangle < 10^{13}$ we find for $\theta_{12} < \theta_{13}, \theta_{23}$ that

$$f_{NL} \sim \frac{1}{6}\frac{\mathfrak{p}_3(\boldsymbol{\theta}_1, \boldsymbol{\theta}_2, \boldsymbol{\theta}_3)}{2\mathfrak{p}_2(\boldsymbol{\theta}_1, \boldsymbol{\theta}_2)\mathfrak{p}_2(\boldsymbol{\theta}_2, \boldsymbol{\theta}_3)} = \frac{1}{18}\frac{-\frac{1}{2}A_{motif} + \frac{3}{2}A_{1\cap 2}}{N_P N_Q A_{1\cap 2} A_{2\cap 3}}, \tag{43}$$

while for $\theta_{12} \sim \theta_{23} \sim \theta_{13}$ we find that

$$f_{NL} \sim \frac{1}{6}\frac{\mathfrak{p}_3(\boldsymbol{\theta}_1, \boldsymbol{\theta}_2, \boldsymbol{\theta}_3)}{3\mathfrak{p}_2(\boldsymbol{\theta}_1, \boldsymbol{\theta}_2)\mathfrak{p}_2(\boldsymbol{\theta}_2, \boldsymbol{\theta}_3)} = \frac{1}{27}\frac{\frac{\sqrt{3}}{4}\theta_{12}^2 - \frac{1}{2}A_{motif} + \frac{3}{2}A_{1\cap 2}}{N_P N_Q A_{1\cap 2}^2}. \tag{44}$$

For small θ_{12} the term $(\sqrt{3/4})\theta_{12}^2$ is subdominant, so for both cases we need only compare $A_{1\cap 2}$ with A_{motif}. We present the results for the former case, with those for the latter following straightforwardly. For $\theta < \theta_{eq}$, where θ_{eq} is the distance such that $(3/2)A_{1\cap 2} = (1/2)A_{motif}$, we find that $(3/2)A_{1\cap 2} > (1/2)A_{motif}$ and hence that

$$f_{NL} \sim \frac{1}{18}\frac{A_{1\cap 2}}{N_P N_Q A_{1\cap 2} A_{2\cap 3}} = \frac{1}{18}\frac{1}{N_P N_Q A_{2\cap 3}}. \tag{45}$$

For $\theta_{eq} < \theta_{12} < 2\theta_{circ}$ the contribution of $(\sqrt{3/4})\theta_{12}^2$ becomes more important but the dominant contribution is that of $(1/2)A_{motif}$. Moreover since $A_{1\cap 2}$ is a decreasing function of θ_{12} except for θ_{12} very close to $2\theta_{circ}$, then $A_{1\cap 2}A_{2\cap 3} > A_{2\cap 3}^2$ and consequently $-1/A_{1\cap 2}A_{2\cap 3} > -1/A_{2\cap 3}^2$. Thus

$$f_{NL} \gtrsim \frac{1}{18}\frac{-\frac{1}{2}A_{motif}}{N_P N_Q A_{1\cap 2} A_{2\cap 3}} > \frac{1}{18}\frac{-\frac{1}{2}A_{motif}}{N_P N_Q A_{2\cap 3}^2}. \tag{46}$$

A similar calculation would follow for the case $\theta_{12} \sim \theta_{23} \sim \theta_{13}$.

Since these distances must be such that $0 \leq \theta_{12}, \theta_{13}, \theta_{23} \leq 2\theta_{circ}$, then $-\infty \leq f_{NL} < \infty$, i.e. we can generate arbitrary non-Gaussianity. We need, however, to be able to constrain f_{NL} to a finite interval. This requires the use of the foreground maps to constrain the parameters for the case of the uniform model, as well as the functional form of the probability functions for more realistic models.

5. Constraining the Model

Constraints on the parameters of the model can be extracted from the properties of foreground maps. Spatial templates from the WMAP data produced for synchrotron, free-free and dust emission show a temperature distribution strongly dependent on the latitude. In order to account for this observation we must consider a non-uniform probability density that can capture the qualitative aspects of these dependence. The simplest case of a non-uniform distribution of motifs in the sky is that where $Q(\boldsymbol{\theta})$ is a slowly varying function across the scale of the motif, so that the probabilities are changed to

$$\mathfrak{p}_{1,1}(\boldsymbol{\theta}_1,\boldsymbol{\theta}_2) = \frac{1}{2!}N_P^2 A_{motif}^2 Q(\boldsymbol{\theta}_1)Q(\boldsymbol{\theta}_2) \quad (47)$$

$$\mathfrak{p}_2(\boldsymbol{\theta}_1,\boldsymbol{\theta}_2) = \frac{1}{2!}N_P^2 A_{1\cap 2} Q\left(\frac{\boldsymbol{\theta}_1+\boldsymbol{\theta}_2}{2}\right) \quad (48)$$

and to

$$\mathfrak{p}_{1,1,1}(\boldsymbol{\theta}_1,\boldsymbol{\theta}_2,\boldsymbol{\theta}_3) = \frac{1}{3!}N_P^3 A_{motif}^3 Q(\boldsymbol{\theta}_1)Q(\boldsymbol{\theta}_2)Q(\boldsymbol{\theta}_3) \quad (49)$$

$$\mathfrak{p}_{2,1}(\boldsymbol{\theta}_1,\boldsymbol{\theta}_2,\boldsymbol{\theta}_3) = \frac{1}{2!}N_P A_{motif} \times\Big[\mathfrak{p}_2(\boldsymbol{\theta}_1,\boldsymbol{\theta}_2)Q(\boldsymbol{\theta}_3)+\mathfrak{p}_2(\boldsymbol{\theta}_2,\boldsymbol{\theta}_3)Q(\boldsymbol{\theta}_1)+\mathfrak{p}_2(\boldsymbol{\theta}_3,\boldsymbol{\theta}_1)Q(\boldsymbol{\theta}_2)\Big] \quad (50)$$

$$\mathfrak{p}_3(\boldsymbol{\theta}_1,\boldsymbol{\theta}_2,\boldsymbol{\theta}_3) = \frac{1}{3!}N_P^3 A_{1\cap 2\cap 3} Q\left(\frac{\boldsymbol{\theta}_1+\boldsymbol{\theta}_2+\boldsymbol{\theta}_3}{3}\right). \quad (51)$$

The changes in the two and three-point correlation functions, as well as in f_{NL}, are straightforward. Assuming that the contribution of the overlapping motifs add in the intersection region (what we called the "optically thin" approximation), then the value of Q in the mid distance between the centers of two overlapping motifs is the sum of the value of Q at the center of each motif, i.e.

$$Q\left(\frac{\boldsymbol{\theta}_i+\boldsymbol{\theta}_j}{2}\right) = Q(\boldsymbol{\theta}_i)+Q(\boldsymbol{\theta}_j). \quad (52)$$

Since the scale on which $Q(\boldsymbol{\xi})$ varies is large compared to A_{motif}, then

$$Q(\boldsymbol{\xi}) = \frac{\langle T(\boldsymbol{\xi})\rangle}{T_{motif}}\frac{1}{A_{sky}} \quad (53)$$

so that the same normalization condition holds. Possible functional forms for $Q(\boldsymbol{\xi})$ will be explored in Ref. [6]. For the purpose of this section, however, we will not need to be more specific. The conditional probability becomes $P(\boldsymbol{\theta},\boldsymbol{\xi}) = T(\boldsymbol{\theta})\Theta\left[\theta_{circ} - |\boldsymbol{\xi}-\boldsymbol{\theta}|\right]$ which will be simply the average temperature at the point $\boldsymbol{\xi}$ where the motif on which $\boldsymbol{\theta}$ lies is centered.

By quantifying the clean regions of these maps, we can also set bounds on the foreground temperature. From a careful examination of the five-year data maps [5] we extracted the minimum temperature T_0 that a foreground source should have in order to saturate the map with as close to a single temperature value as possible. Taking the minimum value of the temperature of the three foregrounds for each of the five frequency bands analyzed, we found by eye inspection that saturation of the maps was achieved for T_0 a few μK. Hence the temperature at each point of the motif distribution must be such that $T(\boldsymbol{\xi}) > T_0$. On the other hand, the temperature of a single spot must be highly constrained so as not to stand out. This observation is captured by the condition that the power spectrum generated by the motif distribution does not outshine the background radiation. If the motif distribution produced a large power spectrum, as that produced by small and bright sources, then non-Gaussianity would be obvious. We thus set $\mathcal{P}(\boldsymbol{\theta}_i,\boldsymbol{\theta}_j) < (T_0/T_{sky})^2$.

The three cases discriminated in the previous section will now be combined with the constraints on the temperature obtained from the foreground maps. Apart from factors of order $O(1)$, the results will be the same for both a uniform and a non-uniform probability distribution of motifs in the sky. For simplicity we will analyze the results from the uniform distribution.

5.1. *Case* $\theta_{12}, \theta_{13}, \theta_{23} > 2\theta_{circ}$

Here for any two pair of motif centers i, j, the constraints on the temperature yield that

$$\mathcal{P}(\boldsymbol{\theta}_i, \boldsymbol{\theta}_j) = \frac{1}{2} N_P^2 N_Q^2 A_{motif}^2 > \frac{1}{2} \left(\frac{T_0}{T_{sky}} \right)^2 A_{motif}^2 N_Q^2 \tag{54}$$

and simultaneously that

$$\mathcal{P}(\boldsymbol{\theta}_i, \boldsymbol{\theta}_j) < \left(\frac{T_0}{T_{sky}} \right)^2 \tag{55}$$

from which it follows that

$$\frac{1}{2} \langle N \rangle^2 \left(\frac{A_{motif}}{A_{sky}} \right)^2 < 1. \tag{56}$$

This relation constrains the size of the motifs given their average number $\langle N \rangle$ distributed on the sky. Using the relations above in the expression for f_{NL} we find that

$$f_{NL} = \frac{1}{27} \frac{1}{N_P A_{motif} N_Q} < \frac{1}{27} \frac{1}{\langle N \rangle (A_{motif}/A_{sky})(T_0/T_{sky})}. \tag{57}$$

Given the constraint from the power spectrum, we find that the magnitude of f_{NL} is further determined by (T_0/T_{sky}) only. Here f_{NL} is thus a measure of the deviation of T_0 from the average sky temperature.

5.2. *Case* $\theta_{12} < 2\theta_{circ}$ *and* $\theta_{13}, \theta_{23} > 2\theta_{circ}$

Here for $\langle N \rangle > 10^{13}$, or $\langle N \rangle < 10^{13}$ and $\theta_{12} \sim 2\theta_{circ}$

$$\mathcal{P}(\boldsymbol{\theta}_1, \boldsymbol{\theta}_2) \sim \mathcal{P}(\boldsymbol{\theta}_1, \boldsymbol{\theta}_3) = \mathcal{P}(\boldsymbol{\theta}_2, \boldsymbol{\theta}_3) = \frac{1}{2} N_P^2 A_{motif}^2 N_Q^2 > \frac{1}{2} \left(\frac{T_0}{T_{sky}} \right)^2 A_{motif}^2 N_Q^2, \tag{58}$$

which reduces to the previous case. For $\langle N \rangle < 10^{13}$ and $\theta_{12} \ll \theta_{circ}$, however,

$$\mathcal{P}(\boldsymbol{\theta}_1, \boldsymbol{\theta}_2) \sim \frac{1}{2} N_P^2 A_{1\cap 2} N_Q > \left(\frac{T_0}{T_{sky}} \right)^2 A_{1\cap 2} N_Q \tag{59}$$

which together with Eqn. (55) yields

$$\frac{1}{2} \langle N \rangle \frac{A_{1\cap 2}}{A_{sky}} < 1 \tag{60}$$

This relation constrains the overlapping between any two motifs given the average number of motifs. Then for f_{NL} we find that

$$f_{NL} \sim \frac{1}{12} \frac{1}{N_P A_{motif} N_Q} < \frac{1}{12} \frac{1}{\langle N \rangle (A_{motif}/A_{sky})(T_0/T_{sky})}. \tag{61}$$

This relation sets a constraint on f_{NL} of the same order of magnitude as the previous case.

5.3. Case $\theta_{12}, \theta_{13}, \theta_{23} < 2\theta_{circ}$

Here for $\langle N \rangle > 10^{13}$ we find the same result as in the first case where the constraint on the power spectrum is given by Eqn. (56). For $\langle N \rangle < 10^{13}$ the constraint on the power spectrum is the same as in the second example of the second case and given by Eqn. (60). We further discriminate between two cases, namely $\theta_{12} < \theta_{13}, \theta_{23}$ and $\theta_{12} \sim \theta_{13} \sim \theta_{23}$, and each case for two regimes of the parameter θ_{12}. Thus for $\theta_{12} < \theta_{13}, \theta_{23}$ we find that

$$f_{NL} \sim \frac{1}{18} \frac{A_{1\cap 2\cap 3}}{N_P A_{1\cap 2} A_{2\cap 3} N_Q} < \frac{1}{18} \frac{-\frac{1}{2} A_{motif} + \frac{3}{2} A_{1\cap 2}}{\langle N \rangle A_{1\cap 2} (A_{2\cap 3}/A_{sky})(T_0/T_{sky})}, \tag{62}$$

which for $\theta_{12} < \theta_{eq}$ becomes

$$f_{NL} < \frac{1}{18} \frac{1}{\langle N \rangle (A_{2\cap 3}/A_{sky})(T_0/T_{sky})}, \tag{63}$$

whereas for $\theta_{eq} < \theta_{12} < 2\theta_{circ}$

$$f_{NL} \gtrsim \frac{1}{18} \frac{-\frac{1}{2} A_{motif}}{\langle N \rangle A_{1\cap 2} (A_{2\cap 3}/A_{sky})(T_0/T_{sky})} > \frac{1}{18} \frac{-\frac{1}{2} A_{motif}}{\langle N \rangle A_{2\cap 3} (A_{2\cap 3}/A_{sky})(T_0/T_{sky})}. \tag{64}$$

A similar calculation would follow straightforwardly for the case $\theta_{12} \sim \theta_{13} \sim \theta_{23}$. Eqn. (63) sets a weaker constraint on f_{NL} than Eqn. (61) by the order of magnitude of $A_{2\cap 3}/A_{motif}$. Since $A_{1\cap 2}$ is predominantly a decreasing function of θ_{12}, Eqn. (64) sets a weaker still constrain while allowing for a negative correlation among the three motifs.

6. Discussion

We propose a simple family of models for mimicking foreground sources that could contaminate the non-Gaussian signal of the CMB. This contamination could lead to the misidentification of a detection of a non-Gaussian signal for primordial when in fact we would be looking at the spurious signal from late-time, non-linear sources. Qualitatively non-Gaussian aspects of foregrounds that are likely to give a significant signal of $f_{NL} \neq 0$ result from the modulation of the small-scale power by the large-scale power.

Our model allows to generate foreground maps by distributing motifs in the sky according to a probability density $Q(\boldsymbol{\xi})$ and correlating them according to a conditional probability $P(\boldsymbol{\theta}, \boldsymbol{\xi})$ of finding one, two or three points inside the same motif. The statistical properties of the resulting maps are determined by the two and three-point correlation functions, which are calculated by simply evaluated tree-level Feynman diagrams. We find the expression for f_{NL} in terms of the parameters of $Q(\boldsymbol{\xi})$ and $P(\boldsymbol{\theta}, \boldsymbol{\xi})$, namely the mean number of motifs and the intersection areas among two and three motifs. We suggest a prescription for introducing temperature in order to interpret the statistical properties of the motif ensemble as statistical properties of the temperature anisotropies. We also indicate how to use the foreground maps to constrain the normalization factors and accordingly we constrained the values for f_{NL}.

In the forthcoming paper we will use the model to generate concrete mock foreground maps consistent with the level of foreground contamination observed and make detailed analysis of the impact on f_{NL} for *Planck* and other experiments.

Acknowledgments

The author thanks Martin Bucher and Bartjan van Tent for useful discussions and insightful comments. The author was supported by the Fundação para a Ciência e a Tecnologia under the fellowship /BPD/18236/2004. The author also acknowledges the use of the Legacy Archive for Microwave Background Data Analysis (LAMBDA) and the HEALPix package [16].

References

[1] Komatsu E et al. 2008 Five-Year Wilkinson Microwave Anisotropy Probe (WMAP) Observations: Cosmological Interpretation (*Preprint* arXiv:0803.0547)

[2] Yadav A P S and Wandelt B D 2008 Evidence of Primordial Non-Gaussianity f_{NL} in the Wilkinson Microwave Anisotropy Probe 3-Year Data at 2.8σ *Phys.Rev.Lett.* **100** 181301 (*Preprint* arXiv:0712.1148)

[3] Babich D 2005 Optimal estimation of non-Gaussianity *Phys.Rev. D* **72** 043003 (*Preprint* astro-ph/0503375)

[4] Creminelli P, Senatore L and Zaldarriaga M 2007 Estimators for local non-Gaussianities *JCAP* **0703** 019 (*Preprint* astro-ph/0606001)

[5] Gold B et al. 2008 Five-Year Wilkinson Microwave Anisotropy Probe (WMAP) Observations: Galactic Foreground Emission (*Preprint* arXiv:0803.0715)

[6] Bucher M, Carvalho C S, van Tent B J W (in preparation)

[7] Komatsu E and Spergel D N 2001 Acoustic Signatures in the Pimary Microwave Background Bispectrum *Phys.Rev. D* **63** 063002 (*Preprint* asto-ph/0005036)

[8] Maldacena J M 2003 Non-Gaussian features of primordial fluctuations in single field inflationary models *JHEP* **0305** 013 (*Preprint* astro-ph/0210603)

[9] Acquaviva V, Bartolo N, Matarrese S and Riotto A 2002 Second order cosmological perturbations from inflation *Nucl.Phys. B* **667** 119 (*Preprint* astro-ph/0209156)

[10] Bartolo A, Matarrese S and Riotto A 2002 Nongaussianity from inflation. *Phys.Rev. D* **65** 103505 (*Preprint* hep-ph/0112261)

[11] Rigopoulos G I, Shellard E P S and van Tent B J W 2005 Large non-Gaussianity in Multiple-Field Inflation *Phys.Rev. D* **73** 083522 (*Preprint* astro-ph/0506704)

[12] Komatsu E 2004 The Pursuitof Non-Gaussian Fluctuations in the Cosmic Microwave Background (*Preprint* astro-ph/0206039)

[13] Komatsu E, Spergel D N and Wandelt B D 2003 Measuring primordial non-Gaussianity in the cosmic microwave background (*Preprint* astro-ph/0305189)

[14] Bennet C L et al. 2003 First Year Wilkinson Microwave Anisotropy Probe (WMAP) Observations: Foregroud Emission *ApJS* **148** 97 (*Preprint* astro-ph/0302208)

[15] Peebles P J E *The Large Scale Structure of the Universe* (Princeton Series in Physics)

[16] Grski K M, Hivon E, Banday A J, Wandelt B D, Hansen F K, Reinecke M and Bartelmann M 2005 HEALPIX – a Framework for High Resolution Discretization, and Fast Analysis of Data Distributed on the Sphere *ApJ* **622** 759 (*Preprint* astro-ph/0409513)

Properties of codimension-2 braneworlds in six-dimensional Lovelock theory

Christos Charmousis[1,2], **Antonios Papazoglou**[3]

[1]Laboratoire de Physique Théorique, Université de Paris-Sud, Bât. 210, 91405 Orsay CEDEX, France

[2]Laboratoire de Mathematiques et Physique Théorique, Université Francois Rabelais, Parc de Grandmount, 37200 Tours, France

[3]Institute of Cosmology and Gravitation, University of Portsmouth, Portsmouth PO1 3FX, UK

E-mail: `christos.charmousis@th.u-psud.fr`, `antonios.papazoglou@port.ac.uk`

Abstract. We consider maximally symmetric 3-branes embedded in a six-dimensional bulk spacetime with Lovelock dynamics. We study the properties of the solutions with respect to their induced curvature, their vacuum energy and their effective compactness in the extra dimensions. Some simple solutions are shown to give rise to self-accelerating braneworlds, whereas several others solutions have self-tuning properties. For the case of geometric self-acceleration we argue that the cross-over scale in between four-dimensional and higher-dimensional gravity and the scale of late-time geometric acceleration, fixed by the present horizon size, are related via the conical deficit angle of the six-dimensional bulk solution, which is a free parameter.

1. Introduction

Cosmological and astrophysical data indicate that more than two thirds of the content of the Universe is of the form of dark energy. The best fit to the data indicates that this component has the form of a small cosmological constant. This results to a significant theoretical problem since its natural value, from the point of view of Quantum Field Theory, is of the order of the ultraviolet cutoff we would impose for our quantum field theory (anything from SUSY breaking scale to the Planck scale). In addition, one has to understand why this cosmological constant is of the order of the dark matter energy density now, taking into account the completely different cosmological evolution of these components of the Universe.

The above, of course, are correct statements given that we have a homogeneous Universe described by the Einstein's field equations. Taking into account the above difficulties, modifying gravity in the infrared is a legitimate theoretical hypothesis that should be taken seriously. In the context of braneworld models, there have been some interesting ideas. Firstly, it was shown that five-dimensional models with an induced gravity term on the brane (DGP models) can have a *self-accelerating* phase [2], where the current acceleration of the Universe is not due to a component of the Universe energy density, but rather a geometrical effect. An even more ambitious proposal, that of the *self-tuning* [3, 4], was to find models where the vacuum energy of the brane can be large without affecting the curvature of the brane and without fine-tuning of it with other brane or bulk parameters. Both proposals change the way that we see the interplay between vacuum energy and curvature in gravity. There were not, however, without problems. Self-accelerating braneworlds were typically infested by ghosts [5] and self-tuning ones had hidden fine-tunings or curvature singularities [6].

In our recent work [1], we examined a completely novel possibility of obtaining acceleration due to geometry as well as certain self-tuning properties. The modified gravity theory that we will study is Lovelock theory [7] in six dimensions, which is the natural extension of General Relativity in higher dimensions. The Lovelock theory in six dimensions has, in addition to the Einstein-Hilbert term, the Gauss-Bonnet combination (for a recent review on Lovelock theory see [8]). Although the latter is a topological invariant in four dimensions, it becomes dynamical for higher dimensions and modifies the gravitational theory. This theory is special because, on the one hand it provides geometric novel solutions which are absent in Einstein theory, and on the other hand, it gives an induced gravity term on the brane. In the context of this theory, we found examples of both self-accelerating and self-tuning cases. These examples open new possibilities for consistent self-acceleration and effective self-tuning which need to be considered in more detail in the future.

2. Lovelock braneworlds

Let us consider the six-dimensional dynamics of Lovelock gravity with a bare cosmological constant Λ and a Gauss-Bonnet term. The action of the system reads

$$S = \int d^6x \sqrt{-g}\left[\frac{1}{16\pi G_6}(R + \hat{\alpha}\mathcal{L}_{GB}) - 2\Lambda\right] , \tag{1}$$

where

$$\mathcal{L}_{GB} = R_{MNK\Lambda}R^{MNK\Lambda} - 4R_{MN}R^{MN} + R^2 , \tag{2}$$

is the Gauss-Bonnet Lagrangian density, G_6 the six-dimensional Newton's constant and $\hat{\alpha}$ the Gauss-Bonnet coupling.

The procedure we use to generate brane world solutions, is to doubly Wick rotate black hole solutions (first studied by Boulware and Deser [9]). In this procedure, the positions of the horizons r_h will be the endpoints of the internal space where there codimension-2 branes are in principle located. The brane tension is in fact related to the temperature of the black hole horizon, and a warped two brane setup will correspond to black hole solutions with double horizons. In fact, via a generalized version of Birkhoff's theorem [10], it can be shown that the most general axisymmetric solutions of (1) with maximally symmetric four-dimensional subspaces are

$$ds^2 = V(r)d\theta^2 + \frac{dr^2}{V(r)} + r^2 h^{(\kappa)}_{\mu\nu} dx^\mu dx^\nu , \tag{3}$$

where the potential is given by

$$V(r) = \kappa + \frac{r^2}{2\alpha}\left[1 + \epsilon\sqrt{1 + 4\alpha\left(a^2 - \frac{\epsilon\mu}{r^5}\right)}\right] , \tag{4}$$

with $\alpha = 6\hat{\alpha}$, $16\pi G_6\Lambda = 20a^2$ the positive cosmological constant[1]. The four-dimensional metric brane metric $h^{(\kappa)}_{\mu\nu}$ is parametrized by $\kappa = 0, -1, 1$, for four-dimensional Minkowski, AdS_4 and dS_4 respectively with curvature $R[h] = 12\kappa$. Finally, $\epsilon = \pm 1$, giving rise to two distinct branches

[1] Here we have omitted the bulk charge parameter for simplicity.

of solutions. The branch $\epsilon = +1$ (*Gauss-Bonnet branch*) does not have an Einstein theory limit as $\alpha \to 0$ (more recently the vacuum in this branch was shown to be unstable [11]). This limit is regular for the other branch with $\epsilon = -1$ (*Einstein branch*). The case where $1 + 4\alpha a^2 = 0$ is special, because the theory can be written in a Born-Infeld (BI) form [12]. It does not have an Einstein theory limit, however, this is the only case that we have a unique vacuum.

Defining the Gaussian Normal radial coordinate $\rho = \sqrt{4(r - r_h)/V'_{r_h}}$ and expanding around e.g. one root r_h of V, we get that the internal space is locally conical

$$ds_2^2 \approx \left(\frac{1}{4}V'^2_{r_h}\right)\rho^2 d\theta^2 + d\rho^2 \ . \tag{5}$$

If the angular coordinate has periodicity $\theta \in [0, 2\pi c)$, then the deficit angle which is induced at the brane position is $\delta = 2\pi(1 - \beta)$ with $\beta = \frac{1}{2}|V'_h|c$. Note that the conical deficit is related as usual to the temperature of the Wick rotated black hole horizon. From the Lovelock equations supplemented by a brane tension term, one can separate the distributional Dirac parts and write down induced Einstein equations for the brane. These brane junction conditions are [13]

$$2\pi(1-\beta)\left(-\gamma_{\mu\nu} + 4\hat{\alpha}G^{ind}_{\mu\nu}\right) = 8\pi G_6 T^{brane}_{\mu\nu} \ , \tag{6}$$

where $\gamma_{\mu\nu} = r_h^2 h^{(\kappa)}_{\mu\nu}$ is the induced metric on the brane with curvature $R[\gamma] = 12\kappa/r_h^2 \equiv 12\kappa H^2$, and $G^{ind}_{\mu\nu} = -3\kappa H^2 \gamma_{\mu\nu}$ is the induced Einstein tensor. The brane position $V(r_h) = 0$ depends on the bulk parameters via (4). The induced Newton's constant on the brane can be determined from (6) to be

$$G_4 = \frac{3G_6}{4\pi\alpha(1-\beta)} \ . \tag{7}$$

Note that in order to have positive induced Newton's constant, we should have have angle deficit ($\beta < 1$) for $\alpha > 0$ and angle excess ($\beta > 1$) for $\alpha < 0$.

Substituting the $G^{ind}_{\mu\nu}$ back in (6), we find a relation between the Hubble parameter H on the brane and the action parameters ($T_\mu^{brane\ \nu} = -T\delta_\mu^\nu$)

$$\kappa H^2 = -\frac{1}{2\alpha} + \frac{8\pi G_4}{3}T \ . \tag{8}$$

The above equations (7), (8) are very important since they firstly relate the hierarchy in between the scales of the theory and secondly the curvature on the brane H^2 to its sources, namely the brane tension and the Gauss-Bonnet coupling. We see in particular from (8) that the junction conditions tell us that the effective expansion H is in one part due to the Gauss-Bonnet induced cosmological term and in another part due to the vacuum energy of the brane. On the other hand we expect (7) to be indicating a cross-over scale in between a four-dimensional gravity phase and a six-dimensional one, $r_c^2 = G_6/G_4$. Unlike the codimension-1 DGP model note the appearance of two scales α and β dictating the size of r_c.

3. Self-properties of the solutions

Let us now discuss the physical consequences of the above solutions. In particular, we wish to see whether we can obtain codimension-2 braneworlds exhibiting *self-accelerating* or *self-tuning* behavior. The key relation for picking these solutions is (8).

3.1. Self-acceleration

For self-acceleration we need dS_4 vacua ($\kappa = 1$), where one has a positive geometrical contribution to the curvature, i.e. $\alpha < 0$ coming from the Gauss-Bonnet term in the action (1). In addition, this contribution should be dominant in comparison to the brane tension contribution (in other words the T-term in (8) should be negligible). There are two cases where this can happen.

First, if we have no bulk cosmological constant $a^2 = 0$ and we are in the Einstein branch $\epsilon = -1$. Then the internal space is non-compact and the brane position is bounded from below as $r_h(\mu) > \sqrt{2|\alpha|}$. The limit $r_h(\mu_s) = \sqrt{2|\alpha|}$, with $\mu_s \equiv \sqrt{2}|\alpha|^{3/2}$, is singular since it corresponds to the branch cut singularity of the square root of (4). The solution is self-accelerating when μ is in the neighborhood of μ_s. The bulk solution is simply a Wick rotation of the six-dimensional Schwarzschild black hole.

Second, if the mass parameter vanishes $\mu = 0$ and we are in the neighborhood of the BI point $a^2_{BI} = 1/(4|\alpha|)$ (in both branches $\epsilon = \pm 1$). Then, the brane position is a function of the bulk cosmological constant $r_h = r_h(a^2)$ and at the BI limit $r_h(a^2_{BI}) = \sqrt{2|\alpha|}$. In these cases the internal spaces are compact and since the space has no singularity at $r = 0$, we can extend the radial coordinate to $r < 0$ and consider the region of $-r_h \le r \le r_h$. The internal space is symmetric around $r = 0$, thus we have Z_2 symmetry around the equator of the internal space.

In both the above cases in order for the geometrical acceleration to account for the current acceleration of the Universe, the Gauss-Bonnet coupling appearing in the 6 dimensional action should be enormous, roughly of the order $\alpha \sim 10^{120} M_{Pl}^{-2}$. This hierarchy means that the bulk gravity is essentially dictated by the higher order Lovelock term that in turn gives ordinary four-dimensional gravity on the brane according to (6). The hierarchy in between G_4 and G_6 maybe reduced by sufficiently fine-tuning β to be close to 1^-. In fact combining (7) and (8), for $T \approx 0$, we have that,

$$\delta \sim \frac{3r_c^2}{2r_0^2}, \tag{9}$$

where $r_0 = H_0^{-1}$ is the horizon size and δ the angular excess, $\delta < 0$. Hence the crossover scale, r_c, is now a combination of a purely topological number δ and the horizon size r_0, unlike the situation in codimension-1 DGP where the two scales are the same. We should emphasize however, that strictly speaking the cross-over scale should be obtained from the brane propagator of the above bulk solution (3) and is not necessarily the scale appearing in the junction condition. Clearly this requires further investigation.

3.2. Self-tuning

For self-tuning to operate, one should be able to absorb variations of the brane tension in integration constants like c and μ, with the crucial demand of keeping the curvature of that brane as well as other bulk or brane parameters constant. We will be obviously interested in dS_4 ($\kappa = 1$) or flat ($\kappa = 0$) vacua. It turns out that if there is more than one brane present in the compactification, there are unavoidable fine-tunings in the model. Thus, the only possible self-tuning vacua can be the ones with only one brane present, or when an extra mirror brane is present, as in the Z_2-symmetric model that we mentioned in the previous subsection. A supplementary requirement for the self-tuning to be satisfying, is that the scales of H^2 and T should be dissociated. For the latter to happen, one should have $|\alpha|H^2 \ll 1$ (in other words the H-term in (8) should be negligible). In this case we can see from (8) and (7) that changes to T can be absorbed in c. Moreover, α can have much more natural values, i.e. $|\alpha| \sim M_{Pl}^{-2}$, than the ones for the previous self-accelerating vacua. In all these cases, since c should follow the variation of T, we will obtain a vacuum-energy-dependent Newton's constant G_4 as seen from (7). There are several examples where this can happen.

First, all the non-compact flat vacua, for $\alpha > 0$ or $\alpha < 0$, and with or without bulk cosmological constant, satisfy the above requirements. In this cases, we have exact self-tuning solutions.

For the dS_4 vacua, there are many possibilities with non-compact or compact internal space. A non-compact example is when we have no bulk cosmological constant $a^2 = 0$, we are in the Einstein branch $\epsilon = -1$, for both $\alpha > 0$ and $\alpha < 0$, and for large enough mass $\mu|\alpha|^{-3/2} \gg 1$. On the other hand, the only compact (Z_2-symmetric) case is when the mass parameter vanishes $\mu = 0$, we are in the Einstein branch $\epsilon = -1$, for $\alpha < 0$ and for small enough bulk cosmological constant $|\alpha|a^2 \ll 1$.

4. Discussion and Conclusions

We have obtained several self-accelerating and self-tuning solutions in a Lovelock six-dimensional theory with codimension-2 branes. The new setting of these models may help overcome the obstacles of the previous self-accelerating and self-tuning models in the literature.

An important property of the above models is that the codimension-2 junction conditions have an induced Einstein tensor term. Therefore, one should expect that the theory (even for non-compact or very large volume models) behaves in a four-dimensional way up to some cross-over scale which we have argued to be $r_c^2 = G_6/G_4$. We should emphasize that a characteristic element of the present setting is that this cross-over scale is in principle different from the scale of self-acceleration r_0 because the deficit angle δ which comes also into play. In principle the scale r_c could be much lower than the horizon size r_0 thus reducing the hierarchy between G_6 and G_4. How low we can set the crossover scale r_c, compared to r_0 depends on the modified higher dimensional gravity and how much it will modify large scale cosmological observables.

What is, therefore, the next step in the analysis of the codimension-2 brane models presented here, is the linear perturbation of these solutions as well as the cosmology analysis. The former, will tell us of the stability and the precise gravitational spectrum. For the latter, although one expects that the introduction of matter on the brane introduces singularities which need to be regularized, the fact that there exists an induced Einstein equation on the brane, may allow for the cosmology to be studied without the need of an explicit regularization. Furthermore, knowledge of the modified Friedmann equations on the brane will tell us a lot more on the cosmological scales that we have hinted upon here and on the validity of codimension-2 braneworlds.

Acknowledgments

A.P. is supported by the Marie Curie Intra-European Fellowship EIF-039189.

References

[1] C. Charmousis and A. Papazoglou, JHEP **0807** (2008) 062 [arXiv:0804.2121 [hep-th]].

[2] G. R. Dvali, G. Gabadadze and M. Porrati, Phys. Lett. B **485** (2000) 208 [arXiv:hep-th/0005016]; C. Deffayet, Phys. Lett. B **502** (2001) 199 [arXiv:hep-th/0010186].

[3] N. Arkani-Hamed, S. Dimopoulos, N. Kaloper and R. Sundrum, Phys. Lett. B **480** (2000) 193 [arXiv:hep-th/0001197]; S. Kachru, M. B. Schulz and E. Silverstein, Phys. Rev. D **62** (2000) 045021 [arXiv:hep-th/0001206].

[4] S. M. Carroll and M. M. Guica, arXiv:hep-th/0302067; I. Navarro, JCAP **0309** (2003) 004 [arXiv:hep-th/0302129]; Y. Aghababaie, C. P. Burgess, S. L. Parameswaran and F. Quevedo, Nucl. Phys. B **680** (2004) 389 [arXiv:hep-th/0304256].

[5] D. Gorbunov, K. Koyama and S. Sibiryakov, Phys. Rev. D **73** (2006) 044016 [arXiv:hep-th/0512097]; K. Koyama, Phys. Rev. D **72** (2005) 123511 [arXiv:hep-th/0503191]; C. Charmousis, R. Gregory, N. Kaloper and A. Padilla, JHEP **0610** (2006) 066 [arXiv:hep-th/0604086].
[6] S. Forste, Z. Lalak, S. Lavignac and H. P. Nilles, Phys. Lett. B **481** (2000) 360 [arXiv:hep-th/0002164]; S. Forste, Z. Lalak, S. Lavignac and H. P. Nilles, JHEP **0009** (2000) 034 [arXiv:hep-th/0006139].
[7] D. Lovelock, J. Math. Phys. **12** (1971) 498; B. Zumino, Phys. Rept. **137** (1986) 109 .
[8] C. Charmousis, Lect. Notes Phys. **769** (2009) 299 [arXiv:0805.0568 [gr-qc]].
[9] D. G. Boulware and S. Deser, Phys. Rev. Lett. **55** (1985) 2656.
[10] R. Zegers, J. Math. Phys. **46** (2005) 072502 [arXiv:gr-qc/0505016].
[11] C. Charmousis and A. Padilla, JHEP **0812** (2008) 038 [arXiv:0807.2864 [hep-th]].
[12] J. Crisostomo, R. Troncoso and J. Zanelli, Phys. Rev. D **62** (2000) 084013 [arXiv:hep-th/0003271].
[13] P. Bostock, R. Gregory, I. Navarro and J. Santiago, Phys. Rev. Lett. **92** (2004) 221601 [arXiv:hep-th/0311074]; C. Charmousis and R. Zegers, JHEP **0508** (2005) 075 [arXiv:hep-th/0502170]; C. Charmousis and R. Zegers, Phys. Rev. D **72** (2005) 064005 [arXiv:hep-th/0502171].

Recent Developments in Gravity (NEB XIII) IOP Publishing
Journal of Physics: Conference Series **189** (2009) 012008 doi:10.1088/1742-6596/189/1/012008

Canonical quantization of some midi-superspace models in 2+1 and 3+1 dimensions

T. Christodoulakis[1], G. Doulis[1], P. A. Terzis[1], E. Melas[2], Th. Grammenos[3], G. O. Papadopoulos[4], and A. Spanou[5]

[1] Nuclear and Particle Physics Section, Physics Department, University of Athens, GR 157–71 Athens
[2] Technological Educational Institution of Lamia, Electrical Engineering Department, GR 35–100, Lamia
[3] Department of Mechanical Engineering, University of Thessaly, GR 383–34 Volos
[4] Department of Mathematics and Statistics, Dalhousie University, Halifax, Nova Scotia, Canada B3H 3J5
[5] School of Applied Mathematics and Physical Sciences, National Technical University of Athens, GR 157–80, Athens

E-mail: `tchris@phys.uoa.gr`, `gdoulis@phys.uoa.gr`, `pterzis@phys.uoa.gr`, `evangelosmelas@yahoo.co.uk`, `thgramme@uth.gr`, `gopapad@mathstat.dal.ca`, `aspanou@central.ntua.gr`

Abstract. A proposal is put forward which enables the canonical quantization of a family of axially symmetric geometries in 2+1 dimensions and a corresponding spherically symmetric family in 3+1 dimensions. The proposal consists of a particular renormalization *assumption* and an accompanying *requirement* and results in a Wheeler–DeWitt equation which is based on a renormalized manifold parametrized by three smooth scalar functionals. The aforementioned equation is analytically solved for both the 2+1 and 3+1 case.

1. Introduction

After Dirac's pioneering work on the treatment of systems with constraints [1], [2], [3], [4] the way was open for a systematic treatment of constrained dynamics. Some of the landmarks in the study of constrained systems have been the connection between constraints and invariances [5], the extension of the formalism to describe fields with half-integer spin through the algebra of Grassmann variables [6] and the introduction of the BRST formalism [7]. All these results have helped the quantization of gauge theories. There exist several excellent reviews studying constraint systems with a finite number of degrees of freedom [8] or constraint field theories [9], as well as more general presentations [10], [11], [12], [13], [14], [15]. In particular, the conventional canonical analysis approach of quantum gravity has been initiated by B.S. DeWitt [16] based on earlier work of P.G. Bergmann [17].

Since a full theory of quantum gravity does not today exist, it is reasonably important to address the quantization of (classes of) simplified geometries. An elegant way to achieve a degree of simplification is to impose some symmetry. For example, the assumption of a G_3

symmetry group acting simply transitively on the surfaces of simultaneity, i.e. the existence of three independent space-like Killing vector fields, leads to classical and subsequently quantum homogeneous cosmology (see, e.g., [18], [19]). The imposition of lesser symmetry, e.g. fewer Killing vector fields, results in the various inhomogeneous cosmological models [20]. The canonical analysis under the assumption of spherical symmetry, which is a G_3 group acting multiply transitively on two-dimensional space-like subsurfaces of the three-slices, has been first considered in [21], [22].

This work consists in two parts. In the first part, we deal with the simplified case of 2+1 dimensions (see, e.g., [26], [27], [28] and references there in), under the assumption of one spacelike Killing vector field, while in the second part, the canonical quantization of all geometries admitting two-dimensional surfaces of maximal symmetry, i.e. spheres (constant positive curvature), planes (zero curvature) and Gauss-Bolyai-Lobachevsky (henceforth GBL) spaces (constant negative curvature) is considered. The structure of the paper is as follows: In Section 2 we give the reduced metrics, the space of classical solutions and the Hamiltonian formulation of the reduced Einstein-Hilbert action principle, resulting in one (quadratic) Hamiltonian and two (linear) momentum constraints, all being first class. In Section 3, we consider the quantization of this constraint system following Dirac's proposal of implementing the quantum operator constraints as conditions annihilating the wave-function [4]. Our guideline is a conceptual generalization of the quantization scheme developed in [30], [31] for the case of constraint systems with finite degrees of freedom, to the present case. Even though after the symmetry reduction the system still represents a field theory (all remaining metric components depend on time and the radial coordinate), we manage to extract and subsequently completely solve a Wheeler-DeWitt equation in terms of three unique smooth scalar functionals of the appropriate components of the reduced spatial metric. This is achieved through an appropriate re-normalization assumption we adopt. Finally, some concluding remarks are included in the discussion.

2. Possible Metrics, Classical Solutions and Hamiltonian Formulation

2.1. The 2+1 Case

Our starting point is the action principle:

$$I = \int d^3x\sqrt{-g}(R - 2\Lambda). \tag{2.1}$$

The equations of motion arising upon variation of this action are

$$R_{IJ} - \frac{1}{2}g_{IJ}R + \Lambda g_{IJ} = 0, \tag{2.2}$$

where $I, J = 0, 1, 2$. Of course, since in three dimensions the Riemann curvature tensor is expressible in terms of both the Ricci tensor and scalar, the space of solutions to (2.2) consists simply of all maximally symmetric 3D metrics (AdS3). If topological considerations are taken into account, the above space might be enriched containing, for example, the stationary BTZ "black" hole [32], [33]

$$ds^2 = -(M - \Lambda r^2)dt^2 - Jdtd\phi + \left(M - \Lambda r^2 + \frac{J^2}{4r^2}\right)^{-1} dr^2 + r^2 d\phi^2, \tag{2.3}$$

or the "cosmological" solutions [34], [35]

$$ds^2 = -\frac{1}{4t^2\Lambda}dt^2 + \frac{1}{2t\sqrt{\Lambda}}(dr^2 + d\phi^2), \tag{2.4}$$

$$ds^2 = -\left(\frac{4}{16t^2 - \Lambda}\right)^2 dt^2 + \frac{4}{16t^2 - \Lambda}dr^2 + \frac{4e^{-4r}}{16t^2 - \Lambda}d\phi^2. \tag{2.5}$$

The above metrics admit a manifest G_2 symmetry group. Here, we consider a generalization consisting in the imposition of a G_1 (axial) symmetry only, i.e we impose one Killing vector field, say $\xi = \frac{\partial}{\partial\phi}$. Subsequently, all components of the metric become functions of only the time and the radial coordinate. The canonical decomposition of such a metric is given in terms of the spatial metric $g_{ij}(t,r)$, the lapse function $N^o(t,r)$ and the shift "vector" $N_i(t,r)$ [10]:

$$ds^2 = \left(-(N^o)^2 + g^{ij}N_iN_j\right)dt^2 + 2N_idtdx^i + g_{ij}dx^idx^j, \tag{2.6}$$

where

$$g_{ij} = \begin{pmatrix} \dfrac{\rho^2+\sigma^2\chi^2}{\sigma} & \sigma\chi \\ \sigma\chi & \sigma \end{pmatrix}, \qquad g^{ij} = \begin{pmatrix} \dfrac{\sigma}{\rho^2} & -\dfrac{\sigma\chi}{\rho^2} \\ -\dfrac{\sigma\chi}{\rho^2} & \dfrac{\rho^2+\sigma^2\chi^2}{\sigma\rho^2} \end{pmatrix}, \tag{2.7}$$

with $i,j = 1,2$, and $x^i = (r,\phi)$.

Utilizing the Gauss-Codazzi equation (see, e.g., [36]), we eliminate all second time-derivatives from the Einstein-Hilbert action, by subtracting an appropriate divergence, and arrive at an action quadratic in the velocities. The application of the Dirac algorithm results firstly in the three primary constraints $P_o \equiv \frac{\delta L}{\delta \dot{N}^o} \approx 0$, $P^i \equiv \frac{\delta L}{\delta \dot{N}_i} \approx 0$ and the Hamiltonian

$$H = \int \left(N^o\mathcal{H}_o + N^i\mathcal{H}_i\right)dr, \tag{2.8}$$

where $\mathcal{H}_o$, $\mathcal{H}_i$ are given by

$$\mathcal{H}_o = \frac{1}{2}G^{\alpha\beta}\pi_\alpha\pi_\beta + V, \tag{2.9a}$$

$$\mathcal{H}_1 = \sigma'\pi_\sigma - \rho\pi'_\rho - \chi\pi'_\chi, \tag{2.9b}$$

$$\mathcal{H}_2 = -\pi'_\chi, \tag{2.9c}$$

the indices (α,β) take the values (ρ,σ,χ) and $' \equiv \dfrac{\partial}{\partial r}$. The Wheeler-DeWitt midi-superspace metric $G^{\alpha\beta}$ reads

$$G^{\alpha\beta} = \begin{pmatrix} -\rho & -\sigma & 0 \\ -\sigma & 0 & 0 \\ 0 & 0 & \dfrac{\rho}{\sigma^2} \end{pmatrix}, \tag{2.10}$$

while the potential V is

$$V = 2\Lambda\rho + \left(\frac{\sigma'}{\rho}\right)'. \tag{2.11}$$

The requirement for preservation, in time, of the primary constraints leads to the secondary constraints

$$\mathcal{H}_o \approx 0, \qquad \mathcal{H}_1 \approx 0, \qquad \mathcal{H}_2 \approx 0. \tag{2.12}$$

At this stage, a tedious but straightforward calculation produces the following "open" Poisson bracket algebra of these constraints:

$$\begin{aligned}
\{\mathcal{H}_o(r), \mathcal{H}_o(\tilde{r})\} &= [g^{1j}(r)\mathcal{H}_j(r) + g^{1j}(\tilde{r})\mathcal{H}_j(\tilde{r})]\delta'(r,\tilde{r}),\\
\{\mathcal{H}_1(r), \mathcal{H}_o(\tilde{r})\} &= \mathcal{H}_o(r)\delta'(r,\tilde{r}),\\
\{\mathcal{H}_2(r), \mathcal{H}_o(\tilde{r})\} &= 0,\\
\{\mathcal{H}_1(r), \mathcal{H}_1(\tilde{r})\} &= \mathcal{H}_1(r)\delta'(r,\tilde{r}) - \mathcal{H}_1(\tilde{r})\delta(r,\tilde{r})',\\
\{\mathcal{H}_1(r), \mathcal{H}_2(\tilde{r})\} &= \mathcal{H}_2(r)\delta'(r,\tilde{r}),\\
\{\mathcal{H}_2(r), \mathcal{H}_2(\tilde{r})\} &= 0,
\end{aligned} \tag{2.13}$$

indicating that they are first class and also signaling the termination of the algorithm. Thus, our system is described by (2.12); the "dynamical" Hamilton-Jacobi equations $\frac{d\,\pi_\rho}{d\,t} = \{\pi_\rho, H\}$, $\frac{d\,\pi_\sigma}{d\,t} = \{\pi_\sigma, H\}$, $\frac{d\,\pi_\chi}{d\,t} = \{\pi_\chi, H\}$ are satisfied by virtue of the time derivatives of (2.12). These three equations, when expressed in the velocity phase-space with the help of the definitions $\frac{d\,\rho}{d\,t} = \{\rho, H\}$, $\frac{d\,\sigma}{d\,t} = \{\sigma, H\}$, $\frac{d\,\chi}{d\,t} = \{\chi, H\}$, are completely equivalent to the three independent Einstein's field equations satisfied by (2.6).

Under changes of the radial variable r of the form $r \to \tilde{r} = h(r)$, it can easily be inferred from (2.6) and (2.7) that:

$$\begin{aligned}
&\tilde{\rho}(\tilde{r}) = \rho(r)\,\frac{d\,r}{d\,\tilde{r}}, \qquad \tilde{\sigma}(\tilde{r}) = \sigma(r), \qquad \tilde{\chi}(\tilde{r}) = \chi(r)\,\frac{d\,r}{d\,\tilde{r}},\\
&\frac{d\,\tilde{\sigma}(\tilde{r})}{d\,\tilde{r}} = \frac{d\,\sigma(r)}{d\,r}\,\frac{d\,r}{d\,\tilde{r}}, \qquad \frac{d}{d\,\tilde{r}}\left(\frac{\tilde{\chi}(\tilde{r})}{\tilde{\rho}(\tilde{r})}\right) = \frac{d}{d\,r}\left(\frac{\chi(r)}{\rho(r)}\right)\frac{d\,r}{d\,\tilde{r}},
\end{aligned} \tag{2.14}$$

where the t-dependence has been omitted for the sake of brevity. Thus, under the above coordinate transformations, σ, $\frac{\chi}{\rho}$ are scalars, while ρ, χ and the derivatives of σ, $\frac{\chi}{\rho}$ are covariant rank 1 tensors (one-forms), or, equivalently in one dimension, scalar densities of weight -1. Therefore, the scalar derivative is not $\frac{d}{d\,r}$ but rather $\frac{d}{\rho\,d\,r}$ or $\frac{d}{\chi\,d\,r} \equiv \frac{\rho}{\chi}\frac{d}{\rho\,d\,r}$. Finally, if we consider an infinitesimal transformation $r \to \tilde{r} = r - \eta(r)$, it is easily seen that the corresponding changes induced on the basic fields are:

$$\delta\,\rho(r) = (\rho(r)\,\eta(r))', \qquad \delta\,\sigma(r) = \sigma'(r)\,\eta(r), \qquad \delta\,\chi(r) = (\chi(r)\,\eta(r))', \tag{2.15}$$

i.e., nothing but the one-dimensional analogue of the appropriate Lie derivatives.

With the use of (2.15), we can reveal the nature of the action of $\mathcal{H}_1$ on the basic configuration space variables as that of the generator of spatial diffeomorphisms:

$$\begin{aligned}
\left\{\rho(r)\,,\int d\tilde{r}\,\eta(\tilde{r})\,\mathcal{H}_1(\tilde{r})\right\} &= (\rho(r)\,\eta(r))',\\
\left\{\sigma(r)\,,\int d\tilde{r}\,\eta(\tilde{r})\,\mathcal{H}_1(\tilde{r})\right\} &= \sigma'(r)\,\eta(r),\\
\left\{\chi(r)\,,\int d\tilde{r}\,\eta(\tilde{r})\,\mathcal{H}_1(\tilde{r})\right\} &= (\chi(r)\,\eta(r))'.
\end{aligned} \tag{2.16}$$

Thus, we are justified to consider $\mathcal{H}_1$ as the representative, in phase-space, of an arbitrary infinitesimal reparametrization of the radial coordinate. As far as $\mathcal{H}_2$ is concerned, the situation is a little more complicated: the imposition of the symmetry generated by the Killing vector field $\xi = \partial/\partial\phi$ has left all configuration variables without any ϕ dependence; subsequently we can not expect $\mathcal{H}_2$ to generate arbitrary infinitesimal reparametrization of ϕ. Nevertheless, we can identify a property of $\mathcal{H}_2$ which links its existence to the existence of ξ. This property is described by the relation: $\{\mathcal{H}_2(r), \{\mathcal{H}_2(r), \{\mathcal{H}_2(r), g_{ij}(\tilde{r})\}\}\} = 0 \underset{ndence}{\overset{correspo}{\Longleftrightarrow}} \mathcal{L}_\xi g_{ij} = 0$.

2.2. The 3+1 Case

Our starting point is the two-dimensional spaces of positive, zero or negative constant curvature. Their line elements are respectively:

$$ds^2 = d\theta^2 + \sin^2\theta\, d\phi^2, \qquad ds^2 = d\theta^2 + \theta^2\, d\phi^2, \qquad ds^2 = d\theta^2 + \sinh^2\theta\, d\phi^2, \tag{2.17}$$

with an obvious range of the coordinates for each case. The corresponding (maximal) symmetry groups are generated by the following KVF's:

$$\xi_1 = \frac{\partial}{\partial\phi}, \quad \xi_2 = -\cos\phi\frac{\partial}{\partial\theta} + \cot\theta\sin\phi\frac{\partial}{\partial\phi}, \quad \xi_3 = \sin\phi\frac{\partial}{\partial\theta} + \cot\theta\cos\phi\frac{\partial}{\partial\phi}, \tag{2.18}$$

$$\xi_1 = \frac{\partial}{\partial\phi}, \quad \xi_2 = -\cos\phi\frac{\partial}{\partial\theta} + \frac{\sin\phi}{\theta}\frac{\partial}{\partial\phi}, \quad \xi_3 = \sin\phi\frac{\partial}{\partial\theta} + \frac{\cos\phi}{\theta}\frac{\partial}{\partial\phi}, \tag{2.19}$$

$$\xi_1 = \frac{\partial}{\partial\phi}, \quad \xi_2 = -\cos\phi\frac{\partial}{\partial\theta} + \coth\theta\sin\phi\frac{\partial}{\partial\phi}, \quad \xi_3 = \sin\phi\frac{\partial}{\partial\theta} + \coth\theta\cos\phi\frac{\partial}{\partial\phi}. \tag{2.20}$$

Trivially prolonging these KVF's to four-dimensional space-time, we arrive at the three classes of metrics described by the following line element:

$$\begin{aligned} ds^2 = &\left(-\alpha(t,r)^2 + \frac{\beta(t,r)^2}{\gamma(t,r)^2}\right) dt^2 + 2\,\beta(t,r)\,dt\,dr + \gamma(t,r)^2\,dr^2 \\ &+ \psi(t,r)^2\,d\theta^2 + \psi(t,r)^2\,f(\theta)^2\,d\phi^2, \end{aligned} \tag{2.21}$$

where $f(\theta) = \sin\theta$ in the case of spherical symmetry, $f(\theta) = \theta$ in the case of plane symmetry, and $f(\theta) = \sinh\theta$ in the case of the GBL symmetry.

In order to attain the classical space of solution for these geometries one can, exploiting the freedom to change coordinates in the (t, r) subspace, bring the upper left block of the metric in conformally flat form and readily solve the vacuum Einstein's field equations. The result is given, in the light-cone coordinates $u = \dfrac{t+r}{2}$, $v = \dfrac{-t+r}{2}$, by the following two line elements:

$$ds^2 = 2\,\epsilon\,\frac{A + 2\,\psi(u,v)}{4\,\psi(u,v)}\,du\,dv + \psi(u,v)^2\,d\theta^2 + \psi(u,v)^2\,f(\theta)^2\,d\phi^2, \tag{2.22}$$

where

$$\psi(u,v) = -\frac{A}{2}\left(1 + \mathrm{ProductLog}(-\frac{\exp^{-\frac{\lambda(A+u+\epsilon\,v)}{A}}}{A})\right),$$

$\epsilon = \pm 1$, $\lambda = \pm 1$ for $f(\theta) = \sin\theta$, $\lambda = \pm i$, for $f(\theta) = \sinh\theta$ and ProductLog(z) is the principal solution for w to the equation $z = w \exp^w$, and

$$ds^2 = 2\,\epsilon \frac{1}{\psi(u,v)} du\, dv + \psi(u,v)^2 d\theta^2 + \psi(u,v)^2 \theta^2 d\phi^2, \tag{2.23}$$

where

$$\psi(u,v) = \pm\sqrt{2u + 2\epsilon\, v}.$$

The canonical analysis proceeds similarly as the previous case and the reduced Hamiltonian reads

$$H = \int \left(N^o \mathcal{H}_o + N^i \mathcal{H}_i\right) dr, \tag{2.24}$$

where

$$N^o = \alpha(t,r), \quad N^1 = \frac{\beta(t,r)}{\gamma(t,r)^2}, \quad N^2 = 0, \quad N^3 = 0,$$

and $\mathcal{H}_o$, $\mathcal{H}_i$ are given by

$$\mathcal{H}_o = \frac{1}{2} G^{\alpha\beta} \pi_\alpha \pi_\beta + V, \tag{2.25a}$$

$$\mathcal{H}_1 = -\gamma\, \pi'_\gamma + \psi'\, \pi_\psi, \qquad \mathcal{H}_2 = 0, \qquad \mathcal{H}_3 = 0, \tag{2.25b}$$

while the indices $\{\alpha, \beta\}$ take the values $\{\gamma, \psi\}$ and $' = \dfrac{d}{dr}$. The reduced Wheeler-DeWitt super-metric $G^{\alpha\beta}$ reads

$$G^{\alpha\beta} = \begin{pmatrix} \dfrac{\gamma}{4\,\psi^2} & -\dfrac{1}{4\,\psi} \\ -\dfrac{1}{4\,\psi} & 0 \end{pmatrix}, \tag{2.26}$$

while the potential V is

$$V = -2\,\epsilon\,\gamma + 2\Lambda\gamma\,\psi^2 - 2\frac{\psi'^2}{\gamma} + 4\left(\frac{\psi\,\psi'}{\gamma}\right)', \tag{2.27}$$

with $\epsilon = \{1, 0, -1\}$ for the families of two-dimensional subspaces with positive, zero or negative constant curvature, respectively. The requirement for preservation, in time, of the primary constraints leads to the secondary constraints

$$\mathcal{H}_o \approx 0, \qquad \mathcal{H}_1 \approx 0, \tag{2.28}$$

The "open" Poisson bracket algebra satisfied by these constraints is:

$$\begin{aligned} \{\mathcal{H}_o(r), \mathcal{H}_o(\tilde{r})\} &= \left[\frac{1}{\gamma^2(r)}\mathcal{H}_1(r) + \frac{1}{\gamma^2(\tilde{r})}\mathcal{H}_1(\tilde{r})\right]\delta'(r,\tilde{r}), \\ \{\mathcal{H}_1(r), \mathcal{H}_o(\tilde{r})\} &= \mathcal{H}_o(r)\delta'(r,\tilde{r}), \\ \{\mathcal{H}_1(r), \mathcal{H}_1(\tilde{r})\} &= \mathcal{H}_1(r)\delta'(r,\tilde{r}) - \mathcal{H}_1(\tilde{r})\delta(r,\tilde{r})', \end{aligned} \tag{2.29}$$

indicating that they are first class and also signaling the termination of the algorithm. Thus, our system is described by (2.28) and the dynamical Hamilton-Jacobi equations $\frac{d\,\pi_\gamma}{d\,t} = \{\pi_\gamma, H\}$, $\frac{d\,\pi_\psi}{d\,t} = \{\pi_\psi, H\}$. These equations, when expressed in the velocity phase-space with the help of the definitions $\frac{d\,\gamma}{d\,t} = \{\gamma, H\}$, $\frac{d\,\psi}{d\,t} = \{\psi, H\}$, are completely equivalent to the independent Einstein's field equations satisfied by (2.21).

Under changes of the radial variable r of the form $r \to \tilde{r} = h(r)$, it can easily be inferred from (2.21) that

$$\tilde{\gamma}(\tilde{r}) = \gamma(r)\,\frac{d\,r}{d\,\tilde{r}}, \qquad \tilde{\psi}(\tilde{r}) = \psi(r), \qquad \frac{d\,\tilde{\psi}(\tilde{r})}{d\,\tilde{r}} = \frac{d\,\psi(r)}{d\,r}\,\frac{d\,r}{d\,\tilde{r}}, \tag{2.30}$$

where the t-dependence has again been omitted. Thus, under the above coordinate transformations, ψ is a scalar, while γ, ψ' are covariant rank 1 tensors (one-forms), or, equivalently in one dimension, scalar densities of weight -1. Therefore, the scalar derivative is not $\frac{d}{d\,r}$ but rather $\frac{d}{\gamma\,d\,r}$. Finally, if we consider an infinitesimal transformation $r \to \tilde{r} = r - \eta(r)$, it is easily seen that the corresponding changes induced on the basic fields are:

$$\delta\,\gamma(r) = (\gamma(r)\,\eta(r))', \qquad \delta\,\psi(r) = \psi'(r)\,\eta(r), \tag{2.31}$$

i.e., nothing but the one-dimensional analogue of the appropriate Lie derivatives.
With the use of (2.31), we can reveal the nature of the action of $\mathcal{H}_1$ on the basic configuration space variables as that of the generator of spatial diffeomorphisms:

$$\begin{aligned} \left\{\gamma(r)\,,\,\int d\tilde{r}\,\eta(\tilde{r})\,\mathcal{H}_1(\tilde{r})\right\} &= (\gamma(r)\,\eta(r))', \\ \left\{\psi(r)\,,\,\int d\tilde{r}\,\eta(\tilde{r})\,\mathcal{H}_1(\tilde{r})\right\} &= \psi'(r)\,\eta(r). \end{aligned} \tag{2.32}$$

Thus, we are justified to consider $\mathcal{H}_1$ as the representative, in phase-space, of an arbitrary infinitesimal reparametrization of the radial coordinate.

3. Quantization

We are now interested in attempting to quantize these two Hamiltonian systems following Dirac's general spirit of realizing the classical first class constraints as quantum operator constraint conditions annihilating the wave function. The main motivation behind such an approach is the justified desire to construct a quantum theory manifestly invariant under the "gauge" generated by the constraints. To begin with, let us first note that, despite the simplification brought by the imposition of the symmetry $\xi = \partial/\partial\phi \Leftrightarrow \mathcal{L}_\xi g_{IJ} = 0$, the system is still a field theory in the sense that all configuration variables and canonical conjugate momenta depend not only on time (as is the case in homogeneous cosmology), but also on the radial coordinate r. Thus, to canonically quantize the system in the Schrödinger representation, we first realize the classical momenta as functional derivatives with respect to their corresponding conjugate fields

$$\hat{\pi}_\rho(r) = -i\,\frac{\delta}{\delta\,\rho(r)}, \qquad \hat{\pi}_\sigma(r) = -i\,\frac{\delta}{\delta\,\sigma(r)}, \qquad \hat{\pi}_\chi(r) = -i\,\frac{\delta}{\delta\,\chi(r)},$$

$$\hat{\pi}_\gamma(r) = -i\,\frac{\delta}{\delta\,\gamma(r)} \qquad \hat{\pi}_\psi(r) = -i\,\frac{\delta}{\delta\,\psi(r)}.$$

We next have to decide on the initial space of state vectors. To elucidate our choice, let us consider the action of a momentum operator on some function of the configuration field variables, say

$$\hat{\pi}_\rho(r)\rho(\tilde{r})^2 = -2i\rho(\tilde{r})\delta(\tilde{r},r).$$

The Dirac delta-function renders the outcome of this action a distribution rather than a function. Also, if the momentum operator were to act at the point at which the function is evaluated, i.e. if $\tilde{r} = r$, then its action would produce a $\delta(0)$ and would therefore be ill-defined. Both of these unwanted features are rectified, as far as expressions linear in momentum operators are concerned, if we choose as our initial collection of states all *smooth functionals* (i.e., integrals over r) of the configuration variables $\rho(r), \sigma(r), \chi(r), \gamma(r), \psi(r)$ and their derivatives of any order. Indeed, as we infer from the previous example,

$$\hat{\pi}_\rho(r)\int d\tilde{r}\rho(\tilde{r})^2 = -2i\int d\tilde{r}\rho(\tilde{r})\delta(\tilde{r},r) = -2i\rho(r);$$

thus the action of the momentum operators on all such states will be well-defined (no $\delta(0)$'s) and will also produce only local functions and not distributions. However, even so, $\delta(0)$'s will appear as soon as local expressions quadratic in momenta are considered, e.g.,

$$\hat{\pi}_\rho(r)\,\hat{\pi}_\rho(r)\int d\tilde{r}\rho(\tilde{r})^2 = \hat{\pi}_\rho(r)(-2i\int d\tilde{r}\rho(\tilde{r})\delta(\tilde{r},r)) = \hat{\pi}_\rho(r)(-2i\rho(r)) = -2i\delta(r,r).$$

Another problem of equal, if not greater, importance has to do with the number of derivatives (with respect to r) considered: A momentum operator acting on a smooth functional of degree n in derivatives of $\rho(r), \sigma(r), \chi(r), \gamma(r), \psi(r)$ will, in general, produce a function of degree $2n$, e.g.,

$$\hat{\pi}_\rho(r)\int d\tilde{r}\rho''(\tilde{r})^2 = -2i\int d\tilde{r}\rho''(\tilde{r})\delta''(\tilde{r},r) = -2i\rho^{(4)}(r).$$

Thus, clearly, more and more derivatives must be included if we desire the action of momentum operators to keep us inside the space of integrands corresponding to the initial collection of smooth functionals; eventually, we have to consider $n \to \infty$. This, in a sense, can be considered as the reflection to the canonical approach, of the non-renormalizability results existing in the so-called covariant approach. The way to deal with these problems is, loosely speaking, to regularize (i.e., render finite) the infinite distribution limits, and re-normalize the theory by, somehow, enforcing n to terminate at some finite value.

In the following, we are going to present a quantization scheme of our systems which: (a) avoids the occurrence of $\delta(0)$'s, (b) reveals the value $n = 1$, as the only possibility to obtain a closed space of state vectors, and (c) extracts a finite-dimensional Wheeler-DeWitt equation governing the quantum dynamics. The scheme closely parallels, conceptually, the quantization developed in [30],[31] for finite systems with one quadratic and a number of linear first class constraints. Therefore, we deem it appropriate and instructive to present a brief account of the essentials of this construction.

To this end, let us consider a system described by a Hamiltonian of the form

$$\begin{aligned} H &\equiv \mu X + \mu^i \chi_i \\ &= \mu\left(\frac{1}{2}G^{AB}(Q^\Gamma)P_A P_B + U^A(Q^\Gamma)P_A + V(Q^\Gamma)\right) + \mu^i\,\phi_i^A(Q^\Gamma)P_A, \end{aligned} \tag{3.1}$$

where $A, B, \Gamma \ldots = 1, 2 \ldots, M$ count the configuration space variables and $i = 1, 2, \ldots, N < (M-1)$ numbers the super-momenta constraints $\chi_i \approx 0$, which along with the super-Hamiltonian constraint $X \approx 0$ are assumed to be first class:

$$\{X, X\} = 0, \qquad \{X, \chi_i\} = XC_i + C_i^j \chi_j, \qquad \{\chi_i, \chi_j\} = C_{ij}^k \chi_k, \tag{3.2}$$

where the first (trivial) Poisson bracket has been included only to emphasize the difference from the first of (2.13).

The physical state of the system is unaffected by the "gauge" transformations generated by (X, χ_i), **but also** under the following three changes:

(I) Mixing of the super-momenta with a non-singular matrix

$$\bar{\chi}_i = \lambda_i^j (Q^\Gamma) \chi_j,$$

(II) Gauging of the super-Hamiltonian with the super-momenta

$$\bar{X} = X + \kappa^{(Ai}(Q^\Gamma)\phi_i^{B)}(Q^\Gamma) P_A P_B + \sigma^i(Q^\Gamma)\phi_i^A(Q^\Gamma) P_A,$$

(III) Scaling of the super-Hamiltonian

$$\bar{X} = \tau^2(Q^\Gamma) X.$$

Therefore, the geometrical structures on the configuration space that can be inferred from the super-Hamiltonian are really equivalence classes under actions (I), (II) and (III); for example (II), (III) imply that the super-metric G^{AB} is known only up to conformal scalings and additions of the super-momenta coefficients $\bar{G}^{AB} = \tau^2(G^{AB} + \kappa^{(Ai}\phi_i^{B)})$. It is thus mandatory that, when we Dirac-quantize the system, we realize the quantum operator constraint conditions on the wave-function in such a way as to secure that the whole scheme is independent of actions (I), (II), (III). This is achieved by the following steps:

(1) Realize the linear operator constraint conditions with the momentum operators to the right

$$\hat{\chi}_i \Psi = 0 \leftrightarrow \phi_i^A(Q^\Gamma) \frac{\partial \Psi(Q^\Gamma)}{\partial Q^A} = 0,$$

which maintains the geometrical meaning of the linear constraints and produces the $M - N$ independent solutions to the above equations $q^\alpha(Q^\Gamma)$, $\alpha = 1, 2, \ldots, M-N$ called physical variables, since they are invariant under the transformations generated by χ_i.

(2) Define the induced structure $g^{\alpha\beta} \equiv G^{AB} \dfrac{\partial q^\alpha}{\partial Q^A} \dfrac{\partial q^\beta}{\partial Q^B}$ and realize the quadratic in momenta part of X as the conformal Laplace-Beltrami operator based on $g_{\alpha\beta}$.

We are now ready to proceed with the quantization of our systems (2.8-2.12), (2.24-2.28) respectively, in close analogy to the scheme above outlined.

3.1. The 2+1 Case

In order to realize the equivalent to step 1, we first define the quantum analogue of $\mathcal{H}_1(r) \approx 0$ as

$$\hat{\mathcal{H}}_1(r)\Phi = 0 \leftrightarrow -\rho(r)\left(\frac{\delta \Phi}{\delta \rho(r)}\right)' + \sigma'(r) \frac{\delta \Phi}{\delta \sigma(r)} - \chi(r)\left(\frac{\delta \Phi}{\delta \chi(r)}\right)' = 0. \tag{3.3}$$

As explained in the beginning of the section, the action of $\hat{\mathcal{H}}_1(r)$ on all smooth functionals is well defined, i.e., produces no $\delta(0)$'s. It can be proven that, in order for such a functional to be annihilated by this linear quantum operator, it must be scalar, i.e. have the form

$$\Phi = \int \rho(\tilde{r})\, f\left(\Sigma^{(0)}, \Sigma^{(1)}, \ldots, \Sigma^{(n)}, X^{(0)}, X^{(1)}, \ldots, X^{(n)}\right) d\tilde{r}, \tag{3.4a}$$

$$\Sigma^{(0)} \equiv \sigma(\tilde{r}), \quad \Sigma^{(1)} \equiv \frac{\sigma'(\tilde{r})}{\rho(\tilde{r})}, \ldots, \quad \Sigma^{(n)} \equiv \frac{1}{\rho(\tilde{r})}\frac{d}{d\tilde{r}}\left(\underbrace{\cdots}_{n-1}\sigma(\tilde{r})\right), \tag{3.4b}$$

$$X^{(0)} \equiv \frac{\chi(\tilde{r})}{\rho(\tilde{r})}, \quad X^{(1)} \equiv \frac{1}{\rho(\tilde{r})}\left(\frac{\chi(\tilde{r})}{\rho(\tilde{r})}\right)', \ldots, \quad X^{(n)} \equiv \frac{1}{\rho(\tilde{r})}\frac{d}{d\tilde{r}}\left(\underbrace{\cdots}_{n-1}\frac{\chi(\tilde{r})}{\rho(\tilde{r})}\right), \tag{3.4c}$$

where f is any function of its arguments. We note that, as it is discussed at the end of the previous section, $\frac{\sigma'}{\rho}$ is the only scalar first derivative of σ, and likewise for the higher derivatives. The proof of this statement is analogous to the corresponding result concerning full gravity [37]: consider an infinitesimal r-reparametrization $\tilde{r} = r - \eta(r)$. Under such a change, the left-hand side of (3.4a), being a number, must remain unaltered. If we calculate the change induced on the right-hand side we arrive at

$$0 = \int \left[f\delta\rho + \rho\frac{\delta f}{\delta\sigma}\delta\sigma + \rho\frac{\delta f}{\delta(\chi/\rho)}\delta\left(\frac{\chi}{\rho}\right)\right] dr = \int [\rho\hat{\mathcal{H}}_1(f)]\eta(r)dr, \tag{3.5}$$

where use of (2.15) and a partial integration has been made. Since this must hold for any $\eta(r)$, the result sought for is obtained.

We now turn to the second linear constraint and try to see what are the restrictions it brings into our space of state vectors. We define

$$\hat{\mathcal{H}}_2(r)\Phi = 0 \leftrightarrow (\frac{\delta\,\Phi}{\delta\,\chi(r)})' = 0 \leftrightarrow \frac{\delta\,\Phi}{\delta\,\chi(r)} = k, \tag{3.6}$$

where k is any constant (with respect to r) independent of the basic fields and their derivatives, and Φ is given by (3.4a)−(3.4c). As we argued before, the functional derivative $\frac{\delta}{\delta\chi(r)}$ acting on $X^{(n)}$ will produce, upon partial integration of the nth derivative of the Dirac delta function, a term proportional to $X^{(2n)}$. Since the arguments of f in (3.4a) reach only up to $X^{(n)}$, it is evident that f must be such that the coefficient of $X^{(2n)}$ vanishes; more precisely

$$\begin{aligned}
\frac{\delta\,\Phi}{\delta\,\chi(r)} = k \leftrightarrow \quad & \ldots \quad + \int \rho(\tilde{r})\frac{\partial f}{\partial X^{(n)}(\tilde{r})}\frac{\delta X^{(n)}(\tilde{r})}{\delta\chi(r)}\,d\tilde{r} = k \leftrightarrow \\
& \ldots \quad + \int \rho(\tilde{r})\frac{\partial f}{\partial X^{(n)}(\tilde{r})}\frac{1}{\rho(\tilde{r})}\frac{d}{d\tilde{r}}\left(\underbrace{\cdots}_{n-1}\frac{\delta(r,\tilde{r})}{\rho(\tilde{r})}\right) d\tilde{r} = k \leftrightarrow \\
& \ldots \quad +(-1)^n \int \frac{\partial^2 f}{\partial(X^{(n)}(\tilde{r}))^2} X^{(2n)}(\tilde{r})\delta(r,\tilde{r})\,d\tilde{r} = k \leftrightarrow \\
& \ldots \quad +(-1)^n \frac{\partial^2 f}{\partial(X^{(n)})^2} X^{(2n)} = k.
\end{aligned}$$

Thus, since all the terms hidden in ... do not involve $X^{(2n)}$, then $\frac{\partial^2 f}{\partial (X^{(n)})^2} = 0$ in order for this equation to have a possibility to be satisfied. Subsequently:

$$f = f_1 \left(\Sigma^{(0)}, \ldots, \Sigma^{(n)}, X^{(0)}, \ldots, X^{(n-1)}\right) X^{(n)} + f_2 \left(\Sigma^{(0)}, \ldots, \Sigma^{(n)}, X^{(0)}, \ldots, X^{(n-1)}\right).$$

Now, the term in Φ corresponding to f_1 is, up to a surface term, equivalent to a general term depending on $X^{(0)}, \ldots, X^{(n-1)}$ only:

$$\Phi_1 = \int \rho(\tilde{r}) f_1 \frac{1}{\rho(\tilde{r})} \frac{d}{d\tilde{r}} X^{(n-1)} d\tilde{r},$$

which upon subtraction of the surface term

$$A = \int d\tilde{r} \frac{d}{d\tilde{r}} \left(\int dX^{(n-1)} f_1 \right),$$

produces a smooth functional with arguments up to $X^{(n-1)}$ only. Thus, only f_2 is important for the local part of Φ. The entire argument can be repeated successively for $n-1$, $n-2, \ldots, 1$; Therefore all $X^{(n)}$'s are suppressed from f except when $n = 0$. Thus, finally, upon inserting into (3.6) the resulting functional:

$$\Phi = \int \rho(\tilde{r}) \, h \left(\Sigma^{(0)}, \ldots, \Sigma^{(n)}, X^{(0)}\right) d\tilde{r},$$

we obtain

$$\begin{aligned} \frac{\delta \Phi}{\delta \chi(r)} &= k \leftrightarrow \int \rho(\tilde{r}) \frac{\partial h}{\partial X^{(0)}} \frac{\delta(r, \tilde{r})}{\rho(\tilde{r})} d\tilde{r} = k \leftrightarrow \frac{\partial h}{\partial X^{(0)}} = k \leftrightarrow \\ h &= k \frac{\chi(r)}{\rho(r)} + L \left(\Sigma^{(0)}, \ldots, \Sigma^{(n)}\right). \end{aligned}$$

We have thus reached the conclusion that the imposition of both linear quantum operators $\hat{\mathcal{H}}_1$ and $\hat{\mathcal{H}}_2$ dictates the form of the smooth functional to be:

$$\Phi = k \int d\tilde{r} \, \chi(\tilde{r}) + \int d\tilde{r} \, \rho(\tilde{r}) \, L \left(\Sigma^{(0)}, \ldots, \Sigma^{(n)}\right). \tag{3.7}$$

We now try to realize step 2 of the program previously outlined. We have to define the equivalent of Kuchař's induced metric on the so far "physical" space of states described by (3.7). Let us start our investigation by considering **one** initial candidate of the above form. Then the induced metric will be given by

$$g^{\Phi\Phi} = G^{\alpha\beta} \frac{\delta \Phi}{\delta x^\alpha} \frac{\delta \Phi}{\delta x^\beta}, \tag{3.8}$$

where $(x^\alpha, x^\beta) = (\rho, \sigma, \chi)$ and $G^{\alpha\beta}$ is given by (2.10). Note that this metric is well defined since it contains only first functional derivatives of the state vectors, as opposed to any second order functional derivative operator that might have been considered as a quantum analogue of the kinetic part of $\mathcal{H}_o$. Nevertheless, $g^{\Phi\Phi}$ will turn out to be a local function and not a smooth functional. It is thus clear that, if we want the induced metric $g^{\Phi\Phi}$ to be composed out of the "physical" states annihilated by $\hat{H}_1$, $\hat{H}_2$, we must establish a correspondence between local functions and smooth functionals. A way to achieve this is to adopt the following ansatz:

Assumption: *We assume that, as part of the re-normalization procedure, we are permitted to map local functions to their corresponding smeared expressions e.g.,* $\chi(r) \leftrightarrow \int d\tilde{r}\chi(\tilde{r})$.

Let us be more specific, concerning the meaning of the above Assumption. Let $\mathcal{F}$ be the space which contains all local functions, and define the equivalence relations

$$\sim: \{f_1(r) \sim f_2(\tilde{r}), \tilde{r} = g(r)\}, \quad \approx: \{h_1(r) \approx h_2(\tilde{r}) \frac{d\tilde{r}}{dr}, \tilde{r} = g(r)\}, \tag{3.9}$$

for scalars and densities respectively.

Now let $\mathcal{F}_o = \{f \in \mathcal{F}, \mod (\sim, \approx)\}$ and $\mathcal{F}_I$ the space of the smeared functionals. We define the one to one maps $\mathfrak{G}$, $\mathfrak{G}^{-1}$

$$\mathfrak{G} : \mathcal{F}_o \mapsto \mathcal{F}_I : \quad \chi(r) \mapsto \int \chi(\tilde{r})\, d\tilde{r}, \quad \mathfrak{G}^{-1} : \mathcal{F}_I \mapsto \mathcal{F}_o : \quad \int \chi(\tilde{r})\, d\tilde{r} \mapsto \chi(r). \tag{3.10}$$

The necessity to define the maps $\mathfrak{G}, \mathfrak{G}^{-1}$ on the equivalence classes and not on the individual functions, stems out of the fact that we are trying to develop a quantum theory of the geometries (2.6) and not of their coordinate representations. If we had tried to define the map $\mathfrak{G}$ from the original space $\mathcal{F}$ to $\mathcal{F}_I$ we would end up with states which would not be invariant under spatial coordinate transformations (**r**-reparametrizations). Indeed, one can make a correspondence between local functions and smeared expressions, but smeared expressions *must* contain another arbitrary smearing function, say $s(r)$. Then the map between functions and smeared expressions is one to one (as is also the above map) and is given by multiplying by $s(r)$ and integrating over r; while the inverse map is given by varying w.r.t. $s(r)$. However, this would be in the opposite direction from that which led us to the states (3.4a) by imposition of the linear operator constraint. As an example consider the action of this operator on one particular case of the states (3.4a), containing the structure $s(r)$:

$$\hat{\mathcal{H}}_1(r) \int s(\tilde{r})\, \rho(\tilde{r})\, \sigma(\tilde{r})\, d\tilde{r} = -s'(r)\, \rho(r)\, \sigma(r) \neq 0 \quad \text{for arbitrary}\, s(r), \tag{3.11}$$

$$\hat{\mathcal{H}}_2(r) \int s(\tilde{r})\, \chi(\tilde{r})\, d\tilde{r} = s'(r) \neq 0 \quad \text{for arbitrary}\, s(r). \tag{3.12}$$

Thus, every foreign to the geometry structure $s(r)$ is not allowed to enter the physical states.

Now, after the correspondence has been established, we can come to the basic property the induced metric must have. In the case of finite degrees of freedom the induced metric depends, up to a conformal scaling, on the physical coordinates q^α by virtue of (3.2). In our case, due to the dependence of the configuration variables on the radial coordinate r, the above property is not automatically satisfied; e.g. the functional derivative $\frac{\delta}{\delta\sigma(r)}$ acting on $\Sigma^{(n)}$ will produce, upon partial integration of the nth derivative of the Dirac delta function, a term proportional to $\Sigma^{(2n)}$. Therefore, since L in (3.7) contains derivatives of $\sigma(r)$ up to $\Sigma^{(n)}$, the above mentioned property must be *enforced.* The need for this can also be traced to the substantially different first Poisson bracket in (2.13), which signals a non trivial mixing between the dynamical evolution generator $\mathcal{H}_o$ and the linear generators $\mathcal{H}_1, \mathcal{H}_2$.

Journal of Physics: Conference Series **189** (2009) 012008 doi:10.1088/1742-6596/189/1/012008

Thus, according to the above reasoning, in order to proceed with the generalization of Kuchař's method, we have to demand that:

Requirement: $L\left(\Sigma^{(0)},\ldots,\Sigma^{(n)}\right)$ *must be such that* $g^{\Phi\Phi}$ *becomes a general function, say* $F\left(k\chi(r)+\rho(r)\,L(\Sigma^{(0)},\ldots,\Sigma^{(n)})\right)$ *of the integrand of* Φ*, so that it can be considered a function of this state:* $g^{\Phi\Phi} \overset{Assumption}{\equiv} F\left(k\int\chi(\tilde{r})d\tilde{r}+\int\rho(\tilde{r})\,L(\Sigma^{(0)},\ldots,\Sigma^{(n)})d\tilde{r}\right)=F(\Phi)$.

At this point, we must emphasize that the application of the **Requirement** in the subsequent development of our quantum theory will result in very severe restrictions on the form of (3.4a). Essentially, all higher derivatives of $\sigma(r)$ (i.e $\Sigma^{(2)}\ldots\Sigma^{(n)})$) are eliminated from Φ. This might, at first sight, strike as odd; indeed, the common belief is that all the derivatives of the configuration variables should enter the physical states. However, before the imposition of both the linear *and* the quadratic constrains there are no truly physical states. Thus, no physical states are lost by the imposition of the **Requirement**; ultimately the only true physical states are the solutions to the Wheeler–DeWitt equation, which will be constructed at a later stage.

Having clarified the way in which we view the **Assumption** and **Requirement** above, we now proceed to the restrictions implied by their use.

In the case of finite degrees of freedom the induced metric depends, up to a conformal scaling, on the physical coordinates q^α by virtue of (3.2). The reason for this not being the case for our system might be traced to the substantially different first Poisson bracket in (2.13). A first consequence of the requirement that $g^{\Phi\Phi}=F\left(k\,\chi(r)+\rho(r)\,L(\Sigma^{(0)},\ldots,\Sigma^{(n)})\right)$ is the vanishing of k. This follows from (a) the property that $g^{\Phi\Phi}$ is homogenous in the functional derivative $\frac{\delta}{\delta\chi(r)}$, (b) that $G^{\alpha\beta}$ in (2.10) does not contain any $\chi(r)$; namely

$$g^{\Phi\Phi}=\ldots+G^{33}\frac{\delta\Phi}{\delta\chi(r)}\frac{\delta\Phi}{\delta\chi(r)}\leftrightarrow g^{\Phi\Phi}=\underbrace{\ldots}_{no\ \chi}+\frac{\rho}{\sigma^2}\,k^2\equiv F(k\,\chi+\rho\,L).$$

Since ... are terms not involving $\chi(r)$, the final identification is possible iff $k=0$. Thus, Φ is reduced to:

$$\Phi=\int d\tilde{r}\,\rho(\tilde{r})\,L\left(\Sigma^{(0)},\ldots,\Sigma^{(n)}\right). \tag{3.13}$$

We now turn to the degree of derivatives (n) of $\sigma(r)$. The situation is similar to the corresponding case with $X^{(n)}$ considered before; again the functional derivative $\frac{\delta}{\delta\sigma(r)}$ acting on Φ will bring a maximum term $\Sigma^{(2n)}$ while $\frac{\delta}{\delta\rho(r)}$ a corresponding term $\Sigma^{(2n-1)}$. More precisely

$$g^{\Phi\Phi}=\ldots+2\,G^{12}\frac{\delta\Phi}{\delta\rho(r)}\frac{\delta\Phi}{\delta\sigma(r)}\,.$$

Journal of Physics: Conference Series **189** (2009) 012008 doi:10.1088/1742-6596/189/1/012008

Where the functional derivatives are:

$$\begin{aligned}
\frac{\delta\Phi}{\delta\sigma} &= \ldots + \int \rho \frac{\partial L}{\partial \Sigma^{(n)}} \frac{\delta \Sigma^{(n)}}{\delta\sigma}\, d\tilde{r} = \ldots + \int \rho \frac{\partial L}{\partial \Sigma^{(n)}} \frac{1}{\rho}\frac{d}{d\tilde{r}} \left(\underbrace{\ldots}_{n-1} \delta(r,\tilde{r}) \right) d\tilde{r} = \\
&= \ldots - \int \frac{d}{d\tilde{r}} \left(\frac{\partial L}{\partial \Sigma^{(n)}} \right) \frac{1}{\rho}\frac{d}{d\tilde{r}} \left(\underbrace{\ldots}_{n-2} \delta(r,\tilde{r}) \right) d\tilde{r} = \\
&= \ldots - \int \rho \frac{\partial^2 L}{\partial \left(\Sigma^{(n)}\right)^2} \Sigma^{(n+1)} \frac{1}{\rho}\frac{d}{d\tilde{r}} \left(\underbrace{\ldots}_{n-2} \delta(r,\tilde{r}) \right) d\tilde{r} = \\
&= \ldots + (-1)^n \int \rho(\tilde{r}) \frac{\partial^2 L}{\partial \left(\Sigma^{(n)}\right)^2} \Sigma^{(2n)}\, \delta(r,\tilde{r})\, d\tilde{r} = \\
&= \ldots + (-1)^n \rho \frac{\partial^2 L}{\partial \left(\Sigma^{(n)}\right)^2} \Sigma^{(2n)},
\end{aligned}$$

and

$$\begin{aligned}
\frac{\delta\Phi}{\delta\rho} &= \ldots + \int \rho \frac{\partial L}{\partial \Sigma^{(n)}} \frac{\delta \Sigma^{(n)}}{\delta\rho}\, d\tilde{r} = \ldots + \int \rho \frac{\partial L}{\partial \Sigma^{(n)}} \frac{1}{\rho}\frac{d}{d\tilde{r}} \left(\underbrace{\ldots}_{n-2} - \frac{\delta(r,\tilde{r})}{\rho(\tilde{r})^2} \sigma'(\tilde{r}) \right) d\tilde{r} = \\
&= \ldots + \int \rho \frac{\partial L}{\partial \Sigma^{(n)}} \frac{1}{\rho}\frac{d}{d\tilde{r}} \left(\underbrace{\ldots}_{n-2} - \frac{\delta(r,\tilde{r})}{\rho(\tilde{r})} \Sigma^{(1)} \right) d\tilde{r} \\
&= \ldots - \int \frac{d}{d\tilde{r}} \left(\frac{\partial L}{\partial \Sigma^{(n)}} \right) \frac{1}{\rho}\frac{d}{d\tilde{r}} \left(\underbrace{\ldots}_{n-3} - \frac{\delta(r,\tilde{r})}{\rho(\tilde{r})} \Sigma^{(1)} \right) d\tilde{r} = \\
&= \ldots + (-1)^{n-1} \int \frac{\partial^2 L}{\partial \left(\Sigma^{(n)}\right)^2} \Sigma^{(2n-1)}\, \Sigma^{(1)}\, \delta(r,\tilde{r})\, d\tilde{r} = \\
&= \ldots + (-1)^{n-1} \frac{\partial^2 L}{\partial \left(\Sigma^{(n)}\right)^2} \Sigma^{(2n-1)}\, \Sigma^{(1)}.
\end{aligned}$$

Therefore

$$g^{\Phi\Phi} = \ldots - 2\,\sigma\,(-1)^{2n-1} \left(\frac{\partial^2 L}{\partial \left(\Sigma^{(n)}\right)^2} \right)^2 \Sigma^{(1)}\, \Sigma^{(2n-1)}\, \Sigma^{(2n)},$$

where the $\ldots$ stand for all other terms, not involving $\Sigma^{(2n)}$. Now, according to the aforementioned **Requirement** we need this to be a general function, say $F(\rho L)$, and for this to happen the coefficient of $\Sigma^{(2n)}$ must vanish, i.e.

$$\frac{\partial^2 L}{\partial \left(\Sigma^{(n)}\right)^2} = 0 \Leftrightarrow L = L_1\left(\Sigma^{(0)}, \ldots, \Sigma^{(n-1)}\right) \Sigma^{(n)} + L_2\left(\Sigma^{(0)}, \ldots, \Sigma^{(n-1)}\right).$$

Again the term of Φ corresponding to L_1 is, up to a total derivative, equivalent to a local smooth functional containing $\Sigma^{(0)}, \ldots, \Sigma^{(n-1)}$. The argument can be repeated for $(n-1), (n-2), \ldots, 2$.

The case $n = 1$ needs separate consideration since, upon elimination of the linear in $\Sigma^{(2)}$ term we are left with a local function of $\Sigma^{(1)}$, and thus the possibility arises to meet the **Requirement** by solving a differential equation for L. In more detail, if

$$\Phi \equiv \int \rho(\tilde{r}) L\left(\sigma, \Sigma^{(1)}\right) d\tilde{r}, \tag{3.14}$$

$g^{\Phi\Phi}$ reads

$$\begin{aligned} g^{\Phi\Phi} &= -\rho\left(L - \Sigma^{(1)} \frac{\partial L}{\partial \Sigma^{(1)}}\right)\left[L - \Sigma^{(1)} \frac{\partial L}{\partial \Sigma^{(1)}} + 2\,\sigma\left(\frac{\partial L}{\partial \sigma} - \Sigma^{(1)} \frac{\partial^2 L}{\partial \sigma\, \partial \Sigma^{(1)}}\right)\right] + \\ &+2\,\rho\,\sigma\left(L - \Sigma^{(1)} \frac{\partial L}{\partial \Sigma^{(1)}}\right) \frac{\partial^2 L}{\partial (\Sigma^{(1)})^2}\, \Sigma^{(2)}. \end{aligned} \tag{3.15}$$

Through the definition

$$H \equiv L - \Sigma^{(1)} \frac{\partial L}{\partial \Sigma^{(1)}}, \tag{3.16}$$

we obtain

$$\begin{aligned} \frac{\partial H}{\partial \sigma} &= \frac{\partial L}{\partial \sigma} - \Sigma^{(1)} \frac{\partial^2 L}{\partial \sigma\, \partial \Sigma^{(1)}}, \\ \frac{\partial H}{\partial \Sigma^{(1)}} &= -\Sigma^{(1)} \frac{\partial^2 L}{\partial (\Sigma^{(1)})^2}. \end{aligned}$$

Thus $g^{\Phi\Phi}$ assumes the form

$$g^{\Phi\Phi} = -\rho\left(H^2 + 2\,\sigma\, H \frac{\partial H}{\partial \sigma} + \frac{2\,\sigma}{\Sigma^{(1)}} H \frac{\partial H}{\partial \Sigma^{(1)}} \Sigma^{(2)}\right),$$

which upon addition, by virtue of the **Assumption**, of the surface term

$$A = \frac{d}{dr}\left(\int \frac{2\,\sigma}{\Sigma^{(1)}} H \frac{\partial H}{\partial \Sigma^{(1)}}\, d\Sigma^{(1)}\right),$$

gives

$$g^{\Phi\Phi} = -\rho\left(H^2 + 2\,\sigma\, H \frac{\partial H}{\partial \sigma} - \Sigma^{(1)} \frac{\partial}{\partial \sigma} \int \frac{2\,\sigma}{\Sigma^{(1)}} H \frac{\partial H}{\partial \Sigma^{(1)}}\, d\Sigma^{(1)}\right). \tag{3.17}$$

Since in the last expression we have only a multiplicative $\rho(r)$, it is obvious that the **Requirement**

$$g^{\Phi\Phi} = F(\rho\, L),$$

can be satisfied only by

$$g^{\Phi\Phi} = -\kappa\, \rho\, L, \tag{3.18}$$

with $g^{\Phi\Phi}$ given by (3.17). Upon differentiation of this equation with respect to $\Sigma^{(1)}$ we get

$$-\frac{\partial}{\partial \sigma} \int \frac{2\,\sigma}{\Sigma^{(1)}} H \frac{\partial H}{\partial \Sigma^{(1)}} = \kappa \frac{\partial L}{\partial \Sigma^{(1)}}.$$

Multiplying the last expression by $\Sigma^{(1)}$ and subtracting it from (3.18) we end up with the autonomous necessary condition for $H(\sigma,\, \Sigma^{(1)})$:

$$H\left(H + 2\,\sigma\,\frac{\partial H}{\partial \sigma} - \kappa\right) = 0,$$

where the definition of H was also used. The above equation can be readily integrated giving

$$\begin{aligned} H &= 0, \\ H &= \kappa + \frac{a(\Sigma^{(1)})}{\sqrt{\sigma}}, \end{aligned}$$

where $a(\Sigma^{(1)})$ is an arbitrary function of its argument. The first possibility gives according to the definition of H, $L = \lambda\,\Sigma^{(1)}$ which, however, contributes to Φ a surface term, and can thus be ignored. Inserting the second solution into the definition of H we construct a partial differential equation for L, namely

$$L - \Sigma^{(1)}\,\frac{\partial L}{\partial \Sigma^{(1)}} = \kappa + \frac{a(\Sigma^{(1)})}{\sqrt{\sigma}},$$

which upon integration gives

$$L = \kappa - \frac{\Sigma^{(1)}}{\sqrt{\sigma}} \int \frac{a(\Sigma^{(1)})}{\Sigma^{(1)2}}\; d\Sigma^{(1)} + c_1(\sigma)\,\Sigma^{(1)}\,.$$

Since this form of L emerged as a necessary condition, it must be inserted (along with H) in (3.18). The result is that $c_1(\sigma) = 0$. Thus L reads

$$L = \kappa - \frac{\Sigma^{(1)}}{\sqrt{\sigma}} \int \frac{a(\Sigma^{(1)})}{\Sigma^{(1)2}}\; d\Sigma^{(1)}\,. \tag{3.19}$$

By assuming that the $\Sigma^{(1)}$–dependent part of L equals $b(\Sigma^{(1)})$, i.e.

$$-\Sigma^{(1)} \int \frac{a(\Sigma^{(1)})}{\Sigma^{(1)2}}\; d\Sigma^{(1)} = b(\Sigma^{(1)}),$$

we get, upon a double differentiation with respect to $\Sigma^{(1)}$, the ordinary differential equation

$$-\frac{a'(\Sigma^{(1)})}{\Sigma^{(1)}} = b''(\Sigma^{(1)}),$$

with solution

$$a(\Sigma^{(1)}) = b(\Sigma^{(1)}) + \kappa_1 - \Sigma^{(1)}\,b'(\Sigma^{(1)}),$$

where κ_1 is a constant. Substituting this equation into (3.19) and performing a partial integration we end up with

$$L = \kappa + \frac{\kappa_1}{\sqrt{\sigma}} + \frac{b(\Sigma^{(1)})}{\sqrt{\sigma}}\,. \tag{3.20}$$

κ, κ_1 and $b(\Sigma^{(1)})$ being completely arbitrary and to our disposal; the two simpler choices $\kappa_1 = 0$, $b(\Sigma^{(1)}) = 0$ and $\kappa = 0$, $b(\Sigma^{(1)}) = 0$ lead respectively to the following two basic local smooth functionals:

$$q^1 = \int d\tilde{r}\rho(\tilde{r}), \qquad q^2 = \int d\tilde{r}\frac{\rho(\tilde{r})}{\sqrt{\sigma(\tilde{r})}}. \tag{3.21}$$

The next simplest choice $\kappa = 0$, $\kappa_1 = 0$ and $b(\Sigma^{(1)})$ arbitrary leads to a generic $q^3 = \int d\tilde{r}\rho(\tilde{r})\,\frac{b(\Sigma^{(1)})}{\sqrt{\sigma(\tilde{r})}}$. However, it can be proven that, for any choice of $b(\Sigma^{(1)})$, the corresponding renormalized induced metric

$$g^{AB} = G^{\alpha\beta}\frac{\delta q^A}{\delta x^\alpha}\frac{\delta q^B}{\delta x^\beta} \qquad where \quad A, B = 1, 2, 3$$

is singular. The calculation of g^{AB} gives:

$$\begin{aligned}
g^{11} &= G^{\alpha\beta}\frac{\delta q^1}{\delta x^\alpha}\frac{\delta q^1}{\delta x^\beta} = -\rho \overset{Assumption}{\Longleftrightarrow} g^{11}_{ren} = -q^1,\\
g^{12} &= G^{\alpha\beta}\frac{\delta q^1}{\delta x^\alpha}\frac{\delta q^2}{\delta x^\beta} = -\frac{\rho}{2\sqrt{\sigma}} \overset{Assumption}{\Longleftrightarrow} g^{12}_{ren} = -\frac{q^2}{2},\\
g^{22} &= G^{\alpha\beta}\frac{\delta q^2}{\delta x^\alpha}\frac{\delta q^2}{\delta x^\beta} = 0 \overset{Assumption}{\Longleftrightarrow} g^{22}_{ren} = 0,\\
g^{13} &= G^{\alpha\beta}\frac{\delta q^1}{\delta x^\alpha}\frac{\delta q^3}{\delta x^\beta} = \rho\left(-\frac{b}{2\sqrt{\sigma}} + \frac{\Sigma^{(1)}}{2\sqrt{\sigma}}b' + \sqrt{\sigma}\,\Sigma^{(2)}\,b''\right) \overset{Assumption}{\Longleftrightarrow}\\
g^{13}_{ren} &= \int dr\rho\left(-\frac{b}{2\sqrt{\sigma}} + \frac{\Sigma^{(1)}}{2\sqrt{\sigma}}b' + \sqrt{\sigma}\,\Sigma^{(2)}\,b''\right) - \int dr\frac{d}{dr}\left(\int d\Sigma^{(1)}\sqrt{\sigma}\,b''\right) =\\
&= -\int dr\rho\,\frac{b}{2\sqrt{\sigma}} = -\frac{q^3}{2},
\end{aligned}$$

$$\begin{aligned}
g^{23} &= G^{\alpha\beta}\frac{\delta q^2}{\delta x^\alpha}\frac{\delta q^3}{\delta x^\beta} = \rho\,\Sigma^{(2)}\,b'' = \frac{d}{dr}\,b' \overset{Assumption}{\Longleftrightarrow} g^{23}_{ren} = 0,\\
g^{33} &= G^{\alpha\beta}\frac{\delta q^3}{\delta x^\alpha}\frac{\delta q^3}{\delta x^\beta} = 2\,\rho\left(b - \Sigma^{(1)}\,b'\right)\Sigma^{(2)}\,b'' \overset{Assumption}{\Longleftrightarrow}\\
g^{33}_{ren} &= 2\int dr\rho\left(b - \Sigma^{(1)}\,b'\right)\Sigma^{(2)}\,b'' - 2\int dr\,\frac{d}{dr}\left[\int d\Sigma^{(1)}\left(b - \Sigma^{(1)}\,b'\right)b''\right] = 0,
\end{aligned}$$

where by $'$ we denote differentiation with respect to $\Sigma^{(1)}$. Thus the renormalized induced metric reads

$$g^{AB}_{ren} = -\frac{1}{2}\begin{pmatrix} 2\,q^1 & q^2 & q^3 \\ q^2 & 0 & 0 \\ q^3 & 0 & 0 \end{pmatrix}.$$

Effecting the transformation $(\tilde{q}^1, \tilde{q}^2, \tilde{q}^3) = \left(q^1, q^2, f(\frac{q^3}{q^2})\right)$ we bring g^{AB}_{ren} into a manifestly degenerate form:

$$g^{AB}_{ren} = -\frac{1}{2}\begin{pmatrix} 2\,q^1 & q^2 & 0 \\ q^2 & 0 & 0 \\ 0 & 0 & 0 \end{pmatrix}.$$

So, it seems as though the relevant part of the renormalized metric is described by the upper 2×2 block. This fact is consistent with the form of the renormalized potential $V = 2\,\Lambda\,q^1$ which indeed does not contain any $\Sigma^{(1)}$ term.

However, this is not the end of our investigation for a suitable space of state vectors: the argument leading to q^1, q^2 depends upon the original choice of <u>one</u> initial candidate smooth scalar functional (3.14); to complete the search we must close the circle by starting with the two already secured smooth functionals (q^1, q^2), and a <u>third</u> of the general form

$$q^3 = \int dr\,\rho\,L(\Sigma^{(1)}),$$

since the σ dependence has already been fixed to either 1 or $\frac{1}{\sqrt{\sigma}}$. The calculation of the, related to q^3, components of the induced metric g^{AB} gives:

$$\begin{aligned}
g^{13} &= \rho\left(-L+\Sigma^{(1)}\,L'+\sigma\,\Sigma^{(2)}\,L''\right) \overset{Assumption}{\Longleftrightarrow} \\
g^{13}_{ren} &= \int dr\rho\left(-L+\Sigma^{(1)}\,L'+\sigma\,\Sigma^{(2)}\,L''\right) - \int dr\frac{d}{dr}\left(\int d\Sigma^{(1)}\,\sigma\,L''\right) = -\int dr\rho\,L = \\
&= -q^3, \\
g^{23} &= \rho\left(-\frac{L}{2\sqrt{\sigma}}+\frac{\Sigma^{(1)}}{2\sqrt{\sigma}}L'+\sqrt{\sigma}\,\Sigma^{(2)}\,L''\right) \overset{Assumption}{\Longleftrightarrow} \\
g^{23}_{ren} &= \int dr\rho\left(-\frac{L}{2\sqrt{\sigma}}+\frac{\Sigma^{(1)}}{2\sqrt{\sigma}}L'+\sqrt{\sigma}\,\Sigma^{(2)}\,L''\right) - \int dr\frac{d}{dr}\left(\int d\Sigma^{(1)}\,\sqrt{\sigma}\,L''\right) = \\
&= -\int dr\rho\,\frac{L}{2\sqrt{\sigma}} \overset{Assumption}{=} -\frac{q^2\,q^3}{2\,q^1}, \\
g^{33} &= -\rho\left(L-\Sigma^{(1)}L'\right)^2 + 2\,\rho\,\sigma\left(L-\Sigma^{(1)}L'\right)\Sigma^{(2)}L''.
\end{aligned}$$

By following the procedure presented between (3.15) and (3.17) we end up with the expression

$$g^{33} = -\rho\left[\left(L-\Sigma^{(1)}L'\right)^2 - \Sigma^{(1)}\int\frac{d\Sigma^{(1)}}{\Sigma^{(1)}}\frac{\partial}{\partial\Sigma^{(1)}}\left(L-\Sigma^{(1)}L'\right)^2\right],$$

which is a generic function of $\Sigma^{(1)}$ and therefore, also of L: let this function be parameterized as $L\left(\Sigma^{(1)}\right)^2 - \frac{1}{H'\left(L(\Sigma^{(1)})\right)}$; this "peculiar" parametrization of the arbitrariness in $L\left(\Sigma^{(1)}\right)$ has been chosen in order to facilitate the subsequent proof that this freedom is a pure general coordinate transformation (gct) of the induced renormalized metric. Indeed, let us first take the simple choice $L\left(\Sigma^{(1)}\right) \equiv {\Sigma^{(1)}}^2$ which results in the renormalized metric

$$g^{AB}_{ren} = -\frac{1}{2}\begin{pmatrix} 2\,q^1 & q^2 & 2\,q^3 \\ q^2 & 0 & \frac{q^2 q^3}{q^1} \\ q^3 & \frac{q^2 q^3}{q^1} & -\frac{2\,(q^3)^2}{3\,q^1} \end{pmatrix}, \qquad g_{AB\,ren} = \frac{1}{2}\begin{pmatrix} \frac{3}{2\,q^1} & -\frac{4}{q^2} & -\frac{3}{2\,q^3} \\ -\frac{4}{q^2} & \frac{8\,q^1}{(q^2)^2} & 0 \\ -\frac{3}{2\,q^3} & 0 & \frac{3\,q^1}{2(q^3)^2} \end{pmatrix}. \tag{3.22}$$

Considering a generic $L\left(\Sigma^{(1)}\right)$, i.e. $x^3 = \int dr\,\rho\,L(\Sigma^{(1)})$ (along with (3.21)) we are led to

$$g^{13} = -x^3, \qquad g^{23} = -\frac{q^2\,x^3}{2\,q^1},$$

and

$$g^{33} = -\rho\left[L^2 - \frac{1}{H'(L)}\right] = -\frac{(\rho\,L)^2}{\rho} + \frac{\rho}{H'\left(\frac{\rho L}{\rho}\right)} \overset{Assumption}{\Longleftrightarrow} g^{33}_{ren} = -\frac{(x^3)^2}{q^1} + \frac{q^1}{H'\left(\frac{x^3}{q^1}\right)}.$$

Remarkably enough, the new induced renormalized metric can be put in gct equivalence with the metric (3.22) through the transformation

$$(q^1, q^2, x^3) = (q^1, q^2, q^1 H^{-1}(\frac{\sqrt{3}}{2}\ln\frac{q^3}{q^1})),$$

with H^{-1} denoting the function inverse to H.

We can therefore consider, without loss of generality, the reduced renormalized manifold to be parameterized by the following three smooth scalar functionals:

$$q^1 = \int d\tilde{r}\rho(\tilde{r}), \qquad q^2 = \int d\tilde{r}\frac{\rho(\tilde{r})}{\sqrt{\sigma(\tilde{r})}}, \qquad q^3 = \int d\tilde{r}\,\frac{\sigma'(\tilde{r})^2}{\rho(\tilde{r})}. \tag{3.23}$$

Any other functional, say $q^4 = \int d\tilde{r}\,\rho(\tilde{r})\,L(\sigma(\tilde{r}), \Sigma^{(1)}(\tilde{r}))$, can be considered as a function of q^1, q^2, q^3; indeed, since the scalar functions appearing in the integrands of q^2, q^3 form a base in the space of $\sigma, \Sigma^{(1)}$, we can express the generic L in q^4 as $F(\frac{\rho}{\rho\sqrt{\sigma}}, \frac{\rho\Sigma^{(1)2}}{\rho})$), which (through the **Assumption**) gives $q^4 = q^1 F(\frac{q^2}{q^1}, \frac{q^3}{q^1})$.

The geometry of this space is described by the induced renormalized metric (3.22). Any function $\Psi(q^1, q^2, q^3)$ on this manifold is of course annihilated by the quantum linear constraints, i.e.

$$\hat{\mathcal{H}}_1\Psi(q^1, q^2, q^3) = \frac{\partial\Psi(q^1, q^2, q^3)}{\partial q^1}\hat{\mathcal{H}}_1\,q^1 + \frac{\partial\Psi(q^1, q^2, q^3)}{\partial q^2}\hat{\mathcal{H}}_1\,q^2 + \frac{\partial\Psi(q^1, q^2, q^3)}{\partial q^3}\hat{\mathcal{H}}_1\,q^3 = 0,$$

$$\hat{\mathcal{H}}_2\Psi(q^1, q^2, q^3) = \frac{\partial\Psi(q^1, q^2, q^3)}{\partial q^1}\hat{\mathcal{H}}_2\,q^1 + \frac{\partial\Psi(q^1, q^2, q^3)}{\partial q^2}\hat{\mathcal{H}}_2\,q^2 + \frac{\partial\Psi(q^1, q^2, q^3)}{\partial q^3}\hat{\mathcal{H}}_2\,q^3 = 0,$$

since the derivatives with respect to r are transparent to the partial derivatives of Ψ (which are, just like the q^A's, r-numbers).

The final restriction on the form of Ψ will be obtained by the imposition of the quantum analog of the quadratic constraint $\mathcal{H}_o$. According to the above exposition we postulate that the quantum gravity of the geometries given by (2.6), (2.7) will be described by

$$\hat{\mathcal{H}}_o\Psi \equiv [-\frac{1}{2}\,\Box_c + 2\,\Lambda\,q^1]\,\Psi(q^1,q^2,q^3) = 0, \tag{3.24}$$

with

$$\Box_c = \Box + \frac{d-2}{4\,(d-1)}\,R, \tag{3.25}$$

being the conformal Laplacian based on $g_{AB\,ren}$, R the Ricci scalar, and d the dimensions of $g_{AB\,ren}$. Metric (3.22) is conformally flat with Ricci scalar $R = \frac{1}{2\,q^1}$, its dimension is $d = 3$, and therefore the Wheeler–DeWitt equation becomes

$$\begin{aligned}&2\,q^1\,\Lambda\,\Psi(q^1,q^2,q^3) - \frac{1}{32\,q^1}\Psi(q^1,q^2,q^3) + \frac{q^3}{12\,q^1}\frac{\partial\Psi(q^1,q^2,q^3)}{\partial q^3} + \frac{q^2}{8\,q^1}\frac{\partial\Psi(q^1,q^2,q^3)}{\partial q^2} +\\ &\frac{3}{4}\frac{\partial\Psi(q^1,q^2,q^3)}{\partial q^1} + \frac{q^2q^3}{2\,q^1}\frac{\partial^2\Psi(q^1,q^2,q^3)}{\partial q^2\,\partial q^3} + q^3\,\frac{\partial^2\Psi(q^1,q^2,q^3)}{\partial q^1\,\partial q^3} + \frac{q^2}{2}\frac{\partial^2\Psi(q^1,q^2,q^3)}{\partial q^1\,\partial q^2} +\\ &\frac{q^1}{2}\frac{\partial^2\Psi(q^1,q^2,q^3)}{\partial\,(q^1)^2} - \frac{(q^3)^2}{6\,q^1}\frac{\partial^2\Psi(q^1,q^2,q^3)}{\partial\,(q^3)^2} = 0.\end{aligned} \tag{3.26}$$

The change to new coordinates (x^1, x^2, x^3) described by

$$(q^1,\,q^2,\,q^3) = \left(e^{x^2},\,e^{\frac{1}{2}(x^1+x^2)},\,e^{x^2+\frac{2}{\sqrt{3}}x^3}\right),$$

transforms the metric into the manifestly conformally flat form $diag\{e^{x^2}, -e^{x^2}, -e^{x^2}\}$ and brings (3.26) into the form

$$\begin{aligned}&2\,e^{2\,x^2}\,\Lambda\,\Psi(x^1,x^2,x^3) - \frac{1}{32}\,\Psi(x^1,x^2,x^3) + \frac{1}{4}\frac{\partial\Psi(x^1,x^2,x^3)}{\partial x^2} - \frac{1}{2}\frac{\partial^2\Psi(x^1,x^2,x^3)}{\partial\,(x^1)^2} +\\ &\frac{1}{2}\frac{\partial^2\Psi(x^1,x^2,x^3)}{\partial\,(x^2)^2} + \frac{1}{2}\frac{\partial^2\Psi(x^1,x^2,x^3)}{\partial\,(x^3)^2} = 0.\end{aligned} \tag{3.27}$$

This equation is readily solved by the method of separation of variables: assuming $\Psi(x^1,x^2,x^3) = X^1(x^1)\,X^2(x^2)\,X^3(x^3)$ and dividing (3.27) by Ψ we get the three ordinary differential equations:

$$\frac{1}{2\,X^1(x^1)}\frac{d^2X^1(x^1)}{d\,(x^1)^2} + \frac{1}{32} = m + n, \tag{3.28a}$$

$$\frac{1}{2\,X^2(x^2)}\frac{d^2X^2(x^2)}{d\,(x^2)^2} + \frac{1}{4\,X^2(x^2)}\frac{dX^2(x^2)}{dx^2} + 2\,e^{2\,x^2}\,\Lambda = n, \tag{3.28b}$$

$$\frac{1}{2\,X^3(x^3)}\frac{d^2X^3(x^3)}{d\,(x^3)^2} = m, \tag{3.28c}$$

where m and n are separation constants. Their solutions are:

$$
\begin{aligned}
X^1(x^1) &= c_1\, e^{\frac{1}{4}\sqrt{32\,m+32\,n-1}\; x^1} + c_2\, e^{-\frac{1}{4}\sqrt{32\,m+32\,n-1}\; x^1}\,,\\
X^2(x^2) &= c_3\, e^{-x^2/4}\,\Gamma\left(1-\frac{1}{4}\sqrt{32\,n+1}\right) J_{-\frac{1}{4}\sqrt{32\,n+1}}\left(2\,e^{x^2}\sqrt{\Lambda}\right)+\\
&+\; c_4\, e^{-x^2/4}\,\Gamma\left(1+\frac{1}{4}\sqrt{32\,n+1}\right) J_{\frac{1}{4}\sqrt{32\,n+1}}\left(2\,e^{x^2}\sqrt{\Lambda}\right)\\
X^3(x^3) &= c_5\, e^{\sqrt{2\,m}\; x^3} + c_6\, e^{-\sqrt{2\,m}\; x^3}\,,
\end{aligned}
$$

where $\Gamma\left(1 \pm \frac{1}{4}\sqrt{32\,n+1}\right)$ are Gamma functions and $J_{\pm\frac{1}{4}\sqrt{32\,n+1}}\left(2\,e^{x^2}\sqrt{\Lambda}\right)$ Bessel functions of the first kind and of non-integer order.

3.2. The 3+1 Case

The quantization of the Hamiltonian system (2.25a), (2.25b) proceeds in exact analogy to the 2+1 case described above. In order to realize the equivalent to step 1, we first define the quantum analogue of $\mathcal{H}_1(r) \approx 0$ as

$$\hat{\mathcal{H}}_1(r)\Phi = 0 \leftrightarrow -\gamma(r)\,(\frac{\delta\,\Phi}{\delta\,\gamma(r)})' + \psi'(r)\,\frac{\delta\,\Phi}{\delta\,\psi(r)} = 0. \tag{3.29}$$

The general solution to this equation is given by the following family of smooth scalar functionals

$$\Phi = \int \gamma(\tilde{r})\, L\left(\Psi^{(0)}, \Psi^{(1)}, \ldots, \Psi^{(n)}\right) d\tilde{r}, \tag{3.30a}$$

$$\Psi^{(0)} \equiv \psi(\tilde{r}), \quad \Psi^{(1)} \equiv \frac{\psi'(\tilde{r})}{\gamma(\tilde{r})}, \ldots, \Psi^{(n)} \equiv \frac{1}{\gamma(\tilde{r})}\frac{d}{d\tilde{r}}\left(\underbrace{\cdots}_{n-1}\psi(\tilde{r})\right). \tag{3.30b}$$

In trying to realize step 2 of the program, we have to define the equivalent of Kuchař's induced metric on the so far space of "physical" states Φ described by (3.30a) which are the analogues, in our case, of Kuchař's physical variables q^α. The imposition of the **Assumption** and the **Requirement** described in detail in the previous section results in the following three smooth scalar functionals which parametrize the reduced renormalized manifold:

$$y^1 = \int \gamma(\tilde{r})d\tilde{r}, \qquad y^2 = \int \gamma(\tilde{r})\psi(\tilde{r})^2 d\tilde{r}, \qquad y^3 = \int \gamma(\tilde{r})L(\Psi^{(1)})d\tilde{r}. \tag{3.31}$$

Any other functional, say $y^4 = \int d\tilde{r}\,\gamma(\tilde{r})\,K\left[\psi(\tilde{r}), \Psi^{(1)}(\tilde{r})\right]$, can be considered as a function of y^1, y^2, y^3; indeed, since the scalar functions appearing in the integrands of y^2, y^3 form a base in the space spanned by $\psi, \Psi^{(1)}(r)$, we can express the generic K in y^4 as $K\left[\sqrt{\frac{\gamma\psi^2}{\gamma}}, \Psi^{(1)}\right]$, which (through the **Assumption**) gives $y^4 = y^1\,K\left[\sqrt{\frac{y^2}{y^1}}, L^{-1}\left(\frac{y^3}{y^1}\right)\right]$.

The geometry of this space is described by the induced renormalized metric

$$g^{AB}_{ren}(y^1,y^2,y^3) = -\frac{1}{4}\begin{pmatrix} -\frac{(y^1)^2}{y^2} & y^1 & -\frac{y^1y^3}{y^2} \\ y^1 & 3\,y^2 & y^3 \\ -\frac{y^1y^3}{y^2} & y^3 & -\frac{(y^3)^2}{y^2}+\frac{4\,(y^1)^2F\left(\frac{y^3}{y^1}\right)^2}{3\,y^2F'\left[F\left(\frac{y^3}{y^1}\right)\right]^2} \end{pmatrix},$$

$$g_{AB_{ren}}(y^1,y^2,y^3) = \begin{pmatrix} \frac{3\,y^2}{(y^1)^4}\left((y^1)^2-\frac{(y^3)^2F'\left[F\left(\frac{y^3}{y^1}\right)\right]^2}{F\left(\frac{y^3}{y^1}\right)^2}\right) & -\frac{1}{y^1} & \frac{3\,y^2y^3F'\left[F\left(\frac{y^3}{y^1}\right)\right]^2}{(y^1)^3F\left(\frac{y^3}{y^1}\right)^2} \\ -\frac{1}{y^1} & -\frac{1}{y^2} & 0 \\ \frac{3\,y^2y^3F'\left[F\left(\frac{y^3}{y^1}\right)\right]^2}{(y^1)^3F\left(\frac{y^3}{y^1}\right)^2} & 0 & -\frac{3\,y^2F'\left[F\left(\frac{y^3}{y^1}\right)\right]^2}{(y^1)^2F\left(\frac{y^3}{y^1}\right)^2} \end{pmatrix}. \quad (3.32)$$

Any function $\Psi(y^1,y^2,y^3)$ on this manifold is of course annihilated by the quantum linear constraint, i.e.

$$\hat{\mathcal{H}}_1\Psi(y^1,y^2,y^3) = \frac{\partial\Psi(y^1,y^2,y^3)}{\partial y^1}\,\hat{\mathcal{H}}_1\,y^1 + \frac{\partial\Psi(y^1,y^2,y^3)}{\partial y^2}\,\hat{\mathcal{H}}_1\,y^2 + \frac{\partial\Psi(y^1,y^2,y^3)}{\partial y^3}\,\hat{\mathcal{H}}_1\,y^3 = 0,$$

since the derivatives with respect to r are transparent to the partial derivatives of Ψ (which are, just like the y^A's, r-numbers).

The covariant metric (3.32) describes a three dimensional conformally flat geometry, since the Cotton-York tensor vanishes. The Ricci scalar is $R = \frac{3}{8\,y^2}$, indicating that the arbitrariness in F (and thus also in L) is a *pure gauge*. The change of coordinates

$$(y^1,y^2,y^3) = (e^{-\frac{1}{8}(5\,Y^1+3\,Y^3)}, e^{Y^1+Y^2+Y^3}, e^{-\frac{1}{8}(5\,Y^1+3\,Y^3)}F^{-1}(e^{\frac{1}{24}(-9\,Y^1+8\,Y^2-15\,Y^3)})), \quad (3.33)$$

(where F^{-1} denotes the function inverse to F, i.e $F^{-1}(F(x)) = x$) brings the metric to the manifestly conformally flat form:

$$g_{AB_{ren}}(Y^1,Y^2,Y^3) = \begin{pmatrix} e^{Y^1+Y^2+Y^3} & 0 & 0 \\ 0 & -\frac{4}{3}\,e^{Y^1+Y^2+Y^3} & 0 \\ 0 & 0 & -e^{Y^1+Y^2+Y^3} \end{pmatrix}, \quad (3.34)$$

in which all the F dependence has indeed disappeared.

The final restriction on the form of Ψ will be obtained by the imposition of the quantum analog of the quadratic constraint $\mathcal{H}_o$. According to the above exposition we postulate that the quantum gravity of the geometries given by (2.21) will be described by the following partial differential equation (in terms of the Y^A's)

$$\hat{\mathcal{H}}_o\Psi \equiv [-\frac{1}{2}\Box_c + V_{ren}]\,\Psi(Y^1,Y^2,Y^3) = 0, \tag{3.35}$$

with

$$\Box_c = \Box + \frac{d-2}{4\,(d-1)}\,R, \tag{3.36}$$

being the conformal Laplacian based on $g_{AB\,ren}(Y^1,Y^2,Y^3)$, R the Ricci scalar, and d the dimensions of $g_{AB\,ren}$. The metric (3.34) is conformally flat with Ricci scalar $R = \frac{3}{8}\,e^{-Y^1-Y^2-Y^3}$, and its dimension is $d = 3$. The renormalized form of the potential (2.27) offers us the possibility to introduce, in a dynamical way, topological effects into our wave functional: Indeed, under our **Assumption**, the first two terms become $-2\,\epsilon\,y^1$ and $2\,\Lambda\,y^2$, respectively, while the last, being a total derivative, becomes $A_T \equiv 4\,\frac{\psi\,\psi'}{\gamma}\,|_\alpha^\beta$ $(if\;\alpha < r < \beta)$. In the spirit previously explained we should drop this term, however one could also keep it. The renormalized form of the remaining, third, term of the potential can be obtained as follows

$$y^3 \;=\; \gamma\,L(\Psi^{(1)}) \Leftrightarrow L(\Psi^{(1)}) = \frac{y^3}{\gamma} \overset{Assumption}{\Longleftrightarrow} L(\Psi^{(1)}) = \frac{y^3}{y^1} \Leftrightarrow \Psi^{(1)} = L^{-1}\left(\frac{y^3}{y^1}\right),$$

thus finally

$$\frac{\psi'}{\gamma} = L^{-1}\left(\frac{y^3}{y^1}\right),$$

and the third term becomes $-2\,y^1\left[L^{-1}\left(\frac{y^3}{y^1}\right)\right]^2$. Finally, effecting the transformation (3.33) the form of the renormalized potential is

$$\begin{aligned} V_{ren} \;=\;& -2\,\epsilon\,e^{-\frac{1}{8}(5\,Y^1+3\,Y^3)} - 2\,e^{-\frac{1}{8}(5\,Y^1+3\,Y^3)}\left[L^{-1}\left(F^{-1}(e^{\frac{1}{24}(-9\,Y^1+8\,Y^2-15\,Y^3)})\right)\right]^2 + \\ & 2\,\Lambda\,e^{Y^1+Y^2+Y^3} + A_T, \end{aligned} \tag{3.37}$$

and the Wheeler-deWitt equation is given as

$$\begin{aligned} & -2\,\epsilon\,e^{\frac{1}{8}(3\,Y^1+5\,Y^3)+Y^2}\,\Psi(Y^1,Y^2,Y^3) + 2\,\Lambda\,e^{2(Y^1+Y^2+Y^3)}\,\Psi(Y^1,Y^2,Y^3) - \\ & 2\,e^{\frac{1}{8}(3\,Y^1+5\,Y^3)+Y^2}\left[L^{-1}\left(F^{-1}(e^{\frac{1}{24}(-9\,Y^1+8\,Y^2-15\,Y^3)})\right)\right]^2\Psi(Y^1,Y^2,Y^3) + \\ & A_T\,e^{Y^1+Y^2+Y^3}\,\Psi(Y^1,Y^2,Y^3) - \frac{3}{128}\Psi(Y^1,Y^2,Y^3) - \frac{1}{4}\frac{\partial\Psi(Y^1,Y^2,Y^3)}{\partial Y^1} + \\ & \frac{3}{16}\frac{\partial\Psi(Y^1,Y^2,Y^3)}{\partial Y^2} + \frac{1}{4}\frac{\partial\Psi(Y^1,Y^2,Y^3)}{\partial Y^3} - \frac{1}{2}\frac{\partial^2\Psi(Y^1,Y^2,Y^3)}{\partial\,(Y^1)^2} + \frac{3}{8}\frac{\partial^2\Psi(Y^1,Y^2,Y^3)}{\partial\,(Y^2)^2} + \\ & \frac{1}{2}\frac{\partial^2\Psi(Y^1,Y^2,Y^3)}{\partial\,(Y^3)^2} = 0. \end{aligned}$$

Since F^{-1} is an arbitrary function of its arguments, we may contemplate the choice:

$$F^{-1}\left(e^{\frac{1}{24}(-9\,Y^1+8\,Y^2-15\,Y^3)}\right) = L\left(\sqrt{e^{\frac{1}{24}(-9\,Y^1+8\,Y^2-15\,Y^3)} - \epsilon}\right).$$

Of course, there is a question of existence for such a choice: any demand that F has a specified form (much more in terms of L) constitutes a restriction on the form of L. Consequently, one has to prove that such an L indeed exists. The appropriate form of L is implicitly given by

$$L(\omega) = m + \int \frac{(\epsilon+\omega^2)^{1/4}}{\omega^{5/8}} e^{k-\frac{3\epsilon}{16\omega^2}}\, d\omega,$$

where $\omega \equiv \sqrt{\Psi^{(1)2} - \epsilon}$ and m, k are related by

$$c_1 m + c_2 + c_3 e^k = 0.$$

For a detailed elaboration of the results, see [39].

Anyway, the above choice of F reduces the Wheeler–DeWitt equation to the final separable form:

$$\begin{aligned}&\left[2\Lambda\, e^{2(Y^1+Y^2+Y^3)} - 2\,e^{\frac{4}{3}Y^2} + A_T\, e^{Y^1+Y^2+Y^3} - \frac{3}{128}\right]\Psi(Y^1,Y^2,Y^3) - \\ &\frac{1}{4}\frac{\partial\Psi(Y^1,Y^2,Y^3)}{\partial Y^1} + \frac{3}{16}\frac{\partial\Psi(Y^1,Y^2,Y^3)}{\partial Y^2} + \frac{1}{4}\frac{\partial\Psi(Y^1,Y^2,Y^3)}{\partial Y^3} - \\ &\frac{1}{2}\frac{\partial^2\Psi(Y^1,Y^2,Y^3)}{\partial\,(Y^1)^2} + \frac{3}{8}\frac{\partial^2\Psi(Y^1,Y^2,Y^3)}{\partial\,(Y^2)^2} + \frac{1}{2}\frac{\partial^2\Psi(Y^1,Y^2,Y^3)}{\partial\,(Y^3)^2} = 0. \end{aligned} \tag{3.38}$$

This equation is readily solved, for $\Lambda = 0$ and $A_T = 0$, by the method of separation of variables: assuming $\Psi(Y^1,Y^2,Y^3) = \Psi^1(Y^1)\,\Psi^2(Y^2)\,\Psi^3(Y^3)$ and dividing (3.38) by Ψ we get the three ordinary differential equations:

$$\begin{aligned}&\frac{1}{4\,\Psi^1(Y^1)}\frac{d\Psi^1(Y^1)}{dY^1} + \frac{1}{2\,\Psi^1(Y^1)}\frac{d^2\Psi^1(Y^1)}{d\,(Y^1)^2} = m,\\ &\frac{3}{16\,\Psi^2(Y^2)}\frac{d\Psi^2(Y^2)}{dY^2} + \frac{3}{8\,\Psi^2(Y^2)}\frac{d^2\Psi^2(Y^2)}{d\,(Y^2)^2} - 2\,e^{\frac{4}{3}Y^2} = n,\\ &\frac{1}{4\,\Psi^3(Y^3)}\frac{d\Psi^3(Y^3)}{dY^3} + \frac{1}{2\,\Psi^3(Y^3)}\frac{d^2\Psi^3(Y^3)}{d\,(Y^3)^2} - \frac{3}{128} = m-n,\end{aligned}$$

where m and n are separation constants. Their solutions are:

$$\begin{aligned}\Psi^1(Y^1) &= c_1\, e^{\frac{1}{4}\left(-1-\sqrt{1+32\,m}\right)Y^1} + c_2\, e^{\frac{1}{4}\left(-1+\sqrt{1+32\,m}\right)Y^1},\\ \Psi^2(Y^2) &= c_3\, e^{-Y^2/4}\, I_{-\frac{\sqrt{3}}{8}\sqrt{3+128\,n}}\left(2\sqrt{3}\,e^{2Y^2/3}\right) + c_4\, e^{-Y^2/4}\, I_{\frac{\sqrt{3}}{8}\sqrt{3+128\,n}}\left(2\sqrt{3}\,e^{2Y^2/3}\right),\\ \Psi^3(Y^3) &= c_5\, e^{\frac{1}{8}\left(-2-\sqrt{7+128\,m-128\,n}\right)Y^3} + c_6\, e^{\frac{1}{8}\left(-2+\sqrt{7+128\,m-128\,n}\right)Y^3},\end{aligned}$$

where $I_{\pm\frac{\sqrt{3}}{8}\sqrt{3+128\,n}}\left(2\sqrt{3}\,e^{2Y^2/3}\right)$ are modified Bessel functions of the first kind and of non-integer order.

4. Discussion

We have considered the canonical analysis and subsequent quantization of two midi-superspace models in 2+1 and 3+1 dimensions. The simplifying assumption was the existence of appropriate Killing fields, which reduced the components of the metrics to functions of time and a radial coordinate. Thus, the emanating field theory is effectively one-dimensional. The quantum linear constraints reduced the initial Hilbert space consisting of all smooth functionals to the space of all smooth scalar functionals. The appropriate imposition of the quadratic constraint leads us to the **Assumption** and the **Requirement**, which strongly further restrict the space of functionals to the three functionals (3.23) and (3.31) for the 2+1 and 3+1 system, respectively. It is quite interesting that, although the systems considered are different, the resulting renormalized manifolds, their geometry, described by (3.22) and (3.32), and the corresponding Wheeler–DeWitt equations are strikingly similar. This points to the fact that the imposition of the **Assumption** and the **Requirement** may not be simply an elegant way to reduce the number of spatial derivatives of the configuration fields, contained in the scalar functional, but is rather a tool for unraveling the underlying geometrical structure of Quantum Gravity (in the approximation considered). We have reasons to believe that our method can be applied to more general models, even in full pure gravity, and we hope to return with relevant investigations in the very near future.

4.1. Acknowledgments

One of the authors (G. O. Papadopoulos) is a Killam Postdoctoral Fellow and acknowledges the relevant support from the Killam Foundation.

References

[1] Dirac P A M 1950 *Can. J. Math.* **2** 129
[2] Dirac P A M 1951 *Can. J. Math.* **3** 1
[3] Dirac P A M 1958 *Proc. R. Soc.* (London) A **246** 326
[4] Dirac P A M 1964 *Lectures on Quantum Mechanics* (New York: Yeshiva University, Academic Press)
[5] Bergmann P G 1961 *Rev. Mod. Phys.* **33** 510
[6] Berezin F A and Marinov M S 1977 *Ann. Phys.* (N.Y.) **104** 336
[7] Becchi C, Rouet C and Stora R 1976 *Ann. Phys.* (N.Y.) **98** 287
[8] Sudarshan E C G and Mukunda N 1974 *Classical Dynamics: A Modern Perspective* (New York: Wiley)
[9] Hanson A J, Regge T and Teitelboim C 1976 *Constrained Hamiltonian Systems* (Vatican: Accademia Nazionale dei Lincei)
[10] Sundermeyer K 1982 *Lecture Notes in Physics* **169** *Constrained Dynamics* (Heidelberg: Springer-Verlag)
[11] Gitman D M and Tyutin I V 1991 *Quantization of Fields with Constraints* Springer Series in Nuclear and Particle Physics
[12] Govaers J 1991 *Hamiltonian Quantization and Constrained Dynamics* (Leuven: Leuven University Press)
[13] Henneaux M and Teitelboim C 1992 *Quantization of Gauge Systems* (Princeton:Princeton University Press)
[14] Wipf A *Hamilton's Formalism for Systems with Constraints* (*Preprint* hep-th/9312078)
[15] Thiemann Th 2007 *Modern Canonical Quantum General Relativity* (Cambridge: Cambridge University Press)
[16] DeWitt B S 1967 *Phys. Rev.* **160**(5) 1113
[17] Bergmann P G 1966 *Phys. Rev.* **144** 1078
[18] Ryan M P Jr and Shepley L C 1975 *Homogeneous Relativistic Cosmologies* (Princeton: Princeton University Press)
[19] Christodoulakis T 2002 *Quantum Cosmology Lect.Notes Phys.* **592** Springer Verlag 318-350
[20] Krasinski A 1997 *Inhomogeneous Cosmological Models* (Cambridge: Cambridge University Press)
[21] Thomi P, Isaak B and Hajicek P 1984 *Phys. Rev.* D **30** 1168
[22] Hajicek P 1984 *Phys. Rev.* D **30** 1178
[23] Rodrigues L M C S, Soares I D and Zanelli J 1989 *Phys. Rev. Lett.* **62** 989
[24] Kiefer C 2007 *Quantum Gravity* 2nd ed (Oxford: Oxford University Press) ch 7
[25] Kiefer C, Müller-Hill J and Vaz C 2006 *Phys. Rev.* D **73** 044025
[26] Carlip S 2003 *Quantum Gravity in 2+1 dimensions* (Cambridge: Cambridge University Press)
[27] Carlip S 2005 *Class. Quantum Grav.* **22** R85
[28] Carlip S 2005 *Living Rev. Rel.* **8** 1
[29] Schroers B J, 1990 *Lessons from (2+1)-dimensional quantum gravity* (*Preprint* gr-qc/0710.5844)

[30] Hajicek P and Kuchař K V 1990 *Phys. Rev.* D **41** 1091
[31] Hajicek P and Kuchař K V 1990 *J. Math.Phys.* **31** 1723
[32] Bañados M, Teitelboim C, and Zanelli J 1992 *Phys. Rev. Lett.* **69** 1849
[33] Bañados M, Henneaux M, Teitelboim C, and Zanelli J 1993 *Phys. Rev.* D **48** 1506
[34] Christodoulakis T and Papadopoulos G O *Preprint* gr-qc/0405144v2
[35] Barrow J, Shaw D and Tsagas C 2006 *Class. Quantum Grav.* **23** 5291
[36] Eisenhart L P 1964 *Riemannian Geometry*, (Princeton: Princeton University Press, New Jersey) 5th printing pp.146
[37] Thomas T Y 1991 *The Differential Invariants of Generalized Spaces*, (New York: Chelsea Publishing Company) pp.140
[38] Bojowald M and Swiderski R 2006 *Class. Quantum Grav.* **23** 2129
[39] Christodoulakis T et al. *Towards a Canonical Quantum Gravity for Geometries Admitting Maximally Symmetric Two-Dimensional Surfaces* work in progress

BTZ solutions on codimension-2 braneworlds

B. Cuadros-Melgar[1], E. Papantonopoulos[2], M. Tsoukalas[2], V. Zamarias[2]

[1]Departamento de Física, Universidad de Santiago de Chile, Casilla 307, Santiago, Chile
[2]Department of Physics, National Technical University of Athens, GR 157 73 Athens, Greece

E-mail: berthaki@gmail.com, lpapa@central.ntua.gr, minasts@central.ntua.gr, zamarias@central.ntua.gr

Abstract. We consider five-dimensional gravity with a Gauss-Bonnet term in the bulk and an induced gravity term on a 2-brane of codimension-2. We show that this system admits BTZ-like black holes on the 2-brane which are extended into the bulk with regular horizons.

1. Introduction

Codimension-2 braneworlds, i.e, a brane embedded in a bulk with two extra dimensions, have recently stimulated a growing interest. The most attractive feature of these models is that the vacuum energy of the brane instead of curving the brane world-volume, merely induces a deficit angle in the bulk around the brane [1].This property was used to solve the cosmological constant problem [2]. However, soon it was realized [3] that one can only find nonsingular solutions if the brane stress tensor is proportional to its induced metric. To obtain the Einstein equation on the brane one has to introduce a cut-off (brane thickness) [4], losing the predictability of the theory, or alternatively, one can add to the gravitational action a Gauss-Bonnet (GB) term [5] or a scalar curvature term (induced gravity) on the brane [6].

Time dependent cosmological solutions in this scenario are also a complex topic. In the thin brane limit, since the brane and bulk energy momentum tensors are related, we cannot get the standard cosmology on the brane [7]. Thus, one has to regularize the codimension-2 branes by introducing some thickness and then consider matter on them [8], arriving to a time-dependent bulk solution. Or alternatively, one can conceive a codimension-1 brane moving in the regularized static background [9]. However, the resulting cosmology is unrealistic having a negative Newton's constant.

Moreover, the issue of localization of a black hole on the brane and its extension to the bulk is not fully understood. In codimension-1 braneworlds, the most natural generalization of the four-dimensional Schwarzschild metric was to consider a black string infinite in the fifth dimension [10]. However, although the curvature scalars are everywhere finite, the Kretschmann scalar diverges at the AdS horizon at infinity, which turns the above solution into a physically unsuitable object. It has been argued that there exists a localized black cigar solution with a finite extension along the extra dimension due to a Gregory-Laflamme [11] type of instability near the AdS horizon, although no analytical solution has been found until the present time. Further attempts deal with the Einstein equations projected on the brane [12], which include an unknown bulk dependent term, the Weyl tensor projection. Due to this reason the system is not closed, and some assumptions have to be made either in the form of the metric or in the Weyl term [13]. The stability and thermodynamics of these solutions were worked out in [14].

A lower dimensional version of a black hole living on a (2+1)-dimensional braneworld was considered in [15]. The authors based their analysis on the C-metric [16] modified by a cosmological constant term. After some coordinate transformations, the resulting geometry corresponds to a BTZ black hole [17] on the brane. This object can be extended as a BTZ string in a four-dimensional anti-de Sitter (AdS) bulk, which has been cut in order to avoid the conical singularity along the symmetry semi-axis. Their thermodynamical stability analysis showed that the black string remains a stable configuration when its transverse size is comparable to the four-dimensional AdS radius, being destabilized by the Gregory-Laflamme instability above that scale, breaking up to a BTZ black hole on a 2-brane.

In the codimension-2 scenario, a six-dimensional black hole on a 3-brane was proposed in [18]. This is a generalization of the $4D$ Aryal, Ford, Vilenkin [19] black hole pierced by a cosmic string adjusted to the codimension-2 branes with a conical structure in the bulk and deformations fitting the deficit angle. However, it is not clear how to realize these solutions in the thin 3-brane limit.

In this work we study black holes on an infinitely thin conical 2-brane and their extension into the five-dimensional bulk.

2. The Setup

We consider the following gravitational action in five dimensions with a Gauss-Bonnet (GB) term in the bulk and an induced three-dimensional curvature term on the brane

$$\begin{aligned} S_{\rm grav} &= \frac{M_5^3}{2}\left\{\int d^5x\sqrt{-g^{(5)}}\left[R^{(5)}+\alpha\left(R^{(5)2}-4R^{(5)}_{MN}R^{(5)MN}+R^{(5)}_{MNKL}R^{(5)MNKL}\right)\right]\right. \\ &+ \left. r_c^2\int d^3x\sqrt{-g^{(3)}}\,R^{(3)}\right\}+\int d^5x\mathcal{L}_{bulk}+\int d^3x\mathcal{L}_{brane}\,, \end{aligned} \tag{1}$$

where $\alpha\,(\geq 0)$ is the GB coupling constant, $r_c = M_3/M_5^3$ is the induced gravity "cross-over" scale, which marks the transition from 3D to 5D gravity, and M_5, M_3 are the five and three-dimensional Planck masses, respectively.

The above induced term has been written in the particular coordinate system in which the metric is

$$ds_5^2 = g_{\mu\nu}(x,\rho)dx^\mu dx^\nu + a^2(x,\rho)d\rho^2 + b^2(x,\rho)d\theta^2\,. \tag{2}$$

Here $g_{\mu\nu}(x,0)$ is the brane metric, whereas x^μ denotes three dimensions, $\mu = 0,1,2$, and ρ,θ denote the radial and angular coordinates of the two extra dimensions (the ρ direction may or may not be compact, and the θ coordinate ranges form 0 to 2π). Capital M, N indices will take values in the five-dimensional space. Note that we have assumed that there exists an azimuthal symmetry in the system, so that both the induced three-dimensional metric and the functions a and b do not depend on θ.

The Einstein equations resulting from the variation of the action (1) are

$$G^{(5)N}_M + r_c^2 G^{(3)\nu}_\mu g^\mu_M g^N_\nu\frac{\delta(\rho)}{2\pi b} - \alpha H^N_M = \frac{1}{M_5^3}\left[T^{(B)N}_M + T^{(br)\nu}_\mu g^\mu_M g^N_\nu\frac{\delta(\rho)}{2\pi b}\right]\,, \tag{3}$$

where

$$\begin{aligned} H^N_M &= \left[\frac{1}{2}g^N_M(R^{(5)\,2}-4R^{(5)\,2}_{KL}+R^{(5)\,2}_{ABKL})-2R^{(5)}R^{(5)N}_M\right. \\ &\quad \left.+4R^{(5)}_{MP}R^{NP}_{(5)}\quad+4R^{(5)\quad N}_{KMP}R^{KP}_{(5)}-2R^{(5)}_{MKLP}R^{NKLP}_{(5)}\right]\,. \end{aligned} \tag{4}$$

To obtain the braneworld equations we expand the metric around the brane as

$$b(x,\rho) = \beta(x)\rho + O(\rho^2)\ . \tag{5}$$

At the boundary of the internal two-dimensional space where the 2-brane is situated the function b behaves as $b'(x,0) = \beta(x)$, where a prime denotes derivative with respect to ρ. We also demand that the space in the vicinity of the conical singularity is regular, which imposes the supplementary conditions, $\partial_\mu\beta = 0$ and $\partial_\rho g_{\mu\nu}(x,0) = 0$ [5].

The extrinsic curvature in the particular gauge $g_{\rho\rho} = 1$ that we are considering is given by $K_{\mu\nu} = g'_{\mu\nu}$. The above decomposition will be helpful in the following for finding the induced dynamics on the brane. We will now use the fact that the second derivatives of the metric functions contain δ-function singularities at the position of the brane. The nature of the singularity then gives the following relations [5]

$$\frac{b''}{b} = -(1-b')\frac{\delta(\rho)}{b} + \text{non} - \text{singular terms}\ , \tag{6}$$

$$\frac{K'_{\mu\nu}}{b} = K_{\mu\nu}\frac{\delta(\rho)}{b} + \text{non} - \text{singular terms}\ . \tag{7}$$

From the above singularity expressions and using the Gauss-Codazzi equations we can match the singular parts of the Einstein equations (3) and get the following "boundary" Einstein equations

$$G^{(3)}_{\mu\nu} = \frac{1}{M^3_{(5)}(r_c^2 + 8\pi(1-\beta)\alpha)}T^{(br)}_{\mu\nu} + \frac{2\pi(1-\beta)}{r_c^2 + 8\pi(1-\beta)\alpha}g_{\mu\nu}\ . \tag{8}$$

The effective three-dimensional Planck mass and cosmological constant are simply

$$M^2_{Pl} = M^3_{(5)}(r_c^2 + 8\pi(1-\beta)\alpha)\ , \tag{9}$$

$$\Lambda_{(3)} = \lambda - 2\pi M^3_{(5)}(1-\beta)\ , \tag{10}$$

where λ is the brane tension. Note that the Planck mass can depend on the deficit angle. This is an effect of solely the bulk Gauss-Bonnet term.

As a result of the Gauss-Codazzi reduction, the above boundary Einstein equations will also contain terms proportional to the extrinsic curvature and additional terms coming from the bulk GB term. However, if we allow only regular conical singularities there is no contribution from these terms [5].

3. Black String Solutions

We assume that there is a localized (2+1) black hole on the brane described by the following metric

$$ds_3^2 = \left(-n(r)^2dt^2 + n(r)^{-2}dr^2 + r^2d\phi^2\right)\ , \tag{11}$$

where $0 \le r < \infty$ is the radial coordinate, and ϕ has the usual periodicity $(0, 2\pi)$. We will look for black string solutions of the Einstein equations (3) using the five-dimensional metric (2) in the form

$$ds_5^2 = f^2(\rho)\left(-n(r)^2dt^2 + n(r)^{-2}dr^2 + r^2d\phi^2\right) + d\rho^2 + b^2(r,\rho)d\theta^2\ , \tag{12}$$

where we have chosen $a(r,\rho) = 1$ without loss of generality.

The space outside the conical singularity is regular, therefore, we demand that the warp function $f(\rho)$ is also regular everywhere. We assume that there is only a cosmological constant Λ_5 in the bulk. Then, the bulk Einstein equations take the form

$$G^{(5)}_{MN} - \alpha H_{MN} = -\frac{\Lambda_5}{M_5^3}g_{MN}\ . \tag{13}$$

By combining the (rr) and $(\phi\phi)$ equations we get

$$\left(\dot{n}^2 + n\ddot{n} - \frac{n\dot{n}}{r}\right)\left(1 - 4\alpha\frac{b''}{b}\right) = 0 \ , \tag{14}$$

while a combination of the $(\rho\rho)$ and $(\theta\theta)$ equations gives

$$\left(f'' - \frac{f'b'}{b}\right)\left[3 - 4\frac{\alpha}{f^2}\left(\dot{n}^2 + n\ddot{n} + 2\frac{n\dot{n}}{r} + 3f'^2\right)\right] = 0 \ , \tag{15}$$

where a dot implies derivatives with respect to r. The solutions of the equations (14) and (15) are summarized in the following table [20].

$n(r)$	$f(\rho)$	$b(\rho)$	$-\Lambda_5$	Constraints
BTZ	$\cosh\left(\frac{\rho}{2\sqrt{\alpha}}\right)$	$\forall b(\rho)$	$\frac{3}{4\alpha}$	$L_3^2 = 4\,\alpha$
BTZ	$\cosh\left(\frac{\rho}{2\sqrt{\alpha}}\right)$	$2\,\beta\,\sqrt{\alpha}\,\sinh\left(\frac{\rho}{2\sqrt{\alpha}}\right)$	$\frac{3}{4\alpha}$	-
BTZ	$\cosh\left(\frac{\rho}{2\sqrt{\alpha}}\right)$	$2\,\beta\,\sqrt{\alpha}\,\sinh\left(\frac{\rho}{2\sqrt{\alpha}}\right)$	$\frac{3}{4\alpha}$	$L_3^2 = 4\,\alpha$
BTZ	±1	$\frac{1}{\gamma}\,\sinh\left(\gamma\,\rho\right)$	$\frac{3}{l^2}$	$\gamma = \sqrt{-\frac{2\Lambda_5}{3+4\alpha\Lambda_5}}$
$\forall n(r)$	$\cosh\left(\frac{\rho}{2\sqrt{\alpha}}\right)$	$2\,\beta\,\sqrt{\alpha}\,\sinh\left(\frac{\rho}{2\sqrt{\alpha}}\right)$	$\frac{3}{4\alpha}$	-
Corrected BTZ	$\cosh\left(\frac{\rho}{2\sqrt{\alpha}}\right)$	$2\,\beta\,\sqrt{\alpha}\,\sinh\left(\frac{\rho}{2\sqrt{\alpha}}\right)$	$\frac{3}{4\alpha}$	$L_3^2 = 4\,\alpha$
Corrected BTZ	±1	$2\,\beta\,\sqrt{\alpha}\,\sinh\left(\frac{\rho}{2\sqrt{\alpha}}\right)$	$\frac{1}{4\alpha}$	$\Lambda_5 = -\frac{1}{4\alpha} = -\frac{3}{L_3^2}$

Table 1. BTZ String-Like Solutions in Five-Dimensional Braneworlds of Codimension-2

where L_3 is the length of three-dimensional AdS space. The BTZ solution is given by [17]

$$n^2(r) = -M + \frac{r^2}{L_3^2} \ . \tag{16}$$

This black hole for positive mass has a horizon at $r = L_3\sqrt{M}$, and the radius of curvature of the AdS_3 space $L_3 = (-\Lambda_3)^{-1/2}$ provides the length scale necessary to define this horizon. For the mass $-1 < M < 0$, which is dimensionless, the BTZ black hole has a naked conical singularity while for $M = -1$ the vacuum AdS_3 space is recovered.

Moreover, the "corrected" version of the BTZ black hole is found to be

$$n^2(r) = -M + \frac{r^2}{L_3^2} - \frac{\zeta}{r} \ . \tag{17}$$

This black hole corresponds to the BTZ black hole conformally coupled to a massless scalar field [21]. The scalar field does not introduce an independent conserved charge, for this reason it does not show up in the form of the metric. It merely modifies the energy-momentum tensor of the three-dimensional Einstein equations as we will see in the next section.

Although $n(r)$ remains undetermined in one of our solutions, it exists a connecting relation between this function and the matter on the brane $T_1^1 = T_2^2 = nn'/r + \Lambda_3$, $T_3^3 = n'^2 + nn'' + \Lambda_3$ from (8).

All these solutions extend the black hole on the brane into the bulk. In the BTZ case if we calculate the curvature invariants, we find no $r = 0$ curvature singularity for the BTZ string-like solution as expected. The warp function $f^2(\rho)$ gives the shape of the horizon of the string-like solution while the size of the horizon is defined by the scale $\sqrt{\alpha}$.

From our solutions we can notice that consistency of the five-dimensional bulk equations requires a fine-tuned relation between the GB coupling constant and the AdS_5 length (through the five-dimensional cosmological constant). The use of this fine-tuning gives to the non-singular horizon the shape of a throat up to the boundary of the AdS space.

4. Localization of the Black Hole on the Brane

In order to complete our solution with the introduction of the brane we must solve the corresponding junction conditions given by the "boundary" Einstein equations (8) using the induced metric on the brane given by (11). For the case when $n(r)$ corresponds to the BTZ black hole (16), and the brane cosmological constant is given by $\Lambda_{(3)} = -1/L_3^2$, we found that the energy-momentum tensor is null. Therefore, the BTZ black hole is localized on the brane in vacuum.

When $n(r)$ is of the form given in (17), we found the following traceless energy-momentum tensor

$$T_\alpha^\beta = \mathrm{diag}\left(\frac{\zeta}{2r^3}, \frac{\zeta}{2r^3}, -\frac{\zeta}{r^3}\right), \tag{18}$$

which is conserved on the brane, $\nabla_\beta T_\alpha^\beta = 0$. This property has been studied in [22] for codimension-2 braneworlds in six dimensions; however, it seems to be a property independent of the bulk dimension for Einstein-Gauss-Bonnet codimension-2 models having an axially symmetric metric like (2) with $a^2(x,\rho) = 1$, and obeying junction conditions analogous to (8). If we consider the energy-momentum tensor in [21] necessary to sustain the solution (17), and we take the limit $r/L_3 << 1$, we get the unexpected result that it reduces to (18) which is necessary to localize the black hole on the conical 2-brane. A way to understand this result is that because in this limit r is very small, the black hole will be localized around the conical singularity and therefore, any matter will take a distributional form around this singularity. Note also that this solution is a result of the presence of the GB term in the bulk. If we switch off the GB coupling, then from relations (14) and (15) it can be seen that only the BTZ black hole is a solution.

5. Conclusions

We discussed black hole localization on an infinitely thin 2-brane of codimension-2 and its extension into a five-dimensional AdS bulk. To have a three-dimensional gravity on the brane we introduced a five-dimensional Gauss-Bonnet term in the bulk and an induced gravity term on the 2-brane. We showed that the (2+1) BTZ black hole and its short-distance-corrected extension can be localized on the 2-brane, while in the bulk these solutions describe BTZ strings. Consistency of the five-dimensional bulk equations requires a fine-tuned relation between the Gauss-Bonnet coupling constant and the length of the five-dimensional AdS space. The use of this fine-tuning gives to the non-singular horizon the shape of a throat up to the boundary of the AdS space.

We did not allow more severe singularities than conical. This assumption has fixed the deficit angle to a constant value.

This analysis has also been extended to a conical 3-brane embedded in a six-dimensional bulk as reported in [23]. This study also showed the relevance of the GB term as it dictates the kind of brane and bulk matter necessary to sustain a black hole on the brane.

Acknowledgments
This work was supported by the NTUA research program PEVE07. The work of B.C-M. is supported by Fondo Nacional de Desarrollo Científico y Tecnológico (FONDECYT), Chile, under grant 3070009.

References
[1] Chen J W, Luty M A and Ponton E, 2000, *JHEP* **0009**, 012.
[2] Carroll S M, Guica M M, *Preprint* arXiv:hep-th/0302067; Aghababaie Y *et al.*, 2004, *Nucl. Phys.* B **680**, 389; Nilles H P, Papazoglou A, Tasinato G, 2004, *Nucl. Phys.* B **677**, 405; Kehagias A, 2004, *Phys. Lett.* B **600**, 133.
[3] Cline J M, Descheneau J, Giovannini M, Vinet J, 2003, *JHEP* **0306**, 048.
[4] Kanno S, Soda J, 2004, *JCAP* **0407**, 002. Vinet J, Cline J M, 2004, *Phys. Rev.* D **70**, 083514; Navarro I, Santiago J, 2005, *JHEP* **0502**, 007.
[5] Bostock P, Gregory R, Navarro I and Santiago J, 2004, *Phys. Rev. Lett.* **92**, 221601.
[6] Papantonopoulos E, Papazoglou A, 2005, *JCAP* **0507**, 004.
[7] Kofinas G, 2006, *Phys. Lett.* B **633**, 141. Papantonopoulos E, Papazoglou A, 2005, *JHEP* **0509**, 012.
[8] Carter B, Battye R A and Uzan J P, 2003, *Commun. Math. Phys.* **235**, 289; Kolanovic M, Porrati M and Rombouts J W, 2003, *Phys. Rev.* D **68**, 064018; Peloso M, Sorbo L and Tasinato G, 2006, *Phys. Rev.* D **73**, 104025; Papantonopoulos E, Papazoglou A and Zamarias V, 2007, *JHEP* **0703**, 002. Himmetoglu B and Peloso M, 2007, *Nucl. Phys.* B **773**, 84; Kobayashi T and Minamitsuji M, 2007, *Phys. Rev.* D **75**, 104013; Burgess C P, Hoover D and Tasinato G, 2007, *JHEP* **0709**, 124; Kobayashi T and Takamizu Y I, 2007, *Class. Quant. Grav.* **25**, 015007; Copeland E J and Seto O, 2007, *JHEP* **0708**, 001.
[9] Papantonopoulos E, Papazoglou A and Zamarias V, 2008, *Nucl. Phys.* B **797**, 520-536; Minamitsuji M and Langlois D, 2007, *Phys. Rev.* D **76**, 084031.
[10] Chamblin A, Hawking S W, Reall H S, 2000, *Phys. Rev.* D **61**, 065007.
[11] Gregory R and Laflamme R, 1993, *Phys. Rev. Lett.* **70**, 2837.
[12] Shiromizu T, Maeda K I and Sasaki M, 2000, *Phys. Rev.* D **62**, 024012.
[13] Dadhich N, Maartens R, Papadopoulos P and Rezania V, 2000, *Phys. Lett.* B **487**, 1; Bruni M, Germani C and Maartens R, 2001, *Phys. Rev. Lett.* **87**, 231302; Casadio R, Fabbri A and Mazzacurati L, 2002, *Phys. Rev.* D **65**, 084040; Kofinas G, Papantonopoulos E and Pappa I, 2002, *Phys. Rev.* D **66**, 104014; Kofinas G, Papantonopoulos E and Zamarias V, 2002, *Phys. Rev.* D **66**, 104028.
[14] Abdalla E, Cuadros-Melgar B, Pavan A B and Molina C, 2006, *Nucl. Phys.* B **752**, 40.
[15] Emparan R, Horowitz G T and Myers R C, 2000, *JHEP* **0001**, 007; Emparan R, Horowitz G T and Myers R C, 2000, *JHEP* **0001**, 021.
[16] Kinnersley W and Walker M, 1970, *Phys. Rev.* D **2**, 1359.
[17] Bañados M, Teitelboim C and Zanelli J, 1992, *Phys. Rev. Lett.* **69**, 1849.
[18] Kaloper N and Kiley D, 2006, *JHEP* **0603**, 077. Kiley D, 2007, *Phys. Rev.* D **76**, 126002.
[19] Aryal M, Ford L H and Vilenkin A, 1986, *Phys. Rev.* D **34**, 2263; Achucarro A, Gregory R and Kuijken K, 1995, *Phys. Rev.* D **52**, 5729.
[20] Cuadros-Melgar B, Papantonopoulos E, Tsoukalas M and Zamarias V, 2008, *Phys. Rev. Lett.* **100**, 221601.
[21] Martinez C and Zanelli J, 1996, *Phys. Rev.* D **54**, 3830.
[22] Kofinas G, 2005, *Class. Quant. Grav.* **22**, L47.
[23] Cuadros-Melgar B, Papantonopoulos E, Tsoukalas M and Zamarias V, 2009, *Nucl. Phys.* B **810**, 246-265.

Self-dual Maxwell field in 3D gravity with torsion and dynamical role of central charges

Milutin Blagojević and Branislav Cvetković

Institute of physics, P. O. Box 57, 11001 Belgrade, Serbia

E-mail: cbranislav@phy.bg.ac.yu, mb@phy.bg.ac.yu

Abstract. We show that general structure of the field equations for the system of a self-dual Maxwell field coupled to 3D gravity with torsion implies that classical central charges have a new dynamical role. By using simple correspondence between self-dual solutions with torsion and their Riemannian counterparts we construct two exact solutions, corresponding to the sectors with a massless and massive Maxwell field, and calculate their conserved charges.

1. Introduction

The three-dimensional (3D) gravity has been actively studied for nearly three decades as an interesting model which shares the same conceptual features with general relativity (GR). Among many outstanding results [1], the discovery of the BTZ black hole [2] was particularly important, since it improved our basic understanding of the classical and quantum black hole dynamics.

Mielke and Baekler (MB) [3] proposed a new geometric framework for 3D gravity by replacing the traditional Riemannian geometry of GR by the Riemann-Cartan geometry, in which the gravitational dynamics is characterized by both the curvature and the torsion. Such an approach led to a number of interesting results [4, 5, 6, 7], which confirm that the topological MB model has a rich dynamical structure. The most recent results are related to the electrically charged sector of the theory [8, 9]. By analyzing *static* configurations of the MB model, we found an exact solution with azimuthal electric field [8]. Here, we continue the investigation of the charged sector by studying the *self-dual* Maxwell field as a source.

The layout of the paper is as follows. In section 2 we give a brief account of 3D gravity with torsion, with Maxwell field modified by a topological mass term (μ) [10] as a source. In section 3, we impose a restriction on the Maxwell field by requesting it to be self-dual. As a consequence, we find that two possible values of the classical central charge of 3D gravity with torsion [5] are directly related to the two self-duality sectors of the Maxwell field. We establish a simple correspondence between self-dual solutions with torsion, and their Riemannian counterparts. In section 4, we construct two exact, stationary and spherically symmetric solutions, characterized by $\mu = 0$ and $\mu \neq 0$ and calculate their conserved charges—energy, angular momentum and electric charge. Section 5 is devoted to concluding remarks.

We use the same conventions as in [9]: the Latin indices refer to the local Lorentz frame, the Greek indices refer to the coordinate frame; the middle alphabet letters $(i, j, k, ...; \mu, \nu, \lambda, ...)$ run over 0,1,2, the first letters of the Greek alphabet $(\alpha, \beta, \gamma, ...)$ run over 1,2; the metric components in the local Lorentz frame are $\eta_{ij} = (+, -, -)$; totally antisymmetric tensor ε^{ijk} and the related tensor density $\varepsilon^{\mu\nu\rho}$ are both normalized so that $\varepsilon^{012} = 1$; the Hodge-star operation is $\star$.

2. Lagrangian and field equations

Theory of gravity with torsion can be naturally described as a Poincaré gauge theory (PGT). Basic gravitational variables in PGT are the triad b^i and the Lorentz connection $A^{ij} = -A^{ji}$ (1-forms), and the corresponding field strengths are the torsion T^i and the curvature R^{ij}. In 3D, one can introduce the notation $A^{ij} =: -\varepsilon^{ij}{}_k\omega^k$ and $R^{ij} =: -\varepsilon^{ij}{}_k R^k$, which yields:

$$T^i = db^i + \varepsilon^i{}_{jk}\omega^j \wedge b^k\,, \qquad R^i = d\omega^i + \frac{1}{2}\varepsilon^i{}_{jk}\omega^j \wedge \omega^k\,. \tag{2.1}$$

PGT is characterized by a useful identity:

$$\omega^i \equiv \tilde{\omega}^i + K^i\,, \tag{2.2a}$$

where $\tilde{\omega}^i$ is the Levi-Civita (Riemannian) connection, and K^i is the contortion 1-form, defined implicitly by $T^i =: \varepsilon^i{}_{mn}K^m \wedge b^n$. Using this identity, one can express the curvature $R_i = R_i(\omega)$ in terms of its *Riemannian* piece $\tilde{R}_i = R_i(\tilde{\omega})$ and the contortion K_i:

$$2R_i \equiv 2\tilde{R}_i + 2\tilde{\nabla}K_i + \varepsilon_{imn}K^m \wedge K^n\,. \tag{2.2b}$$

The covariant derivative $\nabla = \nabla(\omega)$ acts on a tangent-frame spinor/tensor in accordance with its spinorial/tensorial structure.

The antisymmetry of the Lorentz connection A^{ij} implies that the geometric structure of PGT corresponds to Riemann-Cartan geometry.

The *topological* MB model for 3D gravity [3] represents a natural generalization of GR with a cosmological constant (GR_Λ):

$$L_0 = 2ab^iR_i - \frac{\Lambda}{3}\varepsilon_{ijk}b^ib^jb^k + \alpha_3\left(\omega^i d\omega_i + \frac{1}{3}\varepsilon_{ijk}\omega^i\omega^j\omega^k\right) + \alpha_4 b^iT_i\,, \tag{2.3a}$$

where $a = 1/16\pi G$ and the wedge product sign $\wedge$ is omitted for simplicity. The complete dynamics of the topologically massive Maxwell field coupled to 3D gravity with torsion is described by:

$$L = L_0 + L_M\,, \qquad L_M := -\frac{1}{2}F\,{}^\star F - \frac{\mu}{2}AF\,, \tag{2.3b}$$

where $F = dA$.

The gravitational field equations are obtained by varying L with respect to b^i and ω^i. In the nondegenerate sector with $\Delta := \alpha_3\alpha_4 - a^2 \neq 0$, they take the following simple form [8, 9]:

$$T_i = \varepsilon_{imn}K^mb^n\,, \qquad K^m = \frac{1}{2}(pb^m + ut^m)\,, \tag{2.4a}$$

$$2R_i = q\varepsilon_{imn}b^mb^n - v\varepsilon_{imn}t^mb^n\,, \tag{2.4b}$$

where $p := (\alpha_3\Lambda + \alpha_4 a)/\Delta$, $u := \alpha_3/\Delta$, $q := -(\alpha_4^2 + a\Lambda)/\Delta$, $v := a/\Delta$, and the 1-form t^i is expressed in terms of the energy-momentum tensor $\mathcal{T}^i{}_k$:

$$t^i := -\left(\mathcal{T}^i{}_k - \frac{1}{2}\delta^i_k\mathcal{T}\right)b^k\,, \qquad \mathcal{T}^i{}_k := -F^{im}F_{km} + \frac{1}{4}\delta^i_kF^2\,,$$

with $\mathcal{T} = \mathcal{T}^k{}_k$ and $F^2 = F^{mn}F_{mn}$. These equations, together with the modified Maxwell equations

$$d\,{}^\star F + \mu F = 0\,, \tag{2.4c}$$

define the complete dynamics of our system. The Cartan curvature R_i is calculated using the identity (2.2b),

$$2R_i = 2\tilde{R}_i + u\tilde{\nabla}t_i + \varepsilon_{imn}\left(\frac{p^2}{4}b^mb^n + \frac{up}{2}t^mb^n + \frac{u^2}{4}t^mt^n\right)\,, \tag{2.5}$$

where the Maxwell field contribution is compactly represented by the 1-form t^i.

3. General dynamical characteristics of self-dual solutions

General structure of the field equations (2.4) depends essentially on the form of the Maxwell field. In this section, we discuss some important dynamical characteristics of exact solutions corresponding to the self-dual Maxwell field as a source.

We are looking for a *spherically symmetric* and *stationary* solution. Choosing the local coordinates $x^\mu = (t, r, \varphi)$, we make the following ansatz for the triad and for the Maxwell field:

$$b^0 = N dt\,, \qquad b^1 = B^{-1} dr\,, \qquad b^2 = K(d\varphi + C dt)\,, \tag{3.1}$$

$$F = E b^0 b^1 - H b^1 b^2\,. \tag{3.2}$$

Here, N, B, C, K and E, H are six unknown functions of the radial coordinate r.

We assume a generalized *self-duality* of the Maxwell field:

$$E = \epsilon H\,, \qquad \epsilon^2 = 1\,. \tag{3.3}$$

After introducing a new radial coordinate $\rho = \rho(r)$, defined by $d\rho = dr/B$, we find that the first integrals of the modified Maxwell equations are given as

$$\begin{aligned} EK &= -Q_e \exp(-\epsilon\mu\rho)\,, \\ -HN &= (Q_m - Q_e C)\exp(-\epsilon\mu\rho)\,, \end{aligned} \tag{3.4}$$

where Q_e, Q_m are integration constants, and $\epsilon Q_e N = K(Q_m - Q_e C)$, as a consequence of (3.3).

The self-duality condition implies $F^2 = 0$ and simplifies the form of the electromagnetic energy-momentum tensor,

$$\mathcal{T}^i{}_j = \begin{pmatrix} E^2 & 0 & EH \\ 0 & 0 & 0 \\ -EH & 0 & -H^2 \end{pmatrix},$$

as well as the form of t^i:

$$t^0 = -E(Eb^0 + Hb^2)\,, \qquad t^1 = 0\,, \qquad t^2 = -\epsilon t^0\,.$$

This defines the contortion as in (2.4a), and exhausts the content of the first field equation.

Going now to the second field equation (2.4b), we calculate the Cartan curvature (2.5). Then, by substituting this result into (2.4b), we obtain:

$$\begin{aligned} 2\tilde{R}_i &= \Lambda_{\text{eff}}\, \varepsilon_{imn} b^m b^n - V \varepsilon_{imn} t^m b^n\,, \\ V &:= v + \frac{pu}{2} + \epsilon\frac{u}{\ell}\,. \end{aligned} \tag{3.5}$$

Note that the factor $V = V(\epsilon)$ is proportional to the *classical central charge* $c(\epsilon)$, characterizing the asymptotic conformal structure of 3D gravity with torsion [5]:

$$V(\epsilon) = \frac{1}{\Delta}\frac{1}{24\pi\ell}c(\epsilon)\,, \qquad c(\mp 1) = 24\pi\left[a\ell + \alpha_3\left(\frac{p\ell}{2} \mp 1\right)\right].$$

- Equation (3.5) reveals a new feature of the central charges $c(\mp 1)$, showing that they have a direct dynamical influence on the self-duality modes $\epsilon = \mp 1$ of the system.

The set of equations (3.5) is overdetermined. However, there is a consistent solution of this system, which can obtained either directly [9], or by noting that in the limit $u \to 0$, or equivalently, $V \to v = -1/a$, equation (3.5) becomes equivalent to the Einstein equation for the self-dual Maxwell field in *Riemannian* 3D gravity. This property can be formulated as the following constructive statement:

- Starting with any self-dual solution in Riemannian 3D gravity, one can generate the related self-dual solution with torsion by making the replacement $v \to V$.

In what follows, we shall illustrate the power of this theorem by constructing two different solutions of (3.5), belonging to the sectors with $\mu = 0$ and $\mu \neq 0$, respectively.

4. Exact self-dual solutions

4.1. Solution with $\mu = 0$

The condition $\mu = 0$ refers to the standard, massless Maxwell field. The field equations define the form of K^2. To realize the construction of the solution, we first conveniently fix g_2 by demanding $g_2 = r_0^2$. After choosing the radial coordinate r by $B = (r^2 - r_0^2)/\ell r$ and imposing the boundary condition $C \to 0$ for $r \to \infty$, we find the general solution for $\mu = 0$ [9]:

$$B = \frac{r^2 - r_0^2}{\ell r}, \qquad N = \frac{r^2 - r_0^2}{\ell K}, \qquad K^2 = r^2 + r_0^2 \ln\left|\frac{r^2 - r_0^2}{r_0^2}\right| + h_1\,,$$
$$C = \frac{\epsilon}{\ell}\left(1 - \frac{r^2 - r_0^2}{K^2}\right), \qquad E = \epsilon H = -\frac{Q_e}{K}\,, \tag{4.1a}$$

where $h_1 = g_1 - r_0^2$. To complete the solution, we display also the electromagnetic potential:

$$A = \left(\frac{Q_e}{2}\ln\left|\frac{r^2 - r_0^2}{r_0^2}\right| + h_2\right)(dt + \epsilon\ell d\varphi)\,. \tag{4.1b}$$

In the limit $u \to 0$, or $V \to -1/a$, the solution (4.1) coincides with the Kamata-Koikawa self-dual solution, found in Riemannian GR$_\Lambda$ [11] (see also [12]). Thus, the field configuration (4.1) represent a generalization of the Kamata-Koikawa solution to the self-dual solution with torsion. In the limit $Q_e \to 0$, the solution (4.1) reduces asymptotically to the vacuum state of the BTZ-like black hole with torsion [5].

The scalar Cartan curvature of the self-dual solution (4.1) is $R = -6q$, while $\tilde{R} = -6\Lambda_{\text{eff}}$. However, the form of $\tilde{R}_i$ implies that the metric of the solution is not maximally symmetric. To gain a deeper insight into the nature of the self-dual solution (4.1), we now turn our attention to its conserved charges.

The family of solutions (4.1) is parametrized by (Q_e, h_1, h_2). Considering the neutral limit $Q_e \to 0$ in Riemannian GR$_\Lambda$, Kamata and Koikawa [11] concluded that one should fix the parameter h_1 to zero, as it leads to $K^2 \to r^2$ in this limit. However, the choice $h_1 = 0$ leads to divergent values of energy and angular momentum [11, 9].

An interesting attempt to handle these divergences by a suitable *regularization* procedure was proposed in [15]. In this procedure, we enclose the system in a circle C_{as} having a large, but finite radius r_{as}. This circle represents a regularized spatial boundary at infinity, and the asymptotic region is defined by $r \to r_{\text{as}}$. Then, we make a choice of the boundary conditions by fixing the values of dynamical variables at C_{as}. Finally, we take the limit $r_{\text{as}} \to \infty$.

In our example, after introducing C_{as} we choose the boundary conditions $K^2 = r_{\text{as}}^2$ and $A = 0$ [9]; in this way, we effectively eliminate the logarithmic terms from the boundary C_{as}. The standard canonical procedure yields the following expressions for the energy E, the angular momentum M and the electric charge Q of our self-dual solution [9]:

$$\ell E = \epsilon Q_e^2 \frac{u}{12\ell}\, c(-\epsilon) = \epsilon M\,,$$
$$Q = 2\pi Q_e\,. \tag{4.2}$$

Although the regularized self-dual solution has the same *leading* asymptotic terms as the black hole with torsion, its conserved charges E and M are quite different. The reason for this lies in the fact that E and M are basically determined by the *sub-leading* asymptotic terms. This is clearly seen in the canonical formalism, where the analysis of the relevant surface terms shows that the non-vanishing E and M are generated by a combined effect of the Maxwell field (t^i) and the gravitational Chern-Simons term $(u \sim \alpha_3)$.

The regularized solution and the one with $h_1 = h_2 = 0$, are two different members of the same family (4.1), which have *different conserved charges*. Another difference is found in their geometric properties, since the regularized solution is valid only in the asymptotic region [9].

4.2. Solution with $\mu \neq 0$

The topological mass μ is introduced to regularize the asymptotic behavior of the Maxwell field. As shown in [9], the solution for K^2 in the case $\mu \neq 0$ (with $\epsilon\mu\ell \neq -1$), takes the form

$$K^2 = g_1 + g_2 \exp(2\rho/\ell) - \frac{r_0^2}{\mu^2\ell^2 + \epsilon\mu\ell} \exp(-2\epsilon\mu\rho)\,, \tag{4.3a}$$

while the general self-dual solution with torsion (for $\mu \neq 0$, $\epsilon\mu\ell \neq -1$) is given by:

$$N = \frac{g_3 \exp(2\rho/\ell)}{K}\,, \qquad C = \frac{Q_m}{Q_e} - \epsilon\frac{N}{K}\,, \qquad E = \epsilon H = -\frac{Q_e}{K}\exp(-\epsilon\mu\rho)\,, \tag{4.3b}$$

where g_1, g_2 and g_3 are constants of integration.

For a fixed ϵ, the solution is characterized by eight parameters. This number can be significantly reduced by imposing various physical/geometric requirements. A convenient normalization of the time coordinate leads to $g_3 = \ell$. Demanding that C vanishes at spatial infinity, we obtain $Q_m/Q_e = \epsilon/\ell$. The choice $g_2 = \ell^2$ ensures that in the limit $Q_e \to 0$, $g_1 \to 0$, the solution (4.3) reduces to the black hole vacuum. Finally, we note that solutions with vanishing Maxwell field at spatial infinity are characterized by $\epsilon\mu > 0$. Hence, we impose the requirement $\epsilon\mu = 1/\ell$ to simplify the calculations [13].

Since the solution (4.3) is defined only up to a choice of the radial coordinate, we are free to choose $K = r$. Expressed in terms of r, the solution (4.3) reads:

$$\begin{aligned} &K = r\,, \qquad B = \frac{\sqrt{(r^2 - g_1)^2 + 2\ell^2 r_0^2}}{\ell r}\,, \qquad N = \frac{(r^2 - g_1 + \ell r B)}{2\ell r}\,, \\ &C = \epsilon\left(\frac{1}{\ell} - \frac{N}{r}\right)\,, \qquad E = -\frac{Q_e}{r}\sqrt{\frac{\ell}{Nr}}\,, \qquad A = -Q_e\sqrt{\frac{\ell}{Nr}}\,(dt + \epsilon\ell d\varphi)\,. \end{aligned} \tag{4.4}$$

This result represents a natural generalization of the self-dual solution found by Fernando and Mansouri [13], in the context of Riemannian theory. In the limit $Q_e \to 0$, the solution reduces to the extreme black hole with torsion [5].

From the above expressions, we conclude that (4.4) is a perfectly regular solution, since $B > 0$ and $N > 0$ over the whole range of r, while the Cartan curvature is constant, $R = -6q$.

To calculate the conserved charges we first fix the asymptotic form of the fields in (4.4):

$$N \sim \frac{r}{\ell} - \frac{g_1}{\ell r}\,, \qquad B \sim \frac{r}{\ell} - \frac{g_1}{\ell r}\,, \qquad C \sim \frac{\epsilon g_1}{\ell r^2}\,, \qquad E \sim \frac{Q_e}{r^2}\,, \qquad A_\mu \sim \frac{Q_e}{r}\,. \tag{4.5}$$

Note that the electromagnetic potential has the same asymptotic behavior as in 4D, due to the presence of the topological mass μ, which acts as an infrared regulator and modifies the asymptotics of all the fields.

Because of the fast asymptotic decrease of the Maxwell potential, the electromagnetic contribution to the energy and angular momentum vanishes. Moreover, the asymptotics defined by (4.5) coincides with that of the extreme BTZ black hole with torsion [5], so that the energy and angular momentum of our self-dual solution (4.4) are given by:

$$\ell E = \frac{g_1}{6\ell}c(-\epsilon) = \epsilon M\,. \tag{4.6a}$$

In the Riemannian limit $u \to 0$, these expressions agree with those obtained in the quasilocal formalism by Fernando and Mansouri [13], and by Dereli and Obukhov [14].

The electric charge of the solution vanishes:

$$Q = 0\,. \tag{4.6b}$$

Thus, despite the existence of the radial electric field, its asymptotic fall off is too fast to produce a nonvanishing electric charge. The constant Q_e can not be interpreted as the electric charge, and our Fernando-Mansouri-like solution should be called a *neutral* self-dual solution.

5. Concluding remarks

(1) Studying dynamical properties of the self-dual Maxwell field modified by a topological mass term and coupled to 3D gravity with torsion, we found that the dynamical evolution is directly influenced by the classical central charges.

(2) We constructed two exact, stationary and self-dual solutions with torsion, characterized by $\mu = 0$ and $\mu \neq 0$. They represent respective generalizations of the Kamata-Koikawa and Fernando-Mansouri solutions, found earlier in Riemannian GR_Λ. For each of these solutions, we calculated its conserved charges.

Acknowledgments
This work was supported by the Serbian Science Foundation.

References

[1] For a review of the subject and an extensive list of references, see: Carlip S 1998 *Quantum Gravity in 2+1 Dimensions* (Cambridge: Cambridge University Press)
[2] Bañados M, Teitelboim C and Zanelli J, The Black Hole in Three-Dimensional Spacetime, *Phys. Rev. Lett.* **16** 1849
[3] Mielke E W, Baekler P 1991 Topological gauge model of gravity with torsion *Phys. Lett. A* **156** 399
Baekler P, Mielke E W, Hehl F W 1992 Dynamical symmetries in topological 3D gravity with torsion *Nuovo Cim. B* **107** 91
[4] García A, Hehl F W, Heinecke C and Macías A 2003 Exact vacuum solutions of (1+2)-dimensional Poincaré gauge theory: BTZ solution with torsion *Phys. Rev. D* **67** 124016
Mielke E W and Maggiolo A A R 2003 Rotating black hole solution in a generalized topological 3D gravity with torsion *Phys. Rev. D* **68** 104026
Blagojević M and Vasilić M 2003 3D gravity with torsion as a Chern-Simons gauge theory, *Phys. Rev. D* **68** 104023
[5] Blagojević M and Cvetković B 2006 Canonical structure of 3D gravity with torsion in *Progress in General Relativity and Quantum Cosmology*, vol. 2, ed. Ch. Benton (New York: Nova Science Publishers) 103-123;
Blagojević M and Cvetković B 2006 Black hole entropy in 3D gravity with torsion Class. *Quantum Grav.* **23** 4781-4795
[6] Obukhov Y 2003, New solutions in 3D gravity, *Phys. Rev. D* **68** 124015
Cacciatori S, Caldarelli M, Giacomini A, Klemm D and Mansi D 2006 Chern-Simons formulation of three-dimensional gravity with torsion and nonmetricity, *J. Geom. Phys.* **56** 2623-2643
Klemm D, Tagliabue G 2008 The CFT dual of AdS gravity with torsion, *Class. Quant. Grav.* **25** 035011
[7] Giacomini A, Troncoso R and Willison S 2007 Three-dimensional supergravity reloaded, *Class. Quantum Grav.* **24** 2845-2860
Cvetković B and Blagojević M 2007 Supersymmetric 3D gravity with torsion: asymptotic symmetries, *Class. Quantum Grav.* **24** 3933-3950
[8] Blagojević M and Cvetković B 2008 Electric field in 3D gravity with torsion *Phys. Rev. D* **78** 044036
[9] Blagojević M and Cvetković B 2008 Self-dual Maxwell field in 3D gravity with torsion *Phys. Rev. D* **78** 044037
[10] Deser S, Jackiw R and Templeton S 1982 Topologically Massive Gauge Theories *Ann. Phys.* **140** 372-411
Deser S, Jackiw R and Templeton S 1982 Three-Dimensional Massive Gauge Theories *Phys. Rev. Lett.* **48** 975-978
[11] Kamata M and Koikawa T 1995 The electrically charged BTZ black hole with self (anti-self) dual Maxwell field, *Phys. Lett. B* **353** 196-200
Chan K 1996 Comment on the calculation of the angular momentum and mass for the (anti-)self dual charged spinning BTZ black hole *Phys. Lett. B* **373** 296-298
[12] Clément G 1996 Spinning charged BTZ black holes and self-dual particle-like solutions *Phys. Lett. B* **367** 70-74
[13] Fernando S and Mansouri F 1998 Rotating Charged Solutions to Einstein-Maxwell-ChernSimons Theory in 2+1 Dimensions *Comm. Math. Theor. Phys.* **1** 14
[14] Dereli T and Obukhov Yu N 2000 General analysis of self-dual solutions for the Einstein-Maxwell-Chern-Simons theory in (1+2) dimensions *Phys. Rev. D* **62** 024013
[15] Martinez C, Teitelboim C and Zanelli J 2000 Charged Rotating Black Hole in Three Spacetime Dimensions *Phys. Rev. D* **61** 104013

Dark energy and inhomogeneity

G F R Ellis

Mathematics Department, University of Cape Town, South Africa

Abstract. The standard understanding that distant supernova data implies existence of either dark energy or a cosmological constant Λ relies on the assumption of spatial homogeneity of the universe. A large–scale smoothed–out model of the universe ignores small–scale inhomogeneities, but the averaged effects of those inhomogeneities may alter both observational and dynamical relations at the larger scale, mimicking the effect of a cosmological constant. Alternatively large scale spatial inhomogeneity with zero Λ might mimic the effect of a positive cosmological constant. Whether this is the case or not is open to direct observational testing.

1. Inhomogeneity or Dark energy?

The explanation of dark energy is a central preoccupation of present day cosmology. Its presence is indicated by the recent apparent speeding up of the expansion of the universe indicated by supernova observations [51, 67, 2], which is usually taken to imply the existence of a cosmological constant or quintessence causing acceleration of the universe at recent times [59], and is confirmed by other observations such as those of cosmic background radiation anisotropies and Large Scale Structure studies[1]. Like dark matter, its existence was discovered, not predicted.

The astronomical observations are being refined in many sophisticated ways and used to confirm the acceleration and test the equations of state of the hypothetical dark energy. Whether it is constant or varying, its existence is a major problem for theoretical physics. It is therefore crucial to pursue the possibility of other theoretical explanations.

The deduction of the existence of dark energy is based on the assumption that the universe has a Robertson-Walker (RW) geometry - spatially homogeneous and isotropic on a large scale. However the interpretation of these observations is ambiguous. The observations can at least in principle be accounted for without the presence of any dark energy, if we consider the possibility of inhomogeneity. This can happen in two ways: locally via backreaction and observational effects, and globally via large scale inhomogeneity. I will consider these options in turn.

2. Different scale descriptions

Any mathematical description of a physical system depends on an *averaging scale* characterizing the nature of the envisaged model. This averaging scale is usually hidden from view: it is taken to be understood. Thus, when a fluid is described as a continuum, this assumes one is using an averaging scale large enough that the size of individual molecules is negligible. If the averaging scale is close to molecular scale, small changes in the position or size of the averaging volume lead to large changes in the measured density and velocity of the matter, as individual

[1] For full discussion, see the *Journal of General Relativity and Gravitation* special issue on dark energy: GERG **40**: 219-660, 2008.

molecules are included or excluded from the reference volume. Then the fluid approximation is no longer applicable; rather one is using a detailed description of the fluid where individual molecules are represented. Usual work referring to the fluid density and velocity assumes a medium–size averaging scale: not so small that molecular effects matter, but not so large that spatial gradients in the properties of the fluid are significant ([3], p.5). The actual averaging scale, or rather the acceptable range of averaging scales, is not explicitly stated but is in fact a key–feature underlying the description used, and hence the effective macroscopic dynamical laws investigated. Indeed, different types of physics (particle physics, atomic physics, molecular physics, macroscopic physics, astrophysics) correspond to different assumed averaging scales. Thus, instead of referring to a density function ϱ, one should really refer to a function ϱ_L: the density averaged over volumes characterized by scale length L. The key–point about the fluid approximation is that, provided this length scale is in the appropriate domain, then its actual value does not matter; i.e. when it is in this range, then changing L by a factor of 10, 100, or even much more makes no difference: the measured density and average velocity will not change. But if you change L by a very large amount until outside this range, this is no longer true. Hence, there is a range of validity $L_1 < L < L_2$ where the fluid approximation holds [3] and explicit mention of the associated averaging scale may be omitted.

In electromagnetic theory, polarization effects result from a large–scale field being applied to a medium with many microscopic charges. The macroscopic field E differs from the point–to–point microscopic field, which acts on the individual charges because of a fluctuating internal field E_i, the total internal field at each point being $D = E + E_i$ ([45], p.116). Spatially averaging, one regains the average field because the internal field cancels out: $E = \langle D \rangle$, indeed this is how the macroscopic field is defined (implying invariance of the background field under averaging: $E = \langle E \rangle$). On a microscopic scale, however, the detailed field D is the effective physical quantity, and so is the field "measured" by electrons and protons at that scale. Thus, the way different test objects respond to the field crucially depends on their scale (a macroscopic device will measure the averaged field).

Now, exactly the same issue arises with regard to the gravitational field. Applications such as the solar system tests of general relativity theory are at solar system scales. We apply gravitational theory, however, at many other scales: to star clusters, galaxies, clusters of galaxies, and large–scale structures (walls and voids), as well as to black holes (occurring at solar system and star cluster scales, and possibly at much smaller scales).

2.1. Coarse–graining the gravitational field

A range of scales of description are relevant to cosmology. There are levels of approximation in modeling the universe, each with a hidden averaging scale. One can have a description in which every star is represented, or every galaxy (the stars averaged over), or only the largest scale cosmological structures (even galaxies averaged over, as in the fluid approximation).

The General Relativistic cosmological perturbation solutions used to study structure formation embody two interacting levels: the background (zero–order) model, almost always a Robertson–Walker metric, and the perturbed (first–order) model representing the growth of inhomogeneities, represented by a perturbed Robertson–Walker metric. This however may not represent accurately major inhomogeneities such as massive voids on scales bigger than clusters of galaxies. The question then is how do models on two or more different scales relate to each other in Einstein's gravitational theory [31]. This is a difficult issue both because of the non–linearity of Einstein's equations, and because of the lack of a fixed background spacetime – one of the core features of Einstein's theory. This causes major problems in defining suitable averaging processes as needed in studying these processes. While there have been many analyses of this problem, there are still issues to be resolved in relation both to observations and dynamics, and in how this relates to gravitational entropy and the arrow of time. We will not consider the latter here.

2.2. Non–commutativity of averaging and dynamics

The key–point in considering dynamical effects is that the two processes involved in relating the field equations at different scales do not commute [31]. These processes are:

E: calculating the Einstein tensor $G_{1ab} := R_{1ab} - \frac{1}{2}R_1 g_{1ab}$ from a metric tensor g_{1ab}, and, hence, determining the quantity $E_{1ab} := G_{1ab} - \kappa T_{1ab}$ for g_{1ab}, where T_{1ab} is the matter tensor appropriate to the scale represented by g_{1ab};

A: averaging the metric tensor g_{1ab} to produce a smoothed metric tensor $g_{2ab} : g_{2ab} = \langle g_{1ab}\rangle$ and the matter tensor T_{1ab} to produce a corresponding smoothed matter tensor T_{2ab} : $T_{2ab} = \langle T_{1ab}\rangle$.

Now in general the averaging process does not commute with taking derivatives: for a function g, usually $\partial_i\langle g\rangle \neq \langle\partial_i g\rangle$. Furthermore the inverse metric g_2^{ab} (non–linearly dependent on the metric tensor components g_{1ab}) is not the smoothed version of g_1^{ab}. The resulting Christoffel terms Γ^a_{2bc} are therefore not the smoothed version of Γ^a_{1bc}, hence the Ricci tensor components R_{2ab}, non–linearly dependent on Γ^a_{2bc}, are not the smoothed versions of R_{1ab}. Extra non–linearities occur in calculating the Einstein tensor $G_{2ab} = R_{2ab} - \frac{1}{2}R_2 g_{2ab}$ from the Ricci tensor R_{2ab}. Thus, if you smooth first and then calculate the field equations, you get a different answer than if you calculate the field equations first and then smooth; symbolically $\mathbf{A}(\mathbf{E}(g_{1ab})) \neq \mathbf{E}(\mathbf{A}(g_{1ab}))$.

Suppose the field equations are true at the first scale:

$$E_{1ab} := G_{1ab} - \kappa T_{1ab} = 0, \tag{1}$$

then they will not be true at the second scale:

$$E_{2ab} := G_{2ab} - \kappa T_{2ab} \neq 0. \tag{2}$$

Thus, there will be an extra term in the equations at the smoother scale. We can either regard it as an extra term on the left–hand–side,

$$G_{2ab} - E_{2ab} = \kappa T_{2ab} \ , \tag{3}$$

representing a modified curvature term, or as an extra term on the right–hand–side,

$$G_{2ab} = \kappa T_{2ab} + E_{2ab} \ , \tag{4}$$

where it is regarded as an extra contribution to the matter tensor. Which is the more appropriate interpretation depends on the context.

Szekeres [77] developed a polarization formulation for a gravitational field acting in a medium, in analogy to electromagnetic polarization. He showed that the linearized Bianchi identities for an almost flat spacetime may be expressed in a form that is suggestive of Maxwell's equations with magnetic monopoles. Assuming the medium to be molecular in structure, it is shown how, on performing an averaging process on the field quantities, the Bianchi identities must be modified by the inclusion of polarization terms resulting from the induction of quadrupole moments on the individual "molecules". A model of a medium whose molecules are harmonic oscillators is discussed and constitutive equations are derived. This results in the form:

$$E^{2ab} = Q^{abcd}{}_{;cd} \ , \tag{5}$$

that is E_{2ab} is expressed as the double divergence of an effective quadrupole gravitational polarization tensor Q^{abcd} with suitable symmetries:

$$Q^{abcd} = Q^{[ab][cd]} = Q^{cdab} . \tag{6}$$

Gravitational waves are demonstrated to slow down in such a medium.

The problem with such averaging procedures is that they are not covariant. They can be defined in terms of the background unperturbed space, usually either flat spacetime or a Robertson–Walker geometry, and so will be adequate for linearized calculations where the perturbed quantities can be averaged in the background spacetime (although even here the gauge problem arises, see below). But the procedure is inadequate for non–linear cases, where the integral needs to be done over a generic lumpy (non–linearly perturbed) spacetime that are not "perturbations" of a high–symmetry background. However, it is precisely in these cases that the most interesting effects will occur.

The only tensor integrals that are well–defined over a generic spacelike surface or spacetime region (and one interesting issue is which of these one should use in the averaging process) are for scalars [9, 10], unless one uses the bitensors associated with Synge's world function [76], based on parallel propagation along geodesics, to compare tensors at different points in a normal neighbourhood. The problem is that they cannot be used for averaging the metric tensor, for it is the metric tensor itself that defines the parallel propagation used in this process, and so is left invariant by it (since $g_{ab;c} = 0$). So, one has to devise a procedure in which either the field equations are represented only in terms of scalars, possible for example if one takes components relative to a covariantly uniquely defined tetrad, or else bitensors are used to define averages of quantities other than the metric.

Zalaletdinov has taken this issue seriously, and provided the only sustained such attempt based on bitensors [88]. He proposes a macroscopic description of gravitation based on a covariant spacetime averaging procedure. The geometry of the macroscopic spacetime follows from averaging Cartan's structure equations, leading to a definition of correlation tensors. Macroscopic field equations (averaged Einstein equations) can be derived in this framework. It is claimed that use of Einstein's equations with a hydrodynamic stress–energy tensor means neglecting all gravitational field correlations, and a system of macroscopic gravity equations is given when the correlations are taken into consideration. This approach has not won many adherents, but is nevertheless a systematic and coherent attempt to set up the problem generically.

3. Small scale inhomogeneity: backreaction

The cosmological application is to understand the nature of the backreaction of perturbations in cosmology. The large-scale solution used is the Friedmann-Lemaître-Robertson-Walker ('FLRW') exactly spatially homogeneous and isotropic universe models. The real universe is not like this, it has vast inhomogeneities: walls of clusters of galaxies surrounding voids, forming a soap-bubble like structure. The FLRW models are only a smoothed approximation to this reality. All the issues discussed above arise.

Varied methods have been used to study this problem, including straightforward perturbation approaches. Isaacson's method of averaging in the gravitational radiation case has been used in the cosmological context [37]. When Zalaletdinov's approach to the averaging problem is applied to cosmology, the Einstein field equations on cosmological scales are modified by appropriate gravitational correlation terms [24]; for a spatially homogeneous and isotropic macroscopic spacetime, the correlation tensor is of the form of a spatial curvature term.

One sustained attack is that by Buchert [9, 10, 11]. Here one first thinks of averaging over scalar quantities like the density or the rate of expansion in an inhomogeneous universe model. As long as one works with exact equations for the evolution of those fields in a given foliation of spacetime, such an averaging procedure is covariant. Cosmological parameters like the rate of expansion or the mass density are to be considered as volume–averaged quantities, so the relevant parameters are intrinsically scale–dependent unlike the situation in a Friedmannian cosmology. Averaging scalar characteristics on a Riemannian spatial domain delivers the effective dynamical sources that an observer would measure, but although he measures within the lumpy spacetime, he is going to interpret his observations within a Friedmannian fitting model. This suggests a logical division of the averaging problem into 1) calculating averages in the real manifold, and 2) determining the mapping between averages in the real manifold and averages in the Friedmannian model. The first averaging is straightforward for scalars, and encounters *non–commutativity of averaging and time–evolution*: this is a purely kinematical property that can be expressed, for a scalar field ψ, through the rule

$$\partial_t \langle \psi \rangle - \langle \partial_t \psi \rangle \;=\; \langle \theta \psi \rangle - \langle \theta \rangle \langle \psi \rangle \ . \tag{7}$$

The fluctuation part on the right–hand–side of this rule produces the *kinematical backreaction*. The result of this averaging is the Buchert modified Friedmann and Raychaudhuri equations [11]:

$$3(\dot{a}_D/a_D)^2 = \Lambda + 8\pi G \rho_D - (Q_D + \langle R \rangle_D)/2, \tag{8}$$

$$3(\ddot{a}_D/a_D) = \Lambda - 4\pi G \langle \rho \rangle_D + Q_D, \tag{9}$$

where $a_D(t)$ is the volume averaged scale factor, $\langle R \rangle_D$ the averaged spatial curvature, and

$$Q_D := 2\langle II \rangle_D - \frac{2}{3} \langle I \rangle_D^2, \tag{10}$$

$$II := \theta^2/3 - \sigma^2, \ I := \Theta. \tag{11}$$

This shows that averaging in principle allows acceleration terms to arise from the averaging process. The derivation of these equations, and their numerical implementation, has been studied by many others, see for example [4, 8].

The second "averaging" is more adequately thought of as a rescaling of the tensorial geometry. A (Lagrangian) smoothing as opposed to (Eulerian) rescaling of the metric on regional spatial domains has been proposed by Buchert and Carfora [12, 14], using a global Ricci deformation flow for the metric initially proposed by Carfora and Piotrkowska [17]. They introduced real-space renormalization group (RG) methods, based on properties of the Ricci-Hamilton flow. The smoothing of geometry implies a renormalization of averaged spatial variables, determining the effective cosmological parameters as they appear in the Friedmannian fitting model. Two effects that quantify the difference between background and real parameters were identified: *curvature backreaction* and *volume effect* [13]. Both are the result of an inherent non–commutativity of averaging and spatial rescaling. In this way we look at the averaging problem in two directions in function space: time–evolution (as a deformation in direction of the extrinsic curvature of the space sections encoding the kinematical variables), and scale–"evolution" (as a deformation in direction of the intrinsic 3–Ricci curvature).

Thus there are varied approaches to the averaging issue. However, it is not clear how they relate to the gauge problem.

3.1. The gauge issue

The gauge problem is the following: when you perturb a smooth background cosmological metric $\bar{g}_{1ab}$ to obtain a perturbed metric $g_{ab} = \bar{g}_{1ab} + h_{1ab}$, the inverse relation is not unique: there is no agreed averaging or fitting process that will give back a unique background metric $\bar{g}_{ab}$ back from the "lumpy" metric g_{ab} [35]. Some other smooth metric $\bar{g}_{2ab}$ could have been chosen as the background metric instead, leading to a different definition of the perturbations: $h_{2ab} := g_{ab} - \bar{g}_{2ab}$, instead of $h_{1ab} := g_{ab} - \bar{g}_{1ab}$. The choice of background metric $\bar{g}_{ab}$ for a specific "lumpy" metric g_{ab} is called a 'gauge choice'. The backreaction problem will look very different if described in terms of different gauges.

The best way to look at this is to think of a gauge choice as a mapping of a smooth background metric $\bar{g}_{1ab}$ into the lumpy universe with metric g_{ab} [34]. At each point in the real spacetime the density perturbation $\delta\varrho$ is then defined by $\delta\varrho := \varrho - \bar{\varrho}$, where ϱ is the actual density at that point, and $\bar{\varrho}$ the background density at the same point. The key–issue is the choice of surfaces of constant time in the perturbed spacetime, conventionally taken to represent the image of surfaces of constant density of the background spacetime. It then becomes clear that one can for example set the density perturbation to zero by choosing the mapping so that the surfaces of constant background density $\bar{\varrho}$ are the same as the surfaces of constant real density: for then at each point $\varrho = \bar{\varrho} \Rightarrow \delta\varrho = 0$. However, with this choice, the fluid flow lines will not be orthogonal to the surfaces of constant density, so there will still be a non–zero density variation measured by comoving observers. Gauge issues arising in treating multi–component fluids raise extra issues because of the multiple possible choices of reference velocity field [26].

The remedy to this disconcerting behaviour is to choose gauge invariant variables, for example a set of covariantly defined variables that vanish in the background spacetime [34]. While many studies have been carried out for quantifying backreaction effects in cosmology, where the smoothed–out effect of the small–scale perturbations causes extra terms in the Friedmann equations for the background metric, none have been done that both fully and clearly take the gauge issue into account and go beyond linear order. This is a key–issue waiting to be resolved. One certainly wants to go at least to second order in understanding the effects of non–linear perturbations. Many of the crucial results at linear order no longer hold, for example scalar, vector and tensor perturbations are no longer independent of each other at second order [19], and then the backreaction in turn affects the perturbations themselves [61].

3.2. The fitting issue

Given the difficulty of handling averaging in a covariant and gauge invariant way, the conclusion may be that one should as far as possible use covariant variables, but choose a specific gauge when doing averaging. This means one needs to solve the *Fitting Problem*: one must decide which is the best way to fit a smooth background model to the lumpy real universe [35]. This decides what the magnitude of the backreaction effects is, because it determines the size of the "perturbation" away from the background at each point. One way of trying to do this is by averaging, for example one can fit the background model by averaging densities and pressures, or by fitting observable relations such as the (m, z) and (N, z) relations in order to determine H_0 and q_0 for the best-fit background model. These procedures will in general give different results, particularly because the former will be based on a spacelike averaging but the latter on a null cone averaging (because observations are made on the past null cone). Most observational studies (e.g. the classic study by Sandage [74]) implicitly use a null fitting procedure, without making this explicit.

A proper fitting procedure underlies detailed backreaction studies. If the wrong background is fitted then it may appear to create a back reaction which vanishes if a better fit is chosen. Thus a gauge choice is inherent in the backreaction issue; this will be apparent in any detailed studies, even if it is not clearly made manifest.

It is important to note the following: if you consider a FLRW model G_1 and perturb it to get an almost-FLRW model H_1, it is very tempting to assume that H_1 is the best-fit model to G_1: indeed this assumption is effectively hidden in most perturbation studies. But this is not necessarily the case. For example, if you add to G_1 a Gaussian distribution of density fluctuations suitably distributed over a spacelike surface, you will have raised the density on that surface: so a process of fitting the background model by averaging densities will no longer result in determining G_1 as the best fit; it will rather yield a model G_2, with a higher average density than G_1. To preserve the best-fit model you started with, one needs to add a compensated set of perturbations (e.g. top-hat models) where an underdensity surrounds every overdensity, so that the average density is unchanged by the perturbation procedure. This is rarely done; but when it is not done, a change in the average properties will result in a spurious apparent back-reaction effect, which is in reality due to a failure to use the correct background model for the perturbed spacetime. A proper fitting procedure will lead to integral constraints related to the fitting procedure [78].

3.3. Importance of backreaction effects?

There is no doubt that interesting effects occur [36, 15, 68, 54]. The key issue is whether these backreaction effects are significant in cosmology. On the one hand, they might play a significant role in the inflationary era [63, 39]. On the other, it can possibly help explain the apparent dynamical existence either of dark energy and/or of dark matter as effective terms in the macroscopic dynamics at recent times. Various papers suggest the effect may indeed be significant, for example the observed acceleration of the universe could possibly be the result of the backreaction of cosmological perturbations rather than the effect of a negative–pressure dark energy fluid [82, 54, 62, 53] if astronomical parameters are restricted in specific ways [71]. However, other studies obtain different results [73, 15, 65, 66, 68, 69]. Gauge effects are problematic [38], and many doubt the effect is significant (e.g. [44]).

An interesting new development is the realisation that there are really large voids in the universe (see e.g. [72]) and these have important significance for averaging in cosmology, as emphasized by Wiltshire [85]. The geometry and dynamics in and out of a void are quite different: within voids where structures such as isolated galaxies are established, the geometry is locally quasi-static and vacuum dominated, completely different from the FLRW spatially homogeneous matter–dominated and evolving region outside the voids. They have to be joined with suitable junction conditions so that internal and external geometries agree.

One can model this situation by exact inhomogeneous models (next section), but often it is done by using standard perturbation methods based on a "Newtonianly perturbed FLRW metric", see Ishibashi and Wald [44]:

$$ds^2 = -(1+2\Psi)dt^2 + a^2(t)(1-2\Psi)\gamma_{ij}dx^i dx^j \tag{12}$$

with γ_{ij} the metric of a 3-space of constant curvature $k = +1, 0$, or -1, and Ψ satisfies

$$|\,\Psi\,| \ll 1, \quad |\,\partial\Psi/\partial t\,|^2 \ll (1/a^2)D^i\Psi D_i\Psi, \tag{13}$$

$$(D^i\Psi D_i\Psi)^2 \ll (D^iD^j\Psi)(D_iD_j\Psi), \tag{14}$$

where D_i denotes the derivative operator associated with γ_{ij}. Note that the matter present is not assumed to be comoving in this frame. The assumption made in analyses such as that of Ishibashi and Wald is that one can use such coordinates globally. This is probably correct either in a perturbed FLRW domain, or in a locally isolated quasi-static domain, but it is not obvious such coordinates can be simultaneously used in both for any extended length of time. Such coordinates are probably only locally valid in a realistic description of the universe taking both expanding matter-filled domains and locally static vacuum domains into account; but it is the assumption that these coordinates can be used globally that leads to the conclusion that back-reaction effects are negligible (cf. [86, 52]). Its domain of validity in realistic situations needs to be investigated.

Thus a key issue is, *How large in space and time is the domain where such quasi-Newtonian coordinates (12)-(14) can be used in a realistic model of an expanding universe with voids?* One may note here that the timelike reference congruence associated with such coordinates is shear-free, conformally mapping the 3-spaces onto each other. Assuming such a congruence exists is a major assumption, placing strong limits on the Weyl tensor, inter alia excluding occurrence of vector and tensor perturbations. It may be acceptable in the circumstances considered because matter is not comoving in this frame, and metric perturbations are small despite density perturbations being large; but this assumption certainly needs investigation.

An alternative to using these perturbation techniques is looking for suitable exact inhomogeneous solutions of the field equations. The first important set often used are the Lemaître-Tolman-Bondi ('LTB') spherically symmetric pressure-free models [7], discussed further below. The second ones are Swiss-cheese models, with spherically symmetric static vacuum domains (hence Schwarschild solutions) imbedded in a FLRW universe G_1 (Einstein and Strauss [29], Harwit [40]). These can two sets can be blended by imbedding LTB models in a FLRW universe. One can then study averaging and observational issue in these models. If the fitting is done properly, matching metrics and masses at the junctions, no strange matter exists and the junctions and the existence of the lumps does not influence the background model FLRW G_1 first considered; however that may not then be the best fit model G_2, depending on the fitting procedure used (cf. section 3.2 above). However G_2 and G_1 are not likely to be very different; so the dynamical effect is not likely to be large. However the observational effect is a different matter, as we consider in the next section.

Before doing so, I remark on a rather different way of treating homogeneity: Lindquist and Wheeler [60] examined joining many Schwarzschild cells together, with boundaries subject to equations of motion that make the whole expand as an approximation to a $k = +1$ FLRW universe. They found that regular lattices of this type are possible only for $n = 5, 8, 16, 124$, or 600 vertices. The paper compares these lattice universe with closed Friedmann universes. This differs from the Swiss Cheese models, because in this case there are no FLRW domains used in addition to Schwarzschild cells: rather the composite of Schwarzschild cells IS the approximation to a FLRW universe model. locally static but globally expanding. The Schwarzschild cell method predicts the same relation between radius of the universe and proper time, except that the connecti9n between maximum radius and mass is different. This is the way averaging effects lead to a different FLRW model than an exactly smooth distribution of the same matter. What would be interesting would be extension of such models to irregular lattices; but this might be technically difficult. Again the issue will arise of Best-Fitting of the background model derived in this way.

A tentative interim conclusion is that the dynamical effects are certainly there, but are small [4, 8]. In any case, a detailed investigation of backreaction effects helps to improve fitting models on regional scales for a better interpretation of observational data.

4. Small scale inhomogeneity: observations

The effect on observational relations of introducing local inhomogeneities into a given background spacetime is double: it alters redshifts, and it changes area distances.

4.1. Redshift effects

When a void is intervening between the source and the observer, photons drop into a potential well and then climb out, and they exit when the universe is larger than when they went in. If all is static inside the void the redshift changes in and out cancel, but when structure is forming the potential well is changing with time so one gets the Rees-Sciama effect [70]: a change in redshift due to change in the potential well as the photon traverses it. A non-zero cosmological constant will also lead to such an effect.

When the source is in the void, the photon has to climb out of the void, giving a contribution to observed redshifts. Wiltshire emphasizes [83] that time runs at different rate in and out of voids in a void dominated universe, and the difference is cumulative. This can be thought of as the source of an extra redshift contribution, changing the observable relations between cosmological variables.

4.2. Area distance effects

The usual analysis of cosmological observations is based on the equations relating apparent magnitude and redshift in Robertson–Walker spacetimes. The Sachs optical scalar equations for hypersurface–orthogonal null geodesics in a general space-time are:

$$\frac{d\theta}{dv} = -R_{ab}K^aK^b - 2\sigma^2 - \frac{1}{2}\theta^2; \qquad \frac{d\sigma_{mn}}{dv} = -E_{mn}\,, \tag{15}$$

where θ is the rate of expansion of the null geodesics with tangent vector $K^a = dx^a/dv$ and affine parameter v, σ_{mn} is their shear, R_{ab} the Ricci tensor, and E_{mn} a matrix of Weyl tensor components ([41], p.88; [75], pp.108–9). In the idealized Robertson–Walker case, the Weyl tensor C_{abcd} vanishes, but the Ricci tensor is non–zero, being given via the Einstein field equations from the matter present. Thus, $C_{abcd} = 0 \Rightarrow E_{mn} = 0$, and the relevant solutions are shear–free:

$$\sigma^2 = 0 \quad \Rightarrow \quad \frac{d\theta}{dv} = -R_{ab}K^aK^b - \frac{1}{2}\theta^2\ . \tag{16}$$

Integration gives the Mattig relations applicable to Friedmann universe models [74, 30, 75].

However, in the real universe, as pointed out by Bertotti [5], observations take place via null geodesics lying in the empty spacetime between galaxies. Thus, the real situation in a universe with no intergalactic medium (all the matter is concentrated in galaxies) is the opposite of that above: in the region of spacetime traversed by the geodesics, the Ricci tensor vanishes, so

$$\frac{d\theta}{dv} = -2\sigma^2 - \frac{1}{2}\theta^2\ , \qquad \frac{d\sigma_{mn}}{dv} = -E_{mn}. \tag{17}$$

The Weyl tensor (the tidal gravitational field caused by nearby matter) generates shear that then causes focussing. Thus, the microscopic description of the focussing caused by local inhomogeneities in the vacuum gravitational field ($\sigma \neq 0$, $R_{ab} = 0$, $E_{mn} \neq 0$) is radically different from the macroscopic one where focussing is caused by a smooth matter distribution only ($\sigma = 0$, $R_{ab} \neq 0$, $E_{mn} = 0$), and the area distance–redshift relation may be expected to be different on microscopic scales (i.e. the small solid angle bundles of null geodesics actually used in observations of individual objects), as compared with macroscopic scales (averaging over large solid angles).

Thus Ricci focusing on the large scale is replaced by Weyl focusing on small scales. This micro-focussing due to the Weyl tensor must give the FRW equations when averaged over whole sky, but how this happens is not obvious! It has been suggested that energy conservation will imply this FRW all-sky average [81], but this assumes that areas of a bundle of null geodesics are the same in the perturbed and background models, which will not be the case when one takes the effect of caustics into account [33]. Indeed, areas increase slower than in a Robertson–Walker model in the empty spaces between matter, where the Ricci term is zero, and faster in the high–density regions where matter is concentrated, so one might think these effects cancel out. However, the strongly lensed rays soon go through a caustic and emerge highly divergent, so that areas are rapidly increasing again. It is plausible that on average the overall effect is always an increase in area, that is a lesser area distance than in the smooth background model. But in any case supernova observations are preferentially in directions where there is no intervening matter: they will not correspond to an all-sky average.

Various proposals have been made to deal with this. The most popular is the Dyer–Roeder distance [27, 28], obtained by assuming only a fraction f of the total mass density is encountered

Journal of Physics: Conference Series **189** (2009) 012011 doi:10.1088/1742-6596/189/1/012011

by the light–rays but ignoring the shear. Thus, in (16) one replaces $R_{ab}K^aK^b$ by $fR_{ab}K^aK^b$ and works out the corresponding area distance ([75], pp.138–143; [25]). This may be a good approximation if galaxies are embedded in a fairly uniform intergalactic medium of dark matter, but clearly does not take shear effects and caustics properly into account. How good it is will depend on the nature of clustering in the universe and how the averaged distribution impacts along the line of sight [58]. If the dark matter is uniform, Dyer-Roeder is good; if dark matter is clustered, it is not so good.

One can approach the topic in other ways: for example by using stochastic methods [5], or detailed examination of geodesics in Swiss–Cheese universe models with FLRW regions joined to vacuum regions. In these exact inhomogeneous solutions, the null geodesic equations can be exactly integrated in each domain and matched; this includes shear effects.

Kantowski investigated this in a series of papers, and gave a nice geometrical confirmation of the effect on observational relations [46]. He obtained analytic expressions for distance–redshift relations that have been corrected for the effects of inhomogeneities in the density [48]. The values of the density parameter and cosmological constant inferred from a given set of observations depends on the fractional amount of matter in inhomogeneities and can significantly differ from those obtained by using the Mattig relations for the FLRW universes. As an example, a determination of Ω_0 made by applying the homogeneous distance–redshift relation to SN 1997ap at z = 0.83 could be as much as 50% lower than its true value [48]. When $\Lambda = 0$ the equation is an associated Legendre equation and hence distance-redshift can be written using associated Legendre functions. When Λ is not zero the equation is a Lame' equation which is a special type of the Heun equation. The distance-redshift using Heun functions is in [48, 50], solutions depending on the fraction of the universe's mass density that is transparent to the observing beams. The flat space case is special, even when Λ is not zero; for this case distance-redshift can be written using associated Legendre functions [49].

These results make clear that the apparent acceleration term detected could be at least partly due to this optical effect: focussing of null geodesics is different in a lumpy universe than in a smooth one.

4.3. The combination of effects

Putting this together, averaging processes when there are many local inhomogeneities leads both to dynamical back reaction effects, and optical effects due both to altered redshifts and area distances. This combination of effects seems to have the potential to significantly influence interpretation of observational relationships such as the supernova data.

Is this influence of inhomogeneities sufficient to fully explain the apparent acceleration indicated by the supernova data? It is claimed by some that it is, see for example Leith, Ng and Wiltshire [55], utilizing Wiltshire's solution of the Buchet equations [84]. However the modeling used is controversial, and not yet universally accepted. Investigations such as those by Kantowski [46, 47, 48] and Clifton et al [23] suggest it is certainly large enough to significantly affect the the Hubble diagram: "It is found that intervening voids, between the observer and source, have no noticeable effect, while sources inside voids can be effected considerably. By averaging observable quantities over many randomly generated distributions of voids we find that the presence of these structures has the effect of displacing the average magnitude from its background value, and introducing a dispersion around that average" [23]. A contrasting view, suggesting the effect is much smaller, is in [80]. Thus while it is very debatable if it is enough to fully account for the apparent acceleration, it may well be enough to significantly influence the concordance model parameter values, and hence the present understanding of inflationary effects in those models. Clearly, this topic needs further careful investigation.

5. Large scale inhomogeneity

Perhaps the Copernican assumption of large-scale spatial homogeneity is not true. Maybe there is a large scale inhomogeneity of the observable universe, such as that described by the Lemaître-Tolman-Bondi (LTB) pressure-free spherically symmetric models [7], and we are near the centre of a void. The redshift-distance relation for distant supernova may be measuring spatial inhomogeneity, rather than acceleration of a FLRW universe.

5.1. Can we fit the observations?

The LTB models have comoving coordinates

$$ds^2 = -dt^2 + B^2(r,t) + A^2(r,t)(d\theta^2 + \sin^2\theta d\phi^2) \tag{18}$$

where

$$B^2(r,t) = A'(r,t)^2(1-k(r))^{-1} \tag{19}$$

and the evolution equation is

$$(\dot{A}/A)^2 = F(r)/A^3 + 8pG\rho_\Lambda/3 - k(r)/A^2 \tag{20}$$

with the energy density given by

$$F'(A'A^2)^{-1} = 8\pi G\rho_M. \tag{21}$$

There are two arbitrary functions of the spatial coordinate r: namely $k(r)$ (curvature) and $F(r)$ (matter). This freedom enables us to fit the supernova observations with no dark energy or other exotic physics (this is a theorem, see Mustapha *et al* [64]). The idea that such models can explain the supernova observations without any dark energy is discussed by Celériér [18]. One can also fit the CBR observations because they refer to much larger values of r [1, 22]. One should note here that at least some of the observed CBR dipole can then arise because we are a bit off-centre in the void, so one can re-evaluate the great attractor analysis in this context and the alignment of the dipole and quadrupole. Nucleosynthesis data can also be fitted; what is a bit more problematic is the Baryon Acoustic Oscillations ('BAO'). The key comment to make is that different scales are probed by different observations and can in principle all be fitted by adjusting the free spatial functions at different distances.

A typical observationally viable model is one in which we live roughly centrally (within 10% of the central position) in a large void: a compensated underdense region stretching to $z \simeq 0.08$ with $\delta\rho/\rho \simeq -0.4$ and size $160/h$ Mpc to $250/h$ Mpc, a jump in the Hubble constant of about 1.20 at that distance, and no dark energy or quintessence field present (Biswas *et al* [6], Ishak *et al* [43], Yoo *et al* [87]). The details are being tested by many further models, but the basic point is clear: it is indeed possible at first order to fit the observations by such inhomogeneous models with no dark energy present. In an era of precision cosmology, the present debate is about more accurate modeling, and obtaining a better model-dependent better fit to ever more detailed observations.

5.2. Dynamic evolution?

Given we can fit the observations by such a model, can we find dynamics (inflation followed by a HBB era) that can lead to such a model? It has the same basic dynamics as the standard model (evolution along individual world lines governed by the Friedmann equation) but with distant dependent parameters. Will inflation prevent it? This depends on the initial data, the amount of inflation, and the details of the unknown inflaton. If we are allowed the usual tricks of fiddling the inflationary potential and initial data, and adding in multiple fields as desired, then there is sufficient flexibility that it should certainly be possible.

5.3. Improbability

Many dismiss these models on probability grounds: It is improbable the universe is like this, and it is improbable we are near the centre of such a model. But there is always improbability in cosmology. We can can shift it around, but it is always there. It might be in the nature of a Robertson-Walker geometry (the old view), in the inflationary potential and initial conditions (the current mainstream position), which specific universe domain we are in within a multiverse, or the spatial position in an inhomogeneous universe (the present proposal). Note that we are competing with a probability of 10^{-120} for Λ in a FLRW universe; we do not have to get very high probabilities to outdo that improbability, which is what multiverse proposal aim to handle.

Three comments are in order. First, a key feature of cosmology is that there is only one universe; and the very concept of probability does not apply to a single object, even though we can make many measurements of that single object to determine its detailed nature. Probability applies to the multiple measures we can make of the single universe, but not to issues doing with the existence of the universe itself (Ellis [32]). There is no physically realised ensemble to apply that probability to, unless a multiverse exists in physical reality which is not proven. In essence, there simply is no proof the universe is probable; that is a philosophical assumption, which may not be true. The universe may be improbable!! Secondly, there is no well-justified measure for any such probability proposal even if we ignore the first problem. This is still an issue of debate.

And thirdly, a study by Linde *et al* [57] shows that (given a particular choice of measure) this kind of inhomogeneity actually *is* a probable outcome of inflationary theory, with ourselves being located near the centre! One cannot dismiss such models out of hand for probability reasons.

6. Observational tests of spatial homogeneity

Given that we can find inhomogeneous models that reproduce the observations without any exotic energy, as well as homogeneous models that explain the same observations through some form of dark energy, how can we distinguish between the two? Ideally we need a model-independent test of the basic assumption of most present day cosmology: is a RW geometry the correct metric for the observed universe region? Is the Copernican principle correct? Four kinds of tests are possible, as I now discus.

6.1. CBR based tests

Some tests use scattered CBR photons to check spatial homogeneity (Goodman [42]; Caldwell and Stebbins [16]). If the CBR radiation is anisotropic around distant observers (as will be true in inhomogeneous models), Sunyaev-Zeldovich scattered photons have a distorted spectrum that reflects the spatial inhomogeneity. However this test is somewhat model dependent - it is good for void models but misses, e.g., conformally stationary spacetimes. It also has to take into account other possible causes of spectral distortion.

6.2. Direct observational tests: behaviour near origin

The universe must not have a geometric cusp at the origin, as this implies a singularity there. Thus there are centrality conditions that must be fulfilled in the inhomogeneous models (Vanderveld et al [79]). The distance modulus behaves as

$$\Delta dm(z) = -(5/2)q_0 z \tag{22}$$

in standard ΛCDM models for small z, but if this were true in a LTB void model without Λ this implies a singularity (Clifton *et al* [21]). Observational tests of this requirement will be possible through intermediate redshift supernovae in the future.

6.3. Direct observational tests: constancy of curvature

There are two geometric effects on distance measurements: curvature Ω_k bends null geodesics, expansion $H(z)$ changes radial distances. These are coupled in RW models, as expressed in the relation

$$d_L(z) = \frac{(1+z)}{H_0\sqrt{-\Omega_k}} \sin\left(\sqrt{-\Omega_k}\int_0^z dz' \frac{H_0}{H(z')}\right). \tag{23}$$

While these effects are strictly coupled in RW geometries, they are decoupled in LTB geometries.

In RW geometries, we can combine the Hubble rate and distance data to find the curvature today:

$$\Omega_k = \frac{[H(z)D'(z)]^2 - 1}{H_0 D(z)]^2} \tag{24}$$

This relation is independent of all other cosmological parameters, including dark energy model and theory of gravity. It can be used at single redshift to determine Ω_k. The exciting result of Clarkson *et al* ([20]) is that since Ω_k is independent of z, we can differentiate to get the consistency relation

$$C(z) := 1 + H^2(DD'' - D'^2) + HH'DD' = 0, \tag{25}$$

which depends only on a RW geometry: it is independent of curvature, dark energy, nature of matter, and theory of gravity. Thus it gives the desired consistency test for spatial homogeneity. In realistic models we should expect $C(z) \simeq 10^{-5}$, reflecting perturbations about the RW model related to structure formation. Errors may be estimated from a series expansion

$$C(z) = \left[q_0^{(D)} - q_0^{(H)}\right] z + O(z^2) \tag{26}$$

where $q_0^{(D)}$ is measured from distance data and $q_0^{(H)}$ from the Hubble parameter. It is simplest to measure $H(z)$ from BAO data. It is only as difficult carrying out this test as carrying out dark energy measurements of $w(z)$ from Hubble data, which requires $H'(z)$ from distance measurements or the second derivative $D''(z)$.

This is the simplest direct test of spatial homogeneity, and its implementation should be regarded as a high priority: for if it confirms spatial homogeneity, that reinforces the evidence for the standard view in a satisfying way; but if it does not, it has the possibility of undermining the entire project of searching for a physical form of dark energy.

6.4. Indirect Observational tests

If the standard inverse analysis of the supernova data to determine the required equation of state, as discussed at this meeting, shows there is any redshift range where $w := p/\rho < -1$, this may well be a strong indication that one of these geometric explanations is preferable to the Copernican (Robertson-Walker) assumption, for otherwise the matter model indicated by these observations is non-physical (it has a negative kinetic energy).

There is already data suggesting this may be be the case, see e.g. Lima *et al* [56]. There are some attempts to generate matter models that will give this kind of behaviour without negative kinetic energies, for example through dark energy- dark matter interactions, but they are very speculative physically and with no tested physical basis, supposing multiple unknown forms of matter or energy with arbitrarily proposed interactions between them. It all seems rather reminiscent of the Ptolemaic epicycles for describing the solar system, with more and more complex mechanisms added in as the simpler ones failed to describe the phenomena. The physically most conservative approach is to assume no unusual dark energy or exotic interacting fields, but rather that an inhomogeneous geometry might be responsible for the observed apparent acceleration; this should be seriously considered as an alternative.

7. Conclusion

The issue of what is testable and what is not testable in cosmology is a key issue. The acceleration indicated by supernova data could possibly be due to small scale inhomogeneity that definitely exists, but may not be sufficiently significant to do the job. It could be due to large scale inhomogeneity that can probably do the job, but may not exist. Observational tests of the latter possibility are as important as pursuing the dark energy (exotic physics) option in a homogeneous universe. Theoretical prejudices as to the universe's geometry, and our place in it, must bow to such observational tests.

We should stand firm and insist that genuine science is based on observational testing of plausible hypotheses. There is nothing wrong with physically motivated philosophical explanation: but it must be labeled for what it is. Overall: theory must be subject to experimental and/or observational test; this is the central feature of science. Whatever position we may have on the issue of probability, in the end our philosophy on this question will have to give way to any such possible observational tests.

References

[1] Alexander S, Biswas T, Notari A and Vaid A 2007 Local void vs dark energy: confrontation with WMAP and Type IA supernovae *Preprint* arXiv:0712.0370
[2] Barausse E, Matarrese S and Riotto A 2005 *Phys. Rev.* D **71** 063537
[3] Batchelor G K 1967 *An Introduction to Fluid Dynamics* (Cambridge, Cambridge University Press)
[4] Behrend J, Brown I A and Robbers G 2008 *JCAP* **08** 013
[5] Bertotti B 1966 *Proc. Roy. Soc. London* A **294** 195
[6] Biswas T, Monsouri R and Notari A 2007 *JCAP* **12** 017
[7] Bondi H 1947 *MNRAS* **107** 910
[8] Brown I A, Robbers G and Behrend J 2008 Averaging Robertson-Walker Cosmologies *Preprint* arXiv:0811.4495
[9] Buchert T 2000 *Gen. Rel. Grav.* **32** 105
[10] Buchert T 2001 *Gen. Rel. Grav.* **33** 1381
[11] Buchert T 2008 *Gen. Rel. Grav.* **40**, 467
[12] Buchert T and Carfora M 2002 *Class. Quant. Grav.* **19** 6109
[13] Buchert T and Carfora M 2003 *Phys. Rev. Lett.* **90** 031101
[14] Buchert T and Carfora M 2008 *Class. Quant. Grav.* **25** 195001
[15] Buchert T, Kerscher M and Sicka C 2000 *Phys. Rev.* D **62** 043525
[16] Caldwell R and Stebbins A 2008 *Phys Rev Lett.* **100** 191302
[17] Carfora M and Piotrkowska K 1995 *Phys. Rev.* D **52** 4393
[18] Célérier M-N 2007 *New Advances in Physics* **1** 29
[19] Clarkson C 2004 *Phys. Rev.* D **70** 129902
[20] Clarkson C, Bassett B and Lu T H-C 2008*Phys. Rev. Lett.* **101** 011301
[21] Clifton T, Ferreira P G and Land K 2008 *Phys. Rev. Lett.* **101** 131302
[22] Clifton T, Ferreira P.G and Zuntz J 2009 What the small angle CMB really tells us about the curvature of the Universe *Preprint* arXiv:0902.1313
[23] Clifton T and Zuntz J 2009 Hubble Diagram Dispersion From Large-Scale Structure *Preprint* arXiv:0902.0726 (2009).
[24] Coley A A, Pelavas N and Zalaletdinov R M 2005 *Phys. Rev. Lett.* **95** 151102
[25] Demianski M, de Ritis R, Marino A A and Piedipalumbo E 2003 *Astron. Astrophys.* **411** 33
[26] Dunsby P K S, Bruni M and Ellis G F R 1992' *Astrophys. J.* **395** 54
[27] Dyer C C and Roeder R 1974 *Astrophys. J.* **189** 167
[28] Dyer C C and Roeder R 1975 *Astrophys. J.* **196** 671
[29] Einstein A and Strauss E G 1945 *Rev. Mod. Phys.* **17** 120
[30] Ellis G F R 1971 in *General Relativity and Cosmology*, Proc. Int. School of Physics "Enrico Fermi" (Varenna), Course XLVII, Ed R K Sachs (Academic Press) 104–179
[31] Ellis G F R 1984 in *General Relativity and Gravitation*, Ed B Bertotti et al (Reidel) 215
[32] Ellis G F R 2006 in *Handbook in Philosophy of Physics* Ed J Butterfield and J Earman (Elsevier) 1183
[33] Ellis G F R, Bassett B A C C and Dunsby P K S 1998 *Class. Quant. Grav.* **15** 2345
[34] Ellis G F R and Bruni M 1989 *Phys. Rev.* D **40** 1804
[35] Ellis G F R and Stoeger W R 1987 *Class. Quant. Grav.* **4** 1697
[36] Futamase T 1991 *Prog. Theor. Phys.* **86** 389

[37] Futamase T 1996 *Phys. Rev.* D **53** 681
[38] Geshnizjani G and Brandenberger R H 2002 *Phys. Rev.* D **66** 123507
[39] Geshnizjani G and Brandenberger R H 2005 *JCAP* **4** 006
[40] Harwit M 1992 *Astrophys. J.* **392** 394
[41] Hawking S W and Ellis G F R 1973 *The Large Scale Structure of Space Time* (Cambridge, Cambridge University Press)
[42] Goodman J 1995 *Phys. Rev.* D **52** 1821
[43] Ishak M, Richardson J, Whittington D and Garred D 2008 *Phys. Rev.* D **78** 123531
[44] Ishibashi A and Wald R M 2006 *Class. Quant. Grav.* **23** 235
[45] Jackson J D 1967 *Classical Electrodynamics* (New York, Wiley)
[46] Kantowski R 1969 *Astrophys. J.* **155** 89
[47] Kantowski R 1995 *Astrophys. J.* **447** 35
[48] Kantowski R 1998 *Astrophys. J.* **507** 483
[49] Kantowski R 2001 *Astrophys. J.* **561** 491
[50] Kantowski R 2003 *Phys. Rev.* D **68** 123516
[51] Kantowski R and Thomas R C 2000 Distance–Redshift in inhomogeneous $\Omega_0 = 1$ Friemann-Lemaître-Robertson-Walker cosmology *Preprint* astro–ph/0011176
[52] Kolb E W, Marra V and Matarrese S 2008 *Phys. Rev.* D **78** 103002
[53] Kolb E W, Marra V and Matarrese S 2009 Cosmological background solutions and cosmological back-reactions *Preprint* arXiv:0901.4566
[54] Kolb E W, Matarrese S and Riotto A 2005 '*New J. Phys.* **8** 322
[55] Leith B M, Ng S C C and Wiltshire D L 2008 *Astrophys. J.* **672** L91
[56] Lima P M, Vitenti S and Reboucas M J 2008 *Phys. Rev.* D **77** 083518
[57] Linde A, Linde D and Mezhlumian A 1995 *Phys. Lett.* B **345** 203
[58] Linder E V 1998 Averaging inhomogeneous universes: volume, angle, line of sight *Preprint* astro–ph/9801122
[59] Linder E V 2004 *Phys. Rev.* D **70** 023511
[60] Lindquist R W and Wheeler J A 1957 *Rev. Mod. Phys.* **29** 432
[61] Martineau P and Brandenberger R H 2005 *Phys. Rev.* D **72** 023507
[62] Mattsson T 2007 Dark energy as a mirage *Preprint* arXiv:0711.4264
[63] Mukhanov V F, Abramo L R and Brandenberger R H 1997 *Phys. Rev. Lett.* **78** 1624
[64] Mustapha N, Hellaby C and Ellis G F R 1999 *Mon. Not. R. Astron. Soc.* **292** 817
[65] Nambu Y 2002 *Phys. Rev.* D **65** 104013
[66] Notari A 2005 Late time failure of Friedmann equation *Preprint* astro-ph/0503715
[67] Pyne T and Birkinshaw M 2004 *Mon. Not. R. Astron. Soc.* **348** 581
[68] Räsänen S 2006 *Class. Quant. Grav.* **23** 1823
[69] Räsänen S 2008 *JCAP* **04** 026
[70] Rees M J and Sciama D W 1968 *Nature* **217** 511
[71] Rosenthal E and Flanagan E E 2008 Cosmological backreaction and spatially averaged spatial curvature *Preprint* arXiv:0809.2107
[72] Rudnick L, Brown S and Williams L R 2007 Extragalactic Radio Sources and the WMAP Cold Spot *Preprint* arXiv:0704.0908
[73] Russ H, Soffel M H, Kasai M and Börner G 1997 *Phys. Rev.* D **56** 2044
[74] Sandage A 1961 *Astrophys. J.* **133** 355
[75] Schneider P, Ehlers J and Falco E E 1992 *Gravitational Lenses* (Berlin, Springer-Verlag)
[76] Synge J L 1960 *Relativity: The General Theory* (North Holland)
[77] Szekeres P 1971 *Ann. Phys.* **64** 599
[78] Traschen J 1985 *Phys. Rev.* D **31** 283
[79] Vanderveld R A, Flanagan E E and Wasserman I 2006 *Phys. Rev.* D **74** 023506
[80] Vanderveld R A, Flanagan E E and Wasserman I 2007 *Phys. Rev.* D **76** 083504
[81] Weinberg S 1976 *Astrophys. J.* **208** L1
[82] Wetterich C 2003 *Phys. Rev.* D **67** 043513
[83] Wiltshire D L 2007 *New J. Phys.* **9** 377
[84] Wiltshire D L 2007 *Phys. Rev. Lett.* **99** 251101
[85] Wiltshire D L 2008 in *Dark Matter in Astroparticle and Particle Physics: Proceedings of the 6th International Heidelberg Conference*, Eds H V Klapdor-Kleingrothaus and G F Lewis (Singapore, World Scientific) 565
[86] Wiltshire D L 2008 *Phys. Rev.* D **78** 084032
[87] Yoo C-M, Kai T and Nakao K-I 2008 *Prog. Theor. Phys.* **120** 937
[88] Zalaletdinov R M 1997 *Bull. Astron. Soc. India* **25** 401

Geometric properties of magnetized black hole event horizons and ergosurfaces

E. P. Esteban

Physics Department, University of Puerto Rico-Humacao, Humacao, PR 00792
Laboratory of Theoretical Physics, Physics Department, P.O. Box 23343, University of Puerto Rico-Rio Piedras, Rio Piedras, PR 00931-3343

Abstract. In this paper we focus in the geometric properties of the magnetized Kerr-Newman metric. Three applications are considered. First, the event horizon surface area is calculated and from there we derive the first law of thermodynamics for magnetized black holes. We have obtained analytical expressions for the surface gravity, angular velocity, electric potential, and magnetic moment at the magnetized Kerr-Newman black hole event horizon. An approximate expression for the surface area of the magnetized black hole ergosurface was also obtained. Second, we study the magnetized Kerr-Newman black hole's circumferences. We found that for small values of the angular momentum ($\mid \frac{J}{M^2} \mid < 0.1$) the event horizon has a prolate spheroid shape. Increasing the value of the angular momentum will change the event horizon shape from a prolate ellipsoid to an oblate spheroid. For small values of the angular momentum and charge the ergosurface shape is an oblate spheroid. Increasing these two parameters will change the ergosurface shape from a oblate spheroid to a prolate spheroid. Third, analytical expressions for the magnetized Kerr-Newman event horizon and ergosurface Gaussian curvatures were obtained although not explicitly shown. Instead a graphical analysis was carried out to visualize regions where Gaussian curvatures take negative or positive values. We found that the Gaussian curvature at the event horizon poles has negative values and do not satisfy Pelavas condition. Therefore, these regions can not be embedded in E^3. However, the magnetized Kerr-Newman ergosurface can be embedded in E^3 regardless the negative Gaussian curvature values in some regions of the ergosurface.

1. Introduction

As is well known, in rotating black holes there are two surfaces of astrophysical interest. The first surface (black hole's event horizon) is a one way membrane which play an extremely important role in black hole physics. The second surface (stationary limit or ergosurface) bound a region called the ergosphere where objects can not be stand still, and Penrose processes could take place.

With regarding the geometry of both surfaces, three problems are usually considered. First, it is the calculation of the surfaces areas. In doing so, Smarr [1], Beckenstein [2], Bardeen, Carter, and Hawking [3] established a connection between the black hole's event horizon surface area and thermodynamics. Recently, the properties of a Kerr black hole's ergosurface area were revisited by Pelavas et al [4]. They found an exact form of the Kerr black hole's ergosurface area in terms of elliptical integrals, and also proved that the Kerr's ergosurface remains topologically spherical.

The second problem is to calculate the intrinsic distortion of both surfaces from sphericity. In

Journal of Physics: Conference Series **189** (2009) 012012 doi:10.1088/1742-6596/189/1/012012

fact, a conclusive study of the Kerr-Newman black hole's event horizon and ergosurface shapes were done by Smarr [5] and Kokkotas [6], respectively. Smarr [5] found that increasing the value of the black hole's angular momentum means an increase of the oblateness of the Kerr Newman event horizon shape. In the case of the Kerr-Newman black hole ergosurface, Kokkotas showed that increasing the black hole's angular momentum it will bring a change in the Kerr-Newman's ergosurface shape from an oblate spheroid to a prolate spheroid.

The third problem in the study of the black holes surfaces geometry is to investigate whether or not the surfaces under consideration can be isometrically embedded in E^3. With regard to this problem, Smarr [5] found negative values for the Gaussian curvature at the black hole event horizon's poles when the black hole's angular momentum increases of value. He concluded that polar zones at the Kerr-Newman's event horizon can not be isometrically embedded in E^3. Kokkotas [6] obtained for the Kerr-Newman black hole's ergosurface the same qualitative results as Smarr [5], as long as the black hole's angular momentum keeps low. However, he also found that even in regions where the Gaussian curvature becomes negative the embedding in E^3 of the black hole's ergosurface is always possible.

Although the above studies in the case of isolated standard black holes are conclusive, in more realistic astrophysical scenarios we may have to relax the assumptions of flatness and/or isolateness. For example, in AGN's galaxies the supermassive black hole engine is not isolated but surrounded by strong magnetic fields and accretion disks. In this case, the no-hair theorem fails and the standard Kerr-Newman family can not used to describe the exterior magnetized black hole spacetime.

Fortunately, in the seventies, an exact non-flat solution to the combined Einstein-Maxwell equations was found by Ernst-Wild [7]. The Ernst-Wild electrovac solution [7] is a generalization of the Kerr-Newman metric and can be interpreted as providing a model for the exterior spacetime due to a charged and rotating black hole which is placed in an external magnetic field. From now on, we shall call this electrovac metric as the magnetized Kerr-Newman metric.

Earlier efforts to study the geometrical properties of particular cases of the magnetized Kerr-Newman metric were carried out by several authors. For instance, Wild W. J. et al [8] and Krori and Barua [9] attempted to calculate the Gaussian curvature of the event horizon of the magnetized Kerr metric (a particular case of the magnetized Kerr-Newman metric). Unfortunately, a misunderstanding in the interpretation of the magnetized Kerr metric parameters made the results and conclusions given in Refs. [8, 9] unreliable and in need to be revisited. We shall also mention that the Gaussian curvature of the event horizon of uncharged static black holes immersed in external magnetic fields were obtained (using different approaches), investigated, and discussed by Wild and Kerns [10], and S.K. Bose and E. P. Esteban [11].

In this paper, we would like to investigate how the conclusions obtained by Smarr [5] and Kokkotas [6] regarding the Kerr-Newman's event horizon and ergosurface surfaces geometry are modified when external magnetic fields are present. We did not find in the literature any attempt to study the geometry of the magnetized Kerr-Newman metric.

This paper is organized as follows. In Sec. 1, we establish the notation to be used, and discuss briefly the interpretation of the magnetized Kerr-Newman metric parameters. In Sec. 2, the two-dimensional metrics of the magnetized Kerr-Newman event horizon and ergosurface are explicitly written down. In Sec. 3, we calculated the surface areas of the magnetized Kerr-Newman event horizon and ergosurface, and explore the thermodynamics of magnetized black holes. In Sec. 4, the equatorial and polar circumferences of the magnetized Kerr-Newman black hole event horizon and ergosurface are obtained and compared to determine the shape of the surfaces due to the black hole's charge, angular momentum, and the influence of external magnetic fields. Finally, in Sec. 5 we calculated in an analytical form the Gaussian curvature of both black hole's surfaces and study its embedding in E^3.

2. The Magnetized Kerr-Newman Metric

The magnetized Kerr-Newman metric is generated by a Harrison type transformation applied to the familiar Kerr-Newman solution. The line element can be written as (in units such that c=G=1)

$$ds^2 = f_{mkn}^{-1}(-2P^{-2}d\zeta\, d\zeta^* + \rho^2 dt^2) - f_{mkn}(|\Lambda_0|^2 d\varphi - \omega_{mkn}dt)^2, \tag{1}$$

where

$$d\zeta = \frac{1}{\sqrt{2}}\left(\frac{dr}{\sqrt{\Delta}} + i\, d\theta\right), \tag{2}$$

$$\rho = \sqrt{\Delta}\, \sin\theta, \tag{3}$$

$$P = \frac{1}{\sqrt{A}\sin\theta}, \tag{4}$$

with

$$\Delta = r^2 - 2\, M\, r + a^2 + e^2, \tag{5}$$

$$A = (r^2 + a^2)^2 - \Delta\, a^2 \sin^2\theta, \tag{6}$$

$$\Sigma = r^2 + a^2\cos^2\theta. \tag{7}$$

In Eq. (1) ω_{mkn} and f_{mkn} can be obtained from the following

$$\nabla\omega_{mkn} = |\Lambda|^2\, \nabla\omega + \frac{\rho}{f}(\Lambda^*\nabla\Lambda - \Lambda\nabla\Lambda^*), \tag{8}$$

$$f_{mkn} = \frac{f}{|\Lambda|^2} = -\frac{A\, \sin^2\theta}{|\Lambda|^2\Sigma}, \tag{9}$$

where $\nabla = \sqrt{\Delta}\frac{\partial}{\partial r} + i\frac{\partial}{\partial\theta}$, $\omega = \frac{a\,(2\,M\,r - e^2)}{A}$, and the asterisk (*) means complex conjugate. Also, the complex functions $(\Lambda, \Phi, \varepsilon)$ can be explicitly written as $\Lambda = \Lambda_r + i\,\Lambda_i$, $\Phi = \Phi_r + i\,\Phi_i$, and $\varepsilon = \varepsilon_r + i\,\varepsilon_i$, with

$$\varepsilon_r = -\left(r^2 + a^2\right)\sin^2\theta - e^2\cos^2\theta - \frac{2\,a^2\,\sin^2\theta}{\Sigma}(M\,r\,\sin^2\theta + e^2\cos^2\theta), \tag{10}$$

$$\varepsilon_i = 2\,M\,a(3 - \cos^2\theta)\,\cos\theta - \frac{2\,a\,\cos\theta\,\sin^2\theta}{\Sigma}(\,r\,e^2 - a^2 M\,\sin^2\theta), \tag{11}$$

$$\Lambda_r = 1 + B\,(\Phi_r - \frac{1}{4}\,B\,\varepsilon_r), \tag{12}$$

$$\Lambda_i = B\,(\Phi_i - \frac{1}{4}\,B\,\varepsilon_i\,), \tag{13}$$

$$\Phi_r = \frac{e\,a\,r}{\Sigma}\sin^2\theta, \tag{14}$$

$$\Phi_i = -e\,\cos\theta\,(1 + \frac{a^2\sin^2\theta}{\Sigma}). \tag{15}$$

Notice in Eq. (1) that the term Λ_0, which is Λ calculated at $\theta = 0$ or π, is not present in Ernst-Wild original derivation [7]. It was introduced later by Aliev and Gal'tsov [12], to eliminate the existence of a conical singularity. From Eqs. (12) – (13), we obtain

$$|\Lambda_0|^2 = (\frac{e^4}{16} + a^2M^2)B^4 + 2aeMB^3 + \frac{3}{2}e^2B^2 + 1. \tag{16}$$

Journal of Physics: Conference Series **189** (2009) 012012 doi:10.1088/1742-6596/189/1/012012

The magnetized Kerr-Newman metric has four parameters (B, M, a, e). The parameter B represents the asymptotically external magnetic field. The remaining three parameters M, a, e, were originally assumed to have the same interpretation as in the Kerr-Newman metric. However, a decade later, Aliev and Gal'tsov [13] found an alternative interpretation in which the parameters of the magnetized Kerr-Newman metric, M, a and e, are related but not equal to the respective parameters in the Kerr-Newman metric. Other attempts to obtain a physical interpretation of the magnetized Kerr-Newman metric's parameters have been carried out by Dockuchaev [14], and Karas and Vokrouhlicky [15]. All of these authors gave the same interpretation for the e and B parameters, although somewhat disagree with the analytical expressions derived for the remaining parameters. For this reason, and also to make a comparison between the results obtained for the magnetized Kerr-Newman metric with those for the standard Kerr-Newman metric, most derived expressions presented in this paper will be written in terms of the a and M parameters.

An invariant quantity of particular interest is the magnetized Kerr-Newman total charge (Q_{mkn}). It is given in the Ernst formalism by the following expression

$$Q_{mkn}= -|\Lambda_0|^2\ \Phi_i^{'}(r_+,0)\,, \tag{17}$$

where $r_+ = M + (M^2 - a^2 - e^2)^{1/2}$ gives the location of the black hole event horizon, and $\Phi_i^{'}$ is the imaginary part of $\frac{2\Phi - B\varepsilon}{2\Lambda}$. We obtain

$$Q_{mkn}= \ -\frac{e^3}{4}B^2 + e + 2\ aMB. \tag{18}$$

Eq. (18) is valid for any value of the B parameter. This expression was also derived by Aliev and Gal'tsov [12] and Karas and Vokrouhlicky [15]. A quick inspection of Eq.(18) shows that indeed the e parameter does not represent the total charge associated to the magnetized Kerr-Newman metric. Notice that the e parameter associated to the magnetized Kerr metric can be defined by setting $Q_{mkn} = 0$, in Eq.(18), and clearly it is not $e = 0$, as it was assumed in Refs. [8, 9]. Therefore, in Eq. (18) given $a \neq 0$, $B \neq 0$ and $Q_{mkn} = 0$, the value of the e parameter associated to the magnetized Kerr metric must be found by solving the resulting cubic equation. In the case $(\beta = BM << 1)$, an approximate expression of the e parameter for the magnetized Kerr-Newman metric is $e = Q_{mkn} - 2aMB$. The term $2aMB$ can be interpreted as an accretion charge (q_{acc}),and e as the bare charge. Thus, when $\beta = BM << 1$ the e parameter for the magnetized Kerr metric is given by $e = -2aMB$. The values of $e = 0$ and $a = 0$ satisfy trivially Eq. (18), and thus define the magnetized Schwarzschild metric.

With regarding to the value given to the β dimensionless parameter in this paper, we shall point out that in AGN galaxies supermassive black holes have estimate masses from $10^5 M_\odot - 10^9 M_\odot$, and we are assuming a very strong magnetic field of $10^{10}G$. Thus, β can be calculated from

$$\beta= 4.25\ 10^{-9}(\frac{M}{M_\odot})\ (\frac{B}{10^{12}G}). \tag{19}$$

Therefore, in Secs. 3-4 we have set $\beta = BM = 0.0425$ for most calculations.

3. Two Dimensional Metric of the Magnetized Kerr-Newman

As usual the magnetized Kerr-Newman black hole's event horizon and ergosurface can be described by a two dimensional metric. Namely

$$ds^2 = E^2 d\theta^2 + G^2 d\varphi^2, \tag{20}$$

which in the case of the magnetized Kerr-Newman black hole's event horizon can be written as

$$ds^2 = |\Lambda_+|^2\Sigma_+ d\theta^2 + |\Lambda_0|^4 \frac{(r_+^2 + a^2)^2 \sin^2\theta}{|\Lambda_+|^2\Sigma_+} d\varphi^2, \tag{21}$$

where the subscript + means the respective variable evaluated at r_+.

In the case of the magnetized Kerr-Newman ergosurface its two-dimensional metric is given by

$$ds^2 = |\Lambda_{sls}|^2\Sigma_{sls}(\frac{M^2 - e^2}{M^2 - e^2 - a^2\cos^2\theta})d\theta^2 + |\Lambda_0|^4 \frac{A_{sls}\sin^2\theta}{|\Lambda_{sls}|^2\Sigma_{sls}} d\varphi^2, \tag{22}$$

where the subscript sls means that the given variable must to be calculated at the static limit surface ($r_{sls} = M + (M^2 - e^2 - a^2\cos^2\theta)^{1/2}$).

4. Surface Areas of the Event Horizon and Ergosurface

The surface areas (A) can be obtained by solving the following integral

$$A = \int_0^{2\pi}\int_0^{\pi} EG\ d\theta d\varphi, \tag{23}$$

which for the area of the magnetized Kerr-Newman's event horizon (A_h) gives

$$A_h = 4\pi|\Lambda_0|^2(r_+^2 + a^2). \tag{24}$$

Eq. (24) shows that in general the area surface of a magnetized Kerr-Newman black hole event horizon it will be always greater a Kerr-Newman black hole event horizon. Eq. (24) can also be written as follows

$$M^2 = \frac{A_h}{16\pi} + \frac{4\pi J^2}{A_h} + \frac{\pi e^4}{A_h} + \frac{e^2}{2} - \frac{3\pi q_{acc}^4}{A_h} - \frac{q_{acc}^2}{2} - \frac{6\pi q_{acc}^2 e^2}{A_h} - \frac{8\pi q_{acc}^3 e}{A_h}. \tag{25}$$

The first term on the right of Eq. (25) is the "irreducible" part of M^2. The second term is the contribution of the black hole's rotation energy. The third and fourth terms are the contributions of the electrostatic energy due to bare charge, while the fifth and six terms are the contributions of the electrostatic energy due to accretion charge. The last two terms of Eq. (25) can be regarded as the contribution to the energy due to the coupling of the bare and accretion charges.

To explore the connection between magnetized black holes and thermodynamics Eq. (25) can also be re-written as

$$M^2 = \frac{A_h}{16\pi} + \frac{Q_{mkn}^2}{2} + \frac{\pi Q_{mkn}^4}{A_h} + \frac{4\pi J^2}{A_h} - 2JBQ_{mkn} - \frac{8\pi JBQ_{mkn}^3}{A_h}, \tag{26}$$

where in Eq. (26) we assumed β<<1, and thus $|\Lambda_0|^2$->1. The difference in mass(dM) between two black holes states differing by dA, dJ, dQ_{mkn}, and dB in area,angular momentum,charge, and external magnetic field can be written as

$$dM = \frac{\kappa}{8\pi}dA + \Omega_H dJ + \Phi_H dQ_{mkn} - \Psi_H dB, \tag{27}$$

where the last term in Eq.[27] represent the magnetic energy. We obtain

$$\kappa = 8\pi \left(\frac{\partial M}{\partial \mathcal{A}}\right)_{J,B,Q_{mkn}} = \frac{r_+ - M}{\mathcal{A}}, \tag{28}$$

$$\Omega_H = \left(\frac{\partial M}{\partial J}\right)_{\mathcal{A},B,Q_{mkn}} = \frac{a}{\mathcal{A}} - \frac{2Br_+Q_{mkn}}{\mathcal{A}}, \tag{29}$$

$$\Phi_H = \left(\frac{\partial M}{\partial Q_{mkn}}\right)_{\mathcal{A},J,B} = \frac{r_+ e}{\mathcal{A}}, \tag{30}$$

$$\Psi_H = \left(\frac{\partial M}{\partial B}\right)_{J,\mathcal{A},Q_{mkn}} = aQ_{mkn}(1 + \frac{4\pi Q_{mkn}^2}{\mathcal{A}}). \tag{31}$$

Eq. (28) defines the magnetized Kerr-Newman's black hole's surface gravity (κ). The magnetized Kerr-Newman black hole's angular velocity at the event horizon (Ω_H) is given by Eq. (29). This expression can also be derived from Eq. (2.4) in Ref. [7]. The electric potential at the event horizon (Φ_H) is given by Eq. (30). Notice that the right side of Eq. (31), could be interpreted as the magnetic dipole moment of the magnetized Kerr-Newman black hole multiplied for a dimensionless constant. A more detailed study of the thermodynamics of magnetized black holes is underway and it will be submitted elsewhere.

Coming back to the ergosurface, from Eq. (23) it can be easily show that the surface area of the magnetized Kerr-Newman ergosurface (A_{sls}) can be written as

$$\mathcal{A}_{sls} = 2\pi(M^2 - e^2)^{1/2} \mid \Lambda_0 \mid^2 \int_0^{\pi} \sqrt{f_{sls}} \sin\theta \, d\theta, \tag{32}$$

where

$$f_{sls} = 8M^2 + \frac{8M^3 - 4Me^2 + 4Ma^2\sin^2\theta}{(M^2 - e^2 - a^2\cos^2\theta)^{1/2}} + \frac{4Ma^2 + e^2(e^2 - 2a^2\sin^2\theta)}{(M^2 - e^2 - a^2\cos^2\theta)}. \tag{33}$$

The magnetized Kerr-Newman ergosurface area to second order in a and e can be calculated from Eqs. (32) – (33). Thus

$$\frac{\mathcal{A}_{sls}}{M^2} = 4\pi(4+\frac{a^2}{M^2}-\frac{2e^2}{M^2})-\frac{11\pi a^2 e^2}{6M^4}+6\pi\frac{e^2}{M^2}\beta^2(4+\frac{a^2}{M^2})+32\pi\beta^3\frac{a}{M}\frac{e}{M}+8\pi\beta^4\frac{a^2}{M^2}(2-\frac{e^2}{M^2}). \tag{34}$$

To first order in β, Eq. (34) reduces further to

$$\frac{\mathcal{A}_{sls}}{M^2} = 4\pi(4 + \frac{a^2}{M^2} - \frac{2Q_{mkn}^2}{M^2}(1 + \frac{11}{48}\frac{a^2}{M^2})) + 32\pi\frac{a}{M}\frac{Q_{mkn}}{M}\beta. \tag{35}$$

Notice that for $Q_{mkn} = \beta = 0$, Eq. (35) reduces to $16\pi M^2 + 4\pi a^2$. This result gives the Kerr ergosurface area which found earlier by Pelavas [4]. For a Kerr-Newman black hole, in Eq. (35) we shall set $\beta- > 0$, and $Q_{mkn}- > e$. This result $\frac{\mathcal{A}_{sls}}{M^2} = 4\pi(4 + \frac{a^2}{M^2} - \frac{2e^2}{M^2}(1 + \frac{11}{48}\frac{a^2}{M^2}))$ is new in the literature.. From Eq. (35) it is easy to check that for slow rotation the Kerr-Newman's ergosurface area will always be less than the magnetized Kerr-Newman's ergosurface area. This fact it seems to suggest that in a magnetized Kerr-Newman black hole, Penrose processes are more likely to occur as compared to the standard Kerr-Newman black hole.

5. Circumferences

Next we shall calculate the equatorial and polar circumferences of the magnetized Kerr-Newman black hole's event horizon and ergosurface. The equatorial circumference (C_e) can be obtained by solving the following integral

$$C_e = \int_0^{2\pi} G(\frac{\pi}{2}) d\varphi. \tag{36}$$

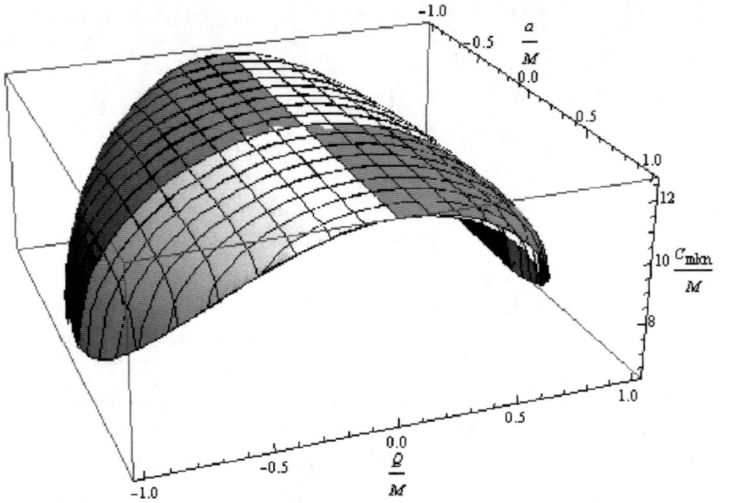

Figure 1. A comparison between the equatorial circumferences of the magnetized Kerr-Newman (gold) and Kerr-Newman (grey) black holes.

If we keep only linear terms in B, we obtain from Eq. (36) the magnetized Kerr-Newman's black hole event horizon equatorial circumference (C_{eh}). Thus

$$C_{eh} = \frac{2\pi(r_+^2 + a^2)}{r_+ + aBQ_{mkn}}. \tag{37}$$

Fig. 1 shows a superposition of two plots obtained from Eq. (37). The first plot (in orange color) corresponds to $\beta = 0.0425$, while the gray one corresponds to $\beta = 0$. In this case, provided $aQ_{mkn} > 0$ ($aQ_{mkn} < 0$), the effect of the external magnetic field is to increase (decrease) the magnetized Kerr-Newman black hole circumference as compared to the Kerr-Newman's equatorial circumference. In a similar way, the equatorial circumference for the magnetized Kerr-Newman's ergosurface can be written as

$$\frac{C_{esls}}{M} = \frac{2\pi\frac{e}{M}(2(1+\frac{a^2}{M^2}) + 2\sqrt{1-\frac{e^2}{M^2}} - \frac{e^2}{M^2})}{(\frac{e}{M} + \frac{a}{M}\beta(1-\sqrt{1-\frac{e^2}{M^2}}))}. \tag{38}$$

Fig. 2 shows a superposition of two plots obtained from Eq. (38). The first plot (in green color) corresponds to $\beta = 0.0425$, while the gray one corresponds to $\beta = 0$. In this case, provided $aQ_{mkn} > 0$ ($aQ_{mkn} < 0$), the effect of the external magnetic field is to increase (decrease) the magnetized Kerr-Newman black hole circumference as compared to the Kerr-Newman's equatorial circumference. A comparison between the event horizon and ergosurface equatorial circumferences is showed in Fig. 3. Similarly, the black hole's polar circumference (C_p) can be obtained from Eq. (20). Thus

$$C_p = \int_0^{2\pi} F(\theta)d\theta. \tag{39}$$

For the magnetized Kerr-Newman black hole event horizon the polar circumference (C_{ph}) given by Eq. (39) can be rewritten as

$$C_{ph} = \int_0^{2\pi} ((r_+^2 + a^2)^{1/2}(1 - k_1\sin^2\theta)^{1/2} + \frac{k_2\sin^2\theta}{(1-k_1)^{1/2}})d\theta, \tag{40}$$

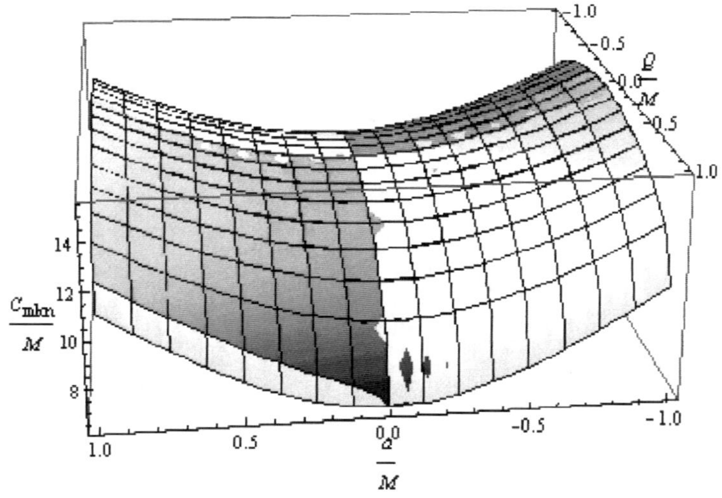

Figure 2. The equatorial circumferences of the ergosurfaces of the magnetized Kerr-Newman (in green) and Kerr-Newman (in gray) black holes.

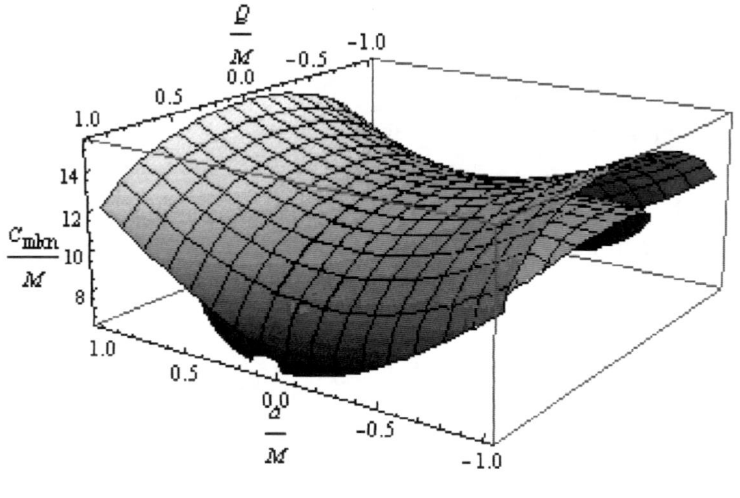

Figure 3. The equatorial circumferences (event horizon (gold), ergosurface (green)) in the magnetized Kerr-Newman black hole.

where $k_1 = \frac{a^2}{(r_+^2+a^2)}$ and $k_2 = \frac{r_+ aBQ_{mkn}}{(r_+^2+a^2)^{1/2}}$. It turns out that Eq. (40) can be solved in terms of complete elliptical integrals of the second kind E and K. Namely

$$C_{ph} = 4(r_+^2 + a^2)^{1/2}\mathsf{E}(k_1) - 4k_2(\frac{(1-k_1)^{1/2}}{k_1}\mathsf{E}(\frac{k_1}{k_1-1}) - \frac{\mathsf{K}(\frac{k_1}{k_1-1})}{k_1(1-k_1)^{1/2}}). \tag{41}$$

Notice that in order to use the standard definitions of the elliptical integrals of the second kind, a square root must be taken to all the the arguments of the elliptical integrals given in Eq. (41). In Fig. 4 we show two superimposed plots obtained from Eq. (41). In the first

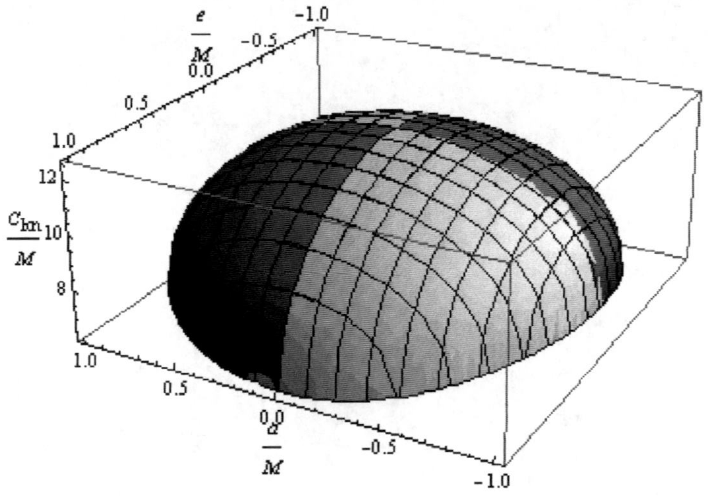

Figure 4. A comparison between the polar circumferences of the magnetized Kerr-Newman (gold) and Kerr-Newman (gray) black holes.

plot (orange), $\beta = 0.0425$, and in the second one $\beta = 0$. It is immediately obviously that for $aQ_{mkn} > 0$, the effect of the external magnetic field is to increase the value of the magnetized Kerr-Newman's event horizon polar circumference as compared to the Kerr-Newman's event horizon polar circumference. The opposite happens when $aQ_{mkn} < 0$.

Regarding the magnetized Kerr-Newman ergosurface polar circumference (C_{psls}), Eq. (39) can be written as

$$\frac{C_{psls}}{M} = \int_0^{2\pi} \sqrt{\frac{(2r_{sls} - e^2)(1 - e^2)}{r_{sls} - 1}} (1 + \frac{aBer_{sls}\sin^2\theta}{2r_{sls} - e^2}) d\theta. \tag{42}$$

Keeping only terms up to B, a^2 and e^2 in Eq. (42), reduces to

$$\frac{C_{psls}}{M} \approx \pi(4 + \frac{3}{4}\frac{a^2}{M^2} + 5\frac{a}{M}\beta\frac{Q_{mkn}}{M} - \frac{Q_{mkn}^2}{M^2}). \tag{43}$$

Thus in particular, increasing the value of a and/or β in Eq. (43) will mean an increase of the polar circumference value. As expected, when $\beta \to 0$, Eq. (43) goes to Eq. (14) in Ref. [8].

To estimate the deviation from sphericity in any of the two surfaces under consideration we shall calculate δ_{mkn} defined as

$$\delta_{mkn} = \frac{C_e - C_p}{C_e}. \tag{44}$$

Fig. 5 shows for the magnetized Kerr-Newman black hole's event horizon the superposition of the plot given by Eq. (44), and a plane $z = 0$. The intersection of both plots allow to easily visualize two regions. The first region ($z > 0, \delta_{mkn} > 0$) defines oblate spheroids and the second region ($z < 0, \delta_{mkn} < 0$) prolate spheroids.

Fig. 6 is similar to Fig. 5 but for the Kerr-Newman black hole. Contrary to Fig. 5 both plots do not intersect and thus there is only one region ($z > 0, \delta_{kn} > 0$). This means that the Kerr-Newman horizon will have always a shape as an oblate spheroid. Coming back to the magnetized Kerr-Newman ergosurface, Eq. (39) can not be solved in closed form.

Table 1 shows for the magnetized Kerr-Newman ergosurface the values of Eq. (44) for different values of $\frac{a}{M}$ and $\frac{Q_{mkn}}{M}$. The last two columns are useful to compare Eq. (44) for the magnetized

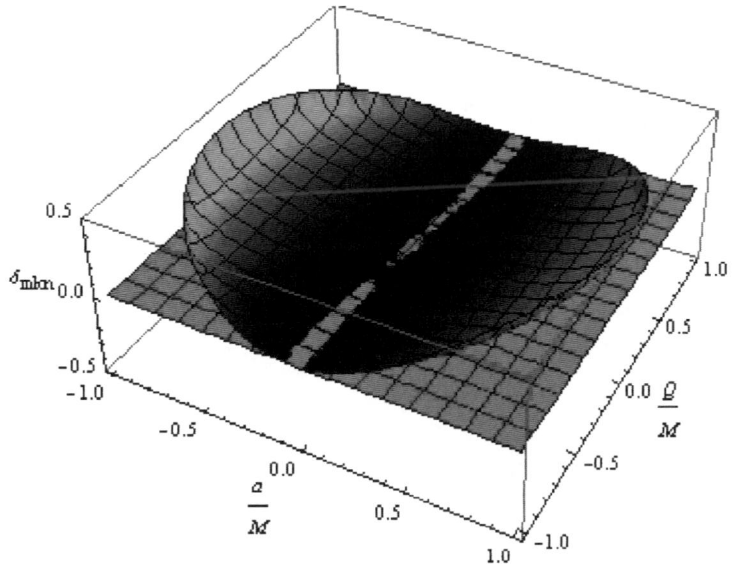

Figure 5. The plot's intersection correspond to prolate spheroids. The remaining regions (in gold) correspond to oblate spheroids.

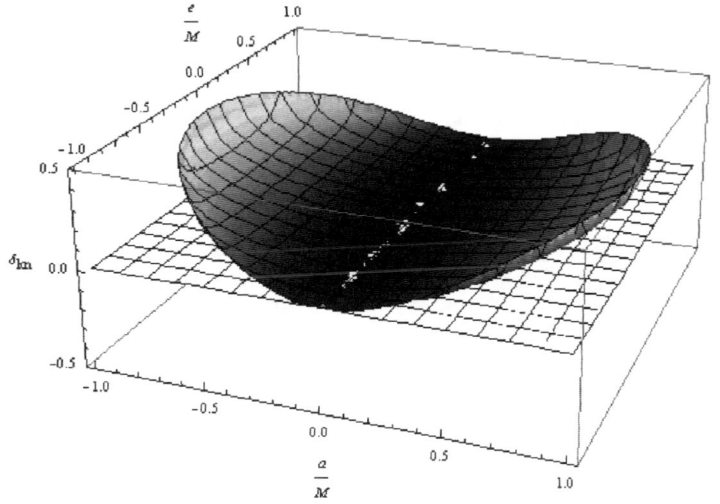

Figure 6. The deviation of sphericity of the Kerr-Newman black hole event horizon. There are no negative values for δ_{kn}.

Table 1. The magnetized Kerr-Newman ergosurface polar circumferences for different values of a andQ_{mkn}. The last column shows the value of δ_{kn} for the Kerr-Newman ergosurface polar circumferences.

$\frac{a}{M}$	$\frac{Q_{mkn}}{M}$	$\frac{C_p}{M}$	$\frac{C_e}{M}$	δ_{mkn}	δ_{kn}
0.1	0.1	12.565	12.569	P	P
0.1	0.2	12.477	12.476	N	P
0.1	0.3	12.323	13.318	N	P
0.1	0.4	12.097	12.087	N	P
0.1	0.5	11.791	11.776	N	P
0.1	0.6	11.390	11.368	N	P
0.1	0.7	10.871	10.840	N	N
0.1	0.8	10.187	10.141	N	N
0.1	0.9	9.231	9.150	N	N
0.5	0.1	13.249	13.311	P	P
0.5	0.2	13.202	13.233	P	P
0.5	0.3	13.104	13.093	N	P
0.5	0.4	12.957	12.887	N	N
0.5	0.5	12.763	12.610	N	N
0.5	0.6	12.532	12.249	N	N
0.5	0.7	12.302	11.786	N	N
0.5	0.8	12.254	11.187	N	N
0.5	0.9	15.492	10.377	N	N
0.9	0.1	16.743	14.887	N	N
0.9	0.2	16.869	14.820	N	N
0.9	0.3	17.220	14.698	N	N
0.9	0.4	18.127	14.519	N	N
0.9	0.5	23.027	14.279	N	N

Kerr-Newman and Kerr-Newmán ergosurfaces. Notice in Table 1, that "P" means $\delta_{mkn} > 0$ or $\delta_{kn} > 0$, and "N" negative values for the same dimensionless constants.

Gaussian Curvature
The Gaussian curvature (K) is a black hole intrinsic invariant and can be evaluated from the following expression

$$K = -\frac{1}{2EG}\frac{d}{d\theta}(\frac{1}{EG}\frac{d}{d\theta}G^2). \tag{45}$$

Thus, the Gaussian curvature (K_h) of the magnetized Kerr-Newman event horizon can be derived from Eq. (21) and Eq. (45). It can be written as

$$K_h = -\frac{1}{2\sin\theta}\frac{d}{d\theta}(\frac{1}{\sin\theta}\frac{d}{d\theta}(\frac{\sin^2\theta}{|\Lambda_+|^2\Sigma_+})). \tag{46}$$

Eq. (46) gives a very large analytical expression which will not be shown here for obvious reasons. However, for a particular case ($e = a = 0$) which correspond to the Gaussian curvature of the magnetized Schwarzschild event horizon (K_{ms}), Eq. (46), can be simplified to

$$K_{ms} = \frac{1 + 8\beta^2\cos^2\theta + \beta^4(3\sin^2\theta - 4)\sin^2\theta}{4M^2(1 + \beta^2\sin^2\theta)^4}. \tag{47}$$

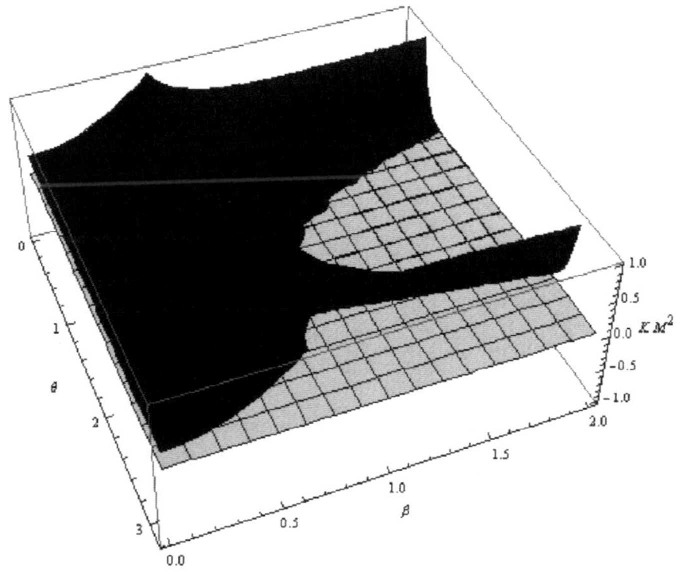

Figure 7. The Gaussian curvature of the event horizon of a magnetized Schwarzschild black hole.

The above value of K_{ms} was obtained first by Wild and Kerns [10], and later using a different approach by S.K. Bose and E.P. Esteban [11] (their definition of the Gaussian curvature was twice the value as given in Eq. 45). A Plot of Eq. (47) intersected by a plane $z = 0$ is given in Fig. 7. In such a figure a "parabolic" curve defines the boundary of a region with negative Gaussian curvature values. This happens when the dimensionless parameter satisfies $\beta \succeq \frac{1}{\sin\theta}$. The equality sign in the previous equation gives the zero Gaussian curvature.

By keeping only linear terms to B, Eq. (46) reduces to the following

$$K_{mkn} = \frac{(a^2 + r_+^2)(r_+^4 - 2a^2r_+^2\cos^2\theta - 3a^4\cos^4\theta + 8aBQ_{mkn}r_+(a^2 + r_+^2)\cos^2\theta)}{\Sigma_+^3(\Sigma_+ + 4aQ_{mkn}r_+B\sin^2\theta)}. \tag{48}$$

The expression for the Gaussian curvature obtained by Smarr [5] for the Kerr-Newman event horizon can be obtained by setting $B = 0$, in Eq. (48). Eq. (48) also shows that the Gaussian curvature of the magnetized Kerr metric event horizon has the same value as in the vacuum Kerr metric if only linear terms in B are kept. At the Equatorial plane ($\theta = \frac{\pi}{2}$), Eq. (48) reduces to

$$K_{mkn}(\frac{\pi}{2}) = \frac{a^2 + r_+^2}{r_+^3(4aBQ_{mkn} + r_+)}, \tag{49}$$

and at the poles ($\theta = 0$), Eq. (48) can be re-written as

$$K_{mkn}(0) = \frac{r_+^2 + 8aBQ_{mkn}r_+ - 3a^2}{(r_+^2 + a^2)^2}. \tag{50}$$

In a similar calculation, we can use again Eq. (45) to obtain the Gaussian curvature for the magnetized Kerr-Newman ergosurface. It happens that the analytical form of the magnetized Kerr-Newman is very long and cumbersome. For instance, its Gaussian curvature at the Equatorial plane and at the poles can be written as

$$K_{sls}(\frac{\pi}{2}) = \frac{f_1 + xf_2 + aQ_{mkn}\beta(f_3 + 4f_4x)}{4f_5 + f_6x + 4aQ_{mkn}\beta(f_7 - f_8x)}, \tag{51}$$

where

$$f_1 = -32 - 40a^2 + 64Q_{mkn}^4 + 56a^2Q_{mkn}^2 - 38Q_{mkn}^4 - 17a^2Q_{mkn}^4 + 6Q_{mkn}^6, \tag{52}$$

$$f_2 = -32 - 40a^2 + 48Q_{mkn}^4 + 36a^2Q_{mkn}^2 - 18Q_{mkn}^4 - 4a^2Q_{mkn}^4 + Q_{mkn}^6, \tag{53}$$

$$f_3 = -256 - 248a^2 + 304Q_{mkn}^4 + 160a^2Q_{mkn}^2 - 72Q_{mkn}^4 - 2a^4Q_{mkn}^2 - 3a^2Q_{mkn}^4, \tag{54}$$

$$f_4 = -48 - 42a^2 + 36Q_{mkn}^2 + 11a^2Q_{mkn}^2 - 3Q_{mkn}^4, \tag{55}$$

$$f_5 = -32 - 16a^2 + 80Q_{mkn}^2 + 32a^2Q_{mkn}^2 - 68Q_{mkn}^4 - 19a^2Q_{mkn}^4 + 22Q_{mkn}^6 + 3a^2Q_{mkn}^6 - 2Q_{mkn}^8, \tag{56}$$

$$f_6 = -128 - 64a^2 + 260Q_{mkn}^2 + 96a^2Q_{mkn}^2 - 182Q_{mkn}^4 - 36a^2Q_{mkn}^4 + 84Q_{mkn}^6 + 2a^2Q_{mkn}^6 - 70Q_{mkn}^8 + 56Q_{mkn}^{10} - 28Q_{mkn}^{12} + 8Q_{mkn}^{14}, \tag{57}$$

$$f_7 = -384 - 160a^2 + 688Q_{mkn}^2 + 208a^2Q_{mkn}^2 - 368Q_{mkn}^4 - 62a^2Q_{mkn}^4 + 57Q_{mkn}^6 + 2a^2Q_{mkn}^6 - Q_{mkn}^8, \tag{58}$$

$$f_8 = 324 - 476Q_{mkn}^2 + 16a^2(8 - 7Q_{mkn}^2 + Q_{mkn}^4) + Q_{mkn}^4(308 + Q_{mkn}^2(-287 - 8Q_{mkn}^2(-35 + 21Q_{mkn}^2 - 7Q_{mkn}^4 + Q_{mkn}^6))), \tag{59}$$

with $x = \sqrt{1 - Q_{mkn}^2 + 4a\beta Q_{mkn}^4}$.

At the poles, the Gaussian curvature of the magnetized Kerr-Newman black hole ergosurface (up to first order in B) can be written as

$$K_{sls}(0) = \frac{g_1 + g_2 y - 4a\beta Q_{mkn}(g_3 + g_4 y)}{y(Q_{mkn}^2 - 1 - 4Q_{mkn}aB)(a^2 + (1+y)^2)^3}, \tag{60}$$

where

$$g_1 = 4(2(Q_{mkn}^2 - 1) + a^2(a^2 - 2)(3a^2 + 3Q_{mkn}^2 - 5)), \tag{61}$$

$$g_2 = 3a^4(Q_{mkn}^2 - 8) + 8(Q_{mkn}^2 - 1) + a^2(36 - 23Q_{mkn}^2 + 3Q_{mkn}^4, \tag{62}$$

$$g_3 = 4(4 + 5a^4 - 2Q_{mkn}^2 + 2a^2(Q_{mkn}^2 - 5)), \tag{63}$$

$$g_4 = 7a^4 - 8(Q_{mkn}^2 - 2) + 5a^2(2Q_{mkn}^2 - 7), \tag{64}$$

with $y = \sqrt{1 - a^2 - Q_{mkn}^2 + 4a\beta Q_{mkn}^4}$.

Although Eqs. (51) – (60) are particular cases of Eq. (45) they look awkward. The only reason that we wrote Eqs. (51) – (60) is that the reader could verify that indeed these equations reduce when $\beta = 0$ and $Q_{mkn} = 0$ (Kerr black hole) to Eqs. 5 and 7 of Ref. [6]. Of course, by setting $\beta = 0$ in Eqs. (51) – (60) we shall obtain the Gaussian curvatures at the Equator and poles of the Kerr-Newman black hole ergosurface. Thus, because the analytical expressions of the Gaussian curvatures of the magnetized Kerr-Newman event horizon and ergosurface are mathematically very long and cumbersome, we have decided to perform a graphical analysis of Eq. (45) for the magnetized Kerr-Newman event horizon and ergosurface to visualize in which regions of both surfaces the Gaussian curvature is positive or negative. For this purpose, we shall present four plots (Fig. 8 – 11) in which we had superimposed graphs of the Gaussian curvature of the event horizon, ergosurface, and a horizontal plane located at $z = 0$. The rational of this

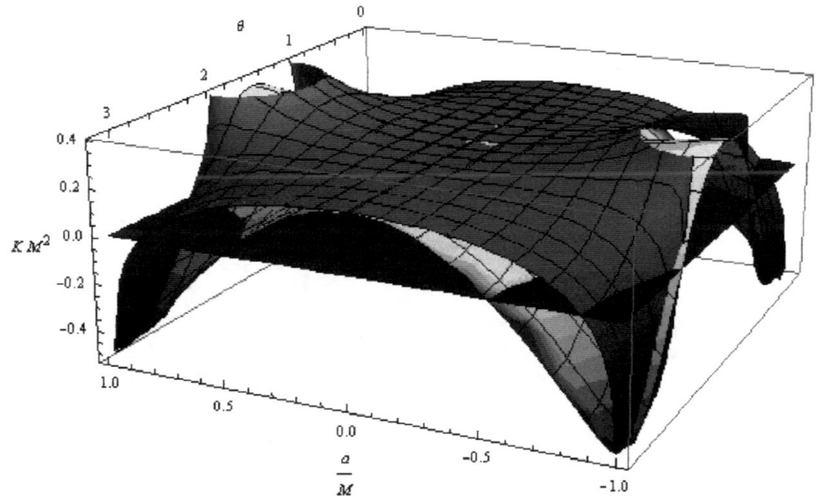

Figure 8. For $\frac{Q_{mkn}}{M}= 0.1$, a comparison of the Gaussian curvatures of the magnetized Kerr-Newman event horizon (brown) and the ergosurface (green).

undertaking is because we know that regions with positive Gaussian curvature values can be always embedded in E^3 . However, regions with negative Gaussian values can only be embedded in E^3 if they satisfy Pelavas condition ($g_{\theta\theta} > (\frac{d\sqrt{g_{\phi\phi}}}{d\theta})^2$).

Fig. 8 shows that for a small charge ($\frac{Q_{mkn}}{M} = 0.1$) the Gaussian curvature of the magnetized Kerr-Newman black hole event horizon (brown) will be always greater than the Gaussian curvature of the ergosurface. Notice that both surfaces have four regions around the poles with negative values while in the Equator the Gaussian curvatures will be always negative. If we increase the charge, Fig. 9 shows that the regions with negative values in the ergosurface shall increase of size as can be seen by comparing Figs. 8 – 9. In this case the event horizon always will always have Gaussian curvature positive values and forbidden regions. Now, we examine the case with small angular momentum ($\frac{a}{M} = 0.2$). In this case the Gaussian curvature at the event horizon will be always positive. The ergosurface will present very small regions in which the Gaussian curvature values are negative.

Finally, for $\frac{a}{M} = 0.9$ (see Fig. 10), the Gaussian curvature at the event horizon (gold) shows negative values and is always greater than the Gaussian curvature at the ergosurface (green). The ergosurface also presents regions with negative values for all possible values of $\frac{Q_{ml}}{M}$ except around the equatorial region.

Regarding to the embedding of both surfaces in E^3 we have noticed that the Pelavas condition [4] fails for the event horizon and holds for the ergosurface.

6. Conclusions

In this contribution we have investigated to first order in B the geometric properties of the magnetized Kerr-Newman event horizon and ergosurface. The surface area of the event horizon was obtained and from there we derived the first law of thermodynamics for magnetized black holes. As a consequence, the surface gravity, angular velocity, electric potential and magnetic moment at the magnetized Kerr-Newman black hole' event horizon were easily obtained. Further analysis and physical consequences of the thermodynamics of magnetized black holes are under study and will be published elsewhere.

The surface area of the magnetized Kerr-Newman ergosurface was not possible to obtained in

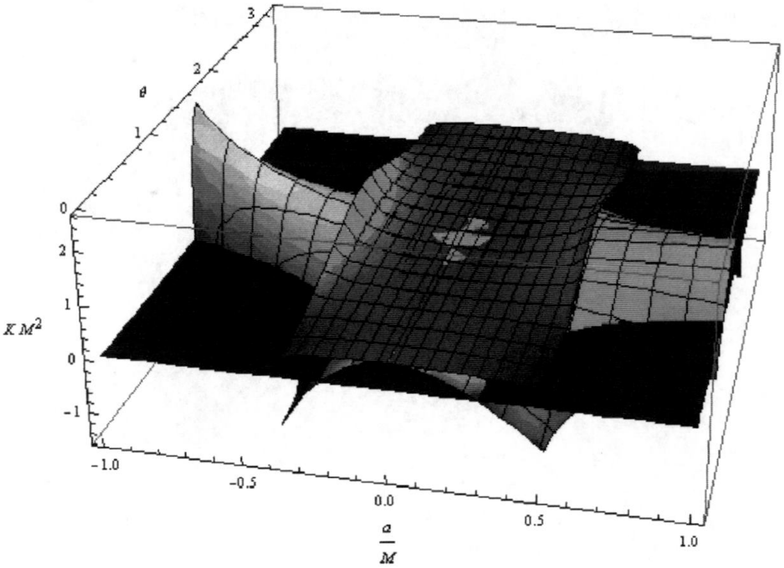

Figure 9. For $\frac{Q_{mkn}}{M} = 0.9$, a comparison of the Gaussian curvatures of the magnetized Kerr-Newman event horizon (brown) and the ergosurface (green).

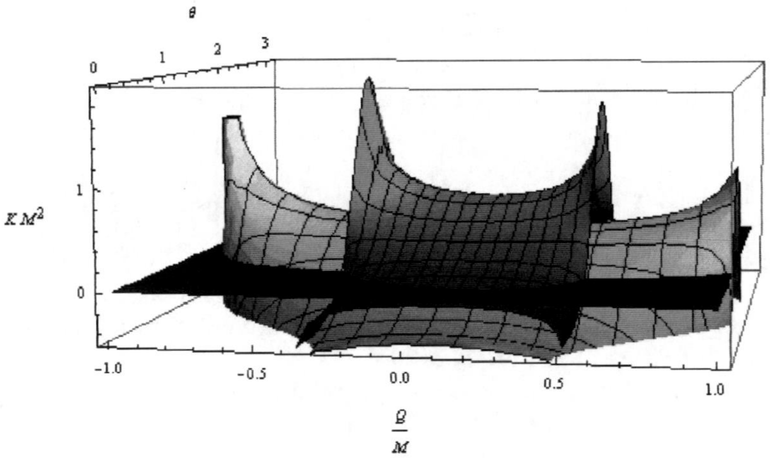

Figure 10. For $\frac{a}{M} = 0.9$, the Gaussian curvatures of the magnetized Kerr-Newman black hole event horizon (brown) and ergosurface (green).

a closed form but to second order in a, Q_{mkn}, and first order in B. This result in the appropriate limit ($Q_{mkn} \to 0$, $B \to 0$) agrees with Eq. (15) in Ref. [4]. The remaining results for the Kerr-Newman($B = 0$), magnetized Kerr-Newman, and magnetized Kerr ($Q_{mkn} = 0$) ergosurface areas are new in the literature.

Next, we studied the deviation from sphericity due to the external magnetic field at the magnetized Kerr-Newman event horizon and ergosurface. We calculated in a closed form the equatorial circumferences of both surfaces. The polar circumference was calculated in terms of elliptical integrals at the event horizon, and in an approximate way at the ergosurface. It turn out that the effect of the magnetic field is to induce a spheroidal prolate shape of the event horizon for low values of the angular momentum ($| \frac{J}{M^2} | < 0.1$). Greater values of the angular momentum will imply that the shape of the event horizon will take a spheroidal oblate. These results must be compared with those found by Smarr for the standard Kerr-Newman black hole in which the event horizon has always a spheroidal oblate shape. Regarding the ergosurface shape the gravitomagnetic effect is also to induce a prolate spheroidal shape to the magnetized Kerr-Newman ergosurface.

Finally, we derived in a closed form the Gaussian curvatures of the magnetized Kerr-Newman metric event horizon and ergosurface. However, these results were not show explicitly in the paper because were too long and cumbersome. Nonetheless we have check that our analytical expressions for the magnetized Kerr-Newman Gaussian curvature at the event horizon and ergosurface reduces when $B = 0$ to those found by Smarr [5] and Kokkotas [6]. Only particular cases of the Gaussian curvature at the magnetized Kerr-Newman Equatorial and polar regions were written down.

For small values of the black hole's charge regions with Gaussian curvature negative values develop at the poles of the event horizon and ergosurface. Increasing the charge value will imply that Gaussian curvature values at the black hole's event horizon has only positive values. However, at the ergosurface the size of the regions with negative Gaussian curvature values will increase. For low angular momentum there will be only positive values of the Gaussian curvature at the event horizon and very small regions with Gaussian curvature negative values around the ergosurface's poles. Increasing the angular momentum will increase the Gaussian curvature negative regions around the event horizon and ergosurface poles. We also have notice that Gausssian curvature positive values exist around the Equator of the event horizon and ergosurface.

The embedding problem was examined using the Pelavas condition [4]. We found that polar regions with Gaussian curvature negative values exist in the event horizon and do not satisfy the Pelavas condition [4], and therefore can not be embedded in E^3. However, regions with Gaussian curvature negative values around the ergosurface poles satisfy the Pelavas condition [4] and therefore can be embedded in E^3. An analogous result was found by Smarr [5] and Kokkotas [6] at the Kerr-Newman's event horizon and ergosurface, respectively.

Acknowledgments

The support given to this research by NASA-PR Consortium, and the University of Puerto Rico-Humacao is greatly appreciated. I also thank undergraduate students K. Rivera, S. Garcia, D. Delgado, and V. de la Rosa for checking via computer algebra most of the analytical expressions presented in this paper.

References

[1] Smarr L., Phys. Rev. Letters 30, 71-73, (1973).
[2] Beckenstein J, Ph.D thesis, Princeton University, (1972).
[3] Bardeen J.M., Carter B, and Hawking S. W., Commun. Math. Phys., 31, 161-170, (1973).
[4] Pelavas N., Neary N., Lake K., Class. Quantum Grav. 18, 1319 (2001).
[5] Smarr L., Phys. Rev. D 7, (1973).
[6] Kokkotas K. D., GRG, 20, 8, 829 (1988).
[7] Ernst F.J. and Wild, W. J.,J. Math. Phys. 17, 182 (1976).
[8] Wild W. J., Kerns R. M., and Drish W. F., Phys. Rev. D 23, 4, 829 (1981).
[9] Krori K. D. and Barua M., Phys. Rev. D 35, 1171 (1987).
[10] Wild W. and Kerns R., Phys. Rev. D 21, 332 (1980).
[11] Bose S. K. and Esteban E. P., J. Math. Phys., 22, 3006, (1981).
[12] Aliev A.N. and Gal'tsov, D. V., Sov. Phys. Usp., 32,75 (1989).
[13] Aliev A. N. and Gal'tsov, D.V., Russian Physics Journal, Vol 32, 10, 790-795 (1989).
[14] Dockuchaev V.I., Sov. Phys. JETP 65,1079 (1987).
[15] Karas W. and Vokrouhlicky, D., J. Math. Phys., 32, 714, (1991).

Evolution of supermassive black hole spins in the ΛCDM cosmology

N. Fanidakis, C. M. Baugh, S. Cole and C. S. Frenk

Institute for Computational Cosmology, Department of Physics, University of Durham, Science Laboratories South Road, Durham DH1 3LE, United Kingdom

E-mail: nikolaos.fanidakis@dur.ac.uk

Abstract. Over the last years, several observations suggest that SMBHs likely reside at the centres of all spheroid galaxies. Even more interestingly, their properties seem to correlate with the bulge luminosity – or stellar velocity dispersion – and the bulge mass, suggesting a single mechanism for assembling BHs and forming spheroids in galaxies. We have been investigating the cosmological co-evolution of galaxies and their central SMBH in hierarchical models of galaxy formation, using the semi-analytic model developed in Durham University. We focus on the spin of the SMBHs and study how does it evolve during accretion of gas and mergers with other SMBHs. We conclude that the global SMBH spin distribution in the present universe may be bimodal, provided that the gas is fed into the hole through a self-gravity limited accretion disk.

1. Introduction

Astrophysical supermassive blacks holes (SMBHs) are expected to possess angular momentum $J_{\rm bh} = aGM_{\rm bh}^2/c$, where $M_{\rm bh}$ is the mass of the black hole (BH) and a is the spin parameter ($0 \leq a \leq 1$). The spin is believed to have a significant impact in the close vicinity of the BH. For example, it determines the efficiency of converting matter into radiation in an accretion disk and it is believed to influence the formation and direction of the radio jets in active galactic nuclei (AGN) [11], [27], [8]. In addition, it is of special interest in the modelling of BH binaries in gravitational wave astronomy [33], [1], [2], [15], [14].

The evolution of spins with time is closely related to the channels that contribute to the growth of SMBHs. It is known that the spin changes due to accretion of gas and mergers with other BHs [3], [41], [42], [21]. Each mechanism of BH growth has varying effects on the spin evolution. For example, accretion of gas that co-rotates with the BH should spin up the hole [3], whereas, capturing of small BHs with randomly oriented spins should spin down the hole [21]. Thus, the spin could be used to study and determine which mechanism most influences BH growth.

Motivated by the above discussion we present a model of BH spin evolution in the context of the ΛCDM cosmology, by performing a series of simulations using the GALFORM semi-analytic model [16]. The paper is organised as following. In Section 2 we review the different mechanisms of BH growth in the ΛCDM cosmology. In Sections 3 and 4 we detail how accretion and mergers influence the spin of supermassive back holes. Finally, in Section 5 we present our simulations and main results.

2. The growth of SMBHs in hierarchical cosmologies

The evolution of BH mass fits naturally in the ΛCDM model of galaxy formation [44], [6], where structures grow hierarchically: the small structures form first, then evolve through mergers into the largest ones observed. In the event of a galaxy merger, the dark matter halo and baryonic matter of the less massive galaxy (satellite) sink in the gravitational potential of the massive central galaxy as a result of dynamical friction [7]. If the mass of the satellite galaxy is comparable to the central galaxy, the galaxies after the merger completely disrupt and form an elliptical galaxy (major merger). The remnant galaxy is characterised by intense star formation as the available cold gas from both progenitors is transformed rapidly into stars. In addition, part of the cold gas reservoir feeds the central SMBH. It has been argued for a long time that AGN must be powered by accretion onto these SMBHs through a disk of cold gas [26]. For highly efficient accretion activity the disk/BH system dominates the energetics of the nucleus and the galaxy becomes visible as a quasar. SMBHs at the centres of galaxies observed in the local universe are regarded as the remnants of those quasar phases at early epochs.

SMBHs accumulate part of their mass via binary mergers with other SMBHs. The formation of a BH-BH binary and the subsequent coalescence of it is a natural evolutionary stage for a SMBH if the host galaxy has experienced multiple mergers. As the galaxies in the dark-matter haloes merge the SMBHs sink to the centre via dynamical friction from distant stars. The transition to a bound binary state after the galaxy merger is an open issue in theoretical astrophysics, however, it is well known that once the separation of the two BHs becomes small enough, gravitational radiation carries away the remaining angular momentum of the binary. The removal of energy from a SMBH binary due to gravitational wave emission leads to a gradual shrinkage of the relative separation of the two members and circularises the orbit. There is a point where the eccentricity reaches zero and the orbit circularises. At that time the two BHs are very close to each other, and strong gravitational waves are emitted copiously. The radiated energy is so large, that two SMBHs which are a few AU apart will lose all their potential energy within a couple of minutes and inevitably coalesce. After the coalescence is completed the binary enters the *ringdown* phase, where the merged members settle into a quiescent remnant hole.

Diffuse gas in haloes undergoing quasi-hydrostatic cooling is also believed to contribute to the growth of SMBHs. As the baryonic content of the satellite halo collapses during galaxy mergers it gets shocked and heated up to its virial temperature with a subsequent phase characterised by cooling and the formation of cooling gas flows towards the core of the central halo. In massive halos at late redshifts the shocking of the gas occurs at a radius comparable to the radius of the dark matter halo. These haloes have cooling times longer than the free-fall time of the gas and, thus, the gas is added to a quasi-static atmosphere surrounding the galaxy rather than simply fall towards the centre. This atmosphere – the *hot halo* regime – is in pressure supported hydrostatic equilibrium and extends up to the virial radius of the dark matter halo. The galaxy is then supplied by the hot halo with fresh gas via cooling flows, which also feed the central SMBH.

Bower et al. [13] adopt a BH growth model where during a galaxy merger the BH accretes a fixed fraction of the available cold gas in the galaxy. The amount of gas deposited into the BH is set by an efficiency factor, which determines what fraction of the available gas reservoir for star formation accretes into the hole. The value of that parameter is chosen to fit the local $M_{\rm bh} - M_{\rm bulge}$ relation [28], [32], [29], [18]. An additional process of cold gas accretion that contributes to the BH mass in the Bower et al. model is the *disk instabilities* triggered by dynamical instabilities (see [16] and references therein). When the mass of the galactic disk becomes sufficiently gravitating, gas collapses towards the centre when the disk equilibrium is distorted, e.g., by an encounter with a satellite galaxy. A fraction of that gas is directly accreted into the BH, while the rest undergoes star formation. We hereafter refer to the accretion of cold gas triggered by disk instabilities and galaxy mergers as *quasar mode* and to the accretion from quasi-hydrostatic haloes as *radio mode*, following the terminology adopted by Bower et al. [13].

3. Evolution of spin due to accretion

Once the SMBH seed forms at the centre of a galaxy, accretion usually initiates the growth era. We assume that the accretion disk is formed in the equatorial plane of the hole. As proposed by Lynden-Bell [26], the gas particles in the disk gradually lose angular momentum due to viscous torques exerted by magnetic fields and radially drift inwards until they reach the inner edge of the disk. The inner edge of an accretion disk is usually taken as the location of the *last stable orbit* (LSO) around the BH. The LSO is a function of the hole's angular momentum and can be written as [4]:

$$R_{\rm lso} = R_{\rm Sch}\{3 + Z_2 \mp [(3 - Z_1)(3 + Z_1 + 2Z_2)]^{1/2}\}, \quad (1)$$

with $R_{\rm Sch} = 2GM_{\rm bh}/c^2$ and

$$\begin{aligned} Z_1 &\equiv 1 + \left(1 - a^2\right)^{1/3}\left[(1+a)^{1/3} + (1-a)^{1/3}\right], \\ Z_2 &\equiv \left(3a^2 + Z_1^2\right)^{1/2}. \end{aligned} \quad (2)$$

Once the particles reach the LSO, they are dumped directly into the BH. In this way, the accretion carries to the BH the specific energy $\tilde{e}$ and angular momentum $\tilde{\ell}$ that the particles have at the LSO. Thus, accretion of a rest mass dM_0 leads to a change in the hole's total mass $M_{\rm bh}$ and angular momentum $J_{\rm bh}$ equal to,

$$dM_{\rm bh} = (\tilde{e}_{\rm lso}/c^2)dM_0, \quad dJ_{\rm bh} = \tilde{\ell}_{\rm lso}dM_0. \quad (3)$$

The change in the hole's spin induced by the accretion of dM_0 is governed by the differential equation

$$\frac{da}{d\ln M_{\rm bh}} = \frac{1}{M_{\rm bh}}\frac{c^3}{G}\frac{\tilde{\ell}_{\rm lso}}{\tilde{e}_{\rm lso}} - 2a. \quad (4)$$

This was integrated by Bardeen [3] using the explicit expressions for $\tilde{e}$ and $\tilde{\ell}$, resulting in the following solution,

$$\begin{aligned} &\text{for} \quad 1 \le M_{\rm bh}^f/M_{\rm bh} \le R_{\rm lso}^{1/2}: \quad a^f = \frac{1}{3}R_{\rm lso}^{1/2}\frac{M_{\rm bh}}{M_{\rm bh}^f}\left[1 - \left(3R_{\rm lso}\frac{M_{\rm bh}}{M_{\rm bh}^f} - 2\right)^{1/2}\right], \\ &\text{for} \quad M_{\rm bh}^f/M_{\rm bh} \ge R_{\rm lso}^{1/2}: \quad a^f = 1, \end{aligned} \quad (5)$$

where, $M_{\rm bh}^f$ and a^f are the final mass and spin of the BH. The expressions in (5) govern the evolution of a during accretion from an initial state of a co-rotating or counter-rotating thin disk (we define a disk to be thin if its semi-thickness is much smaller than its radius, i.e., $H << R$). According to Bardeen's law, a non-rotating BH ($R_{\rm lso} = 6R_{\rm Sch}$) will be spun up to a maximum rotation ($a = 1$, $R_{\rm lso} = R_{\rm Sch}$) after increasing its mass by $(\sqrt{6}-1)M_{\rm bh}$. For a maximally rotating BH in a counter-rotating accretion disk ($a = -1$, $R_{\rm lso} = 9R_{\rm Sch}$), an increase in mass of $2M_{\rm bh}$ is required in order for the BH to be spun down to $a = 0$ and subsequently spun up to maximum rotation. As implied by Eq. (5), further accretion into the BH keeps a equal to one.

Bardeen's calculations are limited to the process of angular momentum transport between the accreted matter and BH without taking into account other processes that may influence the spin evolution. In fact, Thorne [40] argued that the accretion disk radiates and part of this radiation will be accreted by the hole. Capturing of photons with angular momentum opposite to that of the BH will then produce a counteracting torque that prevents spin up beyond the limiting value of $a = 0.998$.

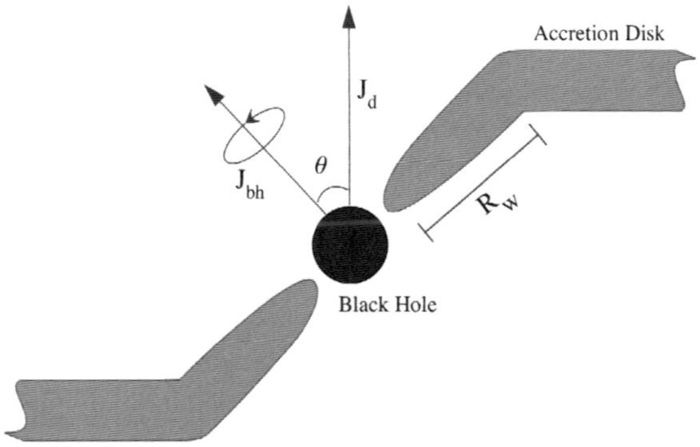

Figure 1. Schematic representation of a warped accretion disk.

3.1. The case of misaligned accretion disks

In the general case where the accretion disk does not lie on the equatorial plane of the BH but has a random orientation relative to the angular momentum of the BH, the evolution of the spin turns out to be a complicated process. Following the discussion in King et al. [22], we assume that a thin disk is inclined at some random angle θ relative to the orientation of the hole's angular momentum vector $\mathbf{J}_{\rm bh}$. We denote the angular momentum of the disk as $\mathbf{J}_{\rm d}$ (see [22] and [42] for a discussion on the nature of $\mathbf{J}_{\rm d}$), and define the total angular momentum vector $\mathbf{J}_{\rm tot}$ as

$$\mathbf{J}_{\rm tot} = \mathbf{J}_{\rm bh} + \mathbf{J}_{\rm d}, \tag{6}$$

with $\hat{\mathbf{J}}_{\rm bh} \cdot \hat{\mathbf{J}}_{\rm d} = \cos\theta$. The angle θ is defined such that $0 \leq \theta \leq \pi$; the values $\theta = 0$ and $\theta = \pi$ correspond to full alignment and anti-alignment. It is obvious that, $\mathbf{J}_{\rm tot}$ is a constant vector and its orientation in space is fixed. Its algebraic expression is given by

$$J_{\rm tot}^2 = J_{\rm bh}^2 + J_{\rm d}^2 + 2J_{\rm bh}J_{\rm d}\cos\theta. \tag{7}$$

When the vectors $\mathbf{J}_{\rm bh}$ and $\mathbf{J}_{\rm d}$ are misaligned, the tilted orbits of the particles in the accretion disk experience a torque due to the Lense-Thirring effect [25], which causes the plane of the accretion disk to precess about the rotational axis of the BH [43], [5], [38]. If the torque is strong enough compared to the internal viscosity, the inner parts of the disk will be forced to rotate in the equatorial plane of the hole resulting in a *warped* disk (see figure 1). The end effect of the Lense-Thirring precession is a BH aligned or anti-aligned with the accretion disk. The expression for the magnitude of $\mathbf{J}_{\rm tot}$ allows us to examine the final configuration of the system. If $J_{\rm bh}^2 > J_{\rm tot}^2$ anti-alignment occurs, which requires

$$\cos\theta < -\frac{J_{\rm d}}{2J_{\rm bh}}. \tag{8}$$

Hence, a BH with $\theta > \pi/2$ and $2J_{\rm bh} > J_{\rm d}$, eventually anti-aligns with the accretion disk.

Further investigation of the warped disks requires the assumption of a thin disk model. We use the solution from the Shakura-Sunyaev disk [39] for our analysis. In the standard Shakura-Sunyaev disk model the analytic expression for the warp radius depends on the values of $J_{\rm bh}$, $M_{\rm bh}$, and the accretion rate $\dot{M}$ of the BH. In terms of the Schwarzschild radius, $R_{\rm warp}$ is written as [42]

$$\frac{R_{\rm warp}}{R_{\rm Sch}} = 3.6 \times 10^3 a^{5/8} \left(\frac{M_{\rm bh}}{10^8 {\rm M}_\odot}\right)^{1/8} f_{\rm Edd}^{-1/4} \left(\frac{\nu_2}{\nu_1}\right)^{-5/8} \alpha^{-1/2}. \tag{9}$$

Here α is the Shakura-Sunyaev viscosity parameter, ν_1 and ν_2 are the accretion and warp propagation viscosity respectively and $f_{\rm Edd} = L/L_{\rm Edd}$ is the Eddington ratio. The accretion luminosity L and Eddington luminosity $L_{\rm edd}$ are defined as

$$L = \epsilon \dot{M} c^2, \tag{10}$$

with ϵ denoting the accretion efficiency, and

$$L_{\rm Edd} = \frac{4\pi G M_{\rm bh} c}{\kappa} = 1.4 \times 10^{46} \left(\frac{M_{\rm bh}}{10^8 {\rm M}_\odot} \right) \ {\rm erg\ s^{-1}}, \tag{11}$$

where κ is the electron scattering opacity. The mass of the disk inside the radius $R_{\rm warp}$ is then

$$M_d(R_{\rm warp}) = \dot{M} t_{\nu_1}(R_{\rm warp}), \tag{12}$$

where the accretion timescale is given by

$$t_{\nu_1} = \frac{R^2_{\rm warp}}{\nu_1} = 3 \times 10^6 \ a^{7/8} \left(\frac{M_{\rm bh}}{10^8 {\rm M}_\odot} \right)^{11/8} f_{\rm Edd}^{-3/4} \left(\frac{\nu_2}{\nu_1} \right)^{-7/8} \alpha^{-3/2} \ {\rm yr}. \tag{13}$$

We can define the total angular momentum $J_{\rm d}$ passing through $R_{\rm warp}$ as,

$$J_{\rm d}(R_{\rm warp}) \lesssim M_d(R_{\rm warp})(G M_{\rm bh} R_{\rm warp})^{1/2}. \tag{14}$$

In terms of the anti-alignment criterion this gives,

$$\frac{J_{\rm d}}{2J_{\rm bh}} = \frac{M_d}{a M_{\rm bh}} \left(\frac{R_{\rm warp}}{R_{\rm Sch}} \right)^{1/2} = 10^{-9} f_{\rm Edd} \left(\frac{t_{\nu_1}}{1 \ {\rm yr}} \right) \left(\frac{R_{\rm warp}}{R_{\rm Sch}} \right)^{1/2} a^{-1}. \tag{15}$$

Derivation of the above quantity shows whether the anti-alignment condition in a misaligned disk is satisfied.

3.2. Self-gravity limited accretion disks

The BH growth process in AGN environments involves vast amounts of accreted gas, often comparable to the initial mass of the accreting hole. This amount of mass settles to the accretion disk around the hole, which often extends in radius to several thousands of gravitational radii, $R_{\rm Sch}$. It is usually assumed that the available mass fuel, $M_{\rm acc}$, is consumed in a single accretion episode, thus, providing a supply of constant angular momentum. In this case, the amount of mass consumed is usually enough to spin up the hole up to a Kerr parameter close to 0.99, even if the BH is maximally spinning in a counter-rotating disk ($a = -1$). Inevitably, subsequent accretion episodes during major galaxy mergers act to spin up the BH to maximum rotation.

Recently, however, King et al. [23] argued that the end effect of the accretion growth channel might be completely different if we take into account the fact that an accretion disk becomes self gravitating at some radius $R_{\rm sg}$ where its mass exceeds $M_{\rm sg} \sim (H/R) M_{\rm bh}$. As a result the mass of the disk is limited by its self gravity to $\Delta M_{\rm episode} \ll M_{\rm acc}$, which gives rise to a series of $N \simeq M_{\rm acc}/\Delta M_{\rm episode}$ well separated accretion episodes. These accretion episodes are assumed to be randomly oriented around the BH since observations suggest that there is no apparent relation between the accretion disk (or radio jet) orientation and the host galaxy disks [24]. Outside $R_{\rm sg}$ the gas is believed to undergo intense star formation which could provide a qualitative explanation for the existence of the ring of stars seen in the near vicinity of the central BH in the Milky Way [17].

We note here that in a large population of accreting BHs with $J_{\rm d} < 2J_{\rm bh}$, we should expect, in principle, counteralignment to occur in a fraction [22]

$$f = \frac{1}{2}\left(1 - \frac{J_{\rm d}}{2J_{\rm bh}}\right), \tag{16}$$

of the accretion episodes. Thus, a succession of episodes that comprise a small fraction of the total mass accreted must have $J_{\rm d} \ll J_{\rm bh}$, which implies that $f \simeq 1/2$. Therefore, counter- and co-alignment are equally possible outcomes of the Lense-Thirring effect. Such a succession of chaotic accretion episodes should then systematically spin down the BH, resulting in a global spin distribution oscillating around zero.

The physics of self-gravitating accretion disks is described in Pringle [35]. Briefly, the criterion that the self-gravity of a disk is negligible is the requirement that the gravitational force along the $\hat{z}$ direction be only due to the central BH. This can be expressed as the surface density of the disk, Σ, being negligible compared to the quantity $M_{\rm bh}H/R^3$, or in terms of the disk mass,

$$M_{\rm d}(< R) \ll \frac{H}{R} M_{\rm acc}. \tag{17}$$

The self-gravity becomes marginally important when $M_{\rm d} \simeq H/RM_{\rm acc}$, which condition gives the mass of a self-gravitating disk [23],

$$M_{\rm sg} = \left(\frac{H}{R} M_{\rm acc}\right)_{R=R_{\rm sg}} = 2.13 \times 10^5 \, \epsilon^{-5/27} \left(\frac{M_{\rm bh}}{10^8 {\rm M}_\odot}\right)^{23/27} f_{\rm Edd}^{5/27} \, \alpha^{-2/17}. \tag{18}$$

As we have described in Section 3.1 the disk once formed with a non-zero misalignment angle about the spin axis of the hole, it will be subject to the Lense-Thirring precession. If the radius, $R_{\rm sg}$, is greater than $R_{\rm warp}$ then,

$$J_{\rm d} = J_{\rm d}(R_{\rm warp}). \tag{19}$$

If, however, $R_{\rm sg} < R_{\rm warp}$, then the entire disk will be subject to Lense-Thirring precession, thus, aligning itself on the equatorial plane of the BH.

4. Spin evolution under binary coalescences

We now examine the evolution of spin during BH-binary coalescences. Following the discussion in Section 2, at the moment of the final merger the smaller BH adds its angular momentum at the LSO to the spin of the larger one,

$$\tilde{l}_{\rm lso} = \pm \frac{2}{3\sqrt{3}} \left[1 + 2\left(3\frac{R_{\rm lso}}{M_{\rm bh}} - 2\right)^{1/2}\right]. \tag{20}$$

Note that, ($\pm$) corresponds to co-rotating and counter-rotating equatorial orbits respectively. Even if the progenitor holes do not possess any angular momentum, the final remnant will preserve the residual orbital momentum of the binary (i.e., the angular momentum that has not been radiated away), thus, the remnant BH will belong to the Kerr family. We note here, that in the field of numerical simulations, recent remarkable breakthroughs in numerical relativity have provided robust simulations of BH mergers by solving directly the Einstein equations in a fully relativistic frame [15], [20], [31]. These simulations have been exploring the parameter space for different configurations of initial masses and spins, allowing accurate measurements of the final spin. For example, a merger of two equal mass BHs with $a = 0$ results in a Kerr BH with $a^f \simeq 0.69$. Analytic fits extend the predictions for a^f to the entire space of parameters and reproduce all the available numerical data. In the case where the masses are unequal, the final spin can be estimated by the analytic expression

$$a^f \simeq 2\sqrt{3}\frac{q}{(1+q)^2} - 2.029\frac{q^2}{(1+q)^4}, \tag{21}$$

where q expresses the mass ratio M_2/M_1 [9].

A more generic semi-analytic fitting formula for predicting the final spin of any given binary configuration in the parameter space has been provided by Rezzolla et al. [36], [37]. In their analysis, they assume that the final spin vector can be expressed as

$$\boldsymbol{a}^f = \frac{1}{(1+q)^2}\left[(\boldsymbol{a_1} + \boldsymbol{a_2})q^2 + \boldsymbol{\ell} q\right]. \tag{22}$$

Here $\boldsymbol{a_{1,2}} = c\mathbf{J}_{1,2}/(GM^2_{1,2})$ and $\boldsymbol{\ell} = \tilde{\boldsymbol{\ell}}/(M_1 M_2)$, with $\boldsymbol{\ell}$ defining the difference between the orbital angular momentum when the binary is widely separated, and the angular momentum radiated away in gravitational waves before the merger,

$$\tilde{\boldsymbol{\ell}} = \mathbf{L} - \mathbf{J}_{\rm rad}. \tag{23}$$

The vector $\boldsymbol{\ell}$ is taken to be parallel to the orbital momentum vector throughout the evolution of the binary, an assumption not strictly valid since the system could radiate away angular momentum in a non-symmetric way. However, the error imposed by this assumption is relatively small for the binary configurations studied in their simulations. A further assumption is that the mass radiated away in gravitational waves is neglected in the calculations, as it accounts for only a minor fraction ($5-7\%$) of the total mass-energy of the binary configurations analysed. Under these assumptions, Rezzolla et al. proposed an analytic expression for the magnitude of the final spin, given by

$$|\boldsymbol{a}_{\rm fin}| = \frac{1}{(1+q)^2}\left[|\boldsymbol{a}_1|^2 + |\boldsymbol{a}_2|^2 q^4 + 2|\boldsymbol{a}_2||\boldsymbol{a}_1|q^2\cos\alpha + 2\left(|\boldsymbol{a}_1|\cos\beta + |\boldsymbol{a}_2|q^2\cos\gamma\right)|\boldsymbol{\ell}|q + |\boldsymbol{\ell}|^2 q^2\right]^{1/2}, \tag{24}$$

with the cosine angles α, β and γ defined as

$$\cos\alpha \equiv \hat{\boldsymbol{a}}_1 \cdot \hat{\boldsymbol{a}}_2, \quad \cos\beta \equiv \hat{\boldsymbol{a}}_1 \cdot \hat{\boldsymbol{\ell}}, \quad \cos\gamma \equiv \hat{\boldsymbol{a}}_2 \cdot \hat{\boldsymbol{\ell}}. \tag{25}$$

The final spin as given by expression (24) is in a good agreement when compared with numerical data, with residuals being restricted to less than 3%.

5. Evolution of spin in hierarchical cosmologies: numerical simulations

5.1. Method

We perform a series of numerical simulations in which we explore the effect of gas accretion and mergers on the spin of SMBHs. The basic tool for our predictions is the `GALFORM` semi-analytic model [16], which implements the Bower et al. [13] model for galaxy evolution in the ΛCDM cosmology. BHs evolve their mass in accordance with the model developed by Malbon et al. [30]. We study the cosmological spin evolution of SMBH seeds for a total number of $\sim 2.1 \times 10^7$ galaxies in a co-moving volume of 1.35×10^8 (Mpc/h)^3 from redshift 127 to redshift zero. Here h is defined in $H_0 = h \times 100$ km s^{-1}Mpc^{-1}, where H_0 is the Hubble constant at redshift zero.

During the evolution of the BHs we take into account spin changes due to accretion of gas and mergers, as described in Sections 3 and 4. The initial population of seeds is assumed to be non-rotating and "pre-galactic" – it could represent the end product of massive Population III stars

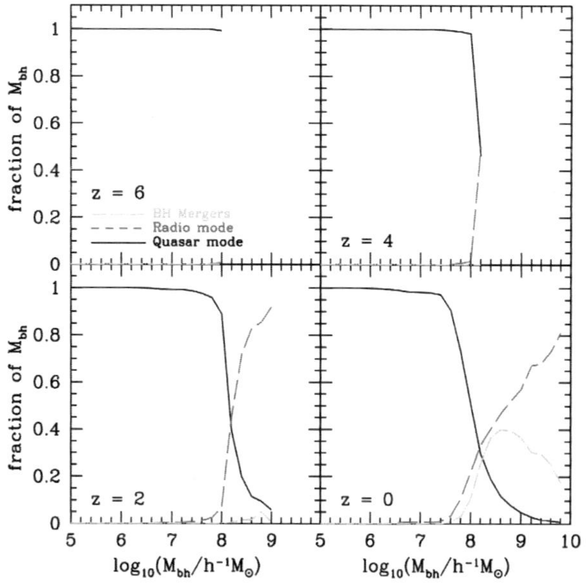

Figure 2. Contribution to the final BH mass from the quasar mode (solid black line), radio mode (dashed red line) and BH-BH mergers (dashed-dotted green line) at different redshifts.

or collapsed clumps of gas in the early universe. These seeds grow via accretion of gas during galaxy mergers and disk instabilities, the radio mode accretion, and BH-BH mergers. Figure 2 shows the contribution from the different growth channels to the final BH mass at different redshifts. As it is illustrated, at high redshifts SMBHs with masses up to $\sim 10^8$ $M_\odot$ built their mass almost exclusively though the quasar mode. On the contrary, accretion of gas from the hot halo in massive galaxies (radio-mode accretion) and BH mergers contribute significantly only to BHs with masses $\gtrsim 10^8$ $M_\odot$ at low redshifts.

To model the physics of the accreted gas, we assume that prior to its collapse it forms an accretion disk whose physics is adequately described by the standard Shakura-Sunyaev disk model. We assume that the available gas to be fed into the BH after a galaxy merger or disk instability accretes over a timescale equal to the dynamical timescale of the galactic bulge that hosts the BH.

The disk is chosen to be randomly oriented relative to the spin axis of the hole. The inclination angle θ is chosen from a flat distribution of cosine values between -1 and 1. If $\theta \neq 0°, 180°$, we assume that the disk precesses around the BH as described in Section 3.1. In brief, for a given BH mass $M_{\rm bh}$ we determine the numerical values of the warp parameters, $R_{\rm warp}$, $M_{\rm d}(R_{\rm warp})$, and the angular momenta $J_{\rm d}(R_{\rm warp})$ and $J_{\rm bh}$. The ratio $J_{\rm d}/2J_{\rm bh}$ allows us through the criterion in relation (8) to check if anti-alignment or alignment occurs. We then allow the BH to accrete an amount of gas equal to $M_{\rm d}(R_{\rm warp})$ and evolve the spin using relation (5). Finally, we update the masses of the BH and we estimate the hole's new spin and the disk precession by calculating the new angle θ. This process is repeated until: (a) the disk aligns or counteraligns ($\theta = 0°$ or $180°$) with the hole's spin, in which case we just consume the rest of the available gas in a single step and evolve the spin according to (5), (b) the accretion disk is entirely consumed, without being able to align or counteralign itself with the spin. In almost al of our cases we find that the accretion disks ultimately align (or anti-align) themselves on the equatorial plane of the BH.

In addition to the standard accretion model, we decide to test the model proposed by King

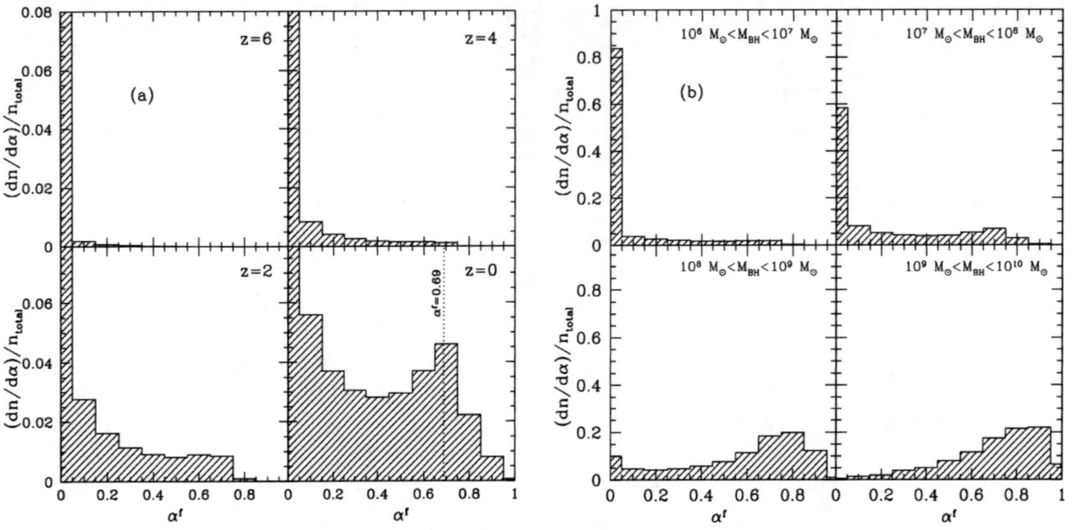

Figure 3. The distribution of spins at different redshifts (a) and in different mass ranges (b) for aligned binary mergers.

et al. [23] where the mass of the accretion disk is limited by its self-gravity. We repeat the same steps as previously taking, however, $M_\mathrm{d} = M_\mathrm{sg}$. Once M_sg is consumed we update the mass of the hole, determine M'_sg computed from the updated hole mass, and settle a new disk in a random orientation around the BH with $M_\mathrm{d} = M'_\mathrm{sg}$.

Finally, we consider the spin changes due to binary coalescences. Following the merger of two galaxies in GALFORM, each harbouring a SMBH, we expect the one hosted by the satellite to sink toward the SMBH of the central galaxy and eventually form a binary. In our mode BH mergers occur in gas-rich environments, where we assume that torques from the accreting gas suffice to align the spin of the BHs with the orbital axes prior the collapse [12]. We, therefore, take $\alpha = \beta = \gamma = 1$ in relation (24). Furthermore, we constrain our results to BH masses greater than $\sim 10^6\mathrm{M}_\odot$ in accordance with the current $M_\mathrm{bh} - M_\mathrm{bulge}$ observations. In addition, we assume that all binaries will ultimately coalesce on a very short timescale, which effectively allows us to coalesce the binary right after the host galaxies merge.

During the merger, the smaller BH is taken to plunge into the larger one carrying along its angular momentum at the LSO. At this point we neglect any mass loss due to gravitational emission, as it corresponds to a small fraction of the total mass-energy of the system. The spin of the final remnant is estimated using the semi-analytic fitting formulae form Rezzolla et al. [36], [37], as described in Section 4.

5.2. Results

In order to establish clearly the spin evolution under binary coalescences and elucidate the contribution of mergers to the global spin distributions, we first allow the BHs to evolve in our model ignoring the effects of accretion. The histograms in figure 3 (a) show the distributions of spins at different redshifts. As demonstrated in these plots, at high redshifts there are almost no mergers occurring in the universe. As redshift decreases, BHs gradually form binaries and merge with each other and eventually at redshift zero nearly 19% of the BHs have experienced at least one merger. What is apparent in the distribution of spins at redshift zero is that it has a strong peak centred around $a^f \sim 0.69$. We recall that $a^f \sim 0.69$ is the maximum spin

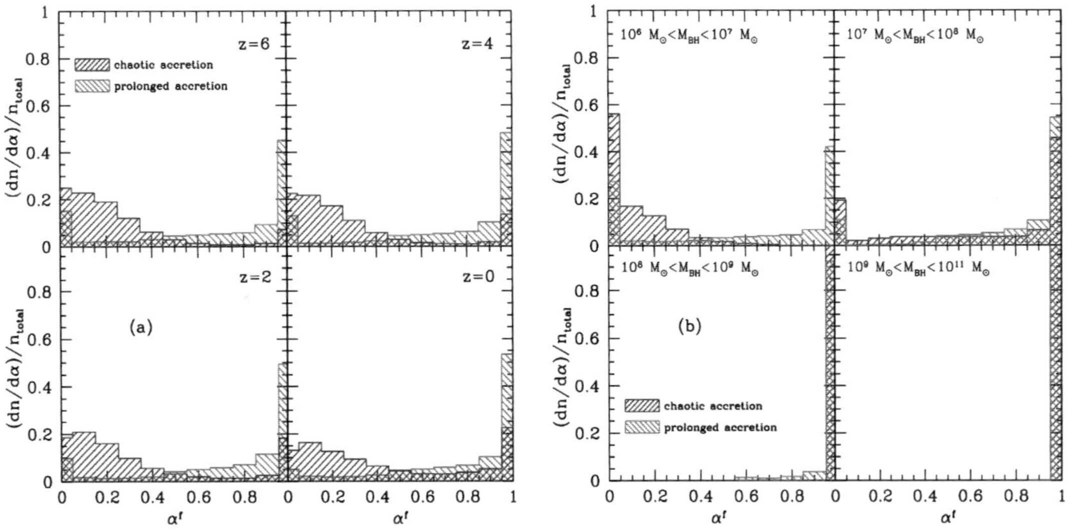

Figure 4. (a) The spin evolution with redshift due to BH-BH mergers and accretion. Different colours represent different accretion scenarios. (b) The distribution of BH spins at $z = 0$ for different mass bins.

that a remnant BH can have when its progenitors are non-spinning and corresponds to the case where the members of the binary have equal masses ($q = 1$). Thus, the region of the histogram with $a^f \leq 0.69$ is dominated by BHs with non-spinning progenitors. BHs with final spins higher than 0.69 are the end product of aligned binaries with spinning members since mergers of holes with parallel spins have the tendency to spin up the remnant (see [10]). Finally, a strong peak appears at the low-spin end of the distribution since most of the BH seeds never experience any merger.

A clear prediction of figure 3 (a) is that BH mergers in the ΛCDM universe are not a very common phenomenon. For example at redshift zero about 6% of the BHs have experienced more than one merger and only $\sim 8\%$ have had a recent major merger ($q > 0.1$). The rare nature of mergers is a consequence of the dominance of accretion as a growth channel (see figure 2).

When we consider the post-merger spins for different BH masses we find that the distribution of spins depends strongly on the mass range chosen. This is demonstrated for $z = 0$ in figure 3 (b), where we show the distribution of spins for different back hole mass ranges. As can be seen, high spin values are common only for massive BHs ($M_{\rm bh} \gtrsim 10^8 M_\odot$). These BHs are hosted by massive elliptical galaxies and have experienced at least one major merger in their past history. Hence, they acquire high spins since aligned mergers with comparable BH masses act to spin BHs up quite efficiently and, thus, account for the high-spin end of the distribution at redshift zero. On the contrary, small BHs built their mass mainly through accretion (see figure 2) and repeated minor mergers. Therefore, BHs with $M_{\rm bh} \lesssim 10^7 M_\odot$ do not acquire very high spins. BHs in the range $10^7 M_\odot \lesssim M_{\rm bh} \lesssim 10^8 M_\odot$ attribute a small fraction of their mass to mergers which allow them to spun up to $a^f \sim 0.7 - 0.8$. The mild bimodality of spins seen here is reflected in the global spin distribution at redshift zero as BHs in this mass range dominate the mass function of BHs in the present universe.

We now simulate the effect of accretion on the evolution of BH spins. The accretion is modelled as in Volonteri [42] where we assume that the available gas is accreted during a single accretion event, and alternatively as in King et al. [23] where we constrain the amount of accretion to the self-gravity limited mass (see Section 3.2). We henceforth refer to these two different scenarios as *prolonged* and *chaotic* accretion models respectively.

Figure 4 (a) reports how the spin evolves under the chaotic and prolonged accretion models. In both cases it is evident that accretion dominates the spin evolution, resulting in spin distributions different from those obtained under mergers only. If we assume chaotic feeding of gas to the BH we find a bimodal distribution of spins. The bimodality is an expression of the nature of accretion in our model. In order to illustrate this better we plot in figure 4 (b) the spin distributions at $z = 0$ in different mass ranges. When BHs acquire their mass through chaotic accretion episodes, the spin evolution depends strongly on the hole's mass. Low mass BHs have moderate spins, very close to $a^f = 0.1$, and account for the low-spin peak in the bottom panel of figure 4. More massive BHs have much higher spins and populate the high-spin end of the global spin distribution at redshift zero.

In our model, low mass BHs ($M_{\rm bh} \lesssim 10^8 \rm M_\odot$) built up their mass mainly through the quasar mode of accretion. For example, for a BH with mass $10^6 \rm M_\odot$ a typical accretion event contains $10^6 \rm M_\odot$ of gas. However, if we assume that the size of the disk is limited by its self gravity, the mass of the gas that settles itself on the disk cannot be more than $10^4 \rm M_\odot$. Thus, a $10^6 \rm M_\odot$ accretion event gives rise to $\sim$ 100 accretion episodes with $\Delta M_{\rm episode} \ll M_{\rm bh}$, all randomly oriented around the BH. As a consequence, these repeated episodes produce a net spin-down to modest values centred around $a^f \sim 0.1$.

For more massive BHs the mass of a self-gravity limited disk can be much higher (remember $M_{\rm sg} \propto \rm M_\odot^{0.85}$ from Eq. (18)). For example a BH of $\sim 10^8 \rm M_\odot$ can sustain a disk of up to $10^6 \rm M_\odot$. Such massive BHs accumulate their mass primarily via accretion during the radio mode. However, the radio mode is characterised by low mass accretion episodes; typical amounts of gas consumed during the radio mode for a $10^8 \rm M_\odot$ BH are of the order of $10^5 \rm M_\odot$. This amount of gas can be accreted in a single episode, resulting in high final spins. This is a common feature for BHs with high masses as above that mass limit the radio mode is the dominant growth channel. Thus, all BHs with $\gtrsim 10^8 \rm M_\odot$ are found to rotate rapidly.

When we consider the prolonged accretion model we find that accretion is a very efficient mechanism for spinning up BHs even at very high redshifts. A great fraction of the back holes have accreted enough gas to be spun-up up to $a^f \approx 1$, even if those BHs originally accrete via retrograde accretion. This results in a significant population of rapidly rotating remnants in the present universe for any given BH mass. If however, there is not enough gas available to the hole and the disk is counter-rotating the BH are on average spun down. As a result, a long tail extending down to $a^f \simeq 0$ is also present.

6. Discussion - Conclusions

We have presented a model for studying the evolution of SMBH spins in the ΛCDM cosmology. We found that the different astrophysical processes that influence the evolution of SMBHs may have a significant effect on the global spin distribution in the present universe depending on the accretion scenario we assume. For example, in the prolonged model, accretion of gas dominates the evolution of spins through cosmic time, resulting in rapidly rotating BHs. However, if we assume that gas is fed into the BH through a self-gravity limited disk we obtain a bimodal spin distribution. In this case, high spin values is a property of BHs with masses $\gtrsim 10^8 \rm M_\odot$. These BHs are hosted in our cosmological model by massive elliptical galaxies. On the contrary, spiral galaxies usually host small BHs ($M_{\rm bh} \lesssim 10^8 \rm M_\odot$) which according to the chaotic accretion-scenario should possess moderate spins.

This is an important result of our model as it could provide insight into the radio loudness of AGNs. If the spin is the BH parameter that determines the strength of the jet in an AGN environment [11], then according to our model since elliptical galaxies have rapidly rotating SMBHs they should host AGNs with powerful radio jets. On the contrary, AGNs in spiral galaxies must have quiescent jets for the reason that their central SMBHs have on average very low spins. This is indeed what observation suggest (see [34] and references therein), since radio jets in AGNs hosted by elliptical galaxies can be about 10^3 times more radio powerful than those in AGNs hosted by spiral galaxies. The relation between the SMBH spin and the radio properties of AGNs will be studied in a forthcoming paper.

Acknowledgments
This work was supported by the Marie Curie Early Stage Training fellowship.

References

[1] Baker, J. G., Centrella, J., Choi, D.-I., Koppitz, M., & van Meter, J. *Phys. Rev. Lett.*, 96, 111102, 2006.
[2] Baker, J. G., McWilliams, S. T., van Meter, J. R., Centrella, J., Choi, D.-I., Kelly, B. J., & Koppitz, M., *Phys. Rev. D*, **75**, 124024, 2007
[3] Bardeen, J. M. 1970, *Nature*, 226, 64
[4] Bardeen, J. M., Press, W. H., & Teukolsky, S. A. 1972, *Astrophys. J.*, 178, 347
[5] Bardeen, J. M., & Petterson, J. A. 1975, *Astrophys. J. Lett.*, 195, L65
[6] Baugh, C. M., Lacey, C. G., Frenk, C. S., Granato, G. L., Silva, L., Bressan, A., Benson, A. J., & Cole, S. 2005, *Mon. Not. R. astr. Soc.*, 356, 1191
[7] Binney, J., & Tremaine, S. 1987, Princeton, NJ, Princeton University Press, 1987
[8] Begelman, M. C., Blandford, R. D., & Rees, M. J. 1984, Reviews of Modern Physics, 56, 255
[9] Berti, E., Cardoso, V., Gonzalez, J. A., Sperhake, U., Hannam, M., Husa, S., & Brügmann, B. 2007, *Phys. Rev. D.*, 76, 064034
[10] Berti, E., & Volonteri, M. 2008, *Astrophys. J.*, 684, 822
[11] Blandford, R. D., & Znajek, R. L. 1977, *Mon. Not. R. astr. Soc.*, 179, 433
[12] Bogdanović, T., Reynolds, C. S., & Miller, M. C. 2007, *Astrophys. J. Lett.*, 661, L147
[13] Bower, R. G., Benson, A. J., Malbon, R., Helly, J. C., Frenk, C. S., Baugh, C. M., Cole, S., & Lacey, C. G. 2006, *Mon. Not. R. astr. Soc.*, 370, 645
[14] Buonanno, A., Cook, G. B., & Pretorius, F. 2007, *Phys. Rev. D.*, 75, 124018
[15] Campanelli, M., Lousto, C. O., Zlochower, Y., Krishnan, B., & Merritt, D. 2007, *Phys. Rev. D.*, 75, 064030
[16] Cole, S., Lacey, C. G., Baugh, C. M., & Frenk, C. S. 2000, *Mon. Not. R. astr. Soc.*, 319, 168
[17] Genzel, R., et al. 2003, *Astrophys. J.*, 594, 812
[18] Häring, N., & Rix, H.-W. 2004, *Astrophys. J. Lett.*, 604, L89
[19] Hawley, J. F., Gammie,
[20] Herrmann, F., Hinder, I., Shoemaker, D. M., Laguna, P., & Matzner, R. A. 2007, *Phys. Rev. D.*, 76, 084032
[21] Hughes, S. A., & Blandford, R. D. 2003, *Astrophys. J. Lett.*, 585, L101
[22] King, A. R., Lubow, S. H., Ogilvie, G. I., & Pringle, J. E. 2005, *Mon. Not. R. astr. Soc.*, 363, 49
[23] King, A. R., Pringle, J. E., & Hofmann, J. A. 2008, *Mon. Not. R. astr. Soc.*, 385, 1621
[24] Kinney, A. L., Schmitt, H. R., Clarke, C. J., Pringle, J. E., Ulvestad, J. S., & Antonucci, R. R. J. 2000, *Astrophys. J.*, 537, 152
[25] Lense, J., & Thirring, H. 1918, Phys. Z., 19, 156
[26] Lynden-Bell, D. 1969, *Nature*, 223, 690
[27] MacDonald, D., & Thorne, K. S. 1982, *Mon. Not. R. astr. Soc.*, 198, 345
[28] Magorrian, J., et al. 1998, *Astr. J.*, 115, 2285
[29] Marconi, A., & Hunt, L. K. 2003, *Astrophys. J. Lett.*, 589, L21
[30] Malbon, R. K., Baugh, C. M., Frenk, C. S., & Lacey, C. G. 2007, *Mon. Not. R. astr. Soc.*, 382, 1394
[31] Marronetti, P., Tichy, W., Brügmann, B., González, J., Hannam, M., Husa, S., & Sperhake, U. 2007, Classical and Quantum Gravity, 24, 43
[32] McLure, R. J., & Dunlop, J. S. 2002, *Mon. Not. R. astr. Soc.*, 331, 795
[33] Merritt D., and Milosavljević M., Living Rev. Relativity **8**, 2005.
[34] Sikora, M., Stawarz, L., & Lasota, J.-P. 2007, *Astrophys. J.*, 658, 815
[35] Pringle, J. E. 1981, *Ann. Rev. Astr. Astrophys.*, 19, 137
[36] Rezzolla, L., Diener, P., Dorband, E. N., Pollney, D., Reisswig, C., Schnetter, E., & Seiler, J. 2008a, *Astrophys. J. Lett.*, 674, L29
[37] Rezzolla, L., Barausse, E., Dorband, E. N., Pollney, D., Reisswig, C., Seiler, J., & Husa, S. 2008b, *Phys. Rev. D.*, 78, 044002
[38] Scheuer, P. A. G., & Feiler, R. 1996, *Mon. Not. R. astr. Soc.*, 282, 291
[39] Shakura, N. I., & Syunyaev, R. A. 1973, *Astr. Astrophys.*, 24, 337

Journal of Physics: Conference Series **189** (2009) 012013 doi:10.1088/1742-6596/189/1/012013

[40] Thorne, K. S. 1974, *Astrophys. J.*, 191, 507
[41] Volonteri, M., Madau, P., Quataert, E., & Rees, M. J. 2005, *Astrophys. J.*, 620, 69
[42] Volonteri, M., Sikora, M., & Lasota, J.-P. 2007, *Astrophys. J.*, 667, 704
[43] Wilkins, D. C. 1972, *Phys. Rev. D.*, 5, 814
[44] White, S. D. M., & Frenk, C. S. 1991, *Astrophys. J.*, 379, 52

Recent Developments in Gravity (NEB XIII) IOP Publishing
Journal of Physics: Conference Series **189** (2009) 012014 doi:10.1088/1742-6596/189/1/012014

On the Tolman and Møller mass-energy formulae in general relativity

P. S. Florides
School of Mathematics, Trinity College Dublin, Dublin 2, Ireland

Abstract. Two of the most satisfactory descriptions of mass–energy and its localization in general relativity are given in terms of the Tolman and Møller formulae. The form of these two equations, as well as their derivation, are so utterly different that for thirty–five years, there was never even the slightest suspicion that there was any relationship between them. Yet, as it will be shown in this talk, for time–independent fields the two formulae are in fact identical.

1. Introduction

The problem that has bedeviled the general theory of relativity most, from the early days of the theory to the present, is the problem of mass-energy and of its localization. Despite the immense effort of many relativists there is still no satisfactory, universally accepted formula for the total (that is, matter plus field) mass-energy of either the whole or part of a physical system.

As is well known there is an infinite number of widely different formulae purporting to give the total mass-energy of a physical system. The one thing that all these formulae have in common is that, for an isolated physical system which is time–independent and which, asymptotically, is strictly the Minkowski space-time expressed in *Cartesian coordinates*, they all give the correct value for the total mass-energy of the *whole* of the system. Unless these conditions are strictly satisfied almost all the formulae give values for the mass-energy of the *whole* system which are, more often than not, totally absurd. The most striking absurdity is that for the Minkowski space–time, when expressed in *spherical Polar coordinates*, the Einstein formula (given by equations (4–6) below), gives an *infinite* amount of energy for the whole of the 3–space $t = constant$. Exactly the same absurd result is obtained for the Schwarzschild (exterior) space–time when it is expressed in the usual *curvature coordinates*.

The next, and related problem, concerns the *localization* of mass-energy. That is the problem of finding the (total) mass-energy within a given region of some 3-space $x^o \equiv t =$constant. Again, almost all of the mass-energy formulae are unable to deal with this problem.

To the best of my knowledge the two mass-energy formulae which, for time independent fields at any rate, can deal satisfactorily with both the total mass-energy of the *whole* system and the *localization* of mass-energy are the Tolman and Møller formulae. These formulae are discussed briefly in section two where the main steps of the derivation of the Møller formula is also given. Despite the totally different derivations of the two formulae and their totally different form, it is shown in section three that, for time independent fields, the two formulae are, in fact, completely equivalent. This equivalence is discussed briefly in section four. The equivalence of the two formulae was first derived by the author in references [1, 2]. The present talk was prompted by Dr E. C. Vagenas whose work [3], and references therein, showed that the Møller formula is still a topic of active research.

2. Tolman and Møller formulae

We shall denote by m_T the Tolman mass-energy of a physical system. It is given by the Tolman formula [4, 5]

$$m_T = \int_V (T_1^1 + T_2^2 + T_3^3 - T_0^0)\sqrt{-g}dV \equiv \int_V (T_\alpha^\alpha - T_0^0)\sqrt{-g}dV\,. \tag{1}$$

Here, T_j^i is the energy-momentum-stress tensor of the system, g the determinant of the space-time metric g_{ij}, V the region of the 3-space x^o = constant occupied by matter or electromagnetic fields, and dV is the coordinates 3-volume element. {Throughout this paper Latin indices take the values, 0,1,2,3 and Greek indices the values 1,2,3 with Einstein's summation and range conventions applied in each case. Relativistic units are used so that $c = G = 1$, and the signature of the matrix is +2. Comma (,) denotes partial differentiation $(f_{,i} = \dfrac{\partial f}{\partial x^i})$.} This formula has, as Tolman remarked, "the great advantage that it can be evaluated by integrating over the region occupied by matter or electromagnetic energy".
Since Tolman's formula is widely known it should suffice to say that Tolman's derivation of formula (1) is based on Einstein's energy pseudotensor, and so too is the much simpler derivation due to Papapetrou [6]. By far the simplest derivation, and one which circumvents the use of pseudotensors, was given by Landau and Lifshitz in their book on Classical Fields [7]. Another derivation, for *static* fields, was given by Whittaker [8].
We come now to the Møller formula. Since this is not as well-known as Tolman's formula it will be helpful to outline the main steps by which Møller arrived at his equation. Møller starts with the Einstein energy complex Θ_b^a which is defined by

$$\Theta_b^a = \sqrt{-g}(T_b^a + t_b^a)\,, \tag{2}$$

t_b^a being Einstein's energy pseudotensor. This complex satisfies the divergence equation

$$\Theta_{b,a}^a = 0\,, \tag{3}$$

which leads, in the usual way, to the definition of the energy-momentum (four) vector of the system

$$(-E, P_\alpha) = \int_V \Theta_a^o dV\,. \tag{4}$$

A simpler and more useful form of the Einstein complex is in terms of the so–called Freud [9] superpotentials h_b^{ac} defined by

$$h_b^{ac} = -h_b^{ca} = \frac{g_{bi}}{2\kappa\sqrt{-g}}[(-g)(g^{ai}g^{cj} - g^{ci}g^{aj})]_{,j}\,, (\kappa = 8\pi)\,, \tag{5}$$

that is,

$$\Theta_b^a = h_{b,c}^{ac}\,. \tag{6}$$

We note that because of the antisymmetry of h_b^{ac} in a and c, equation (3) is satisfied automatically.
It must be emphasised again that Einstein's equation (4) is meaningful only when V is the *whole* 3-space x^o=constant of a physical system which is time–independent and, asymptotically, its field is strictly the Minkowski space-time expressed in *Cartesian coordinates.* It was precisely the attempt to remove this last restriction, the adherence to Cartesian coordinates, that lead Møller to his mass-energy formula.
Møller argued [10] that equation (3) determines Θ_b^a only to within a quantity S_b^a with identically vanishing divergence, i.e. $S_{b,a}^a = 0$. He uses this freedom to define a new energy complex $\mathcal{J}_b^a$ by

$$\mathcal{J}_b^a = \Theta_b^a + S_b^a \,, \tag{7}$$

where S_b^a behaves like a tensor density of weight -1 under all linear transformations, and is chosen so that

(i) $S_{b,a}^a = 0$

(ii) $\int_V S_b^a dx^1 dx^2 dx^3 = 0$ when the coordinates x^1, x^2, x^3 are quasi-Galilean and V is the *whole* of the 3-space x^o=constant,

(iii) $\mathcal{J}_o^a$ behaves like a 4-vector density under all spatial coordinate transformations

$$x^\alpha \to \overline{x}^\alpha = \overline{x}^\alpha(x^1, x^2, x^3), \qquad x^o \to \overline{x}^o = x^o \quad , \qquad (\alpha = 1,2,3) \quad , \tag{8}$$

These three requirements lead Møller to the expression

$$S_b^a = h_{b,c}^{ac} - \delta_b^a h_{i,j}^{ij} + \delta_b^i h_{j,i}^{ja} \,, \tag{9}$$

where h_b^{ac} are defined by equation (5). Furthermore, Møller showed that $\mathcal{J}_b^a$ can be written in the form

$$\mathcal{J}_b^a = \chi_{b,c}^{ac} \,, \tag{10}$$

where the (Møller) superpotentials χ_b^{ac} are given by

$$\chi_b^{ac} = -\chi_b^{ca} = \frac{\sqrt{-g}}{\kappa} g^{ai} g^{cj} (g_{bj,i} - g_{bi,j}) \,. \tag{11}$$

Because of the antisymmetry of χ_b^{ac} in a and c we get

$$\mathcal{J}_{b,a}^a = 0 \,,$$

from which follows the usual interpretation that the energy-momentum is given by

$$(-m_M, P_\alpha) = \int_V \mathcal{J}_a^o dV \,, \tag{12}$$

having written m_M for the Møller mass-energy of the system under consideration.
This is, in essence, Møller's theory. The advantages of Møller's formula (12) over Einstein's formula (4) are obvious. Because of condition (ii) above, Møller's formula agrees with Einstein's whenever the latter is applicable. In contradistinction to Einstein's formula,

Møller's formula can, in view of condition (iii), be used in *any* spatial coordinate system. Furthermore Møller's formula can be used to find the total energy and momentum within any region of a 3-space x^o=constant. Thus, unlike Einstein's theory, it is meaningful to talk about the localization of energy in Møller's theory.
Over the years the author applied Møller's formula on many occasions to many different physical systems. On all occasions the results were meaningful and interesting. It was, for example, on the basis of Møller's formula that the author discussed, for the first time, quantitatively the contribution of the electromagnetic energy of a charged sphere to the total gravitational mass of the sphere. For further details and applications of Møller's formula see [11, 12].
Let us concentrate on the zeroth component of (12), namely the Møller mass-energy m_M. Remembering that $\chi_b^{ac} = -\chi_b^{ca}$ equations (10) and (12) give

$$m_M = -\int_V \mathcal{J}_o^o dV = -\int \chi_{o,\alpha}^{o\alpha} dV \qquad (\alpha = 1, 2, 3), \tag{13}$$

where

$$\chi_o^{o\alpha} = \frac{\sqrt{-g}}{\kappa} g^{oi} g^{\alpha j} (g_{oj,i} - g_{oi,j}). \tag{14}$$

Recall, also, that the Tolman mass-energy m_T is given by equation (1), i.e.

$$m_T = \int_V (T_\alpha^\alpha - T_o^o)\sqrt{-g} dV\,. \tag{15}$$

The expressions for m_T and m_M, as well as their derivation, are so utterly different that, for thirty-five years, there was never even the slightest suspicion that there was any relationship between them. Yet, whenever the two formulae were applied by the author to time-independent spherically symmetric fields, the results were always identical. In view of these results the author ventured to ask the question: "Is it possible that these totally different looking formulae are in fact identical for *general* time-independent fields?".
Once the question was asked it was easy to give the answer. Surprisingly, almost in disbelief, it is proved in the following section that the two formulae are indeed identical.

3. The equivalence of the Tolman and Møller formulae

The proof of the equivalence of the two formulae (13) and (15) is based on the following simple formula due to Landau and Lifshitz [7]:

$$R_o^o = -\frac{1}{\sqrt{-g}} (\sqrt{-g} g^{io} \Gamma_{oi}^\alpha)_{,\alpha}\,, \tag{16}$$

R_o^o being the (0,0) component of the Ricci tensor R_j^i. The formula holds for *arbitrary* time independent fields. Now, from Einstein's field equations, written in the form $R_j^i = -\kappa(T_j^i - \frac{1}{2}\delta_j^i T_a^a)$, $\kappa = 8\pi$, we get

$$R_o^o = \frac{\kappa}{2}(T_\alpha^\alpha - T_o^o). \tag{17}$$

Using this in (16) we get

$$(T_\alpha^\alpha - T_o^o)\sqrt{-g} = -\frac{2}{\kappa}(\sqrt{-g} g^{io} \Gamma_{oi}^\alpha)_{,\alpha}. \tag{18}$$

Now, from the definition of the Christoffel symbols, we get

$$g^{io}\Gamma^{\alpha}_{oi} = g^{io}g^{\alpha j}[oi, j] = \frac{1}{2}g^{io}g^{\alpha j}(g_{oj,i} + g_{ij,o} - g_{oi,j}),$$

or, since by assumption g_{ij} is time-independent,

$$g^{io}\Gamma^{\alpha}_{oi} = \frac{1}{2}g^{io}g^{\alpha j}(g_{oj,i} - g_{oi,j}). \tag{19}$$

Substituting this in equation (18) we get

$$(T^{\alpha}_{\alpha} - T^{o}_{o})\sqrt{-g} = -\frac{1}{\kappa}[\sqrt{-g}g^{io}g^{\alpha j}(g_{oj,i} - g_{oi,j})]_{,\alpha}$$

or, by equation (14),

$$(T^{\alpha}_{\alpha} - T^{o}_{o})\sqrt{-g} = -\chi^{o\alpha}_{o,\alpha}, \tag{20}$$

χ^{ac}_{b} being the Møller superpotential. We have thus arrived, quite unexpectedly, at the result that the integrands of the Tolman and Møller mass-energy formulae are identical. The two formulae are therefore completely equivalent. However different the dresses of the two formulae, they cover one and the same body.

4. Discussion

It is, of course, satisfying to see the number of mass-energy formulae reduced even by one; even if the original number is infinite! The equivalence of the Tolman (m_T) and Møller (m_M) formulae, however, is also important in at least two further aspects: (i) Because the integrand of m_M is in the form of plain divergence it is often simpler to calculate the mass-energy of a system using the formula m_M (and Gauss's theorem) rather than m_T. (ii) The mass-energy of a system should appear as the zeroth component of an energy-momentum 4-vector. Up until now the Tolman mass-energy m_T stood entirely on its own. From the equivalence $m_T = m_M$ established in the paper, m_T (just as m_M) is now the zeroth component of the Møller 4-vector $(-m_M, P_\alpha)$ defined by equation (12); strictly speaking $(-m_M, P_\alpha)$ is a 4-vector under the transformation (8).

For a critical re-appraisal of Tolman's (and, therefore, Møller's) formula the reader is referred to papers [13]-[15].

References

[1] Florides P S 1994 *Proceedings of the Cornelius Lanczos International Centenary Conference* SIAM 503
[2] Florides P S 1994 *Gen. Rel. Grav.* **26** 1145.
[3] Vagenas E C 2006 *Mod. Phys. Letters* **A21**
[4] Tolman R C 1930 *Phys. Rev.* **35** 875.
[5] Tolman R C 1962 *Relativity, Thermodynamics and Cosmology* (Clarendon, Oxford)
[6] Papapetrou A 1947 *Proc. Roy. Irish Acad* **XV** 191
[7] Landau L D and Lifshitz E M 1975 *The Classical Theory of Fields* (Pergamon Press, Oxford)
[8] Whittaker E T 1935 *Proc. Roy. Soc.* A **149** 384.
[9] Von Freud P 1939 *Annals of Mathematics* **40** 417
[10] Møller C. 1958 *Annals of Physics* **4** 347
[11] Florides P S 1961 *Proc. Cambr. Phil. Soc.* **58** 102
[12] Florides P S 1961 *Proc. Cambr. Phil. Soc.* **58** 110.
[13] Cooperstock F I and Sarracino R S 1978 *J. Phys.* A **11** 877.
[14] Cooperstock F I, Sarracino R S and Bayin S S 1981 *J. Phys.* A **14** 181
[15] Devitt J and Florides P S 1989 *Gen. Rel. Grav.* **21** 585

Recent Developments in Gravity (NEB XIII) IOP Publishing
Journal of Physics: Conference Series **189** (2009) 012015 doi:10.1088/1742-6596/189/1/012015

Hidden symmetries and black holes

Valeri P. Frolov

Theoretical Physics Institute, University of Alberta, Edmonton, Alberta, Canada T6G 2G7

E-mail: frolov@phys.ualberta.ca

Abstract. The paper contains a brief review of recent results on hidden symmetries in higher dimensional black hole spacetimes. We show how the existence of a principal CKY tensor (that is a closed conformal Killing-Yano 2-form) allows one to generate a 'tower' of Killing-Yano and Killing tensors responsible for hidden symmetries. These symmetries imply complete integrability of geodesic equations and the complete separation of variables in the Hamilton-Jacobi, Klein-Gordon, Dirac and gravitational perturbation equations in the general Kerr-NUT-(A)dS metrics. Equations of the parallel transport of frames along geodesics in these spacetimes are also integrable.

1. Introduction

There are several reasons why a subject of higher dimensional black holes becomes so popular recently. The idea that the spacetime can have more than four dimensions is very old. Kaluza and Klein used this idea about 80 years ago in their attempts to unify electromagnetism with gravity. The modern superstring theory is consistent (free of conformal anomalies) only if the spacetime has a fixed number (10) of dimensions. Usually it is assumed that extra dimensions are compactified. The natural size of compactification in the string theory is of order of the Planckian scale. In recently proposed models with large extra-dimensions it is also assumed that the spacetime has more than 3 spatial dimensions. A new feature is that the size of the extra dimensions can be much larger than the Planckian size, 10^{-33}cm (up to 0.1mm) . In order to escape contradictions with observations it is usually assumed that the matter and fields (except the gravitational one) are confined to a four-dimensional brane representing our world, while the gravity can propagate in the bulk. In so called ADD models [1, 2] the extra dimensions are flat. In the Randall-Sundrum models [3, 4] the bulk 5D spacetime is curved and it has anti-deSitter asymptotics. Black holes in the string theory and in the models with larger extra dimensions play an important role serving as probes of extra dimension. Study of higher dimensional black holes is a very important problem of the modern theoretical and mathematical physics.

One of the main features of the models with large extra dimensions is a prediction that gravity becomes strong at small distances. This conclusion implies that for the particle collision with the energy of the order of TeV the gravitational channel would be as important as the electroweak channel of the interaction. Under these conditions two qualitatively new effects are possible: (1) bulk emission of gravitons, and (2) mini-black-hole production. These effects have been widely discussed in the connection with the expected new data at the Large Hadronic Collider (LHC) (see e.g. [5] and references therein). Hawking radiation produced by such mini black holes has several observable features which allows one, in principle, to single out such events in observations [6, 7, 8].

When the gravitational radius of a black hole is much smaller than the size of extra dimensions and the tension of the brane is neglected one can consider a the black hole as an isolated one. Such a black hole is described by a solution of the higher dimensional Einstein equations which is asymptotically flat or has (A)dS asymptotics. It was shown that besides 'standard' black holes with a spherical topology of the horizon, in higher dimensions there exist black objects with a different horizon topology. Black rings [9] and black saturns [10] are examples of such solutions in the 5D spacetime. For a review of higher dimensional black objects and their properties see, e.g., [11].

The most general known higher dimensional black hole solution of the Einstein equations with the spherical topology of the horizon is a *Kerr-NUT-(A)dS metric* which was discovered recently [12]. The metric belongs to the special algebraic type **D** [13] of the higher-dimensional algebraic classification [14, 15, 16]. This makes these metrics quite different from the black ring and black saturn type solutions, which are of type I_i [16]. In this paper we review properties of the higher dimensional black hole solutions and, especially, their hidden symmetries. We demonstrate that the properties of the higher dimensional black holes, in many aspects are similar to the properties of the four dimensional black holes.

An isolated stationary black hole in 4-dimensional asymptotically flat spacetime is uniquely specified by two parameters, its mass and angular momentum. The corresponding Kerr metric possesses a number of properties, which was called by Chandrasekhar 'miraculous'. In particular the Kerr metric allows the separation of variables in the geodesic Hamilton-Jacobi equation and a massless field equations. These properties look 'miraculous' since the spacetime symmetries of the Kerr metric are not sufficient to explain them. Really, spacetime symmetries are 'responsible' for two integrals of motion, the energy and the azimuthal component of the angular momentum. This, together with the conservation of $\boldsymbol{p}^2$, gives only 3 integrals of motion. Carter [17, 18] constructed the fourth required integral of motion, which is quadratic in momentum and is connected with the Killing tensor [19]. Penrose and Floyd [20] showed that this Killing tensor is a 'square' of an antisymmetric Killing-Yano tensor [21].

In many aspects a Killing-Yano tensor is more fundamental than a Killing tensor. Namely, its 'square' is always Killing tensor, but the opposite is not generally true (see, e.g., [22]). It was shown by Collinson [23] a 4–dimensional vacuum spacetime which admits a non–degenerate Killing-Yano tensor is of the type **D**. All the vacuum type **D** solutions were obtained by Kinnersley [24]. Demianski and Francaviglia [25] showed that in the absence of the acceleration these solutions admit Killing and Killing-Yano tensors. It should be also mentioned that if a spacetime admits a non–degenerate Killing-Yano tensor it always has at least one Killing vector [26].

Separability of massless and massive field equations (including the gravitational perturbations) in the 4D Kerr metric [27, 28, 29, 30, 31] is a direct consequence of the hidden symmetries. The separability property plays a key role in study of properties of rotating black holes, including the proof of their stability and the calculation of Hawking radiation.

In 2003 is was discovered that five-dimensional Myers-Perry black holes possess a Killing tensor [32, 33]. This makes geodesic equations completely integrable (see e.g [34] where the cross-section of particles and photons for 5D black holes was studied.) After this a lot of attempts were made to study the hidden symmetries of the higher dimensional black holes. In the most of these publications an additional assumption was made. Namely special conditions were imposed on the rotation parameters of a black hole which enhance the symmetry of these solutions. In 2007 it was discovered that Myers-Perry metrics with arbitrary rotation parameters, as well as a general Kerr-NUT-(A)dS metrics possesses a closed conformal Killing-Yano 2-form [35, 36]. This result was used to demonstrate that the general high dimensional black hole solutions in many aspects are similar to their four dimensional 'cousin'. This paper contains a brief review of the recent development of the theory of higher dimensional black holes and hidden symmetries (see also [37, 38, 39]).

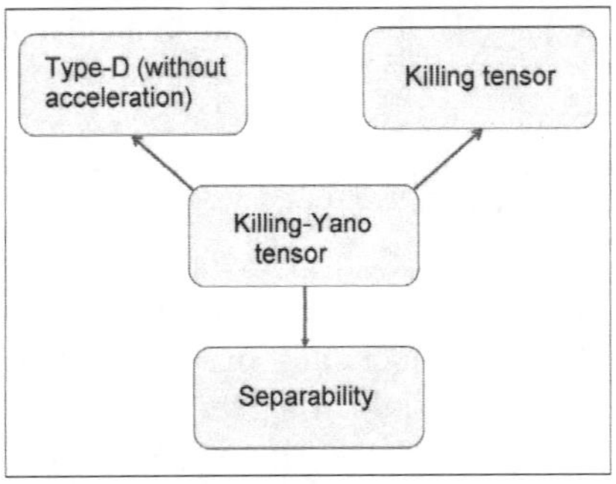

Figure 1. Killing-Yano tensors in 4D spacetime and their properties.

2. Hidden symmetries

The concept of symmetries is one of the most powerful tools of modern theoretical physics. Noether's theorem relates continuous symmetries to conservation laws. The most fundamental of them are connected with the symmetries of the background spacetime. A curved spacetime possesses a symmetry if there exists a diffeomorphism preserving the geometry. Consider a one-parameter family of continuous transformations

$$x^a \to \bar{x}^a = F^a(x,t)\,. \tag{1}$$

Denote by $\boldsymbol{\xi}$ a vector field generating these transformations

$$\xi^a = \left.\frac{dF^a}{dt}\right|_{t=0}\,. \tag{2}$$

Invariance of the metric $\boldsymbol{g}$ under the transformations Eq.(1) implies that

$$\mathcal{L}_{\boldsymbol{\xi}} g_{ab} = 0\,, \tag{3}$$

where $\mathcal{L}_{\boldsymbol{\xi}}$ is the Lie derivative. A generator $\boldsymbol{\xi}$ of the continuous symmetry transformation (isometry) is called a *Killing vector*. The equation Eq.(3) can be identically rewritten in the form

$$\xi_{(a;b)} = 0\,. \tag{4}$$

In the presence of the symmetry generated by the Killing vector $\boldsymbol{\xi}$ geodesic equations possess an integral of motion

$$P_\xi = \xi_a u^a\,, \tag{5}$$

where $u^a = dx^a/d\lambda$ is a tangent vector to a geodesic and λ is an affine parameter. Really,

$$\frac{dP_\xi}{d\lambda} = u^b u^a \xi_{a;b} + u^b u^a{}_{;b}\xi_a = 0\,. \tag{6}$$

The first term in the right hand side vanishes because of the Killing equation Eq.(4), while the second one vanishes because of the geodesic equation of motion $u^b u^a{}_{;b} = 0$.

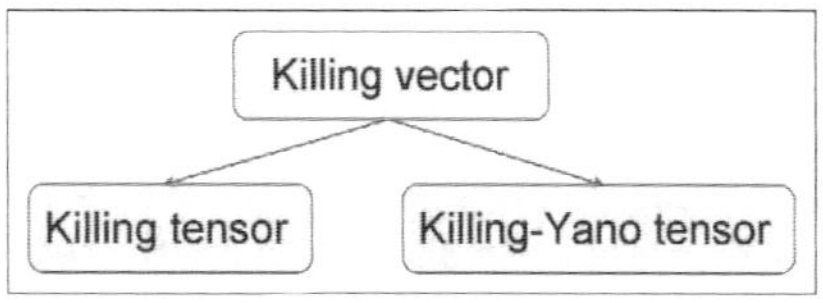

Figure 2. Possible generalizations of the Killing vector.

A Killing tensor $K_{a_1\ldots a_k}$ is a natural *symmetric generalization* of the Killing vector. Let us assume that for any geodesic with a tangent vector u^a the following object

$$P_K = K_{a_1\ldots a_k} u^{a_1} \ldots u^{a_k} \tag{7}$$

is conserved

$$\frac{dP_K}{d\lambda} = 0\,. \tag{8}$$

For a geodesic motion

$$\frac{dP_K}{d\lambda} \equiv u^b u^{a_1} \ldots u^{a_k} K_{a_1\ldots a_k;b} \tag{9}$$

Since this relation is valid for an arbitrary $\boldsymbol{u}$ one has

$$K_{(a_1\ldots a_k;b)} = 0\,. \tag{10}$$

A symmetric tensor $K_{(a_1\ldots a_k)}$ obeying the relation Eq.(10) is called a *Killing tensor.*

A Killing-Yano tensor $f_{a_1\ldots a_k}$ is an *antisymmetric generalization* of the Killing vector. Let us assume that for any geodesic with a tangent vector u^a the following object

$$P_{a_1\ldots a_{k-1}} = f_{a_1\ldots a_k} u^{a_k} \tag{11}$$

is parallel propagated along the geodesic

$$u^b P_{a_1\ldots a_{k-1};b} = 0\,. \tag{12}$$

Using the geodesic equation one obtains

$$f_{a_1\ldots a_k;b} u^{a_k} u^b = 0\,. \tag{13}$$

Since this relation is valid for an arbitrary u^a it implies

$$f_{a_1\ldots (a_k;b)} = 0\,. \tag{14}$$

A skewsymmetric tensor $f_{a_1\ldots a_k}$ obeying this relation is called a *Killing-Yano tensor.*

It is easy to show that if $f_{a_1\ldots a_k}$ is a Killing-Yano tensor then

$$K_{ab} = f_{aa_1\ldots a_{k-1}} f_b{}^{a_1\ldots a_{k-1}} \tag{15}$$

is a Killing tensor.

An important generalization of the symmetry Eq.(3) is a *conformal symmetry.* A conformal Killing vector $\boldsymbol{\xi}$ generating such a transformation obeys the equation

$$\mathcal{L}_{\boldsymbol{\xi}} g_{ab} = \beta(x) g_{ab}\,, \tag{16}$$

or, equivalently

$$\xi_{(a;b)} = \tilde{\xi} g_{ab}\,, \qquad \tilde{\xi} = D^{-1}\xi^{b}{}_{;b}\,. \tag{17}$$

The corresponding expression Eq.(5) is conserved for null geodesics. The conformal generalizations of the Killing and Killing-Yano tensors are defined as follows. A *symmetric* tensor $K_{a_1a_2\ldots a_p} = K_{(a_1a_2\ldots a_p)}$ is called a conformal Killing tensor if it obeys the equation

$$K_{(a_1a_2\ldots a_p;b)} = g_{b(a_1}\tilde{K}_{a_2\ldots a_p)}\,. \tag{18}$$

It is easy to show that a symmetrized tensor product of two conformal Killing tensors is again a conformal Killing tensor. Similarly, a tensor product of two Killing tensors is again a Killing tensor. A (conformal) Killing tensor is called *reducible* if it can be written as a linear combination of tensor products of lower rank (conformal) Killing tensor. If a Killing tensor of rank q is reducible, the corresponding conserved quantity for a geodesic motion, which is a polynomial of rank q in momentum, can be written as a linear combination of products of conserved quantities of lower than q powers in momentum. It means that a reducible Killing tensor does not generate any new *independent* conserved quantities.

An *antisymmetric* generalization of the conformal Killing vector is known as a *conformal Killing-Yano tensor.* An antisymmetric tensor $h_{a_1a_2\ldots a_p} = h_{[a_1a_2\ldots a_p]}$ is called a *conformal Killing-Yano tensor* (or, briefly, CKY tensor) if it obeys the following equation [40]

$$\nabla_{(a_1}h_{a_2)a_3\ldots a_{p+1}} = g_{a_1a_2}\tilde{h}_{a_3\ldots a_{p+1}} - (p-1)g_{[a_3(a_1}\tilde{h}_{a_2)\ldots a_{p+1}]}\,. \tag{19}$$

By tracing the both sides of this equation one obtains the following expression for $\tilde{\boldsymbol{h}}$

$$\tilde{h}_{a_2a_3\ldots a_p} = \frac{1}{D-p+1}\nabla^{a_1}h_{a_1a_2\ldots a_p}\,. \tag{20}$$

3. Conformal Killing-Yano tensors

Let us discuss properties of the conformal Killing-Yano (CKY) tensors in more details. The CKY tensors are forms and operations with them are greatly simplified if one uses the "language" of differential forms. We just remind some of the relations we use in the present paper. If $\boldsymbol{\alpha}_p$ and $\boldsymbol{\beta}_q$ are p- and q-forms, respectively, the external derivative (d) of their external product ($\wedge$) obeys a relation

$$d(\boldsymbol{\alpha}_p \wedge \boldsymbol{\beta}_q) = d\boldsymbol{\alpha}_p \wedge \boldsymbol{\beta}_q + (-1)^p \boldsymbol{\alpha}_p \wedge d\boldsymbol{\beta}_q\,. \tag{21}$$

A Hodge dual $*\boldsymbol{\alpha}_p$ of the p-form $\boldsymbol{\alpha}_p$ is $(D-p)$-form defined as

$$*\boldsymbol{\alpha}_p \Leftrightarrow (*\alpha)_{a_1\ldots a_{D-p}} = \frac{1}{p!}\alpha^{b_1\ldots b_p}e_{b_1\ldots b_p a_1\ldots a_{D-p}}\,, \tag{22}$$

where $e_{a_1\ldots a_D}$ is a totally anti-symmetric tensor. The exterior *co-derivative* δ is defined as follows

$$\delta\boldsymbol{\alpha}_p = (-1)^p\epsilon_p * d * \boldsymbol{\alpha}_p\,, \;\; \epsilon_p = (-1)^{p(D-p)}\frac{\det(g)}{|\det(g)|}\,. \tag{23}$$

One also has $* * \boldsymbol{\alpha}_p = \epsilon_p\boldsymbol{\alpha}_p$.

If $\{\boldsymbol{e}_a\}$ is a basis of vectors, then dual basis of 1-forms $\boldsymbol{\omega}^a$ is defined by the relations $\boldsymbol{\omega}^a(\boldsymbol{e}_b) = \delta^a_b$. We denote $\eta_{ab} = g(\boldsymbol{e}_a, \boldsymbol{e}_b)$ and by η^{ab} the inverse matrix. Then the operations with the indices enumerating the basic vectors and forms are performed by using these matrices. In particular, $\boldsymbol{e}^a = \eta^{ab}\boldsymbol{e}_b$, and so on. We denote a covariant derivative along the vector $\boldsymbol{e}_a$ by $\nabla_a = \nabla_{\boldsymbol{e}_a}$. One has

$$d = \boldsymbol{\omega}^a \wedge \nabla_a\,, \;\; \delta = -\boldsymbol{e}^a \lrcorner \nabla_a\,. \tag{24}$$

In the tensor notations the 'hook' operator applied to a p-form $\boldsymbol{\alpha}_p$ corresponds to a contraction

$$\boldsymbol{X} \lrcorner \boldsymbol{\alpha}_p \Leftrightarrow X^{a_1} \alpha_{a_1 a_2 \dots a_p} \,. \tag{25}$$

For a given vector $\boldsymbol{X}$ one defines $\boldsymbol{X}^\flat$ as a corresponding 1-form with the components $(X^\flat)_a = g_{ab}X^b$. In particular, one has $\eta^{ab}(\boldsymbol{e}_b)^\flat = \boldsymbol{\omega}^a$.

The definition Eq.(19) of the CKY tensor $\boldsymbol{h}$ (which is a p-form) is equivalent to the following equation (see e.g. [41, 42])

$$\nabla_{\boldsymbol{X}} \boldsymbol{h} = \frac{1}{p+1} \boldsymbol{X} \lrcorner d\boldsymbol{h} - \frac{1}{D-p+1} \boldsymbol{X}^\flat \wedge \delta \boldsymbol{h} \,. \tag{26}$$

If $\delta\boldsymbol{h} = 0$ this is an equation for the Killing-Yano tensor.

The CKY tensors possess the following properties:

(i) If $\boldsymbol{\alpha}$ is a CKY tensor then $*\boldsymbol{\alpha}$ is also a CKY tensor;

(ii) If $\boldsymbol{\alpha}$ is a closed CKY tensor ($d\boldsymbol{\alpha} = 0$) then $*\boldsymbol{\alpha}$ is a Killing-Yano (KY) tensor;

(iii) If $\boldsymbol{\alpha}$ and $\boldsymbol{\beta}$ are closed CKY tensors then $\boldsymbol{\alpha} \wedge \boldsymbol{\beta}$ is also a closed CKY tensor.

The first two properties can be proved by using a relation

$$\boldsymbol{X} \lrcorner * \boldsymbol{\omega} = *(\boldsymbol{\omega} \wedge \boldsymbol{X}^\flat) \,. \tag{27}$$

Applying this relation to Eq.(26) one has

$$\nabla_{\boldsymbol{X}} (*\boldsymbol{h}) = \frac{1}{p_*+1} \boldsymbol{X} \lrcorner d(*\boldsymbol{h}) - \frac{1}{D-p_*+1} \boldsymbol{X}^\flat \wedge \delta(*\boldsymbol{h}) \,, \quad p_* = D - p \,. \tag{28}$$

It means that a Hodge dual $*\boldsymbol{h}$ of a CKY tensor $\boldsymbol{h}$ is again a CKY tensor. Moreover, if the CKY is closed, $d\boldsymbol{h} = 0$, then its dual $(D-p)$-form $f = *\boldsymbol{h}$ is a Killing-Yano tensor. The proof of the third property can be found in [37, 43]. This property means that the closed CKY tensors form an algebra.

4. Principal CKY tensor and Killing-Yano tower

Consider a D-dimensional spacetime. We write

$$D = 2n + \varepsilon \,, \tag{29}$$

where $\varepsilon = 1$ for the odd number of dimensions and and $\varepsilon = 0$ for the even number. Consider a 2-form h_{ab} which is a CKY tensor. We assume that it is closed, $d\boldsymbol{h} = 0$, and non-degenerate, that is it has a matrix rank $2n$. We call such a tensor a *principal CKY tensor*. The principal CKY tensor obeys the following equation

$$\nabla_X \boldsymbol{h} = \boldsymbol{X}^\flat \wedge \boldsymbol{\xi}^\flat \,, \quad \xi_b = \frac{1}{D-1} \nabla_d h^d{}_b \,, \tag{30}$$

where $\boldsymbol{X}$ is an arbitrary vector field.

Starting with the principal CKY tensor one can construct a a Killing-Yano tower of tensors [43, 44]. The following diagram illustrates this construction.

The first row of this diagram contains external powers $\boldsymbol{h}^{\wedge k}$ of the principal CKY tensor, which again are closed CKY tensors. Taking the Hodge dual of these tensors one obtains a set of Killing-Yano tensors $\boldsymbol{k}_i$. 'Squares' of $\boldsymbol{k}_i$ give a set of second rank Killing tensors $\boldsymbol{K}^i$. For even D the last column can be omitted since $\boldsymbol{h}^{\wedge n}$ is proportional to a totally antisymmetric tensor,

$$h \Rightarrow \quad h^{\wedge 2} = h \wedge h \quad \Rightarrow \ldots \Rightarrow h^{\wedge j} = \underbrace{h \wedge h \wedge \ldots \wedge h}_{j\ \text{times}} \quad \Rightarrow h^{\wedge n} = \underbrace{h \wedge h \wedge \ldots \wedge h}_{n\ \text{times}}$$

$$2 \qquad 4 \qquad 2j \qquad 2n$$

$$k_1 = {*h} \qquad k_2 = {*h^{\wedge 2}} \qquad k_j = {*h^{\wedge j}} \qquad k_n = {*h^{\wedge n}}$$

$$D-2 \qquad D-4 \qquad D-2j \qquad D-2n = \varepsilon$$

$$K^1 = k_1 \bullet k_1 \quad K^2 = k_2 \bullet k_2 \qquad K^j = k_j \bullet k_j \qquad K^n = k_n \bullet k_n$$

Figure 3. Killing-Yano tower

and hence it does not produce a non-trivial Killing tensor. For the odd case, the Killing tensor $\boldsymbol{K}^n$ is a product of two Killing vectors and hence it is reducible. The first $n-1$ Killing tensors are irreducible. The metric itself is also a Killing tensor, $\boldsymbol{K}^0 = \boldsymbol{g}$. We call the Killing-Yano tower *extended* if the metric is included into it. Thus the extended Killing-Yano tower allows one to construct n independent quadratic in momenta conserved quantities for a geodesic motion.

Besides a tower of the rank 2 Killing tensors, the principal CKY tensor generates a set of the Killing vectors. To demonstrate this let us notice that for a CKY tensor $\boldsymbol{h}$ of rank-2 the vector

$$\xi^{(0)a} = \frac{1}{D-1} \nabla_b h^{ab}, \tag{31}$$

obeys the following equation [40, 45]

$$\xi^{(0)}_{(a;b)} = -\frac{1}{D-2} R_{c(a} h_{b)}{}^c . \tag{32}$$

Thus, in an Einstein space, that is when $R_{ab} = \Lambda g_{ab}$, $\boldsymbol{\xi}^{(0)}$ is the Killing vector. It can be shown that even if the Einstein equations with the cosmological constant are not imposed the vector $\xi^{(0)a}$ constructed for the principal CKY tensor is always the Killing vector[46]. We call it a *primary Killing vector*. Acting by the Killing tensors $\boldsymbol{K}^i$ on the primary Killing vector one obtains a set of the independent commuting Killing vectors[43, 46, 47]

$$\xi^i_a = K^i_{ab} \xi^{(0)b} . \tag{33}$$

There Killing vectors (with an additional Killing vector $\boldsymbol{k}_n$ for $\varepsilon = 1$) give $n + \varepsilon$ conserved first order in momentum quantities for the geodesic motion. These conserved quantities together with $n-$ conserved quantities connected with the Killing tensors (including the metric $\boldsymbol{g}$) form a set of $D = 2n + \varepsilon$ conserved quantities.

5. Darboux basis

Let us consider an eigenvalue problem for a conformal Killing tensor $\boldsymbol{H}$ associated with $\boldsymbol{h}$

$$H_{ab} \equiv h_{ac} h_b{}^c . \tag{34}$$

It is easy to show that in the Euclidean domain its eigenvalues x^2,

$$H^a_{\ b} v^b = x^2 v^a , \tag{35}$$

are real and non-negative. Using a modified Gram–Schmidt procedure it is possible to show that there exists such an orthonormal basis in which the operator $\boldsymbol{h}$ has the following structure:

$$\mathrm{diag}(\mathbf{0}, \mathbf{\Lambda}_1, \dots, \mathbf{\Lambda}_p)\,, \tag{36}$$

where $\mathbf{\Lambda}_i$ are matrices of the form

$$\Lambda_i = \begin{pmatrix} 0 & -x_i\boldsymbol{I} \\ x_i\boldsymbol{I} & 0 \end{pmatrix}, \tag{37}$$

$\mathbf{0}$ is a zero matrix and $\boldsymbol{I}$ are unit matrices. For a non-degenerate principal CKY tensor $\boldsymbol{h}$ the eigenspaces corresponding to the eigenvalues x_i are two dimensional (*Darboux subspaces*). Since the matrix rank of $\boldsymbol{h}$ is $2n$ the eigenvalue 0 is present only for $\varepsilon = 1$, and the corresponding eigenspace in this case is one dimensional. We assume also that $\boldsymbol{h}$ is non-degenerate, that is $p = n$ and the eigenspaces of $\boldsymbol{H}$ are two-dimensional so that the matrices $\mathbf{\Lambda}$ has the form

$$\Lambda_i = \begin{pmatrix} 0 & -x_i \\ x_i & 0 \end{pmatrix}. \tag{38}$$

(See Section 9 for a discussion of a degenerate case.)

We denote by $\boldsymbol{e}_{\hat{\mu}}$ and $\tilde{\boldsymbol{e}}_{\hat{\mu}} \equiv \boldsymbol{e}_{\hat{n}+\hat{\mu}}$, where $\mu = 1, \dots, n$, mutually orthogonal unit vectors in the 2D Darboux space corresponding to the eigenvalue x_μ. For $\varepsilon = 1$ we introduce also a unit vector $\boldsymbol{e}_{\hat{0}}$ in the subspace corresponding to zero eigenvalue. The basis of the dual forms we denote by $\boldsymbol{\omega}^{\hat{\mu}}$ and $\tilde{\boldsymbol{\omega}}^{\hat{\mu}} \equiv \boldsymbol{\omega}^{\hat{n}+\hat{\mu}}$ (and $\boldsymbol{\omega}^{\hat{0}}$ if $\varepsilon = 1$). The metric $\boldsymbol{g}$ and the principal CKY tensor $\boldsymbol{h}$ in this basis take the form

$$\boldsymbol{g} \;=\; = \sum_{\mu=1}^{n} (\boldsymbol{\omega}^{\hat{\mu}}\boldsymbol{\omega}^{\hat{\mu}} + \tilde{\boldsymbol{\omega}}^{\hat{\mu}}\tilde{\boldsymbol{\omega}}^{\hat{\mu}}) + \varepsilon \boldsymbol{\omega}^{\hat{0}}\boldsymbol{\omega}^{\hat{0}}\,, \tag{39}$$

$$\boldsymbol{h} \;=\; \sum_{\mu=1}^{n} x_\mu \boldsymbol{\omega}^{\hat{\mu}} \wedge \tilde{\boldsymbol{\omega}}^{\hat{\mu}}\,. \tag{40}$$

Starting with the principal CKY tensor $\boldsymbol{h}$ written in the *canonical form*, one can find the explicit expressions for the other associated tensors of the Killing tower. In particular, for the associated Killing tensors one has

$$\boldsymbol{K}^{(j)} \;=\; \sum_{\mu=1}^{n} A^{(j)}_{\mu} (\boldsymbol{\omega}^{\hat{\mu}}\boldsymbol{\omega}^{\hat{\mu}} + \tilde{\boldsymbol{\omega}}^{\hat{\mu}}\tilde{\boldsymbol{\omega}}^{\hat{\mu}}) + \varepsilon A^{(j)} \boldsymbol{\omega}^{\hat{0}}\boldsymbol{\omega}^{\hat{0}}, \tag{41}$$

$$A^{(j)} \;=\; \sum_{\nu_1<\dots<\nu_j} x^2_{\nu_1} \dots x^2_{\nu_j}\,, \quad A^{(j)}_{\mu} = \sum_{\substack{\nu_1<\dots<\nu_j \\ \nu_i \neq \mu}} x^2_{\nu_1} \dots x^2_{\nu_j}\,. \tag{42}$$

It should be emphasized that Killing tensors are defined up to an arbitrary constant factor. In order to obtain the above expressions this factors were specially chosen, so that $\boldsymbol{K}^{(j)}$ in Eq.(41) contains an extra constant factors with respect to $\boldsymbol{K}^{j}$ from the diagram 3 for the Killing tower.

6. Canonical form of the metric

In the presence of the principal CKY tensor, one can use its independent eigenvalues x_i as n coordinates. Moreover, the principal CKY tensor generates $n + \varepsilon$ Killing vectors, one primary $\xi^{(0)}$, Eq.(31), and $n - 1 + \varepsilon$ secondary $\xi^{(i)}$, defined by Eq.(33) with $\boldsymbol{K}^{i}$ substituted by $\boldsymbol{K}^{(i)}$, ones. Denote by ψ_0 and ψ_i the corresponding Killing parameters,

$$\xi^{(0)} = \partial_{\psi_0}\,, \quad \xi^{(i)} = \partial_{\psi_i}\,. \tag{43}$$

A set of n *essential* (Darboux) coordinates x_i and $n+\varepsilon$ *Killing* coordinates $\{\psi_0, \psi_i\}$ can be used as $D = 2n+\varepsilon$ coordinates associated with the principal CKY tensor. We call these coordinates *canonical*. It can be shown that the metric of a spacetime, which admits a principal CKY tensor, can be written in the canonical coordinates in the form[1]

$$\boldsymbol{\omega}^{\hat{\mu}} = \frac{1}{\sqrt{Q_\mu}}dx_\mu\,, \quad \tilde{\boldsymbol{\omega}}^{\hat{\mu}} = \sqrt{Q_\mu}\sum_{k=0}^{n-1} A_\mu^{(k)}\,d\psi_k\,, \quad \boldsymbol{\omega}^0 = (-c/A^{(n)})^{1/2}\sum_{k=0}^{n} A^{(k)}\boldsymbol{d}\psi_k\,, \tag{44}$$

$$Q_\mu = X_\mu/U_\mu\,, \quad U_\mu = \prod_{\nu\neq\mu}(x_\nu^2 - x_\mu^2)\,. \tag{45}$$

This result was first proved in [49] assuming the the following two conditions

$$\mathcal{L}_\xi \boldsymbol{g} = 0\,, \qquad \mathcal{L}_\xi \boldsymbol{h} = 0\,, \tag{46}$$

where $\boldsymbol{\xi}$ is defined by Eq.(30), are satisfied. Later it was shown [46] that these conditions follow from the very existence of the principal CKY tensor and the canonical form was derived in a general case for a spacetime with the principal CKY tensor [46, 50]. Here $A_\mu^{(k)}$ and $A^{(k)}$ are given by Eq.(42), and X_μ are functions of a single argument x_μ. It should be emphasized that in the derivation of this form of the metric the Einstein equations were not used. In this sense, this is an *off-shell* result.

By imposing the D-dimensional Einstein equations with a cosmological constant one obtains that the functions X_μ are of the form[2]

$$X_\mu = \sum_{k=\varepsilon}^{n} c_k x_\mu^{2k} - 2b_\mu x_\mu^{1-\varepsilon} + \frac{\varepsilon c}{x_\mu^2}\,. \tag{47}$$

'Time' is denoted by ψ_0, azimuthal coordinates by ψ_k, $k = 1, \ldots, m = n-1+\varepsilon$, and x_μ, $\mu = 1, \ldots, n$, stand for 'radial' and latitude coordinates. The physical metric with proper signature is recovered when standard radial coordinate $r = -ix_n$ and new parameter $M = (-i)^{1+\epsilon}b_n$ are introduced. The total number of constants which enter the solution is $2n+1$: ε constants c, $n+1-\varepsilon$ constants c_k and n constants b_μ. The form of the metric is invariant under a 1-parameter scaling coordinate transformations, thus a total number of independent parameters is $D-\varepsilon$. These parameters are related to the cosmological constant, mass, angular momenta, and NUT parameters. One of them may be used to define a scale, while the other $D-1-\varepsilon$ parameters can be made dimensionless. (For more details see [12].) In the absence of NUT parameters and for vanishing cosmological constant the metric Eq.(44)-Eq.(47) reduces to the Myers-Perry metric describing an isolated rotating higher dimensional black hole in an asymptotically flat spacetime. The existence of the closed CKY tensor for Myers-Perry [51] and Kerr-NUT-(A)dS was established first in [35, 36], where it was also demonstrated that the corresponding principal CKY tensor does not depend on the parameters of the solution (*universality property*).

7. Separation of variables

The Hamilton-Jacobi equation for geodesic motion

$$\frac{\partial S}{\partial\lambda} + g^{ab}\partial_a S \partial_a S = 0\,, \tag{48}$$

[1] A similar canonical form of a $4D$ metric possessing a Killing-Yano tensor was obtained by Carter [48]

[2] The components of the curvature tensor for the metric Eq.(44) were calculated in [13].

in the Kerr-NUT-(A)dS spacetime allows a complete separation of variables [52]

$$S = -w\lambda + \sum_{k=0}^{n+\varepsilon-1} \Psi_k \psi_k + \sum_{\mu=1}^{n} S_\mu(x_\mu) \,. \tag{49}$$

The functions S_μ obey the first order ordinary differential equations

$$S_\mu'^{\,2} = \frac{V_\mu}{X_\mu} - \frac{W_\mu^2}{X_\mu^2} \,, \tag{50}$$

Here

$$W_\mu = \sum_{k=0}^{n+\varepsilon-1} \Psi_k (-x_\mu^2)^{n-1-k} \,, \quad V_\mu = \sum_{k=0}^{n+\varepsilon-1} \kappa_k (-x_\mu^2)^{n-1-k} \,. \tag{51}$$

For $\varepsilon = 1$ we put $\kappa_n = \Psi_n^2/c$. The parameters κ_k and Ψ_k are separation constants. The existence of D independent conserved constants which enter these functions implies a complete integrability of the geodesic motion equations in the spacetimes admitting the principal CKY tensor (for more details, see [43, 44]).

Similarly, the massive scalar field equation

$$\Box\Phi - \mu^2\Phi = 0 \,, \tag{52}$$

in the Kerr-NUT-(A)dS metric allows a complete separation of variables [52]. Namely, the solution can be decomposed into modes

$$\Phi = \prod_{\mu=1}^{n} R_\mu(x_\mu) \prod_{k=0}^{n+\varepsilon-1} e^{i\Psi_k \psi_k} \,. \tag{53}$$

Substitution of Eq.(53) into the equation Eq.(52) results in the following *second order ordinary differential equations* for functions $R_\mu(x_\mu)$

$$(X_\mu R_\mu')' + \varepsilon \frac{X_\mu}{x_\mu} R_\mu' + \left(V_\mu - \frac{W_\mu^2}{X_\mu} \right) R_\mu = 0 \,. \tag{54}$$

Here W_μ and V_μ are given by Eq.(51) and $\kappa_0 = -\mu^2$

In [53] it was shown that the following operators

$$\hat{\xi}_{(k)} = -i\xi^{(k)a}\partial_a \,, \quad k = 0, \dots, m \,, \tag{55}$$

$$\hat{K}_{(j)} = -\frac{1}{\sqrt{|g|}} \partial_a \Big(\sqrt{|g|} K^{(j)ab} \partial_b \Big) \,, \quad j = 0, \dots, n-1 \,, \tag{56}$$

determined by a principal CKY tensor, form a complete set of commuting operators for the Klein–Gordon equation in the Kerr-NUT-(A)dS background.

Using Eq.(56) one has $\hat{K}_{(0)} = -\Box$. Since all the operators Eq.(55)–Eq.(56) commute with one another, their common eigenvalues can be used to specify the modes. It is possible to show [53] that the eigen-vectors of these commuting operators are the modes Eq.(53) and one has

$$\hat{\xi}_{(k)}\Phi = \Psi_k \Phi \,, \quad \hat{K}_{(j)}\Phi = \kappa_j \Phi \,. \tag{57}$$

Later it was shown that the massive Dirac equation in the Kerr-NUT-(A)dS spacetime also allows the separation of variables [54]. More recently Oota and Yasui [55] demonstrated separability of the tensor type gravitational perturbations in a (generalized) Kerr-NUT-(A)dS spacetime. It was also proved that the stationary test string equations in the Kerr-NUT-(A)dS spacetime are completely integrable [56] .

8. Parallel transport of frames along geodesics

One of the additional remarkable properties of the 4D Kerr metric, discovered by Marck in 1983 [57, 58, 59], is that the equations of parallel transport can be integrated. This result allows a generalization: In a higher dimensional spacetime which admits a principal CKY tensor equations of a parallel-propagated frame along a geodesic can be solved explicitly [60, 61].

The main idea of this construction is the following. Let $\boldsymbol{u}$ be a tangent vector to a geodesic. Consider a 2-form $\boldsymbol{F}$ which is obtained by projecting $\boldsymbol{h}$ onto a subspace V orthogonal to $\boldsymbol{u}$

$$F_{ab} = P_a^c P_b^d h_{cd} = h_{ab} + u_a u^c h_{cb} + h_{ac} u^c u_b \,, \tag{58}$$

where $P_a^b = \delta_a^b + u^b u_a$ is the projector onto V. This form $\boldsymbol{F}$ can be also written as follows

$$\begin{aligned} \boldsymbol{F} &= -\boldsymbol{u} \lrcorner (\boldsymbol{u}^\flat \wedge \boldsymbol{h}) = \boldsymbol{h} + \boldsymbol{u}^\flat \wedge \boldsymbol{s} \,, && (59) \\ \boldsymbol{s} &= \boldsymbol{u} \lrcorner \boldsymbol{h} \,, \qquad s_b = u^a h_{ab} \,. && (60) \end{aligned}$$

Let us demonstrate that $\boldsymbol{F}$ is parallel propagated along a geodesic. The definition Eq.(30) of the principal CKY tensor implies

$$\nabla_u \boldsymbol{h} = \boldsymbol{u}^\flat \wedge \boldsymbol{\xi}^\flat \,. \tag{61}$$

Hence

$$\nabla_u (\boldsymbol{u}^\flat \wedge \boldsymbol{h}) = \boldsymbol{u}^\flat \wedge \nabla_u \boldsymbol{h} = \boldsymbol{u}^\flat \wedge \boldsymbol{u}^\flat \wedge \boldsymbol{\xi}^\flat = 0 \,. \tag{62}$$

Thus for a geodesic motion, $\nabla_u \boldsymbol{u} = 0$, the 2-form $\boldsymbol{F}$ is parallel propagated along the geodesic [44].

The 2-form $\boldsymbol{F}$ has its own Darboux basis, which is called *comoving*. For any geodesic the comoving basis is determined along its trajectory. Since $\boldsymbol{F}$ is parallel propagated, its eigenvalues and its Darboux subspaces, which are called the *eigenspaces* of $\boldsymbol{F}$, are parallel-transported. For a *generic* timelike (or spacelike) geodesic the eigenspaces of $\boldsymbol{F}$ are at most 2-dimensional. In fact, the eigenspaces with non-zero eigenvalues are 2-dimensional, and the zero-value eigenspace is 1-dimensional for odd number of spacetime dimensions and 2-dimensional for even. So, the comoving basis is defined up to rotations in each of the 2D eigenspaces. The *parallel-propagated* basis is a special comoving basis. It can be found by solving a set of the first order ordinary differential equations for the angles of rotation in the 2D eigenspaces. It is possible to show that these ordinary differential equations can be solved by means of separation of variables [60]. A modification of this procedure can be used to construct parallel propagated frames along *null geodesics* [61].

9. Degenerate case

Consider a closed rank-2 CKY tensor $\boldsymbol{h}$ and denote by x_μ ($\mu = 1, \cdots, n$) and ξ_i ($i = 1, \cdots, N$) its non-constant eigenvalues and the non-zero constant ones, respectively. Suppose the eigenvalues of the "square of the CKY tensor" $H = (H^a{}_b) = (-h^a{}_c h^c{}_b)$ have the following multiplicities:

$$\{\underbrace{x_1^2, \cdots, x_1^2}_{2\ell_1}, \cdots, \underbrace{x_n^2, \cdots, x_n^2}_{2\ell_n}, \underbrace{\xi_1^2, \cdots, \xi_1^2}_{2m_1}, \cdots, \underbrace{\xi_N^2, \cdots, \xi_N^2}_{2m_N}, \underbrace{0, \ldots, 0}_{K}\}, \tag{63}$$

where $D = 2(|\ell| + |m|) + K$. Here $|\ell| = \sum_{\mu=1}^n \ell_\mu$ and $|m| = \sum_{i=1}^N m_i$. In the previous consideration we assumed that the principal CKY tensor $\boldsymbol{h}$ is non-degenerate, that is

$$\ell_1 = \ldots = \ell_n = 1 \,, \quad m_1 = \ldots = m_N = 0 \,, \quad K = \varepsilon \,. \tag{64}$$

If these conditions are violated we call a principal CKY tensor *degenerate*. Recently there was obtained a general canonical form of a spacetime which admits a degenerate principal CKY tensor [47, 62]. These papers also contain a *generalized Kerr-NUT-(A)dS metrics* which are solutions of the higher dimensional Einstein equations with a cosmological constant with a arbitrary degenerate principal CKY tensor. It is interesting to notice that a 4D Taub-NUT metric belongs to this class of the solutions. An example of a generalized Kerr-NUT-(A)dS with a degenerate principal CKY tensor is a special subclass of Kerr-(A)dS solutions [63, 64]. In a general case these solutions contain $[(D-1)/2]$ different angular momenta. In the case when some of them are equal, the principal CKY tensor has constant eigenvalues [55].

At the moment not so much is known about properties and interpretation of the generalized Kerr-NUT-(A)dS solutions in higher dimensions.

10. Concluding remarks

During past 2-3 years a lot of important and interesting results concerning higher dimensional black hole solutions with the spherical topology of the horizon were obtained. It was discovered that a key role in this study is played by the principal CKY tensor. The very existence of this tensor in a spacetime restricts the form of the metric. In the non-degenerate case this canonical metric obeying the Einstein equations coincides with the Kerr-NUT-(A)dS solution. The class of metrics admitting a principal CKY tensor allows separation of variables in the Hamilton-Jacobi, Klein-Gordon, Dirac, and tensor perturbation. It is still an open question whether other massless fields equations, e.g. the Maxwell field, allow separation of variables in the (generalized) Kerr-NUT-(A)dS spacetime. Another interesting question concerns a possibility to separate variables in the higher dimensional spacetimes in the absence of the principal CKY tensor. It may happen when such a space admits a sufficient number of commuting Killing vectors and Killing tensors [65]. Especially, this problem of separation of variables is important for study black ring and black saturn solutions[3]. Separation of variables for the gravitational perturbations is important for the analysis of the stability of these higher dimensional solutions. Another interesting direction of research is study of the hidden symmetries in supergravity black holes (see e.g. [67]) and, more generally, study of the relation between hidden symmetries and supersymmetry.

Acknowledgments

The author is grateful to the Natural Sciences and Engineering Research Council of Canada and the Killam Trust for partial support.

[1] Arkani-Hamed N, Dimopoulos S and Dvali G.R. 1998 *Phys. Lett.* B **429** 263
[2] Antoniadis I, Arkani-Hamed N, Dimopoulos S and Dvali G.R. 1998 *Phys. Lett.* B **436** 257
[3] Randall L and Sundrum R 1999 *Phys. Rev. Lett.* **83** 3370
[4] Randall L and Sundrum R 1999 *Phys. Rev. Lett.* **83** 4690
[5] Giddings, S B 2007 *AIP Conf.Proc.* **957** 69 e-Print: arXiv:0709.1107
[6] Kanti P 2004 *Int. J. Mod. Phys.* A **19** 4899
[7] Kanti P 2007 e-Print: arXiv:0802.2218
[8] Park S C 2008 e-Print: arXiv:0809.2571
[9] Emparan R and Real H S 2002 *Phys. Rev. Lett.* **88** 101101
[10] Elvang H and Figueras P 2007 *JHEP* **0705** 050
[11] Emparan R and Real H S 2008 e-Print arXiv:0801.3471
[12] Chen W, Lü H and Pope C N 2006 *Class. Quantum Grav.* **23** 5323
[13] Hamamoto N, Houri T, Oota T and Yasui Y 2007 *Journ. Phys.* A **40** F177
[14] Milson R, Coley A, Pravda V and Pravdová A 2005 *Int. J. Geom. Meth. Mod. Phys.* **2** 41
[15] Coley A, Milson R, Pravda V and Pravdová A 2004 *Class. Quantum Grav.* **21** L35
[16] Coley A 2008 *Class. Quantum Grav.* **25** 033001
[17] Carter B 1968 *Phys. Rev.* **174** 1559
[18] Carter B 1968 *Commun. Math. Phys.* **10** 280

[3] See e.g. discussion of the integrability problem for geodesic motion near black rings in [66]

[19] Walker M and Penrose R 1970 *Commun. Math. Phys.* **18** 265
[20] Penrose R 1973 *Ann. N. Y. Acad. Sci.* **224** 125; Floyd R 1973 *The dynamics of Kerr fields*, PhD Thesis, London
[21] Yano K 1952 *Ann. Math.* **55** 328
[22] Collinson C D 1976 *Int. J. Theor. Phys.* **15** 311; Stephani H 1977 *Gen. Rel. Grav.* **9** 789; Ferrando J J and Sáez J A 2002 *Gen. Rel. Grav.* **35** 1191
[23] Collinson C D 1974 *Tensor* **28** 173
[24] Kinnersley W 1969 *J. Math. Phys.* **10** 1195
[25] Demianski M and Francaviglia M 1980 *Int. J. Theor. Phys.* **19** 675
[26] Dietz W and Rüdiger R 1981 *Proc. R. Soc.* A **375** 361; Dietz W and Rüdiger R 1982 *Proc. R. Soc.* A **381** 315
[27] Teukolsky S A 1972 *Phys. Rev. Lett.* **29** 1114
[28] Unruh W 1973 *Phys. Rev. Lett.* **31** 1265
[29] Teukolsky S A 1973 *Astrophys. J.* **185** 635
[30] Chandrasekhar S 1976 *Proc. Roy. Soc. London* **A349** 571
[31] Page D N 1976 *Phys. Rev.* D **14** 1509
[32] Frolov V P and Stojkovic D 2003 *Phys. Rev.* D **67** 084004
[33] Frolov V P and Stojkovic D 2003 *Phys. Rev.* D **68** 064011
[34] Gooding and Frolov A V 2008 *Phys. Rev.* D **77** 104026
[35] Frolov V P and Kubizňák D *Phys. Rev. Lett.* 982007011101
[36] Kubizňák D and Frolov V P *Class. Quantum Grav.* 242007F1
[37] Frolov V 2008 *Prog. Theor. Phys. Suppl.* **172** 210
[38] Frolov V and Kubizňák D 2008 *Class. Quantum Grav.* **25** 154005
[39] Kubizňák D 2008 e-Print: arXiv:0809.2452
[40] Tachibana S 1969 *Tôhoku Math. J.* **21** 56; Kashiwada T 1968 *Nat. Sci. Rep., Ochanomizu University* **19** 67
[41] Benn I M, Charlton P and Kress J 1997 *J. Math. Phys.* **38** 4504
[42] Kress J 1997 *Generalized Conformal Killing-Yano Tensors: Applications to Electrodynamics*, PhD Thesis, University of Newcastle
[43] Krtouš P, Kubizňák D, Page D N and Frolov V P 2007 *J. High Energy Phys.* **02** 004
[44] Page D N, Kubizňák D, Vasudevan M and Krtouš P 2007 *Phys. Rev. Lett.* **98** 061102
[45] Jezierski J 1997 *Class. Quantum Grav.* **14** 1679
[46] Krtouš P, Frolov V P , Krtouš P 2008 e-Print: arXiv:0804.4705
[47] Houri T, Oota T and Yasui Y 2008 *J. Phys.* A **41** 025204
[48] Carter B 1968 *Phys. Lett.* A **26** 399
[49] Houri T, Oota T and Yasui Y 2007 *Phys. Lett.* B **656** 214
[50] Houri T, Oota T and Yasui Y 2008 *Phys. Lett.* B **666** 391
[51] Myers R C and Perry M J 1986 *Ann. Phys. (N.Y.)* **172** 304
[52] Frolov V P, Krtouš P and Kubizňák D 2007 *J. High Energy Phys.* **02** 005
[53] Sergyeyev A and Krtouš P 2008 *Phys. Rev.* D **77** 044033
[54] Oota T and Yasui Y 2008 *Phys. Lett.* B **659** 688
[55] Oota T and Yasui e-Print: arXiv:0812.1623 (2008)
[56] Kubizňák D and Frolov V P 2008 *JHEP* **0802** 007
[57] Marck J A 1983 *Proc. R. Soc.* A **385** 431
[58] Marck J A 1983 *Phys. Lett.* B **97** 140
[59] Kamran N and Marck J A 1986 *J. Math. Phys.* **27** 1589
[60] Connell P, Frolov V and Kubizňák D 2008 *Phys. Rev.* D **78** 024042
[61] Kubizňák D, Frolov V, Krtouš, and Connell P arXiv:0809.2452v1 [gr-qc], 2008
[62] Houri T, Oota T and Yasui Y 2008 e-Print: arXiv:0805.3877
[63] Gibbons G W, Lu H, Page D, Pope C N 2004 *Phys. Rev. Lett.* **93** 171102
[64] Gibbons G W, Lu H, Page D, Pope C N 2005 *J.Geom.Phys.* **53** 49
[65] Benenti S and Francaviglia M 1979 *Gen. Rel. Grav.* **10** 79
[66] Hoskisson J 2008 *Phys. Rev.* D **78** 064039
[67] Chow D D K 2008 e-Print: arXiv:0811.1264

Recent Developments in Gravity (NEB XIII) IOP Publishing
Journal of Physics: Conference Series **189** (2009) 012016 doi:10.1088/1742-6596/189/1/012016

Oscillations and instabilities of fast rotating neutron stars

Erich Gaertig[1,2] **and Kostas D. Kokkotas**[1,3]

[1] Theoretical Astrophysics, University of Tuebingen, Tuebingen 72076, Germany
[2] MPI für Gravitationsphysik, Albert-Einstein-Institut, Hannover 30167, Germany
[3] Department of Physics, Aristotle University of Thessaloniki, Thessaloniki 54124, Greece

E-mail: gaertig@tat.physik.uni-tuebingen.de

Abstract. We present our results of our study of the oscillations and instabilities of rapidly rotating relativistic stars in the Cowling approximation. This is the first study of its kind and provides information on the effect of fast rotation on the oscillation spectra while it offers the possibility in studying the complete problem by including spacetime perturbations. Our study includes both axisymmetric and non-axisymmetric perturbations and provides limits for the onset of the secular bar mode rotational instability. We also present approximate formulae for the dependence of the oscillation spectrum from rotation. The results suggest that it is possible to extract the relativistic star's parameters from the observed gravitational wave spectrum.

1. Introduction

During their evolution relativistic stars may undergo oscillations which can become unstable under certain conditions. Rotation strongly affects these oscillations and perturbed stars can become unstable if they rotate faster than some critical velocities. During these oscillatory phases of their lives compact stars emit copious amounts of gravitational waves which together with viscosity tend to suppress the oscillations. The oscillations are divided into different families according to the restoring force [1, 2]. If pressure is the main restoring force then these modes are called (pressure) *p-modes*, buoyancy is the restoring force of another class of oscillation mode, the *g-modes* while Coriolis force is the restoring force for the *inertial modes*. Spacetime induces another family of oscillations which couple only weakly to the fluid, these are the so-called *w-modes* [3]. There are more families of modes if one assumes the presence of crust [4, 5, 1, 6] or magnetic fields [7]. For a complete description of the relativistic star perturbation theory one may refer to [2, 8, 9].

During the last two decades, these studies became even more important due to the relations of the oscillations and instabilities to the emission of gravitational waves and the possibility of getting information about the stellar parameters (mass, radius, equation of state) by the proper analysis of the oscillation spectrum [10, 11, 12, 13, 14, 15, 16]. Still, all these studies were mainly dealing with non-rotating stars, because the combination of rotation and general relativity made both the analytic and numerical studies extremely involved. This led to certain approximations in studying rotating stars in GR. The most obvious of them include the so called "slow-rotation" and the "Cowling" approximation. Actually, both approximations were known and have been used extensively in Newtonian theory of stellar oscillations [17]. Here, we use the *Cowling approximation*, where one typically neglects the perturbations of the Newtonian potential or the spacetime in the case of GR. This is a quite good approximation, both qualitatively and quantitatively, for the higher order *p-modes*, for the *g-modes* and the inertial modes while it is only qualitatively good for the *f-modes*, see for example [18].

This article is a short description of the work presented in [19] and [20].

2. The Perturbation Equations

The normal mode analysis of rotating neutron stars involves the solution of the full nonlinear set of Einstein's equations of General Relativity together with the equations of motion for the fluid (we set $G = c = 1$ here)

$$G_{\mu\nu} = 8\pi T_{\mu\nu}, \tag{1}$$

$$\nabla^{\mu} T_{\mu\nu} = 0, \tag{2}$$

where $G_{\mu\nu}$ is the Einstein-tensor and $T_{\mu\nu}$ is the energy-momentum–tensor. For a first step in solving this complicated problem, we will introduce some useful approximations.

First of all, we linearize equations (1) and (2), which means that we constrain our study to small perturbations around the equilibrium. Secondly, we will work in the so-called Cowling-approximation, which means that we will neglect all metric perturbations. This simplifies significantly the equations one has to solve since the space-time is considered as "frozen"; and we only have to solve the linearized version of (2). Under this assumption equation (2) will be written as

$$\nabla^{\mu}(\delta T_{\mu\nu}) = g^{\mu\kappa}(\partial_{\mu}\delta T_{\kappa\nu} - \Gamma^{\lambda}{}_{\kappa\mu}\delta T_{\lambda\nu} - \Gamma^{\lambda}{}_{\mu\nu}\delta T_{\kappa\lambda}) = 0, \tag{3}$$

where $g^{\mu\nu}$ is the metric describing neutron star's spacetime, $\Gamma^{\lambda}{}_{\kappa\mu}$ are the Christoffel symbols and $\delta T_{\mu\nu}$ is the Eulerian perturbation of the energy-momentum–tensor. We assume that the matter has no viscosity or shear stresses, i.e. that it can be described by a perfect fluid. Thus $\delta T_{\mu\nu}$ has the form

$$\delta T_{\mu\nu} = (\epsilon + p)(u_{\mu}\delta u_{\nu} + u_{\nu}\delta u_{\mu}) + (\delta p + \delta\epsilon)u_{\mu}u_{\nu} + \delta p g_{\mu\nu}\,. \tag{4}$$

Here ϵ is the energy-density, p is the pressure, u_{μ} is the 4-velocity and δu_{μ} are its perturbation. Energy density and pressure are not independent quantities but are connected via an equation of state (EoS) which we assume to be polytropic, i.e.

$$p \;=\; K\tilde{\rho}^{1+1/N} \quad \text{where} \quad \epsilon = \tilde{\rho} + Np\,. \tag{5}$$

Here $\tilde{\rho}$ is the rest-mass density, K the polytropic constant, N the polytropic exponent and $\Gamma = 1 + 1/N$ the polytropic index. For barotropic oscillations, both the unperturbed background and the perturbations are described by the same equation of state. In this case the pressure variation can be replaced by the corresponding energy-density variation via $\delta p = c_s^2 \delta\epsilon$, where c_s is the speed of sound

$$c_s^2 = \frac{\partial p}{\partial \epsilon}\,. \tag{6}$$

For polytropic EoS of the form (5) it is given by

$$c_s^2 = \frac{\Gamma p}{\epsilon + p}\,. \tag{7}$$

The background model is a compact relativistic star that rotates uniformly up to its Kepler-limit, i.e. the point were it is torn apart by centrifugal forces. In this work we will adopt the metric in a comoving frame of reference as it is described in [21]. In Lewis-Papapetrou coordinates $(\rho, \zeta, \varphi, t)$ the metric reads

$$ds^2 = e^{-2U}\left[e^{2k}\left(d\rho^2 + d\zeta^2\right) + W^2 d\varphi^2\right] - e^{2U}(dt + a d\varphi)^2, \tag{8}$$

where U, k, W and a are depending on ρ and ζ. In general, the (φ, t)-metric component is proportional to the function a and vanishes in the absence of rotation.

In contrast to nonrotating or slowly rotating objects, fast rotating neutron stars are deformed due to centrifugal forces. This means that it is no longer possible to decompose the angular part of the perturbation quantities in spherical harmonics as it was usually done for the non-rotating case (see [22]) or in the slow-rotation-approximation (for example [23] and [24]). Instead we can only separate the azimuthal dependence and the perturbation functions that we use will be written as

$$\begin{aligned}
(\epsilon + p) W e^U \,\delta u_\rho &= f_1(\rho, \zeta, t) e^{im\varphi},\\
(\epsilon + p) W e^U \,\delta u_\zeta &= f_2(\rho, \zeta, t) e^{im\varphi},\\
(\epsilon + p)\,\delta u_\varphi &= f_3(\rho, \zeta, t) e^{im\varphi},\\
c_s^2 e^U \,\delta\epsilon &= H(\rho, \zeta, t) e^{im\varphi},
\end{aligned} \tag{9}$$

and we end up with the following set of four equations:

$$\begin{aligned}
\frac{\partial f_1}{\partial t} &= -We^U \frac{\partial H}{\partial \rho} - \frac{e^{5U}}{W}\frac{\partial a}{\partial \rho} f_3 - \frac{W}{c_s^2}\frac{\partial U}{\partial \rho} e^U H,\\
\frac{\partial f_2}{\partial t} &= -We^U \frac{\partial H}{\partial \zeta} - \frac{e^{5U}}{W}\frac{\partial a}{\partial \zeta} f_3 - \frac{W}{c_s^2}\frac{\partial U}{\partial \zeta} e^U H,\\
\frac{\partial f_3}{\partial t} &= \frac{im}{F}\left(a c_s^2 e^{4U} f_3 + W^2 H\right) + \frac{W a c_s^2 e^{3U-2k}}{F}\left(\frac{\partial f_1}{\partial \rho} + \frac{\partial f_2}{\partial \zeta}\right),\\
&\quad - \frac{e^{3U-2k}}{F} W \frac{\partial a}{\partial \rho} f_1 - \frac{e^{3U-2k}}{F} W \frac{\partial a}{\partial \zeta} f_2,\\
\frac{\partial H}{\partial t} &= \frac{im}{F}\left(a c_s^2 e^{4U} H + c_s^2 e^{4U} f_3\right) + \frac{W c_s^2 e^{3U-2k}}{F}\left(\frac{\partial f_1}{\partial \rho} + \frac{\partial f_2}{\partial \zeta}\right),\\
&\quad - \frac{c_s^2 e^{7U-2k}}{F}\frac{a}{W}\frac{\partial a}{\partial \rho} f_1 - \frac{c_s^2 e^{7U-2k}}{F}\frac{a}{W}\frac{\partial a}{\partial \zeta} f_2,
\end{aligned} \tag{10}$$

where

$$F := a^2 c_s^2 e^{4U} - W^2. \tag{11}$$

Keep in mind that since a comoving frame of reference is used here, there is no explicit dependence on the angular velocity Ω in this set of equations.

Together with the correct boundary conditions and initial data, this system forms an appropriate description for linear perturbations on rotating compact objects; see [19] for more details.

3. Numerical Implementation

The utilization of a comoving frame of reference helped us to keep the formulation of the time evolution equations rather simple. The next step for a successful numerical treatment of the problem is the use of surface-fitted coordinates. Since the shape of a rapidly rotating equilibrium configuration is an oblate spheroid, the standard choice of spherical coordinates for covering the physical domain of interest usually leads to problems at the surface of the compact object. In this case the surface now lies somewhere in between different $r = const$ slices, so one has to find the correct way of implementing the boundary conditions there. With surface-fitted coordinates on the other hand there is not such a problem; for all rotation rates, the coordinate system keeps attached to the surface. The following Figure 1 depicts this situation for a fast rotating star.

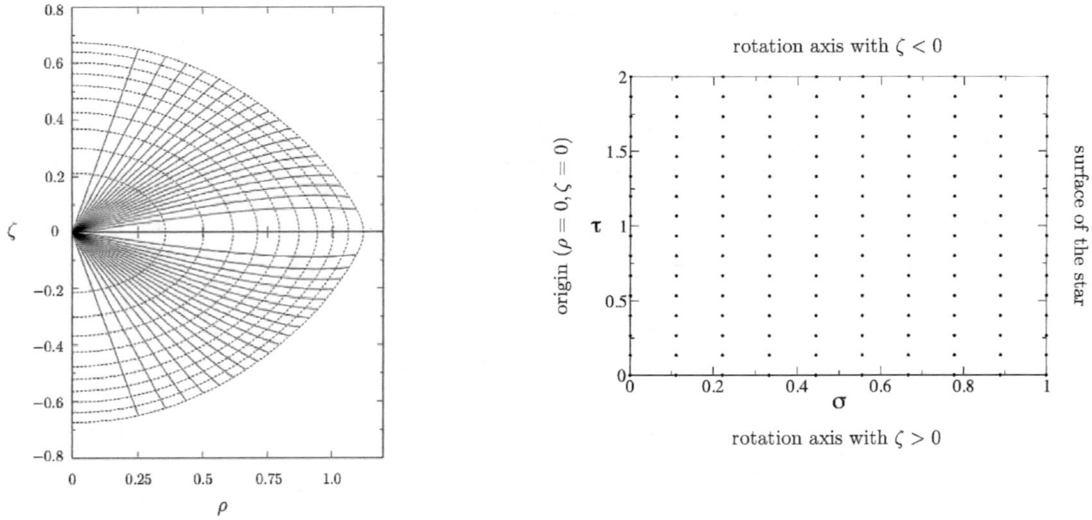

Figure 1. *Left panel:* A cut through a rapidly rotating neutron star and how it is covered by the coordinate system. $\sigma = const.$ lines are shown dotted while $\tau = const.$ lines are drawn solid *Right panel:* The transformation on a rectangular domain and the physical meaning of the boundaries

The other ingredients are standard in numerical relativity; we use second order finite-difference stencils for spatial discretization in both dimensions and an Iterated Crank-Nicholson scheme with swapped weights for time-evolution, see [25, 26] for more details.

As it turned out, the numerical evolution of the equations was unstable i.e. after the first few time steps, high frequency oscillations of exponentially growing modes developed near the center of the star (i.e. at $\sigma = 0$). This is most likely due to the presence of a coordinate singularity at the origin but also because some of the coefficients on the right-hand-sides of our evolution system get nearly singular close to the center. It should be noted that the occurrence of singular terms in general doesn't necessarily lead to a failure of the numerical scheme.

The solution to this problem was the utilization of additional viscous terms in the evolution equations. Since they do not represent any physical effect in the initial setup of the problem, they are commonly referred to as *artificial viscosity.* Here, to each of the four equations we added a Kreiss-Oliger like term of the form (see [27] for details)

$$\mathcal{V}(f) = \alpha \left(D^+_\sigma D^-_\sigma + D^+_\tau D^-_\tau \right) f \,, \tag{12}$$

where f is a perturbation variable, $\alpha = const.$ is the dissipation coefficient, (σ, τ) are the spatial coordinates and D^+, D^- are the standard forward- and backward–difference operators. By using dissipation coefficients $\alpha_i\,, i = 1 \ldots 4$ ranging from $10^{-3} - 10^{-4}$ it become possible to stabilize the time-evolution code against exponentially growing instabilities.

After a successful simulation, the real part of the complex solution is taken and a discrete Fourier-transformation at several points inside the star is performed on these data to extract the oscillation frequencies. If N is the number of points in this time series and Δt is the temporal resolution, the corresponding frequency resolution and the Nyquist-frequency are given by

$$\Delta f = \frac{1}{N\Delta t} \quad \text{and} \quad f_c = \frac{1}{2\Delta t}. \tag{13}$$

Since an explicit numerical scheme has been used for time-evolution, there are certain restrictions on the absolute value of Δt. The timestep cannot be arbitrarily large and depends on the spatial resolution of our grid (CFL-criterion). For most of the simulations, a timestep of the order of $\Delta t \approx 10^{-6}$ sec and $N \approx 10^4$ was used. The total evolution time then is $t_{max} \approx 30 - 40$ ms with $\Delta f \approx 15 - 30$ Hz and $f_c \approx 8 - 12$ kHz.

4. Results

We study first axisymmetric perturbations mainly to test the code against published results. Later we will proceed by studying non-axisymmetric perturbations for several polytropic equations of state.

4.1. Axisymmetric Perturbations

Following the procedure for the non-linear problem presented in [28], a density perturbation of the form

$$\delta\rho = A\rho_c \sin\left(\frac{\pi r}{r_s(\theta)}\right), \tag{14}$$

is used as initial data. Here A is the perturbation amplitude, ρ_c is the central density, (r, θ) denote spherical coordinates and $r_s(\theta)$ is the coordinate radius of the star. The next Figure 2 shows the variation of certain mode frequencies with rotation rate and compares it to the values found in [28]; for more details see [19].

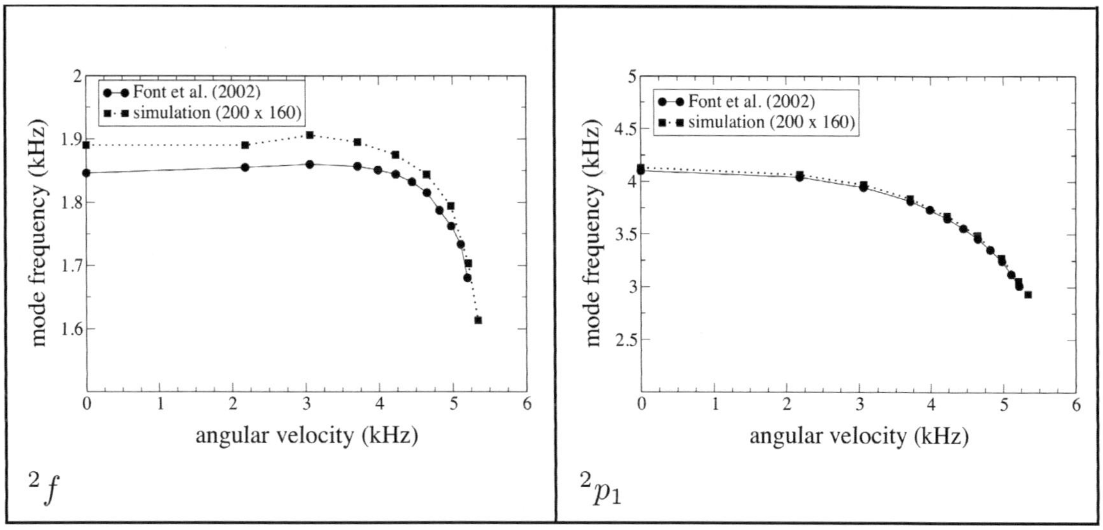

Figure 2. *Left panel:* The fundamental ($l = 2, m = 0$) frequency is plotted as function of the rotation rate *Right panel:* The same plot for the first overtone. For comparison, the results by [28] are shown as well

The agreement between these two different methods is extremely good; the relative difference of the various results is lower than 3%.

Additionally, it is also possible to extract the shape of the eigenfunction which belongs to a certain oscillation. In fact this is how one usually identifies frequency peaks in the power spectrum. In Figure 3 we show the eigenfunctions for the ${}^2f$-mode and its first overtone in the nonrotating limit. Keep in mind, that in the left panel the phase of the eigenfunction is displayed correctly while in the right panel the absolute value of the eigenfunction is depicted, i.e. the phase is not adjusted there. Again, more examples can be found in [19, 20].

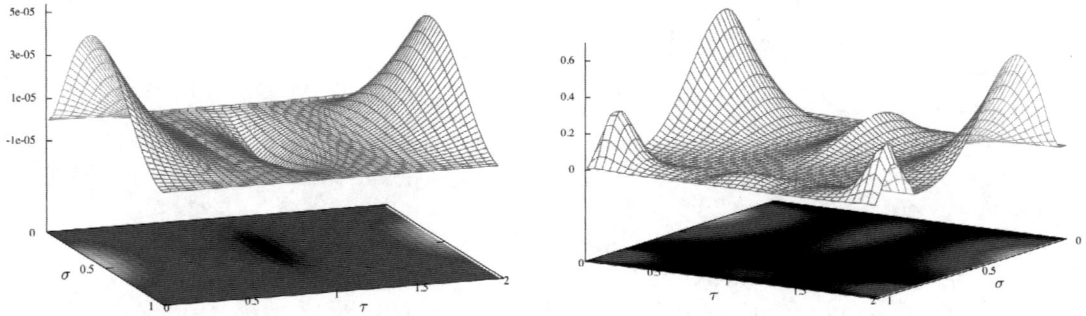

Figure 3. *Left panel:* Eigenfunction of the ${}^2f$-mode with corrected phase *Right panel:* The shape of the ${}^2p_1$-eigenfunction; the node in radial direction is clearly visible

Low-frequency modes pose a special challenge to the computer program. In order to accurately determine these frequencies one needs long evolution times, see equation (13), and therefore the code has to be long-term stable as well. Once rotation sets in, a new class of modes restored by the Coriolis force appears with frequencies $\omega \sim \Omega$. We again tested our implementation against other published results for these so-called inertial modes. Figure 4 shows a comparison between results obtained by this linear code in the Cowling approximation and a nonlinear code using the CFC-approach, see [29].

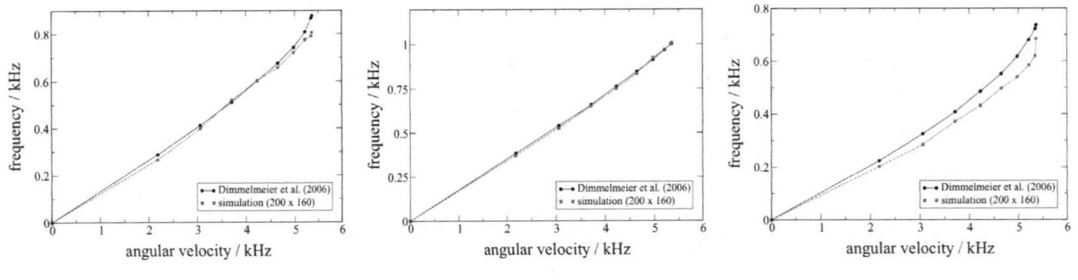

Figure 4. Change of the i_1, i_2 and i_{-2} inertial mode frequency with rotation rate. Note, that these modes are degenerate in the nonrotating limit

As in the case for pressure driven modes discussed above, the agreement between these two methods is very good. This gives us enough confidence to trust our handling of the artificial viscosity.

4.2. Nonaxisymmetric Perturbations

We will now turn to the study of non-axisymmetric oscillations on rotating compact objects with emphasis on the $m = \pm 2$ perturbations. In the nonaxisymmetric case, the frequencies of co- and counter-rotating modes split in the presence of rotation and there is a critical rotation rate for which these frequencies vanish for an inertial observer. This signals the onset of a rotational instability, the so-called CFS-instability. For higher rotation rates this oscillation will go unstable due to the emission of gravitational waves.

We study the onset of the f-mode instability for several polytropic equations of state. All results presented so far make use of $\Gamma = 2$, $K = 100$ and a fixed central rest-mass density $\rho_c = 1.28 \times 10^{-3}$ in units where $G = c = M_\odot = 1$. In the nonrotating case, this leads to a stellar model with a gravitational mass of $M = 1.4\,M_\odot$ and a circumferential radius of $R = 14.15$ km. In addition to this polytropic model we have used two realistic equations of state for which polytropic fits are available. More specifically we take $\Gamma = 2.46$ and $K = 0.00936$ for EoS A and $\Gamma = 2.34$ and $K = 0.0195$ for EoS II; these values are given in geometric units and with [km] as length scale, see also [16]. The two new configurations that were chosen are very close to their maximum mass limit; in particular the background model for EoS A has a mass of $M = 1.61\,M_\odot$ and a radius of $R = 9.51$ km while the EoS II model has a mass of $M = 1.91\,M_\odot$ and a radius of $R = 11.68$ km.

It is well known that the f-mode frequencies in the nonrotating case scale with the mean density of the background star. We therefore expect the highest frequency for the EoS A model followed by EoS II. The next Figure 5 shows our results for the different equations of state used here and confirms the prediction about the relative amplitude of the fundamental quadrupolar mode frequency.

As one can see from Figure 5, the CFS-instability is actually reached for all three different background models although the precise rotation rate where the modes become unstable varies from one EoS to another. The model labeled BU gets unstable just at its Kepler limit while the other two models only need $85 - 90\%$ of their maximum angular velocity to enter the instability.

Despite the obvious differences in the variation of the f-mode frequencies for the distinct background models, there is a nice common feature that becomes apparent in a comoving coordinate system. If we normalize the oscillation frequencies with its value in the nonrotating limit and similarly the rotation frequency with the Kepler-limit, we arrive at Figure 6.

For a large range of rotation frequencies, the three different frequency distributions for co- and counter-rotating modes can be fitted by a single 2nd-order polynomial which is drawn with green dots in Figure 6. If we write for all rotation parameters

$$\frac{f}{f_0} = 1.0 + C^{(1)}_{lm}\left(\frac{\nu}{\nu_K}\right) + C^{(2)}_{lm}\left(\frac{\nu}{\nu_K}\right)^2, \qquad (15)$$

and make a least-square fit through the data points, one arrives at $C^{(1)}_{22} = -0.25 \pm 0.02$, $C^{(2)}_{22} = -0.25 \pm 0.02$ for $m = 2$ and $C^{(1)}_{2-2} = 0.48 \pm 0.03$, $C^{(2)}_{22} = -0.55 \pm 0.04$ in the $m = -2$ case.

5. Conclusions

In this work we presented a study of the oscillation properties of fast rotating relativistic stars for both axisymmetric and non-axisymmetric perturbations. The study was based on the Cowling approximation using a 2D version of the perturbation equations. The results for axisymmetric perturbations are in excellent agreement with earlier ones while the non-axisymmetric results are the first of their kind in the literature. We demonstrated the neutral points for the onset of the CFS instability and suggested possible normalizations which can be used in order to extract the parameters of the rotating star.

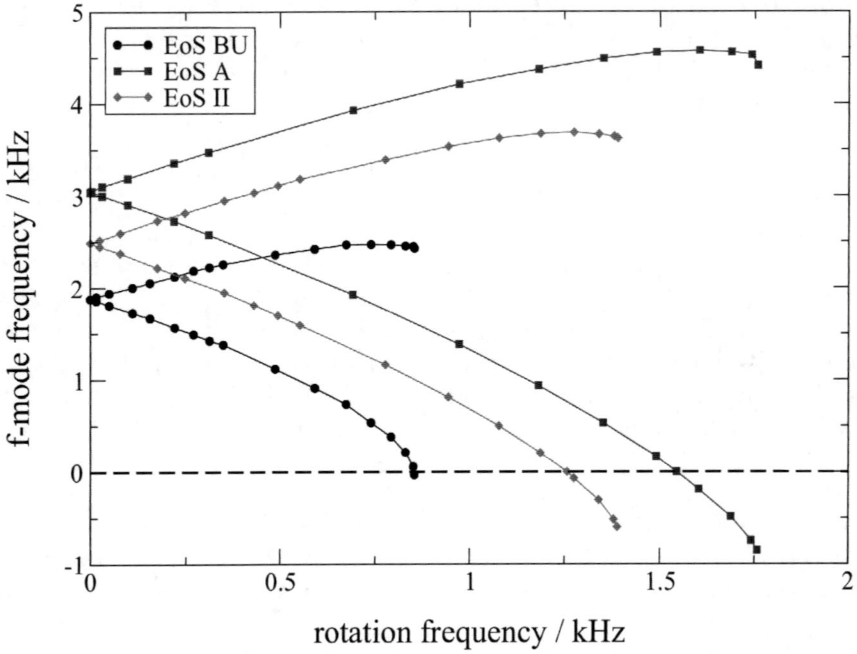

Figure 5. Splitting of the $(l = 2, |m| = 2)$ fundamental mode for the three different EoS as seen by an inertial observer.

The method presented here (evolution of the 2D linearized equations) can be extended by including the perturbations of the spacetime. This will offer the possibility in testing the earlier results [30] for the onset of the secular instability in fast rotating stars. Finally, the effect of differential rotation on the spectra can be studied both for testing earlier, mainly Newtonian, results [31, 32] but more importantly in finding the correct dependence of the frequencies on rotation as well as the neutral points for the onset of the secular instabilities. This is actually an important step since newly born neutron stars are expected to rotate differentially at least during the stages that they will be in an oscillatory phase or the even when they are secularly unstable.

Acknowledgments: This work was supported by the German Science Foundation (DFG) via SFB/TR7.

References

[1] McDermott P N, van Horn H M and Hansen C J 1988 *ApJ* **325** 725–748
[2] Kokkotas K D and Schmidt B G 1999 *Living Reviews in Relativity* **2**
[3] Kokkotas K D and Schutz B F 1992 *MNRAS* **255** 119–128
[4] Schumaker B L and Thorne K S 1983 *MNRAS* **203** 457–489
[5] McDermott P N, van Horn H M, Hansen C J and Buland R 1985 *ApJ* **297** L37–L40
[6] Vavoulidis M, Kokkotas K D and Stavridis A 2008 *MNRAS* **384** 1711–1724

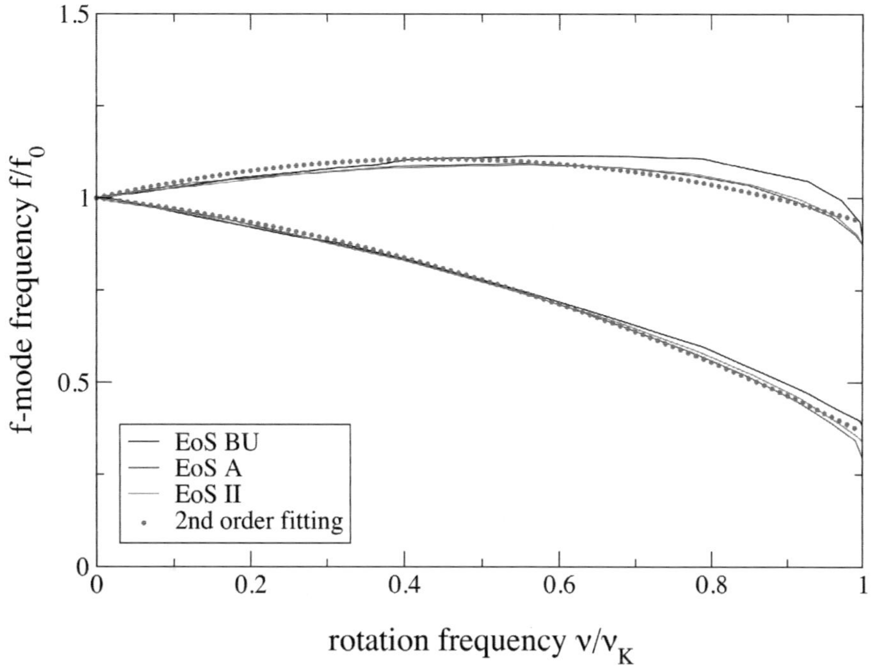

Figure 6. Same as Figure 5 but normalized and in the comoving frame.

[7] Carroll B W, Zweibel E G, Hansen C J, McDermott P N, Savedoff M P, Thomas J H and van Horn H M 1986 *ApJ* **305** 767–783
[8] Andersson N and Kokkotas K D 2001 *International Journal of Modern Physics D* **10** 381–441
[9] Stergioulas N 2003 *Living Reviews in Relativity* **6**
[10] Andersson N and Kokkotas K D 1996 *Phys. Rev. Lett.* **77** 4134–4137
[11] Andersson N and Kokkotas K D 1998 *MNRAS* **297** 493–496
[12] Benhar O, Berti E and Ferrari V 1999 *Mon. Not. Roy. Astron. Soc.* **310** 797–803
[13] Kokkotas K D, Apostolatos T A and Andersson N 2001 *MNRAS* **320** 307
[14] Benhar O, Ferrari V and Gualtieri L 2004 *Phys. Rev.* **D70** 124015
[15] Sotani H, Kohri K and Harada T 2004 *Phys. Rev. D* **69** 084008
[16] Sotani H and Kokkotas K D 2004 *Phys. Rev. D* **70** 084026
[17] Unno W, Osaki Y, Ando H, Saio H and Shibahashi H 1989 *Nonradial oscillations of stars* (Tokyo: University of Tokyo Press, 1989, 2nd ed.)
[18] Robe H 1968 *Annales d'Astrophysique* **31** 549
[19] Gaertig E and Kokkotas K D 2008 *Phys. Rev. D* **78** 064063
[20] Gaertig E 2008 *Linear Oscillations of Compact Stars in the Cowling Approximation* Ph.D. thesis Eberhard-Karls-Universität Tübingen
[21] Ansorg M, Kleinwächter A and Meinel R 2003 *A&A* **405** 711–721
[22] Thorne K S and Campolattaro A 1967 *ApJ* **149** 591
[23] Kojima Y 1992 *Phys. Rev.* **D46** 4289
[24] Stavridis A and Kokkotas K D 2005 *International Journal of Modern Physics D* **14** 543–571
[25] Teukolsky S A 2000 *Phys. Rev. D* **61** 087501
[26] Leiler G and Rezzolla L 2006 *Phys. Rev. D* **73** 044001
[27] Gustafsson B, Kreiss H O and Oliger J 1995 *Time dependent problems and difference methods* (Wiley)
[28] Font J A, Dimmelmeier H, Gupta A and Stergioulas N 2001 *MNRAS* **325** 1463–1470
[29] Dimmelmeier H, Stergioulas N and Font J A 2006 *MNRAS* **368** 1609–1630

[30] Stergioulas N and Friedman J L 1998 *Astrophys. J.* **492** 301
[31] Karino S, Yoshida S and Eriguchi Y 2001 *Phys. Rev. D* **64** 024003
[32] Yoshida S, Rezzolla L, Karino S and Eriguchi Y 2002 *ApJ* **568** L41–L44

Journal of Physics: Conference Series **189** (2009) 012017 doi:10.1088/1742-6596/189/1/012017

Scalar-field cosmological and collapse models with general self-interaction potentials

Roberto Giambò[1], Fabio Giannoni[1], Giulio Magli[2]
[1]Dipartimento di Matematica e Informatica, Università di Camerino, Italy
[2]Dipartimento di Matematica, Politecnico di Milano, Italy
E-mail: roberto.giambo@unicam.it, fabio.giannoni@unicam.it, magli@mate.polimi.it

Abstract.
We present the results of the investigation of a wide class of self-interacting, self-gravitating homogeneous scalar fields models, characterized by quite general conditions on the scalar field potential, and including both asymptotically polynomial and exponential behaviors. We show that the generic evolution is always divergent in a finite time, and this result is used to construct cosmological models as well as radiating collapsing star models of the Vaidya type – for the latter it turns out that black holes are generically formed.

1. Introduction

It is well known that scalar fields as sources of self-gravitating models play an important role as models of the early universe, but it has been one of the first test bed model for gravitational collapse also. However, there are crucial differences in the way the two situations has been treated. First, consider the cosmological models. Here, the scalar field is always assumed as *self-interacting*; in other words, the cosmological expansion is "driven" by a non-vanishing scalar field potential, and the "free" case (i.e. the case of the quadratic field Lagrangian) corresponds, as is well known, just to the case of a dynamical cosmological constant. Thus, in cosmology, the presence of a non vanishing field potential is a key point, and indeed many efforts have been made to study the possible dynamical behaviors of the models in dependence of the choice of the potential, and even to try to put in evidence possible large-scale observable effects (see e.g.[19, 20] and references therein). The situation is even more intriguing when issues from string theories come into play; indeed, here one is lead to test simple models which are asymptotically *anti*-De Sitter, and the potential for the scalar field might have, as a consequence, "non-standard" behaviors (we shall come back on this interesting issue later on). All in all, self-interacting fields with a non trivial potential are a key ingredient in cosmological models and therefore, of course, in the nature of the cosmological singularities.

Consider, now, the "astrophysical counterpart" of the cosmological solutions, that is, gravitational collapse. Here, as is well known, there still exist, unsolved, the issue of the cosmic censorship, that is whether the singularities which form at the end state of the collapse are always covered by an event horizon, or not [6, 12]. The scalar field model of course proposes itself a good test-bed also in this different scenario. However, in this context the scalar fields sources have usually been considered as *free*, i.e. minimally coupled to gravity via equivalence principle only. For this very special case the existence of naked singularities has been shown,

Journal of Physics: Conference Series **189** (2009) 012017 doi:10.1088/1742-6596/189/1/012017

but it has been also shown that such singularities are in a sense non generic with respect to the choice of the initial data [3, 4]. Only in recent years, a few results have been added to this scenario with the study of gravitational collapse of *homogeneous* self-interacting scalar fields. In these papers formation of naked singularities has been found [7, 10], but with the choice of very special potentials.

All in all, in both applications (cosmology, and gravitational collapse) it would be especially welcome a treatment of self-interacting scalar field dynamics able to treat the models in a unified manner, and to predict the qualitative behavior in dependence of the choice of the potential, viewed as an element of a space of admissible "equations of state" for the matter source. It is the aim of the present paper to provide this unified framework, at least in the special case of *homogeneous* scalar fields. Indeed, we study here the dynamical behavior of such fields for a very general class of potentials, which satisfies simple physical requirements - the potentials are bounded from below and the weak energy condition is satisfied - as well as a few more technical assumptions to be discussed below.

In the case of cosmological models, our results widely extend the recent relevant results obtained by Rendall ([17, 18]) and by Miritzis ([13]). In the case of collapsing models, we rely on slightly more general assumptions on the potential than the ones used by Foster [5], that however, using a dynamical system approach, actually imposes further conditions on some coordinate transformation of the unknown variables, that we do not need here. Of course, whether this result can actually be shown to hold also in the much more difficult case of both inhomogeneous and self-interacting scalar fields remains an open problem.

2. Expanding and collapsing models

We consider homogeneous, spatially flat spacetimes

$$g = -\mathrm{d}t \otimes \mathrm{d}t + a^2(t)\left[\mathrm{d}x^1 \otimes \mathrm{d}x^1 + \mathrm{d}x^2 \otimes \mathrm{d}x^2 + \mathrm{d}x^3 \otimes \mathrm{d}x^3\right], \tag{1}$$

where gravity is coupled to a scalar field ϕ, self-interacting with a scalar potential $V(\phi)$. We introduce the following conditions on V:

(i) $V(\phi)$ is a $\mathcal{C}^2$ function bounded from below;

(ii) The critical points of V are isolated. They are either minimum points or non degenerate maximum points.

Denoting the energy density of the scalar field by

$$\epsilon = \frac{1}{2}\dot{\phi}^2 + V(\phi), \tag{2}$$

we will consider the following system in the unknown functions $a(t), \phi(t)$:

$$\left(\frac{\dot{a}}{a}\right)^2 = \frac{\epsilon}{3}. \tag{3a}$$

$$\ddot{\phi} + V'(\phi) = -3\frac{\dot{a}}{a}\dot{\phi}, \tag{3b}$$

Introducing the sign function

$$\chi(t) := \mathrm{sgn}(\dot{a}(t)),$$

describing an expanding (respectively, collapsing or re-collapsing) model at time t, our final system is composed by

$$\frac{\dot{a}}{a} = \chi\sqrt{\frac{\epsilon}{3}}, \tag{4a}$$

$$\ddot{\phi} + V'(\phi) = -\chi\sqrt{3}\sqrt{\epsilon}\,\dot{\phi}. \tag{4b}$$

Let us observe that, using (2) and (4b), the following relation can be seen to hold:

$$\dot{\epsilon} = -\chi\sqrt{3}\sqrt{\epsilon}\,\dot{\phi}^2, \tag{5}$$

that is $\epsilon(t)$ is monotone in each interval in which a is.

We shall divide our discussion of the properties of the solutions into the two following sections, the first dealing with the expanding case and the other with the collapsing case. Of course, the two cases are not disconnected, because in the expanding case, where $\epsilon(t)$ is decreasing, there might be the possibility of reaching a vanishing ϵ in a finite time t_0. If this happens, the model will be ruled from t_0 onward by the equations for the collapsing situation, where $\epsilon(t)$ is increasing, and it will actually be proved to be divergent at some finite time for almost every choice of the initial data, thus yielding a singularity. Moreover this fact, and of course the particular form of system (4a)–(4b) allows us to perform the same analysis to study also *backwards* qualitative behavior of the solution, concluding that the expanding model originates from a big–bang singularity.

We conclude the section stating a couple of results – see [8] for the proof – that will be used throughout all the paper. The first essentially states that the velocity of all (finite energy) solutions which extend infinitely in the future asymptotically vanishes: if $\phi(t)$ a solution of (4b) that can be extended for all $t > 0$, then

$$\epsilon(t),\ V'(\phi(t)) \text{ bounded } \Rightarrow \lim_{t\to+\infty} \dot{\phi}(t) = 0, \tag{6}$$

and, in addition,

$$V''(\phi(t)) \text{ bounded } \Rightarrow \lim_{t\to+\infty} V'(\phi(t)) = 0. \tag{7}$$

The second result shows local existence/uniqueness of solutions with initial zero–energy: let ϕ_0, v_0 such that $v_0^2 + 2V(\phi_0) = 0$. Then, there exists $t_* > 0$ such that the Cauchy problem

$$\begin{cases} \ddot{\phi}(t) = -V'(\phi(t)) + \sqrt{\frac{3}{2}(\dot{\phi}(t)^2 + 2V(\phi(t)))}\,\dot{\phi}(t), \\ \phi(0) = \phi_0, \\ \dot{\phi}_0 = v_0, \end{cases} \tag{8}$$

has a unique solution $\phi(t)$ defined in $[0, t_*]$ with the property

$$\epsilon(t) = \frac{3}{4}\left(\int_0^t \dot{\phi}(s)^2\,\mathrm{d}s\right)^2, \ \forall t \in]0, t_*]. \tag{9}$$

Moreover if $(\phi_{0,m}, v_{0,m}) \to (\phi_0, v_0)$, $(v_{0,m})^2 + 2V(\phi_{0,m}) = 0$ and ϕ_m is the solution of (8) with initial data $(\phi_{0,m}, v_{0,m})$ satisfying condition (9), then $\phi_m \to \phi$ with respect to the C^2-norm in the interval $[0, t_*]$.

3. Qualitative analysis of the expanding models

In this section we analyze the global behavior of the solutions of (4a), (4b) in the expanding ($\chi = 1$) case. Throughout the present section we shall assume the following further condition on V:

$$\lim_{|\phi|\to\infty} V(\phi) = +\infty. \tag{10}$$

We begin our study considering a solution ϕ of (4b) with $\chi = 1$ such that $\epsilon(0) = \frac{1}{2}\dot{\phi}(0)^2 + V(\phi(0)) > 0$. Due to (5), there is $T > 0$ such that $\sqrt{3}\int_0^T \dot{\phi}(s)^2\,\mathrm{d}s = 2\sqrt{\epsilon(0)}$ or

$$\int_0^t \dot{\phi}(s)^2\,\mathrm{d}s < 2\sqrt{\frac{\epsilon(0)}{3}}, \qquad \forall t > 0.$$

In the latter case, $\epsilon(t) \in]0, \epsilon_0]$, $\forall t > 0$. Since V is bounded from below we obtain that the solution is defined for all $t > 0$. Then, by (10) we have that $\phi(t)$ is bounded and (6) implies that

$$\lim_{t\to+\infty} \dot{\phi}(t) = 0, \quad \lim_{t\to+\infty} V'(\phi(t)) = 0.$$

By the hypotheses made on $V(\phi)$ we immediately see that ϕ must be convergent to a critical point of V and we conclude that either there $\exists T = T(\phi(0), \dot{\phi}(0)) > 0$ such that $\epsilon(T) = 0$, or $\phi(t)$ is defined in $\mathbb{R}^+$ and there exists ϕ_* critical point of V such that:

$$\lim_{t\to+\infty} \phi(t) = \phi_*, \quad \lim_{t\to+\infty} \dot{\phi}(t) = 0.$$

We now proceed to study the two possible cases described above starting from the behavior of the solutions for which ϵ vanishes in a finite time $T > 0$. Observe first that this situation may happen only if $V(\phi(T)) \leq 0$. However, we can show that if $\inf V < 0$ this situation is generic. Indeed, reversing time direction in (8), the C^2–norm convergence result of (unique) solutions of (8) with initial zero–energy stated before allows for genericity of expanding solutions such that the energy $\epsilon(t)$ vanishes at some finite time T, in the sense that solutions of this kind exist, and that the set of initial data leading to solutions such that $V(\phi(T)) < 0$ is open. Of course, for $t > T$ the system will be ruled by the equations for the collapsing situation discussed in Section 4 below.

Now, here remains to see what happens if $\phi(t)$ converges to a critical point for $V(\phi)$. In the following, we shall study first the behavior of the solution near a local minimum point of V.

Let ϕ_* be a local minimum point of V (not necessarily non degenerate). Let us consider a sublevel of the potential $\{V(\phi) \leq \bar{V}\}$, with $V(\phi_*) < \bar{V}$, such that the connected component containing ϕ_* does not have critical points other that ϕ_*. It is easy to show that a solution $\phi(t)$ with $\phi(0)$ in this connected component and $\epsilon(0) \leq \bar{V}$ is defined on all $\mathbb{R}^+$ and

$$\lim_{t\to+\infty} \phi(t) = \phi_*, \quad \lim_{t\to+\infty} \dot{\phi}(t) = 0. \tag{11}$$

To prove the above fact, it is crucial to observe that $\epsilon(t)$ is is decreasing for expanding solutions, to show that the solution satisfies hypotheses of (6).

Whenever $V(\phi_*) = 0$ further information about the asymptotic behavior of the solution near the minimum point ϕ_* can be obtained. In [18] it is indeed proved the oscillation of ϕ around the minimum point using asymptotic analysis. If ϕ_* is a non degenerate minimum point we can prove the oscillatory character also using an alternative argument as follows. Suppose, by contradiction, that ϕ does not oscillate around ϕ_*, this means that there exists $t_1 > 0$ such that

$$\phi(t) \neq \phi_*, \, \forall t \geq t_1.$$

To fix our ideas suppose $\phi(t) > \phi_*\, \forall t \geq t_1$. Then, if $t \geq t_1$ and $\dot{\phi}(t) = 0$ it is $\ddot{\phi}(t) = -V'(\phi(t)) < 0$ (since $V' > 0$ in a right neighborhood of ϕ_*). This implies that there exists $t_2 > 0$ such that

$$\dot{\phi}(t) < 0, \, \forall t \geq t_2.$$

Moreover there are not sequences $t_k \to +\infty$ where $\ddot{\phi}(t_k) = 0$. Otherwise we should have $\dddot{\phi}(t_k) = -(V''(\phi(t_k) + \sqrt{3}\frac{\dot{\epsilon}}{\sqrt{\epsilon}})\dot{\phi}(t_k) = -(V''(\phi(t_k) - 12(\dot{\phi}(t_k))^2)\dot{\phi}(t_k))$, so $\dddot{\phi}(t_k) < 0$ for any k sufficiently large, getting a contradiction. Therefore there exists $t_3 > 0$ such that

$$\ddot{\phi}(t) > 0, \ \forall t \geq t_3.$$

Then, by (4b),

$$-V'(\phi) - \sqrt{3\epsilon}\dot{\phi} > 0, \ \forall t \geq t_3,$$

that implies (recalling that $\dot{\phi} < 0$)

$$3(\dot{\phi})^4 + 6V(\dot{\phi})^2 - (V')^2 \geq 0, \ \forall t \geq t_3,$$

and

$$(\dot{\phi})^2 \geq \sqrt{V^2 + \frac{(V')^2}{3}} - V, \ \forall t \geq t_3.$$

Now $V(\phi) = a(\phi - \phi_*)^2 + o(\phi - \phi_*)^2$ with $a > 0$, so there exists $b > 0$ and $t_4 > 0$ such that

$$|\dot{\phi}| \geq b\sqrt{\phi - \phi_*}, \ \forall t_4,$$

(recall that $\phi(t) > \phi_*$). Finally $\forall t \geq t_4$

$$\dot{\phi} \leq -b\sqrt{\phi - \phi_*},$$

and integrating we obtain,

$$\sqrt{\phi(t) - \phi_*} - \sqrt{\phi(t_4) - \phi_*} \leq -\frac{b}{2}t, \forall t \geq t_4,$$

in contradiction with $\phi(t) \geq \phi_*$ for any $t \geq t_4$.

Now let us consider the case when ϕ_* is a local non degenerate maximum point of V, with nonnegative critical value. Therefore, it follows immediately by classical theory on stability of nonlinear dynamical system in a neighborhood of an equilibrium point (see e.g. [16]) that there are only two solutions ϕ_1 and ϕ_2 of (4b) such that

$$\lim_{t \to +\infty} \phi_i(t) = \phi_*, \ \lim_{t \to +\infty} \dot{\phi}_i(t) = 0, \ i = 1, 2. \tag{12}$$

The proof relies on the study of the first order system associated to (4b)

$$\begin{cases} \dot{x} = -V'(y) - \sqrt{3E(x,y)}x, \\ \dot{y} = x, \end{cases} \tag{13}$$

where $E(x,y) = \frac{1}{2}x^2 + V(y)$. Note that the function $\sqrt{3E(x,y)}x$ is not of class C^1 in a neighborhood of $(0, y_*)$ whenever $V(y_*) = 0$. However the quantity $\sqrt{3E(x,y)}x$ is an infinitesimal of order greater than 1 in $(0, y_*)$, therefore the classical theory can be adapted to this case, studying the linear system

$$\begin{cases} \dot{\alpha} = -V''(y_*)\beta, \\ \dot{\beta} = \alpha, \end{cases}$$

obtaining the same conclusion of the case $V(y_*) > 0$.

The growth condition (10) on the potential has been introduced to avoid cases when $V(\phi)$ admits a flat plateau at infinity; on such situations the function $\phi(t)$ may possibly also diverge as $t \to \infty$, with $\epsilon(t)$ approximating the plateau of $V(\phi)$. In this case, the critical point at infinity is non hyperbolic and arguments leading for instance to (12) do not apply anymore.

With an argument that uses of the Center Manifold Theorem, a result of local asymptotical instability for critical point at infinity is found in [14, 15], where the potential is supposed to be $V_\infty(1 - e^{-\sqrt{3/2}\phi})^2$. It can be seen that the Center Manifold Theorem applies to extend results of [14, 15] to a more general situation where, as $\phi \to +\infty$, $V(\phi)$ has a finite limit, $V'(\phi) \to 0^+$, and

$$\lim_{\phi \to +\infty} \frac{V''(\phi)}{V'(\phi)},$$

exists finite (and negative).

We can summarize the results of this section as follows: let be ϕ a solution of (4b) where $\chi = 1$, with initial data such that $\epsilon(0) > 0$. Then, one of the two mutually exclusive situations occur:

either

$$\exists T = T(\phi(0), \dot\phi(0)) > 0 \text{ such that } \epsilon(T) = \frac{1}{2}(\dot\phi(T))^2 + V(\phi(T)) = 0, \tag{14}$$

and the set of initial data such that (14) is satisfied with $V(\phi(T)) < 0$ is open and not empty, while there are only a finite number of solutions satisfying (14) with $V(\phi(T)) = 0$,

or

$$\exists \phi_* \text{ critical point of } V \text{ with } V(\phi_*) \geq 0 \text{ such that } \lim_{t \to +\infty} \phi(t) = \phi_*, \text{ and } \lim_{t \to +\infty} \dot\phi(t) = 0, \tag{15}$$

and the measure of the initial data set such that $\phi(t) \to \phi_*$ maximum point is zero, while the set of initial data such that $\phi(t) \to \phi_*$ minimum point is open and not empty.

Moreover, situation (14) occurs only if $\inf V < 0$.

4. Qualitative analysis of the collapsing models

The aim of the present section is to study the qualitative behavior of the solution of (4a)–(4b) in the collapsing case ($\chi = -1$), for which we will require that $V(\phi)$ satisfies some further conditions. We assume that V is a potential satisfying assumptions stated at the beginning of Section 2 such that $V'(\phi)$ is eventually positive (respectively, negative) as $\phi \to +\infty$ (resp., $\phi \to -\infty$). Moreover, defining the function

$$u(\phi) := \frac{V'(\phi)}{\sqrt{6}V(\phi)}, \tag{16}$$

we assume that V satisfies the growth conditions

$$\limsup_{\phi \to \pm\infty} |u(\phi)| < 1, \tag{17}$$

$$\exists \lim_{\phi \to \pm\infty} u'(\phi) \, (= 0). \tag{18}$$

The main result concerning the collapsing behavior of the scalar field under the action of V as above is stated as follows:

Under the above assumptions made on $V(\phi)$, except at most for a measure zero set of initial data satisfying weak energy condition, there exists $t_s \in \mathbb{R}$ such that the scalar field solution becomes singular at $t = t_s$, that is $\lim_{t\to t_s^-} \epsilon(t) = +\infty$, and $\lim_{t\to t_s^-} a(t) = 0$. Moreover $\lim_{t\to t_s^-} \dot{\phi}(t) = +\infty$.

The proof of this result, that can be found in its full details in [8], consists of several steps. Here we shall give a sketch.

- Since we consider only data which satisfy the weak energy condition and thus, it must be $\epsilon_0 = \epsilon(0) = \frac{1}{2}\dot{\phi}^2(0) + V(\phi(0)) \geq 0$.
- Using existence/uniqueness results of solutions of (8) with initial zero–energy, $\epsilon_0 > 0$ up to shifting initial time of observation. Note that $\epsilon(t)$ is increasing, and therefore, defined
$$t_s = \sup \mathbb{I} \in \mathbb{R}^+ \cup \{+\infty\},$$
where $\mathbb{I}$ is the maximal interval of definition of the solution, the limit $\lim_{t\to t_s^-} \epsilon(t)$ exists.
- Up to shifting initial time of observation again, and up to a zero–measured set of initial data, all critical points of $V(\phi)$ can be thought as contained in the sublevel $V(\phi) < \epsilon_0$. Indeed, if that is not the case, and $\phi(t)$ is a solution such that $\epsilon_0 \leq \epsilon(t) \leq V(\phi_*)$ where ϕ_* is a critical point for V, then (6) applies, to find $\lim_{t\to+\infty} \dot{\phi}(t) = 0$, $\lim_{t\to+\infty} V'(\phi(t)) = 0$. By the assumptions made on the potential, $\phi(t)$ converges to a critical point of V, that we will call ϕ_* again for sake of simplicity. Whenever ϕ_* is a (non degenerate) maximum point we can study the linearization of the first order system equivalent to (4b) in a neighborhood of the equilibrium point $(0, \phi_*)$ obtaining a result totally analogous to (12). Moreover using again the existence/uniqueness result of solutions of (8) with initial zero–energy (in a "time-reversed" version) we see if ϕ_* is a minimum point and ϕ starts with initial data close to $(0, \phi_*)$, then ϕ moves far away from ϕ_*. To sum up, if the evolution is constrained in a potential well below some critical point, we obtain either a non generic situation, or a contradiction.
- $\phi(t)$ is unbounded. This can be seen by a contradiction argument – if ϕ was bounded, it can be proven that $\dot{\phi}$ would also be, and it would be $t_s = +\infty$. That would give, in view of (6), that $\lim_{t\to+\infty} \dot{\phi}(t) = 0$, which would imply $\lim_{t\to+\infty} V(\phi(t)) = \lim_{t\to+\infty} \epsilon(t) > \epsilon_0$, namely $\phi(t)$ moves in a region where V is invertible, and so it converges to some ϕ_*. Then by (4b), $\ddot{\phi}(t) \to -V'(\phi_*) \neq 0$, that is a contradiction.
- $\lim_{t\to t_s^-} |\phi(t)| = +\infty$. This can be proved with a contradiction argument, where is crucial to observe that the function
$$\rho(t) := \frac{2V(\phi(t))}{\dot{\phi}(t)^2}, \tag{19}$$
satisfies the equation
$$\dot{\rho} = \sqrt{6}\dot{\phi}\, \rho\sqrt{(1+\rho)}\left(u(\phi)\sqrt{1+\rho} - \operatorname{sgn}(\dot{\phi})\right). \tag{20}$$
- $\lim_{t\to t_s^-} \rho(t) = 0$. Indeed, supposing ϕ positively diverging, (20) allows to see ρ as a function of ϕ, and to study late time behavior of ρ it suffices to study the solutions of the ODE
$$\frac{\mathrm{d}\rho}{\mathrm{d}\phi} = \rho\sqrt{6(1+\rho)}\left(u(\phi)\sqrt{1+\rho} - 1\right). \tag{21}$$
which are indefinitely extendable to the right, and except for a zero–measured set of initial data, $\rho(t) \to 0$.
- $t_s < +\infty$, and $\lim_{t\to t_s^-} \epsilon(t) = +\infty$, $\lim_{t\to t_s^-} a(t) = 0$. That the solutions cannot be indefinitely extended follows from the relation $\dot{\epsilon}(t) = 2\sqrt{3}\epsilon^{3/2}\frac{1}{1+\rho(t)}$, and the other results are consequence of (4a), (5).

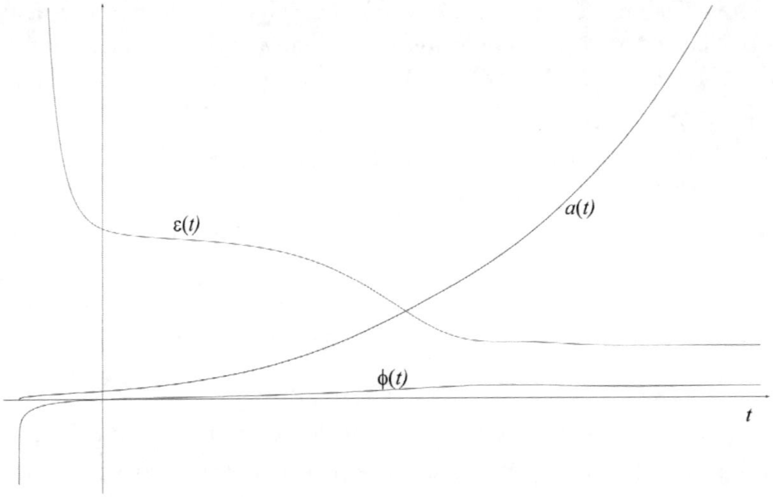

Figure 1. Behavior of the function $\phi(t), \epsilon(t)$, and $a(t)$ with potential given by $V(\phi) = 2x^4 - x^3 - 6x^2 + 10$. The initial conditions are $\phi_0 = 0$, $\dot{\phi}_0 = 1$, $a_0 = 1$, $\chi_0 = 1$.

5. Examples

Let us now review some examples, where the potential satisfies the conditions assumed at the beginning of Section 4. We stress that this class contains the potentials which are usually considered in the physical applications, and it obviously suffices for V to diverge at infinity to be also in the class studied in Section 3.

To begin, let us consider potentials with polynomial leading term at infinity (i.e. $\lambda^2\phi^{2n}$). For instance, for a quartic potential

$$V(\phi) = -\frac{1}{2}m^2\phi^2 + \lambda^2\phi^4, \tag{22}$$

with $\lambda, m \neq 0$, the function $u(\phi)$ goes as $\sqrt{3/2}\,\phi^{-1}$ for $\phi \to +\infty$, and all conditions listed above hold.

Figure 1 shows the behavior of the scalar, under the action of a potential from this class, that is expanding at initial time. Backwards time analysis amounts to study a collapsing situation, as we have already pointed out, and a singularity forms in the past in a finite amount of comoving time. On the other side, since the absolute minimum is at a strictly positive critical value, we expect an indefinite expansion forward in time, as it actually happens. The scalar field approaches one of the local minimum of $V(\phi)$, say ϕ_*, and the energy ϵ converges to $V(\phi_*)$.

When a local minimum ϕ_* such that $V(\phi_*) = 0$ exists, the oscillating behavior of an expanding solution discussed in Section 3 generically takes place; in this case, the energy $\epsilon(t)$ converges to zero. A situation of this kind is represented in Figure 2. Of course, time backwards analysis in this case is completely similar to the previous case, and a big bang singularity exists again.

Finally, when the potential also attains negative values, as explained Section 3, recollapse generically happens, in the sense that this situation is stable under perturbation of the initial data. Figure 3 shows the behavior in this case: the energy of an expanding solution vanishes at some $T > 0$, and for $t > T$ the solution recollapses, to form a singularity in the future – a singularity in the past is also formed, of course, as in the other cases.

In the case of exponential potentials, the results hold for asymptotic behaviors with leading

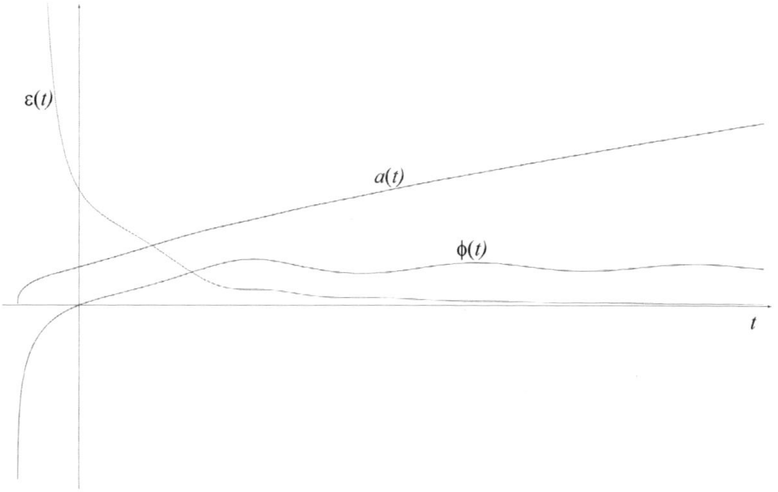

Figure 2. The oscillating behavior occurs when minima at zero critical value exist. In this case, $V(\phi) = (x^2 - 1)^2$, and the initial conditions are $\phi_0 = 0$, $\dot{\phi}_0 = 1$, $a_0 = 1, \chi_0 = 1$.

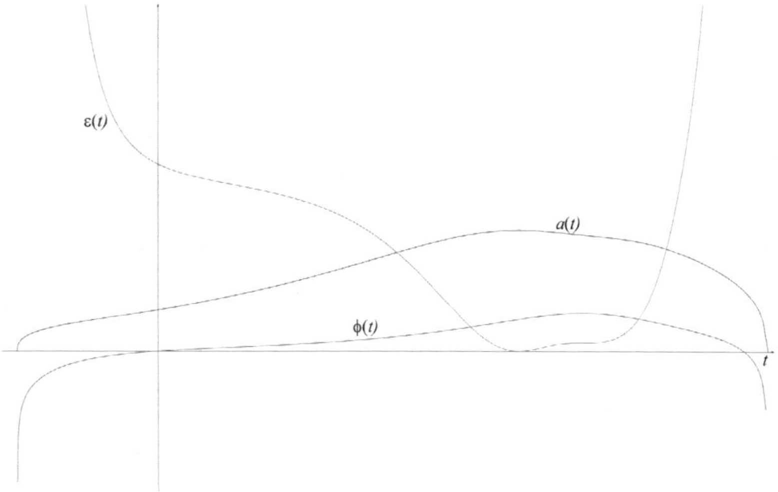

Figure 3. When the potential allows for negative value, expanding solutions may recollapse. In this case, $V(\phi) = 2x^4 - x^3 - 6x^2 + 4$, and the initial conditions are $\phi_0 = 0$, $\dot{\phi}_0 = 1$, $a_0 = 1, \chi_0 = 1$.

term at infinity of the form $V_0 e^{\sqrt{6}\lambda|\phi|}$ with $\lambda < 1$. For instance for

$$V(\phi) = V_0 \cosh(\sqrt{6}\lambda\phi), \tag{23}$$

where $V_0, \lambda > 0$, the quantity $u(\phi)$ goes like λ and so (17) is verified if $\lambda < 1$.

6. Gravitational collapse models

Homogeneous scalar fields may be used as sources for collapsing models, provided that we can find a suitable matching with an exterior spacetime. The natural choice for the exterior is the *generalized Vaidya solution* [21]

$$\mathrm{d}s^2_{\text{ext}} = -\left(1 - \frac{2M(U,Y)}{Y}\right)\mathrm{d}U^2 - 2\,\mathrm{d}Y\,\mathrm{d}U + Y^2\,\mathrm{d}\Omega^2, \tag{24}$$

where M is an arbitrary (positive) function, and the junction is performed along a hypersurface $\Sigma = \{r = r_b\}$, where r_b is some positive constant. Israel–Darmois junction conditions – whose explicit form can be found in [7, Proposition 4.1 and Remark 4.2] – are equivalent to require that Misner–Sharp mass M is continuous and $\frac{\partial M}{\partial U} = 0$ on the junction hypersurface.

The end state of the collapse of these "homogeneous scalar field stars" can be analyzed making use of the result in [7, Theorem 5.2], that states that if, in the approach to the singularity, the quantity $\dot{a}^2 = a^2\epsilon$ is bounded, then the apparent horizon cannot form, and the singularity is naked. Otherwise, if $\dot{a}^2$ is unbounded, the trapped region forms and the collapse ends into a black hole. Now, we know the existence of $\rho_\infty = \lim_{\phi\to\infty}\rho(\phi)$ (which is also equal to $\lim_{t\to t_s^-}\rho(t)$), and that, actually, ρ_∞ vanishes except for a zero–measured set of initial data. On the other hand, field equations also imply

$$\ddot{a} = -\frac{\dot{a}^2}{a}\left(\frac{2-\rho}{1+\rho}\right), \tag{25}$$

and therefore $\ddot{a} \leq -\frac{\dot{a}^2}{a}$ eventually holds in a left neighborhood of t_s, say $[t_0, t_s[$, where $\dot{a}$ is decreasing. It follows that $\forall t \in]t_0, t_s[$, $\dot{a}(t) \leq \dot{a}(t_0)$, which is negative by hypotheses. Then

$$\dot{a}(t) - \dot{a}(t_0) = \int_{t_0}^{t} \ddot{a}(\tau)\,\mathrm{d}\tau \leq -\int_{t_0}^{t} \frac{\dot{a}(\tau)^2}{a(\tau)}\,\mathrm{d}\tau \leq -\dot{a}(t_0)\int_{t_0}^{t} \frac{\dot{a}(\tau)}{a(\tau)}\,\mathrm{d}\tau,$$

that diverges to $-\infty$, and so $\dot{a}(t)$ is unbounded. Then we can conclude the following.

Except at most for a measure zero set of initial data, the scalar field model matched with the generalized Vaidya solution (24) *collapses to a black hole.*

6.1. *Non-generic situations*

In the analysis of collapsing case, the function ρ (19) has been observed to vanish, in the late time behavior, for almost every choice of the initial data. Now we want to reconsider the non generic situation where the limit $\lim_{t\to t_s^-} u(\phi(t))\sqrt{1+\rho(\phi(t))}$ equals 1. Actually, if the potential $V(\phi)$ satisfies the condition

$$\liminf_{\phi\to\pm\infty} |u(\phi)| > 0,$$

the same argument can be adapted also to this (non generic) case, to prove the formation of the singularity in a finite amount of time. Indeed, in this case ρ is bounded above, so that one can find a constant $\kappa > 0$ such that $\dot{\epsilon} > \kappa\epsilon^{3/2}$ in a left neighborhood of $\sup\mathbb{I}$.

Moreover, recalling equation (25), if

$$\liminf_{\phi\to\pm\infty} |u(\phi)| > \tfrac{\sqrt{3}}{3},$$

the argument seen above also applies here, and the collapse ends into a black hole. In particular, observe that in this case $2-\rho$ is bounded away from zero, and positive. On the other side, if

$$\limsup_{\phi\to\pm\infty} |u(\phi)| < \tfrac{\sqrt{3}}{3},$$

then $\ddot{a}$ is eventually positive, and this implies that $\dot{a}$ is bounded, so that the formation of the apparent horizon is forbidden. As a consequence, the collapse, in this non–generic situation, ends into a naked singularity.

6.2. Examples

The potential in exponential form of Example (23) is such that $u(\phi)$ behaves like a positive constant λ, and then it always gives rise to a black hole. The – non generic – data producing the solution such that $\lim_{\phi\to\infty}\rho(\phi) = \frac{1}{\lambda^2} - 1$ are discussed in [7, 10]; a choice of $\lambda < \frac{\sqrt{3}}{3}$ makes the limit ρ_∞ greater than 2, and the apparent horizon cannot form, resulting in a naked singularity. In [11] it is argued that cosmic censorship is restored for these models when loop quantum gravity modifications are taken into account – however, by the aforesaid, it should be stressed that even at a classical level these naked singularities are the outcome of a special choice of initial data.

Let V a potential in polynomial form, as the ones treated in Example (22). We already know that almost every choice of initial data forms a singularity, which happens to be a black hole. Note that, in the general case of a polynomial potential with leading term $\lambda^2\phi^{2n}$, with $n > 2$, it can be shown that *every* initial data gives rise to a singularity, although the function $u(\phi)$ diverges and therefore the above arguments cannot be applied. Indeed, if $\sqrt{\rho+1}$ goes like $u(\phi)^{-1}$ – that in this case behaves like $\frac{n}{\sqrt{3}\phi}$, then using (19) it can be seen that $\dot{\phi} \cong \phi^{n-1}$, and then if $n > 2$ the solution must diverge in a finite amount of time. The non genetically data yielding this particular situation, moreover, forbid apparent horizon formation, and so the resulting singularity is naked.

6.3. Collapse in higher order gravity theories

Scalar fields arise also when higher order gravity theories, described by Lagrangians of the form $L = f(R)\sqrt{-g}$, are considered. Indeed [1, 2], under the conformal transformation $g_{\mu\nu} \mapsto f'(R)g_{\mu\nu}$, fields equations reduce to Einstein field equations with a scalar field (interacting with a potential) as a matter source. The function f determines the form of the potential. In paper [9] this situation is analyzed where, before performing the conformal transformation, a perfect fluid matter contribution $p = (\gamma - 1)\rho$ $(0 \leq \gamma < 2)$ is added to the Lagrangian. Interestingly enough, it is proved that for a wide class of potentials – including the potential corresponding to the conformal frame of the $R + \beta R^2$ theory – the qualitative picture is the same as in the case studied in the present paper, i.e. the addition of a fluid does not affect the general results stated before: the singularity generically forms in a finite amount of comoving time, and it is generically covered. Perfect fluid energy may diverge at the singularity – and this fact generically takes place when $\gamma > 2/3$ – yet producing a lower order curvature divergence with respect to the energy of the scalar field.

References

[1] J.D. Barrow and S. Cotsakis, *Phys. Lett.* **B214** 515 (1988).
[2] S. Capozziello and M. Francaviglia, *Gen Rel. Grav.* **40** 357-420 (2008).
[3] D. Christodoulou, Ann. Math. **140** 607 (1994).
[4] D. Christodoulou, Ann. Math. **149** 183 (1999).
[5] Foster, S. Class.Quant.Grav. 15 (1998) 3485-3504
[6] R. Giambò, F. Giannoni, G. Magli, P. Piccione, Comm. Math. Phys. **235(3)** 545-563 (2003)
[7] R. Giambò, Class. Quantum Grav. **22** (2005) 1-11
[8] R. Giambò, F. Giannoni, G. Magli, J. Math. Phys. **49** 042504 (2008)
[9] R. Giambò, *Gravitational collapse of homogeneous perfect fluids in higher-order gravity theories*, preprint.
[10] R. Goswami, P. S. Joshi, gr-qc/0410144
[11] R. Goswami, P. S. Joshi, P. Singh, Phys. Rev. Lett. **96**, 031302 (2006)
[12] P. S. Joshi, Modern Phys. Lett. A **17** 1067–1079 (2002)
[13] J. Miritzis, Class. Quantum Grav. **20** (2003), no. 14, 2981–2990
[14] J. Miritzis, J. Math. Phys. **44** (2003) 3900-3910
[15] J. Miritzis, J. Math. Phys. **46** (2005) 082502
[16] L. Perko, *Differential Equations and Dynamical Systems*, Springer–Verlag, New York, 1991
[17] A.D. Rendall Class.Quant.Grav. **21** (2004) 2445-2454

[18] A.D. Rendall Class.Quant.Grav. **24** (2007) 667-678
[19] C. Rubano, J. D. Barrow, Phys.Rev. **D64** (2001) 127301
[20] C. Rubano, P. Scudellaro Gen.Rel.Grav. **34** (2002) 307-328
[21] A. Wang and Y. Wu, 1999 Gen. Rel. Grav. **31** 107

Recent Developments in Gravity (NEB XIII) IOP Publishing
Journal of Physics: Conference Series **189** (2009) 012018 doi:10.1088/1742-6596/189/1/012018

Prompt gamma-ray burst emission from gradual energy dissipation

Dimitrios Giannios
Max Planck Institute for Astrophysics, Box 1317, D-85741 Garching, Germany
Department of Astrophysical Sciences, Peyton Hall, Princeton University, Princeton, NJ 08544, USA

E-mail: giannios@astro.princeton.edu

Abstract. I describe the main aspects of a model for the gamma-ray burst (GRB) emission in which energy is dissipated gradually in a Poynting-flux-dominated flow. In this picture, the energy of the radiating particles is determined by heating and cooling balance. Detailed radiative transfer calculations show that, at Thomson optical depths of order of unity, the dominant radiative process is inverse Compton scattering. Synchrotron-self-absorbed emission and inverse Compton dominate in the Thomson thin parts of the flow. The electrons stay thermal throughout the dissipation region because of Coulomb collisions (Thomson thick part of the flow) and exchange of synchrotron photons (Thomson thin part). The resulting spectrum naturally explains the observed sub-MeV break of the GRB emission and the spectral slopes above and below the break. In this model, the Amati relation indicates a tendency for the more luminous bursts to have more energy per baryon. If this tendency also holds for individual GRB pulses, the model predicts the observed narrowing of the width of pulses with increasing photon energy. The model also predicts that the γ-ray power-law tail has a high-energy cutoff typically in the ~GeV energy range that should be observable with *FERMI*. A prompt emission component in the optical and UV is predicted to be associated with the GeV emission.

1. Introduction

The GRB emission is the likely result of internal energy release in an ultrarelativistic flow. The dissipative and radiative mechanisms for the GRB largely remain uncertain. A popular model for the energy dissipation invokes internal shocks in an unsteady flow (Paczynski & Xu 1994; Rees & Mészáros 1994). An alternative proposal is magnetic dissipation (Thompson 1994; Spruit et al. 2001) in a strongly magnetized flow (Usov 1992).

Internal shocks are efficient in dissipating a large fraction of the kinetic energy of the flow provided that it is highly variable, i.e., composed of distinct ejection events with strong variation in their bulk Lorentz factor γ (e.g. Kobayashi et al. 1997). Energy is dissipated by the shocks at the location of the collision of the shells. Particles are assumed to be accelerated on a very short timescale at the shock front to ultrarelativistic speeds and non-thermal distributions. Subsequently, they radiate a fraction of the dissipated energy via synchrotron and inverse Compton processes. The relevant radiative mechanisms and the emitted spectra depend, to a large extent, on the shock microphysics and the corresponding Thomson optical depth of the flow at the radius of the collision.

On the other hand, the energy dissipation that powers the GRB emission may be gradual and distributed over a large part of the volume of the flow. The energy of the radiating particles is determined by the heating/cooling equilibrium (slow heating model; Ghisellini & Celotti 1999; Stern & Poutanen 2004). Such an energy balance is expected to lead to sub-relativistic or mildly relativistic temperatures in the flow. Magnetic dissipation in a strongly magnetized flow can provide a possible physical setup where gradual dissipation is realized. As shown in Drenkhahn (2002) and Drenkhahn & Spruit (2002; hereafter DS02), dissipation though reconnection takes place over several decades in radius; typically in both Thomson thick and thin conditions. Dissipation in the reconnection model is responsible for both the acceleration of the flow and the prompt emission.

In this talk, I summarize our work on the spectral and temporal properties predicted by the reconnection model in the context of GRB flows (Giannios 2006; Giannios & Spruit 2007; Giannios 2008; hereafter G06, GS07, and G08 respectively). The detailed Monte Carlo calculations have shown that, due to the released energy, the flow develops a hot photosphere with comoving electron temperatures of tens of keV. Upscattering of photons that are produced deeper in the flow by those hot electrons leads to a powerful photospheric emission; it accounts for $\sim 3-30\%$ of the luminosity of the flow. The resulting $E \cdot f(E)$ spectrum has a characteristic ~ 1 MeV peak followed by a flat high-energy power-law emission.

Dissipation of magnetic energy and associated emission typically continues in the Thomson thin region of the flow as well. Because of the strong magnetic fields and the mildly relativistic electrons, synchrotron self absorption results in efficient energy exchange of the electrons, keeping their distribution thermal (Ghisellini et al. 1998). Under these conditions, synchrotron-self-absorbed (SSA) emission is an important radiative mechanism in addition to inverse Compton dominating the observed emission in the soft X-rays and softer bands.

The efficient thermalization of the emitting particles throughout the flow reduces the dependence of the model on the, poorly understood, mechanisms of particle acceleration that operate in magnetic reconnection. The model is defined by just 3 main parameters (luminosity, baryon loading and a reconnection-rate parameter). In contrast to the internal shock model, no quantities have to added to parameterize the particle distributions and the amplification of magnetic fields; the field strength is an integral part of the reconnection model. In the following, I summarize how this emitted spectrum is computed in G06, GS07, G08.

2. Magnetic reconnection versus internal shocks

The internal shock model for the GRB emission invokes high variability in the bulk Lorentz factor of the flow that leads to internal collisions. The location where the collision of two shells takes place depends on their initial separation and bulk Lorentz factors. These collisions can dissipate a substantial fraction of the kinetic energy of the flow. Internal shocks are assumed to lead to particle acceleration and magnetic field amplification at the shock front. If electrons receive a large fraction of the dissipated energy then they are accelerated to ultrarelativistic speeds. They cool down radiatively by synchrotron and inverse Compton mechanisms. The resulting spectrum and the relevant radiative mechanisms depend on details of the distribution of the accelerated particles, the magnetic field strength and the Thomson optical depth of the flow at the location of the collision. If the collision of two shells takes place at the Thomson thin region, as needed to explain the typical variability properties of the GRB lightcurves (Daigne & Mochkovitch 1998; Nakar & Piran 2002; Mimica et al. 2005), synchrotron self Compton is likely the most promising radiative mechanism (Rees & Mészáros 1994; Katz 1994; Tavani 1996). In this picture optically thin synchrotron emission dominates the observed hard X-ray, $\sim$MeV spectrum. Despite its attractive features, the synchrotron model has theoretical and observational difficulties (as discussed, for example, in Ghirlanda et al. 2003). If, on the other hand, the collisions take place close to the Thomson photosphere, Compton scattering is the dominant radiative mechanism that shapes the spectrum and results in very different emission (Mészáros & Rees 2000).

As an alternative to internal shocks, magnetic dissipation can power the prompt emission provided that the flow is launched Poynting-flux dominated (or with a substantial fraction of its energy in the form of Poynting flux). Magnetic dissipation through, for example, reconnection can release energy smoothly in a large fraction of the volume of the flow. This energy release can take place while the flow expands over several decades in radius (DS02). The energy of the radiating particles is determined by balancing heating and radiative (or adiabatic) cooling at each radius. In this case the slow heating picture described by Ghisellini & Celotti (1999) takes place. The electrons are subrelativistic or mildly relativistic and their synchrotron emission is self absorbed. In a strongly magnetized flow, such SSA emission guarantees efficient energy exchange and thermalization of the electrons on a very short timescale (Ghisellini et al. 1998). The resulting emission does not depend on details of particle acceleration and magnetic field amplification that one faces in internal shock models. The model is defined by just the luminosity, the baryon loading and the reconnection-rate parameter of the flow. This contributes significantly to the predictive power of the magnetic reconnection model. The total observed flux is the integrated emission from the different parts of the flow in which dissipation of energy takes place. It contains both photospheric (Thompson 1994; Stern 1999; G06; GS07) and Thomson thin components (Stern & Poutanen 2004; G08).

2.1. Implications from the observed GRB variability

One additional difference of the internal shock and the magnetic reconnection model is connected to implications from the observed variability of the lightcurves. Internal shocks are efficient only in variable flows. Variability and dissipation are, a priori, unrelated in the magnetic reconnection model in which dissipation takes place, even in a steady outflow. On the other hand, the observed lightcurves are often highly variable showing that the flow does evidently evolve during a GRB. In the context of the reconnection model, the observed variability reflects changes in the luminosity and baryon loading of the flow during the burst. As shown in GS07, the flow can be treated as quasi-stationary for all but the shortest (msec) time scales observed in a burst, the variation of spectral properties during a burst directly reflects variations in the central engine. This is in contrast to models in which the prompt radiation is produced at much larger distances from the source, such as external shock models. It is also in contrast with the internal shock model, since the internal evolution of the flow between the source and the level where radiation is emitted is a key ingredient in this model. Deducing properties of the central engine from observed burst properties is thus a much more direct prospect in the magnetic dissipation model.

2.1.1. Central engine: the stronger the cleaner A interesting deduction for the central engine is that the luminosity of the flow L correlates with the baryon loading $\eta \propto L^{\sim 0.6-0.7}$ in different bursts and *during* the evolution of the bursts (GS07). If this $L-\eta$ correlation holds, then the so-called Amati relation (Amati et al. 2002; Amati 2006) *and* the energy dependence of the width of pulses in a burst (e.g. Fenimore et al. 1995) can be naturally understood.

2.1.2. Reported constraints of the emission radius of the GRB The ~MeV emission in this model mainly comes from moderate Thomson optical depths. The corresponding distance from the source is of order $r_{\rm MeV} \sim 10^{12}$ cm. On the other hand, studies that attribute the steep decay of the prompt emission (that may last for tens of minutes after the burst) to the "high latitude" emission (Kumar & Panaitescu 2000), favor a much larger emission radius of $r_{\rm MeV} \gtrsim 10^{15}$ cm (Lazzati & Begelman 2006; Lyutikov 2006; Kumar et al. 2007). There is evidence, however, that the decay lightcurves do *not* follow the spectra-temporal signatures of high-latitude emission in the majority of the bursts, making this interpretation questionable (O'Brien et al. 2006).

The "high latitude" emission is *not* relevant in determining the decaying part of the lightcurve in the reconnection model. Any high latitude emission lasts for $t \sim r_{\rm Mev}\theta^2/c \sim 1$ sec (assuming

$\theta \sim 0.1$). In this model, the decay part is "just" a reflection of the decreasing luminosity of the flow at the end of the burst. The steep decay reflects changes in the flow in the radial direction and is not result of off-axis emission.

3. Energy release because of magnetic reconnection

Magnetic dissipation can take place gradually in the GRB flow. The rate at which energy is dissipated as a function of radius depends on the magnetic field geometry and the exact mechanism through which magnetic energy dissipates. If the flow is launched with field of large scale, energy dissipation can be a result of global MHD instabilities (such as current-driven instabilities; e.g., Lyutikov & Blandford 2003; Giannios & Spruit 2006). On the other hand, if the flow contains reversing magnetic fields of sufficiently small scale, dissipation can take place directly through reconnection (Drenkhahn 2002; DS02; Thompson 2006). Here, I focus on the reconnection model, which makes clear prediction for the energy dissipation as function of radius; essential for the radiative transfer calculations presented here. Though the results presented here are directly applicable to the DS02 model, qualitatively similar results are expected from other gradual, magnetic dissipation models.

3.1. The reconnection model

An important physical quantity in the reconnection model is the ratio σ_0 of the Poynting flux to kinetic energy flux at the Alfvén radius r_0. It parameterizes the baryon loading of the flow η and determines the terminal bulk Lorentz factor of the flow $\gamma_\infty \sim \eta \simeq \sigma_0^{3/2}$. The flow must start Poynting-flux dominated with $\sigma_0 \gtrsim 30$ for it to be accelerated to ultrarelativistic speeds with $\gamma_\infty \gtrsim 100$ that are relevant for GRB flows.

In the reconnection model, the magnetic field in the flow changes polarity on a scale λ as in the oblique rotator model for pulsar winds where $\lambda \simeq 2\pi c/\Omega$, where Ω is the angular frequency of the rotator (Coroniti 1990; Lyubarsky & Kirk 2001; DS02; Kirk & Skjæraasen 2003). The rate of magnetic reconnection DS02 model is parameterized through the velocity $v_{\rm r}$ with which magnetic fields of opposite direction merge. The $v_{\rm r}$ is assumed to scale with the Alfvén speed, $v_{\rm A}$, i.e. $v_{\rm r} = \varepsilon v_{\rm A}$. A nominal value used for ϵ is 0.1 (see Lyubarsky 2005). For the flows with $\sigma_0 \gg 1$ that are of interest here, the energy density of the magnetic field is larger than the rest mass energy density, hence $v_{\rm A} \approx c$, and the reconnection takes place with subrelativistic speeds.

3.1.1. Properties of the flow In the reconnection model, magnetic dissipation takes place all the way from the initial radius r_0 till the saturation radius r_s. Part of the dissipated energy (approximately half) is directly used to accelerate the flow. The acceleration of the flow is gradual following the $\gamma \sim r^{1/3}$ scaling as function of radius in the regime $r_0 \ll r \ll r_{\rm s}$. To first order approximation, no further acceleration takes place beyond the saturation radius (DS02)

$$\begin{aligned} \gamma &= \gamma_\infty \left(\frac{r}{r_{\rm s}}\right)^{1/3} = 148\; r_{11}^{1/3} (\varepsilon\Omega)_3^{1/3} \sigma_{0,2}^{1/2}, \quad \text{for} \quad \mathrm{r} < \mathrm{r_s}, \\ \gamma &= \gamma_\infty = \sigma_0^{3/2}, \quad \text{for} \quad \mathrm{r} \geq \mathrm{r_s}, \end{aligned} \tag{1}$$

while the saturation radius is

$$r_{\rm s} = \frac{\pi c \gamma_\infty^2}{3\varepsilon\Omega}; \quad \text{or} \quad r_{\rm s,11} = 310\; \frac{\sigma_{0,2}^3}{(\varepsilon\Omega)_3}. \tag{2}$$

The notation $A = 10^x A_x$ is used; the 'reference values' of the model parameters are $\sigma_0 = 100$, $\varepsilon = 0.1$, $\Omega = 10^4$ rad·s^{-1}. The product of ε and Ω parameterizes the reconnection rate. The physical quantities of the flow depend on this product.

In the steady, spherical flow under consideration the comoving number density can be written as

$$n' = \frac{L}{r^2 \sigma_0^{3/2} \gamma m_{\rm p} c^3}, \qquad (3)$$

where L is the luminosity per steradian of the GRB flow. The reference value used is $L = 10^{52}$ erg·s^{-1}·sterad^{-1}. (In this form the expression can be compared with the fireball model, where the baryon loading parameter η replaces the factor $\sigma_0^{3/2}$).

The expression (1) is deviating from the exact numerical solution presented in Drenkhahn (2002) at $r \gtrsim r_s$. The reason is that the dissipation does not stop abruptly at $r_{\rm s}$ but there is modest energy release at a slower rate at larger radii. Though, not important for the global energetics, the remaining dissipation at the radii $r \gtrsim r_{\rm s}$ results in synchrotron emission that dominates the observed radiation in soft bands (such as optical and near ultra violet). This emission is mainly a result of the large emitting surface at these outer parts of the flow. I take into account the residual dissipation, so that to correctly describe the soft emission.

The comoving magnetic field strength B' below the saturation radius is given by setting $L \simeq L_{\rm p} = r^2 B'^2 \gamma^2 c/4\pi$ and solving for B':

$$B' = \left(\frac{4\pi L}{cr^2\gamma^2}\right)^{1/2} \quad \text{for} \quad \mathrm{r} < \mathrm{r_s}. \qquad (4)$$

The rate of energy density release in a comoving frame can be found by the following considerations. The time scale over which the magnetic field decays is that of advection of magnetic field of opposite polarity to the reconnection area. The reconnection speed is $v_r = \varepsilon v_{\rm A} \simeq \varepsilon c$, while the magnetic field changes polarity over a comoving length scale $\lambda' = 2\pi\gamma c/\Omega$. The decay timescale for the magnetic field, therefore, is

$$t_{\rm dec} = \frac{\lambda'}{v_r} = \frac{2\pi\gamma}{\varepsilon\Omega}. \qquad (5)$$

Using the last expression and Eqs. (1) and (4), the rate of dissipation of magnetic energy density in the comoving frame is

$$P_{\rm diss} = \frac{(B')^2/8\pi}{t_{\rm dec}/2} = \frac{\varepsilon\Omega L}{2\pi c r^2 \gamma^3} \quad \text{for} \quad \mathrm{r} < \mathrm{r_s}. \qquad (6)$$

The bulk Lorentz factor γ, the density n', the magnetic field strength B' and rate of dissipation of magnetic energy density $P_{\rm diss}$ of the flow as functions of radius are the quantities needed for the study of the resulting emission.

4. Emission

If all the energy is dissipated at large optical depths, adiabatic expansion converts most of this energy into kinetic at the expense of radiation. Then one essentially deals with a fireball. Gradual dissipation heats the flow continuously and maintains a substantial fraction of the energy in the form of radiation. This radiation is released at the photosphere of the flow. If dissipation takes place further out in the flow it can result in additional emission coming from the Thomson thin region. The total flux received by the observer is the integrated emission from the different parts of the flow where dissipation takes place.

In the case of magnetic dissipation (as well as for other dissipative mechanisms), the fate of the released energy is rather uncertain. An interesting possibility is that dissipation leads to MHD turbulence where particle acceleration can take place by scattering of photons by Alfvén waves (Thompson 1994). On the other hand, the magnetic energy can directly be dissipated to the particles in the flow, most likely to the electrons due to their higher mobility. Following G06, GS07 we assume that a fraction f_e of order of unity of the dissipated energy heats up the electrons. For the results presented here we set $f_e = 0.5$.

The resulting emission does not depend only on the amounts of energy released but also on the distribution of the emitting particles. I assume that the electron distribution is thermal *throughout* the region where dissipation takes place. As I discuss in more detail below, the thermalization of the electrons is result of Coulomb collisions in the inner parts of the flow and of exchange of synchrotron photons at the outer parts. I first summarize the results of G06, GS07 on the photospheric emission from the reconnection model and then turn to the study of the Thomson thin emission (G08).

4.1. Moderate optical depths

In addition to the saturation radius r_s, another characteristic radius of the flow is the Thomson photosphere. The radius of the Thomson thick-thin transition is found by setting $\tau = 1$ in Eq. (7) and solving for r_{ph}:

$$r_{ph,11} = 6\frac{L_{52}^{3/5}}{(\varepsilon\Omega)_3^{2/5}\sigma_{0,2}^{3/2}}. \tag{7}$$

In deriving these expressions, we have assumed that $r_{ph} < r_s$. A similar calculation gives the radius of the photosphere in the $r_{ph} > r_s$ case.

One can check that for a large parameter space relevant for GRB flows, $r_{ph} < r_s$ which means that dissipation proceeds throughout the photospheric region. In terms of the physical properties of the flow, there is a critical value of the magnetization $\sigma_{0,cr}$, for which $r_{ph} = r_s$. For $\sigma_0 > \sigma_{0,cr}$, $r_{ph} < r_s$. Using Eqs. (7) and (2) one finds

$$\sigma_{0,cr} = 42(L_{52}(\varepsilon\Omega)_3)^{2/15}. \tag{8}$$

The critical baryon loading depends weakly on the parameters of the flow: $\eta_{cr} = \sigma_{0,cr}^{3/2} = 270(L_{52}(\varepsilon\Omega)_3)^{1/5}$.

For $\sigma_0 \ll \sigma_{0,cr}$ dissipation ceases deep in the flow (at high optical depths). For $\sigma_0 \gg \sigma_{0,cr}$, dissipation takes place in both Thomson thick and thin conditions with most of the energy released in the outer parts of the flow. Particles and radiation are found to be in thermal equilibrium deep in the flow. There, the comoving temperature T_{th} of the flow is calculated, under the assumption of complete thermalization, by integrating the energy released at different radii in the flow and taking into account adiabatic cooling. Due to the dominance of scattering, the details of radiative transfer become important already at fairly large optical depth in the flow. Equilibrium between radiation and matter holds only at Thomson depths greater than about 50. At smaller optical depths the electron distribution stays thermalized, but is out of equilibrium with the photon field. More discussion on the processes that lead to thermalization of the electron distribution is presented in the next section. Compton scattering of the photons is treated in detail in this region with Monte Carlo Comptonization simulations (G06; GS07). In this region synchrotron emission is negligible since it is strongly self absorbed. Energy dissipation at moderate and low optical depths is shown to lead to emission that has a highly non-thermal appearance.

4.2. The Thomson thin region

When dissipation continues at the optically thin parts of the flow the electron temperature becomes of the order of the electron rest mass. At those temperatures, synchrotron self

Compton emission has important effect on the emitted spectrum (see also Stern & Poutanen 2004). Furthermore, SSA affects the electron distribution in the flow. It is shown in Ghisellini et al. (1998) that electrons with energy of the order of their rest mass can thermalize on a few synchrotron cooling times (as defined for thin synchrotron emission) by emitting and absorbing synchrotron photons. In the magnetically dominated flow under consideration, the thermalization of the electrons takes place on a timescale shorter than the heating/cooling one. One can, therefore, assume that the electrons are approximately thermal when calculating their emission.

One should keep in mind, however, that thermalization of the particles is not achieved if the dissipated energy accelerates a small fraction of the particles in the flow. In this case, particles can be accelerated to relativistic speeds and cool efficiently through optically thin synchrotron emission. In this case, synchrotron self absorption is not an efficient thermalization mechanism. This case has been investigated in Giannios & Spruit (2005). Here, I assume that the dissipated energy is distributed among a large fraction of the particles and hence thermalization is achieved in the electron distribution.

4.2.1. The synchrotron emission The electron temperature becomes mildly relativistic at small optical depths $\tau \sim 0.1$. It increases further at larger radii resulting in substantial synchrotron emission. The synchrotron-self-absorbed emission from mildly relativistic plasma has a characteristic spectrum that consists by a Rayleigh-Jeans part up to the so-called turnover frequency ν_t where the optical depth due to synchrotron absorption of the flow becomes unity. Most of the energy is emitted at the turnover frequency. At higher frequencies, the spectrum is very steep following the exponentially decaying tail of the synchrotron thin emission. The typical model parameters and radii $r \sim r_{\rm s}$, the turnover frequency is $\nu_{\rm t} \sim 10^3\nu_{\rm c}$ which results in $\nu_{\rm t}^{\rm obs} = \gamma_\infty \nu_{\rm t}/(1+z) \lesssim 1$ keV (z is the redshift of the burst). The synchrotron-self-absorbed emission appears mainly in the soft X-rays and softer bands (G08).

The procedure for the numerical radiative transfer calculation is the following. I make a choice for the temperature profile as function of radius of the electron temperature in the flow. Analytical estimates provide a good initial guess. The radiative transfer in the flow is studied using the Monte Carlo Comptonization code described in G06. The calculation includes the thermal radiation field carried with the flow which is injected in the inner numerical boundary at the "equilibrium" radius where radiation and particles drop out of equilibrium. At larger radii the synchrotron emitted flux is also included. Both sources of photons are propagated through the medium and their scattering by electrons is followed. The code calculates the spectrum and the radius-dependent cooling rate of the electrons. Cooling because of inverse Compton, synchrotron emission and adiabatic expansion is taken into account. The adiabatic cooling becomes important at high optical depths and at the very outer parts of the flow where the expansion timescale $r/\gamma c$ is shorter than the radiative cooling one. The outer boundary of the calculations is set at large enough radius so that it does not have an effect on the computed spectra. I iterate the electron temperature until the cooling rate matches the heating rate predicted by the model reasonably well at all radii.

5. Resulting spectra

In the illustrative case of Fig. 1, I set $\sigma_0 = 70$ (which corresponds to flow with baryon loading $\eta \simeq \sigma_0^{3/2} \sim 600$) and the rest of the parameters to their reference values. Spectra are plotted in the central engine frame. The dotted line shows the input thermal radiation at the "equilibrium radius" which is the inner boundary of the computed domain. The thermal flux is advected with the flow and constitutes a large fraction of the seed photons to be Comptonized further out in the flow.

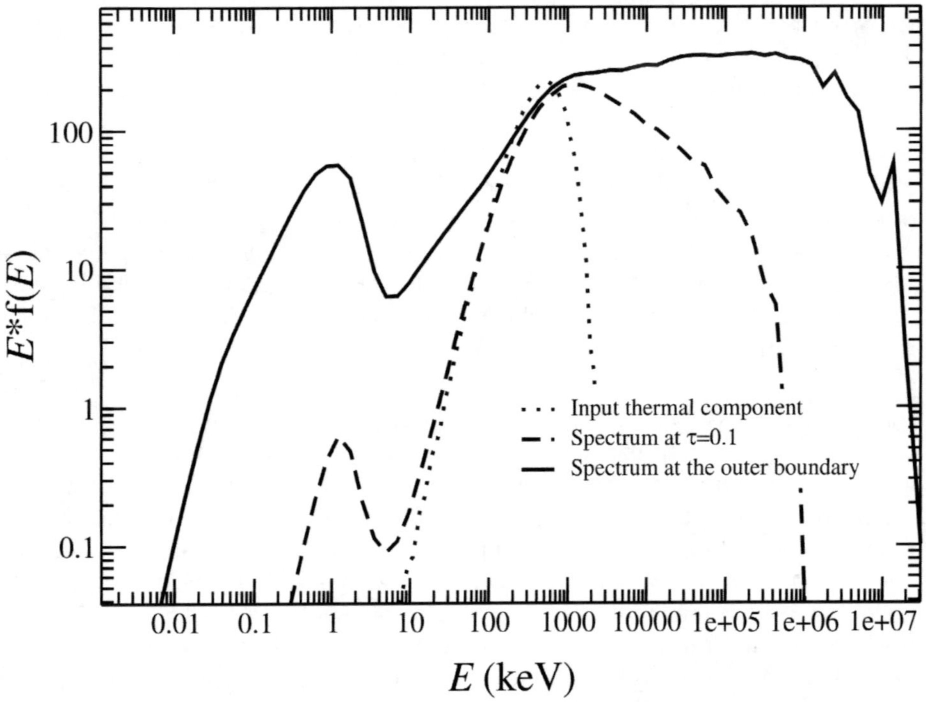

Figure 1. Shape of the photo spectrum at different Thomson optical depths) in the flow. The spectrum is shown in the central engine frame. The dotted line stands for the spectrum at the radius where radiation and electrons decouple. The photospheric emission is shown with the dashed line. The bulk of the MeV emission comes from this region. The overall spectrum (solid line) includes the emission from the Thomson thin region of the flow. Synchrotron-self-absorbed emission from this region dominates the spectrum below $\sim$ 10 keV. Inverse Compton leads to flat γ-ray tail up to $\sim$ 1 GeV.

The appearance of the photon spectrum at radius which corresponds to optical depth $\tau = 0.1$ is shown with dashed line. This radius is the outer boundary used in most of the calculations of G06, GS07. One can clearly see the effect of inverse Compton scattering to the spectrum. The peak of the $E \cdot f(E)$ spectrum increases slightly and the spectrum becomes broader. An important feature is the high-energy tail that is result of unsaturated Comptonization at $\tau \lesssim 1$. Note also a second peak at $\sim$1 keV. This is result of synchrotron emission with the turnover frequency being $\sim$1 keV (in the central engine frame). This component is still weak at $\tau = 0.1$ and has not been included in the calculations of G06, GS07.

The total spectrum is shown with solid line. This includes the emission from the whole volume of the flow where there is dissipation taking place. The overall emission spectrum is much broader. It is characterized by a break at $\sim$ 1 MeV followed by a flat spectrum with photon-number index $\Gamma \simeq -2$ (where $\mathrm{d}N/\mathrm{d}E \sim E^{\Gamma}$), close to the typically observed one. The hard γ-ray tail is extending up to $\sim$1 GeV which corresponds to the Lorentz boosted temperature of the flow at its outer layers (where it reaches its maximum values). Comparing the spectrum at $\tau = 0.1$ and the total one, it is clear that Comptonization proceeds throughout the Thomson thin region strengthening the hard γ-ray component.

An important feature of the emitted spectrum is the powerful component that appears in the soft X-rays and softer bands. This comes from synchrotron-self-absorbed emission. SSA dominates by many orders of magnitude the ultra violet and optical emission. The softer emission originates from the Thomson thin part of the flow that is characterized by the higher electron temperatures and larger emitting surface (see also Stern & Poutanen 2004). This emission is very weak in models where dissipation takes place only below or around the Thomson photospheric region (see, e.g., Mészáros & Rees 2000; Pe'er et al. 2006; Ioka et al. 2007).

As a result of the synchrotron-self-Compton component, the spectrum below the MeV peak softens. The spectrum can be well fitted with a power-law in the $30-300$ keV energy range with photon-number index of $\Gamma \simeq -1.2$ which very close to the one typically observed (e.g. Preece et al. 1998).

The SSA emission spectrum hardens considerably in the ultra violet. At lower energies the Rayleigh-Jeans limit is gradually approached. The location of the hardening is determined by the radius where adiabatic expansion starts to dominate the cooling of the electrons. In this example adiabatic expansion dominates at $r \gtrsim r_{\rm ad} \sim 5 \cdot 10^{14}$ cm. The optical and near UV emission is delayed w.r.t. the $\sim$ MeV emission by $\delta t \sim r_{\rm ad}/\gamma_\infty^2 c \sim 0.05$ s. Radiation in these bands reaches the observer with small but maybe detectable lags.

The relative strength of the synchrotron-self-Compton (SSC) component depends on the fraction of energy dissipated in the Thomson thin region of the flow. For $\sigma_0 \gg \sigma_{0,\rm cr}$ most of the energy is dissipated in the Thomson thin region. Correspondingly the SSC component is pronounced. This is evident in Fig. 2 where the spectrum is shown for different values of σ_0 (the rest of the parameters are kept in their reference values). For $\sigma_0 = 40$, dissipation stops close to the photosphere of the flow. The SSA component is almost absent and the emission above the thermal peak at ~ 1 MeV relatively weak. In more baryon loaded models the emission is quasi-thermal since dissipation stops at high Thomson depths.

With increasing σ_0 both the SSA and inverse Compton components become relatively more powerful. The thermal peak is followed by a flat γ-ray emission that extends up to $\sim$GeV. Most of the models show a high energy cutoff in this energy range. This cutoff corresponds to the highest energies to which photons are upscattered. It is determined by the Lorentz boosted electron temperature at $r \sim r_{\rm s}$. The spectral slope below the $\sim$1 MeV break becomes softer with increasing σ_0. The photon-number index in the $30-300$ keV energy band varies in the range $-1.2 \lesssim \Gamma \lesssim -0.4$ in agreement to that typically observed (e.g. Preece et al. 1998). The high σ_0 models have powerful optical and near ultra violet emission. The flux $f(E)$ that is emitted in these bands is similar to the X-rays one. The optical spectrum is hard with photon number index $0 \lesssim \Gamma \lesssim +1$.

Varying the baryon loading of the flow has moderate effect in the emission in the *BATSE* energy range but profound implications in other bands. The model predicts that flows with low baryon loading (i.e. high σ_0) have powerful optical, UV and GeV emission. More on the comparison of the model with observations is presented in the next section.

5.1. Application to observations

The prompt GRB emission has been typically observed in the hard X-rays up to ~ 1 MeV γ-rays. The spectrum in this energy range shows a characteristic sub-MeV break followed by a flat power-law γ-ray tail (e.g. Band et al. 1993). Below the break the spectrum has typical photon number index of $\Gamma \sim -1$ although much harder spectra have also been observed[1]. The observed

[1] These hard spectra cannot be explained by the thin synchrotron model (e.g. Ghirlanda, Celotti & Ghisellini 2003).

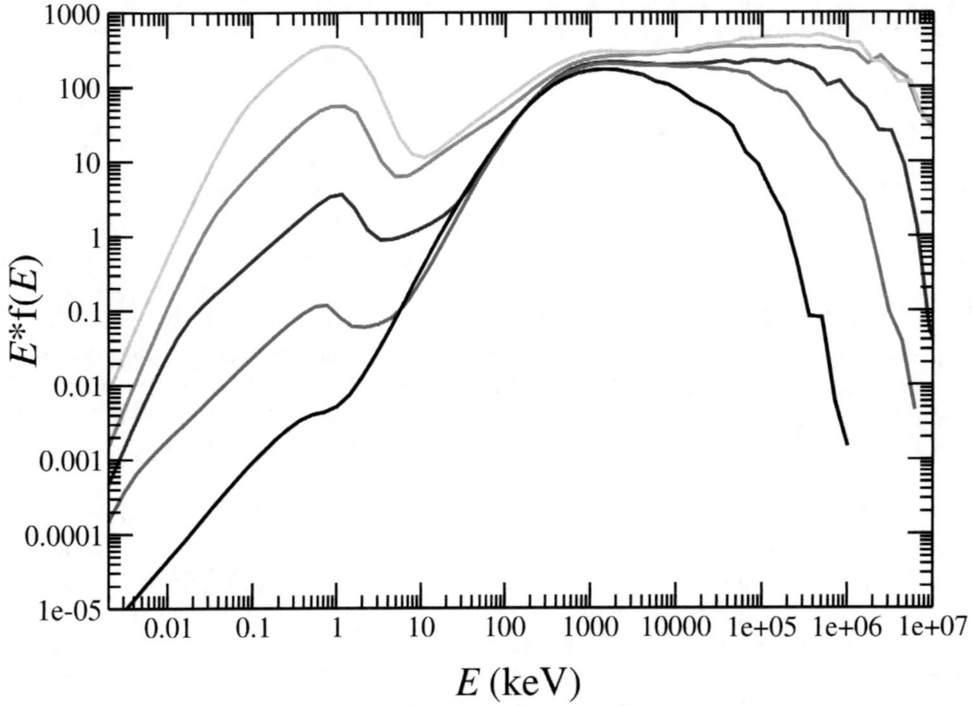

Figure 2. Resulting spectrum (in the central engine frame) for different baryon loadings of the flow. From bottom to top the curves correspond to magnetization σ_0 =40, 50, 60, 70, 100 (or corresponding baryon loading $\eta \simeq$ 250, 350, 460, 590, 1000) respectively. The high σ_0 flows are characterized broader spectra. The model predicts that bright prompt optical and UV emission is accompanied by powerful ~GeV emission. For bright optical emission, the optical spectrum is expected to be hard.

sub-MeV break and the spectral slopes above and below the break are naturally explained by the gradual dissipation model discussed here. Furthermore, the model makes specific predictions on the prompt emission from the optical to GeV; bands that are currently accessible to observations. The model predicts that the flat γ-ray tail extends up to a cutoff that typically appears at ~GeV. It also predicts the prompt optical and UV emission. For low baryon loading, the emission in these bands is powerful with energy flux $f(E)$ similar to that of X-rays. The optical emission is characterized by a hard spectrum. Optically bright bursts have powerful GeV emission and softer spectra below the ~1 MeV break.

In recent years, several observations in softer bands have been made simultaneously with the prompt GRB emission. The *Swift* satellite has observed the prompt emission in the X-rays and ultra violet (e.g. Page et al. 2007) and robotic telescopes in the optical and infra red (e.g. Akerlof 1999; Vestrad 2005; Blake 2005; Boër et al. 2006; Vestrand et al. 2006; Klotz et al. 2006). Furthermore, *FERMI* is expected to probe the emission from GRBs up to ~ 100 GeV.

Here, I compare the model to the very well sampled prompt emission of GRB 061121. This burst has been observed from optical to ~ 1 MeV (Page et al. 2007). The prompt emission

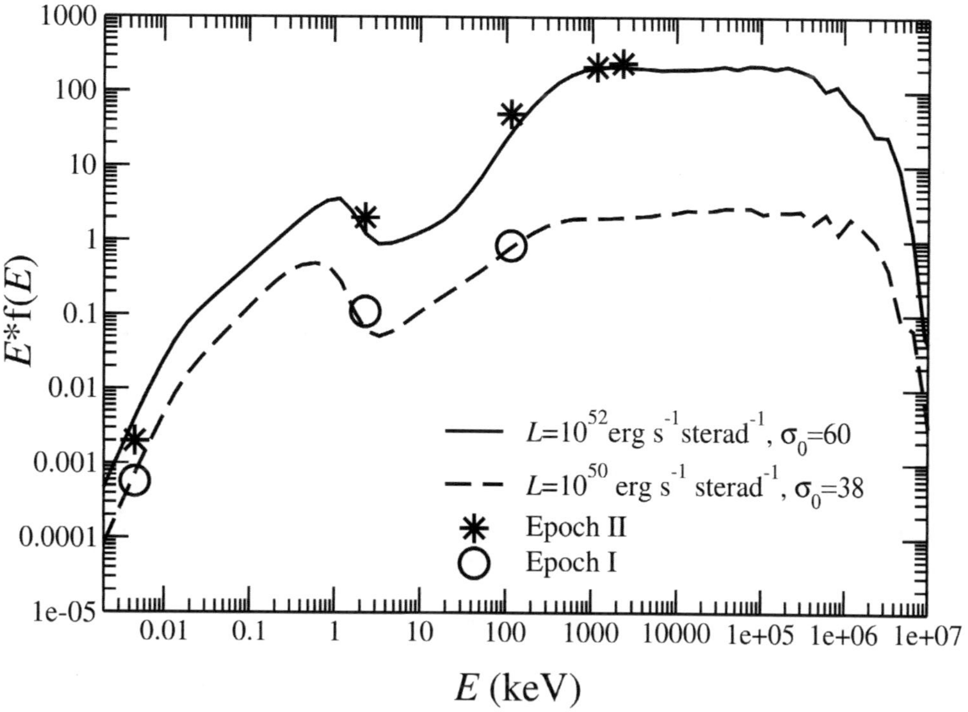

Figure 3. Comparison of the model to multi frequency observations of GRB 061121 (see Fig. 11 in Page et al. 2007). The circles stand for the observations of epoch I (just before the main pulse of the burst) and the stars for those of epoch II (during the pulse). Observations are blue-shifted by $1+z$ to the burst rest frame ($z = 1.131$ for GRB 061121). The solid and dashed curves show spectra for two different sets of the parameters of the flow that illustrate that the model can account for the broad-band prompt spectra.

has been followed with XRT and UVOT on board to *Swift* and in the optical with *ROTSE* simultaneously to γ-ray observations with BAT and *Konus-Wind.* There are two time resolved spectra just before and during the main pulse of the prompt emission that appears $\sim$ 75 sec after the onset of the burst. The pulse is clear in the lightcurves in all observed energy bands. The correlation between the different bands indicates that the optical, UV and X-ray and γ-ray components have common origin (i.e. they are connected to the prompt GRB). This is unlike cases where the optical lightcurves are not tracing the γ-rays (e.g. Akerlof 1999; Boër et al. 2006; Klotz et al. 2006) suggestive of a different physical origin with respect to that of the prompt emission.

In Fig. 3, the data shown with circles refer to the pre-spike emission (epoch I in the Page et al. 2007 terminology) and the stars to the peak observed luminosity of the burst (epoch II). The data span approximately 6 orders of magnitude in frequency from the optical to $\sim$ 1 MeV. Overplotted are the spectral predictions of the model for two different sets of parameters. The two models (not meant to be detailed fits) are reproducing the observations quite closely.

Note that for a given observed luminosity, the baryon loading is essentially the only free parameter of the model. This can be constrained by the ratio of the $\sim$ 1MeV-to-optical flux. Additional constraints can come from observations of the prompt emission in harder bands. In this respect *FERMI* observations in the $\sim$GeV range are going to be of particular importance. The high luminosity model (that describes the epoch II observations) is characterized by higher σ_0 with respect to the lower luminosity one. This is in qualitative agreement with the baryon loading-luminosity correlation during the evolution of the burst needed to explain observed energy-dependent properties of the GRB pulses in the context of the reconnection model (for details see sect. 4 in GS07). However since GS07 do not consider the Thomson thin emission in the calculations, the quantitative results of section 4 in GS07 have to be revisited.

Likewise, the model can be applied to the famous "naked eye" GRB 080319B. GRB 080319B is very luminous in the γ-ray band, exhibiting powerful optical emission that appears to loosely follow the γ-ray lightcurve. The optical emission is well above the extrapolation of the X-ray spectrum (arguing against the synchrotron interpretation of the MeV emission). In the context of the reconnection model, the very bright optical emission of the burst can be understood as coming from a very "clean" (high-η) and luminous flow. I estimate that for $\eta \sim 3000$, $L \sim 10^{54}$ erg/sec/sterad the model can explain the basic features of both the optical luminosity and γ-ray emission and peak energy of the spectrum of the burst. A weak point of the model is that it predicts that this high-η flow should have a rather flat spectrum with photon-number index $\Gamma \simeq -2$ above the $\sim$MeV break, in contrast to the observed spectrum which is very steep with photon-number index $\Gamma \simeq -3.6$ (Racusin et al. 2008).

6. Concluding

As an alternative to internal shocks, magnetic dissipation in a strongly magnetized flow can power the GRB (Thompson 1994). Magnetic dissipation may lead to gradual release of energy over a wide range of radii (e.g. Drenkhahn 2002; DS02). It typically proceeds in both Thomson thick and thin regions of the flow. The released energy can be distributed to a large fraction of the particles of the flow leading to the slow heating scenario for the GRB emission (Ghisellini & Celotti 1999; Stern & Poutanen 2004). The emitting particles (i.e. electrons) are heated up to mildly relativistic speeds. Because of the strong magnetic fields exchange of synchrotron photons provides an efficient mechanism for the thermalization of the electron distribution (Ghisellini et al. 1998). Since the emitting particles are thermal the resulting emission does not depend sensitively on poorly understood physics of particle acceleration in magnetic reconnection. The model is defined by just the luminosity of the flow, its baryon loading and a reconnection-rate parameter and makes direct and stable predictions for the electromagnetic spectrum.

The radiative transfer study was made with Monte Carlo simulations. Those calculations have shown that the flow is characterized by powerful photospheric emission with most of the energy appearing in the hard X-rays and $\sim$ 1 MeV $\gamma-$rays. This emission is the result of photons, produced deep into the flow, that are inverse Compton scattered by sub-relativistic electrons at Thomson optical depths of order of unity. Energy released at large radii leads to mildly relativistic electrons that cool down though emitting synchrotron radiation and inverse Compton scattering soft photons. SSA emission dominates the observed radiation in the soft X-ray and softer bands. Inverse Compton in the Thomson thin region leads to a flat high-energy spectrum that extends up to GeV energies.

The resulting spectra from the radiative transfer calculations naturally explain the observed sub-MeV break of the GRB emission and the spectral slopes below and above the break with photon-number index $\Gamma \sim -1$ and $\Gamma \sim -2$ respectively (Band et al. 1993). Furthermore, the model makes rather robust predictions for the emission in other energy bands. The flat γ-ray spectrum is expected to show a cutoff in the $\sim$GeV energy range that should be observable with *FERMI*. The optical and ultra violet emission can be powerful and the optical spectrum hard with photon number index $0 \lesssim \Gamma \lesssim 1$. Bright prompt optical emission is predicted to be accompanied by powerful $\sim$ GeV emission and rather soft spectra below the sub-MeV break.

Comparison with multi-frequency observations of the the prompt emission from GRB 061121 that span from the optical to the $\sim$ MeV range (Page et al. 2007) supports the model. The model can also explain the optical luminosity and γ-ray emission and peak energy of the spectrum of the "naked eye" GRB 080319B. A weak point of the model is that it predicts a rather flat spectrum above the MeV break with photon-number index $\Gamma \simeq -2$, in contrast to the observed spectrum which has a photon-number index of $\Gamma \simeq -3.6$ (Racusin et al. 2008).

In this model, the observed variability is direct manifestation of modulations of the luminosity (and baryon loading) of the GRB flow. An interesting deduction for the central engine is that the luminosity of the flow L correlates with the baryon loading $\eta \propto L^{\sim 0.6-0.7}$ in different bursts and *during* the evolution of a single burst (GS07). If the $L - \eta$ correlation holds, then the so-called Amati relation (Amati et al. 2002; Amati 2006) *and* the energy dependence of the width of pulses in a burst (e.g. Fenimore et al. 1995) can be naturally understood.

The $\sim$MeV emission in this model mainly comes from moderate Thomson optical depths with corresponding distance from the source of order $r_{\rm MeV} \sim 10^{12}$ cm. On the other hand, studies that attribute the steep decay of the prompt emission (that may last for tens of minutes after the burst) to the "high latitude" emission (Kumar & Panaitescu 2000), favor a much larger emission radius of $r_{\rm MeV} \gtrsim 10^{15}$ cm (Lazzati & Begelman 2006; Lyutikov 2006; Kumar et al. 2007). The "high latitude" emission is *not* relevant in determining the decaying part of the lightcurve in the reconnection model which typically lasts for a few seconds. In this model, the steep decaying part is reflection of the turning-off of the burst and not off-axis emission. The model in principle allows for steeper decline of the prompt emission (due to rapid turn off of the central engine), than that predicted from high-latitude emission models.

Acknowledgments

I thank Henk Spruit for stimulating discussions throughout the development of this work. I acknowledge support from the Lyman Spitzer Jr. Fellowship awarded by the Department of Astrophysical Sciences at Princeton University.

References

Akerlof C, et al. 1999 *Nature* **398** 400
Amati L, et al. 2002 *Astron. Astroph.* **390** 81
Amati L 2006 *Month. Not. Roy. Astron. Soc.* **372** 233
Band D, et al. 1993 *Astrophys. J.* **413** 281
Blake C H, et al. 2005 *Nature* **435** 181
Boër M, Atteia J L, Damerdji Y, Gendre B, Klotz A and Stratta G 2006 *Astrophys. J.* **638** L71
Coroniti F V 1990 *Astrophys. J.* **349** 538
Daigne F and Mochkovitch R 1998 *Month. Not. Roy. Astron. Soc.* **296** 275
Drenkhahn G 2002 *Astron. Astroph.* **387** 714
Drenkhahn G and Spruit H C 2002 *Astron. Astroph.* **391** 1141 (DS02)
Fenimore E E, in 't Zand J J M, Norris J P, Bonnell J T and Nemiroff R J 1995 *Astrophys. J.* **448** L101
Ghirlanda G, Celotti A and Ghisellini G 2003 *Astron. Astroph.* **406** 879
Ghisellini G, Haardt F and Svensson R 1998 *Month. Not. Roy. Astron. Soc.* **297** 348
Ghisellini G and Celotti A 1999 *Astron. Astroph. Suppl.* **138** 527
Giannios D 2006 *Astron. Astroph.* **457** 763 (G06)
Giannios D and Spruit H C 2005 *Astron. Astroph.* **430** 1
Giannios D and Spruit H C 2006 *Astron. Astroph.* **450** 887
Giannios D and Spruit H C 2007 *Astron. Astroph.* **469** 1 (GS07)
Giannios D 2008 *Astron. Astroph.* **480** 305 (G08)
Ioka K, Murase K, Toma K, Nagataki S and Nakamura T 2007 *Astrophys. J.* **670** L77
Katz J I 1994 *Astrophys. J.* **432** L107
Kirk J G and Skjæraasen O 2003 *Astrophys. J.* **591** 366
Klotz A, Gendre B, Stratta, G, Atteia, J L, Boër M Malacrino F, Damerdji Y and Behrend R 2006 *Astron. Astroph.* **451** L39

Kobayashi S, Piran T and Sari, R 1997 *Astrophys. J.* **490** 92
Kumar P and Panaitescu, A 2000 *Astrophys. J.* **541** L51
Kumar P, et al. 2007 *Month. Not. Roy. Astron. Soc.* **376** L57
Lazzati D and Begelman M C 2006 *Astrophys. J.* **641** 972
Lyubarsky Y E 2005, *Month. Not. Roy. Astron. Soc.* **358** 113
Lyubarsky Y and Kirk J G 2001 *Astrophys. J.* **547** 437
Lyutikov M and Blandford R 2003 ArXiv e-prints, arXiv:astro-ph/0312347
Lyutikov M 2006 *Month. Not. Roy. Astron. Soc.* **369** L5
Mahadevan R, Narayan R and Yi I 1996 *Astrophys. J.* **465** 327
Mészáros P and Rees M J 2000 *Astrophys. J.* **530** 292
Mimica P, Aloy M A, Müller E and Brinkmann W 2005 *Astron. Astroph.* **441** 103
Nakar E and Piran T 2002 *Astrophys. J.* **572** L139
O'Brien P T, et al. 2006 *Astrophys. J.* **647** 1213
Page K L, et al. 2007 *Astrophys. J.* **663** 1125
Paczynski B and Xu G 1994 *Astrophys. J.* **427** 708
Pe'er A, Mészáros P and Rees M J 2006 *Astrophys. J.* **642** 995
Petrosian V 1981 *Astrophys. J.* **251** 727
Preece R D, et al. 1998 *Astrophys. J.* **506** L23
Racusin J L, et al. 2008 *Nature* **455** 183
Rees M J and Mészáros P 1994 *Astrophys. J.* **430** L93
Spruit H C, Daigne F. and Drenkhahn G 2001 *Astron. Astroph.* **369** 694
Stern B 1999, in ASP Conf. Ser. 161, *High Energy Processes in Accreting Black Holes*, ed. J. Poutanen, & R. Svensson (ASP San Francisco) 277
Stern B E and Poutanen J 2004 *Month. Not. Roy. Astron. Soc.* **352** L35
Tavani M 1996 *Astrophys. J.* **466** 768
Thompson C 1994 *Month. Not. Roy. Astron. Soc.* **270** 480
Thompson C 2006 *Astrophys. J.* **651** 333
Usov V V 1992 *Nature* **357** 472
Vestrand W T, et al. 2005 *Nature* **435** 178
Vestrand W T, et al. 2006 *Nature* **442** 172

Journal of Physics: Conference Series **189** (2009) 012019 doi:10.1088/1742-6596/189/1/012019

Recent results on short-range gravity experiment

Maki Hata[1], Takashi Akiyama[1], Yuki Ikeda[1], Hirokazu Kawamura[1,2], Keigo Narita[1], Kazufumi Ninomiya[1], Naruya Ogawa[1], Toshiaki Sato[1], Etsuko Seitaibashi[1], Yuta Sekiguchi[1], Ryosuke Tsutsui[1], Kazumasa Yazawa[1] and Jiro Murata[1]

[1]Department of Physics, Rikkyo University, Nishi-Ikebukuro, Tokyo 171-8501, JAPAN
[2] RIKEN Nishina Center, 2-1 Hirosawa, Wako, Saitama 351-0198, JAPAN

htmaki@gmail.com

Abstract. According to the ADD model [1], deviation from Newton's inverse square law is expected at below sub-millimeter scale. Present study is an experimental investigation of the Newton's gravitational law at a short range scale. We have developed an experimental setup using torsion balance bar, and succeeded to confirm the inverse square law at a centimeter scale. In addition, composition dependence of gravitational constant G is also tested at the centimeter scale, motivated to test the weak equivalence principle.

1. Physics Motivation

1.1. Test of Inverse Square Law

Present study is to test the Newton's inverse square law at a short range scale, motivated on one of the most fundamental questions in the modern physics, that why gravitational force is so weak comparing to other three gauge interactions. Based on the ADD model [1], there is a possibility of dilution of gravitational field into bulk space towards directions of the large extra dimensions as illustrated in Figure 1.1

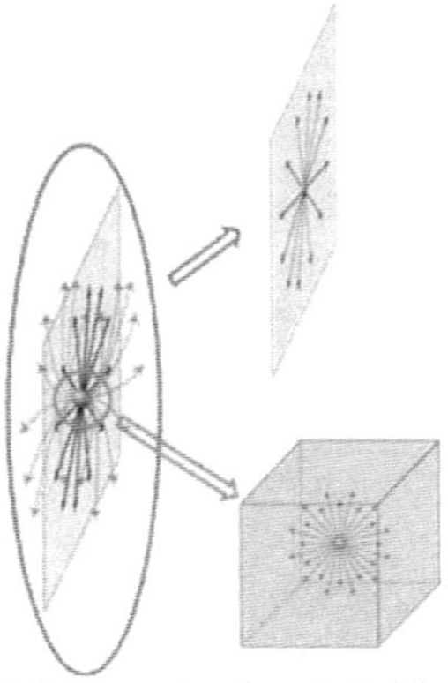

Figure 1.1: Image of dilution of the gravitational field towards extra-dimension world.

If " n " large extra dimensions exist, gravitational potential should be modified because of the Gauss's law, if the size of the extra dimension Λ is as large as the experimental scale. From the naive picture, most simple form of the expected gravitational potential is written as:

$$V_{r\geq\Lambda}(r) = -G\frac{Mm}{r},$$
$$V_{r\leq\Lambda}(r) = -G'\frac{Mm}{r^{1+n}}. \quad (1.1)$$

Here, n is the number of the extra dimensions, and $G' = \Lambda^n G$. In Figure 1.2, modified gravitational force is plotted for a case of $\Lambda = 2cm$ and $n = 2$, being compared with that for pure Newtonian gravity. It can be noticed that the gravitational force is enhanced at distances below $r < \Lambda$.

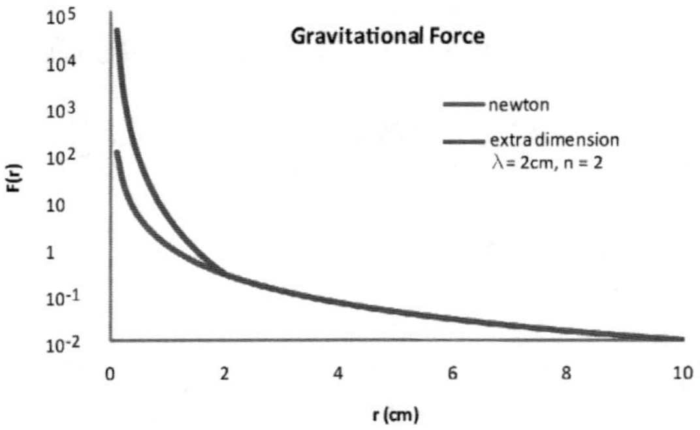

Figure 1.2: Modified gravitational force is plotted as a function of distance. "extra dimension" line is calculated for a case of $\Lambda = 2cm$ and $n = 2$ using formula (1.1), which is compared with the case of pure Newtonian gravity.

The functional form of the modified gravitational force is historically expressed using a Yukawa interaction form instead of (1.1);

$$V(r) = -G\frac{Mm}{r}\left(1 + \alpha \cdot exp\left[-\frac{r}{\lambda}\right]\right). \quad (1.2)$$

Here, λ is the range and α is the coupling constant of the "new" interaction. In this expression, the new Yukawa interaction term is interpreted as a new interaction arising from an exchange of a KK graviton, which has non-zero momentum component of the extra dimensions, therefore, observed as a massive graviton in our three dimensional world. The Compton wave length λ is obtained from the KK graviton mass m_{KK} as $\lambda = \hbar / m_{KK}c$.

For decades, existence of the new Yukawa term is experimentally searched in a wide range from cm to astrophysical scales. In Figure 1.3, experimentally excluded regions are shown. As shown in the Figure 1.3, precision measurement is performed only at around the Earth - Mars distance scale. Therefore, it can be said that there are huge room for the existence of the new Yukawa term at short range scale, which is theoretically predicted as described above.

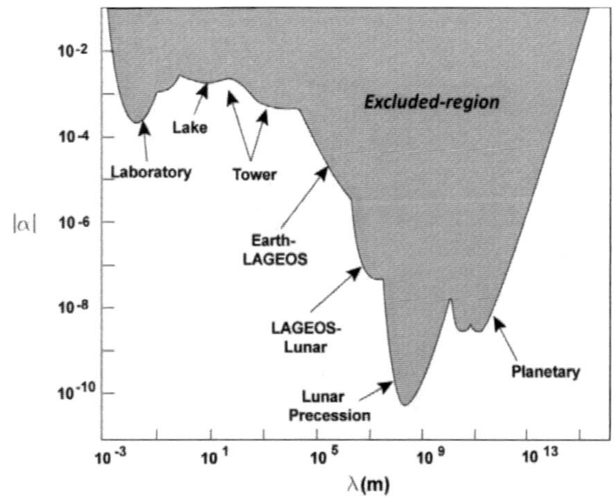

Figure 1.3: Summary of experimental search of the Yukawa term. Excluded region is illustrated as the shaded area. This figure is drawn using summarized data in [2].

In the present study, we are aiming to explore the existence of non-zero α around 4, based on a model of toroidal-compactification which predicts $\alpha = 2n$ [3]. In this sense, we are aiming to perform the highest precision test of the inverse square law at around 0.1mm scale, where $\alpha = 4$ is not excluded experimentally yet. Present study is a first step experiment towards such sub-millimeter test aiming to confirm the performance of our experimental devices in centimeter scale measurement.

1.2. Test of Weak Equivalence Principle

Our device can investigate not only the inverse square law, but also can test weak equivalence principle using the same experimental setup. The weak equivalence principle is expressed as followings; a ratio between inertial mass and gravitational mass is independent of its composition, source interaction of the mass energies. From the definition of the inertial mass $a = F / m_I$ and that of gravitational mass $F \propto m_g$, gravitational acceleration must satisfies $a \propto m_g / m_I$. Therefore, if the ratio m_g / m_I is constant for all the materials, universality of free fall must be expected. Thus, test of the universality of free fall is interpreted as a test of weak equivalence principle. Indeed, numbers of experiments have been performed in order to test the universality of free fall in this sense. However, it should be noticed that a violation of the universality of free fall do not directly implies the violation of weak equivalence principle. It is because existence of not only weak equivalence principle violating gravitational force, but also a new non-gravitational force can violate the universality of free fall, whereas weak equivalence principle can live with such non-gravitational force.

Direct experimental confirmations of the weak equivalence principle have been performed, for example, by measuring the ratio between gravitational force from the Earth/the Sun, and the inertial force by the rotation of the Earth. From these results, weak equivalence principle can be believed with high precision in such long range regions.

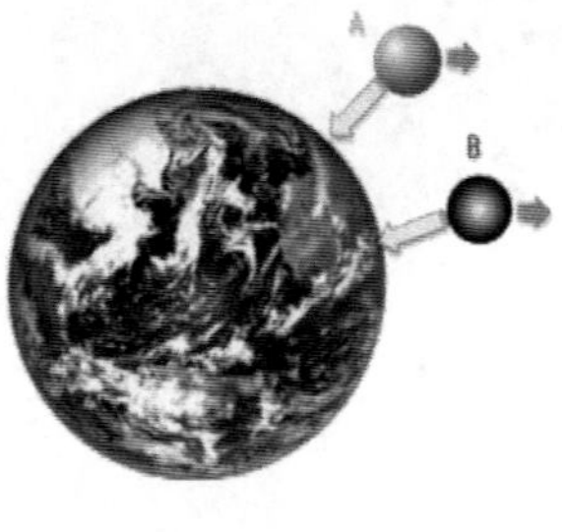

Figure 1.4: Many experiments confirmed that the ratio between gravitational force and inertial force is independent for the different compositions in the Earth system.

On the other hand, there are no experimental investigations at around the short range regions yet. In order to directly perform the test measurement, we have checked the weak equivalence principle by testing universality of the gravitational constant G for different compositions. Numbers of experimental tests confirmed that ratio between inertial mass for different compositions is same for the ratio between gravitational mass.

$$\frac{m_I(A)}{m_I(B)} = \frac{m_{g-long}(A)}{m_{g-long}(B)}. \qquad (1.3)$$

Here, A and B represent material names. m_{g-long} denotes gravitational mass determined using long range gravitational force. This is same as the test of weak equivalence principle at long rage regions. Indeed, (1.3) can be rewritten as

$$\frac{m_I(A)}{m_{g-long}(A)} = \frac{m_I(B)}{m_{g-long}(B)}, \qquad (1.4)$$

which directly implies the weak equivalence principle. Then, if we measure the gravitational mass at short range scales, we can test the new relation

$$\frac{m_{g-long}(A)}{m_{g-long}(B)} = \frac{m_{g-short}(A)}{m_{g-short}(B)}, \qquad (1.5)$$

which leads

$$\frac{m_I(A)}{m_{g-short}(A)} = \frac{m_I(B)}{m_{g-short}(B)}, \qquad (1.6)$$

by assuming (1.3). If we suppose that the gravitational mass is independent of the force range, scale and composition dependence can be included into the gravitational "constant" G. Therefore, short range test of composition dependence of the gravitational constant G can be interpreted as a test of weak equivalence principle for a short range gravitational force.

There is a theoretical model which predicts a violation of the weak equivalence principle. A new interaction which couples to baryon number, instead of gravitational mass is proposed by Lee and Yang [4], motivated for providing existence reason for the observed baryon number conservation law. If such new interaction exists, the weak equivalence principle seems to be violated, because the "gravity-like" force, which is sum of the gravitational force and the new baryon number coupling force, is no more proportional to the gravitational mass. The baryon number coupling force must be similar to the gravitational force because mass density is roughly proportional to the baryon number. However, gravitational mass is slightly different from baryon number time nucleon mass, because of the mass defect and proton/neutron mass difference.

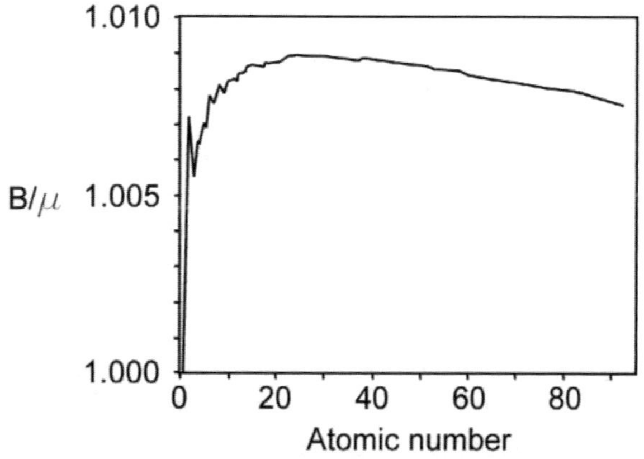

Figure 1.5: Ratio between baryon number and mass is plotted for various materials as function of atomic number. Here, μ is defined as $m = \mu \times m_{hydrogen}$ using hydrogen mass.

As shown in Figure 1.5, ratios between baryon number and gravitational mass are not constant for different materials, therefore, ratio between gravitational force and baryon number coupling force must be slightly different for different materials. Therefore, existence of the new baryon number coupling force can be examined by measuring the "gravity-like" force between different substances.

In this model [4], the new baryon number coupling term is added using new coupling constant f.

$$V(r) = -G\frac{m_A m_B}{r} \rightarrow -G\frac{m_A m_B}{r} + f^2\frac{B_A B_B}{r}. \qquad (1.7)$$

Now, for we are interested in a new interaction which violates weak equivalence principle only at short range, (1.7) should be modified using Yukawa form [2]:

$$V(r) = -G\frac{m_A m_B}{r} \rightarrow -G\frac{m_A m_B}{r}\left(1 - \xi_B \frac{B_A B_B}{\mu_A \mu_B} exp\left[-\frac{r}{\lambda}\right]\right). \qquad (1.8)$$

Here, B_A is baryon number for material A, and μ_A is gravitational mass in hydrogen mass unit. Therefore, violation of the weak equivalence principle at short range scale is appeared as an existence of composition depending α term in the extra dimension model (1.2). This relation is represented as,

$$\alpha \equiv \alpha_{AB} = \xi_B \frac{B_A B_B}{\mu_A \mu_B}. \qquad (1.9)$$

In the above expressions, gravitational constant G is supposed to be constant. Then, in a different expressions where we suppose inverse square law, but do not suppose universality of the gravitational constant G ,i.e.,

$$V = -G_{AB}\frac{m_A m_B}{r}, \qquad (1.10)$$

the new composition depending gravitational constant is defined using ξ_B;

$$G_{AB} = G(1 + \xi_B \frac{B_A B_B}{\mu_A \mu_B} exp\left[-\frac{r}{\lambda}\right]). \qquad (1.11)$$

Therefore, by measuring the composition dependence of the gravitational constant G, we can test the existence of non-zero ξ_B.

Apart from the baryon number coupling force, it can be shown that our test on the gravitational constant has sensitivity on weak equivalence principle, not concerning only for different composition, but also for different type of the mass source interactions. It is because both of the electromagnetic and strong interactions contribute on the nuclear mass, and the contribution fractions of them are varying as function of atomic numbers. For example, Coulomb energy contribution on a nuclear binding energy is about 20% for gold, whereas that for aluminium is only 5%. Therefore, we can say that our test on the composition dependence on the gravitational constant is a model independent test of the universality of free fall, for different source of the mass energy, too.

2. Experiment

In our experiment named Newton-I, we use a torsion balance bar as described in [5]. Sample CCD camera images of the torsion balance bar are shown in Figure 2.1 In Figure 2.1, torsion balance bar is pictured at the centre. Attractor positions are also shown. In this experiment, angular displacement of the torsion balance bar is the gravity signal. For example, expected angular displacement is about 0.1 degrees for the case of Figure 2.1 The tiny angular displacement is extracted by digital image analysis [6].

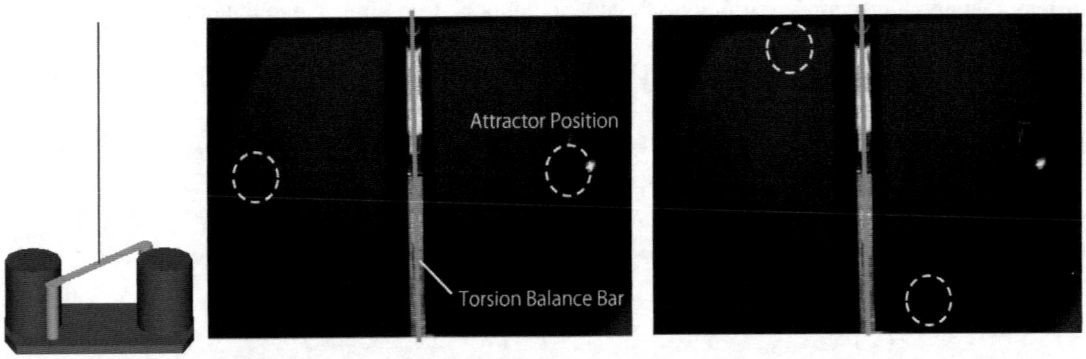

Figure 2.1: Shape of the torsion balance bar and sample CCD images. Tiny angular displacement of the torsion balance bar is extracted by digital image analysis.

Using the torsion balance bar system, we have performed a series of experiment, Newton-I, at around centimeter scales. Typical results from Newton-I experiment are shown in Figure 2.2, where time evolution of torsion balance bar angle is plotted for the cases of lead, copper and aluminum attractors. The expected gravity signal is clearly observed for all the attractor materials, which are in good agreement with the Newton gravity predictions in absolute scale. From this result, density dependence can be also confirmed.

Position dependence is examined by measuring the attractor setting angle dependence on the torsion balance bar angular displacement. The results are shown in Figure 2.3 We can clearly confirm the position dependences, which is also consistent with Newton gravity predictions quantitatively. The error bars are statistical error, which is obtained by standard deviations of multiple measurements. The largest systematic error in this measurement is electric and magnetic influences. They are evaluated by a separate dedicated measurement. From the results, electric influences are negligible thanks to using the electric shield. Although using non-magnetic materials, magnetic influences are observed. Typical magnetic effect is also shown in Figure 2.3

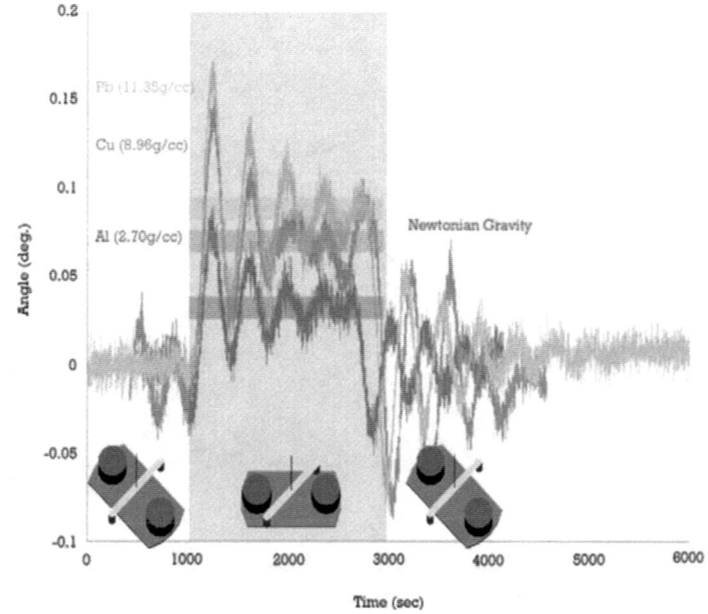

Figure 2.2: Typical results in Newton-I experiment. Time evolution of torsion balance bar angle is plotted for the cases of lead, copper and aluminum attractors. Newton gravity estimations are also shown as references.

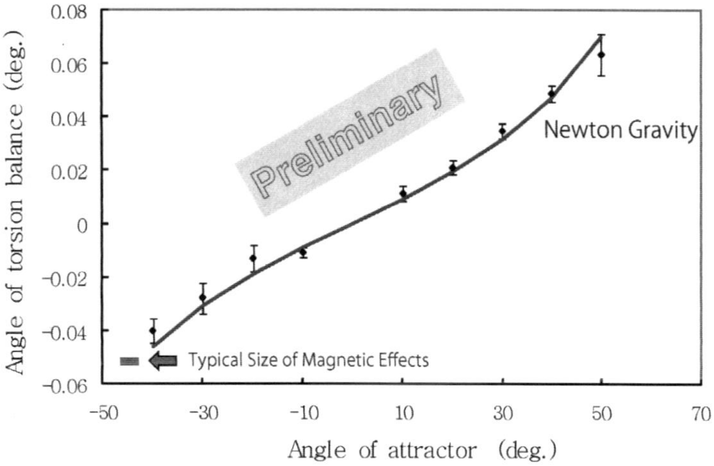

Figure 2.3: Position dependence of the torsion balance bar angle is shown. Solid line represents the Newton gravity prediction. Size of the small square shows a typical size of magnetic effects.

3. Discussion

Obtained results are compared with the previous experiments. In Figure 3.1, current upper limit of the Yukawa term is plotted as function of position scale in short range regions. It should be noticed that there are no precision measurement at sub-millimeter scales. Although our result is not the highest precision data, it is approaching to the current upper limits. In order to perform improved precision measurement, we have built a next generation experimental device, Newton-III, using completely different measurement principle.

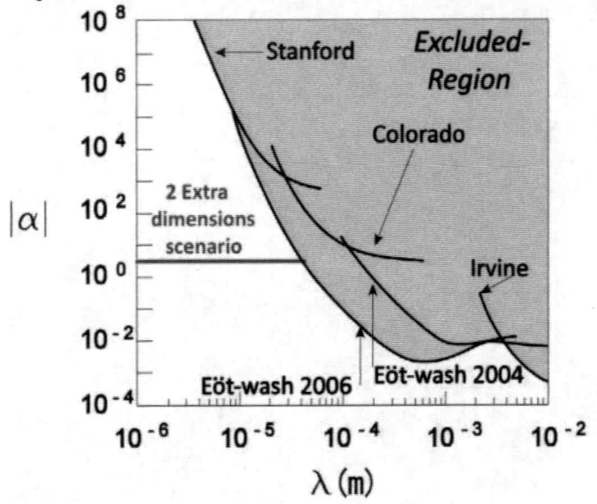

Figure 3.1: Current upper limit on α in short range scales.

As for the substance dependence of gravitation constant G, we obtained them for between brass target and aluminum, copper and lead attractors. The results for $G_{Brass\text{–}Lead} / G_{Brass\text{–}Copper}$ and $G_{Brass\text{–}Aluminum} / G_{Brass\text{–}Copper}$ are shown in Figure 3.2 From the composition dependence on gravitation constant G, we can extract the upper limit on ξ_B. The results are also shown in Figure 3.2. Note that our data from Newton-I experiment gives the most precise limit on the baryon number coupling ξ_B at the shortest scale. In the Newton-I experiment, the largest systematic error on the composition dependence on gravitation constant G comes from attractor replacement. In order to cancel out the systematic effects and to obtain more precise results, we are building next generation experiment Newton-II, which is described in detail in [5].

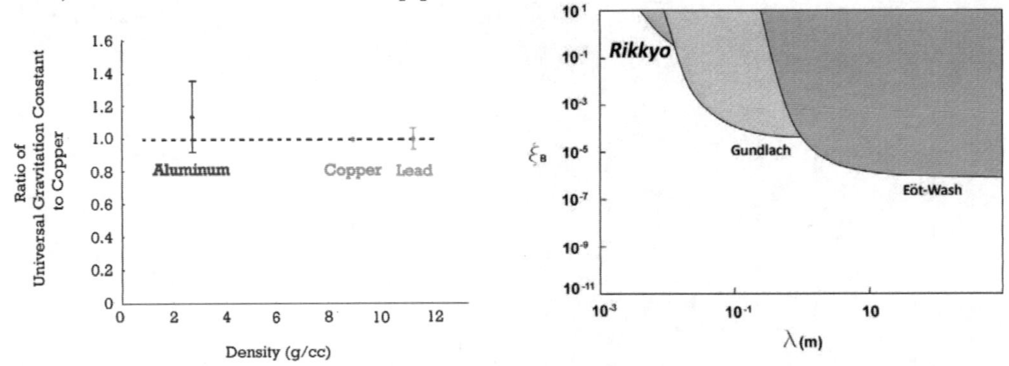

Figure 3.2: Results on composition dependence of the gravitational constant G and the corresponding upper limit on the baryon number coupling force. Data from [2] is used to plot ξ_B.

The results on the test of inverse square law can be compared with the sensitivities of high energy collider search. In the ADD model, Planck-mass in "4+n" dimensional world is modified if scale of extra dimension Λ is large.

$$m_{4+n}{}^{2+n} = \left(\frac{\hbar}{c}\right)^n \frac{m_{pl}{}^2}{\Lambda^n}. \quad (3.1)$$

Here, m_{4+n} is the modified Planck mass in $4+n$ dimensional world, m_{pl} is normal Planck mass. High energy collider can search gravitational effects if their leptonic/partonic energy is reaching the modified Planck energy $m_{4+n}c^2$, by searching creation of small black holes or KK particles. On the other hand, table top experiments which directly examine the inverse square law can see a deviation if the experimental scale is smaller than the size of the extra dimension Λ. Considering (3.1), they can be compared with each others. In Figure 3.3, sensitivities of them are plotted for comparison. It can be noticed that our results are competitive to that of highest energy colliders, including LHC. In Newton-III experiment, we are aiming to reach highest sensitivity, by utilizing the new experimental techniques.

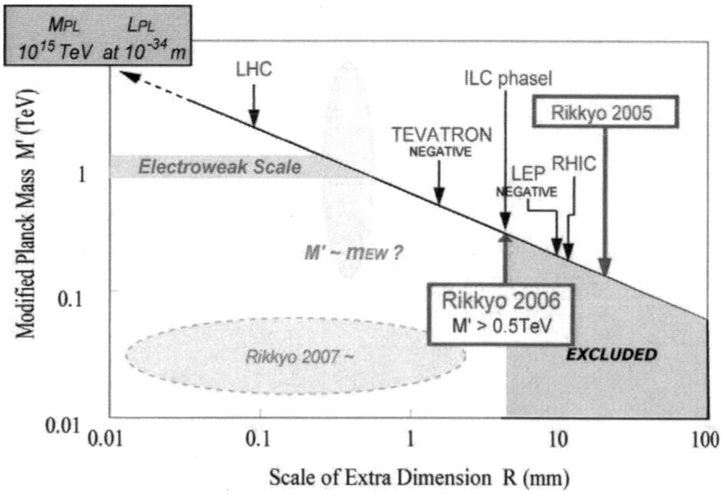

Figure 3.3: Comparing plot of sensitivity on extra dimension in ADD model, for high energy collider and table top experiments.

4. Summary

We have built a prototype apparatus Newton-I, and succeeded to demonstrate its ability to test the Newton's inverse square law in centimeter scale. For the detector resolution is better enough comparing to the current requirement, we will be able to access well below sub-millimeter scale, where even LHC cannot be accessible. We will also be able to test the weak equivalence principle in the shortest range scale. In order to utilize the full ability of our detector system, we are trying to build a new experimental setup, Newton-III, based on a new detection principle without using torsion balance pendulum.

Acknowledgments

Part of the present works are performed as undergraduate student experiments. The author thanks to Y. Miyano, M. Takahashi, T. Tsuneno, T. Amanuma, S. Danbara, T. Iino, S. Mizuno, Y. Araki, T. Ohmori, Y. Sakurai, S. Yamaoka, H. Ooishi for their vital efforts on this work. This work is supported by Grant-in-Aid for Exploratory Research (18654048, 20654024)

Journal of Physics: Conference Series **189** (2009) 012019 doi:10.1088/1742-6596/189/1/012019

References

[1] Nima Arkani-Hamed, Savas Dimopoulos, Gia Dvali, Physics Letters B 429 (1998) 263-272.
[2] Ephraim Fishback, Carrick Talmadge, Nature 356 (1992) 207 – 215; "The Search for Non-Newtonian Gravity", Springer (1998) ISBN 0387984909
[3] E.G. Floratos, G.K. Leontaris, Physics Letters B 465 (1999) 95-100.
[4] T.D. Lee and C.N. Yang, Physical Review 98 (1955) 1501.
[5] Kazufumi Ninomiya *et.al.*,"New Experimental Technique for Short-Range Gravity Measurement" in this volume
[6] J. Murata *et. al.*, Nucl. Instrum. Meth. A 500(2003) 309-317; "Pico-Precision Displacement Sensor using Digital Image Analysis", IEEE Nuclear Science Symposium Conference Record 675 (2005).

Journal of Physics: Conference Series **189** (2009) 012020 doi:10.1088/1742-6596/189/1/012020

Brane-world black holes

Panagiota Kanti

Division of Theoretical Physics, Department of Physics, University of Ioannina, Ioannina GR 45110, Greece

E-mail: pkanti@cc.uoi.gr

Abstract. In this talk, I present and discuss a number of attempts to construct black hole solutions in models with Warped Extra Dimensions. Then, a contact is made with models with Large Extra Dimensions, where black-hole solutions are easily constructed – here the focus will be on the properties of microscopic black holes and the possibility of using phenomena associated with them, such as the emission of Hawking radiation, to discover fundamental properties of our spacetime.

1. Introduction

A breakthrough in string theory [1, 2, 3], that decoupled its fundamental scale from the Planck scale, led to the construction of two higher-dimensional gravitational models a decade ago: the scenario with Large Extra Dimensions [4, 5] and the one with Warped Extra Dimensions [6] (for some early higher-dimensional gravitational models, see [7, 8, 9, 10]). Both models use the idea of a 3-brane that plays the role of our 4-dimensional world and is embedded in a higher-dimensional bulk: n flat, spacelike dimensions in the first case, and a fifth spacelike dimension that is part of an Anti de Sitter spacetime in the second. In both cases, all ordinary matter is localised on the brane, while gravitons have access to the whole spacetime.

The main point in both models is that the introduction of a higher-dimensional spacetime is supplemented by a new fundamental scale. In the Large Extra Dimensions scenario the higher-dimensional theory has a new scale for gravity, M_*, that is related to the effective 4-dimensional Planck scale through the equation $M_P^2 \simeq \mathcal{R}^n\, M_*^{2+n}$ [4, 5]. If the size of the extra dimensions satisfies the inequality $\mathcal{R} \gg l_P$, M_* can be significantly lower than the 4-dimensional one M_P, and the fundamental gravitational constant $G_D = 1/M_*^{n+2}$ much larger than the 4-dimensional one G_D. In the case of Warped Extra Dimensions, it is the Electroweak scale that is derived from the fundamental one through the relation $M_{EW} = e^{-kL}\, M_*$, where L is the distance between our observable brane and a hidden one, and k the curvature scale associated with the negative cosmological constant that fills the 5-dimensional spacetime of the model [6].

In the context of these two scenaria, we are looking for black hole solutions. As we will see, the low scale for gravity of the Large Extra Dimensions will have important consequences for the creation and evaporation process of black holes in these theories. On the other hand, the richer topological structure of the model with Warped Extra Dimensions will prove to be a major obstacle for the construction of black-hole solutions in these theories. We will thus start with the latter class of models and present a number of approaches to this problem, their results, and the new questions emerging. We will see how a natural contact is made between the two scenaria when it comes to black holes, and thus we will move to the Large Extra Dimensions scenario for the study of the much simpler, and easier to construct, black-hole solutions. A brief study of the creation process of the microscopic black holes and their properties will be made, and many current results on the characteristics of their evaporation process will be presented (for a number of reviews where more information is offered on black-hole solutions in both types of models, see [11, 12, 13]).

2. The Warped Case

Despite the presence of additional spacelike dimensions in the Large and Warped Extra Dimensions scenaria, our conception of the creation of a black hole has not changed: whenever ordinary matter, which according to the assumptions of the models is localised on our brane, undergoes gravitational collapse, we expect a black hole to form. We also anticipate that the produced black holes will have the following characteristics:

- *Location :* Being a gravitational object, the black hole will be, at least initially, attached to our brane but it will also extend off our brane
- *Geometry :* Due to the warping along the fifth dimension in the Warped Extra Dimensions Scenario, we expect the black hole to have an additional cylindrical symmetry
- *Asymptotics :* Due to the localization of the source on the brane, the black hole should have a regular asymptotic form both on the brane (Minkowski, de Sitter or Anti de Sitter) and off the brane (Anti de Sitter).

However, in practice, the task of constructing a black hole solution with the above characteristics has proven to be extremely challenging. The attempts to overcome this difficulty have led to different approaches of the problem, some of which are discussed below:

2.1. "The-brane-observer-comes-first" point of view

Starting from a pure 4-dimensional Schwarzschild solution on the brane, the authors of [14] embedded it in a 5-dimensional Anti de Sitter spacetime. The line-element then assumed the form:

$$ds^2 = e^{-2k|y|}\left[-\left(1-\frac{2M}{r}\right)dt^2+\left(1-\frac{2M}{r}\right)^{-1}dr^2+r^2\,d\Omega_2^2\right]+dy^2\,. \tag{1}$$

The above ansatz was an attractive choice as it satisfied the Randall-Sundrum field equations without the need for any extra bulk matter T_{MN} – since the Schwarzschild solution is a vacuum solution and the above line-element was of a factorised type – and its projection on the brane was a well-known black hole solution. However, the five-dimensional background was *not*. By using the above line-element, one easily finds that

$$R_{MNRS}\,R^{MNRS} \sim \frac{48M^2 e^{4k|y|}}{r^6}\,. \tag{2}$$

The above results reveals that the metric (1) describes a black string and not a black hole as the singularity at $r=0$ extends over the whole range of the fifth coordinate. As it was later shown [15], the same solution was plagued by the well-known Gregory-Laflamme [16] instability of black string solutions that turns the black string into an infinitely long row of 'black cigars'.

2.2. "The-brane-observer-is-all-that-matters" point of view

In this approach [17], the view was taken that although it is indeed difficult to find an analytic solution describing a 5-dimensional localised-close-to-the-brane black hole, one may still study its 4-dimensional projection on our brane, which is, in any case, what a brane observer would ever see. For the construction of the 4-dimensional projected line-element, the following brane field equations were used [18]

$$^{(4)}G_{\mu\nu} = 8\pi G_4 T_{\mu\nu} + (8\pi G_5)^2\, \pi_{\mu\nu} + E_{\mu\nu}\,, \tag{3}$$

where $\pi_{\mu\nu}$ is a tensor quadratic to the energy-momentum tensor $T_{\mu\nu}$ – and thus may be ignored in the low-energy limit – and $E_{\mu\nu} =^{(5)} C_{y\mu y\nu}$ is a projected-on-the-brane 5-dimensional Weyl tensor. Although brane black hole solutions could be derived from the above field equations, this could only be done by employing certain assumptions about the form of $E_{\mu\nu}$, whose exact expression can follow only by solving the complete 5-dimensional gravitational problem.

2.3. *"Restore-the-bulk" point of view*

Returning to the approach taken by the authors of [14], an alternative idea was proposed in [19] according to which the 5-dimensional line-element assumed the non-factorised form

$$ds^2 = e^{-2k|y|}\left[-\left(1-\frac{w(y)}{r}\right)dt^2 + \left(1-\frac{w(y)}{r}\right)^{-1} dr^2 + r^2\, d\Omega_2^2\right] + dy^2\,. \tag{4}$$

For the above line-element to satisfy the 5-dimensional bulk equations, an additional bulk energy momentum tensor had to be introduced. Its presence then guaranteed that a localising mechanism for the black-hole singularity was in action if we demanded that

$$w(y) \to 2M \quad (y=0)\,, \qquad w(y) \to 0 \quad (y \to \infty)\,. \tag{5}$$

When the bulk equations were solved for the expression of the energy-momentum tensor, the profile, shown in Fig. 1, was found. The bulk matter had a shell-like distribution along the extra dimension thus engulfing the brane and restricting the extent of the black hole singularity. In addition, it satisfied a stiff equation of state with $p_r \simeq \rho$, and it obeyed the energy conditions on the brane: $\rho > 0$ and $\rho + p_i > 0$. The main drawback of the model was that conventional scalar or gauge field configurations failed to support such an energy-momentum tensor.

2.4. *"The-bulk-comes-first" point of view*

A different approach was taken in [20] where a regular 5-dimensional black hole solution was put at 'the center of the stage' (see also [21]). The line-element was assumed to be of the form

$$ds^2 = -U(r)\, dt^2 + \frac{dr^2}{U(r)} + r^2\,(d\chi^2 + \sin^2\chi\, d\Omega_2^2)\,, \tag{6}$$

describing a black hole in Anti de Sitter spacetime with

$$U(r) = 1 + k^2 r^2 - \frac{\mu}{r^2}\,. \tag{7}$$

In this background, a 3-brane was then introduced through the junction conditions $[K_{\mu\nu} - K\,h_{\mu\nu}]^+_- = \kappa_5\, T_{\mu\nu}$ [22], where $h_{\mu\nu}$ is the induced metric on the brane, $K_{\mu\nu}$ the extrinsic curvature and $T_{\mu\nu}$ the brane energy-momentum tensor of a perfect fluid. Since $U(r)$ was known, the junction conditions formed a system of coupled differential equations for $\chi(t,r)$ (the trajectory of the brane), $\rho(t,r)$ (the brane energy density) and $w(t,r)$ (the brane equation of state). After a detailed analytical study, the following conclusions were drawn: (i) all regular trajectories stayed clear of the bulk black-hole horizon, (ii) no regular black-hole solution with the singularity on the brane was found, and (iii) the most physically interesting solution had ρ and p on the brane resembling those of a star (for a set of brane trajectories, see Fig. 2).

Journal of Physics: Conference Series **189** (2009) 012020 doi:10.1088/1742-6596/189/1/012020

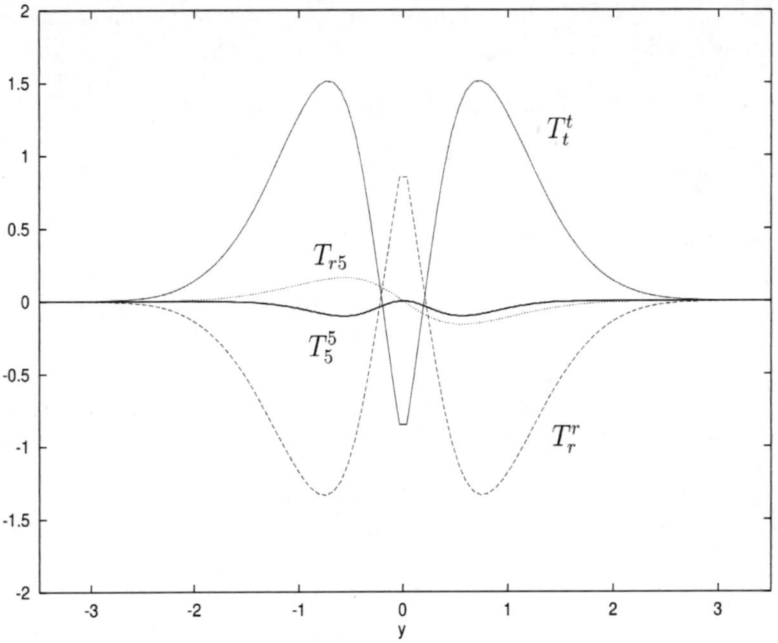

Figure 1. The profile of the bulk energy-momentum tensor T_{MN} along the fifth dimension [19].

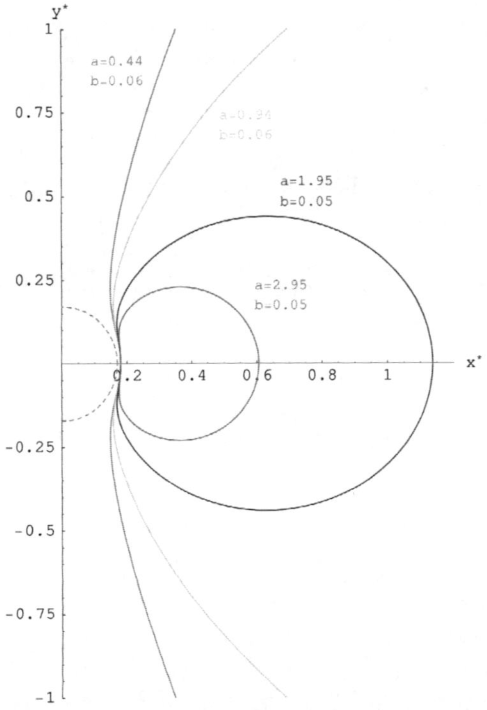

Figure 2. Brane trajectories in a 5D Schwarzschild-AdS background [20].

Journal of Physics: Conference Series **189** (2009) 012020 doi:10.1088/1742-6596/189/1/012020

2.5. "Should-these-BH's-exist-at-all?" point of view

By using the AdS/CFT correspondence, in [23] the action of the Randall-Sundrum model was written as

$$S_{RS} = S_{EH}^{(4)} + 2W_{CFT}\,, \tag{8}$$

where $S_{EH}^{(4)}$ is the Einstein-Hilbert action and W_{CFT} the connected Green function of a Conformal Field Theory in 4 dimensions. According to the above, in the context of the 4-dimensional CFT, if a black hole is formed, it does so in the presence of a large number of conformal fields. This will inevitably induce emission of Hawking radiation of CFT modes with a significant back-reaction to the metric. As a result, a static localised black hole with $r_H \gg \ell_{AdS}$ (the regime in which the AdS/CFT correspondence is valid) may *not exist* at all in a brane-world theory. Similar claims were made around the same time in the context of other works (see, for example, [24, 25]).

With the difficulty in constructing analytically regular black hole solutions in warped models having been clearly established, numerical analyses were undertook next. In [26] the following ansatz for the 5-dimensional black-hole line-element was considered

$$ds^2 = \frac{\ell_{AdS}^2}{z^2}\left[-T^2 dt^2 + e^{2R}\,(d\rho^2 + \frac{\rho^2}{4\xi}\,d\xi^2) + r^2\,e^{2C}\,d\Omega_2^2\right]. \tag{9}$$

The boundary conditions supplementing the desired solution were: (i) the existence of a horizon at a constant value ρ_h, (ii) the demand that the metric functions and their derivatives are finite at ρ_h, and (iii) that at asymptotic infinity, the Anti de Sitter spacetime is recovered with $T \to 1$ and $R, C \to 0$. In the numerical analysis that was performed, small, static, asymptotically AdS black holes were indeed found but with $\rho_h \leq \ell_{AdS}/5$. A similar analysis in 6 dimensions [27] extended that limit up to $\rho_h = 2\ell_{AdS}$. No larger black holes in a warped spacetime have been found numerically so far, a result that seems to support the claims of [23, 24, 25].

A recent numerical analysis performed in [28] followed the construction method of [20] and considered a 5-dimensional topological black hole in Anti de Sitter spacetime with line-element

$$ds^2 = -U(r)\,dt^2 + \frac{dr^2}{U(r)} + r^2\,d\Omega_{3(\beta)}^2\,, \tag{10}$$

with

$$U(r) = \beta + k^2 r^2 - \frac{\mu}{r^2}\,, \qquad (\beta = +1, 0, -1)\,. \tag{11}$$

Then, a brane was introduced through the junction conditions and its trajectory and induced metric were sought for with the condition that an apparent horizon was formed. Once again, no black solutions, i.e. solutions with an apparent horizon, with horizon area larger than that of a black string – which always arises as a solution of the field equations – were found for $A_{AH} \geq \mathcal{O}(\ell_{AdS}^3)$. As soon as the horizon radius of the black object fell below the AdS length, black hole solutions emerged, too.

The analytical studies so far have clearly demonstrated the difficulty in constructing regular, localised black-hole solutions in the context of a warped spacetime, even in the simplest of the models. Numerical studies up to now seem to support claims that the largest class of these objects might not exist at all. Several open questions remain: why the black string solution was so easy to find but not the same happens with the black hole one? Is the stability of the sought solutions the answer? But the black string solution was also unstable yet it was found – and its instability was demonstrated later. If small black holes exist in warped models why we cannot find not even one by analytical means? And why quantum effects are so important for large black holes that they forbid their existence as classical solutions? The community remains divided waiting anxiously for a breakthrough.

3. The Flat Case

From the discussion of the previous section, it becomes clear that, in the context of the warped models, the only type of black holes that arise naturally from our field equations are the ones with $r_H \leq \ell_{AdS}$. If we are then to study the class of these small black holes, a particularly convenient limit is the one where $r_H \ll \ell_{AdS}$. Under this assumption, such a black hole lives, to first approximation, in a 5-dimensional *flat* spacetime with negligible warping, where the fifth dimension enters the line-element on an equal footing as the other three spacelike ones. One might wish to generalise the model by adding a number of additional spacelike dimensions, in which case a contact is made with the Large Extra Dimensions Scenario: here, a flat higher-dimensional line-element can be used under the assumption that $r_H \ll \mathcal{R}$ so that the black hole can not distinguish the finite-sized extra dimensions from the infinite-sized usual ones.

The simplest line-element that we may write that describes a spherically-symmetric, neutral black hole living in a $(4+n)$-dimensional spacetime is the Schwarzschild-Tangherlini one [29, 30]

$$ds^2 = -\left[1-\left(\frac{r_H}{r}\right)^{n+1}\right] dt^2 + \left[1-\left(\frac{r_H}{r}\right)^{n+1}\right]^{-1} dr^2 + r^2 d\Omega_{2+n}^2\,, \tag{12}$$

where $d\Omega_{2+n}^2$ is the line-element of a $(2+n)$-dimensional unit sphere

$$d\Omega_{2+n}^2 = d\theta_{n+1}^2 + \sin^2\theta_{n+1}\left(d\theta_n^2 + \sin^2\theta_n\left(... + \sin^2\theta_2\left(d\theta_1^2 + \sin^2\theta_1\, d\varphi^2\right)...\right)\right). \tag{13}$$

By applying the Gauss law in $D = 4+n$ dimensions, we find for the horizon radius the result [30]

$$r_H = \frac{1}{M_*}\left(\frac{M_{BH}}{M_*}\right)^{\frac{1}{n+1}}\left(\frac{8\Gamma(\frac{n+3}{2})}{(n+2)\sqrt{\pi}^{(n+1)}}\right)^{1/(n+1)}. \tag{14}$$

The well-known linear relation between the mass and the horizon radius of the black hole holding in 4 dimensions arises readily by setting $n = 0$ in the above expression, and substituting the fundamental Planck scale M_* with the 4-dimensional one M_P – as we will shortly see, the presence of the lower scale M_* in the expression of r_H plays an important role on deciding whether black holes may be created at high-energy particle collisions as well as on their properties.

The line-element (12) satisfies the higher-dimensional gravitational equations in the vacuum – the mass of the black hole is assumed to be much larger than M_* so that the black hole can be considered as a classical object. Since both Λ_B and the brane self-energy σ are assumed to be of the same order of the fundamental Planck scale, they also may be ignored and as a result no junction conditions need to be solved, an assumption that simplifies significantly the study of the black hole.

The most optimistic, and also the most phenomenologically interesting, scenario is the one in which M_* is of the order of a few TeV. In that case, collider experiments with $E > M_*$ can access the so-called trans-planckian energy regime and probe strong gravity phenomena! The idea that heavy, extended objects, such as p-branes and black holes, could then produced at ground-based accelerators during high-energy scattering experiments was put forward in [31]. If, during such a process, the impact parameter b between the colliding particles is larger than the Schwarzschild radius $r_H(E)$ that corresponds to the center-of-mass energy of the experiment, elastic and inelastic processes will take place, dominated by the exchange of gravitons. If, on the other hand, $b < r_H(E)$, a black hole will be formed according to the Thorne's Hoop Conjecture [32] (see also [33] for a higher-dimensional formulation of the same argument).

Equivalently, a black hole could be created if the Compton wavelength $\lambda_C = 4\pi/E$ of the colliding particle of energy $E/2$ lies within the corresponding Schwarzschild radius $r_H(E)$ [34].

Table 1. The values of the ratio $x_{min} = E/M_*$, necessary for the creation of a black hole, as a function of n.

$n = 2$	$n = 3$	$n = 4$	$n = 5$	$n = 6$	$n = 7$
$x_{min} = 8.0$	$x_{min} = 9.5$	$x_{min} = 10.4$	$x_{min} = 10.9$	$x_{min} = 11.1$	$x_{min} = 11.2$

By using the expression for the horizon radius (14), the above is written as

$$\frac{4\pi}{E} < \frac{1}{M_*}\left(\frac{E}{M_*}\right)^{\frac{1}{n+1}}\left(\frac{8\Gamma(\frac{n+3}{2})}{(n+2)\sqrt{\pi}^{(n+1)}}\right)^{1/(n+1)}, \tag{15}$$

and leads to the values of $x_{min} = E/M_*$ as a function of the number of extra dimensions n, necessary for the creation of the black hole, shown in Table 1. From these, we conclude that the center-of-mass energy of the collision must be approximately one order of magnitude larger than the fundamental Planck scale M_* – we remind the reader that the maximum center-of-mass energy that can be achieved at the Large Hadron Collider at CERN is 14 TeV.

We should note here that only a Quantum Theory of Gravity could offer a detailed analysis of a trans-planckian collision between two particles leading to the creation of a black hole. As the complete, consistent form of such a theory is still missing, this problem is most often modelled by the collision of two gravitational Aichelburg-Sexl shock waves [35] moving at the speed of light. Solving the boundary value problem of the creation or not of a closed trapped surface, the maximum value of the colliding impact parameter $b_{\rm max}$ and the mass of the produced black hole can be found [36, 37, 38, 39, 40, 41, 42, 43]. The value of the former parameter can also lead to an estimate of the production cross-section of the black holes through the relation $\sigma_{\rm production} \simeq \pi b^2_{\rm max}$, valid for a high-energy collision. The most recent analyses seem to agree quite well with some early estimates [44, 45, 46] according to which the production cross-section takes the values 10^5 fbar and 10^1 fbar for a black hole with mass $M_{BH} = 5\,$TeV and $M_{BH} = 10\,$TeV, respectively, and under the assumption that $M_* = 1\,$TeV and $D = 10$. In both cases, the value of the production cross-section is considered to be significant for beyond the Standard Model processes.

Let us now briefly discuss the properties of the produced black holes [12, 47]. The value of the horizon radius as a function of n may easily follow from Eq. (14) upon particular choices for M_* and M_{BH}. Assuming that $M_* = 1$ TeV and $M_{BH} = 5$ TeV, the corresponding values of r_H are shown in Table 2. In the same Table, the values of the black hole temperature, that for the simple black-hole prototype (12) is given by the equation

$$T_H = \frac{k}{2\pi} = \frac{1}{4\pi}\,\frac{1}{\sqrt{|g_{tt}\,g_{rr}|}}\left(\frac{d|g_{tt}|}{dr}\right)_{r=r_H} = \frac{(n+1)}{4\pi\,r_H}, \tag{16}$$

are also shown. Note that the values of r_H lie in the subnuclear regime while those of T_H in the GeV range - both regimes are easily accessible by present day experiments and detection techniques.

The non-vanishing temperature gives rise to a thermal type of radiation emitted by the black hole in the form of elementary particles. This is the well-known Hawking radiation whose energy spectrum is given by the expression [48, 49]

$$\frac{dE(\omega)}{dt} = \int \frac{|\mathcal{A}(\omega)|^2\,\omega}{\exp\left(\omega/T_H\right) \mp 1}\,\frac{d\omega}{(2\pi)}\,. \tag{17}$$

In the above, ω is the energy of the emitted particles and ± 1 a statistics factor for fermions and bosons, respectively. The origin of the Hawking radiation is quantum-mechanical since classically nothing is allowed to escape through the horizon of the black hole: a virtual pair of particles is produced outside the horizon of the black hole; when the antiparticle falls inside the black hole and the particle is allowed to propagate towards infinity, the black hole emits radiation characterized by its temperature. Its spectrum is of a black-body type except for the factor $|\mathcal{A}(\omega)|^2$ appearing in the numerator of the emission rate, called the Absorption Probability (or, *greybody factor*). Its presence is due to the fact that not all of the outward propagating particles will reach infinity - some of them, will be reflected back due to the strong gravitational field of the black hole.

The emission of Hawking radiation causes the black hole to evaporate and thus to have a finite lifetime. For the same values of M_* and M_{BH} as above, the typical lifetime of the black hole comes out to be $\tau = (1.7-0.5)\times 10^{-26}$ sec for $n = 1-7$. Contrary to the large astrophysical black holes, whose lifetime exceeds that of the universe, these microscopic black holes will evaporate almost instantly after their creation. Another important factor that needs to be taken into account, as it also determines the rate of evaporation, is that a higher-dimensional black hole emits simultaneously radiation in two channels, the 'brane' and the 'bulk' one. The first one involves all the Standard Model Particles living in our 4-dimensional world, fermions, gauge bosons and Higgs-like scalars; the second emits mainly gravitons and possibly other particles, like scalar fields, that are allowed to live in the bulk provided that they carry no quantum charges under the Standard Model gauge group. Although the 'brane' channel is the most phenomenologically interesting to us as it can be directly measured by a brane observer, the 'bulk' one is almost of equal importance as we need to know how much black-hole energy is lost in the non-accessible transverse dimensions.

In what follows, we will therefore start with the study of the emission of Hawking radiation on the brane, and then we will discuss what is known for the emission in the bulk. We will consider the emission during the two intermediate stages of the life of a black hole, the axially-symmetric *spin-down* phase and the spherically-symmetric *Schwarzschild* one. These two phases are preceded by the *balding* phase, in which the black hole first forms and settles down to a state that satisfies the no-hair theorem of General Relativity, and are followed by the *Planck* phase, in which the black hole turns to a quantum object with unknown properties.

Let us start with the Hawking radiation emitting phase with the simplest gravitational background, the Schwarzschild one. Its line-element is given by Eq. (12) and it is the background seen by any particle propagating in the higher-dimensional spacetime. But for a particle localised on the brane, like the Standard Model fields, we need to consider the projection of (12) on the brane - this is done by setting the values of all additional θ_i coordinates, with $i = 2, ..., n+1$, introduced to describe the additional spacelike dimensions, to $\frac{\pi}{2}$. Then, the resulting brane

Table 2. Horizon radius and temperature of the Schwarzschild-Tangherlini black hole as a function of the number of extra dimensions, for $M_* = 1\,\text{TeV}$ and $M_{BH} = 5\,\text{TeV}$ [12]

n	1	2	3	4	5	6	7
r_H (10^{-4} fm)	4.06	2.63	2.22	2.07	2.00	1.99	1.99
T_H (GeV)	77	179	282	379	470	553	629

line-element assumes the form

$$ds_4^2 = -\left[1 - \left(\frac{r_H}{r}\right)^{n+1}\right] dt^2 + \left[1 - \left(\frac{r_H}{r}\right)^{n+1}\right]^{-1} dr^2 + r^2\, d\Omega_2^2 \,. \tag{18}$$

The non-trivial n-dependence of the projected-on-the-brane background will introduce a similar dependence in the field equations. By using the Newman-Penrose method [50, 51], these take the form of a Teukolsky-like [52] "master" equation of motion with the spin s appearing as a parameter [12, 53, 54]. By using also a factorized ansatz for the wavefunction of the field of the form

$$\Psi_s = e^{-i\omega t}\, e^{im\varphi}\, \Delta^{-s}\, R_s(r)\, S^m_{sl}(\theta)\,, \tag{19}$$

the general field equation reduces to two decoupled equations, one for the radial function $R_s(r)$ and one for the spin-weighted spherical harmonics [55] $S^m_{sl}(\theta)$

$$\Delta^s \frac{d}{dr}\left(\Delta^{1-s}\frac{dR_s}{dr}\right) + \left[\frac{\omega^2 r^2}{h} + 2i\omega s r - \frac{is\omega r^2 h'}{h} - \lambda_{sl}\right] R_s(r) = 0\,, \tag{20}$$

and

$$\frac{1}{\sin\theta}\frac{d}{d\theta}\left(\sin\theta \frac{dS^m_{sl}}{d\theta}\right) + \left[-\frac{2ms\cot\theta}{\sin\theta} - \frac{m^2}{\sin^2\theta} + s - s^2\cot^2\theta + \lambda_{sl}\right] S^m_{sl}(\theta) = 0\,, \tag{21}$$

respectively. In the above, we have defined the function $\Delta \equiv r^2 h \equiv r^2 \left[1 - (\frac{r_H}{r})^{n+1}\right]$, while $\lambda_{sl} = l(l+1) - s(s-1)$ is the eigenvalue of the spin-weighted spherical harmonics. By setting the values $s = 0, 1/2$ and 1, one may easily derive from the above the corresponding radial and angular equations for brane-localised scalars, fermions and gauge bosons.

The angular equation carries no useful information as, given the symmetry of the gravitational background, the emitted radiation is evenly distributed over a 4π solid angle. But the radial

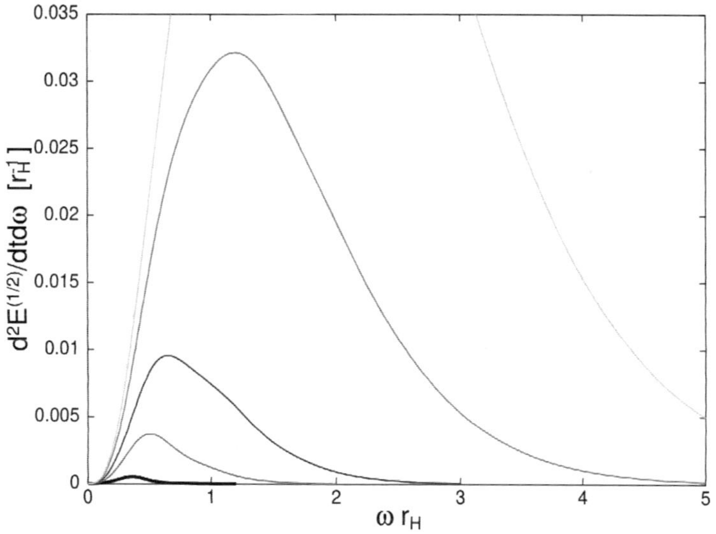

Figure 3. Energy emission rates for fermions on the brane or $n = 0, 1, 2, 4$ and 6 (from bottom to top) [12].

Table 3. Total emissivities for brane-localised scalars, fermions and gauge bosons [57]

n	0	1	2	3	4	5	6	7
Scalars	1.0	8.94	36.0	99.8	222	429	749	1220
Fermions	1.0	14.2	59.5	162	352	664	1140	1830
G. Bosons	1.0	27.1	144	441	1020	2000	3530	5740

equation needs to be explicitly solved as the solution for the radial function $R_s(r)$ determines the Absorption Probability through the formula

$$|\mathcal{A}(\omega)|^2 \equiv 1 - |\mathcal{R}(\omega)|^2 \equiv \frac{\mathcal{F}_{\rm horizon}}{\mathcal{F}_{\rm infinity}}, \tag{22}$$

where $\mathcal{R}(\omega)$ is the Reflection coefficient and $\mathcal{F}$ the flux of energy towards the black hole. The Absorption Probability can be computed both analytically [53, 54, 56] and numerically [57]. Whereas the first method can offer a more qualitative insight into the details of the emission of different species of fields, the second provides us with exact quantitative results for the greybody factor. When this is combined with the expression for the temperature of the black hole in the context of Eq. (17), we obtain the radiation spectra for scalars, fermions, and gauge bosons. The differential emission rate per unit time and frequency, in terms of the number of transverse spacelike dimensions n, for the indicative case of fermions, is shown in Fig. 3. What we may easily observe – with the same holding for scalars and gauge bosons – is that the energy emission rate on the brane is greatly enhanced by the number of additional spacelike dimensions. The same effect is shown in numbers in the entries of Table 3, where the total emissivities have been normalised to the ones for $D=4$: as the value of n increases the enhancement factors reach 3 or 4 orders of magnitude [57].

In the context of the same work, it was also shown that not only the emission rate but also the type of radiation emitted by the black hole depends strongly on n. In Fig. 4, the energy emission rates, for scalars, fermions and gauge bosons, are depicted for the two indicative cases of $n=0$ and $n=6$ [57]. It is clear that, whereas a 4-dimensional spherically-symmetric black hole prefers to emit scalar fields, a 10-dimensional one exhibits a clear preference for gauge bosons. This feature may also be used as an observable signature that will help us determine the number of additional dimensions in nature through the observation of the emitted Hawking radiation from a microscopic black hole.

The gravitational background around a higher-dimensional, rotating black hole is significantly more complicated and takes the form of the Myers-Perry solution [30]

$$\begin{aligned} ds^2 &= \left(1-\frac{\mu}{\Sigma\, r^{n-1}}\right) dt^2 + \frac{2a\mu \sin^2\theta}{\Sigma\, r^{n-1}}\, dt\, d\varphi - \frac{\Sigma}{\Delta} dr^2 \\ &- \Sigma\, d\theta^2 - \left(r^2 + a^2 + \frac{a^2\mu \sin^2\theta}{\Sigma\, r^{n-1}}\right) \sin^2\theta\, d\varphi^2 - r^2\cos^2\theta\, d\Omega_n^2\,, \end{aligned} \tag{23}$$

where

$$\Delta = r^2 + a^2 - \frac{\mu}{r^{n-1}}\,, \qquad \Sigma = r^2 + a^2\cos^2\theta\,. \tag{24}$$

In the above, we have chosen a simple case of the Myers-Perry class of rotating solutions, namely the one where the black hole has only one non-vanishing angular momentum component – this is due to the fact that the black hole has been created by particles that are localised on the brane and have non-zero impact parameters only along a brane spacelike coordinate. The parameters μ and a are then associated to the black hole mass and angular momentum, respectively, through the relations

$$M_{BH} = \frac{(n+2)A_{2+n}}{16\pi G}\,\mu \qquad \text{and} \qquad J = \frac{2}{n+2}\,a\,M_{BH}\,, \tag{25}$$

where A_{2+n} is the area of a $(2+n)$-dimensional unit sphere. The horizon radius is found by setting $\Delta(r_H) = 0$ and is found to be: $r_H^{n+1} = \mu/(1+a_*^2)$, where we have defined the quantity $a_* \equiv a/r_H$. Finally, the temperature and rotation velocity of this black hole are given by

$$T_H = \frac{(n+1)+(n-1)\,a_*^2}{4\pi(1+a_*^2)\,r_H}\,, \qquad \Omega_H = \frac{a}{(r_H^2+a^2)}\,. \tag{26}$$

The derivation of the radiation spectra in the case of the spin-down phase follows along the same lines as in the case of the Schwarzschild phase. We first need to find the line-element seen by the brane-localised Standard Model fields - this follows again by fixing the values of the "extra" angular coordinates in which case the $d\Omega_n^2$ part of the metric (23) disappears leaving the remaining unaltered. By employing again the Newman-Penrose method and a similar factorised ansatz for the spin-s field perturbation – with the spin-weighted spherical harmonics being now replaced by the spin-weighted spheroidal harmonics [58], we derive two decoupled master equations, one for the radial part of the field and one for the angular part, namely [12, 59]

$$\Delta^{-s}\frac{d}{dr}\left(\Delta^{s+1}\frac{dR_s}{dr}\right) + \left[\frac{K^2 - iKs\Delta'}{\Delta} + 4is\omega r + s\left(\Delta''-2\right)\delta_{s,|s|} - \Lambda_{sj}^m\right]R_s = 0 \tag{27}$$

and

$$\frac{1}{\sin\theta}\frac{d}{d\theta}\left(\sin\theta\frac{dS_{sj}^m}{d\theta}\right) + \left[-\frac{2ms\cot\theta}{\sin\theta} - \frac{m^2}{\sin^2\theta} + a^2\omega^2\cos^2\theta \right.$$
$$\left. -2a\omega s\cos\theta + s - s^2\cot^2\theta + \lambda_{sj}\right]S_{sj}^m(\theta) = 0\,, \tag{28}$$

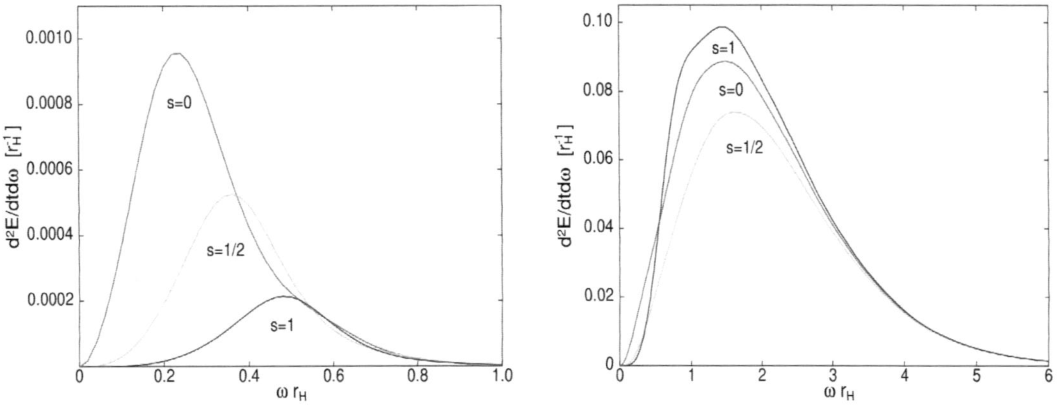

Figure 4. Energy emission rates for brane fields for $n = 0$ (left plot) and $n = 6$ (right plot) [57].

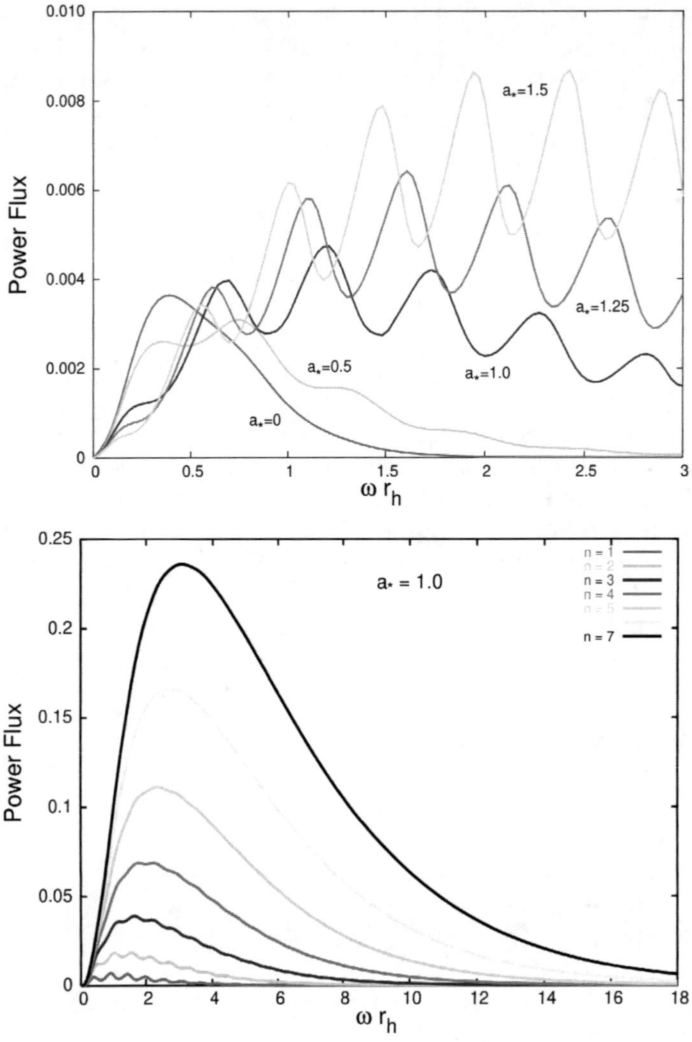

Figure 5. Energy emission rates for brane-localised scalar fields in terms of the angular parameter (top) [64] and gauge bosons in terms of the number of extra dimensions (bottom) [59].

where we have used the definitions

$$K = (r^2 + a^2)\,\omega - am\,, \qquad \Lambda_{sj}^m = \lambda_{sj} + a^2\omega^2 - 2am\omega\,. \tag{29}$$

The angular eigenvalue λ_{sj} connecting the radial and angular equations does not exist in closed form. It may be computed either analytically, through a power series expansion in terms $a\omega$ [60, 61, 62] or numerically [59, 63, 64, 65]. In the same way, the Absorption Probability can be found by solving the radial equation, for all species of fields, either analytically [66, 67] or numerically [59, 63, 64, 65, 68, 69, 70]. As before, the exact spectra follow only by means of numerical analysis by using the derived value of the Absorption Probability together with the

Table 4. Bulk-to-Brane Relative Emissivities Ratio for scalar fields in terms of n [57]

n	0	1	2	3	4	5	6	7
Bulk/Brane	1.0	0.40	0.24	0.22	0.24	0.33	0.52	0.93

ones of T_H and Ω_H in the context of the formula [48, 49, 71]

$$\frac{d^2 E}{dt\, d\omega} = \frac{1}{2\pi} \sum_{j,m} \frac{|\mathcal{A}(\omega)|^2\, \omega}{\exp(\tilde{\omega}/T_H) \mp 1}, \tag{30}$$

where $\tilde{\omega} = \omega - m\Omega_H$. The radiation spectra will now depend on two topological parameters, the angular-momentum parameter a_* of the black hole and the number of additional spacelike dimensions n. In Fig. 5, we present the energy emission rates, for the indicative cases of brane-localised scalars and gauge bosons, in terms of these two parameters. We may easily conclude that the energy emission rates are significantly enhanced by both parameters: the enhancement factor is of order $\mathcal{O}(10)$ in terms of a_* and of order $\mathcal{O}(100)$ in terms of n. As n increases, the gauge bosons are again the species of particles that a rotating black hole increasingly prefers to emit.

Finally, let us briefly add that, contrary to the case of a spherically-symmetric black hole, the emission of Hawking radiation by a rotating black hole has a non-trivial angular distribution in space. As it was shown in detail in [59, 64, 65] by numerically solving the angular equation (28), the centrifugal force causes all particles with intermediate and high frequency to be emitted along the equatorial plane, i.e. vertically to the rotation axis of the black hole. In addition, the emission of particles with non-vanishing spin and low frequency is polarised along the rotation axis with the effect being more prominent the larger the value of s is.

Let us now focus on the emission of particles in the bulk by the higher-dimensional black hole. Although the bulk emission can never be detected by a brane observer, it determines the amount of energy left for emission on the brane. The black hole can emit only particular species of particles in the bulk, namely gravitons and possibly scalar fields. We will first consider the emission of the latter species during the Schwarzschild phase. By following a similar analysis as the one for emission on the brane, the bulk scalar equation can be solved in the background of Eq. (12) and the corresponding greybody factor can be found both analytically [53, 72] and numerically [57]. The exact numerical results allow us to compute the complete radiation spectra that exhibit a similar behaviour to the ones for brane emission, namely a strong enhancement with the number of additional spacelike dimensions. More importantly, they allow us to compute the relative emissivity for scalar fields, i.e. the ratio of the energy emitted by the black hole per unit time over the whole frequency regime in the two emission channels, 'bulk' and 'brane'. The values of this ratio as a function of n is shown in Table 4 [57]. From these, it becomes clear that the brane scalar channel remains the dominant one for all values of n, however, as n takes large values, the emission in the two channels becomes comparable.

Similar analyses, analytical [73, 74] as well as numerical [75, 76], have been performed for the emission of gravitons in the bulk during the Schwarzschild phase. The results produced in [75] have revealed an enhancement factor for the energy emission rate for gravitons with n that surpasses the one for any other species of field, reaching the value of 10^{5-6}. Nevertheless, contrary to our naive expectation this is not enough to render the bulk channel the dominant one: the total number of brane degrees of freedom is larger than the one for gravitons living in the bulk, and overall the brane channel still dominates over the bulk one, thus offering strong support to an early work [77] where a similar claim was made[1].

The study of the emission of fields in the bulk during the spin-down phase has not been completed yet. Until now only the emission of scalar fields has been studied [80, 81, 82] as the field equations for gravitons in an axially-symmetric gravitational background remains largely unknown. In [82], a comprehensive study of the scalar emission in the bulk was made and compared to the one on the brane. It was demonstrated that although the total energy output of the black hole in the scalar channel increases with the angular momentum of the black hole, the bulk emission is actually suppressed especially in the low a_* regime (see Fig. 6). This is due to the fact that the temperature of the black hole decreases with a_*, so it is left to the behaviour of the greybody factor to determine the final form of the spectrum: whereas $|\mathcal{A}(\omega)|^2$ for brane scalars increases significantly with a_* overcoming the decrease in temperature, for bulk scalars its enhancement is not large enough to achieve the same effect. As a result, the brane dominance persists even in the spin-down phase, at least in the scalar channel of emission.

4. Conclusions

In this review talk, we have discussed some recent as well as some less recent attempts to construct and study black-hole solutions in the context of theories postulating the existence of additional spacelike dimensions in nature. These attempts have been proven to be very challenging in some cases and very informative in others.

[1] For some special cases where the bulk channel may dominate over the brane one, see [78, 79].

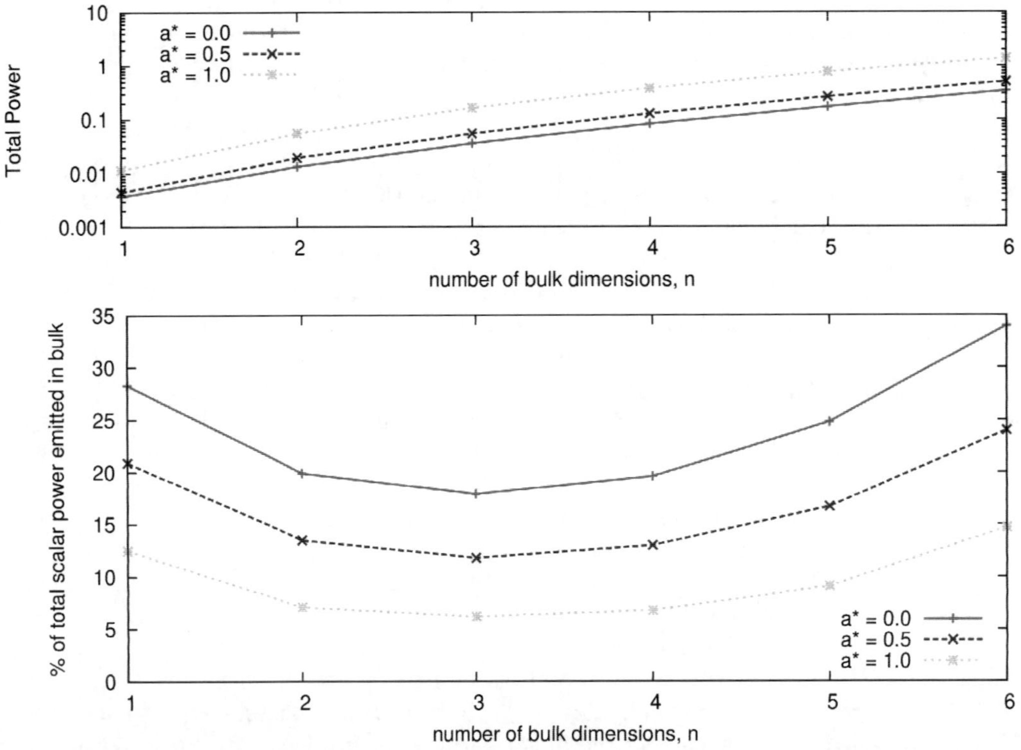

Figure 6. Total power emitted by a rotating black hole in the scalar channel (upper plot) and the % of this power emitted in the bulk (lower plot) [82].

In the context of theories with Warped Extra Dimensions, all efforts to construct analytically a regular black hole solution have so far failed. Different approaches have been developed over the years, where either the bulk or the brane solution played the central role, however none of them managed to achieve a breakthrough. The long-sought regular, localised, static black hole remains elusive whereas its 'twin' solution, the black string, readily emerges although unstable. Theoretical arguments and numerical studies seem even to support that the existence of large, static black holes is highly disfavoured in warped models.

Whereas constructing a regular black hole in a spacetime filled with a negative cosmological constant is highly non-trivial, in higher-dimensional flat spacetime this task is straightforward with regular black-hole solutions being known for decades now. The theoretical challenges of the existence of these black holes are, in the context of the Large Extra Dimensions model, removed – the interest is turned instead to the study of their properties and any phenomenological implications that the potential creation of very small black holes at high-energy particle collisions may have. If strong gravity emerges indeed at energies close to 1 TeV, black holes may even be produced at ground-based colliders such as the Large Hadron Collider (LHC). One of the most distinctive signatures should be the Hawking radiation spectra from which a wealth of information can be derived on the emitted particle properties and the topological parameters of our spacetime.

The frontiers in both areas remain open – the outcome of our efforts to construct regular black hole solutions in the context of warped models will show how realistic these models, inspired from superstring theory, really are. On the other hand, the potential detection of a microscopic black hole, guided by our predictions for their properties and observable signatures, will prove the validity of the theoretical foundation of black hole physics we have built both in 4-dimensional and higher-dimensional spacetime.

Acknowledgments
I am grateful to my collaborators (K. Tamvakis, I. Olasagasti, J. March-Russell, C. Harris, A. Barrau, J. Grain, G. Duffy, E. Winstanley, M. Casals, S. Creek, O. Efthimiou, R. Gregory, B. Mistry and S. Dolan, in chronological order) for our enjoyable and fruitful collaborations. I would also like to thank the organisers of the "NEB XIII – New Developments in Gravity" conference for their kind invitation to give this review talk. I finally acknowledge participation in the RTN Universenet (MRTN-CT-2006035863-1 and MRTN-CT-2004-503369).

References

[1] Antoniadis I 1990 *Phys. Lett.* B **246** 377
[2] Horava P and Witten E 1996 *Nucl. Phys.* B **460** 506; 1996 *Nucl. Phys.* B **475** 94
[3] Lykken J 1996 *Phys. Rev.* D **54** 3693
[4] Arkani-Hamed N, Dimopoulos S and Dvali G 1998 *Phys. Lett.* B **429** 263; 1999 *Phys. Rev.* D **59** 086004
[5] Antoniadis I, Arkani-Hamed N, Dimopoulos S and Dvali G 1998 *Phys. Lett.* B **436** 257
[6] Randall L and Sundrum R 1999 *Phys. Rev. Lett.* **83** 3370; 19991 *Phys. Rev. Lett.* **83** 4690
[7] Akama K 1982 *Lect. Notes Phys.* **176** 267
[8] Rubakov V A and Shaposhnikov M E 1983 *Phys. Lett.* B **125** 139; 1983 *Phys. Lett.* B **125** 136
[9] Visser M 1985 *Phys. Lett.* B **159** 22
[10] Gibbons G W and Wiltshire D L 1987 *Nucl. Phys.* B **287** 717
[11] Majumdar A S and Mukherjee N 2005 *Int. J. Mod. Phys.* D **14** 1095; Gregory R 2009 *Lect. Notes Phys.* **769** 259
[12] Kanti P 2004 *Int. J. Mod. Phys.* A **19** 4899; 2009 *Lect. Notes Phys.* **769** 387 (0802.2218 [hep-th])
[13] Cavaglià M 2003 *Int. J. Mod. Phys.* A **18** 1843; Landsberg G 2004 *Eur. Phys. J.* C **33** S927; Harris C M hep-ph/0502005; Casanova A and Spallucci E 2006 *Class. Quant. Grav.* **23** R45; Winstanley E 0708.2656 [hep-th]
[14] Chamblin A, Hawking S W and Reall H S 2000 *Phys. Rev.* D **61** 065007
[15] Gregory R 2000 *Class. Quant. Grav.* **17** L125
[16] Gregory R and Laflamme R 1993 *Phys. Rev. Lett.* **70** 2837

[17] Dadhich N, Maartens R, Papadopoulos P and Rezania V 2000 *Phys. Lett.* B **487** 1; Casadio R, Fabbri A and Mazzacurati L 2002 *Phys. Rev.* D **65** 084040; Kofinas G, Papantonopoulos E and Zamarias V 2000 *Phys. Rev.* D **66** 104028; Karasik D, Sahabandu C, Suranyi P and Wijewardhana L C R 2004 *Phys. Rev.* D **70** 064007
[18] Shiromizu T, Maeda K i and Sasaki M 2000 *Phys. Rev.* D **62** 024012
[19] Kanti P and Tamvakis K 2002 *Phys. Rev.* D **65** 084010; Kanti P, Olasagasti I and Tamvakis K 2003 *Phys. Rev.* D **68** 124001
[20] Creek S, Gregory R, Kanti P and Mistry B 2006 *Class. Quant. Grav.* **23** 6633
[21] Frolov V P, Snajdr M and Stojkovic D 2003 *Phys. Rev.* D **68** 044002; Galfard C, Germani C and Ishibashi A 2006 *Phys. Rev.* D **73** 064014; A. Flachi, O. Pujolas, M. Sasaki and T. Tanaka, hep-th/0601174
[22] Israel W 1966 *Nuovo Cimento Soc. Ital. Phys.* B **44** 4349
[23] Tanaka T 2003 *Prog. Theor. Phys. Suppl.* **148** 307
[24] Bruni M, Germani C and Maartens R 2001 *Phys. Rev. Lett.* **87** 231302
[25] Emparan R, Fabbri A and Kaloper N 2002 *JHEP* **0208** 043
[26] Kudoh H, Tanaka T and Nakamura T 2002 *Phys. Rev.* D **68** 024035
[27] Kudoh H 2004 *Phys. Rev.* D **69** 104019 [2004 Erratum-ibid. D **70** 029901]
[28] Tanahashi N and Tanaka T 2008 *JHEP* **0803** 041
[29] Tangherlini F R 1963 *Nuovo Cim.* **27** 636
[30] Myers R C and Perry M J 1986 *Annals Phys.* **172** 304
[31] Banks T and Fischler W hep-th/9906038
[32] Thorne K S 1972 *Magic without Magic*, ed. Klauder J R (San Fransisco)
[33] Ida D and Nakao K i 2002 *Phys. Rev.* D **66** 064026
[34] Meade P and Randall L 2008 *JHEP* **0805** 003
[35] Aichelburg P C and Sexl R U 1971 *Gen. Rel. Grav.* **2** 303
[36] Penrose R 1974 presented at the Cambridge University Seminar, *unpublished*
[37] D'Eath P D and Payne P N 1992 *Phys. Rev.* D **46** 658; 1992 *Phys. Rev.* D **46** 675; 1992 *Phys. Rev.* D **46** 694
[38] Cardoso V and Lemos J P S 2002 *Phys. Lett.* B **538** 1; 2003 *Phys. Rev.* D **67** 084005
[39] Sperhake U, Cardoso V, Pretorius F, Berti E and Gonzalez J A, 0806.1738 [gr-qc]
[40] Eardley D M and Giddings S B 2002 *Phys. Rev.* D **66** 044011
[41] Yoshino H and Nambu Y 2002 *Phys. Rev.* D **66** 065004; 2003 *Phys. Rev.* D **67** 024009
[42] Yoshino H and Rychkov V S 2005 *Phys. Rev.* D **71** 104028
[43] Yoshino H and Mann R B 2006 *Phys. Rev.* D **74** 044003
[44] Giddings S B 2003 *Phys. Rev.* D **67** 126001
[45] Giddings S B and Thomas S 2002 *Phys. Rev.* D **65** 056010
[46] Dimopoulos S and Landsberg G 2001 *Phys. Rev.* D **87** 161602
[47] Argyres P C, Dimopoulos S and March-Russell J 1998 *Phys. Lett.* B **441** 96
[48] Hawking S W 1975 *Commun. Math. Phys.* **43** 199
[49] Unruh W G 1974 *Phys. Rev.* D **10** 3194; 1976 *Phys. Rev.* D **14** 3251
[50] Newman E and Penrose R 1962 *J. Math. Phys.* **3** 566
[51] Chandrasekhar S 1983 *The Mathematical Theory of Black Holes* (Oxford University Press, New York)
[52] Teukolsky S A 1972 *Phys. Rev. Lett.* **29** 1114; 1973 *Astrophys. J.* **185** 635
[53] Kanti P and March-Russell J 2002 *Phys. Rev.* D **66** 024023
[54] Kanti P and March-Russell J 2003 *Phys. Rev.* D **67** 104019
[55] Goldberg J N, MacFarlane A J, Newman E T, Rohrlich F and Sudarshan E C 1967 *J. Math. Phys.* **8** 2155
[56] Ida D, Oda K and Park S C 2003 *Phys. Rev.* D **67** 064025 [2004 Erratum-ibid. D **69** 049901]
[57] Harris C M and Kanti P 2003 *JHEP* **0310** 014
[58] Flammer C 1957 *Spheroidal Wave Functions* (Stanford University Press, Stanford, USA)
[59] Casals M, Kanti P and Winstanley E 2006 *JHEP* **0602** 051
[60] Starobinskii A A and Churilov S M 1974 *Sov. Phys.-JETP* **38** 1
[61] Fackerell E D and Crossman R G 1977 *J. Math. Phys.* **18** 1849
[62] Seidel E 1989 *Class. Quant. Grav.* **6** 1057
[63] Harris C M and Kanti P 2006 *Phys. Lett.* B **633** 106
[64] Duffy G, Harris C, Kanti P and Winstanley E 2005 *JHEP* **0509** 049
[65] Casals M, Dolan S R, Kanti P and Winstanley E 2007 *JHEP* **0703** 019
[66] Creek S, Efthimiou O, Kanti P and Tamvakis K 2007 *Phys. Rev.* D **75** 084043
[67] Creek S, Efthimiou O, Kanti P and Tamvakis K 2007 *Phys. Rev.* D **76** 104013
[68] Frolov V and Stojkovic D 2003 *Phys. Rev.* D **67** 084004
[69] Ida D, Oda K y and Park S C 2005 *Phys. Rev.* D **71** 124039

[70] Ida D, Oda K y and Park S C 2006 *Phys. Rev.* D **73** 124022
[71] Ottewill A C and Winstanley E 2000 *Phys. Rev.* D **62** 084018
[72] Frolov V and Stojkovic D 2002 *Phys. Rev.* D 66 084002
[73] Cornell A S, Naylor W and Sasaki M 2006 *JHEP* **0602** 012
[74] Creek S, Efthimiou O, Kanti P and Tamvakis K 2006 *Phys. Lett.* B **635** 39
[75] Cardoso V, Cavaglia M and Gualtieri L 2006 *Phys. Rev. Lett.* **96** 071301 [2006 Erratum-ibid. **96** 219902]; 2006 *JHEP* **0602** 021
[76] Park D K 2006 *Phys. Lett.* B **638** 246
[77] Emparan R, Horowitz G T and Myers R C 2000 *Phys. Rev. Lett.* **85** 499
[78] Grain J, Barrau A and Kanti P 2005 *Phys. Rev.* D **72** 104016
[79] Cho H T, Cornell A S, Doukas J and Naylor W, 2008 *Phys. Rev.* D **77** 016004
[80] Jung E and Park D K 2005 *Nucl. Phys.* B **731** 171
[81] Creek S, Efthimiou O, Kanti P and Tamvakis K 2007 *Phys. Lett.* B **656** 102
[82] Casals M, Dolan S R, Kanti P and Winstanley E 2008 *JHEP* **0806** 071

Gravitational waves versus cosmic strings

Kostas Kleidis[1,2]

[1]Department of Physics, Aristotle University of Thessaloniki, GR-54124 Thessaloniki, Greece
[2]Department of Civil Engineering, Technological Education Institute of Serres, GR-62124 Serres, Greece

E-mail: `kleidis@astro.auth.gr`

Abstract. The equation which governs the temporal evolution of a gravitational wave (GW) in curved space-time can be treated as the Schrödinger equation for a particle moving in the presence of an *effective potential.* When GWs propagate in an expanding Universe with *constant* effective potential, there is a *critical value* (k_c) of the comoving wave-number which discriminates the metric perturbations into *oscillating* ($k > k_c$) and *non-oscillating* ($k < k_c$) modes. The effective potential is reduced to a non-vanishing constant in a cosmological model which is driven by a *two-component fluid,* consisting of *radiation* (dominant) and *cosmic strings* (subdominant). However, the cosmological evolution (gradually) results in the *scaling* of any long-cosmic-string network and, therefore, after some time ($\Delta\tau$) the Universe enters in the *pure-radiation* epoch. The evolution of the non-oscillatory GW modes during $\Delta\tau$, results in the *distortion* of the *low-frequency part* of the *stochastic GW power-spectrum,* which, therefore, departs from *scale invariance* (anticipated in the pure-radiation case). As regards the corresponding *high-frequency part* (which is determined by the evolution of the oscillating modes), we find that the presence of cosmic strings gives rise to the quantum-gravitational creation of *gravitons,* leading to the *amplification* of the GW signal by (almost) two orders of magnitude.

1. Introduction

The so-called *cosmological gravitational waves* (CGWs) represent small-scale perturbations to the Universe metric tensor [1]. Since gravity is the weakest of the four known forces, these metric corrections decouple from the rest of the Universe at very early times, presumably at the Planck epoch [2]. Their subsequent propagation is governed by the space-time curvature, encapsulating in the field equations the inherent coupling between relic GWs and the Universe matter-content; the latter being responsible for the background gravitational field [3].

In this context, we consider the interaction between CGWs and *cosmic strings.* They are one-dimensional objects that can be formed as linear defects at a symmetry-breaking phase transition [4], [5]. If they exist, they may help us to explain some of the large-scale structures seen in the Universe today, such as gravitational lenses [6]. They may also serve as *seeds* for density perturbations [7], [8], as well as potential sources of relic gravitational radiation [9], [10].

In this article we present some recent results concerning another possibility: The presence of cosmic strings in a radiation-dominated Universe being responsible for the constancy of the *effective potential* which drives the temporal evolution of a CGW in curved space-time.

In particular, we shall demonstrate that, a (non-vanishing) constant effective potential is associated to a *critical value* (k_c) of the comoving GW wave-number, which discriminates the metric perturbations into *oscillating* ($k > k_c$) and *non-oscillating* ($k < k_c$) modes, something that is reflected in the CGWs' *power-spectrum* [11]. In this context, the presence of cosmic strings at the early stages of the Universal evolution will modify the expected *profile* of any *relic-GW signal* that can be observed today. In what follows, we shall attempt to illustrate how, in a realistic setting.

We begin with the fact that the cosmological evolution (gradually) results in the *scaling* of any long-cosmic-string network [12] and, therefore, the Universe, if ever experienced a short *radiation-plus-strings stage*, eventually, will enter in the *pure-radiation* epoch. Accordingly, following the evolution of the non-oscillatory GW modes during the radiation-plus-strings stage, we find that the *low-frequency part* of a *relic-GW signal* will depart from *scale-invariance*, in contrast to what is anticipated in the pure-radiation case [13]. On the other hand, as regards the oscillatory GW modes, their evolution from an *early* radiation epoch to a *late* radiation era through the radiation-plus-strings stage leads to the quantum-gravitational creation of *gravitons*, thus resulting in the *amplification* of the *high-frequency part* of the relic-GW signal by (almost) two orders of magnitude.

2. CGWs in a Friedmann model

In the system of units where c (the velocity of light) equals to unity, a weak CGW ($|h_{ij}| \ll 1$), propagating in a spatially flat Friedmann - Robertson - Walker (FRW) cosmological model (as *inflation* advocates), is defined by [9]

$$ds^2 = dt^2 - R^2(t)(\delta_{ij} + h_{ij})dx^i dx^j, \tag{1}$$

and satisfies the Klein-Gordon equation [14]

$$h_{ij\;;\alpha}^{\;\;;\alpha} = 0, \tag{2}$$

In Eqs (1) and (2), Greek indices refer to the four-dimensional space-time (in connection, Latin indices refer to the three-dimensional spatial section), the semicolon denotes covariant derivative, δ_{ij} is the Kronecker symbol and the dimensionless scale factor $R(t)$ is a solution to the Friedmann equations with matter-content in the form of a *perfect fluid*. In terms of the so-called *conformal-time* coordinate, $\tau = \int dt/R(t)$, the solution to Eq (2) is a *linear superposition of plane-wave modes*

$$h_{ij}(\tau, x^r) = \phi_k(\tau)\, \varepsilon_{ij}\, e^{\imath k_r x^r}, \tag{3}$$

where, $\phi_k(\tau)$ is a (complex) function of time and ε_{ij} is the polarization tensor, depending only on the direction of the *comoving* wave-vector k_r. For a fixed wave-number $k^2 = \sum k_r^2$, the time-dependent part of the corresponding GW mode satisfies the second-order differential equation (e.g. see [15])

$$\phi_k'' + 2\frac{R'}{R}\,\phi_k' + k^2\,\phi_k = 0, \tag{4}$$

where, the prime denotes differentiation with respect to τ. Under the further decomposition

$$h_k(\tau) = \phi_k(\tau)\, R(\tau), \tag{5}$$

the equation which governs the temporal evolution of a CGW in a FRW model, is written in the form

$$h_k'' + (k^2 - \frac{R''}{R})\, h_k = 0. \tag{6}$$

Eq (6) can be treated as the *Schrödinger equation* for a particle moving in the presence of the *effective potential*

$$V_{eff} = \frac{R''}{R}. \tag{7}$$

To solve this equation, we need an evolution formula for the cosmological model under consideration.

In terms of the conformal time, the spatially flat FRW model is a solution to the Friedmann equation

$$(\frac{R'}{R^2})^2 = \frac{8\pi G}{3}\,\rho(\tau), \tag{8}$$

(where, G is Newton's constant) with matter-content in the form of a perfect fluid, $T_{\mu\nu} = diag(\rho, -p, -p, -p)$, which obeys the conservation law

$$\rho' + 3\frac{R'}{R}\,(\rho + p) = 0, \tag{9}$$

and the equation of state

$$p = (\frac{m}{3} - 1)\,\rho, \tag{10}$$

where, $\rho(\tau)$ and $p(\tau)$ represent the mass-density and the pressure, respectively.

The linear equation of state (10) covers most of the matter-components considered to drive the evolution of the Universe, such as quantum vacuum ($m = 0$), a network of domain walls ($m = 1$), a gas of cosmic strings ($m = 2$), dust ($m = 3$), radiation ($m = 4$) and Zel'dovich ultra-stiff matter ($m = 6$). For each component, the continuity equation (9) yields

$$\rho = \frac{M_m}{R^m}, \tag{11}$$

where, M_m is an integration constant. Provided that the various components *do not interact* with each other, a mixture of them obeys [16]

$$\rho = \sum_m \frac{M_m}{R^m}, \tag{12}$$

where, now, Eq (9) holds for each matter-constituent separately.

3. Constancy of the effective potential

There is a case of particular interest involved in the temporal evolution of a primordial GW, in which the effective potential is *constant* for every τ, namely

$$\frac{R''}{R} = const = \frac{8\pi G}{3}M, \tag{13}$$

where, M is a non-negative constant of units gr/cm^3. In this case, Eq (6) is written in the form

$$h_k'' + (k^2 - \frac{8\pi G}{3}M)h_k = 0. \tag{14}$$

According to Eq (14), a *critical value* of the comoving wave-number arises, through the condition

$$k > k_c = \sqrt{\frac{8\pi G}{3}M}. \tag{15}$$

This critical value discriminates the primordial GWs in modes, with $k > k_c$, which *oscillate* for every τ,

$$h_{k>k_c}(\tau) \sim e^{\imath\sqrt{k^2-k_c^2}\,\tau}, \tag{16}$$

and modes, with $k < k_c$, which *grow exponentially* for every τ,

$$h_{k<k_c}(\tau) \sim e^{\sqrt{k_c^2 - k^2}\,\tau}, \tag{17}$$

(the corresponding *exponentially-decaying* solutions are neglected).

Now, the question arises on whether there exists a spatially flat FRW cosmological model in which the effective potential is constant. To answer this question, we set $\mathcal{H} = \frac{R'}{R}$. Accordingly, V_{eff} is written in the form

$$\frac{R''}{R} = \mathcal{H}' + \mathcal{H}^2. \tag{18}$$

Upon consideration of Eqs (8) and (18), Eq (13) results in the ordinary differential equation

$$\rho' + 4\rho\frac{R'}{R} = 2M\frac{R'}{R^3}, \tag{19}$$

which admits the solution

$$\rho(\tau) = \frac{M}{R^2} + \frac{C}{R^4}, \tag{20}$$

where, C is an integration constant of units gr/cm^3. For $C = 0$, Eq (20) results in $\rho \sim R^{-2}$, which, by virtue of Eqs (10) and (11), leads to $p = -\frac{1}{3}\rho$. This case, now known as the *string-dominated Universe* [17], was considered by Ford and Parker [18] in their effort for a natural *renormalization* of the energy-momentum tensor of the *gravitational-perturbation field.* Upon consideration of Eqs (11) and (12), we distinguish two cases [in Eq (20)] with respect to M: (i) $M = 0$ and (ii) $M \neq 0$. We consider each one of them separately.

3.1. CGWs in a radiation-dominated Universe

For $M = 0$, in correspondence to Eq (11), Eq (20) reads

$$\rho(\tau) = \frac{M_4}{R^4}. \tag{21}$$

This is the matter-content of a cosmological model filled with relativistic particles - radiation. Now, the effective potential vanishes and Eq (14) is written in the form

$$h_k'' + k^2\,h_k = 0, \tag{22}$$

i.e. it is a simple *harmonic-oscillator* equation. In this case, a *conformally-invariant field* (the metric perturbation) propagates in a *conformally-flat* space-time and there is no particle production [19]. Therefore, the choice of a solution to Eq (22) corresponds to the choice of an *initial (or final) quantum state* for the gravitational-perturbation field. Eq (22) admits the solution

$$h_k^{in}(\tau) = \sqrt{\frac{2}{\pi k}}\,e^{-\imath k\tau}, \tag{23}$$

and the gravitational-perturbation field is in the well-defined *adiabatic-vacuum* state [20].

3.2. CGWs in the presence of cosmic strings

For $M \neq 0$, in correspondence to Eq (12), Eq (20) reads

$$\rho(\tau) = \frac{M_2}{R^2} + \frac{M_4}{R^4}. \tag{24}$$

This type of matter-content corresponds to a radiation model *contaminated* by cosmic strings, in which the two constituents (relativistic particles and linear defects) do not interact with each other, as it should be the case shortly after the *dynamic friction* between them [4] became unimportant [5]. It is worth noting that, the constant M appearing in Eq (13) is related to the initial amount of the linear defects in the mixture (M_2). Now, Eq (14) results in

$$h_k'' + (k^2 - k_c^2)\, h_k = 0, \tag{25}$$

and, as regards the oscillating ($k > k_c$) modes, it can be put in the more convenient form

$$h_k'' + \gamma^2 k^2\, h_k = 0, \tag{26}$$

where, we have introduced the parameter

$$\gamma = \sqrt{1 - (\frac{k_c}{k})^2}, \tag{27}$$

which measures the departure from the *pure-radiation* case. For every $k_c < k < \infty$ $(0 < \gamma < 1)$, the general solution to Eq (26) is written in the form

$$h_k(\tau) = \sqrt{\frac{2}{\pi\gamma k}}\left[c_1\, e^{-\imath\gamma k\tau} + c_2 e^{+\imath\gamma k\tau}\right], \tag{28}$$

where, c_1 and c_2 are complex numbers, satisfying the Wronskian condition

$$|c_1|^2 - |c_2|^2 = 1. \tag{29}$$

We conclude that, in the presence of cosmic strings the effective potential is reduced to a non-vanishing constant and, therefore, oscillation of the metric perturbations is possible only if their comoving wave-number is larger than a critical value, depending on the mass-density of the linear defects

$$k > k_c = \sqrt{\frac{8\pi G}{3} M_2}. \tag{30}$$

In other words, a cosmic-string network discriminates the primordial GWs predicted by inflation into oscillating and non-oscillating modes, something that should be reflected in the power-spectrum of the stochastic GW background. We shall attempt to illustrate how, in a realistic setting.

4. A tale of cosmic strings

The presence of cosmic strings in the Universe is purely a question of *topology* [21]. In particular, after inflation (and reheating) the Universe enters in an *early-radiation* (ER) epoch [22], during which the temperature of the Universe drops monotonically ($T \sim R^{-1}$). This cooling process may have resulted in the breaking of a fundamental U(1) local gauge symmetry, which, in turn, led to the formation of *linear topological defects* (for a detailed analysis see [4] and/or [5]). Within this era, the cosmological scale factor behaves as

$$R_{ER}(t) = (\frac{t}{t_{cr}})^{1/2} \;\Rightarrow\; R_{ER}(\tau) = \frac{\tau}{\tau_{cr}}, \tag{31}$$

where $\tau_{cr} = 2t_{cr}$ marks the time at which the cosmic strings are formed and we have normalized $R(t_{cr})$ to unity.

By the time these linear defects are formed ($t_{cr} \sim 10^{-31}$ *sec* for GUT-scale strings), they are moving in a very dense environment and, hence, their motion is heavily damped due to string-particle scattering [23] - [25]. This *friction* becomes subdominant to expansion damping [23] at

$$t_* = \frac{1}{G\mu}\, t_{cr} \;\;\Rightarrow\;\; \tau_* = \frac{1}{\sqrt{G\mu}}\, \tau_{cr}, \tag{32}$$

where μ is the mass per unit-length of the linear defect. For $\tau \geq \tau_*$, the motion of the long cosmic strings can be considered essentially independent of anything else in the Universe and soon they acquire relativistic velocities [4]. Now, the gas of cosmic strings *does not interact* with the fluid of relativistic particles and, hence, Eq (24) holds. Therefore, we may consider that, for $\tau \geq \tau_*$, the evolution of the curved space-time is driven by a two-component fluid, consisting of radiation (dominant) and cosmic strings (subdominant). Accordingly, τ_* marks the beginning of a *radiation-plus-strings* stage. During this stage, the Friedmann equation (8) admits the solution

$$R(\tau) = \sqrt{\frac{M_4}{M_2}}\; \sinh \sqrt{\frac{8\pi G}{3} M_2}\; \tau. \tag{33}$$

Nevertheless, the scale factor (33) can drive the Universe expansion only for a short period of time after τ_*, since cosmic strings should (at any time) be a small proportion of the Universe energy-content. This means that the equation of state (24) has validity only for a limited time-period, otherwise, cosmic strings would eventually dominate the energy-density (see [17]).

In fact, the radiation-plus-strings stage (if ever existed) does not last very long. Numerical simulations of the early 90's [26] - [29], as well as their present-time counterparts [30] - [32], suggest that, after the friction becomes unimportant, the production of loops smaller than the Hubble radius gradually results in the *scaling* of the long-string network. In other words, the linear defects form a *self-similar configuration*, the density of which, eventually, behaves as R^{-4}. In this way, at some (physical) time $t_{sc} > t_*$ the Universe re-enters in the (*late*) radiation (LR) era

$$R_{LR}(t) = R_{sc}\, (\frac{t}{t_{sc}})^{1/2} \;\;\Rightarrow\;\; R_{LR}(\tau) = R_{sc}\, \frac{\tau}{\tau_{sc}}, \tag{34}$$

before it can become string-dominated. The duration ($\Delta\tau = \tau_{sc} - \tau_*$) of the radiation-plus-strings stage is quite uncertain, mostly due to the fact that numerical simulations (which revealed the scaling of the long-strings network) can be run for relatively limited times. For example, the earliest treatments [28], [29] suggested that $t_{sc} \simeq 30\, t_*$ ($\tau_{sc} \simeq 5.5\, \tau_*$), while, the most recent ones [31] raise this value to $t_{sc} \simeq 300\, t_*$ ($\tau_{sc} \simeq 17\, \tau_*$).

5. Implications on the propagation of metric perturbations

5.1. Evolution of the non-oscillatory modes

5.1.1. GW modes outside of the horizon: CGWs are produced by quantum fluctuations during inflation (e.g. see [13]). Some of them escape from the visible Universe, once their reduced physical wavelength [$\lambda_{ph} = \frac{\lambda}{2\pi} R(\tau)$] becomes larger than the (constant) inflationary horizon [$\ell_H = H_{dS}^{-1}$, H_{dS} being the Hubble parameter of the de Sitter space]. Eventually, every CGW with $k \leq k_{max} = H_{dS} R_{dS}$ is *exiled* from the Hubble sphere and *freezes out*, acquiring a constant amplitude [33], [34]

$$\alpha_k^2 = \left[\frac{h_k(\tau)}{R(\tau)}\right]^2 = \frac{16}{\pi} \left(\frac{H_{dS}}{m_{Pl}}\right)^2 k^{-3}, \tag{35}$$

where $m_{Pl} = G^{-1/2}$ is the Planck mass. After inflation, i.e. within the subsequent radiation epoch, analytic solutions for $\alpha_k(\tau)$ can be expressed in terms of the Bessel function $J_{\frac{1}{2}}(k\tau)$ [11]

$$\alpha_k(\tau) \sim \alpha_k\, \frac{\sqrt{\tau}}{R(\tau)}\, J_{\frac{1}{2}}(k\tau) \sim \alpha_k\, \frac{\sin k\tau}{k\tau}. \tag{36}$$

Accordingly, when $k\tau \ll 1$, the perturbation's amplitude evolves slowly and is approximately constant. Once $k\tau \approx 1$, the amplitude decays away rapidly before entering an oscillatory phase with slowly decreasing amplitude, when $k\tau \gg 1$. Physically, this corresponds to a mode that is (almost) frozen beyond the horizon, until its physical wavelength becomes comparable to the Hubble radius, at which point it enters in the visible Universe (e.g. see [35]).

In other words, as the Universe expands, a fraction of the modes that lie beyond the horizon re-enters inside the Hubble sphere. At the time of re-entry their amplitude is given by Eq (34), while, afterwards, they begin oscillating. The k-dependence of their amplitude implies a *scale-invariant* power-spectrum [13].

However, if the cosmological evolution includes a radiation-plus-strings stage, then, during this stage, the effective potential (in the equation which governs the temporal evolution of a CGW) is a non-vanishing constant. In other words, $k_c \neq 0$ and the modes with $k < k_c$ do not oscillate. As a consequence (in contrast to what previously stated), even if they lie outside the horizon, *they do not freeze out.*

At the beginning of the radiation-plus-strings stage, the GW modes that *fit* inside the visible Universe obey the condition

$$\lambda_{ph}(\tau_*) \leq \ell_H(\tau_*) \;\Rightarrow\; k \geq H(\tau_*)R(\tau_*) = \frac{1}{\tau_*}, \tag{37}$$

while, modes of $k < k_* = \tau_*^{-1}$ lie outside the horizon. In order to examine whether $k_c > k_*$ we need to determine the initial mass-density of the linear defects, since, by definition,

$$M_2 = \rho_{str}(\tau_*)R^2(\tau_*). \tag{38}$$

Let us consider a network of cosmic strings characterized by a *correlation length* $\xi(\tau)$. This may be defined as the length such that the mass within a typical volume ξ^3, is $\mu\xi$ [4]. In this case, at $\tau = \tau_*$, the cosmic strings contribute to the Universe matter-content a mean density

$$\rho_{str}(\tau_*) = \frac{\mu}{\xi^2(\tau_*)} = \gamma_*^2 \frac{\mu}{\ell_H^2(\tau_*)}, \tag{39}$$

where γ_* is a numerical constant of the order of unity (in fact, $\gamma_* \simeq 4-11$, e.g. see [4], reflecting the differences among the various methods of numerical integration in [26] - [29]), representing the number of correlation lengths inside the horizon at τ_*. Accordingly,

$$M_2 = \mu \left(\frac{\gamma_*}{\tau_*}\right)^2, \tag{40}$$

and hence

$$k_c = \sqrt{\frac{8\pi}{3}}\,\sqrt{G\mu}\,\frac{\gamma_*}{\tau_*}. \tag{41}$$

The dimensionless quantity $(G\mu)$ characterizes the *strength* of the gravitational interactions of strings. The observations give an upper limit on the value of this parameter. In particular, for GUT-scale strings, the current CMB bound is [36] $G\mu \leq 1.3 \times 10^{-6}$, while, more recent studies on a cosmic-string contribution in the WMAP data [37] have yielded the tighter bound $G\mu \leq 3.3 \times 10^{-7}$, something that is confirmed also by gravitational lensing observations: $G\mu \geq 4 \times 10^{-7}$ [38], [39]. Admitting that $G\mu \sim 10^{-6}$ and $\gamma_* \simeq 7$, Eq (41) results in

$$k_c \simeq 2 \times 10^{-2}\, k_* \ll k_*. \tag{42}$$

In other words, at the beginning of the radiation-plus-strings stage, the GW modes of $k < k_c$ *do not fit* inside the horizon.

On the other hand, for $\tau_* < \tau \leq \tau_{sc}$, the condition of fitting inside the Hubble sphere is written in the form

$$\frac{k}{k_c} \geq \coth\sqrt{\frac{8\pi G}{3} M_2}\, \tau. \tag{43}$$

Since the *hyperbolic cotangent* is larger than unity for every τ, Eq (43) suggests that during the whole radiation-plus-strings stage the GW modes of comoving wave-numbers $k < k_c$ remain outside the horizon.

Nevertheless, due to the constancy of the effective potential along this stage, these modes (although being outside the horizon) do not freeze out. Indeed, the combination of Eqs (17) and (35) suggests that, for $\tau_* < \tau \leq \tau_{sc}$, their amplitude evolves as

$$\alpha_{k<k_c}(\tau) = \frac{4}{\sqrt{\pi}} \left(\frac{H_{dS}}{m_{Pl}}\right) \frac{1}{k^{3/2}} \left[\frac{R(\tau_*)}{R(\tau)}\right] e^{\sqrt{k_c^2-k^2}\,(\tau-\tau_*)}. \tag{44}$$

This behavior ends at τ_{sc}, when the scaling of the long-string network is completed and the Universe re-enters in the (late) radiation era. For $\tau > \tau_{sc}$ the GW modes of $k < k_c$ are no longer influenced by the radiation-plus-strings stage and therefore, just like the rest of the metric perturbations outside the horizon, *(re)freeze out*. As a consequence, their amplitude acquires the constant value

$$\alpha_{k<k_c} = \frac{4}{\sqrt{\pi}} \left(\frac{H_{dS}}{m_{Pl}}\right) \frac{1}{k^{3/2}} \left[\frac{R(\tau_*)}{R(\tau_{sc})}\right] e^{\sqrt{k_c^2-k^2}\,\Delta\tau}. \tag{45}$$

5.1.2. The distorted power-spectrum: Within the late-radiation era these modes remain frozen until the time τ_c. At that time, the mode k_c enters inside the visible Universe, since its physical wavelength ($\lambda_{c_{ph}} \sim \tau_c$) becomes smaller than the corresponding Hubble radius ($\ell_H \sim \tau_c^2$). In accordance, for $\tau > \tau_c$, GW modes of $k < k_c$ also enter inside the Hubble sphere. After entering inside the horizon, the GW modes under consideration begin oscillating, thus producing a part of the power-spectrum we observe today (or at some time in the future). However, since they have experienced the influence of the radiation-plus-strings (*rps*) stage, their amplitude is no longer given by Eq (35), but by Eq (45), thus resulting in the *distortion* of the GW power-spectrum ($\mathcal{P}_k \sim k^3\,\alpha_k^2$), from what it is anticipated in a *pure-radiation* (*rad*) model, at comoving wave-numbers $k < k_c$

$$\begin{aligned}
\mathcal{P}^{rps}_{k<k_c} &= \frac{16}{\pi} \left(\frac{H_{dS}}{m_{Pl}}\right)^2 \left[\frac{R(\tau_*)}{R(\tau_{sc})}\right]^2 e^{2\sqrt{k_c^2-k^2}\,\Delta\tau} \Rightarrow \\
\mathcal{P}^{rps}_{k<k_c} &= \mathcal{P}^{rad}_{k<k_c} \left[\frac{R(\tau_*)}{R(\tau_{sc})}\right]^2 e^{2\sqrt{k_c^2-k^2}\,\Delta\tau}.
\end{aligned} \tag{46}$$

Upon consideration of Eq (33), we obtain

$$\frac{R(\tau_*)}{R(\tau_{sc})} = \frac{1}{\cosh(k_c\Delta\tau) + \coth(k_c\tau_*)\,\sinh(k_c\Delta\tau)}, \tag{47}$$

and therefore

$$\frac{\mathcal{P}^{rps}_{k<k_c}}{\mathcal{P}^{rad}_{k<k_c}} = \frac{4\,e^{2(1+\sqrt{1-x^2})\,k_c\,\Delta\tau}}{\{[\coth(k_c\tau_*)+1]\,e^{2k_c\Delta\tau} - [\coth(k_c\tau_*)-1]\}^2}, \tag{48}$$

where, we have set

$$0 < \frac{k}{k_c} = x = \frac{f}{f_c} < 1, \tag{49}$$

and f is the frequency attributed to the GW mode denoted by k. By virtue of Eq (42), $\coth(k_c\tau_*) \simeq 5$ and, therefore, Eq (48) results in

$$\frac{\mathcal{P}^{rps}_{k<k_c}}{\mathcal{P}^{rad}_{k<k_c}} = \frac{e^{2(1+\sqrt{1-x^2})\, k_c\, \Delta\tau}}{(3\, e^{2k_c\Delta\tau} - 2)^2}. \tag{50}$$

Clearly, for $\Delta\tau = 0$ (i.e. in the absence of the radiation-plus-strings stage) $\mathcal{P}^{rps}_{k<k_c} = \mathcal{P}^{rad}_{k<k_c}$, while, for $\Delta\tau \neq 0$ the inflationary-GW power-spectrum is no longer *scale-invariant.*

The *spectral function* Ω_{gw}, appropriate to describe the *intensity* of a stochastic GW background [40], is related to the power-spectrum as $\Omega_{gw} \sim k\, \mathcal{P}_k$ (e.g. see [2], [9], [41]), so that, Eq (50) is written in the form

$$\frac{\Omega^{rps}_{gw}(f < f_c)}{\Omega^{rad}_{gw}(f < f_c)} = \frac{e^{2(1+\sqrt{1-x^2})\, k_c\, \Delta\tau}}{\left(3\, e^{2k_c\Delta\tau} - 2\right)^2}. \tag{51}$$

Notice that, for every $0 \leq x \leq 1$, Eq (51) yields $\Omega^{rps}_{gw}(f < f_c) \leq \Omega^{rad}_{gw}(f < f_c)$ with the equality being valid only for $\Delta\tau = 0$. In other words, the involvement of a radiation-plus-strings stage in the evolution of the Universe reduces the stochastic GW intensity at lower levels than those expected in a pure-radiation model. To give some numbers, we take into account the numerical results of [28], as well as those of [29]. Accordingly, a reasonable estimate on the duration of radiation-plus-strings stage would be $\tau_{sc} = 5.5\, \tau_*$ and therefore, $k_c\Delta\tau \simeq 9 \times 10^{-2}$. In this case, Eq (51) is written in the form

$$\frac{\Omega^{rps}_{gw}(f < f_c)}{\Omega^{rad}_{gw}(f < f_c)} \simeq 0.47 \times e^{0.18\sqrt{1-x^2}}, \tag{52}$$

from which, it becomes evident that, for $f < f_c$, the value of Ω_{gw} is no longer 8×10^{-14}, as it is predicted by pure-radiation (e.g. see [2], [9]), but rather

$$\Omega^{rps}_{gw} \simeq 0.5\, \Omega^{rad}_{gw} \sim 4 \times 10^{-14}, \tag{53}$$

(Fig. 1). On the other hand, according to [31], which deals with the scaling of a cosmic-string network in an updated numerical fashion, the duration of a potential radiation-plus-strings stage in terms of the conformal time (the *dynamical range*, as it is referred to) is $\tau_{sc} = 17\tau_*$ (corresponding to a factor of 300 in terms of t). Adoption of this result, would lead to a more evident distortion of the inflationary GW spectrum, modifying Eq (53) to

$$\Omega^{rps}_{gw} \simeq 0.2\, \Omega^{rad}_{gw} \sim 1.6 \times 10^{-14}. \tag{54}$$

Such a distortion reflects a change in the distribution of the GW energy-density among the various frequency intervals, probably due to the coupling between metric perturbations and cosmic strings.

The question that arises now is, whether these results are observable by the detectors currently available. To answer this question, we should determine explicitly both f_c (the critical frequency) and t_c (the physical time at which modes of $f < f_c$ begin entering inside the horizon). In what follows, $c \neq 1$.

During the early-radiation epoch, the *physical time* is defined as

$$t = \int R(\tau) d\tau \;\Rightarrow\; t = \frac{\tau^2}{2\tau_{cr}}. \tag{55}$$

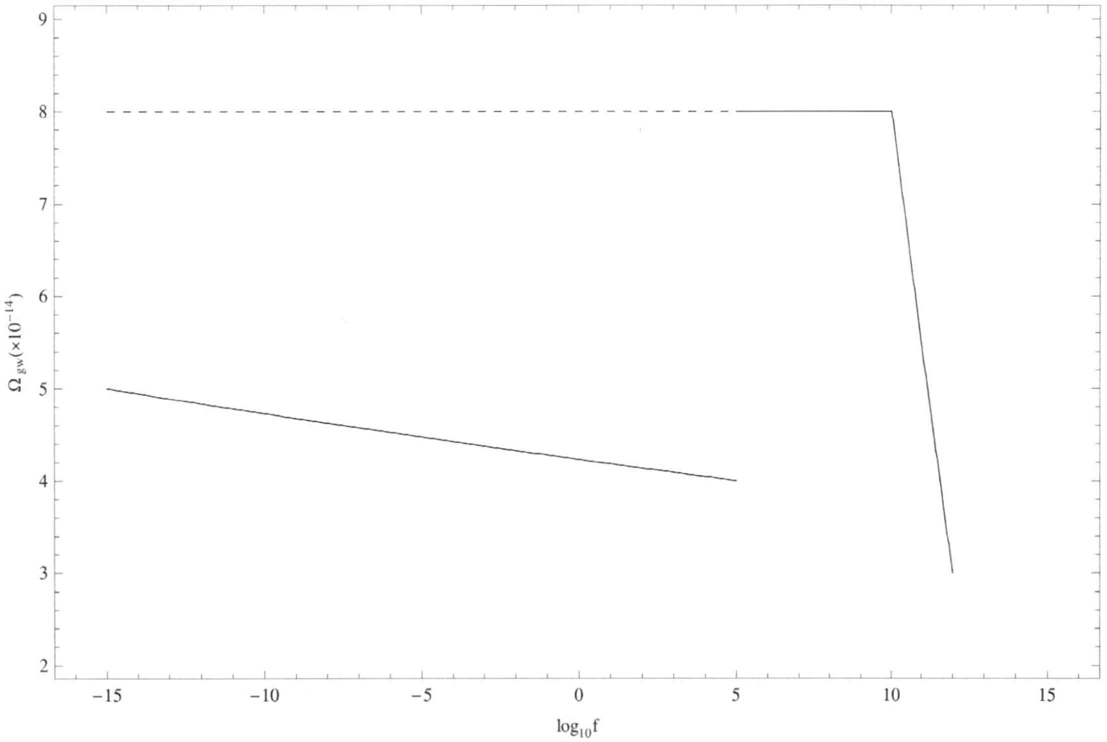

Figure 1. The stochastic GW background from inflation at frequencies within the radiation era, $f \geq 10^{-16}\ Hz$ (dashed line), under the influence of a radiation-plus-strings stage produced by GUT-scale cosmic strings (solid line).

With the aid of Eqs (32) and (55), Eq (41) is written in the form

$$k_c = \sqrt{\frac{2\pi}{3}} \left(\frac{G\mu}{c^2}\right) \frac{\gamma_*}{c\, t_{cr}}, \tag{56}$$

and therefore

$$f_c = \frac{1}{\sqrt{6\pi}} \left(\frac{G\mu}{c^2}\right) \frac{\gamma_*}{t_{cr}}. \tag{57}$$

The first of the GW modes under consideration which enters inside the visible Universe, is the one with the shortest comoving wavelength (λ_c), i.e. the one with the largest frequency, f_c. In terms of the physical time, this process begins at t_c at which $\lambda_{c_{ph}}(t_c) \leq \ell_H(t_c)$.

Within the late-radiation era, the physical time is defined as

$$t = \int R(\tau) d\tau \ \Rightarrow \ t = R_{sc} \frac{\tau^2}{2\tau_{sc}}, \tag{58}$$

and therefore

$$t_c \geq \frac{3\pi}{11\gamma_*^2} \left(\frac{G\mu}{c^2}\right)^{-1/2} R_{sc}\, t_*, \tag{59}$$

where we have used both Eqs (55) and (58), together with Eq (32) and the fact that $\tau_{sc} = 5.5\,\tau_*$. In an expanding Universe

$$R_{sc} > R(t_*) = \left(\frac{G\mu}{c^2}\right)^{-1/2}, \tag{60}$$

and hence

$$t_c > \frac{3\pi}{11\gamma_*^2}\left(\frac{G\mu}{c^2}\right)^{-1} t_* = \frac{3\pi}{11\gamma_*^2}\left(\frac{G\mu}{c^2}\right)^{-2} t_{cr}. \tag{61}$$

Within the Hubble sphere the GW modes of $\lambda > \lambda_c$ correspond to CGWs of frequencies $f < f_c$. Extrapolation of this result into the present epoch ($t_{pr} \simeq 13.7 \times 10^9\ y$), suggests that at frequencies

$$\begin{aligned} f^{pr} \leq f_c^{pr} &= f_c \left(\frac{t_c}{t_{rec}}\right)^{1/2} \left(\frac{t_{rec}}{t_{pr}}\right)^{2/3} \Rightarrow \\ f^{pr} \leq f_c^{pr} &= \frac{1}{\sqrt{22}} \left(\frac{t_{cr}}{t_{rec}}\right)^{1/2} \left(\frac{t_{rec}}{t_{pr}}\right)^{2/3} \frac{1}{t_{cr}}, \end{aligned} \tag{62}$$

(where $t_{rec} = 1.2 \times 10^{13}\ sec$ is the *recombination time*) the inflationary-GW power-spectrum is distorted, departing from scale-invariance.

We note that f_c^{pr} depends only on the (physical) time at which the cosmic strings are formed. These linear defects may have been formed at a grand unification (GUT) transition or, conceivably, much later, at the electro-weak transition or somewhere in between [4]. For GUT-scale strings, $t_{cr} \sim 10^{-31}\ sec$ [5] and therefore $f_c^{pr} \simeq 1.5 \times 10^5\ Hz$. Clearly, this value is far outside of the range where both the ground-based and the space-based laser interferometers may operate. A GW of this frequency could be detected only by a system of *coupled super-conducting microwave cavities* [42], [43].

However, one should have in mind that, this is only the *upper bound* of the distorted GW power-spectrum. In fact, if cosmic strings have contributed to the evolution of the Universe, the GW power-spectrum (that we hope we observe some day) will decline from what it is anticipated by a pure-radiation model at every present-time frequency in the range $10^{-16}\ Hz < f \leq f_c^{pr}$ (see Fig. 1). The lower bound of this range arises from the GWs that began entering inside the horizon after the Universe has become matter-dominated [39]. Clearly, a potential detection of CGWs would give us valuable information on the epoch (and therefore on the physical mechanism, as well) at which the cosmic strings were formed.

5.2. Evolution of the oscillatory modes

The quantum-gravitational creation of gravitons in an expanding FRW Universe was first demonstrated by Grishchuk [14] and Starobinsky [44]. They showed that, in the *linear approximation*, the behavior of a CGW propagating in curved space-time is identical to that of a massless, minimally-coupled scalar field. In particular, each of the two polarization states of the metric perturbation satisfy the Klein-Gordon equation (2). The quantization of primordial GWs and that of the minimally-coupled, massless scalar fields also proceeds along identical lines, as it was demonstrated by Ford and Parker [18], [45].

In what follows, we will show that, if the Universal evolution includes a radiation-plus-strings stage, then, although it could last only for a short period of time, its presence would have resulted also in a measurable effect on the *high-frequency* CGWs' power-spectrum.

Indeed, around τ_* and τ_{sc} (the boundaries of the radiation-plus-strings stage) both $R(\tau)$ and $R'(\tau)$ can be matched to be continuous, but $R''(\tau)$ is essentially discontinuous, acquiring a non-zero value through Eq (13). Consequently, the *scalar curvature* of the space-time changes discontinuously both at τ_* and at τ_{sc}. A discontinuous change in the scalar curvature produces *gravitons* [19], [46].

As a consequence, the evolution of the oscillating GW modes in the transition of the Universe from an early-radiation epoch to the late-radiation era through a radiation-plus-strings stage, is modified as follows:

- For $\tau \leq \tau_*$ the effective potential (13) vanishes. The temporal evolution of the metric perturbations is governed by Eq (22), admitting the solution (23) and the gravitational-perturbation field is in an adiabatic-vacuum state.
- For $\tau_* < \tau < \tau_{sc}$ the effective potential reduces to a non-vanishing constant ($V_{eff} = k_c^2 c^2$). The propagation of the metric perturbations is now driven by Eq (25), for $k > k_c$. As a result of the discontinuous change of the scalar curvature at τ_*, during the whole radiation-plus-strings stage the gravitational-perturbation field is no longer in its vacuum-state and the general solution to Eq (25) is the linear superposition of positive- and negative-frequency solutions (28).
- Finally, for $\tau \geq \tau_{sc}$, $V_{eff} = 0$ and the temporal evolution of a metric perturbation in curved space-time is (once again) governed by Eq (22). However, due to the discontinuity of the scalar curvature at τ_{sc}, this time the general solution to Eq (22) is written in the form ($c \neq 1$)
$$h_k^{out}(\tau) = \sqrt{\frac{2}{\pi k c}} \left[\alpha_k \, e^{-\imath k c \tau} + \beta_k \, e^{+\imath k c \tau} \right], \tag{63}$$
where, the constant (Bogoliubov) coefficients α_k and β_k should satisfy the Wronskian condition
$$|\alpha_k|^2 - |\beta_k|^2 = 1. \tag{64}$$

For $k > k_c$, Eq (63) determines the *final* quantum state of the gravitational-perturbation field. As a result of conformal invariance, this state is also an adiabatic vacuum, but, as long as $\beta_k \neq 0$, it differs from the corresponding state of the early-radiation epoch. Clearly, the occurrence of a non-zero β_k in the late-radiation era would signal that $h_k^{out}(\tau)$ is not a pure *positive-frequency* solution, but contains also a *negative-frequency* component. This means that, if the *in-state* of the gravitational-perturbation field is taken to be vacuum, then, particles are found in the *out-state* [20].

In what follows, we discuss the evolution of the modes (23) with $k > k_c$, in the transition of the Universe from the early-radiation epoch to the late-radiation era through a radiation-plus-strings stage. The coefficients c_1, c_2, α_k and β_k can be determined by the requirement that $h_k(\tau)$ and its first derivative $h_k'(\tau)$ are continuous across the boundaries τ_* and τ_{sc}.

5.2.1. Graviton production by long cosmic strings: Matching the modes (23) and (28), as well as their first derivatives, at $\tau = \tau_*$, we obtain
$$c_1 = \frac{\gamma + 1}{2\sqrt{\gamma}} \, e^{\imath k c (\gamma - 1) \tau_*}, \tag{65}$$
and
$$c_2 = \frac{\gamma - 1}{2\sqrt{\gamma}} \, e^{-\imath k c (\gamma + 1) \tau_*}, \tag{66}$$
for which the Wronskian condition (29) holds. Similarly, matching the modes (28) and (63), as well as their first derivatives, at $\tau = \tau_{sc}$, we obtain
$$\alpha_k = \frac{1}{4\gamma} \left[(\gamma + 1)^2 \, e^{-\imath(\gamma - 1) k c \Delta\tau} - (\gamma - 1)^2 \, e^{+\imath(\gamma + 1) k c \Delta\tau} \right], \tag{67}$$
and
$$\beta_k = \imath \, \frac{\gamma^2 - 1}{2\gamma} \, sin(\gamma k c \Delta\tau) \, e^{-\imath k c (\tau_* + \tau_{sc})}, \tag{68}$$

where we have taken into account Eqs (65) and (66). Using Eq (27), the number of gravitons [20] created in the mode denoted by k, is written in the form

$$N_k = |\beta_k|^2 = \frac{k_c^4}{4\,k^2} \left[\frac{sin(c\Delta\tau\sqrt{k^2 - k_c^2})}{\sqrt{k^2 - k_c^2}} \right]^2 , \tag{69}$$

and the normalization condition $|\alpha_k|^2 - |\beta_k|^2 = 1$, holds. The observable quantity N_k possesses a series of very interesting properties.

Obviously, for $k_c = 0$, N_k vanishes, i.e. no gravitons are created in the absence of cosmic strings. On the other hand, for $k \to k_c$ and/or $\Delta\tau \to 0$, Eq (69) results in

$$N_k \to \frac{k_c^4}{4\,k^2}\, c^2\, \Delta\tau^2. \tag{70}$$

We distinguish the following cases:

(i) $\Delta\tau = 0$: In this case, $N_k = 0$. Although this may look like an unexpected result, it is not. Usually, the instantaneous transition from one epoch to another is a pathological feature of quantum field theory, leading to an infinite number of created particles (e.g. see [20]). However, in our case, as long as $\Delta\tau = 0$ no transition ever takes place. The Universe remains in the radiation era, where no gravitons are produced.

(ii) $k = k_c$: In this case, N_k acquires the constant value

$$N_{k_c} = \frac{k_c^2}{4}\, c^2\, \Delta\tau^2. \tag{71}$$

This is a very important result, suggesting that the number of gravitons in the *lowest-allowed* wave-number is well-defined and finite. Therefore, there should be no *infrared divergences.*

Finally, for $k \gg k_c$ and $k \to \infty$, $N_k \to 0$ and particle production ceases. Consequently, there should be no *ultraviolet divergences*, as well.

On the other hand, according to Eq (69), the *expectation value* of the *number-operator* varies periodically with k. Hence:

(iii) For comoving wave-numbers satisfying the condition

$$c\,\Delta\tau\sqrt{k^2 - k_c^2} = m\pi \;\; (m = 1, 2, ...), \tag{72}$$

N_k admits its absolute minimum value, i.e. $N_k = 0$. The case $m = 0$ is excluded due to the fact that $N_{k_c} \neq 0$. As a consequence, creation of gravitons with comoving wave-numbers

$$k_m^2 = k_c^2 + m^2\, \frac{\pi^2}{(c\Delta\tau)^2} \;\; (m = 1, 2, ...), \tag{73}$$

never occurs by this mechanism. In quantum physics, this case is referred to as *perfect-transmission* [47]. Notice that, the spectrum of these states is *discrete* and their position on the k-axis is determined solely by the choice of $\Delta\tau \neq 0$.

5.2.2. The power-spectrum of these relic gravitons: During the (late) radiation stage of the expansion, the *physical* frequency of a primordial GW is defined as $\omega = ck/R(\tau)$ and the number of states with frequencies in the interval ω and $\omega + d\omega$ is written in the form $dn_\omega = \omega^2 d\omega/2\pi^2 c^3$. Accordingly, the distribution of the gravitational energy-density, summed over the two polarization states of the gravitons present in the late-radiation epoch, is given by

$$d\epsilon_{gw} = \mathcal{P}(\omega)d\omega = 2\hbar\omega \, \frac{\omega^2}{2\pi^2 c^3} \, |\beta_k|^2 \, d\omega, \tag{74}$$

and, in terms of k, is reduced to

$$d\epsilon_{gw} = \mathcal{P}(k)dk = \tag{75}$$
$$\frac{1}{R^4(\tau)} \, \frac{\hbar c}{4\pi^2} \, k_c^4 \left[\frac{sin\left(c\Delta\tau\sqrt{k^2 - k_c^2}\right)}{\sqrt{k^2 - k_c^2}} \right]^2 k \, dk.$$

Therefore, the power-spectrum of the created gravitons [15], [19], [46] reads

$$\begin{aligned} \mathcal{P}(k) &= \frac{d\epsilon_{gw}}{dk} = \\ &= \frac{\hbar c}{4\pi^2} \, k_{c_{ph}}^4 \left[\frac{sin\left(c\Delta\tau\sqrt{k^2 - k_c^2}\right)}{\sqrt{k^2 - k_c^2}} \right]^2 k, \end{aligned} \tag{76}$$

where, we have introduced the (physical) critical wave-number at every τ as

$$k_{c_{ph}}(\tau) = k_c/R(\tau). \tag{77}$$

In terms of the dimensionless parameter

$$x = \frac{k}{k_c} \geq 1, \tag{78}$$

which, in this case, measures the comoving wave-number in units of the corresponding *lowest-allowed* k_c, Eq (76) is written in the form

$$\mathcal{P}(x) = \frac{d\epsilon_{gw}}{dx} = k_c \frac{d\epsilon_{gw}}{dk} = \tag{79}$$
$$(k_c \, c \, \Delta\tau)^2 \, \frac{\hbar c}{4\pi^2} \, k_{c_{ph}}^4 \left[\frac{sin\left(k_c c\Delta\tau\sqrt{x^2 - 1}\right)}{\left(k_c c\Delta\tau\sqrt{x^2 - 1}\right)} \right]^2 x,$$

or else

$$\begin{aligned} \mathcal{P}(x) &= 4\pi \, \frac{1}{4} \, (k_c \, c \, \Delta\tau)^2 \, 2\hbar \, c \, k_{c_{ph}} \, \frac{1}{\lambda_{c_{ph}}^3} \\ &\times \left[\frac{sin\left(k_c c\Delta\tau\sqrt{x^2 - 1}\right)}{\left(k_c c\Delta\tau\sqrt{x^2 - 1}\right)} \right]^2 x, \end{aligned} \tag{80}$$

where, $\lambda_{c_{ph}} = 2\pi/k_{c_{ph}}$ is the *correlation (wave)length*, associated to $k_{c_{ph}}$.

At a fixed $\tau \geq \tau_{sc}$, the gravitational energy contained within the volume $\lambda^3_{c_{ph}}$, is $2\hbar c k_{c_{ph}}$, i.e. the energy of a graviton with physical wave-number $k_{c_{ph}}$ summed over the two polarization states. Therefore, the quantity

$$\epsilon_c(\tau) = N_{k_c}\, 2\hbar\, c\, k_{c_{ph}}\, \frac{1}{\lambda^3_{c_{ph}}}, \tag{81}$$

[where N_{k_c} is given by Eq (71)] represents the overall energy-density of the gravitons created in the state denoted by k_c, at every $\tau_{sc} \leq \tau \leq \tau_c$, where τ_c is the time at which the longest metric perturbation created by this mechanism (i.e. the one with $k = k_c$), enters inside the horizon. At that time its *physical wavelength* $[\lambda_{ph}(\tau_c) = \lambda_c\, R(\tau_c)]$ becomes smaller than the corresponding *Hubble radius* $[\ell_H(\tau_c) = c/H(\tau_c)$, H being the Hubble parameter]. According to Eq (61),

$$\tau_c \geq \left(\frac{G\mu}{c^2}\right)^{-1/2} \tau_*, \tag{82}$$

and, therefore, $\tau_c \sim 10^3\, \tau_* \gg \tau_{sc} \sim 17\, \tau_*$. Since k_c is the lowest-allowed comoving wave-number, after τ_c there are no gravitons (created by this mechanism) to enter inside the Hubble sphere.

Upon consideration of Eqs (71) and (81), the power-spectrum (80) results in

$$\mathcal{P}(x) = 4\pi\, \epsilon_c(\tau) \left[\frac{sin\left(k_c c \Delta\tau \sqrt{x^2-1}\right)}{\left(k_c c \Delta\tau \sqrt{x^2-1}\right)}\right]^2 x. \tag{83}$$

For every $\tau_{sc} \leq \tau \leq \tau_c$, $\mathcal{P}(1)$ is the energy-density of the gravitons created in the state $k = k_c$ (within the solid angle of a sphere - 4π), while, as we depart from the lowest-allowed comoving wave-number, the gravitational power is distributed among the various k-intervals according to Eq (83). In particular, for k very close to k_c ($x \simeq 1$), $\mathcal{P}(k) \sim k$, while, in the large-k limit ($x \to \infty$) it results in a damped oscillation

$$\mathcal{P}(k) \sim \frac{1}{k}\, sin^2(kc\Delta\tau). \tag{84}$$

According to Eq (84), graviton production is suppressed at high frequencies.

On the other hand, the *logarithmic spectrum* is written in the form

$$\frac{d\epsilon_{gw}}{d(\ln k)} = k\, \mathcal{P}(k) = x\, \mathcal{P}(x) = \tag{85}$$

$$4\pi\, \epsilon_c(\tau) \left[\frac{sin\left(k_c c \Delta\tau \sqrt{x^2-1}\right)}{\left(k_c c \Delta\tau \sqrt{x^2-1}\right)}\right]^2 x^2.$$

For $k \simeq k_c$ ($x \to 1$), the logarithmic spectrum is quadratic in k, while, for $k \gg k_c$ ($x \to \infty$), it reduces to a periodic function of the comoving wave-number with constant amplitude (Fig. 2)

$$\frac{d\epsilon_{gw}}{d(lnk)} \sim sin^2(kc\Delta\tau). \tag{86}$$

The importance of this result rests in the fact that the logarithmic spectrum is a re-scaling of the *spectral intensity*

$$\frac{1}{\varepsilon_{cr}} \frac{d\epsilon_{gw}}{d(\ln k)} = \Omega_{gw} = \frac{1}{\varepsilon_{cr}} \frac{d\epsilon_{gw}}{d(lnf)}, \tag{87}$$

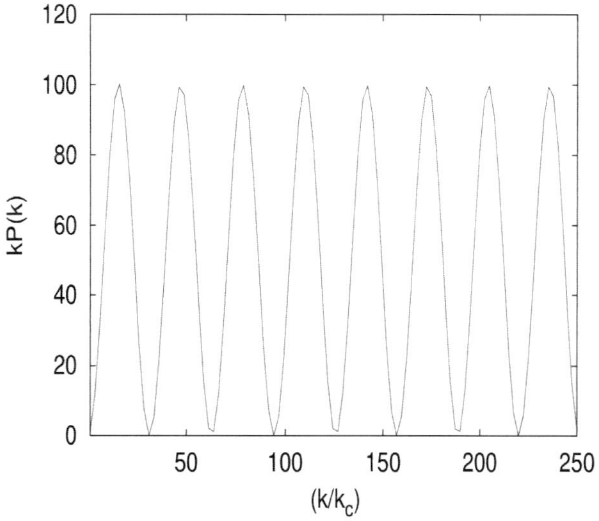

Figure 2. The logarithmic spectrum $k\,\mathcal{P}(k)$ of the metric perturbations created during the scaling of a long-cosmic-string network (normalized over $4\pi\,\epsilon_c$), versus k in units of k_c, for $k_c c\Delta\tau = 0.32$ (corresponding to $\tau_{sc} = 17\tau_*$ [31]).

where ε_{cr} is the *energy-density for closing the Universe.*

Accordingly, if the Universal evolution has undergone through a radiation-plus-strings stage, then, at present-time frequencies *larger* than the critical value (62) (corresponding to $k_{c_{ph}}$ at the present epoch), the relic GW spectrum would no longer be scale invariant (as it is predicted by pure-radiation), but it should have been reduced to a *periodic function* of the frequency (Fig. 2). For $k = k_c$ ($x = 1$), Eq (85) yields

$$k_c\,\mathcal{P}(k_c) = 4\pi\,\epsilon_c(\tau). \tag{88}$$

In this case, from Fig. 2 we observe that, the *amplitude* of the logarithmic spectrum (and therefore of Ω_{gw}, as well) reaches up to 100 times the (initial) value of this function for $k = k_c$. This is a very important result!

It suggests that, after being *filtered* by a long-cosmic-string network, the high-frequency part of the relic GW signal will be *amplified* by (almost) *two orders of magnitude.* To the best of our knowledge, *this is the first time* (since Grishchuk [14]) that an amplification mechanism of a stochastic GW signal is proposed and discussed. In this context, an indirect interaction between relic GWs and cosmic strings could have a major impact on the detection of these waves.

Since the only interference to the propagation of the relic GWs from an early-radiation epoch to the late-radiation era comes from cosmic strings (encapsulated in the cosmological model), it is natural to assume that any *energization* of the gravitational-perturbation field is due to a corresponding *energy-loss* of the linear-defect network. This assumption is compatible to the absence of *high-frequency harmonics* on long cosmic strings [48].

In particular, cosmic-string simulations indicate that, on small scales, the strings are extremely *wiggly.* In other words, there is a substantial amount of small-scale structure on a string network [4], [5]. The small-amplitude perturbations (or waves) on an *infinite* string can become the source of gravitational radiation with the same frequency [49], [50]. However, it has been found that the production of high-frequency gravitational waves from a long cosmic string is negligible, suggesting that the corresponding higher harmonics are strongly suppressed [48].

Our model could provide a physical interpretation for this result, in the sense that, the higher harmonics of the long linear defects might have been *consumed* in the excitation of high-frequency relic GWs (those produced by inflation at earlier epochs and the wavelength of which became less than the horizon during the radiation-plus-strings stage) and not in the production of *new* CGWs. The corresponding physical mechanism probably involves a *resonant interaction* between inflationary GWs and long cosmic strings, which becomes prominent at high frequencies. Such a mechanism is yet unknown and its exploration will be the scope of a future work.

6. Conclusions

The equation which governs the temporal evolution of a CGW in a Friedmann model can be treated as the Schrödinger equation for a particle moving in the presence of the effective potential $V_{eff} = R''/R$. In the present article we show that, if there is a period where the effective potential is constant, this would lead to a critical value (k_c) in the comoving wave-number of the metric fluctuations, discriminating them into oscillating ($k > k_c$) and non-oscillating ($k < k_c$) modes. As a consequence, when the non-oscillatory modes lie outside the horizon do not freeze out, something that should be reflected in the relic-GW power-spectrum.

This property is met in a radiation model contaminated by a fraction of cosmic strings. Therefore, if the cosmological evolution includes a radiation-plus-strings stage, some of the long-wavelength GW modes (although being outside the Hubble sphere) continue to evolve. However, this stage (if ever existed) does not last very long, since, the cosmological evolution gradually results in the scaling of the cosmic-string network and, after some time ($\Delta\tau$), the Universe enters in the late-radiation era.

In a radiation-dominated Universe the metric perturbations of $k < k_c$ can enter the horizon, which now expands faster than their physical wavelength. However, the evolution of the non-oscillatory GW modes during $\Delta\tau$ (while they were outside the horizon) has modified their amplitude and, therefore, oscillation of these modes within the Hubble sphere results in the distortion of the scale-invariant GW power-spectrum at present-time frequencies in the range $10^{-16}\ Hz < f \leq 10^5\ Hz$.

On the other hand, if there was a period in the early Universe in which the matter-content can be modelled by a two-component fluid consisting of radiation and cosmic strings, this would have resulted in a measurable effect on the high-frequency CGWs' power-spectrum, as well. In fact, along the transition from an early-radiation epoch to the late-radiation era through a radiation-plus-strings stage, gravitons with a highly recognizable profile could have been produced. As a consequence, at high (present-time) frequencies ($f \geq 10^5\ Hz$), the GW spectrum would also depart from scale-invariance, becoming a *periodic* function of the frequency, the period of which depends solely on the *duration* of the radiation-plus-strings stage.

But, what's most important for the detection of these waves, is that: The semiclassical interaction between metric perturbations and cosmic strings would lead to the *amplification of the relic GW signal* by (almost) *two orders of magnitude*. To the best of our knowledge, this is the first time since Grishchuk's work, that an amplification mechanism of a stochastic GW signal is proposed and discussed.

Acknowledgments

Financial support by the Greek Ministry of Education under the PYTHAGORAS Program, is gratefully acknowledged.

References

[1] Weinberg S 1972 *Gravitation and Cosmology* (New York: Wiley Interscience)
[2] Maggiore M 2000 *Phys. Rep.* **331** 283
[3] Misner C W, Thorne K S and Wheeler J A 1973 *Gravitation* (San Francisco: Freeman)

[4] Hindmarsh M B and Kibble T W B, 1995 *Rep. Prog. Phys.* **58** 477
[5] Vilenkin A and Shellard E P S 2000 *Cosmic Strings and Other Topological Defects* (Cambridge: Cambridge University Press)
[6] Vilenkin A 1986 *Nature* **322** 613
[7] Zel' dovich Ya B 1980 *MNRAS* **192** 663
[8] Vilenkin A 1981 *Phys. Rev. Lett.* **46** 1169
[9] Allen B 1997 *"The stochastic gravity-wave background: Sources and detection"* in: J A Marck and Lassota J P (eds) *Les Houches School on Astrophysical Sources of Gravitational Waves* (Cambridge: Cambridge University Press)
[10] Battye R A, Caldwell R R and Shellard E P S 1998 *"Gravitational waves from cosmic strings"* in: F Melchiorri and M Signore (eds) *Topological Defects in Cosmology* (Singapore: World Scientific)
[11] Kleidis K, Papadopoulos D B, Verdaguer E and Vlahos L 2008 *Phys. Rev. D* **78** 024027
[12] Kibble T B W 1985 *Nucl. Phys. B* **252** 227
[13] Abbott L F and Wise M B 1984 *Nucl. Phys. B* **244** 541
[14] Grishchuk L P 1975 *Sov. Phys. JETP* **40** 409
[15] Mukhanov V F, Feldman H A and Brandenberger R H 1992 *Phys. Rep.* **215** 203
[16] Bleyer U, Liebscher D E and Polnarev A G 1991 *Class. Quantum Grav.* **8** 477
[17] Vilenkin A 1981 *Phys. Rev. D* **24** 2082
[18] Ford L and Parker L 1977 *Phys. Rev. D* **16** 245
[19] Allen B 1988 *Phys. Rev. D* **37** 2078
[20] Birrell N D and Davies P C W 1982 *Quantum Fields in Curved Space* (Cambridge: Cambridge University Press)
[21] Kibble T B W, Lazarides G and Shafi Q 1982 *Phys. Rev. D* **26** 435
[22] Kolb E W and Turner M S 1990 *The Early Universe* (New York: Addison-Wesley)
[23] Kibble T B W 1976 *J. Phys. A: Math. Gen.* **9** 1387
[24] Kibble T B W 1987 *"Cosmology and Particle Physics"* in *Lectures at GIFT XVIIth Int. Seminar on Theoretical Physics* E Alvarez, R Dominguez-Tenreiro, J M Ibanez-Cabanel and M Quiros (eds) (Singapore: World Scientific)
[25] Vilenkin A 1991 *Phys. Rev. D* **43** 1060
[26] Albrecht A and Turok N 1989 *Phys. Rev. D* **40** 973
[27] Bennett D P and Bouchet F R 1989 *Phys. Rev. Lett.* **63** 2776
[28] Bennett D P and Bouchet F R 1990 *Phys. Rev. D* **41** 2408
[29] Allen B and Shellard E P S 1990 *Phys. Rev. Lett.* **64** 119
[30] Sakellariadou M 2005 *J. Cosmol. Astropart. Phys.* JCAP04(2005)003
[31] Ringeval C, Sakellariadou M and Bouchet F R 2007 *J. Cosmol. Astropart. Phys.* JCAP02(2007)023
[32] Martins C J A P and Shellard E P S 2006 *Phys. Rev. D* **73** 043515
[33] Lidsey J E, Liddle A R, Kolb E W, Copeland E J, Barreiro T and Abney M 1997 *Rev. Mod. Phys.* **69** 373
[34] Grishchuk L P, Lipunov V M, Postnov K A, Prokhorov M E and Sathyaprakash B S 2001 *Phys. Usp.* **44** 1
[35] Pritchard J R and Kamionkowski M 2005 *Ann. of Phys.* **318** 2
[36] Pogosian L, Tye S-H H, Wasserman I and Wyman M 2003 *Phys. Rev. D* **68** 023506
[37] Jeong E and Smoot G F 2004 *"Search for cosmic strings in CMB anisotropies"* Preprint astro-ph/0406432
[38] Sazhin M, Longo G, Capaccioli M, Alcala J M, Silvotti R, Covone G, Khovanskaya O, Pavlov MV, Pannella M, Radovich M and Testa V 2003 *MNRAS* **343** 353
[39] Sazhin M, Khovanskaya O, Capaccioli M, Longo G, Alcala J M, Silvotti R and Pavlov M V 2004 *"Lens candidates in the Capodimonte Deep Field in the vicinity of the CLS1 object"* Preprint astro-ph/0406516
[40] Carr B J 1980 *A&A* **89** 6
[41] Zhao W and Zhang Y 2005 *Phys. Rev. D* **74** 043503
[42] Reece C, Reiner P and Melissinos A 1984 *Phys. Lett. A* **104** 341
[43] Reece C, Reiner P and Melissinos A 1986 *Nucl. Instrum. Methods Phys. Res., Sect. A* **245** 299
[44] Starobinsky A A 1979 *JETP Lett.* **30** 682
[45] Ford L and Parker L 1977 *Phys. Rev. D* **16** 1601
[46] Sahni V 1990 *Phys. Rev. D* **42** 453
[47] Schiff L I 1955 *Quantum Mechanics* (New York: McGraw-Hill) pp 101 - 104
[48] Battye R A and Shellard E P S 1994 *Nucl. Phys. B* **423** 260
[49] Hindmarsh M B 1990 *Phys. Lett. B* **251** 28
[50] Sakellariadou M 1990 *Phys. Rev. D* **42** 354

Journal of Physics: Conference Series **189** (2009) 012022 doi:10.1088/1742-6596/189/1/012022

Cosmological magnetic fields

Kerstin E. Kunze

Departamento de Física Fundamental, Universidad de Salamanca, Plaza de la Merced s/n, 37008 Salamanca, Spain

E-mail: kkunze@usal.es

Abstract. Observations indicate that magnetic fields are associated with most structures in the universe. Upto now the origin of large scale primordial magnetic fields is an open problem. Two mechanisms in particular, taking place during inflation in the very early universe are reviewed.

1. Introduction

Observations indicate that magnetic fields are associated with most structures in the universe. Magnetic fields are observed to be of the order of 10^{13} G in neutron stars, 10^3 G in solar type stars and of order a μG on galactic scale. Magnetic fields of order a few μG also have been detected in radio galaxies at redshifts $z > 2$. The structure of the magnetic field in galaxies depends on the type of galaxy. In spiral galaxies there is typically a large scale component whose coherence length is of order of the size of the visible disk. In addition to this there are small-scale tangled fields. In contrast, magnetic fields in elliptical galaxies seem to be random with a coherence length much smaller than the galactic scale [1]. Basically there are no direct observations of magnetic fields not associated with some kind of structure. However, limits on the strength of truly cosmological magnetic fields can be derived indirectly using for example big bang nucleosynthesis [2] or the anisotropies of the cosmic microwave background [3, 4]. Since magnetic fields enter directly into the initial conditions as well as the evolution equations or, in the case of big bang nucleosynthesis, in the interaction rates, this leads to upper bounds on the magnetic field strength. In order to explain the galactic magnetic field the standard approach is to assume some kind of dynamo mechanism which amplifies an initial "seed" magnetic field. Similarly, magnetic fields in stars are explained by the formation of protostars out of condensed interstellar matter which was pervaded by a preexisting large scale magnetic field [5].

One of the open problems is to explain the origin of this initial magnetic field. There is a multitude of proposals. However, none of them seems to entail the most "natural" model. The proposals can be classified broadly into two groups. On the one hand mechanisms which rely on processes on small scales such as vortical perturbations and phase transitions and on the other hand those which use the potential amplification of perturbations in the electromagnetic field during inflation in the very early universe, see for example, [6, 7]. Inflation provides a mechanism to amplify initially small perturbation amplitudes to cosmologically relevant values. In flat backgrounds in order to reach interesting amplitudes one has to break the conformal invariance of the background. In four dimensions the Maxwell Lagrangian describing linear electrodynamics is conformally invariant. There has been a variety of proposals to break the conformal invariance such as, coupling to a scalar field [8], breaking Lorentz invariance [9], adding extra dimensions [10, 11], coupling to curvature terms [12] or nonlinear electrodynamics [13, 14].

In curved backgrounds with open spatial sections there is an interesting coupling between the geometry and electromagnetism that can lead to the superadiabatic amplification of the field amplitude just within the framework of standard electromagnetism [15].

In particular, here two proposals will be considered in more detail. On the one hand, the generation of magnetic seed fields from extra dimensions will be considered. On the other hand, the way how nonlinear electrodynamics can produce primordial magnetic fields of cosmologically interesting field strengths will be discussed.

2. Primordial magnetic fields from extra dimensions

Space-times with more than four dimensions have been considered since a long time starting with the work of Kaluza and Klein [16]. In this approach the extra dimensions are small enough so that our world is effectively described by physics in four dimensions.

The possibility of having large extra dimensions came about when solutions of this type were found within string/M-theory. Here the observable four dimensional universe is located on a brane which is a four dimensional hypersurface embedded in a higher dimensional space-time. In higher-dimensional gravity the four dimensional Planck scale M_4 is no longer fundamental, but rather it is the higher dimensional Planck scale M_D. Assuming that there are n extra dimensions with a characteristic size R, using Gauss' law the four dimensional and D dimensional Planck masses are related by,

$$M_4^2 = R^n M_D^{n+2}. \tag{1}$$

A lower bound on the ratio M_D/M_4 can be found by experiments of Newtonian gravity. Experiments indicate that Newtonian gravity is valid at least down to length scales of 1 mm [17].

The background space-time is assumed to be homogeneous and anisotropic with a line element,

$$ds^2 = a^2(\eta)\left[d\eta^2 - \delta_{ij}dx^i dx^j\right] - b^2(\eta)\delta_{AB}dy^A dy^B, \tag{2}$$

where $i, j = 1, .., 3$ and $A, B = 4, ..., 3+n$, $n \geq 1$. Furthermore $a(\eta)$ is the scale factor of the "external", 3 -dimensional space and $b(\eta)$ is the scale factor of the internal, n-dimensional space. We assume that an early $(n+4)-$ dimensional vacuum phase of the universe is matched to a standard four dimensional radiation dominated phase. Thus whereas in the higher dimensional phase both scale factors are functions of time, in the succeeding four dimensional phase only the scale factor of the three dimensional space $a(\eta)$ is dynamical. In this stage the extra dimensions are frozen at a constant value of the scale factor $b(\eta)$ =const. The matching of these two phases of the universe, that is the higher-dimensional and the four dimensional stage, takes place at some time $\eta = -\eta_1$. The solutions for the scale factors during these two epochs are given by,

$$a(\eta) = a_1\left(-\frac{\eta}{\eta_1}\right)^{\sigma}, \qquad b(\eta) = b_1\left(-\frac{\eta}{\eta_1}\right)^{\lambda}, \qquad \text{for} \quad \eta < -\eta_1 \tag{3}$$

$$a(\eta) = a_1\left(\frac{\eta+2\eta_1}{\eta_1}\right), \qquad b(\eta) = b_1, \qquad \text{for} \quad \eta \geq -\eta_1 \tag{4}$$

In the following we set $a_1 = 1 = b_1$. Clearly, for $\eta < -\eta_1$ the solution is given by the vacuum Kasner metric, which is determined by the exponents σ and λ which are functions of the number of extra dimensions n. These are related to the Kasner exponents α_E and α_I, which satisfy the Kasner conditions $3\alpha_E + n\alpha_I = 1$ and $3\alpha_E^2 + n\alpha_I^2 = 1$, by

$$\sigma = \frac{\alpha_E}{1-\alpha_E}, \qquad \lambda = \frac{\alpha_I}{1-\alpha_E}. \tag{5}$$

Furthermore, in the case of an expanding, external space and a contracting, internal space the exponents σ and λ are of the form [10],

$$\sigma = -\frac{1}{2}\left(\sqrt{\frac{3n}{n+2}} - 1\right), \qquad \lambda = \sqrt{\frac{3}{n(n+2)}}. \tag{6}$$

Starting with Maxwell's equations in D dimensions, $\nabla_{\tilde{A}} F^{\tilde{A}\tilde{B}} = 0$, where $\tilde{A}, \tilde{B} = 0, ..., n+3$ equations for the gauge potential A_i in 3+1 dimensions can be derived. Imposing the radiation gauge simplifies these equations. In order to proceed from there the canonical field $\Psi_i = b^{\frac{n}{2}} A_i$ is introduced and Fourier expanded. This finally leads to a mode equation, which during the early stage of dynamical extra dimensions is given by, for $\eta < -\eta_1$, [11]

$$\Psi_l'' + \left[k^2 + \left(-\frac{\eta}{\eta_1}\right)^{2\beta} q^2 - \frac{N}{\eta^2}\right]\Psi_l = 0, \tag{7}$$

where $' \equiv \frac{\partial}{\partial\eta}$ and $N \equiv \frac{1}{4}(n\lambda - 1)^2 - \frac{1}{4}$. Furthermore, $\beta \equiv \sigma - \lambda$. During the later radiation dominated epoch where the extra dimensions are frozen to a particular size and only the three spatial dimensions of the observable universe are dynamical, the mode equation takes the form, for $\eta \geq -\eta_1$

$$\Psi_l'' + \left[k^2 + \left(\frac{\eta + 2\eta_1}{\eta_1}\right)^2 q^2\right]\Psi_l = 0. \tag{8}$$

The next step is to solve the mode equations in the two regimes, using the proper matching conditions at the transition time, $\eta = -\eta_1$. This leads to an expression for the Bogoliubov coefficient determining the number of created photons which can be used to calculate the total magnetic energy density in four dimensions. As the universe had a very high conductivity during most of its history, the magnetic field evolves such that its flux is conserved and thus the dimensionless ratio $r \equiv \rho_B/\rho_\gamma$ is approximately constant [7], where ρ_B is the magnetic field energy density and ρ_γ the energy density of the background radiation. r calculated at the galactic scale of order 1 Mpc determines the magnetic field strength at the time of galaxy formation. Assuming that a galactic dynamo has been operating since the time of galaxy formation implies a lower limit on the primordial seed field strength of $B_s \sim 10^{-20}$G [5]corresponding to $r \sim 10^{-37}$. This number can be lowered significantly if the non-vanishing cosmological constant is taken into account. In that case the minimal required magnetic seed field strength is $B_s \sim 10^{-30}$G and $r \sim 10^{-57}$. Imposing in addition that $r(\omega) < 1$, the bound on M_D/M_4 from Newtonian gravity and that standard big bang nucleosynthesis takes place, leads to bounds on the parameters of the model. It is found that if the momenta lying along the extra dimensions are taken into account magnetic fields strong enough to seed the galactic magnetic field can be created. For details see reference [11].

3. Primordial magnetic fields from nonlinear electrodynamics

Nonlinear electrodynamics naturally breaks the conformal invariance of electrodynamics in four dimensions. Thus it seems to be a good candidate to amplify perturbations of the electrodynamic field during inflation in the early universe. Nonlinear electrodynamics can be understood as taking into account the self interaction of the electromagnetic field. Born and later Infeld were the first to treat a particular model of nonlinear electrodynamics in the search for a classical, singularity free theory of the electron [18]. Another class of models of nonlinear electrodynamics is given by the Heisenberg-Euler Lagrangian which describes effectively the propagation of a photon in an external electromagnetic field [19].

The question that arises in the context of generation of primordial magnetic fields is if models of nonlinear electrodynamics can generate primordial magnetic fields of cosmologically interesting strength. As we will see below this is actually the case. The cosmological model to be considered consists of a stage of de Sitter inflation followed by reheating and a standard radiation dominated stage. During inflation quantum fluctuations in the electromagnetic field are excited within the horizon. Being stretched beyond the horizon they become classical perturbations. It is assumed that during inflation electrodynamics is nonlinear which could be motivated by the presence of possible quantum corrections to quantum electrodynamics at high energies. Furthermore, it is assumed that once inflation ends electrodynamics becomes linear so that the subsequent evolution is described by the standard model of cosmology.

In general, electrodynamics coupled minimally to gravity can be written as, [20]

$$S=\frac{1}{16\pi G_N}\int d^4x\sqrt{-g}R+\frac{1}{4\pi}\int d^4x\sqrt{-g}L(X,Y), \tag{9}$$

where $L(X,Y)$ is the Lagrangian of nonlinear electrodynamics. Furthermore, the invariants are denoted by $X\equiv\frac{1}{4}F_{\mu\nu}F^{\mu\nu}$ and $Y\equiv\frac{1}{4}F_{\mu\nu}\,{}^*F^{\mu\nu}$, where ${}^*F^{\mu\nu}$ is the dual bi-vector given by ${}^*F^{\mu\nu}=\frac{1}{2\sqrt{-g}}\epsilon^{\mu\nu\alpha\beta}F_{\alpha\beta}$, and $\epsilon^{\mu\nu\alpha\beta}$ the Levi-Civita tensor with $\epsilon_{0123}=+1$.

The equations of motion are given by

$$\nabla_\mu P^{\mu\nu}=0 \tag{10}$$

where $P_{\mu\nu}=-\left(L_X F_{\mu\nu}+L_Y\,{}^*F_{\mu\nu}\right)$, furthermore L_A denotes $L_A=\partial L/\partial A$, and

$$\nabla_\mu\,{}^*F^{\mu\nu}=0, \tag{11}$$

which implies that $F_{\mu\nu}=\partial_\mu A_\nu-\partial_\nu A_\mu$ where $\mu,\nu..=0...3$. Moreover, the electromagnetic field is treated as a perturbation so that the vacuum Einstein equations apply to the background cosmology. The background metric is chosen to be of the form

$$ds^2=a^2(\eta)\left[-d\eta^2+d\mathbf{x}^2\right]. \tag{12}$$

Writing the Maxwell tensor in terms of the electric and magnetic field leads to a set of highly nonlinear equations coupled in these two fields [13]. Clearly, a Fourier expansion of the electric and magnetic field leads to mode coupling. Therefore, we used the long wave length approximation which is based on the observation that if variations over a characteristic comoving length scale L much larger than the horizon aH are considered then the spatial derivatives of a quantity can be neglected with respect to its time derivatives. In this approximation the equation for the magnetic field is given by

$$\vec{\mathcal{B}}_k''+\frac{L_X'}{L_X}\vec{\mathcal{B}}_k'+\frac{L_Y'}{L_X}\vec{\mathcal{E}}_k'+\frac{L_Y'}{L_X}\left[\frac{L_X'}{L_X}\vec{\mathcal{E}}_k-\frac{L_Y'}{L_X}\vec{\mathcal{B}}_k\right]=0, \tag{13}$$

where $\vec{\mathcal{B}}_k\equiv a^2\vec{B}_k$, $\vec{\mathcal{E}}_k\equiv a^2\vec{E}_k$ and a prime denotes the derivative with respect to conformal time η, that is $'\equiv\frac{d}{d\eta}$. Now in order to make progress one has to choose a particular theory of nonlinear electrodynamics $L(X,Y)$. Our aim is to see if in principle it is possible to amplify sufficiently perturbations in the electromagnetic field during inflation. Thus for simplicity the Lagrangian is chosen to be of the form,

$$L = -\left(\frac{X^2}{\Lambda^8}\right)^{\frac{\delta-1}{2}} X, \tag{14}$$

where δ is a dimensionless parameter and Λ a dimensional constant. This Lagrangian describes the abelian Pagels-Tomboulis model [21]. Originally, the nonabelian theory was proposed as an effective model of low energy QCD [22]. It is obvious that linear electrodynamics is recovered for the choice $\delta = 1$. As emphasized before the Lagrangian (14) is chosen in such a way since it leads to a simplification of the equations, but still allows to study the effects of a strongly nonlinear theory of electrodynamics on the generation of primordial magnetic fields. Using the Lagrangian (14) in the equations of motion in the long wave length approximation leads to an equation for the evolution of the energy density of the magnetic field which can be approximately solved in three different regimes depending on $\vec{\mathcal{B}}_k^2/\vec{\mathcal{E}}_k^2$ (for details see [13]). Basically the cases $\vec{\mathcal{B}}_k^2/\vec{\mathcal{E}}_k^2$ greater, equal or smaller than one are considered. The key quantity in order to decide if the amplification of the initial perturbation in the electromagnetic field during inflation is cosmologically relevant is as before the ratio of magnetic energy density to radiation energy density, $r = \rho_B/\rho_\gamma$. Requiring that r has to be larger than 10^{-37} which corresponds to a seed magnetic field of $B_s \sim 10^{-20}$G it has been shown that depending on the approximation and the value of the Pagels-Tomboulis parameter δ primordial magnetic fields can be generated that are strong enough to seed the galactic magnetic field. For a detailed discussion see [13].

4. Conclusions

We have presented two particular mechanisms to generate primordial magnetic fields. Both of them take place in flat backgrounds and so rely on breaking the conformal invariance of Maxwell's equations in four dimensions. In particular, the first mechanism presented involves a stage with dynamical extra dimensions during inflation. The second mechanism assumes that due to quantum corrections electrodynamics effectively becomes non linear. In both cases it has been found that there is a range of parameters where primordial magnetic fields of cosmologically relevant strength can be created, so as to seed for a example the galactic dynamo resulting in the galactic magnetic field of 10^{-6}G today.

Acknowledgments

I thank the conference organizers for their hospitality and in particular Christos Tsagas. Financial support from Spanish Science Ministry grants FPA2005-04823, FIS2006-05319 and CSD2007-00042 is gratefully acknowledged.

References

[1] L. M. Widrow, Rev. Mod. Phys. **74** (2002) 775; M. Giovannini, Int. J. Mod. Phys. D **13**, 391 (2004).
[2] B. l. Cheng, D. N. Schramm and J. W. Truran, Phys. Rev. D **49** (1994) 5006; P. J. Kernan, G. D. Starkman and T. Vachaspati, Phys. Rev. D **54** (1996) 7207; B. l. Cheng, A. V. Olinto, D. N. Schramm and J. W. Truran, Phys. Rev. D **54** (1996) 4714; M. Giovannini and M. E. Shaposhnikov, Phys. Rev. Lett. **80** (1998) 22.
[3] M. Giovannini, Phys. Rev. D **70** (2004) 123507; M. Giovannini, Phys. Rev. D **74** (2006) 063002; M. Giovannini, PMC Phys. A **1** (2007) 5.
[4] M. Giovannini and K. E. Kunze, Phys. Rev. D **77** (2008) 061301; M. Giovannini and K. E. Kunze, Phys. Rev. D **77** (2008) 063003; M. Giovannini and K. E. Kunze, Phys. Rev. D **77** (2008) 123001; M. Giovannini and K. E. Kunze, Phys. Rev. D **78** (2008) 023010.
[5] M. J. Rees, Quaterly Journal of the Royal Astronomical Society **28** (1987) 197.

[6] D. Grasso and H. R. Rubinstein, Phys. Rept. **348**, 163 (2001).
[7] M. S. Turner and L. M. Widrow, Phys. Rev. D **37**, 2743 (1988).
[8] B. Ratra, Astrophys. J. **391** (1992) L1; D. Lemoine and M. Lemoine, Phys. Rev. D **52** (1995) 1955; M. Gasperini, M. Giovannini and G. Veneziano, Phys. Rev. Lett. **75** (1995) 3796.
[9] O. Bertolami and D. F. Mota, Phys. Lett. B **455** (1999) 96; A. Ashoorioon and R. B. Mann, Phys. Rev. D **71** (2005) 103509.
[10] M. Giovannini, Phys. Rev. D **62** (2000) 123505.
[11] K. E. Kunze, Phys. Lett. B **623** (2005) 1.
[12] F. D. Mazzitelli and F. M. Spedalieri, Phys. Rev. D **52** (1995) 6694.
[13] K. E. Kunze, Phys. Rev. D **77** (2008) 023530.
[14] L. Campanelli, P. Cea, G. L. Fogli and L. Tedesco, Phys. Rev. D **77** (2008) 043001.
[15] C. G. Tsagas and A. Kandus, Phys. Rev. D **71** (2005) 123506.
[16] T. Kaluza, Sitz. Preuss. Akad. Wiss. **33** (1921) 966; O. Klein, Z. Phys. **37** (1926) 895.
[17] C. D. Hoyle, U. Schmidt, B. R. Heckel, E. G. Adelberger, J. H. Gundlach, D. J. Kapner and H. E. Swanson, Phys. Rev. Lett. **86** (2001) 1418.
[18] M. Born, Nature **132** (1933) 282; Proc. Roy. Soc. **A 143** (1934) 410; M. Born and L. Infeld, Nature **132** (1933) 970 and 1004; Proc. Roy. Soc. **A 144** (1934) 425.
[19] W. Heisenberg and H. Euler, Z. Physik **98** (1936) 714; J. Schwinger, Phys. Rev. **82** (1951) 664; Z. Bialynicka-Birula and I. Bialynicki-Birula, Phys. Rev. D **2** (1970) 2341.
[20] G. W. Gibbons and C. A. R. Herdeiro, Class. Quant. Grav. **18** (2001) 1677; Phys. Rev. D **63** (2001) 064006.
[21] H. Arodz, M. Slusarczyk and A. Wereszczynski, Acta Phys. Polon. B **32** (2001) 2155; M. Slusarczyk and A. Wereszczynski, Acta Phys. Polon. B **34** (2003) 2623.
[22] H. Pagels and E. Tomboulis, Nucl. Phys. B **143** (1978) 485.

Recent Developments in Gravity (NEB XIII) IOP Publishing
Journal of Physics: Conference Series **189** (2009) 012023 doi:10.1088/1742-6596/189/1/012023

Gravitational waves from gauge-invariant perturbations of spherically symmetric spacetimes

Paul Lasky

Centre for Astrophysics and Supercomputing, Swinburne University, Hawthorn VIC 3122, Australia
Centre for Stellar and Planetary Astrophysics, Monash University, Melbourne 3800, Australia
Theoretical Astrophysics, Eberhard Karls University, Tübingen 72076, Germany

E-mail: plasky@astro.swin.edu.au

Abstract. One difficulty associated with perturbations of spherical collapse models in General Relativity is attributed to the junction conditions required at the interface of the interior matter-filled region and the exterior vacuum region. This implies extracting information about gravitational waves at spacelike infinity is also a difficult task. In this talk, I present a method which eliminates the need for junction conditions in both the background and perturbed spacetimes, thereby allowing relatively simple modelling of gravitational waveforms. This is achieved by using a recently developed method that enables a single line element to be expressed for the entire spherically symmetric background spacetime. Perturbing this spacetime in a gauge-invariant manner implies junction conditions are not required at any stage of the perturbation. Wave equations are derived for the Newman-Penrose Weyl scalars which hold in both the matter filled regions of the spacetime as well as the vacuum exterior regions.

1. Introduction

The direct detection of gravitational waves is one of the most eagerly anticipated events in physics, and will usher in a new era in gravitational wave astrophysics. The detection of such waves relies on an *a priori* knowledge of the expected signature, which in turn relies on knowledge of the source of the gravitational wave. Whilst in recent years the numerical relativity community has made amazing progress in modeling such sources – which include the coalescence of binary systems of black holes and neutron stars (e.g. [1, 2, 3]) and the gravitational collapse of isolated sources induced by supernovae (e.g. [4, 5, 6]) – the accurate extraction of gravitational waveforms is fraught with difficulties. The primary reason is that weaker gravitational waves have amplitudes of the same order as gauge modes and intrinsic numerical errors [7].

A common solution to this problem is to analyse non-spherical gravitational collapse using linear perturbations of spherically symmetric spacetimes. This was studied in detail by Gundlach and Martín-García [8, 9] who looked at the perturbations of spherically symmetric perfect fluids with an exterior vacuum spacetime. These two papers highlight the difficulties one faces with the perturbations at the interface between the interior, fluid filled region and the exterior vacuum. This point is emphasized with the recent work of Mars, Mena & Vera [10] who presented a critical review of various formulations of the problem. The difficulty is essentially associated with the identification of the various coordinate systems used either side of the boundary in both the background and perturbed spacetimes.

Journal of Physics: Conference Series **189** (2009) 012023 doi:10.1088/1742-6596/189/1/012023

In this work I develop a method that avoids the problem of matching perturbed spacetimes. This is done by establishing the entire background spacetime in a single coordinate system as first achieved by Adler et al. [11] and later developed by Lasky & Lun [12, 13, 14, 15]. Both regions of the background spacetime being in a single coordinate system implies the perturbed manifold is also expressed in a single coordinate system. That is, no matching is required across the interface in either the background or perturbed manifold.

This work is set out as follows; in section 2, I describe the spherically symmetric background spacetime. In the present work this is done for a perfect fluid interior region with vacuum exterior. For the extension of the background formulation to more complicated fluids, including a general decomposition of the energy momentum tensor and a relativistic plasma, see Lasky & Lun [14] and [15] respectively. In section 3 perturbation equations are presented which are derived using the Geroch-Held-Penrose (GHP) formulation [16] of the Newman-Penrose equations [17]. I further use these perturbation equations to derive a single wave equation on the perturbed Ψ_4 component of the Weyl tensor.

2. The Background Spacetime

The critical aspect of the method outlined in this article is the background spacetime. Once the background is established in the correct way the perturbations can be handled with relative ease. Therefore, for clarity, I shall only consider the simplified background spacetime of a spherically symmetric perfect fluid [12]. Note that the method for setting up the background is more robust than a simple perfect fluid; for instance it has been successful with both a completely general decomposition of the energy-momentum tensor [14], as well as adapting to the case of Einstein-Maxwell plasmas [15]. In these two cases the physics is more complicated as the exterior spacetimes are not necessarily Schwarzschild due to the presence of outgoing radiation. However, even with more complicated physics, the basis of the method remains the same.

The metric representing a spherically symmetric spacetime can be expressed in $3+1$ form as

$$\mathrm{d}s = -\alpha^2 \mathrm{d}t^2 + \frac{(\beta \mathrm{d}t + \mathrm{d}r)^2}{1+E} + r^2 \mathrm{d}\Omega^2. \tag{1}$$

Here, $\alpha(t,r) > 0$ is the lapse function, $\beta(t,r)$ is the radial component of the shift vector, $E(t,r) > -1$ and $\mathrm{d}\Omega^2 := \mathrm{d}\theta^2 + \sin^2\theta \mathrm{d}\phi^2$. The function E shall be referred to as the energy function as in the dust case it reduces to the energy function of the Lemaître-Tolman-Bondi spacetime [18, 19, 20], which is a measure of the energy of a shell at a radius r [21].

The energy-momentum tensor for a perfect fluid is given by

$$T^{\mu\nu} = (\rho + P)\, n^\mu n^\nu + P g^{\mu\nu}, \tag{2}$$

where $\rho(t,r)$ and $P(t,r)$ are respectively the energy-density and isotropic pressure as measured by an observer n^μ. It is worth stressing that n^μ is *not* a comoving observer, rather it is the observer whose vector field is everywhere orthogonal to the spacelike hypersurfaces given by the line element (1). The vector field is required to be timelike and normalized, $n^\alpha n_\alpha = -1$, and hypersurface forming, $n_{[\mu}\nabla_\nu n_{\nu]} = 0$ (where square brackets denote antisymmetrization). Using these conditions, one can express the normal vector in component form

$$n_\mu = [-\alpha,\, 0,\, 0,\, 0] \qquad \text{and} \qquad n^\mu = \alpha^{-1}[1,\, -\beta,\, 0,\, 0]\,. \tag{3}$$

Conservation of energy-momentum is given by $\nabla_\alpha T^\alpha{}_\mu = 0$. In spherical symmetry this provides two independent equations, however one of these equations is trivially satisfied by the Einstein equations (for details see [12]). The remaining conservation equation is Euler's equation which relates the pressure gradient to the lapse function

$$\frac{-1}{\rho + P}\frac{\partial P}{\partial r} = \frac{1}{\alpha}\frac{\partial \alpha}{\partial r}, \tag{4}$$

where $\rho \neq -P$. This equation, coupled to a suitable equation of state, implies that the pressure gradient determines the lapse function uniquely (note that integration of equation (4) introduces an arbitrary function of the temporal coordinate, however this can always be scaled away by a suitable coordinate transformation $t \to t'$). The standard method in numerical relativity is to choose a gauge (i.e. choose α and β) and solve for the remaining metric coefficients. However, the method utilized here is to let the physics dictate the choice of gauge. That is, rather than making an arbitrary gauge choice, equation (4) is solved dependent on the equation of state, thus prescribing the lapse function.

The remaining equations determining the background spacetime are found by reducing the Einstein Field Equations, $G_{\mu\nu} = 8\pi T_{\mu\nu}$. I provide here only a brief summary of the results; for a full derivation the interested reader is referred to Lasky & Lun [12]. Begin by defining a function which is referred to as the "mass",

$$M(t,r) := 4\pi \int_0^r \rho r' dr'. \tag{5}$$

It is noted that this is not necessarily a physical mass as it does not properly take into account the volume element associated with the metric. However, in the derivation this quantity arises in a very natural way [12]. Moreover, in the dust limit this function becomes the familiar mass function of the Lemaître-Tolman-Bondi dust solutions, and in the vacuum limit it also becomes the Schwarzschild mass, suggesting that this quantity is of considerable importance.

After much work, an algebraic relation can be found between the mass and energy functions and the lapse and shift functions

$$\frac{\beta}{\alpha} = \left(\frac{2M}{r} + E\right)^{1/2}, \tag{6}$$

where a positive square root has been taken. Taking the negative root gives an expanding model rather than a contracting model which is evidenced by the fact that changing the sign of β in the metric is equivalent to mapping $t \to -t$. This equation implies explicit dependance on the shift function in the metric is no longer required, and everywhere the shift function is substituted using the above relation.

Finally, the remaining Einstein equations reduce to just two equations which contain the time evolution of the energy function

$$\frac{\mathcal{L}_n E}{1+E} = 2\left(\frac{2M}{r} + E\right)^{1/2} \frac{1}{\rho + P}\frac{\partial P}{\partial r}, \tag{7}$$

and the mass function

$$\mathcal{L}_n M = 4\pi P r^2 \left(\frac{2M}{r} + E\right)^{1/2}. \tag{8}$$

Here, $\mathcal{L}_n$ is the Lie derivative operator which, when acting on a scalar, takes the form

$$\mathcal{L}_n = \frac{1}{\alpha}\frac{\partial}{\partial t} - \frac{\beta}{\alpha}\frac{\partial}{\partial r}. \tag{9}$$

Equations (7) and (8), combined with equations (4), (5) and an equation of state provide all the information required to solve for the metric coefficients, and therefore uniquely determine the evolution of the spacetime. Moreover, it can be shown that the only initial conditions required are the initial energy-density and its time rate of change. With these specified, the equation of state gives the initial lapse function, equation (5) implies the initial mass function is known and equation (8) can be rearranged to find the initial energy function.

So far, the equations have described an interior, fluid filled region of the spacetime. The exterior, vacuum region can now be found by setting $P = \rho = 0$ for all $r \geq r_s(t)$, where $r_s(t)$ is the surface of the collapsing star. Moreover, this is only required to be stipulated on the initial spacelike hypersurface, whereby the subsequent evolution is given by the solution of equations (7) and (8). Setting $P = \rho = 0$ implies $M = M_s$ for $r \geq r_s$, where M_s is the Schwarzschild mass. Moreover, equation (4) implies the lapse function is a function of only the temporal coordinate, which can then be scaled to unity by a suitable coordinate transformation. Equation (8) is now trivially satisfied and equation (7) becomes $\mathcal{L}_n E = 0$. Finally, the metric can be expressed as

$$ds^2 = -dt^2 + \frac{\left(\sqrt{2M_s/r + E}\, dt^2 + dr\right)^2}{1+E} + r^2 d\Omega^2. \tag{10}$$

It can be shown that equation (10), together with the condition $\mathcal{L}_n E = 0$, is exactly the Schwarzschild spacetime in coordinates that generalize the Painleve-Gullstrand coordinates (which are recovered with the trivial solution $E = 0$).

The perfect fluid system of equations reducing to the Schwarzschild spacetime in this way implies that all regions of the spacetime are expressed in a single coordinate system. Setting up a physically relevant problem involves specifying an initial energy-density distribution on the initial spacelike hypersurface, as well as an equation of state relating the pressure and the energy-density. At some finite radius on the initial hypersurface, $r_s(t=0)$, the density and pressure tend to zero. Subsequent evolution of the spacetime is achieved by solving equations (4), (7) and (8). Solving these equations will provide a solution for $r_s(t)$ throughout the spacetime, which is the evolution of the boundary of the interior fluid filled region and the exterior vacuum region. The entire spacetime is therefore described by a single line element (1), where the metric coefficients are the solutions of the equations just described.

3. Perturbations

As noted, a single solution of the Einstein Field Equations is now provided for a collapsing spacetime which includes both the interior collapsing region and the exterior vacuum region. Utilizing a single solution, as opposed to matching two distinct solutions together at some interface, implies that junction conditions are not required in the background spacetime. Whilst the benefits of this methods for the background spacetime have been explicitly discussed in Lasky & Lun [12, 13, 14, 15], the discussion of perturbations of such spacetimes has not been discussed prior to this article.

With the background spacetime specified, one is free to choose their favorite perturbation scheme. For the purposes of this discussion, I choose to work with the GHP formalism [16] due to the relative simplicity in the expressions[1]. This formalism is an extension of the Newman-Penrose [17] equations where the following differential operators are defined;

$$\begin{aligned} \text{þ} &:= \ell^\alpha \nabla_\alpha - p\epsilon - q\bar{\epsilon}, & \text{þ}' &:= k^\alpha \nabla_\alpha - p\gamma - q\bar{\gamma}, \\ \eth &:= m^\alpha \nabla_\alpha - p\beta - q\bar{\alpha}, & \eth' &:= \bar{m}^\alpha \nabla_\alpha - p\alpha - q\bar{\beta}. \end{aligned} \tag{11}$$

Here, ℓ^μ, k^μ and m^μ define the null tetrad, α, β, ϵ and γ are spin coefficients and p and q are scalars associated to the spin and boost weights, s and w respectively, according to

$$s = (p-q)/2, \qquad \text{and} \qquad w = (p+q)/2. \tag{12}$$

[1] Another valid perturbation method, given that the background spacetime has been established using a $3+1$ formalism, is to use a $2+1+1$ perturbation scheme [22, 23, 24, 25].

For a consistent description, one must express the background spacetime using this formalism. It is trivial to show that the vanishing spin coefficients, Weyl and energy momentum components on the background are

$$\kappa = \sigma = \tau = \nu = \lambda = \pi = 0, \tag{13}$$

$$\Psi_0 = \Psi_1 = \Psi_3 = \Psi_4 = \Phi_{01} = \Phi_{02} = \Phi_{12} = 0. \tag{14}$$

Moreover, at the interface between the two regions of the background spacetime, the remaining energy-momentum terms (Φ_{00}, Φ_{11}, Φ_{22} and Λ) also tend to zero such that they are equivalently zero for all $r \geq r_\star(t)$.

A linear perturbation can now be performed on the background system of equations such that for any scalar quantity, φ say,

$$\tilde{\varphi} = \varphi + \varphi_\star, \tag{15}$$

where the subscript '$\star$' denotes a perturbed quantity. After much work, one can show that the perturbed system of equations is given by

Perturbed Ricci Identities:

$$
\begin{aligned}
þ\rho_\star + þ_\star\rho - ð'\kappa_\star &= 2\rho\rho_\star + \Phi_{00\star}, && (16)\\
þ'\mu_\star + þ'_\star\mu - ð\nu_\star &= -2\mu\mu_\star - \Phi_{22\star}, && (17)\\
þ\sigma_\star - ð\kappa_\star &= 2\rho\sigma_\star + \Psi_{0\star}, && (18)\\
þ'\lambda_\star - ð'\nu_\star &= -2\mu\lambda_\star - \Psi_{4\star}, && (19)\\
þ\tau_\star - þ'\kappa_\star &= (\tau_\star + \bar{\pi}_\star)\rho + \Psi_{1\star} + \Phi_{01\star}, && (20)\\
þ'\pi_\star - þ\nu_\star &= -(\pi_\star + \bar{\tau}_\star)\mu - \Psi_{3\star} - \Phi_{21\star}, && (21)\\
ð\rho_\star + ð_\star\rho - ð'\sigma_\star &= -\Psi_{1\star} + \Phi_{01\star}, && (22)\\
ð'\mu_\star + ð'_\star\mu - ð\lambda_\star &= \Psi_{3\star} - \Phi_{21\star}, && (23)\\
ð\tau_\star - þ'\sigma_\star &= \mu\sigma_\star + \rho\bar{\lambda}_\star + \Phi_{02\star}, && (24)\\
ð'\pi_\star - þ\lambda_\star &= -\rho\lambda_\star - \mu\bar{\sigma}_\star - \Phi_{20\star}, && (25)\\
þ'\rho_\star + þ'_\star\rho - ð'\tau_\star &= -\rho\bar{\mu}_\star - \rho_\star\mu - \Psi_{2\star} - 2\Lambda_\star, && (26)\\
þ\mu_\star + þ_\star\mu - ð\pi_\star &= \rho\mu_\star + \bar{\rho}_\star\mu + \Psi_{2\star} + 2\Lambda_\star. && (27)
\end{aligned}
$$

Perturbed Bianchi Identities:

$$
\begin{aligned}
(þ - 4\rho)\Psi_{1\star} &= ð'\Psi_{0\star} - 3\kappa_\star\Psi_2 - (ð_\star + \bar{\pi}_\star)\Phi_{00} - ð\Phi_{00\star} + (þ - 2\rho)\Phi_{01\star} + 2\kappa_\star\Phi_{11}, && (28)\\
(þ' + 4\mu)\Psi_{3\star} &= ð\Psi_{4\star} + 3\nu_\star\Psi_2 - (ð'_\star - \bar{\tau}_\star)\Phi_{22} - ð'\Phi_{22\star} + (þ' + 2\mu)\Phi_{21\star} - 2\nu_\star\Phi_{11}, && (29)\\
ð'\Psi_{1\star} &= (þ - 3\rho)\Psi_{2\star} + (þ_\star - 3\rho_\star)\Psi_2 + (þ' + \mu)\Phi_{00\star} - ð'\Phi_{01\star}\\
&\quad + (þ'_\star + \bar{\mu}_\star)\Phi_{00} + 2þ_\star\Lambda + 2þ\Lambda_\star - 2\rho_\star\Phi_{11} - 2\rho\Phi_{11\star}, && (30)\\
ð\Psi_{3\star} &= (þ' + 3\mu)\Psi_{2\star} + (þ'_\star + 3\mu_\star)\Psi_2 + (þ - \rho)\Phi_{22\star} - ð\Phi_{21\star}\\
&\quad + (þ_\star - \bar{\rho}_\star)\Phi_{22} + 2þ'_\star\Lambda + 2þ'\Lambda_\star + 2\mu_\star\Phi_{11} + 2\mu\Phi_{11\star}, && (31)\\
(þ - 2\rho)\Psi_{3\star} &= ð'\Psi_{2\star} + (ð'_\star + 3\pi_\star)\Psi_2 + (þ - 2\rho)\Phi_{21\star} - ð\Phi_{20\star} + 2\mu\Phi_{10\star}\\
&\quad - 2\pi_\star\Phi_{11} + \bar{\kappa}_\star\Phi_{22} + 2ð'\Lambda_\star + 2ð'_\star\Lambda, && (32)\\
(þ' + 2\mu)\Psi_{1\star} &= ð\Psi_{2\star} + (ð_\star - 3\tau_\star)\Psi_2 + (þ' + 2\mu)\Phi_{01\star} - ð'\Phi_{02\star} - 2\rho\Phi_{12\star}\\
&\quad + 2\tau_\star\Phi_{11} - \bar{\nu}_\star\Phi_{00} + 2ð\Lambda_\star + 2ð_\star\Lambda, && (33)\\
(þ - \rho)\Psi_{4\star} &= ð'\Psi_{3\star} - 3\lambda_\star\Psi_2 - (þ' + \mu)\Phi_{20\star} + ð'\Phi_{21\star} + \bar{\sigma}_\star\Phi_{22} - 2\lambda_\star\Phi_{11}, && (34)\\
(þ' + \mu)\Psi_{0\star} &= ð\Psi_{1\star} + 3\sigma_\star\Psi_2 - (þ - \rho)\Phi_{02\star} + ð\Phi_{01\star} - \bar{\lambda}_\star\Phi_{00} + 2\sigma_\star\Phi_{11}. && (35)
\end{aligned}
$$

Perturbed Contracted Bianchi Identities:

$$\begin{aligned}
þ\Phi_{11\star} + þ_\star\Phi_{11} + þ'\Phi_{00\star} + þ'_\star\Phi_{00} - ð\Phi_{10\star} - ð'\Phi_{01\star} + 3þ\Lambda_\star + 3þ_\star\Lambda& \\
=4\rho\Phi_{11\star} + 2\left(\rho_\star + \bar{\rho}_\star\right)\Phi_{11} - 2\mu\Phi_{00\star} - \left(\mu_\star + \bar{\mu}_\star\right)\Phi_{00},& \quad (36)\\
þ'\Phi_{11\star} + þ'_\star\Phi_{11} + þ\Phi_{22\star} + þ_\star\Phi_{22} - ð'\Phi_{12\star} - ð\Phi_{21\star} + 3þ'\Lambda_\star + 3þ'_\star\Lambda& \\
= -4\mu\Phi_{11\star} - 2\left(\mu_\star + \bar{\mu}_\star\right)\Phi_{11} + 2\rho\Phi_{22\star} + \left(\rho_\star + \bar{\rho}_\star\right)\Phi_{22},& \quad (37)\\
þ\Phi_{12\star} + þ'\Phi_{01\star} - ð\Phi_{11\star} - ð_\star\Phi_{11} - ð'\Phi_{02\star} + 3ð\Lambda_\star + 3ð_\star\Lambda& \\
=3\rho\Phi_{12\star} - 3\mu\Phi_{01\star} - 2\left(\tau_\star - \bar{\pi}_\star\right)\Phi_{11} - \bar{\nu}_\star\Phi_{00} - \kappa_\star\Phi_{22},& \quad (38)\\
þ'\Phi_{10\star} + þ\Phi_{21\star} - ð'\Phi_{11\star} - ð'_\star\Phi_{11} - ð\Phi_{20\star} + 3ð'\Lambda_\star + 3ð'_\star\Lambda& \\
= -3\mu\Phi_{10\star} + 3\rho\Phi_{21\star} + 2\left(\pi_\star - \bar{\tau}_\star\right)\Phi_{11} + \bar{\kappa}_\star\Phi_{22} + \nu_\star\Phi_{00}.& \quad (39)
\end{aligned}$$

Perturbed Commutators:

$$\begin{aligned}
\left[þ, þ'_\star\right] + \left[þ_\star, þ'\right] =& \left(\bar{\tau}_\star + \pi_\star\right)ð + \left(\tau_\star + \bar{\pi}_\star\right)ð' - (p+q)\left(\Psi_{2\star} + \Phi_{11\star} - \Lambda_\star\right), & (40)\\
\left[þ, ð_\star\right] + \left[þ_\star, ð\right] =& \rho ð_\star + \bar{\rho}_\star ð + p\left(\mu\kappa_\star - \Psi_{1\star}\right) - q\left(\rho\bar{\pi}_\star - \Phi_{01\star}\right), & (41)\\
\left[ð, ð'_\star\right] + \left[ð_\star, ð'\right] =& \left(\mu_\star - \bar{\mu}_\star\right)þ + \left(\rho_\star - \bar{\rho}_\star\right)þ' - p\left(\rho\mu_\star + \rho_\star\mu - \Psi_{2\star} + \Phi_{11\star} + \Lambda_\star\right) & \\
&+ q\left(\rho\bar{\mu}_\star + \bar{\rho}_\star\mu - \bar{\Psi}_{2\star} + \Phi_{11\star} + \Lambda_\star\right). & (42)
\end{aligned}$$

Whilst the above system of equations is obviously extremely complicated, it does have the nice feature that the same set of equations holds over the entire spacetime manifold. The standard approach to this subject is to have a background spacetime formulated using two matched solutions of the Einstein field equations. Perturbations are then achieved by perturbing both solutions of the field equations, and then matching the perturbed manifolds together. The boundary between the two regions is then found pointwise in the perturbed regions (for example see [8, 9, 10]). The above equations are a single set of perturbation equations which hold over the entire manifold, implying no matching at the interface is required.

Wave Equation
The Bardeen-Press wave equation [26] can be found by operating on equation (29) with $ð'$ and equation (34) with $þ' + 5\mu$. Adding the resulting equations implies, after much algebra

$$\Box\Psi_{4\star} = \mathcal{S}. \quad (43)$$

Here, $\Box$ is a wave operator defined as

$$\Box := \left(þ' + 5\mu\right)\left(þ - \rho\right) - ð'ð - 3\Psi_2 + 2\Phi_{11}, \quad (44)$$

and the source for the wave equation is given by

$$\begin{aligned}
\mathcal{S} :=& 12\left(\lambda_\star þ' - \nu_\star ð'\right)\Lambda - \left(þ' + 5\mu\right)\left(þ' + \mu\right)\Phi_{20\star}\\
&- ð'ð'_\star\Phi_{22} - ð'ð'\Phi_{22\star} + \bar{\sigma}_\star þ'\Phi_{22} + 5\lambda_\star þ\Phi_{22}\\
&+ \left(þ' + 4\mu\right)ð'\Phi_{21\star} + 2\Phi_{22}\left(þ'\bar{\sigma}_\star + 3\bar{\sigma}_\star\mu - 3\lambda_\star\rho\right)\\
&+ \Phi_{22}\Phi_{20\star} - 4\Phi_{11}þ'\lambda_\star. \qquad (45)
\end{aligned}$$

This equation is still highly coupled to both the background and perturbed spin coefficients, and is therefore extremely difficult to solve. Solving it requires stipulating initial conditions for the perturbations. For example, one may wish to initially perturb the interior, fluid filled region of the spacetime. The perturbation will then propagate outwards, through the interface of the interior and exterior regions, and out to radial infinity. This will be done as a single solution of the above equation with no extra matching required at the interface. In principal, the interface between the two regions is found in the same way as is done for the background spacetime; by solving the full set of equations and finding the point at which all matter terms vanish on each spacelike slice.

4. Conclusion

This article has introduced a general technique for handling perturbations of spherically symmetric spacetimes without the need for matching various regions in either the background or the perturbed manifolds. This was done by utilising the recently developed formalism of Adler et al. [11] and Lasky & Lun [12, 13, 14, 15] which expresses a spherically symmetric background spacetime covering both a vacuum exterior region and a fluid filled interior region as a single line element. This has lead to a single set of perturbation equations covering the entire spacetime.

Acknowledgments

I am extremely grateful to the conference organisers, particularly Nick Stergioulas, for the financial support to attend the conference. Many thanks also to Kostas Kokkotas and the Theoretical Astrophysics group at the Eberhard-Karls-University in Tübingen for their kind hospitality. This work was partially supported by the German Foundation (DFG) via SFG/TR7 and also the Australian Research Council's "Commonwealth Cosmology Initiative" (www.thecci.org: DP0665574).

References

[1] Brugmann B, Tichy W and Jansen N 2004 *Phys. Rev. Lett.* **92**(21) 211101
[2] Pretorius F 2005 *Phys. Rev. Lett.* **95**(12) 121101
[3] González J A, Sperhake U, Brugmann B, Hannam M and Husa S 2007 *Phys. Rev. Lett.* **98** 091101
[4] Shibata M, Liu Y T, Shapiro S L and Stephens B C 2006 *Phys. Rev. D* **74**(10) 104026
[5] Ott C D, Dimmelmeier H, Marek A, Janka H, Hawke I, Zink B and Schnetter E 2006 *Phys. Rev. Lett.* **98** 261101
[6] Dimmelmeier H, Ott C D, Janka H, Marek A and Muller E 2007 *Phys. Rev. Lett.* **98** 251101
[7] Sotani H, Yoshida S and Kokkotas K D 2007 *Phys. Rev. D* **75** 084015
[8] Gundlach C and Martín-García J M 2000 *Phys. Rev. D* **61** 084024
[9] Martín-García J M and Gundlach C 2001 *Phys. Rev. D* **64** 024012
[10] Mars M, Mena F C and Vera R 2007 *Class. Quantum Grav.* **24** 3673
[11] Adler R, Bjorken J D, Chen P and Liu J 2005 *Am. J. Phys.* **73**(12) 1148–59
[12] Lasky P D and Lun A W C 2006 *Phys. Rev. D* **74** 084013
[13] Lasky P D, Lun A W C and Burston R B 2007 *the ANZIAM journal* **49** 53
[14] Lasky P D and Lun A W C 2007 *Phys. Rev. D* **75** 024031
[15] Lasky P D and Lun A W C 2007 *Phys. Rev. D* **75** 104010
[16] Geroch R, Held A and Penrose R 1973 *J. Math. Phys.* **14**(7) 874
[17] Newman E and Penrose R 1962 *J. Math. Phys.* **3**(3) 566
[18] Lemaitre G 1933 *Ann. Soc. Sci. Bruxelles A* **53** 51
[19] Tolman R C 1934 *Proc. Nat. Acad. Sci. USA* **20** 169–76
[20] Bondi H 1947 *Mon. Not. Roy. Astron. Soc.* **107** 410–25
[21] Szekeres P and Lun A 1999 *J. Austral. Math. Soc. Ser. B.* **41** 167–79
[22] Betschart and Clarkson 2004 *Class. Quantum Grav.* **21** 5587–607
[23] Clarkson C 2007 *Phys. Rev. D* **76** 104034
[24] Burston R B 2008 *Class. Quantum Grav.* **25** 075003
[25] Burston R B 2008 *Class. Quantum Grav.* **25** 075004
[26] Bardeen J M and Press W H 1973 *J. Math. Phys.* **17** 7

Neutron star equation of state via gravitational wave observations

C Markakis[1], J S Read[2], M Shibata[3], K Uryū[4], J D E Creighton[1], J L Friedman[1], and B D Lackey[1]

[1] Department of Physics, University of Wisconsin–Milwaukee, P.O. Box 413, Milwaukee, WI 53201, USA
[2] Max-Planck-Institut für Gravitationsphysik, Albert-Einstein-Institut, Golm, Germany
[3] Yukawa Institute for Theoretical Physics, Kyoto University, Kyoto 606-8502, Japan
[4] Department of Physics, University of the Ryukyus, 1 Senbaru, Nishihara, Okinawa 903-0213, Japan

Abstract. Gravitational wave observations can potentially measure properties of neutron star equations of state by measuring departures from the point-particle limit of the gravitational waveform produced in the late inspiral of a neutron star binary. Numerical simulations of inspiraling neutron star binaries computed for equations of state with varying stiffness are compared. As the stars approach their final plunge and merger, the gravitational wave phase accumulates more rapidly if the neutron stars are more compact. This suggests that gravitational wave observations at frequencies around 1 kHz will be able to measure a compactness parameter and place stringent bounds on possible neutron star equations of state. Advanced laser interferometric gravitational wave observatories will be able to tune their frequency band to optimize sensitivity in the required frequency range to make sensitive measures of the late-inspiral phase of the coalescence.

1. Introduction

In numerical simulations of the late inspiral and merger of binary neutron-star systems, one commonly specifies an equation of state for the matter, perform a numerical simulation and extract the gravitational waveforms produced in the inspiral. The scope of this talk is to report work on the inverse problem: if gravitational radiation from an inspiraling binary system is detected, can one use it to infer the bulk properties of neutron star matter and, if so, with what accuracy? To answer this question, a number of simulations with systematic variation of the equation of state (EOS) is performed. The simulations use the evolution and initial-data codes of Shibata and Uryū to compute the last several orbits and the merger of neutron stars, with matter described by a parameterized equation of state, previously obtained in [1]. Our analysis uses waveforms from these simulations to examine the accuracy with which detectors with the sensitivity of Advanced LIGO can extract from inspiral waveforms an EOS parameter associated with the stiffness of the neutron star EOS above nuclear density.

One might expect that gravitational wave observations can potentially measure properties of neutron star EOS by measuring departures from the point-particle limit of the gravitational waveform produced during the late inspiral of a binary system. A way to quantify this departure is to make use of the fact that the gravitational wave phase near the end of inspiral accumulates more rapidly for smaller values of the neutron star compactness (the ratio of the mass of a neutron star to its radius). In this way, greater accuracy may be obtained compared to previous work that suggested the use of an effective cutoff frequency to place constraints on the equation of state. Our results based on this approach suggest that realistic equations of state will lead to gravitational waveforms that are distinguishable from point particle inspirals at a distance $\lesssim 100\,\mathrm{Mpc}$ for an optimally oriented system, using Advanced LIGO in a broadband or narrowband configuration. Waveform analysis, that uses the sensitivity curves of commissioned and proposed gravitational wave detectors, allows us to also estimate the accuracy with which neutron star radii, closely linked to the EOS parameter varied, can be extracted. The choice of EOS parameter varied in this work is motivated by the fact that, as Lattimer and Prakash observed [3], neutron-star radius is closely tied to the pressure at density not far above nuclear. Analysis of the numerical waveforms also indicates that optimizing the sensitivity of gravitational wave detectors to frequencies above 700 Hz can lead to improved constraints on the radius and EOS of neutron stars.

2. Equation of state

Difficulties inherent in astrophysically constraining the EOS of neutron star matter are the wide range of different microphysical models, implying the lack of a model-independent set of fundamental observables, and the fact that models often have more free parameters than the number of astrophysical observables. Therefore, an effort to systematize such constraints necessitates a form of EOS (i) parameterized in a model independent way, with a number of EOS parameters (ii) large enough to accurately model the universe of candidate EOS, but (iii) smaller than the number of neutron star observables that have been measured or will have been measured in the next several years.

In the past, for different purposes, Müller and Eriguchi [4] have used non-relativistic piecewise polytropes, with a large number (20-100) of segments, to accurately approximate realistic EOS, in order to construct Newtonian models of differentially rotating stars. More recently, with conditions (i), (ii) and especially (iii) in mind, Read et al [1, 2] have developed a relativistic piecewise polytropic approximation, which reproduces the features of most realistic neutron star EOS with rms error $\lesssim 4\%$, using only 3-4 segments (or 3-4 free parameters), as described below.

2.1. *Astrophysical constraints on a piecewise polytropic equation of state*

The EOS pressure p is specified as a function of rest mass density ρ (rest mass density $\rho = m_{\mathfrak{b}} n$ is proportional to number density n, where $m_{\mathfrak{b}} = 1.66 \times 10^{-24}\,\mathrm{g}$ is the mean baryon rest mass). The relativistic piecewise polytropic EOS proposed in [1, 2] has the form

$$p(\rho) = K_i \rho^{\Gamma_i}, \tag{1}$$

in a set of intervals $\rho_{i-1} \le \rho \le \rho_i$ in rest mass density, with the constants K_i determined by requiring continuity on each dividing density ρ_i:

$$K_i \rho_i^{\Gamma_i} = K_{i+1} \rho_i^{\Gamma_{i+1}}. \tag{2}$$

The energy density ϵ as a function of ρ is then determined by the first law of thermodynamics,

$$d\frac{\epsilon}{\rho} = -p\, d\frac{1}{\rho}, \tag{3}$$

which is integrated, to give

$$\epsilon(\rho) = (1 + a_i)\rho c^2 + \frac{K_i \rho^{\Gamma_i}}{\Gamma_i - 1}, \tag{4}$$

(for $\Gamma_i \neq 1$), where we used the condition $\epsilon/\rho \to c^2$ as $\rho \to 0$. This condition implies $a_0 = 0$ and the other constants a_i are fixed by continuity of $\epsilon(\rho)$ at each ρ_i. In the above equations, ρ is measured in $\mathrm{g\,cm^{-3}}$, ϵ and p have units of $\mathrm{erg\,cm^{-3}} = \mathrm{dyn\,cm^{-2}}$ and c is the speed of light. The specific enthalpy, $\mathfrak{h}$, is dimensionless and is given by

$$\mathfrak{h} = \frac{\epsilon + p}{\rho c^2}. \tag{5}$$

The crust EOS is considered to be known and, for the purposes of this work, can be modelled with a single polytropic region, fitted to a tabulated crust EOS for the region above neutron drip (10^{11}-$10^{12}\,\mathrm{g\,cm^{-3}}$); the numerical simulations considered do not resolve densities below this. This polytrope has $\Gamma_0 \equiv \Gamma_{\rm crust} = 1.3569$, with $K_0 \equiv K_{\rm crust}$ chosen so that $p = 1.5689 \times 10^{31}\,\mathrm{dyn\,cm^{-2}}$ when $\rho = 10^{13}\,\mathrm{g\,cm^{-3}}$. The core EOS is constructed independently of crust behavior. In [1] it was found that three zones within the core, with adiabatic exponents $\Gamma_1, \Gamma_2, \Gamma_3$ as parameters, are needed to model the full range of proposed EOS models with sufficient accuracy. A fourth parameter, $p_1 \equiv p(\rho_1)$, is needed to fix an overall pressure shift, specified at the fiducial density ρ_1. The dividing densities ρ_1, ρ_2 are *not* considered new parameters. Instead, they are fixed by minimizing the error in fitting a large set of candidate EOS, leading [1] to the preferred values $\rho_1 = 10^{14.7}\,\mathrm{g\,cm^{-3}}$ and $\rho_2 = 10^{15.0}\,\mathrm{g\,cm^{-3}}$. The first dividing density ρ_0 between the crust and core varies by EOS, but is also not a new parameter. It is defined as the density for which the $\log P$ vs. $\log \rho$ curves of the crust and core EOS intersect. Fixing the crust EOS and determining the three dividing densities in this way, allows one to model a large set of candidate EOS with only four parameters $\{p_1, \Gamma_1, \Gamma_2, \Gamma_3\}$.

Using this parameterized EOS, one can construct relativistic models of neutron stars and compute astrophysical observables, such as maximum mass, radius, moment of inertia etc. Doing so, while systematically varying the EOS parameters $\{p_1, \Gamma_1, \Gamma_2, \Gamma_3\}$, Read et al [1] have obtained constraints on the allowed parameter space set by causality [Fig. 1] and by present and near-future astronomical observation. They find that two measured properties of a single star can potentially place stringent constraints on the EOS parameter space and that a potential measurement of the moment of inertia of a pulsar can strongly constrain the maximum neutron-star mass [Figs. 1, 2]. A detailed study of these constraints may be found in [1, 2]. The potential combination of such astrophysical constraints with gravitational wave observation can restrict the allowed EOS parameter space region to a surface (thickened by the error bars of the observations) or even a line. The rest of this talk focuses on constraints placed on neutron star EOS by potential detection of binary inspiral waveforms.

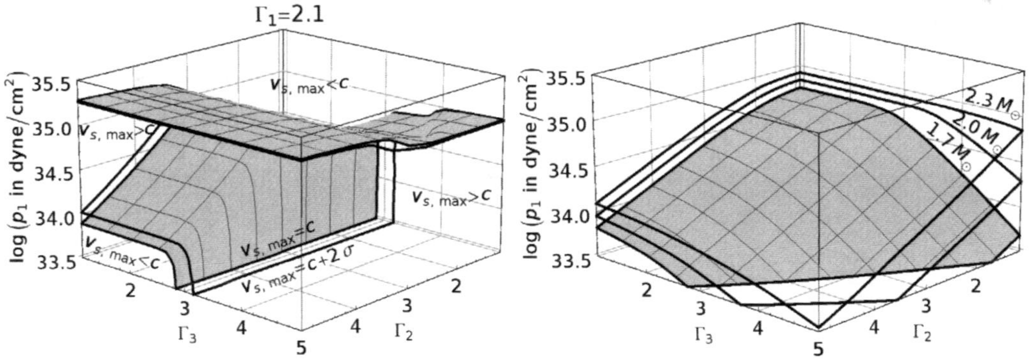

Figure 1. *Left*: Causality constraints are shown for a fixed value of Γ_1. For each EOS, the maximum speed of sound up to the central density of the maximum mass star is considered. The shaded surface separates the EOS parameter space into a region *behind* the surface allowed by causality, labeled $v_s < c$, and a region in which corresponding EOSs violate causality, labeled $v_s > c$. The second, outlined surface shows a weaker constraint to accommodate the expected error (~ 2 standard deviations σ) in the speed of sound associated with a piecewise polytropic approximation to an EOS.
Right: Surfaces representing the set of parameters resulting in a constant maximum mass. A single observation of a massive neutron star constrains the equation of state to lie *above* the corresponding surface. The exponent Γ_1 is set to the least constraining value at each point. Reprinted with permission from: Read J S, Markakis C, Shibata M, Uryu K, Creighton J D E, and Friedman J L, Phys. Rev. D 79, 124033, 2009 ©(2009) by the American Physical Society.

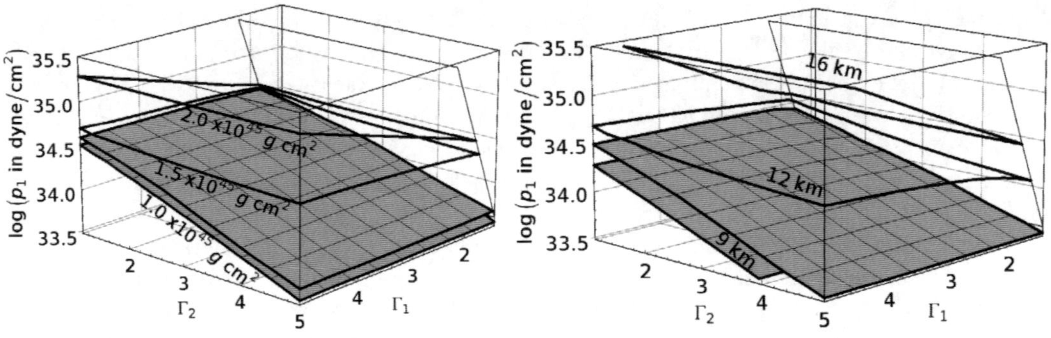

Figure 2. *Left*: Surfaces that represent the set of parameters that result in a star with a mass of $1.338\,M_\odot$ and a given moment of inertia I (from possible near-future observations of PSR J0737-3039A). Observation of moment of inertia I for a known mass M constrains the equation of state to lie *on* the corresponding surface.
Right: Surfaces representing the set of parameters that result in a star with a mass of $1.4\,M_\odot$ and a fixed radius R. Measurement of the radius R of a neutron star with known mass M constrains the equation of state to lie *on* the corresponding surface. The wedge at the back right in both figures corresponds to incompatible values of p_1 and Γ_1. Reprinted with permission from: Read J S, Markakis C, Shibata M, Uryu K, Creighton J D E, and Friedman J L, Phys. Rev. D 79, 124033, 2009 ©(2009) by the American Physical Society.

2.2. *Initial candidate equations of state*

Systematic variation of the EOS parameters also allows us to determine which properties significantly affect the gravitational radiation produced, and thus can be constrained with detection of sufficiently strong gravitational waves.

The models selected for this study use a variation of one EOS parameter in the neutron star core. We vary the core EOS by an overall pressure shift p_1 (specified at the fiducial density $\rho_1 = 10^{14.7}\,\mathrm{g\,cm^{-3}}$), while holding the adiabatic index in the core regions fixed at $\Gamma_1 = \Gamma_2 = \Gamma_3 = 3$. While only a subset of realistic EOS are well-approximated by a single core polytrope, reducing the EOS considered to this single-parameter family allows us to estimate parameter measurability with a reasonable number of simulations.

After fixing the core adiabatic index, increasing the overall pressure scale p_1 produces a family of neutron stars with progressively increasing radius for a given mass; the choice of parameter p_1 is partly motivated by the observation of Lattimer and Prakash [3] that pressure at one to two times nuclear density is closely tied to neutron star radius, with $R \propto p_1^{1/4}$. The radius is less sensitive to variation of the adiabatic index in the neutron star core, for reasonable adiabatic indices [1], as suggested by Fig. 2.

Future work will incorporate additional variations of the EOS within the core. This could involve additional models of varying adiabatic index around a fixed p_1 (with $\Gamma_1 = \Gamma_2 = \Gamma_3$ in all core layers), as well as multiple piecewise-polytrope zones (with $\Gamma_1 \neq \Gamma_2 \neq \Gamma_3$) within the core or EOS parameters yielding neutron stars of the same mass M and radius R but different internal structure. Such work would yield insight into the relative size and correlation of effects on the orbital evolution due to the stellar radius *and* internal structure.

The first set of EOS were chosen to "bracket" the range of existing candidates, seen in Fig. 3. These models are HB with $p_1 = 10^{34.40}\,\mathrm{dyn\,cm^{-2}}$, a standard EOS; 2H with $p_1 = 10^{34.90}\,\mathrm{dyn\,cm^{-2}}$, a stiff EOS; 2B, with $p_1 = 10^{34.10}\,\mathrm{dyn\,cm^{-2}}$, a soft EOS.

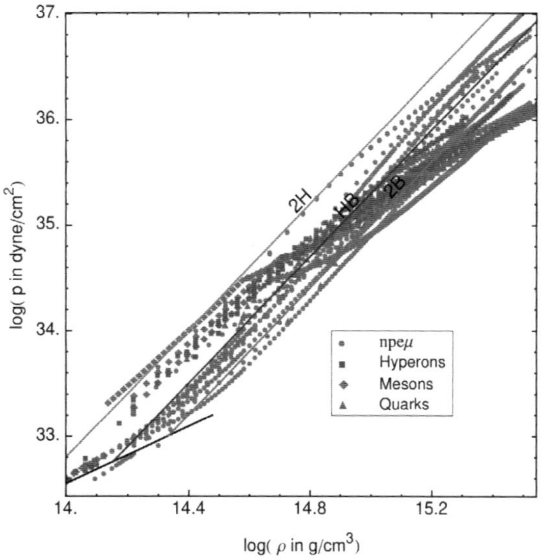

Figure 3. Initial choices of EOS for numerical evolution compared to the set of tabled EOS considered in [1]. Candidates are labeled in order of increasing softness: 2H, HB, 2B. Reprinted with permission from: Read J S, Markakis C, Shibata M, Uryu K, Creighton J D E, and Friedman J L, Phys. Rev. D 79, 124033, 2009 ©(2009) by the American Physical Society.

3. Numerical simulations

For each parameterized EOS considered, the late inspiral and merger of a binary neutron star system is simulated. The gravitational mass of each neutron star in the binary is fixed to $1.35M_{\odot}$, an average value for pulsars observed in binary systems [32, 33]. The significance of tidal effects in this configuration is expected to fairly represent tidal effects over the relatively narrow range of masses and mass ratios in astrophysical binary neutron star systems.

3.1. Initial data

Initial data is generated by constructing quasi-equilibrium sequences of irrotational neutron stars in a binary system, following the methods of [12, 21, 34–37], using the Isenberg-Wilson-Mathews formulation [38, 39]. Since tidal spin-up in neutron star binaries is estimated to be negligible [5, 6], our initial data and quasi-equilibrium sequences are constructed assuming irrotational flow fields and neglecting the neutron star spin. The parameterized EOS of Eq. (1) is incorporated in the code to solve for initial data with a conformally flat spatial geometry coupled to the fluid equations of neutron star matter. The baryon number of each star is equal to that of a spherical isolated star with gravitational mass $M = 1.35\,M_{\odot}$. The initial data for the full numerical simulation is taken from the quasi-equilibrium configuration at a separation such that approximately 3 orbits remain until merger. Relevant quantities of the initial configurations for each parameterized EOS are presented in Tables 1, 2.

3.2. Numerical evolution

The Einstein equations are solved using the Baumgarte-Shapiro-Shibata-Nakamura formulation [40, 41]. The conformal factor of the spatial metric is evolved following [42] and the resulting black hole spacetime is handled using the moving puncture method [43, 44]. The lapse and shift are evolved using a dynamical gauge condition as in [42], while the Einstein equations coupled to the fluid equations are solved using the numerical scheme also described in [42, 56].

Journal of Physics: Conference Series **189** (2009) 012024 doi:10.1088/1742-6596/189/1/012024

Table 1. Properties of initial candidate EOS. These range from the "softest" EOS at the top, which results in a prompt collapse to a black hole upon merger, to the "stiffest" at the bottom. Model HB is considered a typical EOS. The pressure p_1, which is the pressure at density $\rho_1 = 5 \times 10^{14}\,\mathrm{g\,cm^{-3}}$, determines the polytropic EOS for the neutron star core; all candidates have $\Gamma = 3$. Radius R and compactness GM/c^2R are those of a single isolated spherical (TOV) star of gravitational mass $M = 1.35\,M_\odot$, with radius measured in Schwarzschild coordinates. The maximum neutron star mass $M_{\rm max}$ is given in the fifth column.

Model	$\log_{10} p_1$ [dyn cm^{-2}]	R [km]	GM/c^2R	$M_{\rm max}$ [$M_\odot$]
2H	34.90	15.2	0.13	2.83
HB	34.40	11.6	0.17	2.12
2B	34.10	9.7	0.21	1.78

Table 2. Quantities of initial data sets for irrotational binary neutron stars. Each star has a baryon number M_0 equal to that of an isolated star with gravitational mass $M = 1.35\,M_\odot$. The ADM mass $M_{\rm ADM}$ of the initial slice includes the binding energy and J is the total angular momentum of the initial slice.

Model	$\rho_{\rm max}$ [g cm^{-3}]	M_0 [$M_\odot$]	$M_{\rm ADM}$ [$M_\odot$]	$cJ/(GM_{\rm ADM}^2)$	$\Omega/2\pi$ [Hz]
2H	3.73196×10^{14}	1.45488	2.67262	0.993319	324.704
HB	8.27673×10^{14}	1.49273	2.67290	0.995361	309.928
2B	1.38300×10^{15}	1.52509	2.67229	0.987681	321.170

During inspiral, the fluid evolution is essentially free of shocks, whence the cold parameterized EOS specified in Sec. 2 is used in the simulations. During merger, when shocks are developed, we include a hot EOS component with a thermal effective adiabatic index $\Gamma_{\rm eff}$, as described in [46]. Shock heating in the merger can increase the thermal energy up to $\sim$ 20–30% of the total energy [17]. The merger and post-merger evolution is significantly affected by this hot component, since a collapse to a black hole may be delayed by the thermal pressure developed during and after merger.

Gravitational radiation is extracted both by spatially decomposing the metric perturbation about flat spacetime in the wave-zone with spin-2 weighted spherical harmonics and by calculating the outgoing part of the Weyl scalar Ψ_4. For equal mass neutron stars, such as those studied here, the quadrupole ($\ell = 2, m = \pm 2$) mode is much larger than any other mode, so we consider only this mode in this analysis. The waveforms output from the simulations are the cross and plus amplitudes $h_+ c^2 D/GM_{\rm tot}$ and $h_\times c^2 D/GM_{\rm tot}$ of the quadrupole waveform, as would be measured at large distance $D \gg GM_{\rm tot}/c^2$ along the z axis perpendicular to the plane of the orbit, versus the retarded time $t_{\rm ret}$. Here $M_{\rm tot}$ is the sum of the two neutron star masses when they are far apart, $M_{\rm tot} = 2.7 M_\odot$. The early part of the extracted waveforms, for $t_{\rm ret} \lesssim 0$, contains spurious radiation which is discarded in the data analysis.

Each simulation is typically performed for three grid resolutions. In the best-resolution case, the diameter of each neutron star is covered with 60 grid points. Convergence tests with different grid resolutions indicate that, with the best grid resolution, the time duration in the inspiral phase is underestimated by about 1 orbit. This is primarily due to the fact that angular momentum is spuriously lost by numerical dissipation. Thus, the inspiral gravitational waves include a phase error, and as a result, the amplitude of the spectrum for the inspiral phase is slightly underestimated. However, it is found that the waveforms and resulting power spectrum for the late inspiral and merger phases, which we are most interested in for the present work, depend weakly on the grid resolution.

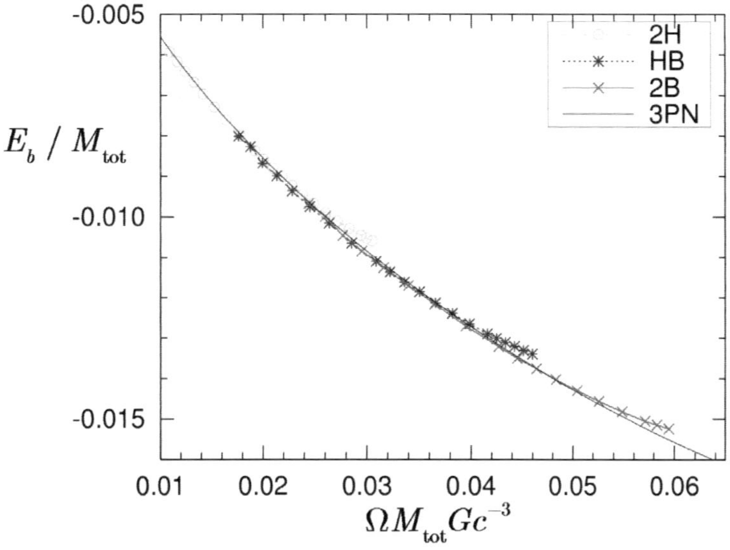

Figure 4. Binding energy $E_b := M_{\rm ADM} - M_{\rm tot}$ as a function of orbital frequency Ω for constant rest mass quasi-equilibrium sequences. The gravitational mass of each neutron star in isolation is $M = M_{\rm tot}/2 = 1.35\,M_\odot$. Departure from the point particle (3PN) curve occurs earlier for the stiffest EOS (candidate 2H – lowest compactness) and later for the softest EOS (candidate 2B – highest compactness).

4. Analysis of inspiral waveforms

From the quadrupole waveform data, we construct the complex quantity

$$h = h_+ - ih_\times. \tag{6}$$

The amplitude and phase of this quantity define the instantaneous amplitude $|h|$ and phase $\phi = \arg h$ of the waveform. The instantaneous frequency f of the quadrupole waveform is then estimated by

$$f = \frac{1}{2\pi}\frac{\Delta\phi}{\Delta t}. \tag{7}$$

The numerical data can be shifted in phase and time by adding a time shift τ to the time series points and multiplying the complex h by $e^{i\phi}$ to shift the overall wave phase by ϕ.

It is useful to define a reference time marking the end of the inspiral and onset of merger. A natural choice for the end of the inspiral portion, considering the behaviour of a PP inspiral waveform, is the time of the peak in the waveform amplitude $|h|$. However, the amplitude of the numerical waveforms oscillates over the course of an orbit. A moving average of the waveform amplitude over 0.5 ms segments is used to average this oscillation; the end of the inspiral is then defined as the time at the end of the maximum amplitude interval. The resulting merger time $t_{\rm M}$ will be marked by solid vertical lines in the plots to follow.

The numerical waveforms begin at different orbital frequencies. To align them for comparison, they are each matched in the early inspiral region to the same post-Newtonian point-particle (or PP) waveform.

4.1. Post-Newtonian point particle

Point-particle inspiral is not well-defined in full general relativity (GR), and one is left with the post-Newtonian point-particle (PP) approximation and fully general relativistic black-hole numerical solutions

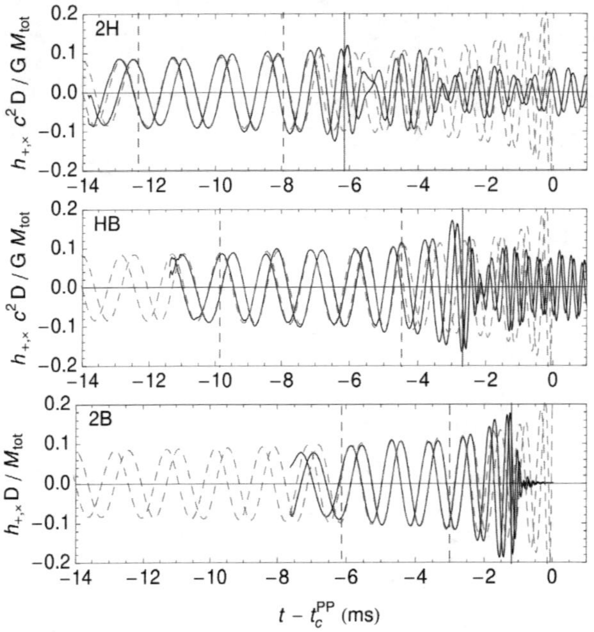

Figure 5. Solid lines show numerical waveforms, scaled by c^2D/GM_{tot}, and aligned in time and phase to the same point-particle post-Newtonian inspiral (dashed line), using a method described in [56]. The two dashed vertical bars indicate the portion of the waveform used for matching; the last vertical bar indicates the end of inspiral time t_{M} for the numerical waveform. The top two simulations, 2H and HB, show the start of post-merger oscillations from a hypermassive neutron star remnant in the simulation. 2B shows quasinormal ringdown from a prompt collapse to a black hole following merger. Reprinted with permission from: Read J S, Markakis C, Shibata M, Uryu K, Creighton J D E, and Friedman J L, Phys. Rev. D 79, 124033, 2009 ©(2009) by the American Physical Society.

as natural substitutes. Fortunately, the Taylor T4 3.0/3.5 post-Newtonian specification, introduced in [47], agrees closely with numerical binary black hole waveforms for many cycles, up to and including the cycle before merger (see also [48]). This empirical agreement allows us to adopt the Taylor T4 waveform as an appropriate PP baseline waveform, compatible with full GR until the last cycles. Quasi-equilibrium sequences [Fig. 4] and time-frequency analysis [Fig. 6] show that the binary neutron star waveforms depart from this waveform 4–8 cycles (200–560 M_{tot}) before the best-fit PP merger, due to finite size effects.

The TaylorT4 waveform is constructed by numerically integrating to obtain a gauge invariant post-Newtonian parameter related to the orbital frequency observed at infinity Ω [56]. The orbital frequency evolution $\Omega(t)$ and orbital phase evolution $\Phi(t)$ are computed to 3.5 post-Newtonian order following [47]. For the numerical integration, one needs to specify the constants of integration by fixing coalescence time $t_{\mathrm{c}}^{\mathrm{PP}}$ and the orbital phase at this time $\Phi(t_{\mathrm{c}}^{\mathrm{PP}}) = \Phi_{\mathrm{c}}^{\mathrm{PP}}$. These two parameters uniquely specify the 3PN waveform for given particle masses. The amplitude of the $(l = 2, m = \pm 2)$ quadrupole waveform is then calculated to 3.0 post-Newtonian order as described in [49, 56].

To match the numerical data to the PP inspiral waveform, the two parameters, $t_{\mathrm{c}}^{\mathrm{PP}}$ and $\Phi_{\mathrm{c}}^{\mathrm{PP}}$ are varied and the best match is obtained [56], to fix a relative time shift and a relative phase shift. The masses of the point particles in the PP waveform are fixed to be the same as the neutron stars in the numerical simulations (the gravitational mass of isolated spherical neutron stars with the same number of baryons) and so masses are not varied in finding the best match. With a goal of signal analysis, we choose the time and phase shift by maximizing a correlation-based match between two waveforms. More details on the matching technique may be found in [56]. We note, however, that longer simulations are required to more precisely fix a post-Newtonian match.

4.2. Comparison of gravitational waveforms

Unlike the case of matching binary black hole simulations to point particle post-Newtonian[47, 48], the binary neutron star simulations show departure from point particle many cycles before the post-Newtonian merger time. Fig. 5 shows the four waveforms shifted so the best-match PP waveforms have the same t_c^{PP} and ϕ_c^{PP}. As the stiffness of the EOS and thus the radius of the neutron stars, increases, the end of inspiral for the binary neutron stars is shifted away from the end of inspiral for post-Newtonian point particle.

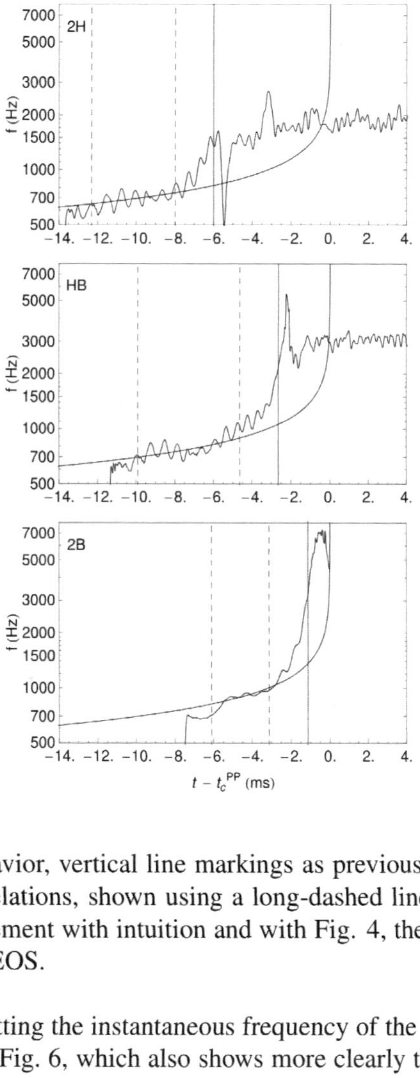

Figure 6. Time-frequency behavior, vertical line markings as previous figure. The departure from the point particle time-frequency relations, shown using a long-dashed line, occurs between 700–1000 Hz depending on the EOS. In agreement with intuition and with Fig. 4, the departure occurs earlier for the stiffest and later for the softest EOS.

This can also be seen by plotting the instantaneous frequency of the numerical simulation waveform with the same time shifts, as in Fig. 6, which also shows more clearly the difference in the post-merger oscillation frequencies of the hypermassive remnants, when present. The larger neutron star produced by the stiff EOS 2H has a lower oscillation frequency than that from the medium EOS HB. The remnant forms with a bar-mode oscillation stable over a longer period than the ~ 10 ms simulated. The signal from such a bar mode may be even stronger when the full lifetime is included. Information that can be extracted from the presence (or absence) and characteristics of a post-merger oscillation signal from

Journal of Physics: Conference Series **189** (2009) 012024 doi:10.1088/1742-6596/189/1/012024

the hyper-massive remnant would complement the information present in the late inspiral. Although such information could probe the equation of state at densities higher than those present in ordinary neutron stars, aspects of the physics neglected in this study will likely come into play. Such aspects may include finite temperature, magnetic field, neutrino cooling and other effects that need to be accurately modeled, thus complicating the parameter extraction. As previously noted, our simulations included a hot component in the equation of state, which strongly affects the post merger behavior (and prompt versus delayed collapse to a black hole in particular), but inclusion of other effects is a subject for future study.

4.3. Spectrum of gravitational waveforms

Given h_+ or $h_\times$, one can construct the discrete Fourier transforms (DFTs) $\tilde{h}_+$ or $\tilde{h}_\times$. Both polarizations yield the same DFT amplitude spectrum $|\tilde{h}|$, with phase shifted by $\pi/2$, if one neglects discretization, windowing, and numerical effects (including eccentricity). The amplitude spectrum $|\tilde{h}|$ is independent of phase and time shifts of the waveform.

The stationary phase approximation is valid for the post-Newtonian waveform up to frequencies of about 1500 Hz (with $\lesssim 10\%$ error), so is used to plot the amplitude of the point particle spectrum. In terms of an instantaneous frequency

$$f(t) = \frac{1}{2\pi}\frac{d\phi}{dt}, \tag{8}$$

the Fourier transform of the waveform has the amplitude

$$|\tilde{h}| \simeq A(f)\left(\frac{df}{dt}\right)^{-1/2}. \tag{9}$$

For binaries comprised of equal mass companions, the gravitational radiation is dominated by quadrupole modes throughout the inspiral, so the analysis includes this mode only, as mentioned earlier. The wave phase ϕ is negligibly different from twice the orbital phase 2Φ until the onset of merger at very high frequencies so we can use the relation

$$\frac{df}{dt} \simeq \frac{1}{\pi}\frac{d\Omega}{dt}; \tag{10}$$

to write the amplitude of the Fourier transform entirely in terms of the functions $d\Omega/dt$ and the amplitude $A(f) = |h|$ of the polarization waveforms.

The translation of emitted waveforms into the strain amplitude measured at a detector involves transformations incorporating the effects of the emitting binary's angle of inclination and sky location. These effects are absorbed into an effective distance D_{eff}, which is equal to the actual distance D for a binary with optimal orientation and sky location, and is greater than the actual distance for a system that is not optimally oriented or located. The detector will detect a single polarization of the waveform, some combination of the plus and cross polarizations of the emitted waveform. The polarizations extracted from the simulation can be used as two estimates of the strain at the detector for a given numerically modelled source, which give very close results and are subsequently averaged.

To compare with noise curves in the usual units, the quantity $f^{1/2}|\tilde{h}(f)|$ is plotted, at a reference distance of $D_{\text{eff}} = 100\,\text{Mpc}$ and rescaling from previously plotted numerical output $h(t)c^2D/GM_{\text{tot}}$ using $M_{\text{tot}} = 2.7M_\odot$.

The full spectra of models 2H and HB, seen in Fig. 7, show peaks at post-merger oscillation frequencies. Waveforms of 2H and HB are truncated while the post-merger oscillation is ongoing; if the simulations were allowed to continue, these peaks would presumably grow further. The simulation of the softest candidate 2B, in contrast, promptly collapses to a black hole upon merger and has a short lived, and relatively small-amplitude, quasinormal mode ringdown.

Note that time-frequency plots like the ones in Fig. 6 showed that numerical waveforms follow the PP waveform at instantaneous frequencies of up to 700–1000 Hz, depending on the EOS. The disagreement

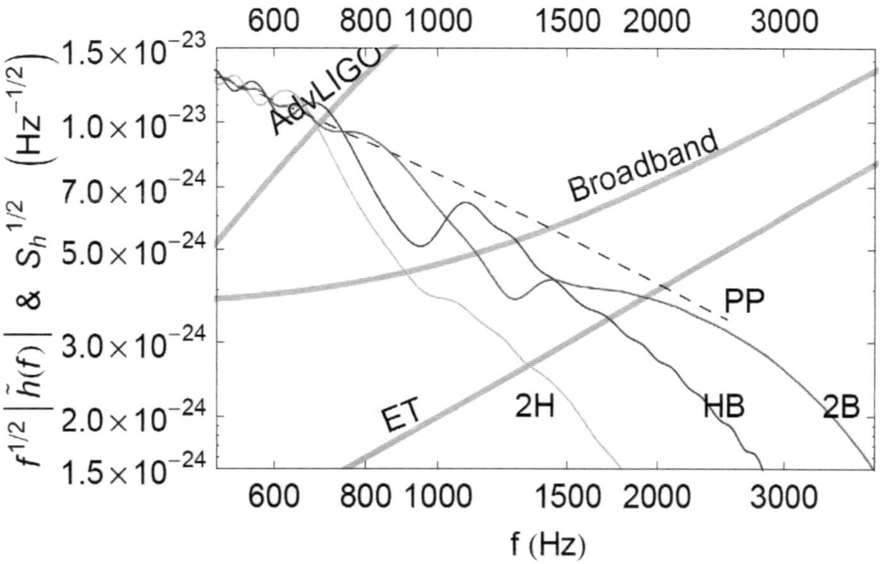

Figure 7. Colored curves show the DFT of full numerical inspiral waveforms, truncated and smoothly joined to best-match PP inspiral time series, at an effective distance $D_{\text{eff}} = 100\,\text{Mpc}$. The stationary-phase point particle is shown by a dashed line for reference. Thick grey lines represent the noise spectra for Advanced LIGO (labeled "AdvLIGO" for the standard configuration and "Broadband" for the broad-band configuration) and the Einstein Telescope (labeled "ET").

in the spectra in Fig. 7 from the PP stationary-phase approximation waveform at frequencies below this is primarily due to the finite starting time of the numerical waveforms. To estimate spectra from the full inspiral, we construct hybrid waveforms. The short-term numerical waveforms are smoothly merged on to long-inspiral PP-PN waveforms, using Hann windowing to smoothly turn on or turn off a signal. To construct hybrid waveforms, these windows are used to turn on the numerical waveform as the matched post-Newtonian is turned off, such that the sum of the two window functions is 1 over the matching range. Because, as mentioned earlier, the dependence on the cold EOS is less straightforward during the merger, due to unmodelled physics that become important at that time, we focus instead on the signal from the waveforms during the inspiral region only, turning off the waveforms after the end of inspiral, t_{M}.

5. Distinguishability and detectability

The question for neutron star astrophysics is whether these differences in the gravitational waveform will be measurable. We consider the possibility of detecting EOS effects with Advanced LIGO style detectors in varying configurations.

We use several detector configurations commonly considered for Advanced LIGO include tunings optimized for 1.4 $M_\odot$ NS-NS inspiral detection ("Standard"), for burst detection ("Broadband"), and for pulsars at 1150 Hz ("Narrowband"). The sensitivity is expressed in terms of the one-sided strain-equivalent amplitude spectral density $S_h(f)$ (which has units of $\text{Hz}^{-1/2}$) of the instrumental noise in Advanced LIGO. We also consider a provisional noise curve for the Einstein Telescope [51]. We consider only a single detector of each type, rather than a combination of detectors, for a preliminary estimate of detectability.

Journal of Physics: Conference Series **189** (2009) 012024 doi:10.1088/1742-6596/189/1/012024

Table 3. $\overline{\varrho}_{\rm diff}$ in standard (NS-NS detection optimized) noise $\times$ $(100\,{\rm Mpc}/D_{\rm eff})$

Model	PP	2B	HB	2H
PP	0	0.32	0.55	0.69
2B		0	0.48	0.63
HB			0	0.60
2H				0

Table 4. $\overline{\varrho}_{\rm diff}$ in broadband (burst-optimized) noise $\times$ $(100\,{\rm Mpc}/D_{\rm eff})$

Model	PP	2B	HB	2H
PP	0	1.86	2.67	2.89
2B		0	2.32	2.54
HB			0	2.37
2H				0

Given two signals h_1 and h_2, and a noise spectrum $S_h(f)$, we define the usual inner product [52]:

$$\langle h_1|h_2\rangle = 4\,{\rm Re}\int_0^\infty \frac{\tilde{h}_1(f)\tilde{h}_2^*(f)}{S_h(f)}\,df. \tag{11}$$

This inner product yields a natural metric on a space of waveforms with distance between waveforms weighted by the inverse of the noise. The detector output s is filtered against an expected waveform h using $\langle s|h\rangle$. Then

$$\varrho = \frac{\langle s|h\rangle}{\sqrt{\langle h|h\rangle}}, \tag{12}$$

is the optimal statistic to detect a waveform of known form h in the signal s. If the detector output contains a particular waveform h that is exactly matched by the template used, then the expectation value of ϱ is the expected signal-to-noise ratio or SNR of that waveform:

$$\overline{\varrho} = \sqrt{\langle h|h\rangle}. \tag{13}$$

Given a signal h_1 to be used as a template, one can ask whether a known *departure* from this signal can be measured. Assuming the modified waveform h_2 is known, the expected SNR of $h_2 - h_1$ is similarly

$$\overline{\varrho}_{\rm diff} = \sqrt{\langle h_2 - h_1|h_2 - h_1\rangle}. \tag{14}$$

We consider the two signals to be marginally distinguishable [53] if the difference between the waveforms h_1 and h_2 has $\overline{\varrho}_{\rm diff} \geq 1$.

To analyze the measurability of finite size effects, we consider differences between hybrid inspiral-only waveforms matched to point particle PN waveforms in the early inspiral. After the numerical waveforms have been matched to the same post-Newtonian point-particle inspiral, the signals will be aligned in time and phase. We then then compare the resulting waveforms to each other, and to a PN-only waveform, using different Advanced LIGO noise spectra. We report results in terms of the SNR measured by a single detector at an effective distance $D_{\rm eff} = 100\,{\rm Mpc}$. These results may be straightforwardly scaled to any distance in the wave zone, since h and $\overline{\varrho}_{\rm diff}$ are inversely proportional to the distance D.

Table 5. $\overline{\varrho}_{\rm diff}$ in narrowband 1150 Hz noise $\times$ (100 Mpc/$D_{\rm eff}$)

Model	PP	2B	HB	2H
PP	0	0.91	3.69	2.12
2B		0	2.92	1.45
HB			0	2.25
2H				0

The difference between waveforms due to finite size effects is not detectable in the NS-NS detection optimized configuration of Advanced LIGO for $\sim$ 100 Mpc effective distances.

However, in both narrowband and broadband the differences can be significant, and waveforms are distinguishable from each other and from the PP waveform. Note that this implies that, even if the details of a NS-NS late inspiral signal are not known, the difference between it and a point particle waveform should be measurable. The quantity $\overline{\varrho}_{\rm diff}$ between the observed waveform and a best fit point particle waveform, limited to differences at high frequency, may be useful in itself to constrain possible EOS independently of waveform details.

5.1. Parameter extraction

Since the early part of the inspiral is well described by a point particle post-Newtonian approximation, we will assume that EOS effects on the waveform impact the late inspiral only. For simplicity, we assume that orbital parameters, such as $M_{\rm tot}$, mass ratio η, point particle post-Newtonian coalescence time $t_{\rm c}^{\rm PP}$, and phase shift $\phi_{\rm c}^{\rm PP}$, are determined from the observations of the earlier inspiral waveform, with sufficient accuracy that their measurement uncertainty will not affect the accuracy to which the late inspiral effects determine the EOS parameters. These measurements would be made by a broad-band instrument, in which the signal-to-noise ratio is expected to be high ($\sim$ 40 at 100 Mpc) and measurement accuracy is expected to be good [52]. Inaccuracies in these measurement could lead to biases in the measured EOS. This will be an important aspect to assess when high-quality binary neutron star simulations with various masses become abundant.

With a one-parameter family of waveforms sampled, we can estimate the accuracy to which this parameter can be measured. There are also other EOS-related parameters which are not considered. In this first analysis we directly estimate the measurability of the EOS parameter p_1, ignoring variations of Γ within the core. In the analysis of radius measurability, variations of the internal structures are respectively neglected. Expanding coverage of the EOS parameter space is underway. However, these initial parameter choices are expected to give the dominant contributions to finite size effects of the waveform. For example, the 1PN tidal effect estimates of [7, 54] predict $\sim$10% variation in the tidal terms contributing to binding energy and luminosity from changing internal structure—varying the apsidal constant—while keeping radius fixed.

We estimate errors in parameter estimation to first order in $1/\overline{\varrho}$ or, equivalently, in $\delta\theta^A$, using the Fisher matrix $\Gamma_{AB} = \langle \partial_A h | \partial_B h \rangle$ [52]. Its inverse, $(\Gamma^{-1})^{AB}$, yields

$$\overline{\delta\theta^A \delta\theta^B} = (\Gamma^{-1})^{AB}, \tag{15}$$

so that the expected error in a given parameter θ^A is

$$\overline{(\delta\theta^A)^2} = (\Gamma^{-1})^{AA}, \tag{16}$$

and the cross terms of the inverse Fisher matrix yield correlations between different parameters.

Table 6. Measurability of radius R (km) with broadband (burst-optimized) and narrowband (1150 Hz) configurations of Advanced LIGO, for a binary system at effective distance 100 Mpc. The values of R here are midpoint values of the radii of candidates shown in Table 1. The error δR (km) is estimated using Eqs. (14), (18) (with $\overline{\varrho}_{\rm diff}$ taken from Tables 4, 5) and scales with effective distance as $\delta R \times (D_{\rm eff}/100\,{\rm Mpc})$.

Configuration:	Broadband	Narrowband (1150 Hz)
$R = 10.65$	± 0.81	± 0.65
$R = 13.40$	± 1.52	± 1.60

Table 7. Measurability of pressure p_1 (dyn cm^{-2}) at the fiducial density $\rho_1 = 10^{14.7}\,{\rm g\,cm^{-3}}$, with broadband (burst-optimized) and narrowband (1150 Hz) configurations of Advanced LIGO, for a binary system at effective distance 100 Mpc. The values of $\log_{10} p_1$ here are midpoint values of $\log_{10} p_1$ from Table 1. The error $\delta\log_{10} p_1$ is estimated using Eqs. (14), (18) (with $\overline{\varrho}_{\rm diff}$ taken from Tables 4, 5) and scales with effective distance as $\delta\log_{10} p_1 \times (D_{\rm eff}/100\,{\rm Mpc})$.

Model	Broadband	Narrowband (1150 Hz)
$\log_{10} p_1 = 34.25$	± 0.13	± 0.10
$\log_{10} p_1 = 34.65$	± 0.21	± 0.22

With a few simulations of varying parameter value, we estimate $\partial h/\partial p_1$ and $\partial h/\partial R$ from two of the sampled waveforms, h_1 and h_2. For a single parameter θ (which can be taken to be either p_1 or R), we have

$$\left.\frac{\partial h}{\partial \theta}\right|_{\theta=(\theta_1+\theta_2)/2} \simeq \frac{h_2 - h_1}{\theta_2 - \theta_1}, \tag{17}$$

where $h_1 = h(\theta_1)$ and $h_2 = h(\theta_2)$, and then, for our one-parameter family where we neglect correlations with other parameters, we have to first order

$$\overline{(\delta\theta)^2} \simeq \frac{(\theta_2 - \theta_1)^2}{\langle h_2 - h_1 | h_2 - h_1 \rangle}. \tag{18}$$

Using adjacent pairs of models to estimate waveform dependence at an average parameter value, we then find estimates of radius measurability as shown in Table 6 and p_1 measurability as shown in Table 7 for the burst-optimized noise and narrowband configurations.

5.2. Sources of error

Many higher order but likely relevant effects have been neglected in this preliminary analysis. Tidal effects may measurably influence the orbital evolution before the start of the numerical simulations, as estimated in [7], slowly enough not to be seen over the few cycles of the waveform matched to PP in this analysis. In one sense this analysis is a worst-case scenario, as it assumes exact PP behavior before the numerical match. Earlier drift away from point particle dynamics would give larger differences between waveforms, and more sensitive radius measurement, but poses a challenge by requiring accurate numerical simulation over many cycles to verify EOS effects. Combining numerical estimation with PN analyses like those of [7, 8], and/or quasiequilibrium sequence information [Fig. 5] may clarify the transition between effectively PP and tidally influenced regimes.

Some residual eccentricity from initial data and finite numerical resolution is present in the waveforms themselves. The eccentricity error may be estimated by comparing the plus polarization to the cross polarization shifted by $\pi/2$ from the same numerical waveform. This results in $\overline{\varrho}_{\rm diff}$ of ~ 0.3 for HB and 2H, rather than the expected quadrupole polarization cross-correlation of zero. The 2B waveform produces $\overline{\varrho}_{\rm diff} \sim 0.05$ between polarizations.

We have only a coarsely sampled family of waveforms; estimates of $\partial h/\partial\theta$ are limited by this. Further, the length of the inspirals (see discussion on PN matching in [56]) limits precision in choosing the best match time. We can estimate these effects on current results by varying the match region considered; this changes $\overline{\varrho}_{\rm diff}$ by up to ~ 0.5 at 100 Mpc in the broadband detector. The resolution from existing numerical simulations is comparable to the difference between parameters of some of the models. The estimated uncertainty in each $\overline{\varrho}_{\rm diff}$ is smaller than (but comparable to) the $\overline{\varrho}_{\rm diff}$ between candidates 2H, HB, 2B and PP.

Finally, we note that use of a Fisher matrix estimate of parameter measurement accuracy is fully valid only in high SNR limit of $\overline{\varrho}_{\rm diff} > 10$ [55], i.e., for distances $\lesssim 20$ Mpc in the broadband detector. The results do not take into account multiple detectors, nor multiple observations, nor parameter correlations. We conclude that, although these are of course first estimates, they should be better than order-of-magnitude. A full estimation of EOS parameter measurability will require more detailed analysis, with a larger set of inspiral simulations sampling a broader region of EOS parameter space, with mass ratios departing from unity, and with more orbits simulated before merger.

6. Summary

This is a first quantitative estimate of the measurability of the EOS of cold matter above nuclear density with gravitational wave detectors. Neutron stars of the same mass but different EOS and thus different radii will emit different gravitational waveforms during a late stage of binary inspiral. We calculated the signal strength of this difference in waveform using the sensitivity curves of commissioned and proposed gravitational wave detectors, and find that there is a measurably different signal from the inspirals of binary neutron stars with different EOS.

Although the standard noise configuration of Advanced LIGO is not sensitive to differences in the waveform from finite size effects, a broadband (burst-optimized) configuration or a narrowband detector configuration of Advanced LIGO can distinguish neutron star EOS from each other and from point particle inspiral, at an effective distance $\lesssim 100$ Mpc. In addition, the provisional standard noise curve of the Einstein Telescope indicates the ability to resolve different EOS at roughly double that distance.

With broadband Advanced LIGO at frequencies between 500 and 1000 Hz, our first estimates show that the radius can be determined to an accuracy of $\delta R \sim 1\,\mathrm{km} \times (100\,\mathrm{Mpc}/D_{\rm eff})$. Related first estimates show that such observations can constrain an EOS parameter parameter p_1, the pressure at a rest mass density $\rho_1 = 10^{14.7}\,\mathrm{g\,cm^{-3}}$, with an accuracy of $\delta p_1 \sim 5 \times 10^{33}\,\mathrm{dyn\,cm^{-2}} \times (100\,\mathrm{Mpc}/D_{\rm eff})$. These estimates neglect correlations between the parameters and other details of internal structure, which are expected to give relatively small tidal effect corrections. Although these results are preliminary, they strongly motivate further work on gravitational wave constraints from binary neutron star inspirals. The accuracy of the estimates will be improved with numerical simulation of more orbits in the late inspiral and a wider exploration of the EOS parameter space.

Acknowledgments

It is a pleasure to thank the organizers of the 13th Conference on Recent Developments in Gravity (NEB-XIII) at Thessaloniki, for providing a very stimulating environment for research and discussion during the meeting. This work was supported in part by NSF grants PHY-0503366, PHY-0701817 and PHY-0200852, by NASA grant ATP03-0001-0027, and by JSPS Grants-in-Aid for Scientific Research(C) 20540275 and 19540263. CM thanks the Greek State Scholarships Foundation for support. Computation was done in part in the NAOJ and YITP systems.

References

[1] Read J S, Lackey B, Friedman J L, and Owen B, `astro-ph/0812.2163`.

[2] Read J S 2008 *PhD Thesis*, University of Wisconsin-Milwaukee
[3] Lattimer J M and Prakash M, *Ap. J.* **550**, 426 (2001), astro-ph/0002232.
[4] Müller E, Eriguchi Y 1985 *Ap. J.* **152**, 325-335.
[5] Kochanek C S 1992 *Ap.J.* **398**, 234.
[6] Bildsten L and Cutler C *Ap.J.* **400**, 175.
[7] Flanagan É É and Hinderer T, *Phys. Rev.* D **77**, 021502 (2008), 0709.1915.
[8] Hinderer T, Lackey B and Read J S 2009 *Work in preparation*
[9] Lai D and Wiseman A G, *Phys. Rev.* D **54**, 3958 (1996), gr-qc/9609014.
[10] Zhuge X, Centrella J M, and McMillan S L W, *Phys. Rev.* D **54**, 7261 (1996), gr-qc/9610039.
[11] Rasio F A and Shapiro S L, *Classical and Quantum Gravity* ***16***, 1 (1999), gr-qc/9902019.
[12] Uryu K, Shibata M, and Eriguchi Y, Phys. Rev. **D62**, 104015 (2000), gr-qc/0007042.
[13] Faber J A, Grandclément P, Rasio F A, and Taniguchi K, *Physical Review Letters* **89**, 231102 (2002), astro-ph/0204397.
[14] Faber J A, Grandclément P, and Rasio F A, *Phys. Rev.* D **69**, 124036 (2004), gr-qc/0312097.
[15] Bejger M, Gondek-Rosińska D, Gourgoulhon E, Haensel P, Taniguchi K, and Zdunik J L, *Astronomy and Astrophysics* **431**, 297 (2005), astro-ph/0406234.
[16] Gondek-Rosińska D, Bejger M, Bulik T, Gourgoulhon E, Haensel P, Limousin F, Taniguchi K, and Zdunik L, Advances in Space Research **39**, 271 (2007).
[17] Shibata M, Taniguchi K, and Uryū K, *Phys. Rev.* D **71**, 084021 (2005), gr-qc/0503119.
[18] Shibata M and Uryū K, *Prog. Theor. Phys.* **107**, 265 (2003).
[19] Oechslin R and Janka H-T, *Phys. Rev. Lett.* **99**, 121102 (2007).
[20] Hughes S A, *Phys. Rev.* D **66**, 102001 (2002), gr-qc/0209012.
[21] Taniguchi K and Gourgoulhon E, *Phys. Rev.* D **68**, 124025 (2003), gr-qc/0309045.
[22] Baumgarte T, Brady P R, Creighton J D E, Lehner L, Pretorius F, and Devoe R, *Phys. Rev.* D **77**, 084009 (2008).
[23] Shibata M, *Phys. Rev.* D **60**, 104052 (1999), gr-qc/9908027.
[24] Shibata M and Uryū K, *Phys. Rev.* D **61**, 064001 (2000), gr-qc/9911058.
[25] Shibata M and Taniguchi K, *Phys. Rev.* D **73**, 064027 (2006), astro-ph/0603145.
[26] Shibata M, Taniguchi K, and Uryū K, *Phys. Rev.* D **68**, 084020 (2003), gr-qc/0310030.
[27] Miller M, Gressman P, and Suen W-M, *Phys. Rev.* D **69**, 064026 (2004), gr-qc/0312030.
[28] Marronetti P, Duez M D, Shapiro S L, and Baumgarte T W, *Phys. Rev. Lett.* **92**, 141101 (2004), gr-qc/0312036.
[29] Liu Y T, Shapiro S L, Etienne Z B, and Taniguchi K, *Phys. Rev.* D **78**, 024012 (2008), 0803.4193.
[30] Yamamoto T, Shibata M, and Taniguchi K, *Phys. Rev.* D **78**, 064054 (2008), 0806.4007.
[31] Baiotti L, Giacomazzo B, and Rezzolla L (2008), gr-qc/0804.0594.
[32] Thorsett S E and Chakrabarty D, *Ap. J.* **512**, 288 (1999), astro-ph/9803260.
[33] Stairs I H 2004, Science **304**, 547.
[34] Uryū K and Eriguchi Y 2000, *Phys. Rev.* D **61**, 124023, gr-qc/9908059.
[35] Bonazzola S, Gourgoulhon E, and Marck J-A 1999, *Phys. Rev. Lett.* **82**, 892, gr-qc/9810072.
[36] Gourgoulhon E, Grandclement P, Taniguchi K, Marck J-A, and Bonazzola S 2001, *Phys. Rev.* D **D63**, 064029, gr-qc/0007028.
[37] Taniguchi K and Gourgoulhon E 2002, *Phys. Rev.* D **66**, 104019, gr-qc/0207098.
[38] Isenberg J A 2008, *Int. J. Mod. Phys.* **D17**, 265, gr-qc/0702113.
[39] Wilson J R and Mathews G J 1995, *Physical Review Letters* **75**, 4161.
[40] Shibata M and Nakamura T, *Phys. Rev.* D **52**, 5428 (1995).
[41] Baumgarte T W and Shapiro S L 1999, *Phys. Rev.* D **59**, 024007, gr-qc/9810065.
[42] Shibata M and Taniguchi K 2008, *Phys. Rev.* D **77**, 084015, 0711.1410.
[43] Campanelli M, Lousto C O, Marronetti P, and Zlochower Y, *Phys. Rev. Lett.* **96**, 111101 (2006), gr-qc/0511048.
[44] Baker J G, Centrella J, Choi D-I, Koppitz M, and van Meter J 2006, *Phys. Rev. Lett.* **96**, 111102, gr-qc/0511103.
[45] Kurganov A and Tadmor E 2000, *J. Comput. Phys.* **160**, 241.
[46] Shibata M, *Phys. Rev.* D **67**, 024033 (2003).
[47] Boyle M, Brown D A, Kidder L E, Mroué A H, Pfeiffer H P, Scheel M A, Cook G B 2007, and Teukolsky S A, *Phys. Rev.* D **76**, 124038, gr-qc/0710.0158.
[48] Gopakumar A, Hannam M, Husa S, and Brügmann B 2007, *Preprint* 0712.3737.
[49] Kidder L E, *Phys. Rev.* D **77**, 044016 (2008).
[50] Allen B, Anderson W G, Brady P R, Brown D A, and Creighton J D E 2005, *Preprint* gr-qc/0509116.
[51] Van Den Broeck C 2008, *Tech. Rep.* ,
(URL https://workarea.et-gw.eu/et/WG4-Astrophysics/base-sensitivity/).
[52] Cutler C and Flanagan É É, *Phys. Rev.* D **49**, 2658 1994, gr-qc/9402014.
[53] Lindblom L, Owen B J, and Brown D A 2008, 0809.3844.
[54] Hinderer T 2008 *Ap. J.* **677**, 1216 , *Preprint* 0711.2420.

[55] Vallisneri M 2008 *Phys. Rev.* D **77**, 042001
[56] Read J S, Markakis C, Shibata M, Uryū K, Creighton J D E, and Friedman J L 2009 *Preprint* `astro-ph/0901.3258`

Journal of Physics: Conference Series **189** (2009) 012025 doi:10.1088/1742-6596/189/1/012025

Homogeneous and isotropic cosmologies in higher order gravity

J. Miritzis

Department of Marine Sciences, University of the Aegean, University Hill, Mytilene 81100, Greece

E-mail: imyr@aegean.gr

Abstract. Homogeneous and isotropic models are studied in the Jordan frame of the second order gravity theory. The late time evolution of the models is analysed with the methods of the dynamical systems. The normal form of the dynamical system has periodic solutions for a large set of initial conditions. This implies that an initially expanding closed isotropic universe may exhibit oscillatory behaviour.

1. Introduction

Quadratic gravitational Lagrangians were proposed shortly after the formulation of general relativity (GR) as alternatives to Einstein's theory. Gravity modifications in the form of higher-order curvature invariants in the Lagrangian are generally known as higher-order gravity (HOG) theories. They arise in string-theoretic considerations, e.g., brane models with Gauss-Bonett terms, or models with a scalar field coupled to the Gauss-Bonett invariant (see [1] for a review) and generally involve linear combinations of all possible second order invariants that can be formed from the Riemann, Ricci and scalar curvatures. A quarter of a century ago there was a resurgence of interest in such theories in an effort to explain inflation. The reasons for considering HOG theories were multiple. Firstly, it was hoped that higher order Lagrangians would create a first approximation to quantum gravity, due to their better renormalisation properties than GR. Secondly, it was reasonable to expect that on approach to a spacetime singularity, curvature invariants of all orders ought to play an important dynamical role. Far from the singularity, when higher order corrections become negligible, one should recover GR. Furthermore, it was hoped that these generalised theories of gravity might exhibit better behavior near singularities. Thirdly, inflation emerges in these theories in a most direct way. In one of the first inflationary models, proposed in 1980 by Starobinsky [2], inflation is due to the R^2 correction term in a gravitational Lagrangian $L = R + \beta R^2$ where β is a constant.

Recently there is a revival of interest in HOG theories in an effort to explain the accelerating expansion of the Universe [3, 4]. The general idea is to add an $1/R$ term to the Einstein-Hilbert Lagrangian or more generally to consider $R + \alpha R^{-n}$ Lagrangians [5]. As the Universe expands, one expects that the inverse curvature terms will dominate and produce the late time accelerating expansion. Most studies are restricted to simple Friedmann-Robertson-Walker (FRW) models because of the complexity of the field equations. At present, the observational viability of these models is a subject of active research. However, it seems that a large class of Lagrangians may fit the observational data, but simple models based on the FRW metric are insufficient to pick the correct Lagrangian (see for example [6] for the reconstruction of the $f(R)$ theory which best reproduces the observed cosmological data). For more general spacetimes it may even be meaningless to say that R is small in some epoch of cosmic evolution and large in some other one (for a thorough critic see [7]).

In this paper we investigate the late time evolution of positively curved FRW models with a perfect fluid in the $R + \beta R^2$ theory. This is the simplest generalization of the Einstein-Hilbert Lagrangian and the addition of the quadratic term represents a correction to general relativity. The simple vacuum case was studied in [8], where oscillatory behaviour of the solutions of closed models was found. Since HOG theories in vacuum are conformally equivalent to GR with a scalar field, it is tempting to say that the R^2 contribution has predictable cosmological consequences. However, this is an oversimplification of the picture. The two frames are mathematically equivalent, but physically they provide different theories. In the Jordan frame, gravity is described entirely by the metric $g_{\mu\nu}$. In the Einstein frame, the scalar field exhibits a non-metric aspect of the gravitational interaction, reflecting the additional degree of freedom due to the higher order of the field equations in the Jordan frame. Inclusion of additional matter fields, further complicates the situation and while the field equations in the Einstein frame are formally the Einstein equations, nevertheless this theory is not physically equivalent to GR. There is no universally acceptable answer to the issue "which conformal frame is physical" [9] (see [10] for a thorough analysis of different views).

The plan of the paper is as follows. In the next section the field equations are written as a constrained four-dimensional polynomial dynamical system. In Section 3 we study the qualitative behaviour of the solutions near the equilibrium points of positively curved models and analyse their late time evolution. The so-called normal form of the dynamical system greatly simplifies the problem, since two of the equations decouple. It is shown that an initially expanding closed FRW universe may exhibit oscillatory behaviour. Section 4 is devoted to the discussion of the results in the context of cyclic cosmology.

2. Field equations

The general gravitational Lagrangian in four-dimensional spacetimes contains curvature invariants of all orders, $R, R^2, R_{\mu\nu}R^{\mu\nu}, \ldots$. The term $R_{\mu\nu\rho\sigma}R^{\mu\nu\rho\sigma}$ is omitted because of the Gauss-Bonnet theorem. A further simplification can be done in homogeneous and isotropic spacetimes where the variation of $R_{\mu\nu}R^{\mu\nu}$ with respect to the metric is proportional to the variation of R^2 [11]. We conclude that for isotropic cosmologies the gravitational Lagrangian contains only powers of the scalar curvature and we may consider HOG theories derived from Lagrangians of the form

$$L = f(R)\sqrt{-g} + L_{\text{matter}},$$

where f is an arbitrary smooth function. It is well-known that the corresponding field equations are fourth-order and take the form

$$f'(R)R_{\mu\nu} - \frac{1}{2}f(R)g_{\mu\nu} - \nabla_\mu\nabla_\nu f'(R) + g_{\mu\nu}\Box f'(R) = T_{\mu\nu}, \quad (1)$$

where $\Box = g^{\alpha\beta}\nabla_\alpha\nabla_\beta$ and a prime ($'$) denotes differentiation with respect to R. The generalised Bianchi identities imply that

$$\nabla^\mu T_{\mu\nu} = 0. \quad (2)$$

Contraction of (1) yields the trace equation

$$3\Box f'(R) + f'(R)R - 2f(R) = T. \quad (3)$$

In contrast to GR where the relation of R to T is algebraic, in HOG theories the trace equation (3) is a differential equation for R, with T as source term [12]. This suggests that in HOG theories both the metric and the scalar curvature are dynamical fields.

In the following, we consider a quadratic Lagrangian without cosmological constant, i.e.

$$f(R) = R + \beta R^2, \ \beta > 0,$$

and confine our attention to cosmologies with a perfect fluid with energy density ρ and pressure p, of the form

$$p = (\gamma - 1)\rho, \quad 0 \leq \gamma \leq 2.$$

For homogeneous and isotropic spacetimes[1] described by the standard FRW metric we need the following useful relations ($i, j = 1, 2, 3$ and $k = 0, \pm 1$)

$$R_{00} = -3\frac{\ddot{a}}{a}, \ R_{ij} = \left(\frac{\ddot{a}}{a} + 2\frac{\dot{a}^2}{a^2} + 2\frac{k}{a^2}\right) g_{ij}, \ R = 6\left(\frac{\ddot{a}}{a} + \frac{\dot{a}^2}{a^2} + \frac{k}{a^2}\right). \tag{4}$$

The 00 component of (1) is

$$H^2 + \frac{k}{a^2} + 2\beta\left[R\left(H^2 + \frac{k}{a^2}\right) + H\dot{R} - \frac{R^2}{12}\right] = \frac{1}{3}\rho, \tag{5}$$

where $H = \dot{a}/a$, is the Hubble function.

Some comments about the choice of variables are necessary. Setting $x = 1/a$, the 00 equation becomes

$$H^2 + kx^2 + 2\beta\left[R\left(H^2 + kx^2\right) + H\dot{R} - \frac{R^2}{12}\right] = \frac{1}{3}\rho, \tag{6}$$

and the evolution equation for x is

$$\dot{x} = -xH. \tag{7}$$

The evolution equation for H comes from the third of (4) and takes the form

$$\dot{H} = \frac{1}{6}R - 2H^2 - kx^2. \tag{8}$$

The conservation equation

$$\dot{\rho} = -3\gamma\rho H, \tag{9}$$

is a consequence of the Bianchi identities (2). With the relation $\Box R = -\ddot{R} - 3H\dot{R}$, equation (3) becomes

$$\ddot{R} + 3H\dot{R} + \frac{1}{6\beta}R = \frac{1}{6\beta}(4 - 3\gamma)\rho. \tag{10}$$

This equation is usually considered as superfluous, since it follows from the differentiation of the equation (6) with respect to t; for example equation (10) was used in [14] as a control of the accuracy of numerical investigations of Bianchi type I and IX models. Thus, one can chose R, x, H and ρ as dynamical variables which obey the evolution equations (6), (7), (8) and (9) and constitute a four-dimensional dynamical system.

The form of equation (6) suggest the choice of expansion normalized variables of the type

$$u \sim \frac{R}{H}, \ \Omega \sim \frac{\rho}{H^2}, \ \frac{d}{d\tau} \sim \frac{1}{H}\frac{d}{dt}, \ldots \tag{11}$$

Detailed studies using this approach for general $f(R)$ theories can be found in [15, 16]. However, even in the simplest case of flat models in vacuum, numerical investigation of the system with

[1] We adopt the metric and curvature conventions of [13]. Here, $a(t)$ is the scale factor, an overdot denotes differentiation with respect to time t, and units have been chosen so that $c = 1 = 8\pi G$.

initial values $H(0) > 0$ shows that $H(t)$ exhibit damped oscillations, with almost zero minima. Therefore although permissible, the transformation (11) may induce fake singularities to the solutions. Moreover, as mentioned in [13], a drawback of this choice of variables is that it does not give a complete description of the evolution for bouncing or recollapsing models. If the Hubble parameter passes through zero the logarithmic time coordinate is ill-defined and the transformation (11) is singular.

In order to circumvent these difficulties, we introduce one more degree of freedom by using (10), thus augmenting the dimension of the dynamical system. The state $(R, \dot{R}, x, \rho, H)$ of the system lies on the hypersurface of $\mathbb{R}^5$ defined by the constraint (6). The presence of $(R, \dot{R})$ in the state vector reflects the fact that there are additional degrees of freedom in HOG theories than in GR (cf. the remark after equation (3)).

We define

$$x_1 = R, \quad x_2 = \dot{R}$$

and use (6) to eliminate ρ, so that equations (10), (7) and (8) constitute a four-dimensional system. The parameter β may be used to define dimensionless variables by rescaling

$$x_1 \to \frac{2}{\beta}x_1, \; x_2 \to \sqrt{\frac{2}{3\beta^3}}\,x_2, \; H \to \frac{H}{\sqrt{6\beta}}, \; x \to \frac{x}{\sqrt{6\beta}}, \; t \to \sqrt{6\beta}\,t,$$

and our system becomes

$$\begin{aligned}
\dot{x}_1 &= x_2, \\
\dot{x}_2 &= -x_1 + (1 - 3\gamma)\,x_2 H + \frac{(4 - 3\gamma)}{4} Z, \\
\dot{x} &= -xH, \\
\dot{H} &= 2x_1 - 2H^2 - kx^2,
\end{aligned} \tag{12}$$

with

$$Z = \left[H^2 + kx^2 - 4x_1^2 + 4x_1\left(H^2 + kx^2\right)\right].$$

Remark. The system (12) is not an arbitrary "free" four-dimensional system. In view of (6) the initial conditions have to satisfy the condition

$$H^2 + kx^2 + 4x_1\left(H^2 + kx^2\right) + 4Hx_2 - 4x_1^2 \geq 0. \tag{13}$$

With a little manipulation of the equations (12) it can be shown that, once we start with initial conditions satisfying (13) at time t_0, the solutions of the system satisfy this inequality for all $t > t_0$. Thus the field equations share the general property of the Einstein equations, namely that the subsequent evolution of the system is such that the solutions respect the constraint.

In the flat, $(k = 0)$, case the dimension of the dynamical system is reduced by one, since the evolution equation for x decouples from the remaining equations. For an analysis of flat models see [17]. We turn our attention to closed models.

3. Positively curved models

In this section we consider an initially expanding closed universe described by the full four-dimensional system (12) with $k = 1$. There are two equilibria:

$\mathcal{M}: x_1 = 0, x_2 = 0, x = 0, H = 0$. This corresponds to the limiting state of an almost empty, slowly varying universe with $H \to 0$, while the scale factor goes to infinity. The point $\mathcal{M}$ which resembles to the Minkowski solution, is located at the boundary of the phase space.

Journal of Physics: Conference Series **189** (2009) 012025 doi:10.1088/1742-6596/189/1/012025

$\mathcal{S}: H=0, x_2=0, x=\bar{x}\equiv\sqrt{\frac{2-3\gamma}{3\gamma-4}}, x_1=\bar{x}_1\equiv\bar{x}^2/2$, with $\frac{2}{3}<\gamma<\frac{4}{3}$. It is a static solution and the eigenvalues of the Jacobian matrix at $\mathcal{S}$ are

$$\pm\frac{8-9\gamma\pm\sqrt{3\gamma\left(36\gamma^2-69\gamma+32\right)}}{2\left(3\gamma-4\right)}.$$

The real parts of the eigenvalues are nonzero for almost all[2] permissible values of γ, and we conclude that the local stable and unstable manifolds through $\mathcal{S}$ are both two-dimensional. The point $\mathcal{S}$ corresponds to the Einstein static universe, where the effective cosmological constant is provided by the curvature equilibrium $\bar{x}_1$. To see this, it is sufficient to write equation (6) in the original variables at the equilibrium point as

$$k\bar{x}^2+\frac{\Lambda}{3}=\frac{1}{3}\bar{\rho},$$

with $\Lambda>0$. The cosmological constant depends on both the parameters β and γ, for example, for $\gamma=1$, $\Lambda=\left(12\beta-1\right)^2/2\beta$. Static solutions have little interest as cosmological models and we turn our attention to the other equilibrium, $\mathcal{M}$.

The point $\mathcal{M}$ is a nonhyperbolic equilibrium, since the eigenvalues of the Jacobian matrix at the origin are $\pm i, 0, 0$ and therefore we cannot infer about stability using the linearisation theorem. For nonhyperbolic equilibrium points there exist no general methods for studying their stability. The normal form of the system may provide some information about the behaviour of the solutions near the equilibrium. The normal form theory consists in a nonlinear coordinate transformation that allows to simplify the nonlinear part of the system (cf. [18] for a brief introduction). This task will be accomplished in three steps in some detail for the convenience of readers with no previous knowledge of the method.

1. Let P be the matrix formed from the eigenvectors which transforms the linear part of the vector field into Jordan canonical form. We write (12) in vector notation (with $\mathbf{x}=(x_1,x_2,x,H)$) as

$$\dot{\mathbf{x}}=A\mathbf{x}+\mathbf{F}\left(\mathbf{x}\right), \tag{14}$$

where A is the linear part of the vector field and $\mathbf{F}\left(\mathbf{0}\right)=\mathbf{0}$.

2. Using the matrix P, we define new variables, $(y_1,y_2,x,y)\equiv\mathbf{y}$, by the equations

$$y_1=x_1,\quad y_2=-x_2,\quad x=x,\quad y=H+2x_2, \tag{15}$$

or in vector notation $\mathbf{y}=P\mathbf{x}$, so that (14) becomes

$$\dot{\mathbf{y}}=P^{-1}AP\mathbf{y}+P^{-1}\mathbf{F}\left(P\mathbf{y}\right).$$

Denoting the canonical form of A by B we finally obtain the system

$$\dot{\mathbf{y}}=B\mathbf{y}+\mathbf{f}\left(\mathbf{y}\right), \tag{16}$$

[2] More precisely, the real parts of the eigenvalues are zero for

$$\frac{2}{3}<\gamma\leq\frac{23-\sqrt{17}}{24}\approx 0.79,$$

and nonzero in the rest of the interval $\left(\frac{2}{3},\frac{4}{3}\right)$.

where $\mathbf{f}(\mathbf{y}) := P^{-1}\mathbf{F}(P\mathbf{y})$. In components system (16) is

$$\begin{bmatrix} \dot{y}_1 \\ \dot{y}_2 \\ \dot{x} \\ \dot{y} \end{bmatrix} = \begin{bmatrix} 0 & -1 & 0 & 0 \\ 1 & 0 & 0 & 0 \\ 0 & 0 & 0 & 0 \\ 0 & 0 & 0 & 0 \end{bmatrix} \begin{bmatrix} y_1 \\ y_2 \\ x \\ y \end{bmatrix} +$$

$$\begin{bmatrix} 0 \\ (4-3\gamma)\,y_1^2 - (3\gamma+2)\,y_2^2 - \frac{4-3\gamma}{4}\left(kx^2+y^2\right) - 3yy_2 - (4-3\gamma)\,Z_1 \\ -2xy_2 - xy \\ -2\,(4-3\gamma)\,y_1^2 + 2\,(3\gamma-2)\,y_2^2 - \frac{3\gamma-2}{2}kx^2 - \frac{3\gamma}{2}y^2 - 2yy_2 + 2\,(4-3\gamma)\,Z_1 \end{bmatrix},$$

with

$$Z_1 = y_1\left[kx^2 + (2y_2+y)^2\right].$$

3. Under the non-linear change of variables

$$\begin{aligned} y_1 &\to y_1 + 3\gamma y_1^2 + (3\gamma-2)\,y_2^2 + \frac{4-3\gamma}{4}\left(x^2+y^2\right) + \frac{3}{4}y_2 y, \\ y_2 &\to y_2 + 4y_1y_2 + \frac{3}{4}y_1 y, \\ x &\to x + 2y_1 x, \\ y &\to y - 2y_1y_2 + 2y_1 y, \end{aligned} \tag{17}$$

and keeping only terms up to second order, the system transforms to

$$\begin{aligned} \dot{y}_1 &= -y_2 - \frac{3}{2}y_1 y + \mathcal{O}(3), \\ \dot{y}_2 &= y_1 - \frac{3}{2}y_2 y + \mathcal{O}(3), \\ \dot{x} &= -xy + \mathcal{O}(3), \\ \dot{y} &= 6\,(\gamma-1)\left(y_1^2+y_2^2\right) - \frac{3\gamma-2}{2}x^2 - \frac{3\gamma}{2}y^2 + \mathcal{O}(3). \end{aligned} \tag{18}$$

Defining cylindrical coordinates $(y_1 = r\cos\theta, y_2 = r\sin\theta, x = x, y = y)$, we obtain

$$\begin{aligned} \dot{r} &= -\frac{3}{2}ry + \mathcal{O}(3), \\ \dot{\theta} &= 1 + \mathcal{O}(2), \\ \dot{x} &= -xy + \mathcal{O}(3), \\ \dot{y} &= 6\,(\gamma-1)\,r^2 - \frac{3\gamma-2}{2}x^2 - \frac{3\gamma}{2}y^2 + \mathcal{O}(3). \end{aligned} \tag{19}$$

We may continue to simplify the third order terms but, since we are interested on the behaviour of the solutions only near the origin, we truncate the vector field at $\mathcal{O}(2)$. We note that the θ dependence of the vector field has been eliminated, so that we can study the system in the (r, x, y) space. The second equation of (19) implies that the trajectory in the $y_1 - y_2$ plane spirals with angular velocity 1. We write the first and third of (19) as a differential equation

$$\frac{dr}{dx} = \frac{3}{2}\frac{r}{x},$$

which has the general solution

$$r = Ax^{3/2}, \quad A > 0. \tag{20}$$

Substitution of (20) into the fourth equation of (19) yields the projection of the fourth-dimensional system on the $x - y$ plane, namely

$$\dot{x} = -xy,$$
$$\dot{y} = b(\gamma - 1)x^3 - \frac{3\gamma - 2}{2}x^2 - \frac{3\gamma}{2}y^2, \quad b > 0. \tag{21}$$

Some general properties of the solutions of (21) follow by inspection. By standard arguments all trajectories are symmetric with respect to the x axis. Note that the new x defined by (17) remains non-negative for initial values $y_1(0)$ sufficiently small and, since the line $x = 0$ is invariant, any trajectory starting at the half plane $x > 0$ remains there for all $t > 0$. System (21) has two equilibrium points, the origin $(0, 0)$ and $(x_*, 0)$, where

$$x_* = \frac{3\gamma - 2}{2b(\gamma - 1)},$$

and therefore $x_* > 0$ for $\gamma < 2/3$ or $\gamma > 1$. The origin is a nonhyperbolic equilibrium point. Computation of the Jacobian matrix at the equilibrium $(x_*, 0)$ shows that for the linearised system, this point is a saddle for $0 \leq \gamma < 2/3$, and a center for $1 < \gamma \leq 2$. For $\gamma > 1$, it is easy to see that x is decreasing in the first quadrant and y is decreasing along the orbits in the strip $0 < x < x_*$. On any orbit starting in the first quadrant with $x < x_*$, y becomes zero at some time and the trajectory crosses vertically the $x-$axis. Once the trajectory enters the second quadrant, x increases and y decreases. For $2/3 < \gamma \leq 1$, all trajectories starting in the first quadrant follow the same pattern.

System (21) has a first integral, *viz.*

$$\phi(x, y) = -\frac{2b}{3}x^{3(1-\gamma)} + x^{2-3\gamma} + \frac{y^2}{x^{3\gamma}}, \quad \gamma \neq 1,$$
$$\phi(x, y) = \frac{1}{x} + \frac{y^2}{x^3}, \quad \gamma = 1.$$

This can be seen by writing (21) as

$$\frac{dy}{dx} = \frac{bx^3 - \frac{3\gamma-2}{2}x^2 - \frac{3\gamma}{2}y^2}{-yx}.$$

Setting $y^2 = z$, we obtain a linear differential equation for z which is easily integrable. The level curves of ϕ are the trajectories of the system.

Proposition. *For the system (21) (i) for every* $\gamma \in (\frac{2}{3}, 2]$ *there are no solutions asymptotically approaching the origin (ii) for* $\gamma \in (\frac{2}{3}, 1]$ *there are no periodic solutions and (iii) for* $\gamma \in (1, 2]$ *there exist periodic solutions and the basin of every periodic trajectory is the set*

$$\left\{(x, y) \in \mathbb{R}^2 : y^2 + x^2 - \frac{2b}{3}x^3 < 0, x > 0\right\}.$$

Proof. The proof mimics that found in [8] for the simple case $\gamma = 4/3$. Let $\phi(x, y) = C$. We have

$$y^2 = x^2\left(-1 + \frac{2b}{3}x + Cx^{3\gamma-2}\right), \tag{22}$$

which implies that the function

$$f(x) = -1 + \frac{2b}{3}x + Cx^{3\gamma-2},$$

must be non-negative. We consider two cases. 1. $C \geq 0$. Then f is strictly increasing for $x \geq 0$ and $f(0) = -1$, thus f has a unique root $x_1 > 0$ depending on C. It follows that for $C \geq 0$ any orbit starting in the first quadrant satisfies

$$x \geq x_1(C) > 0,$$

i.e., there are no solutions approaching the axis $x = 0$. These solutions are not closed since they intersect the x−axis only once at $x_1(C)$. 2. $C < 0$. If $2/3 < \gamma \leq 1$, then f has again a unique root hence the trajectories follow the same pattern as in case 1. If $\gamma > 1$, then f has two zeros, say $x_1(C) < x_2(C)$. This means that $0 < x_1(C) < x < x_2(C)$, i.e., x is bounded, and by (22), so is y. Thus, an orbit of (21) starting in the first quadrant crosses the x−axis at $x_1(C)$ and re-enters in the first quadrant crossing the x−axis at $x_2(C)$ i.e. it is a closed curve and represents a periodic solution. The curve corresponding to $C = 0$ separates the phase space into two disjoint regions I and II. In region II, ($C < 0$), every trajectory corresponds to a periodic solution and we conclude that the basin of every periodic trajectory is the set $y^2 + x^2 - \frac{2b}{3}x^3 < 0$. □

Remark. The mere existence of closed orbits around the equilibrium point $(x_*, 0)$ could be inferred from the following theorem: If an equilibrium point **p** is a center for the linearised system and all trajectories are symmetric with respect to the x axis, then **p** is also a center for the nonlinear system (21) (cf. [18] Theorem 6, page 141). In the above proof we also determine the subset of the phase space which contains all periodic orbits.

Using all this information we may sketch the phase portrait of the system (Figure 1).

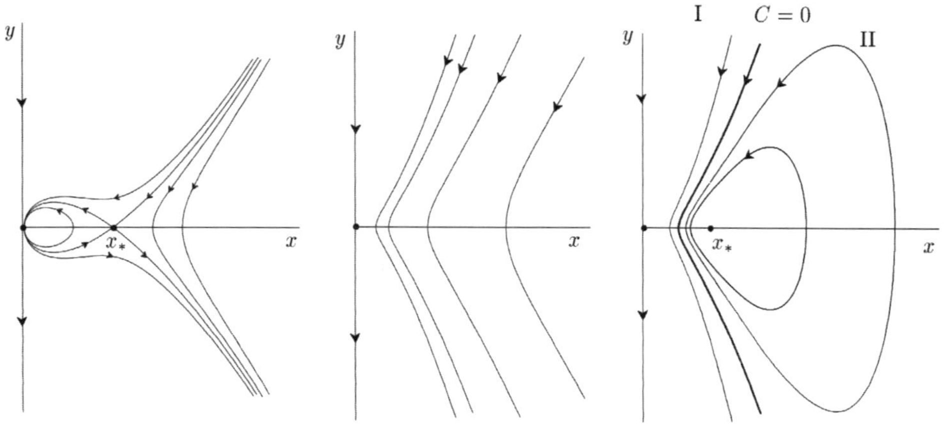

Figure 1. Phase portrait of (21) for $\gamma < \frac{2}{3}$, $\frac{2}{3} < \gamma < 1$ and $\gamma > 1$.

For $\gamma < 2/3$ the homoclinic curves to the origin as well as the curves approaching the origin indicate that an initially expanding closed universe may avoid recollapse. This result is also valid in GR when matter fields violate the strong energy condition [20]. For $\gamma \in (\frac{2}{3}, 1]$ every solution curve becomes unbounded and we may interpret this as an indication that this universe recollapses.

Journal of Physics: Conference Series **189** (2009) 012025 doi:10.1088/1742-6596/189/1/012025

The range $(1,2]$ for γ is the more interesting because of the periodic orbits in region II. The phase portrait in Figure 1 may lead to the conclusion that the periodic orbits are far from the origin. However, the position of the cycles in the phase space depends on the constants b in (21) and C in (22) and therefore, for suitable values of b and C, there exist periodic orbits arbitrary close to the origin. Note that the periodic solutions of (21) induce periodicity to the full four-dimensional system (19) or (18). In fact, if $x(t)$ and $y(t)$ are periodic solutions, then (20) implies that the solutions

$$y_1(t) = r(t)\cos(t+\theta_0), \quad y_2(t) = r(t)\sin(t+\theta_0),$$

oscillate in the $y_1 - y_2$ plane with a periodic amplitude $r(t)$. Note also that the periodic motion in the $x - y$ plane is independent from the rotation of r in the $y_1 - y_2$ plane.

Obviously we cannot assign a physical meaning to the new variables (y_1, y_2, x, y), since the transformations (15) and (17) have "mixed" the original variables of (12) in a nontrivial way. However, the periodic character of the solutions of (18) whatever the physical meaning of the variables be, has the following interpretation. *Close to the equilibrium of the original system (12), there exist periodic solutions.* This result is in agreement with [21] where it is shown that for Lagrangians $R + \beta R^m$, bounces of closed models are allowed for every integer value of m.

Remark. As mentioned in Section 2 numerical experiments show that the solutions of the system have oscillatory behavior. This property is intuitively evident by looking at the harmonic oscillator, equation (10). Using the rescalings of the dynamical variables along with the scaling $\rho \to \beta\rho$, equation (10) becomes

$$\ddot{R} + 3H\dot{R} + R = \frac{(4-3\gamma)}{2}\rho,$$

which is the equation of a forced, damped harmonic oscillator with unit angular frequency. Qualitative arguments supported by numerical solutions indicate that oscillatory motion, possibly slightly damped, is essentially independent of k and γ and becomes the late time behaviour. This is revealed in the normal form of the system (19), where the θ equation shows oscillatory motion with unit angular frequency. We emphasize again that this motion is independent from the periodic motion in the $x - y$ plane. The normal form analysis reveals both kinds of oscillatory behaviour and distinguishes the models which actually exhibit undamped oscillations at late time.

4. Discussion

We analysed the qualitative behaviour of flat and positively curved FRW models filled with ordinary matter described by a perfect fluid in the Jordan frame of the $R + \beta R^2$ theory. We have shown that initially expanding flat models close to the equilibrium $H = 0, R = 0, \rho = 0$ are ever expanding and asymptotically approach flat empty spacetime. Therefore, the late time behaviour of flat models is a common property of quadratic gravity and GR. Closed models can avoid recollapse for $\gamma < 2/3$, but not for γ in the range $\left[\frac{2}{3}, 1\right]$, thereby also behaving similarly to the corresponding general-relativistic cosmologies.

The interesting feature is the existence of periodic solutions near the origin for $\gamma > 1$. This is not revealed in the Einstein frame (see [19]), possibly because in that investigation, the scalar field related to the conformal transformation is not coupled to matter, i.e., the matter Lagrangian is added after performing the conformal transformation. Were the two frames physically equivalent then, a naive physical explanation of the cyclic behavior could rely on the scalar field which behaves like a "cosmological constant" in the high curvature limit. The perfect fluid which is also present dominates in the low curvature regime allowing for a recollapse, but then the effective cosmological constant induces a bounce in the high-curvature regime.

The periodic solutions imply that an initially expanding closed universe can avoid recollapse through an infinite sequence of successive expansions and contractions. The oscillatory open model proposed by Steinhardt and Turok [22] has renewed interest in cyclic universes. However, observations do not exclude Ω_{total} to be slightly larger than one [23]. Oscillatory closed models were considered in the context of Loop Quantum Cosmology [24] and oscillatory (but not periodic) solutions in GR were found in [25] for closed models containing radiation and dust or scalar field. In Figure 1 the basin of all periodic trajectories of the (21) is the domain on the right of the $C = 0$ curve and since it is an open subset of the phase space, we conclude that there is enough room in the set of initial data of (12) which lead to an oscillating scale factor. The $R + \beta R^2$ theory has offered a successful inflationary model, but whether it is capable to explain the acceleration of the universe is an open question that needs to be studied in more detail. In particular, the observed slow acceleration must be compared to the periods of the closed curves of (21).

The normal form theory is a powerful method for determining the qualitative behaviour of a dynamical system near a nonhyperbolic equilibrium point, but does not give any information about the structure of the solutions far from this equilibrium. Our results are based on an analysis of the behaviour of the dynamical system (12) only near the equilibrium solutions. The geometry of the trajectories of a four-dimensional dynamical system may be quite complicated, e.g. strange attractors may be present. For the system (12) the whole picture may come in view only from the investigation of the global structure of the solutions. The study of this question is an interesting challenge for mathematical relativity.

Acknowledgments

I thank Spiros Cotsakis and Alan Rendall for useful comments.

References

[1] Gasperini M and Veneziano G 2003 *Phys. Rept.* **373** 1
[2] Starobinsky A 1980 *Phys. Lett.* B **91** 99
[3] Carroll S M, Duvvuri V, Trodden M and Turner M S 2004 *Phys. Rev.* D **70** 043528
[4] Capozziello S, Carloni S and Troisi A 2003 *Preprint* astro-ph/0303041
[5] Carroll S M, De Felice A, Duvvuri V, Easson D A, Trodden M and Turner M S 2005 *Phys. Rev.* D **71** 063513
[6] Capozziello S, Cardone V F and Troisi A 2005 *Phys. Rev.* D **71** 043503
[7] Sokołowski L M 2007 *Class. Quantum Grav.* **24** 3391
[8] Miritzis J 2008 *Proceedings of the MG11 Meeting on General Relativity (Berlin)*, ed H Kleinert, R T Jantzen and R Ruffini (Singapore: World Scientific) p 2048 (*Preprint* gr-qc/0609025)
[9] Brans C H 1998 *Class. Quantum Grav.* **5** L197; Ferraris M, Francaviglia M and Magnano G 1990 *Class. Quantum Grav.* **7** 261; Cotsakis S 1993 *Phys. Rev.* D **47** 1437; Erratum 1994 *Phys. Rev.* D **49** 1145
[10] Magnano G and Sokołowski L M 1994 *Phys. Rev.* D **50** 5039; Faraoni V, Gunzig E and Nardone P 1999 *Fund. Cosmic Phys.* **20** 121; Faraoni V and Nadeau S 2007 *Phys. Rev.* D **75** 023501
[11] Barrow J D and Ottewill A 1983 *J. Phys.* A **16** 35
[12] Olmo G 2007 *Phys. Rev.* D **75** 023511
[13] Wainwright J and Ellis G F R 1997 *Dynamical Systems in Cosmology* (Cambridge University Press)
[14] Berkin A 1990 *Phys. Rev.* D **42** 1016
[15] Amendola L, Gannouj R, Polarski D and Tsujikawa S 2007 *Phys. Rev.* D **75** 083504
[16] Carloni S, Troisi A and Dunsby P K S 2007 (*Preprint* arXiv/0706.0452)
[17] Miritzis J 2008 *Gen. Rel. Grav.* DOI: 10.1007/s10714-008-0651-3 published online (*Preprint* arXiv:0708.1396)
[18] Perko L 2001 *Differential Equations and Dynamical Systems* (Springer-Verlag)
[19] Miritzis J 2005 *J. Math. Phys.* **46** 082502
[20] Barrow J D, Galloway G and Tipler F 1986 *Mon. Not. R. Astr. Soc.* **223** 835
[21] Carloni S, Dunsby P K S and Solomons D 2006 *Class. Quantum Grav.* **23** 1913
[22] Steinhardt P J and Turok N 2002 Science **296** 1436
[23] Jaffe A H *et al* 2001 *Phys. Rev. Lett.* **86** 3475
[24] Lidsey J E, Mulryne D J, Nunes N J and Tavakol R 2004 *Phys. Rev.* D **70** 063521
[25] Clifton T and Barrow J D 2007 *Phys. Rev.* D **75** 043515

New experimental technique for short-range gravity measurement

Kazufumi Ninomiya[1], Takashi Akiyama[1], Maki Hata[1], Yuki Ikeda[1], Hirokazu Kawamura[1,2], Keigo Narita[1], Naruya Ogawa[1], Toshiaki Sato[1], Etsuko Seitaibashi[1], Yuta Sekiguchi[1], Ryosuke Tsutsui[1], Kazumasa Yazawa[1] and Jiro Murata[1]

[1]Department of Physics, Rikkyo University, Nishi-Ikebukuro, Tokyo 171-8501, JAPAN
[2] RIKEN Nishina Center, 2-1 Hirosawa, Wako, Saitama 351-0198, JAPAN

E-mail:08la013a@rikkyo.ac.jp

Abstract. A new experimental technique for short range gravity measurements using pico-precision displacement sensor using digital image analysis is presented. We have developed a new experimental setup using torsion balance pendulum, aiming to test the Newton's inverse square law at below millimeter scale. Weak equivalence principle can also be tested using the same experimental setup. Detector techniques and the detail of the experimental setup are described.

1. Digital Image Analysis System

Present experimental techniques are originally developed for a high energy collider experiment, as a geometrical detector position monitoring system [1]. For PHENIX experiment at RHIC (Relativistic Heavy Ion Collider) accelerator at Brookhaven National Laboratory, position monitoring resolution of 25microns is required for muon tracking chambers. To satisfy the requirement, a micron precision optical alignment system (OASys), which comprised of fiber optic light source, convex lens and CCD camera, is developed. Position determination resolution of 10nanometers is achieved, which is significantly better than the required resolution of 25 microns.

Triggered by the significant achievement at PHENIX, the authors developed a dedicated position monitoring system using digital video cameras, aiming to obtain better precision than at PHENIX [2]. Schematic diagram of the digital video system is shown in Figure 1.1. A position data is obtained as a video image, captured by the digital video camera. The video data are sent to PC via IEEE1394, then captured and recorded as a movie file in avi file format.

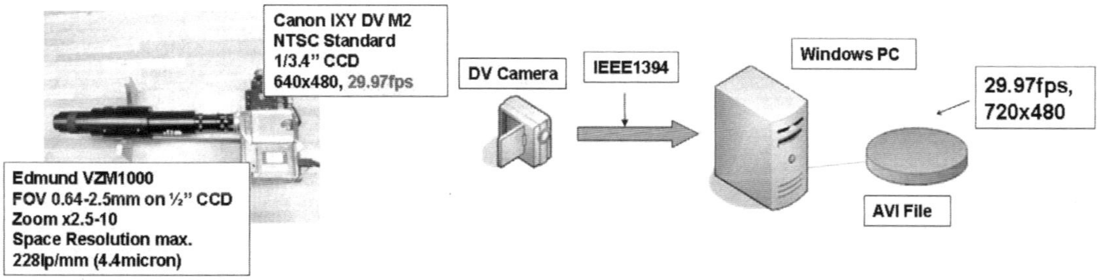

Figure 1.1: Schematic diagram of the digital video system.

In order to extract the position data from the movie files, the avi movie files are converted into sequential static image files in bmp format. The intensity profiles are statistically analyzed in a offline analysis tool developed for this purpose, yielding position information. By using a video lens, position determination resolution of 100 pico-meters is achieved.

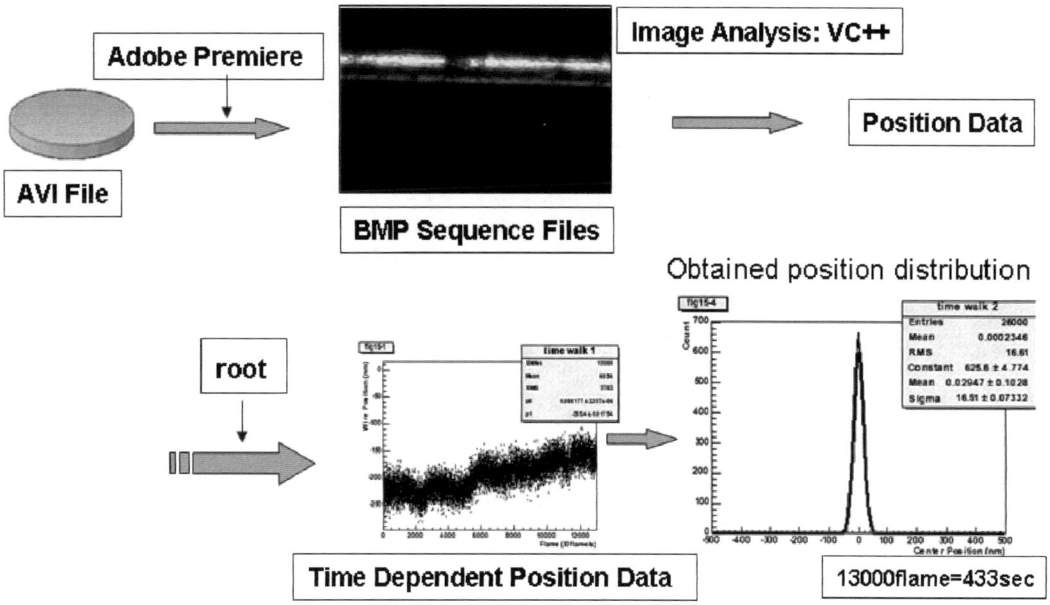

Figure 1.2: Data analysis chain from avi movie files to position data.

In Figure 1.2, data analysis chain is illustrated. At first, intensity histograms are obtained as sliced figures from the bmp static files which are generated by Adobe Premiere software. The object positions are determined as center of gravity positions. If we are interested in positions and angles of line-shaped objects, linear line fitting is applied for the sequence of the center of gravity points, as shown in Figure 1.3. Typical angular resolution of about 1.2×10^{-6} degrees are achieved.

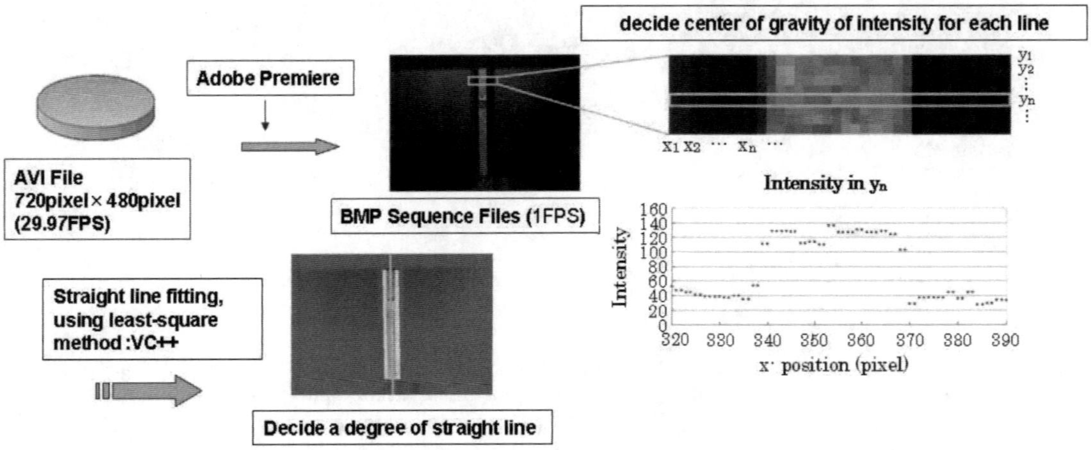

Figure 1.3: Image analysis for determining position and angle of a line-shaped object.

2. Principle of Short Range Gravity Experiment

Utilizing the surprisingly precise profile of the video position monitoring system, we have started to perform a series of gravity experiment at short range scale, where precision displacement measurement is required in order to extract the expected tiny gravity signals. To achieve the measurement, we employ a torsion balance pendulum, which is shown in Figure 2.1.

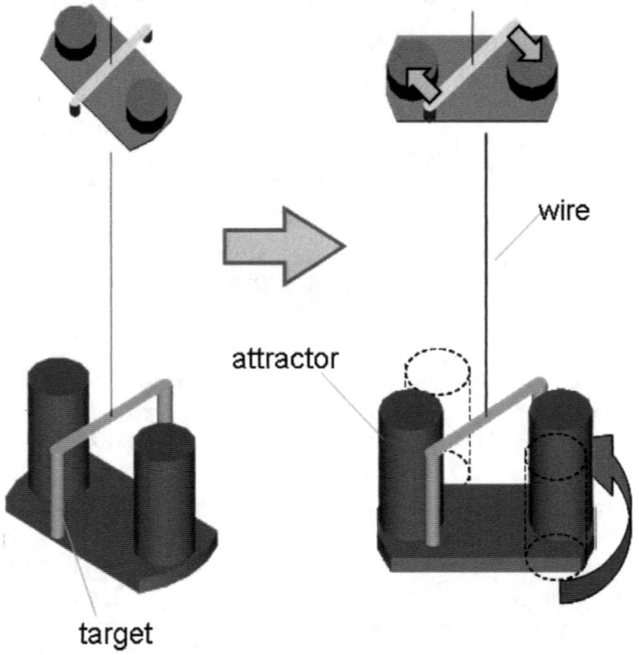

Figure 2.1: Principle of the gravity experiment using torsion balance pendulum.

Principle of the measurement is shown in Figure 2.1. Angular displacements between before and after moving the attractor mass positions are going to be measured as the gravity signal. It is because the balanced position of the torsion balance bar is expected at a position where spring force of the twisting wire and the gravitational force dragged by the attractor mass are getting equals. Therefore, gravitational force can be determined by measuring the displacement of the balanced positions. As a first order approximation, Hooke's law can be assumed;

$$N = \kappa \cdot \Delta\theta \qquad (2.1)$$

Here, N is the torque, κ is the spring constant and $\Delta\theta$ is the angular displacement. Considering that the source of the torque is gravitational force F, (2.1) can be expressed as followings using arm length of the torsion balance bar L and target-attractor distance r as;

$$\kappa \cdot \Delta\theta = 2 \times G \frac{m_{target} m_{attractor}}{r^2} \times L \qquad (2.2)$$

, if we simplify the objects as point mass. From (2.2), we can determine the gravitational force by measuring the angular displacement $\Delta\theta$. In order to extract the gravitational force, spring constant κ needs to be determined. We can experimentally determine the spring constant κ by measuring a periodic oscillation of the torsion balance bar. As shown in Figure 2.2, if we measured the oscillation period T, κ can be determined as $\kappa = 4\pi \frac{I}{T^2}$ using inertia moment I of the torsion balance bar.

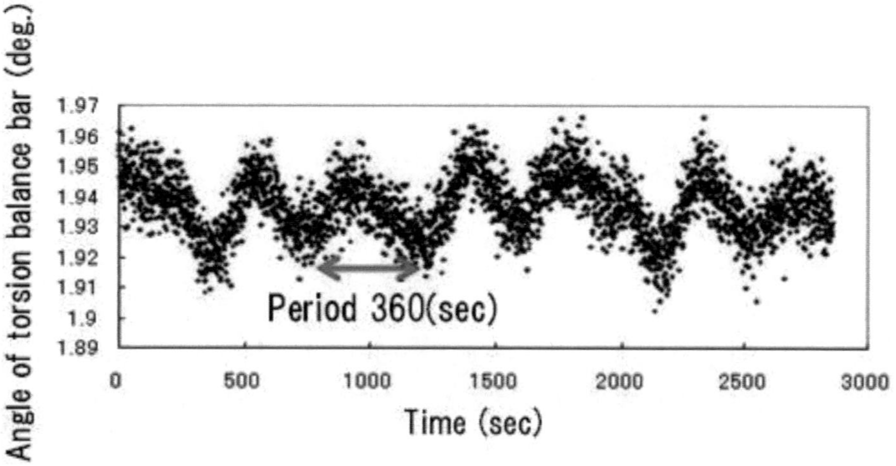

Figure 2.2: Periodic oscillation of the torsion balance bar. The spring constant κ can be determined from the oscillation period T

If we use Tungsten 30microns wire with 80cm length, typical value of the obtained spring constant is

$$\kappa = 60nNm. \qquad (2.3)$$

Typical size of the gravitational force in the first generation setup, Newton-I, is about $F = 1nN$, therefore, angular displacement is about $\Delta\theta = 2 \times 1nN \times 5cm / 60nNm \sim 0.1\deg$.

3. Newton-I Experiment

To perform the measurement, we have built the Newton-I experimental setup. Detailed structure of the Newton-I is illustrated in Figure 3.2. All the components are electrically connected, and made of non-magnetic metals (brass, aluminum, copper and lead). In order to shield a Coulomb force, metal electric shield is set covering the attractors. Position dependence can be examined by rotating the attractor table around the center for the test of inverse square law measurement. Composition dependence of gravitational constant can be measured by replacing the attractor material for the test of weak equivalence principle.

Journal of Physics: Conference Series **189** (2009) 012026 doi:10.1088/1742-6596/189/1/012026

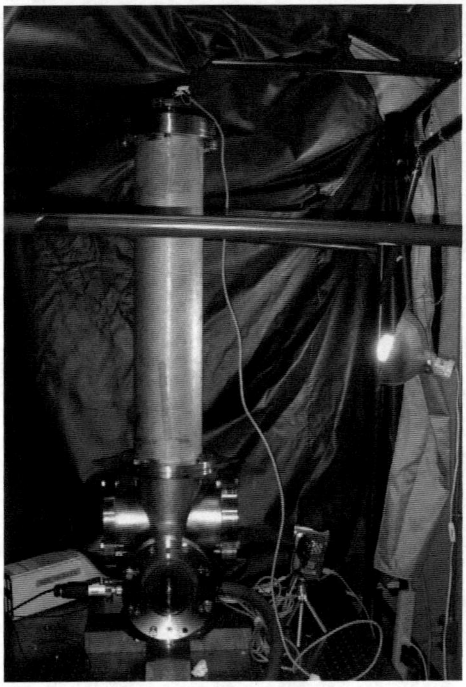

Figure 3.1: Picture of Newton-I. Torsion balance bar is set inside the vacuum chamber.

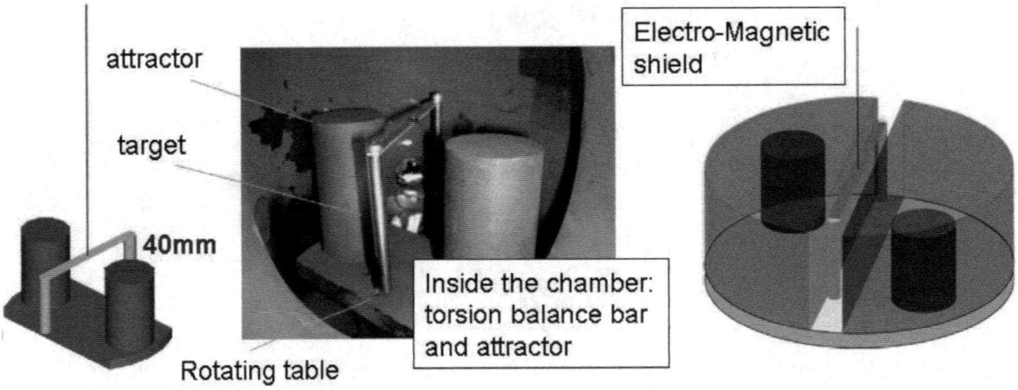

Figure 3.2: Structure of Newton-I setup. Electric shield is set covering the attractors.

4. New Device: Newton-II

As described in the separate contribution [3], Newton-I experiment succeeded to confirm Newton's law at centimeter scale. Especially, we succeeded to set most precise limit on the baryon number coupling force at shortest range scale, which is obtained from measurement of the composition dependence of gravitational constant. In Newton-I, largest error source comes from attractor replacement. In order to suppress the systematic error caused by attractor exchanging, next generation device, Newton-II, is developed. Newton-II is designed to be able to move the attractors rotating outside the torsion balance bar position, without changing the attractor materials by opening the vacuum chamber. As shown in Figure 4.1, we can measure gravity between different materials just by rotating the attractor turning table, without changing the setup.

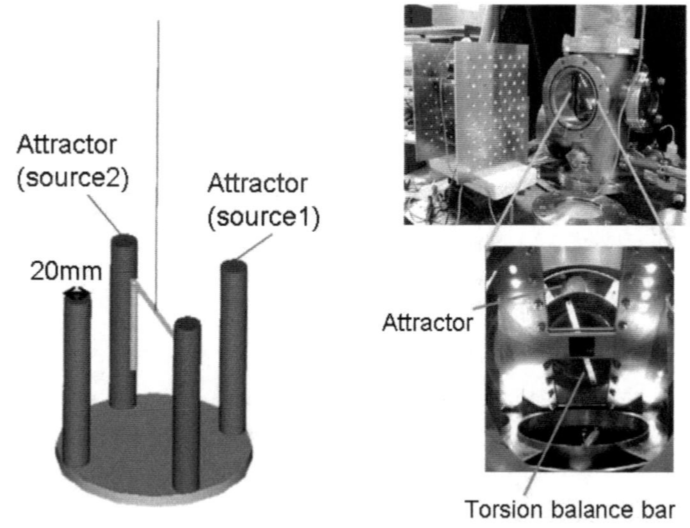

Figure 4.1: Structure of Newton-II. Electric shield set covering the attractors are not drawn.

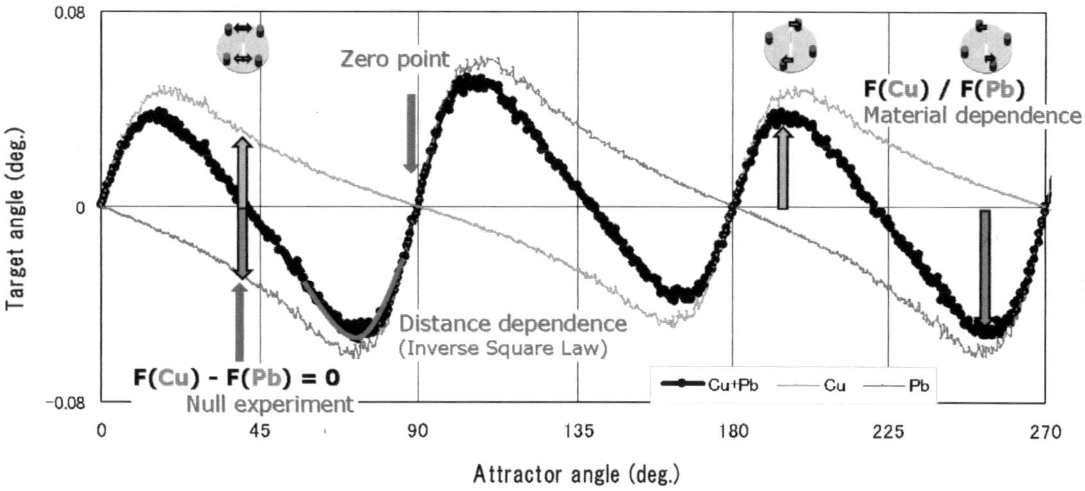

Figure 4.2: Expected signal from Newton-II.

Signal estimation using numerical calculation is performed, as shown in Figure 4.2. Position dependence and the material dependence can be measured at the same time without suffering from the systematic effects. Relative strength between different attractor materials provides precision information about the composition dependence of the gravitational constant. At the same time, curvature around the peak gives us the position dependence. In Newton-II, we will be able to test the weak equivalence principle in the shortest range, and the inverse square law in shorter range than Newton-I experiment.

5. Summary

A new experimental device Newton-I, is developed using digital video displacement sensing techniques with torsion balance pendulum. In order to suppress the systematic error, next generation device Newton-II is also developed, which aims to provide most precise test of weak equivalence principle at shortest range scale. In addition, third generation device, Newton-III is being developed using completely different measuring principle. Data from Newton-II and Newton-III will be available soon.

Acknowledgements

Part of the present works are performed as undergraduate student experiments. The author thanks to Y. Miyano, M. Takahashi, T. Tsuneno, T. Amanuma, S. Danbara, T. Iino, S. Mizuno, Y. Araki, T. Ohmori, Y.Sakurai , S. Yamaoka, H. Ooishi for their vital efforts on this work. This work is supported by Grant-in-And for Exploratory Research (18654048, 20654024).

References

[1] J. Murata *et. al.*, Nucl. Instrum. Meth. A 500(2003) 309-317.
[2] J. Murata, "Pico-Precision Displacement Sensor using Digital Image Analysis", IEEE Nuclear Science Symposium Conference Record 675 (2005).
[3] Maki Hata *et. al.*, "Recent Results on Short Range Gravity Experiment", in this volume

A new anisotropic solution for ellipsoidal spaces

T. Papakostas

TEI of Crete, Branch of Chania, 3 Romanou Street, Chalepa , Chania 731-33, Crete, Greece

E-mail:taxiar@chania.teicrete.gr

Abstract. We use the notion of ellipsoidal spaces in General Relativity and obtain the corresponding Einstein's equations in the case of a perfect fluid with heat flux. We present the integration of the resulting equations and two families of solutions.

1. Introduction

The problem of finding an interior solution in the case of stationary and axially symmetric spaces of General Relativity is still open. The importance of this problem is obvious; it gives an alternative model for a rotating star other than the black hole source of the Kerr metric.

There are very few exact solutions describing the above mentioned situation, which all exhibit unsatisfactory features: The Wahlquist solutions [1] have an unphysical equation of state and a bounding surface prolate rather than oblate, in [2] and [3] the energy conditions are not satisfied and in [4] it is not possible to isolate the hydrostatic pressure from the other stresses in order to define the bounding surface of zero pressure.

In this paper we consider a special case of the stationary and axially symmetric spaces, the ellipsoidal spaces. The notion of an ellipsoid of revolution in curved spaces is given first by Krasinski [6] and presented in a geometric way by [7], [8]. In this context we restrict our study in the family [A] of Carter's spaces and we consider as energy-momentum tensor that of a perfect fluid with heat flux.

In Section 2 we present the Carter's family [A] of spaces and we introduce the notion of curved spaces admitting a foliation by ellipsoids of revolution. In Section 3 we present the energy-momentum space of a perfect fluid with heat flux and we deduce the corresponding Einstein's equations. Finally, in Section 4 we integrate the equations and we present the two families of solutions.

2. Carter's family [A] of solutions and ellipsoidal spaces

We suppose that the space admits an isometry Abelian group of two parameters invertible with non null surface of transitivity [9]. This implies that there is a local coordinate system in which the metric can be written as follows:

$$ds^2 = (Ldt + Mdz)^2 - (Ndt + Pdz)^2 - S^2dx^2 - R^2dy^2, \tag{2.1}$$

where *L, M, N, P, S, R* are real functions of *x* and *y*. The Killing vectors of the Abelian group are:

$\partial/\partial t$ and $\partial/\partial z$, t and z being the ignorable coordinates (time and the polar angle respectively). This form of the metric for the stationary and axially symmetric spaces has been used by many authors [10-12] and has been proved very effective for the calculations. The Carter's spaces are characterized by the fact that the Hamilton-Jacobi (HJ) for the geodesics is solvable by separation of variables. Then the metric (2.1) reduces to the following expression:

$$ds^2 = (x^2+y^2)\left\{\frac{E^2(y)}{(x^2+y^2)^2}(dt-x^2dz)^2 - \frac{H^2(x)}{(x^2+y^2)^2}(dt+y^2dz)^2 - \frac{x^2dx^2}{F^2(x)} - \frac{y^2dy^2}{G^2(y)}\right\}. \tag{2.2}$$

In the case of vacuum Einstein's equations, we obtain the following expressions for the metric (2.2):

$$G^2 = y^2E^2, \qquad F^2 = x^2H^2, \qquad E^2 = \frac{1}{2}a^2 + by + c, \qquad H^2 = -\frac{1}{2}ax^2 + dx + c, \tag{2.3}$$

where a, b, c, d are constants of integration. The Kerr metric is obtained if we set:

$$a=2\ , \ d=0\ , \ b=-2M\ , \ c=-a^2\ , \ y=r\ , \ x=a\cos\theta, \tag{2.4}$$

where M is the mass, a is the angular momentum per unit mass and r, θ are the Boyer-Lindquist coordinates.

The Einstein's equations in the presence of the perfect fluid reduce to two equations in the Newman-Penrose formalism:

$$\Phi_{00}\Phi_{02} = \Phi_{01}{}^2, \quad \text{and} \quad 2\Phi_{11} = \Phi_{00} + \Phi_{02}. \tag{2.5}$$

These equations imply that the energy-momentum tensor admits one simple eigenvalue (the energy-mass density) and triple eigenvalues (the isotropic hydrostatic pressure). The solution of the first of these equations permits to define G^2 and H^2 as follows:

$$W(y) = \frac{G^2(y)}{E^2(y)} = k_4y^4 + k_2y^2 - k_0, \qquad Z(x) = \frac{F^2(x)}{H^2(x)} = -k_4x^4 + k_2x^2 + k_0, \tag{2.6}$$

where k_0, k_2, k_4, are constants of integration. The solution of the second of equations (2.5) permits to obtain a generalization of the Wahlquist solution [5].

The Carter's family [A] of solutions admits a foliation of ellipsoids of revolution if the quotient space of the commoving observers is that of an ellipsoid of revolution in a 3 dimensional curved space [6-8,13]. The vector field of the comoving observers is a linear combination of the Killing fields of Carter's spaces. The existence of the above-mentioned foliation implies that:

$$F^2(x) = x^2(a^2 - x^2), \quad \text{and} \quad G^2(y) = f^2(y)(a^2 + y^2)y^2. \tag{2.7}$$

These expressions of F^2 and G^2 are the sufficient conditions for the existence of a family of ellipsoids in Carter's family [A] of solutions.

3. Perfect fluid with heat flux

The energy-momentum tensor of a perfect fluid with heat flux is given by the following expression:

$$T_{ij} = (\varepsilon + p)u_i u_j - p g_{ij} + q_i u_j + q_j u_i , \tag{3.1}$$

with

$$u_i = \frac{\sqrt{2}}{2}\{\Pi_1(n_i + l_i) - \Pi_2(m_i + \bar{m}_i)\}, \qquad q_i = \frac{\sqrt{2}}{2}[q_1 n_i + q_2 l_i - q_3 \bar{m}_i - q_4 m_i]. \tag{3.2}$$

The fluid moves along the integral curves of u_i the heat flow is described by the space-like vector field q_i which is orthogonal to u_i**:**

$$u_i u^i = 1, \qquad q_i u^i = 0. \tag{3.3}$$

The energy-mass density is ε and the hydrostatic pressure p. In the (NP) formalism the Einstein's equations are:

$$\begin{aligned}
2\Phi_{11} + 6\Lambda &= \frac{1}{2}(\varepsilon + p)\Pi_1^2 + \frac{1}{2}(q_1 + q_2)\Pi_1 - p, \\
2\Phi_{11} - 6\Lambda &= \frac{1}{2}(\varepsilon + p)\Pi_1^2 + \frac{1}{2}(q_3 + q_4)\Pi_2 + p, \\
-2\Phi_{01} &= -\frac{1}{2}(\varepsilon + p)\Pi_1\Pi_2 - \frac{1}{2}(q_1\Pi_2 + q_3\Pi_1), \\
-2\Phi_{10} &= -\frac{1}{2}(\varepsilon + p)\Pi_1\Pi_2 - \frac{1}{2}(q_1\Pi_2 + q_4\Pi_1), \\
-2\Phi_{12} &= -\frac{1}{2}(\varepsilon + p)\Pi_1\Pi_2 - \frac{1}{2}(q_2\Pi_2 + q_4\Pi_1), \\
-2\Phi_{21} &= -\frac{1}{2}(\varepsilon + p)\Pi_1\Pi_2 - \frac{1}{2}(q_2\Pi_2 + q_4\Pi_1) \\
&(q_1 + q_2)\Pi_1 - (q_3 + q_4)\Pi_2 = 0.
\end{aligned} \tag{3.4}$$

In Carter's family [A] of solutions these equations are reduced to the following relations:

$$q_2 = q_1 , \; q_4 = q_3 , \; q_1\Pi_1 = q_3\Pi_2 , \tag{3.5}$$

$$\varepsilon = 3(\Phi_{00} - \Phi_{02}) + 6\Lambda, \tag{3.6}$$

$$p = (\Phi_{00} - \Phi_{02}) - 6\Lambda, \tag{3.7}$$

$$q_1 = \frac{4}{(\Pi_1^2 + \Pi_2^2)}\left[\Phi_{01}\Pi_2 - (\Phi_{00} - \Phi_{02})\Pi_1\Pi_2^2\right], \tag{3.8}$$

$$q_3 = \frac{4}{(\Pi_1^2 + \Pi_2^2)}\left[\Phi_{01}\Pi_1 - (\Phi_{00} - \Phi_{02})\Pi_1^2\Pi_2\right], \tag{3.9}$$

$$\Phi_{00}\Pi_2^2 - 2\Phi_{01}\Pi_1\Pi_2 + \Phi_{02}\Pi_1^2 = 0, \tag{3.10}$$

$$\Pi_1^2 - \Pi_2^2 = 1, \tag{3.11}$$

$$2\Phi_{11} - \Phi_{00} - \Phi_{02} = 0. \tag{3.12}$$

The relation (3.12) is a differential equation for the unknown functions E and H.

4. The integration of equation $2\Phi_{11} - \Phi_{00} - \Phi_{02} = 0$

We write explicitly this equation as:

$$\begin{aligned}
&2\left[x^3y(x^2+y^2)^2\right]WE_{yy}^2 + (x^2+y^2)\left[-2x^3(5y^2+x^2)W + x^3y(x^2+y^2)W_y\right]E_y^2 + \\
&+8x^3y^3(W+Z)E^2 \\
&-2\left[xy^3(x^2+y^2)^2\right]ZH_{xx}^2 - (x^2+y^2)\left[-2y^3(5x^2+y^2)Z + xy^3(x^2+y^2)Z_x\right]H_x^2 - \\
&-8x^3y^3(W+Z)H^2 = 0,
\end{aligned} \tag{4.1}$$

where W and Z are defined as follows:

$$W(y) = \frac{G^2(y)}{E^2(y)}, \quad \text{and} \quad Z(x) = \frac{F^2(x)}{H^2(x)}. \tag{4.2}$$

If we divide Equation (4.1) by x^3y^3 and differentiate five times with respect x and y we can easily show that:

$$W = kE^2 + k_4y^4 + k_3y^3 + k_2y^2 + k_1y + k_0, \tag{4.3}$$

$$Z = kH^2 + l_4x^4 + l_3x^3 + l_2x^2 + l_1x + l_0. \tag{4.4}$$

If we impose now relations (2.7) we have an ellipsoidal space and the solution of (4.1) is split into two families:

I. $k \neq 0$

$$H^2(x) = \frac{a^2 x^2}{l_0},$$

$$E^2(y) = \frac{l_0\sqrt{(l_0^2 + a^4 k)y^4 + (4a^2 k_0 l_0 + 4a^6 k)y^2 + 4a^4 k_0^2}}{2a^2 k l_0} - -\frac{(l_0^2 + a^4 k)y^2 + 2a^2 k_0 l_0}{2a^2 k l_0},$$

II. $k = 0$

$$H^2(x) = \frac{a^4 x^2 (a^2 - x^2)}{l_0 (a^2 - 2x^2)},$$

$$E^2(y) = \frac{1}{4l_0} b(16 l_0 y^4 - a^4 k_2 y^2 - a^4 k_0)^{\frac{1}{8}} \bullet \left(\frac{32 l_0 y^2 - a^2\sqrt{64 k_0 l_0 + a^4 k_2^2} - a^4 k_2}{32 l_0 y^2 + a^2\sqrt{64 k_0 l_0 + a^4 k_2^2} - a^4 k_2}\right)^{\frac{16a^2 l_0 + a^4 k_2}{8a^2\sqrt{64 k_0 l_0 + a^4 k_2^2}}} - \frac{a^4}{4l_0}.$$

The resulting spaces are of type I in the Petrov classification and we intend to make a complete study of them. Relations (3.5) - (3.11) will permit to calculate the heat flux vector, the mass-energy density, the hydrostatic pressure and try to define the surface of zero pressure in order to check if these spaces can be matched to the Kerr metric.

References

[1] Wahlquist H 1968 Phys Rev. **172, 5** 1291.
[2] Herrera L and Jimenez L 1982 J.Math.Phys. **23(12)** 2339.
[3] Gurses M and Gursey F 1979 J.Math.Phys **16(12)** 2385.
[4] Papakostas T 2001 Int.J.Mod.Phys. D **10** 869.
[5] Papakostas T 1998 Int.J.Mod.Phys.D **7** 927
[6] Krasinski A 1978 Ann.of Phys **112** 22.
[7] Racz I 1992 Class.Quantum Grav. **9** L93.
[8] Zsigrai J 2003 Class.Quantum Grav, **20** 2855.
[9] Carter B 1968 Commun.Math.Phys. **10** 280.
[10] Debever R 1971 Bull.Soc.Math.Belgique. **XXIII** 360.
[11] Carter B and McLenaghan R 1979 Phys.Rev. D **19** 1093.
[12] Papakostas T 1983 Bull.Sci.Acad.R.Belgique **LXIX** 495.
[13] Papakostas T 2005 arXiv gr-qc/0511078.

Matching of analytical and numerical solutions for neutron stars of arbitrary rotation

George Pappas

Section of Astrophysics, Astronomy, and Mechanics, Department of Physics, University of Athens, Panepistimiopolis Zografos GR15783, Athens, Greece

E-mail: gpappas@phys.uoa.gr

Abstract. We demonstrate the results of an attempt to match the two-soliton analytical solution with the numerically produced solutions of the Einstein field equations, that describe the spacetime exterior of rotating neutron stars, for arbitrary rotation. The matching procedure is performed by equating the first four multipole moments of the analytical solution to the multipole moments of the numerical one. We then argue that in order to check the effectiveness of the matching of the analytical with the numerical solution we should compare the metric components, the radius of the innermost stable circular orbit (R_{ISCO}), the rotation frequency $\Omega \equiv \frac{d\phi}{dt}$ and the epicyclic frequencies Ω_ρ, Ω_z. Finally we present some results of the comparison.

1. Introduction

Obtaining the geometry of the spacetime that surrounds a rotating distribution of mass in an exact form, is a problem of astrophysical relevance that remains unsolved. When modelling the astrophysical phenomena that we observe around rotating neutron stars, there are only a few choices we can make for the geometry we will use. We can either use one of the very well understood geometries of Schwarzschild or Kerr, which can only be used as very rough approximations, or we have to use a numerically obtained spacetime.

Although there are various groups (see [1], and for an extended list of numerical schemes see [2]), that have expertise in building relativistic models of astrophysical objects with adjustable physical characteristics and can construct the metric inside and outside such objects by solving numerically the full Einstein equations in stationary cases, having the metric of the spacetime in tabulated form with numerical values that correspond to the metric components at a selected grid is not always appropriate for studying the phenomena around the object.

Thus, an analytical solution that could describe the geometry around such an object would be useful. Fortunately, nowadays, there is a large variety of analytical solutions of vacuum Einstein equations, which could be used as candidate metrics to describe well the exterior space-time of axisymmetric astrophysical objects. Ernst [3] formulated the Einstein equations in the case of axisymmetric stationary space-times long time ago, while Manko et al. and Sibgatullin [4, 5, 6, 7, 8, 9] have used various analytical methods to produce such spacetimes parameterized by various parameters that have a different physical context depending on the type of each solution.

Such an analytical solution, appropriately matched to a neutron star, could be used to describe the stationary properties of the spacetime. That is, study the geodesics in the exterior of the neutron star. In other words, from the analytical solution, we could obtain bounds of motion for test particles, the location of the innermost stable circular orbit (ISCO), the rotation frequency of the circular orbits on the equatorial plane as well as the epicyclic frequencies around them.

These properties that are of astrophysical relevance and characteristic of a spacetime, will be used to compare the proposed analytical solution to the numerically produced solutions that describe the spacetime exterior to a neutron star.

The rest of the paper is organized as follows: In Sec. 2 the proposed analytical solution (two-soliton) is briefly presented and some of its properties analyzed. In Sec. 3 we present at first the matching criteria and then the comparison criteria and some results.

2. An Analytical axially symmetric solution: The Two Soliton solution.

The vacuum region of a stationary and axially symmetric spacetime can be described by the Papapetrou line element [10]

$$ds^2 = -f\left(dt - \omega d\phi\right)^2 + f^{-1}\left[e^{2\gamma}\left(d\rho^2 + dz^2\right) + \rho^2 d\phi^2\right], \tag{1}$$

were the functions f, ω and γ are functions of the Weyl-Papapetrou coordinates $(\rho,\ z)$. Using this metric, the Einstein field equations in vacuum reduce to the Ernst equation [3]

$$Re(\mathcal{E})\nabla^2\mathcal{E} = \nabla\mathcal{E}\cdot\nabla\mathcal{E}, \tag{2}$$

were the Ernst potential $\mathcal{E}$, is a complex function of the metric functions.

A general procedure for generating solutions of the Ernst equation was developed by Sibgatullin and Manko [8, 9, 6, 4]. The solution of the Ernst equation is produced from a choice of the Ernst potential on the axis of the form

$$\mathcal{E}(\rho = 0, z) = e(z) = \frac{P(z)}{R(z)}, \tag{3}$$

were the functions $P(z), R(z)$ are polynomials of order n with complex coefficients in general. In [6], one can find a detailed description of the procedure for generating the solutions using the parameters of the Ernst potential. It is shown in [6] that the metric functions are finally expressed in terms of some determinants, which depend on the parameters ξ_n, which are the $2n$ roots of the equation $e(z) + e(z)^* = 0$, the parameters e_n, f_n and β_n, which are defined from the expressions

$$e(z) = 1 + \sum_{k=1}^{n}\frac{e_k}{z-\beta_k},\quad f(z) = \sum_{k=1}^{n}\frac{f_k}{z-\beta_k},$$

and finally the expressions $R_n = \sqrt{\rho^2 + (z-\xi_n)^2}$, which are functions of the coordinates (ρ, z).

The vacuum two-soliton solution (proposed by Manko [4]) is a special case of the previous general axisymmetric solution that is obtained from the ansatz (see also [11])

$$e(z) = \frac{(z - M - ia)(z + ib) - k}{(z + M - ia)(z + ib) - k}, \tag{4}$$

where all the parameters are real while the parameters M, a are the mass and the reduced angular momentum $\frac{J}{M}$ respectively. The first five mass and angular momentum moments of the corresponding spacetime are:

$$
\begin{aligned}
M_0 &= M, \quad M_1 = 0, \quad M_2 = -(a^2 - k)M, \quad M_3 = 0, \\
M_4 &= \left(a^4 - (3a^2 - 2ab + b^2)k + k^2\right) M, \\
J_0 &= 0, \quad J_1 = aM, \quad J_2 = 0, \\
J_3 &= -(a^3 - (2a - b)k)M, \quad J_4 = 0.
\end{aligned} \tag{5}
$$

For the ansatz (4), the equation $e(z) + e(z)^* = 0$ corresponds to:

$$z^4 - (M^2 - a^2 - b^2 + 2k)z^2 + (k - ab)^2 - b^2M^2 = 0, \tag{6}$$

where the coefficients take real values. That means that the roots can either be real, or conjugate pairs. We can also see that the four roots of (6) are of the form

$$\xi_1 = -\xi_3 = \xi_+, \quad \xi_2 = -\xi_4 = \xi_-.$$

Defining the parameters $d = \sqrt{(k - ab)^2 - b^2M^2}$ and $\kappa_\pm = \sqrt{M^2 - a^2 - b^2 + 2k \pm 2d}$, we can express the roots of (6) as

$$\xi_\pm = \frac{1}{2}\left(\kappa_+ \pm \kappa_-\right). \tag{7}$$

Using these parameters we redefine R_n as

$$R_\pm = \sqrt{\rho^2 + (z \pm \xi_+)^2}, \; r_\pm = \sqrt{\rho^2 + (z \pm \xi_-)^2}. \tag{8}$$

Depending on the choice of the parameters used in (4), there can be obtained various types of solutions. For example, if one sets the value of k equal to zero, for arbitrary values of the parameter b, the two-soliton corresponds to the Kerr metric of mass M and reduced angular momentum a (if $|a| < M$) or the hyperextreme Kerr (if $|a| > M$). Further more, if one chooses a equal to zero, then the two-soliton corresponds to the Schwarzschild solution of mass M.

There is also another interesting special case for the two-soliton solution. If the parameter d is real and we have the conditions $M^2 - a^2 - b^2 + 2k + 2d > 0$ and $M^2 - a^2 - b^2 + 2k - 2d = 0$, then the parameter κ_+ is real and the parameter $\kappa_- = 0$. In that case there is a degeneracy and there are only two real roots $\xi_{1,2} = \pm\xi_\pm = \pm\frac{\kappa_+}{2}$. This solution corresponds to the Manko et al. solution [5, 7].

In brief, the two-soliton solution can produce a very rich family of analytical solutions including the classical solutions of Schwarzschild and Kerr as well as the Manko et al. solution [5, 7] that has been used in [12, 13] to approximately match the exterior of a rotating neutron star.

At this point we can point out the advantages of the two-soliton solution against the Manko et al. solution [5, 7]. If we assume a constant mass M, the two-soliton solution has a three dimensional parameter space (a, b, k). In that space the plane $k = 0$ includes all Kerr solutions of mass M while the b axis corresponds to the Schwarzschild solution of mass M.

As we have mentioned earlier, the Manko et al. solution [5, 7] is obtained if we impose the constrain $M^2 - a^2 - b^2 + 2k - 2d = 0$ to the parameters. This is analogous to having a two dimensional surface in the parameter space. There are two things that could be pointed out about that surface, which is drawn in Fig. 1. The first is that the axis of b (the blue line) that corresponds to the Schwarzschild solution doesn't cross the surface and the second is that the Kerr solutions with $M > |a|$, that lie in the parameter space on the plane $k = 0$ between the two red lines parallel to the b axis, do not intersect the surface either.

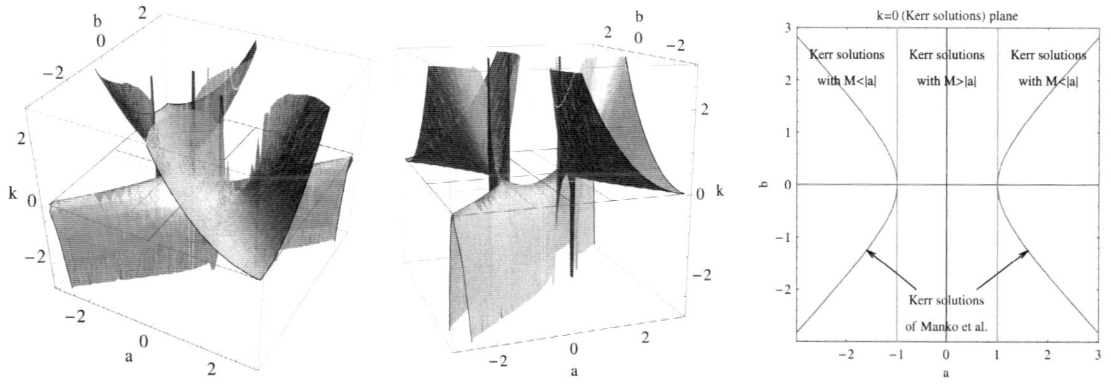

Figure 1. The first two plots show the parameter space (a, b, k) of the two-soliton solution for a specific mass ($M = 1$) from two different view points. The 2D surface plotted is the constrain of k which corresponds to the Manko et al. solution [5, 7] used by Berti and Stergioulas [12]. The third plot is the $k = 0$ plane of the parameter space which corresponds to all Kerr solutions of the two-soliton solution. The plots clearly show that the Manko et al. solution has no set of parameters that describe the case of $k = 0$ which corresponds to the Kerr and the Schwarzschild (for $a = 0$) solutions, since there is no intersection between the constrain of k and the Kerr plane within the appropriate range of parameters. The two hyperbola plotted on the plane $k = 0$ are the points were the constrain of k touches the plane in a tangent manner. As we can see these hyperbola correspond to $|a| > M$.

Thus, this solution can reduce to Schwarzschild or Kerr only formally. These are important drawbacks and were pointed out by Berti and Stergioulas [12]. Particularly important is the fact that it can only be matched with rapidly rotating models of neutron stars, since in the limit of zero rotation it has a non-vanishing quadrupole moment $Q = -\frac{M}{4}\frac{(M^2+b^2)^2}{(M^2-b^2)}$. These problems do not exist for the two-soliton solution, that can represent neutron stars of arbitrary rotation. Another important aspect of the two-soliton is that, as can be seen from the multipole moments (5), the solution can describe spacetimes with an arbitrary quadrupole M_2 and an arbitrary current octupole J_3, since every parameter is introduced linearly in the first moment it appears and thus it can always be determined by the first four non-zero moments.

3. Matching and comparison of the analytical to the numerical metric.

3.1. Matching of the analytical to the numerical solution.

In matching the analytical solution to the numerical one it is desirable to find a criterion that will be characteristic of the whole structure of the spacetime and not of a small region of it. That is, the matching conditions should be global and not local. Berti and Stergioulas [12] have argued that these global conditions should be the matching of the multipole moments. We will work under the same assumption. The multipole moments of a stationary axially symmetric spacetime are connected to the Ernst potential on the axis of symmetry and the spectrum of the moments fully specify it. On the other hand, when the Ernst potential on the axis is given, there exists a spacetime which is fully specified by that Ernst potential [14]. Thus, the multipole moments are characteristic of a spacetime and can be used as a global matching condition.

When a numerical spacetime for a neutron star model is constructed from a numerical algorithm, it is possible to evaluate its mass moments M, Q, ... and angular momentum moments J, S_3, ... (for further discussion see Berti and Stergioulas [12]). These numerically evaluated moments can be used to impose the matching conditions to the analytical spacetime. The first four nonzero multipole moments of the two-soliton solution are:

$$\begin{aligned} M_0 = M \quad &, \quad J_1 = aM, \\ M_2 = -(a^2 - k)M \quad &, \quad J_3 = -(a^3 - (2a - b)k)M, \end{aligned} \tag{9}$$

where we can see that once we specify the mass and the angular momentum of the spacetime, the parameter k is uniquely determined by the quadrupole moment $Q = M_2$ and the parameter b is uniquely determined by the current octupole $S_3 = J_3$. Thus, having specified the parameters of the two-soliton, we have completely determined the analytical spacetime that can be used to describe the exterior of the neutron star. What remains to be seen is how well do the properties of the analytical spacetime compare to those of the numerical one.

3.2. Criteria for the comparison of the analytical to the numerical spacetime.

To compare the analytical to the numerical spacetime, one should use criteria that are characteristic of the geometric structure of the spacetime. The criteria should also be related to properties of the spacetime that could potentially be measured in astrophysical phenomena. A good fit of such properties would mean that they could be used to identify a spacetime through observational data and then associate it with properties of the compact object that generates it.

The first and more obvious criterion one can use is how well the respective components of the analytical and the numerical metric match. If the relative difference of the analytical metric components

$$g_{tt} = -f, \quad g_{t\phi} = f\omega, \quad g_{\phi\phi} = f^{-1}\rho^2 - f\omega^2, \tag{10}$$

to the corresponding numerical is small, one could consider the former as a good substitute of the latter. The comparison of the $g_{\phi\phi}$ component can be performed through the circumferential radius $R_{circ} = \sqrt{g_{\phi\phi}}$ which is as an indicator of how well the measurements of the circumferences in the corresponding spaces (analytical and numerical) with a specific value of the coordinate ρ, compare with each other.

The second criterion is the innermost stable circular orbit (ISCO). Test particles that move on geodesics on the equator follow trajectories that are given by the equation

$$-g_{\rho\rho}\left(\frac{d\rho}{d\tau}\right)^2 = 1 - \frac{\tilde{E}^2 g_{\phi\phi} + 2\tilde{E}\tilde{L}g_{t\phi} + \tilde{L}^2 g_{tt}}{\rho^2} \equiv V(\rho). \tag{11}$$

where $\tilde{E}$ and $\tilde{L}$ are the conserved energy per unit mass and angular momentum (parallel to the axis component) per unit mass and $V(\rho)$ is the effective potential. Since we are interested in circular orbits, the conditions for such orbits are $d\rho/d\tau = 0$ and $d^2\rho/d\tau^2 = 0$ which are equivalent from (11) to the conditions for the effective potential, $V(\rho) = 0, \quad dV(\rho)/d\rho = 0$. The ISCO is determined if we additionally impose the condition that the radius ρ of the circular orbit is also a turning point of the potential, that is $d^2V(\rho)/d\rho^2 = 0$. The location of the ISCO is of astrophysical importance since it indicates the radius at which matter orbiting a compact object can no longer maintain its orbit and plunges into the object. That radius would be the internal radius of accretion disks formed around such a compact object.

The third criterion is the rotation frequency of circular orbits on the equatorial plane. The rotation frequency $\Omega \equiv \frac{d\phi}{dt}$ is determined from (11) and the conditions for circular orbits, $V(\rho) = 0, \quad dV(\rho)/d\rho = 0$, and is given by the equation

$$\Omega(\rho) = \frac{-g_{t\phi,\rho} + \sqrt{(g_{t\phi,\rho})^2 - g_{tt,\rho}g_{\phi\phi,\rho}}}{g_{\phi\phi,\rho}}. \tag{12}$$

The final criterion are the epicyclic frequencies Ω_ρ, Ω_z. These are the frequencies of precession of the periastron and the orbital plane of a perturbed circular orbit, respectively. They can be evaluated by perturbing the equations of motion around a circular orbit

$$-g_{\rho\rho}\left(\frac{d\rho}{d\tau}\right)^2 - g_{zz}\left(\frac{dz}{d\tau}\right)^2 = V(\rho, z), \tag{13}$$

where $V(\rho, z)$ is the effective potential defined in (11) maintaining its z dependence. The final expressions for the epicyclic frequencies are

$$\begin{aligned}\Omega_a &= \Omega - \left\{ -\frac{g^{aa}}{2}\left[(g_{tt} + g_{t\phi}\Omega)^2 \left(\frac{g_{\phi\phi}}{\rho^2}\right)_{,aa} \right.\right. \\ &\quad -2(g_{tt} + g_{t\phi}\Omega)(g_{t\phi} + g_{\phi\phi}\Omega)\left(\frac{g_{t\phi}}{\rho^2}\right)_{,aa} \\ &\quad \left.\left. (g_{t\phi} + g_{\phi\phi}\Omega)^2 \left(\frac{g_{tt}}{\rho^2}\right)_{,aa}\right]\right\}^{1/2},\end{aligned} \tag{14}$$

where a is ρ or z and the expression is evaluated at $z = 0$ and around circular orbits of $\rho = \rho_0$.

These criteria can be related to astrophysical phenomena and in particular, they can be related to properties and observables of accretion disks. One such observable is the quasi-periodic oscillations (QPOs) with modulations in the range of kHz that have been observed in the X-ray flux from accreting X-ray pulsars [15]. The association of QPOs with the above frequencies could be used to determine the properties of the central object from the values of the various frequencies. These properties such as mass, angular momentum and mass quadrupole, could also be used to gain insight on the structure of the central object. Thus, an analytical spacetime that fits well all these parameters could be practically considered equivalent to the numerical one and a useful tool for doing astrophysics.

3.3. Results of the comparison.

In the comparison of the metric components, we have used the two corresponding Manko et al. solutions that were used by Berti and Stergioulas [12], as a reference. The comparison of the metrics is performed on the equatorial plane and on the axis of rotation. Here we will present only one rotating neutron star model, since these results are typical for all models. The chosen example is the one in table 3 of the Berti and Stergioulas [12] work, that corresponds to the model with the fastest rotation of the sequence of maximum mass in the non-rotating limit, for the FPS equations of state (EOS).

In Fig. 2 we have plotted the relative difference of the metric components g_{tt} and $g_{t\phi}$ and of the circumferential radius R_{circ} as a function of coordinate radius ρ on the equatorial plane (the $\mathcal{D}$ operator is defined as $\mathcal{D}$ =(numerical - analytical)/numerical). In the first two graphs we have used for comparison the Two Soliton solution and the two corresponding solutions of Manko et al. which are distinguished by their value of the parameter b of the Ernst potential. We remind here that the Ernst potential of the Manko et al. solution is given by the expression

$$e(z) = \frac{(z - M - ia)(z + ib) + d - \delta - ab}{(z + M - ia)(z + ib) + d - \delta - ab}, \tag{15}$$

$$\delta = \frac{-M^2 b^2}{M^2 - (a-b)^2}, \quad d = \frac{1}{4}\left(M^2 - (a-b)^2\right),$$

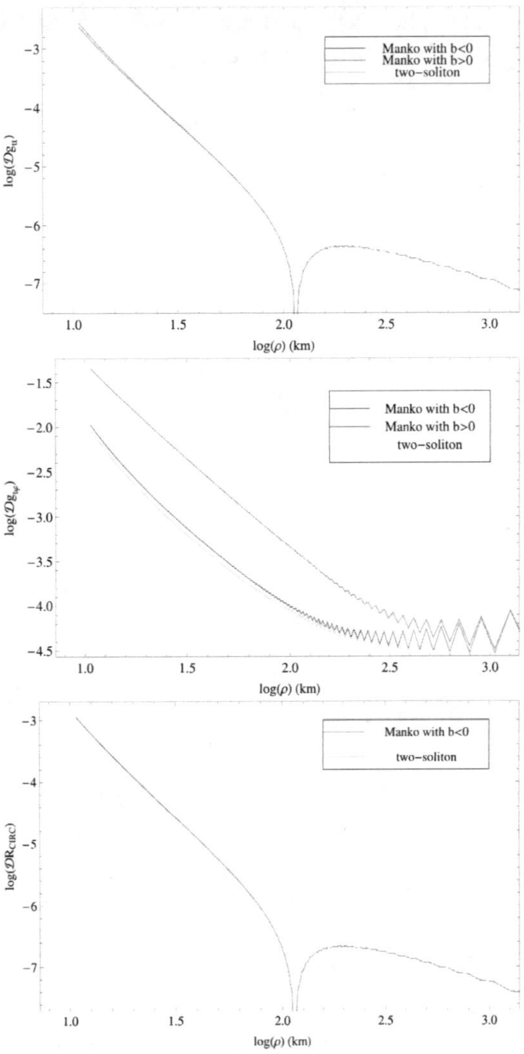

Figure 2. The first two plots show the relative difference of the three analytical metrics with the numerical for the components g_{tt} and $g_{t\phi}$ on the equatorial plane. The analytical metrics used are the two-soliton and the two Manko et al. [5, 7] that correspond to the two values of the parameter b (see Berti and Stergioulas [12]). The third plot shows the relative difference of the circumferential radius on the equatorial plane for the two-soliton and the Manko et al. with the negative value of b (the two solutions almost coincide). As we can see, the errors are well under 1 percent (for the green line) on the surface and fall outwards. The oscillations that appear in the second plot are due to the behavior of the numerical solution.

where there are only three independent parameters M, a, b and for a particular quadrupole moment of the numerical solution (with fixed M, a) there may be no corresponding value of b or there may be two values. For these two values of b we can calculate two metrics which are the ones shown in the graphs. As the second graph shows, the one of the two Manko et al. solutions does not fit well with the numerical metric and thus it describes it poorly [12]. On the other hand, the other solution (corresponding to the negative value of b) is as good as the two-soliton metric. Both metrics are better than 1 percent on the equator and the relative difference falls with increasing radial coordinate ρ. The relative difference shows a kink at about 100km which is an artifact due to a crossing of the numerical and the analytical solutions and has no further physical meaning. In the third graph of Fig. 2 we have used only the one of the Manko et al. metrics (the one with negative b) and the two-soliton. The relative difference for the R_{circ} is under 0.1 percent on the surface and falls with increasing radial coordinate ρ.

In Fig. 3 we have plotted the relative difference of g_{tt} on the axis of symmetry. The plot shows the same good behavior of the analytical metric on the axis as on the equatorial plane. The behavior on the axis of the analytical metric was a matter of confusion that has been clarified in [16].

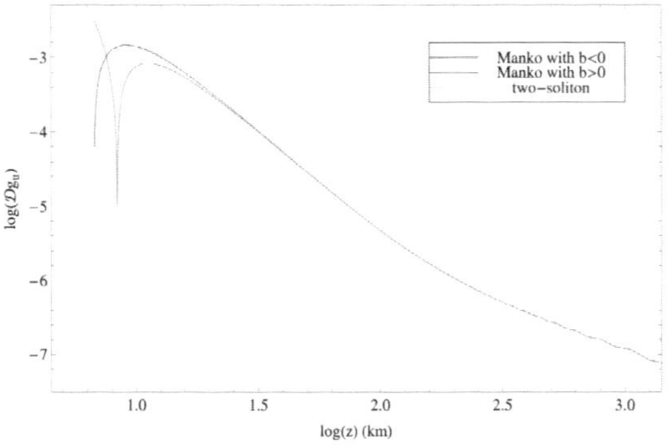

Figure 3. The plot shows the relative difference of the three analytical metrics with the numerical one for the component g_{tt} on the axis of symmetry. The relative difference is under 1 percent on the surface. The behavior is similar to the behavior on the equatorial plane.

As we can see, both the Manko et al. and the two-soliton fit well the numerical metric. But for the Manko et al. there are constrains. If the rotation rate of the neutron star becomes smaller than a critical value (for typical values of the rotation parameter $j = J/M^2$ see [12]), there cannot be found any solutions (based on the first four multipole moments) of the type of Manko et al. that could match to the exterior of the neutron star. On the other hand, the two-soliton can describe arbitrary rotations as was pointed out and the good fit to the numerical metric extends to all rotations.

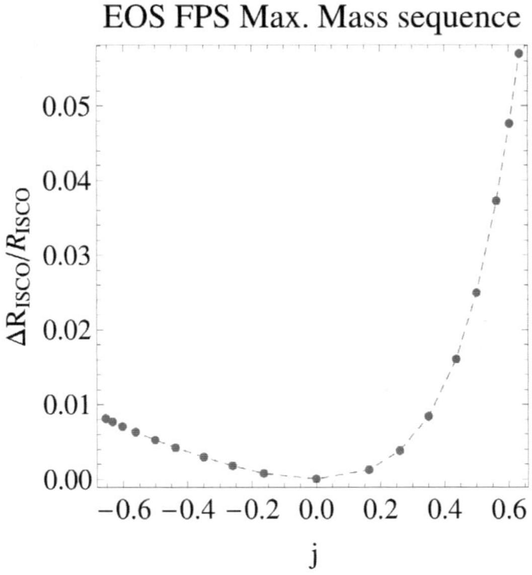

Figure 4. Relative difference of the analytical with the numerical ISCO for the sequence of maximum mass in the non-rotating limit for the equation of state FPS. The maximum difference of the co-rotating ISCO is 5.7 percent for the fastest rotating model that has a co-rotating ISCO and the maximum difference for the counter rotating ISCO is less than 1 percent for the fastest rotating model.

For the comparison of the ISCO we have used a sequence of models of neutron stars with various rotations. The sequence is the one in table 3 of Berti and Stergioulas [12] that corresponds to the model of maximum mass in the non-rotating limit for the FPS equation of state. The comparison of the R_{ISCO} is shown in Fig. 4, where we have plotted the relative difference of the numerical ISCO with the analytical one as a function of the rotation parameter $j \equiv J/M^2$. The negative values of j correspond to the counter-rotating ISCOs which are located further from the star surface. In some cases the co-rotating ISCO is under the surface of the star so, there is no numerical value for it, while that never happens for the counter-rotating ISCO. The co-rotating analytical ISCO is systematically less accurate than the counter-rotating one and that is to be expected since the latter is located further from the star. The difference in the ISCOs is less than 5.7 percent for the co-rotating and less than 1 percent for the counter-rotating orbits. The same general results are true for the Manko et al. metric for rotation parameters above the critical value. We should point out that the results of Fig. 4 are different than the ones shown in Fig. 8 of [12]. The difference is on the counter-rotating branch and it is due to a calculation slip that resulted to overestimation of the error in the counter-rotating ISCOs in [12]. For example the fastest rotating model of fig. 7 in [12] is given an error of 10 percent while the true error is of the order of 1.2 percent. Thus the ISCO comparison is more favorable than it was originally estimated.

The next property that we compare is the rotation frequency of the circular orbits. The relative difference of Ω on the equatorial plane is plotted as a function of the coordinate distance ρ and is shown in Fig. 5. The relative difference is less than 1 percent on the surface of the star and reduces with increasing distance ρ. In this model, since the ISCO is located under the surface of the neutron star, there is no problem to evaluate Ω from the surface outwards. The analytically evaluated Ω, as Fig. 5 shows, fits well the numerical Ω and that conclusion also holds for all rotations.

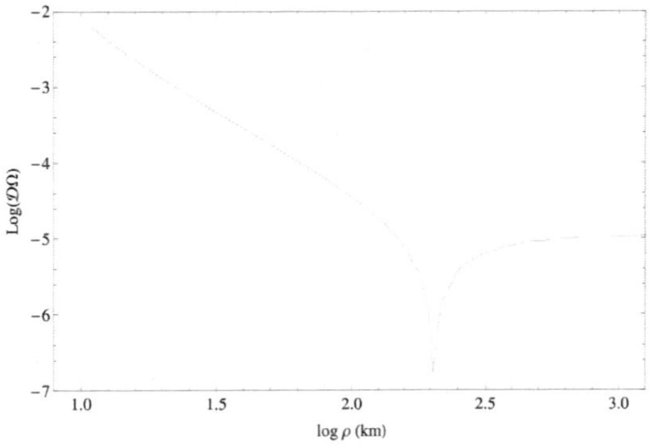

Figure 5. Relative difference in Ω of the two-soliton with the numerical metric. As in the plots of the various metric components, the relative difference is under 1 percent on the surface and falls outwards.

3.4. Conclusions

That was a brief presentation of our results on comparing the analytical two-soliton metric with numerical metrics describing the exterior of neutron stars. We plan to present a more thorough comparison between the two kinds of metrics in a forthcoming paper [17] where the comparison of the epicyclic frequencies will be demonstrated as well.

The results presented here suggest that the two-soliton metric is a very good candidate for describing the exterior spacetime of a neutron star of arbitrary rotation. Therefore, we could use this metric to gain information about the neutron star through observables related to astrophysical processes in the vicinity of the star.

Acknowledgments

I would like to thank Theocharis Apostolatos and Nikolaos Stergioulas for many useful discussions. This work was supported by the research funding program "Kapodistrias" with Grant No 70/4/7672.

References

[1] Stergioulas N and Friedman J L 1995 *ApJ* **444** 306.
[2] Stergioulas N 2003 Rotating Stars in Relativity *Living Rev. Relativity* **6** 3.
[3] Ernst F J 1968 *Phys. Rev.* **167** 1175, Ernst F J 1968, *Phys. Rev.* **168** 1415.
[4] Manko V S, Martin J and Ruiz J E 1995 *J. Math. Phys.* **36** 3063.
[5] Manko V S, Sanabria-Gómez J D and Manko O V 2000 *Phys. Rev.* D **62** 044048.
[6] Ruiz E, Manko V S and Martin J 1995 *Phys. Rev.* D **51** 4192.

[7] Manko V S, Mielke E W and Sanabria-Gómez J D 2000 *Phys. Rev.* D **61** 081501.
[8] Sibgatullin N R 1984 *Oscillations and Waves in Strong Gravitational and Electromagnetic Fields* (Nauka, Moscow, 1984; English translation: Springer-Verlag, Berlin, 1991).
[9] Manko V S and Sibgatullin N R 1993 *Class. Quantum Grav.* **10** 1383.
[10] Papapetrou A 1953 *Ann.Phys.* **12** 309.
[11] Sotiriou T P and Pappas G 2005 *J. Phys.: Conf. Ser.* **8** 23 (*Preprint* gr-qc/0504122).
[12] Berti E and Stergioulas N 2004 *MNRAS* , **350** 1416.
[13] Stute M and Camenzind M 2002 *MNRAS* **336** 831.
[14] Hoenselaers C, Kinnersley W and Xanthopoulos B C 1979 *Phys. Rev. Lett.* **42** 481, Hoenselaers C, Kinnersley W and Xanthopoulos B C 1979 *J. Math. Phys.* **20** 2530, Xanthopoulos B C 1979 *J. Phys. A: Math. Gen.* **12** 1025, Xanthopoulos B C 1981 *J. Math. Phys.* **22** 1254, Hauser I and Ernst F J 1981 *J. Math. Phys.* **22** 1051.
[15] Kluzniak W, Abramowicz M A, Kato S, Lee W H and Stergioulas N 2004 *Astrophys. J.* **603** L89 (*Preprint* astro-ph/0308035).
[16] Pappas G and Apostolatos T A 2008 *Class. Quant. Grav.* **25** 228002 (*Preprint* 0803.0602 [gr-qc]).
[17] Pappas G and Apostolatos T A, *Work in preperation.*

Asymmetrically warped brane models, bulk photons and Lorentz invariance

K Farakos[1], N E Mavromatos[2] and P. Pasipoularides[1]

[a]Department of Physics, National Technical University of Athens, Zografou Campus, 157 80 Athens, Greece.

[b]King's College London, University of London, Department of Physics, Strand WC2R 2LS, London, U.K.

Abstract. We present a brief review of our recent work [1] on asymmetrically warped brane models, where the background metric is characterized by different time and space warp factors. In particular we examine the case of bulk photons and we show that the standard Lorentz invariant dispersion relation is corrected by nonlinear terms which lead to an Energy-dependent speed of light. Stringent constraints on the parameters of our models can be set by comparing the results with recent data from high-energy Gamma-Ray Astrophysics, for instance the MAGIC Telescope.

1. Introduction

Theorists, in an attempt to solve the hierarchy problem, invented new string theory models with relatively large extra dimensions. The early realization [2] that the string scale is an arbitrary parameter, and can be as low as a TeV scale, lead naturally to the consideration of models with large extra dimensions [3, 4], introducing the so-called *brane-world models*. In such constructions, the particle excitations of the Standard Model are assumed to be localized in a 3D brane (our world), embedded in a multi-dimensional manifold (bulk). Subsequently, models in which the bulk space time is warped have been proposed [5, 6], and a much richer phenomenology emerged. In such models, the extra dimensions could be: (i) finite, if a second parallel brane world lies at a finite bulk distance from our world [5], thus providing a new hierarchy of mass scales, or: (ii) infinite, if our world is viewed as an isolated brane, embedded in an (infinite) bulk space [6]. In fact, it is the presence of such warp factors that provides [5] in the case (i) a "resolution" to the hierarchy problem.

Beyond the above standard brane world scenarios, there are models in which all or some of the standard model particles, can live in the bulk. Note, that in the case of the first Randall-Sundrum (RS) model [5] the size of the extra dimension is very small, of the order of the Planck length. In such a case, gauge fields and fermions are not necessarily localized on the brane, see for example [7] and references therein.

There are numerous generalizations of the above generic models, including, for instance, topological defects along the extra dimension(s), or higher-order curvature corrections [8, 9, 10, 11, 12, 13]. Also, there are models in which the standard model particles are localized on the brane dynamically, via a mechanism which is known as *layer-phase mechanism* [14, 15, 16, 17, 18, 19]. Moreover, an effective propagation of standard model particles in the bulk may characterize the so-called "fuzzy" or fluctuating-thick-brane-world scenarios [20], according to which our brane world quantum fluctuates in the bulk. In such a case, there are uncertainties in the bulk position of the brane world, resulting in an "effectively" thick brane [21].

In this talk we focus on the so called asymmetrically warped brane models. We shall review briefly work presented in Ref. [1]. In particular, we consider the following generic ansatz for the metric in five dimensions

$$ds^2 = -\alpha^2(z)dt^2 + \beta^2(z)d\mathbf{x}^2 + \gamma^2(z)dz^2\,, \qquad (1)$$

where z parameterizes the extra dimension, $\alpha(z)$ is the time warp factor and $\beta(z)$ is the three-dimensional-space warp factor. Models which are described by metrics of the form of Eq. (1), in which the time and three-dimensional-space warp factors are different, are often called asymmetrically warped brane models. In these models, although the induced metric on the brane (localized at $z = 0$) is Lorentz invariant upon considering the case $\alpha(0) = \beta(0)$, the metric of Eq. (1) does not preserve 4D Lorentz invariance in the bulk since $\alpha(z) \neq \beta(z)$ for $z \neq 0$.

Models with equal warp factors, such as the RS model [5, 6], have so far attracted the main attention, since 4D Lorentz invariance is assumed as a fundamental symmetry of nature. However, brane models with asymmetrically warped solutions, like that of Eq. (1) above, have also been constructed [22, 23, 24, 25, 26]. The question then arises as to how one can constrain or exclude/falsify brane models with *asymmetric* solutions of the form of Eq. (1), on account of present (or immediate-future) experimental bounds on (local) Lorentz symmetry violation in the bulk.

In the standard brane world scenario, where the bulk is completely unaccessible by the standard model particles, Lorentz violation signals can be observed only by bulk particles in the gravitational sector, like gravitons, or at most particles neutral under the standard model group, e.g. right-handed sterile neutrinos. Bulk fields can "see" the asymmetry between the warp factors in the extra dimension, whilst standard-model particles, which are rigidly "pinned" on the brane, can only "see" equal warp factors $\alpha(0) = \beta(0)$. Gravity effects which could reveal 4D Lorentz violation are described in Ref. [24] (see also Refs. [27, 28, 29]) where superluminous propagation of gravitons is possible for specific models with *asymmetric* solutions. However, since the detection of gravitons is still not an experimental fact, such Lorentz violations are still compatible with the current experiments, both terrestrial and astrophysical, probably awaiting the future detection of gravitational waves in order to be constrained significantly.

However, in the case where some or all of the standard model particles, are allowed to propagate in the bulk, such Lorentz-Invariance-violating effects can be bounded by high precision tests of Lorentz symmetry, since now, 4D Lorentz violation may be revealed even in the standard model sector. In this way, stringent restrictions to *asymmetric* models can be imposed by astrophysical observations and other high-energy experimental tests.

In this paper, we give a brief review of our recent work [1], in which we study the propagation of bulk photons in an *asymmetrically* warped metric background by solving the (classical) equations of motion for a 5D massless U(1) Gauge field in the curved background of Eq. (1). We shall demonstrate that the standard Lorentz invariant dispersion relation for 4D photons possesses nonlinear corrections, which lead to an energy-dependent speed of light on the brane. Specifically, we shall obtain a sub-luminal refractive index for photons $n_{eff}(\omega) = 1 + c_G\ \omega^2$, where ω is the energy of the photon, and the factor c_G is always positive and depends on the free parameters of the models. Finally, comparing these results with astrophysical data, especially from High-energy cosmic Gamma Rays, we can impose stringent constraints on the parameters of our models. As a concrete example, we use the recent findings of the MAGIC Telescope on a delayed arrival of highly energetic photons from the distant galaxy Mk501 [30, 31].

2. 5D U(1) Gauge fields in asymmetrically warped spacetimes

We consider an action which includes 5D gravity, a negative cosmological constant Λ, plus a bulk U(1) Gauge field [24]:

$$S = \int d^5x\sqrt{g}\left(\frac{1}{16\pi G_5}(R^{(5)} - 2\Lambda) - \frac{1}{4}B^{MN}B_{MN}\right) + \int d^4x\sqrt{g_{(brane)}}\mathcal{L}_{matter}, \qquad (2)$$

where G_5 is the five dimensional Newton constant, and $B_{MN} = \partial_M H_N - \partial_N H_M$ is the field strength of the U(1) Gauge field H_M, with $M, N = 0, 1 \ldots 5$. Note, that this additional bulk Gauge field does not interact with charged matter on the brane, so it must not be confused with the usual electromagnetic field A_M, representing a bulk photon, which will be introduced later. The four dimensional term in the action corresponds to matter fields localized on the brane, which is assumed located at $r = r_0$, and are described by a *perfect fluid* with energy density ρ and pressure p. This brane term is necessary for the solution of Eq. (3) to satisfy the junction conditions on the brane (for details see [1] and references there in).

For the metric of the black hole solution we make the ansatz

$$ds^2 = -h(r)dt^2 + \ell^{-2}r^2 d\Sigma^2 + h(r)^{-1}dr^2 , \tag{3}$$

where $d\Sigma^2 = d\sigma^2 + \sigma^2 d\Omega^2$ is the metric of the spatial 3-sections, which in our case are assumed to have zero curvature, in agreement with the current astrophysical phenomenology, pointing towards spatial flatness of the observable Universe. Moreover, ℓ is the AdS radius which is equal to $\sqrt{-\frac{6}{\Lambda}}$. By solving the corresponding Einstein equations we obtain:

$$h(r) = \frac{r^2}{\ell^2} - \frac{\mu}{r^2} + \frac{Q^2}{r^4} , \tag{4}$$

where μ is the mass (in units of the five dimensional Planck scale) and Q the charge of the 5D *AdS-Reissner-Nordstrom* black hole.

Note that, in the case of nonzero charge Q, a non-vanishing component B_{0r} of the bulk field-strength tensor B_{MN}:

$$B_{0r} = \frac{\sqrt{6}}{\sqrt{8\pi G_5}} \frac{Q}{r^3} , \tag{5}$$

is necessary in order for the solution to satisfy the pertinent Einstein-Maxwell equations.

In order to achieve our aim, which is to write the 5D AdS-Reissner-Nordstrom solution as a *linearized perturbation* around the RS metric, we perform the following change of variables $r \to z(r)$ in Eq. (3):

$$\begin{aligned} r &= r_0 e^{-k\,z} , \quad \text{for } z > 0 , \\ r &= r_0 e^{k\,z} , \quad \text{for } z < 0 , \end{aligned} \tag{6}$$

If, in addition, we make the rescaling $x_\mu \to \frac{r_0}{\ell} x_\mu \quad (\mu = 0, \ldots 3)$, we obtain:

$$ds^2 = -a^2(z)h(z)dt^2 + a^2(z)d\mathbf{x}^2 + h(z)^{-1}dz^2 , \tag{7}$$

where $a(z) = e^{-k|z|}$, and $k = \ell^{-1}$ is the inverse AdS_5 radius. For the function $h(z)$ we obtain:

$$h(z) = 1 - \delta h(z), \quad \delta h(z) = \frac{\mu\ell^2}{r_0^4} e^{4k|z|} - \frac{Q^2\ell^2}{r_0^6} e^{6k|z|} . \tag{8}$$

As we describe in detail in Ref. [1] it is not difficult to construct two brane models. Now, the Z_2 symmetry $r \leftrightarrow r_0^2/r$ if it expressed in the frame of the new parameter z, reads $z \to -z$. In addition, the positions of the branes which are located at r_0 and $r_0' = r_0\epsilon$ in the original coordinate system, in the new coordinate system are determined by the equations $z = 0$ and $z = \pi r_c$ correspondingly, where $\epsilon = e^{-k\pi r_c}$ and r_c is radius of the compact extra dimension.

Finally, we assume that $|\delta h(z)| \ll 1$ in the interval $0 < z < \pi r_c$, or equivalently we adopt that $\delta h(z)$ is only a small perturbation around the RS-metric. This implies that r_0, which is the radius that determines the position of the brane in the bulk, is (comparatively) a very large quantity. In particular, we have to satisfy both the following two inequalities $r_0^2\epsilon^2 \gg \sqrt{\mu}\ell$ and $r_0^3\epsilon^3 \gg Q\ell$.

Now we will consider the case of a 5D massless $U(1)$ gauge boson A_N in the background of an asymmetrically warped solution of the form of Eq. (7). This represents a photon propagating in the extra bulk dimensions, which will be of primary interest for our discussion. The corresponding equation of motion for A_N reads:

$$\frac{1}{\sqrt{g}}\partial_M\left(\sqrt{g}g^{MN}g^{RS}F_{NS}\right) = 0 \,, \tag{9}$$

with $F_{NS} = \partial_N A_S - \partial_S A_N$, and $N, S = 0, 1, \ldots 5$. In the background metric of Eq. (7), Eq. (9) gives:

$$-\partial_z(a^2(z)h(z)\partial_z A_j) - \nabla^2 A_j + \frac{1}{h(z)}\partial_0^2 A_j = 0, \quad j = 1, 2, 3 \,, \tag{10}$$

where we have considered the Coulomb gauge condition:

$$\vec{\nabla}\cdot\vec{A} = 0, \quad A_0 = 0, \quad A_z = 0 \,. \tag{11}$$

This is suitable for the case of a Lorentz violating background. On setting in Eq. (9):

$$A_j(x, z) = e^{ip\cdot x}\chi_j(z), \quad p_\mu = (-\omega, \mathbf{p}) \tag{12}$$

and keeping only the linear terms in $\delta h(z)$, we obtain

$$-\partial_z\left\{a^2(z)\left[1 - \delta h(z)\right]\partial_z\chi\right\} + \left\{\mathbf{p}^2 - \left[1 + \delta h(z)\right]\omega^2\right\}\chi = 0 \,, \tag{13}$$

where, for brevity, we have dropped the index j from χ. Note that the *spectrum* of Eq. (13) is *discrete*, due to the orbifold boundary conditions [5], $\chi'(0) = 0$ and $\chi'(\pi r_c) = 0$ (where the prime denotes a z-derivative).

Upon applying the formalism of time-independent perturbation theory to second order (for details see Ref. [1]) after same analytical calculations we obtain the following non-trivial dispersion relation for light:

$$v_{ph} = \frac{\omega}{|\mathbf{p}|} = \frac{1}{\sqrt{(1 + a_G) + b_G\omega^2}} \,, \tag{14}$$

where ω is the energy of the photon. For the computation of the parameters a_G and b_G we have obtained the formulas

$$a_G = \int_0^{\pi r_c} dz \left(\chi_0^{(0)}(z)\right)^2 \delta h(z) \,, \tag{15}$$

$$b_G = \sum_{n\neq 0}\frac{1}{\left(m_n^{(0)}\right)^2}\left(\int_0^{\pi r_c} dz\, \chi_0^{(0)}(z)\chi_n^{(0)}(z)\delta h(z)\right)^2 \,, \tag{16}$$

where $\chi_n^{(0)}(z)$ and $m_n^{(0)}$ (n=0,1,2..) are the eigenfunctions and eigenvalues of the unperturbed Schrodinger equation for bulk photons, which are known analytically [32, 33], see also [1].

If we consider the limit $b_G\omega^2 << 1$:

$$v_{ph} \simeq \frac{1}{\sqrt{(1 + a_G)}} - \frac{b_G}{2(1 + a_G)^{\frac{3}{2}}}\omega^2 \,, \tag{17}$$

On the other hand, for the photon's group velocity we have:

$$v_{gr} \simeq \frac{1}{\sqrt{(1 + a_G)}} - \frac{3\, b_G}{2(1 + a_G)^{\frac{3}{2}}}\omega^2 \,. \tag{18}$$

One can define the constant velocity of light *in standard vacuo* as the low energy limit ($\omega \to 0$) of the phase velocity of Eq. (17)

$$c_{light} = \frac{1}{\sqrt{(1+a_G)}} . \qquad (19)$$

¿From Eqs. (17) and (19) we then obtain an *effective subluminal refractive index* for the *non-standard vacuum* provided by our brane-world constructions:

$$n_{eff}(\omega) = \frac{c_{light}}{v_{ph}} = 1 + \frac{b_G}{2(1+a_G)}\omega^2 . \qquad (20)$$

This is the main result of [1], and of our talk. An issue, we would like to emphasize, is that the phase and group velocities (Eqs. (17) and (18) respectively), as well as the effective refractive index of Eq. (20), depend quadratically on the photon energy ω. Moreover, we note that equation (20) is a perturbative result which is valid only for energies $\omega << 1/\sqrt{b_G}$. For larger energies (for example in the case of ultra high energies cosmic rays $\omega \sim 10^{20}$ eV), Eq. (20) is not valid and a full nonperturbative treatment is necessary. In this limit the subluminal nature of the refractive index may be lost and one may also have birefringence effects. We hope to come to a discussion on such issues in a future work.

3. Instead of Conclusions: Comparison with Data- the MAGIC observations

We wish now to compare the time delays of the more energetic photons, implied by the dispersion relation (20), with experimentally available data. Recently, experimental data from observations of the MAGIC Telescope [30] on photon energies in the TeV range have become available. It therefore makes sense to compare the time delays predicted in our models, due to (20), with such data, thereby imposing concrete restrictions on the free parameters of the models. It should be stressed that the use of the MAGIC data serves only as a concrete example to impose upper bounds on the parameters of our models. From only one set of data one cannot determine with any certainty the physical reasons for the delayed arrival of high energy photons, as compared with their lower-energy counterparts, observed in MAGIC. The source mechanism for the production of high energy photons is still unknown, and certainly one needs many more data and confirmation of certain patterns in the behaviour of high energy photons before reaching conclusions and disentangling source from propagation effects.

For completeness let us first review briefly the relevant observations [30, 31]. MAGIC is an imaging atmospheric Cherenkov telescope which can detect very high energy (0.1 TeV-30 TeV) electromagnetic particles, in particular gamma rays. Photons with very high energy (VHE) are produced from conversion of gravitational energy at astrophysical distances from Earth, when, say, a very massive rotating star is collapsing to form a supermassive black hole. Astrophysical objects like this are called blazars and are active galactic nuclei (AGN).

The observations of MAGIC during a flare (which lasted twenty minutes) of the relatively nearby (red-shift $z \sim 0.03$) blazar of Markarian (Mk) 501 on July 9 (2005), indicated a 4 ± 1 min time delay between the peaks of the time profile envelops of photons with energies smaller than 0.25 TeV and photons with energies larger than 1.2 TeV. Possible interpretations of such delays of the more energetic photons have already been proposed. Conventional (astro)physics at the source may be responsible for the delayed emission of the more energetic photons, as a result, for instance, of some non-trivial versions of the Synchrotron Self Compton (SSC) mechanism [30]. It should be noted at this stage that the standard SSC mechanism, usually believed responsible for the production of VHE photons in other AGN, such as Crab Nebula, fails [30] to explain the results of MAGIC, as a relative inefficient acceleration is needed in order to explain the $\mathcal{O}$(min) time delay. Modified SSC models have been proposed in this respect [34], but the situation is not conclusive. This prompted speculations that new fundamental physics, most likely quantum-gravity effects, during propagation of photons from the source till observation, may be responsible for inducing the observed delays, as a result of an effective refractive index for the vacuum, see for example Ref. [35] and references therein.

Exploring the fact that MAGIC had the ability of observing individual photons, a numerical analysis on the relevant experimental data has been performed in [31], which aimed at the reconstruction of the original electromagnetic pulse by maximizing its energy upon the assumption of a sub-luminal vacuum refractive index with either linear or quadratic quantum-gravity-scale suppression:

$$n_{eff}(\omega) = 1 + \left(\frac{\omega}{M_{QG(n)}}\right)^n, \quad n = 1, 2 \tag{21}$$

The analysis in [31] resulted in the following values for the quantum-gravity mass scale at 95 % C.L.

$$M_{QG(1)} \simeq 0.21 \ \ 10^{18} \text{ GeV}, \quad M_{QG(2)} \simeq 0.26 \ \ 10^{11} \text{ GeV}. \tag{22}$$

We must emphasize at this point that many Quantum Gravity Models seem to predict modified dispersion relations for probes induced by vacuum refractive index effects, which appear to be different for each quantum gravity approach, not only as far as the order of suppression by the quantum gravity scale is concerned, but also its super- or sub-luminal nature. Some models, for instance, entail birefringence effects, which can be severely constrained by astrophysical measurements [36]. There are also alternative models, see for example Refs. [37, 38].

In [39], a non-perturbative mechanism for the observed time delays has been proposed, based on stringy uncertainty principles within the framework of a string/brane theory model of space-time foam. The model entails a brane world crossing regions in bulk space time punctured by point-like D-brane defects (D-particles). As the brane world moves in the bulk, populations of D-particles cross the brane, interact with photons, which are attached on the brane world, and thus affect their propagation. The important feature of the stringy uncertainty delay mechanism is that the induced delays are proportional to a single power of the photon energy ω, thus being linearly suppressed by the string mass scale, playing the quantum gravity scale in this model. It is important to notice that the mechanism of [39] does not entail any modification of the local (microscopic) dispersion relations of photons. In fact the induced delays are also independent of photon polarization, so there are no gravitational birefringence effects. It is also important to notice that, in view of the electric charge conservation, the D-particle-foam defects can interact non trivially only with photons or at most electrically neutral particles and no charged ones, such as electrons, to which the foam looks transparent.

As we have showed in the previous section, a sub-luminal vacuum refractive index may characterize asymmetrically warped brane-world models, assuming propagation of photons in the bulk. Our model, however, predicts a refractive index with quadratic dependence on energy, c.f. (20). Comparing Eqs. (20) with (21), we obtain:

$$\frac{b_G}{2(1+a_G)} \leq M_{QG(2)}^{-2}. \tag{23}$$

The parameters a_G and b_G can be computed numerically by means of Eqs. (15) and (16). Note, that the deviation $\delta h(z)$, the eigenvalues $m_n^{(0)}$ and the eigenfunctions $\chi_n^{(0)}$ are known analytically. We have computed numerically the parameter b_G for two exemplary cases: a) for an AdS-Reissner-Nordstrom Solution (8), and b) for an AdS-Schwarzschild Solution (obtained by setting $Q = 0$ in Eq. (8)) . In the former case we assume for simplicity that μ and Q^2/r_0^2 are of the same order of magnitude (for details see Ref. [29] or the discussion in section 2.2 of our work [1]), and hence we can ignore the contribution of the e^{4kz} term in Eq. (8). In particular we obtain:

$$b_G \simeq 2.95 \, \langle\delta h\rangle^2 \ \text{TeV}^{-2}, \quad \text{AdS} - \text{Reissner} - \text{Nordstrom}, \tag{24}$$

$$b_G \simeq 10 \, \langle\delta h\rangle^2 \ \text{TeV}^{-2}, \quad \text{AdS} - \text{Schwarzschild}, \tag{25}$$

where $\langle\delta h\rangle$ is defined as the average value:

$$\langle\delta h\rangle = \frac{1}{\pi r_c}\int_0^{\pi r_c} \delta h(z) dz \, . \tag{26}$$

We will use $\langle \delta h \rangle$ in order to estimate the degree of violation of Lorentz symmetry in our models. Taking into account Eqs. (22), (23) ,(24) and (25) we find the constraints:

$$| \langle \delta h \rangle | \leq 1.4\; 10^{-8} \quad \mathrm{AdS-Reissner-Nordstrom}\,, \tag{27}$$
$$| \langle \delta h \rangle | \leq 0.75\; 10^{-8} \quad \mathrm{AdS-Schwarzschild}\,. \tag{28}$$

The small values we obtain are consistent with the weak nature of $\langle \delta h \rangle$, as required by treating it as a perturbation. Also we observe that the values for the average deviation of Eqs. (27) and (28) are of the same order of magnitude for both AdS-Schwarzschild and AdS-Reissner-Nordstrom solutions. In the above analysis we ignored the effects due to the Universe expansion, since the latter do not affect the order of magnitude of the above bounds due to the small red shift ($z \sim 0.03$) of Mk501 we restrict our discussion in this section. The inclusion of such effects, which are essential for larger redshifts, is straightforward [31] and does not present any conceptual difficulty.

Note, that although the parameter a_G is not important for our analysis, it is crucial when we have to make comparisons with the velocities of other particles. If we take into account Eq. (15) we see that $a_G = \langle \delta h \rangle$. Hence, the parameter a_G is constrained via the equation

$$| a_G | \leq 10^{-8}\,. \tag{29}$$

This summarizes the constraints to the asymmetrically-warped brane models with bulk photons using the data of the MAGIC experiment. For comparison with other data (for example, recent data from H.E.S.S. experiment [40] and Ultra High Energy Cosmic Rays [41]), we refer the reader to our article [1] for further details. Moreover, in that work we have also made some comparison with the velocity of gravitons (for this latter point, see also [42]).

The induced modification of the refractive index for photons is a result of the breakdown of the higher(five, in our case) -dimensional Lorentz Invariance and photon propagation in the bulk. On the brane world, Lorentz-invariant metrics are assumed, but the asymmetric warp factors in the bulk result in feeding the effects of bulk Lorentz-symmetry breaking back onto our brane world, via the induced refractive indices of four dimensional photons.

Acknowledgements

P.P. wishes to thank the organisers of NEB XIII 2008 (Thessaloniki (Greece)) for the opportunity to present the results of this work, and for providing a stimulating atmosphere during the meeting. The work of N.E.M. is partially supported by the European Union through the FP6 Marie-Curie Research and Training Network *Universenet* (MRTN-2006-035863).

References

[1] Farakos K, Mavromatos N E anf PasipoularidesP 2009 *JHEP* **0901** 057
[2] Antoniadis I 1990 *Phys. Lett. B* **246** 377
[3] Arkani-Hamed N, Dimopoulos S and Dvali G R 1998 *Phys. Lett. B* **429** 263
[4] Antoniadis I, Arkani-Hamed N, Dimopoulos S and Dvali G R 1998 *Phys. Lett. B* **436** 257
[5] Randall L and Sundrum R 1999 *Phys. Rev. Lett.* **83** 3370
[6] Randall L and Sundrum R 1999 *Phys. Rev. Lett.* **83** 4690
[7] Davoudiasl H, Rizzo T G and Soni A 2008 *Phys. Rev. D* **77** 036001
[8] Mavromatos N E and Rizos J 2000 *Phys. Rev. D* **62** (2000) 124004
[9] Mavromatos N E and Rizos J 2003 *Int. J. Mod. Phys. A* **18** 57
[10] Giovannini M 2001 *Phys. Rev. D* **64** 124004; ibid 2006 *Class. Quantum Grav.* **23** L73; ibid 2007 *Phys. Rev. D* **75** 064023; ibid 2007 *Phys. Rev. D* **75** 064023; ibid 2007 *Phys. Rev. D* **76** 124017
[11] Mavromatos N E and Papantonopoulos E 2006 *Phys. Rev. D* **73** 026001
[12] Barbosa-Cendejas N, Herrera-Aguilar A, Nucamendi U and Quiros I 2007 *Preprint* arXiv:0712.309; Barbosa-Cendejas N and Herrera-Aguilar A 2007 *Rev. Mex. Fis.* **S53N4** 5; Barbosa-Cendejas N, Herrera-Aguilar A, Reyes Santos M A and Schubert C 2007 *Preprint* arXiv:0709.3552

[13] Liu Y-X, Zhao Z-H, Wei S-W and Duan Y-S 2009 *Preprint* arXiv:0901.0782; Liu Y-X, Zhang L-D, Zhang L-J and Duan Y-S 2008 *Phys. Rev. D* **78** 065025; Liu Y-X, Zhang L-D, Wei S-W and Duan Y-S 2008 *JHEP* **08** 041; Liu Y-X, Zhang X-H, Zhang L-D and Duan Y-S 2008 *JHEP* **02** 067
[14] Farakos K and Pasipoularides P 2005 *Phys. Lett. B* **621** 224
[15] Farakos K and Pasipoularides P 2006 *Phys. Rev. D* **73** 084012
[16] Farakos K and Pasipoularides P 2007 *Phys. Rev. D* **75** 024018
[17] Farakos K, Koutsoumbas G and Pasipoularides P 2007 *Phys. Rev. D* **76** 064025
[18] Pasipoularides P and Farakos K 2007 *J. Phys. Conf. Ser.* **68** 012041.
[19] Dimopoulos P, Farakos K and Vrentzos S 2006 *Phys. Rev. D* **74** 094506; Farakos K and Vrentzos S 2008 *Phys. Rev. D* **77** 094511; Dimopoulos P, Farakos K and Koutsoumbas G 2002 *Phys. Rev. D* **65** 074505
[20] Campbell-Smith A, Ellis J R, Mavromatos N E and Nanopoulos D V 1999 *Phys. Lett. B* **466** 11; Yu H W and Ford L H 2000 *Preprint* arXiv:gr-qc/0004063
[21] Lizzi F, Mavromatos N E and Szabo R J 1998 *Mod. Phys. Lett. A* **13** 829
[22] Chung D J H, Kolb E W and Riotto A 2002 *Phys. Rev. D* **65** 083516
[23] Visser M 1985 *Phys. Lett. B* **159** 22
[24] Csaki C, Erlich J and Grojean C 2001 *Nucl. Phys. B* **604** 312
[25] Dubovsky S L 2002 *JHEP* **0201** 012
[26] Bowcock P, Charmousis C and Gregory R 2000 *Class. Quant. Grav.* **17** 4745
[27] Cline J M and Firouzjahi H 2002 *Phys. Rev. D* **65** 043501
[28] Cline J M and Firouzjahi H 2003 *AIP Conf. Proc.* **646** 197
[29] Cline J M and Valcarcel L 2004 *JHEP* **0403** 032
[30] Albert J etal 2007 *Astrophys. J.* **669** 862
[31] Albert J etal [MAGIC Collaboration] and Ellis J R, Mavromatos N E, Nanopoulos D V, Sakharov A S and Sarkisyan E K G 2008 *Phys. Lett. B* **668** 253
[32] Davoudiasl H, Hewett J L and Rizzo T G 2000 *Phys. Lett. B* **473** 43
[33] Pomarol A 2000 *Phys. Lett. B* **486** 153
[34] Bednarek W and Wagner R M 2008 *Preprint* arXiv:0804.0619
[35] Ellis J R, Farakos K, Mavromatos N E, Mitsou V A and Nanopoulos D V 2000 *Astrophys. J.* **535** 139; Ellis J R, Mavromatos N E, Nanopoulos D V, Sakharov A S and Sarkisyan E K G 2006 *Astropart. Phys.* **25** 402; ibid 2008 *Astropart. Phys.* **29** 158
[36] Jacobson T, Liberati S and Mattingly D 2006 *Annals Phys.* **321** 150
[37] Gogberashvili M, Sakharov A S and Sarkisyan E K G 2007 *Phys. Lett. B* **644** 179
[38] Carmona J M, Cortes J L, Gamboa J and Mendez F 2003 *Phys. Lett. B* **565** 222
[39] Ellis J R, Mavromatos N E and Nanopoulos D V 2008 *Phys. Lett. B* **665** 412
[40] Aharonian F etal 2008 *Phys. Rev. Lett.* **101** 170402
[41] Galaverni M and Sigl G 2008 *Phys. Rev. D* **78** 063003
[42] Moore G D and Nelson A E 2001 *JHEP* **0109** 023

Time evolution of rapidly rotating stratified neutron stars

Andrea Passamonti
School of Mathematics, University of Southampton, Southampton SO17 1BJ, UK
E-mail: a.passamonti@soton.ac.uk

Abstract. We study the time evolution of non-axisymmetric oscillations of rapidly rotating neutron stars. The rotation is assumed uniform and the stellar models have composition gradients that generate gravity restored oscillation modes. We work in linear Newtonian perturbation theory and in Cowling approximation, i.e. we neglect the gravitational potential perturbations. We study for the first time the behaviour of the inertia-gravity modes in fast spinning stars, and carry out a detailed comparison of non-stratified and stratified models. In particular, we show that each g-mode becomes rotation-dominated, i.e. approaches a particular inertial mode, as the rotation rate of the star is increased.

1. Introduction

The spectral properties of stellar oscillations could help us to clarify many aspects of the neutron star physics, as every physical property and configuration of the star could be related to particular features of the spectrum. In particular, astero-seismology of the electromagnetic and gravitational radiation may be very helpful for constraining the equation of state and understand the state of matter at supernuclear densities. In order to facilitate the signal detection and its interpretation, we must still improve the theoretical models and clarify the nature of various stellar pulsation features.

In this contribution, we report our study of the non-axisymmetric oscillations of rapidly rotating stars [1]. Here, we focus on stratified neutron stars and investigate the properties of the inertia-gravity modes, i.e. pulsations that are restored both by buoyancy and Coriolis force. When buoyancy is dominant, as in the non-rotating limit, inertia-gravity modes exhibit the typical features of gravity modes. Instead in the rapidly rotating limit, the Coriolis force overcomes buoyancy and the modes are very similar to the inertial modes of barotropic stars. It is then important to understand better the influence of these two restoring forces on the mode oscillations and in which regions of the parameter space they dominate. For a more detailed analysis of the inertia-gravity modes see [1]. For an analysis of the spectrum of superfluid neutron stars see [2].

We study the non-axisymmetric pulsations in the time domain by adopting the Newtonian linear perturbation theory on an axisymmetric background. The equilibrium configurations are determined by using the self-consistent Hachisu method [3] for polytropic models. We have found very useful to expand the linear perturbations of an axisymmetric background with respect to the azimuthal number m. In this ways, the time evolution of non-axisymmetric oscillations becomes a two-dimensional problem.

Table 1. This table displays the main quantities of our rotating equilibrium configurations. The stellar models are described by a polytropic equation of state with adiabatic index $\Gamma_\beta = 2$. In the first and second columns we show, respectively, the ratio of polar to equatorial axes and the angular velocity of the star. In the third column, the rotation rate is compared to the Kepler velocity Ω_K that represents the mass shedding limit. The ratio between the rotational kinetic energy and gravitational potential energy $T/|W|$ and the stellar mass are given in the fourth and fifth columns, respectively. All quantities are given in dimensionless units, where G is the gravitational constant, ρ_c represents the central mass density and R_{eq} is the equatorial radius.

R_p/R_{eq}	$\Omega/\sqrt{G\rho_c}$	Ω/Ω_K	$T/\|W\| \times 10^{-2}$	$M/(\rho_c R_{eq}^3)$
1.000	0.000	0.000	0.000	1.273
0.992	0.119	0.105	0.192	1.261
0.950	0.287	0.260	1.170	1.197
0.900	0.403	0.374	2.380	1.118
0.800	0.556	0.550	4.930	0.956
0.700	0.658	0.710	7.570	0.779
0.600	0.717	0.893	9.860	0.579

2. Neutron star model

We consider Newtonian uniformly rotating stars where matter is composed by three degenerates and non-superfluid components: neutrons n, protons p and electrons e. The Coulomb interaction is very efficient to lock protons and electrons together [4], forming a neutral charge conglomerate. Furthermore, we assume that the background star is in β-equilibrium, then the proton fraction $x_{\rm p} \equiv \rho_{\rm p}/\rho$ is a function only of the mass density ρ. With ρ_p we denote the proton mass density. The equilibrium configuration can be then described by a barotropic equation of state $P = P(\rho)$. For simplicity, we consider a simple polytropic model

$$P = K\rho^{\Gamma_\beta}\,, \tag{1}$$

where K is constant and

$$\Gamma_\beta \equiv \frac{d\log P}{d\log \rho}\,, \tag{2}$$

is the adiabatic index.

The axisymmetric background configurations are determined by using the self-consistent Hachisu method [3] for $\Gamma_\beta = 2$ polytropic models. The properties of the rotating models are given in Table 1.

3. Perturbations

In neutron stars, the mode oscillation period is much shorter than the typical time scales of the weak interaction processes. Therefore, a perturbed fluid element does not have time to reach β-equilibrium with the surrounding matter within an oscillation period [4]. The composition of the perturbed fluid element can be considered as frozen, i.e. the Lagrangian perturbation of the proton fraction vanishes:

$$\Delta x_{\rm p} = \delta x_{\rm p} + \xi \cdot \nabla x_{\rm p} = 0\,. \tag{3}$$

The function ξ denotes the Lagrangian displacement of a fluid element, and $\delta x_{\rm p}$ is the Eulerian perturbation of $x_{\rm p}$.

The equation of state of the perturbed star is then a function of two variables, and from equation (3) one can write:

$$\Delta P = \Gamma_{\rm f} \frac{P}{\rho} \Delta\rho \,, \tag{4}$$

where

$$\Gamma_{\rm f} \equiv \left. \frac{\partial \log P}{\partial \log \rho} \right|_{x_{\rm p}} \,, \tag{5}$$

is the adiabatic index for a frozen composition. Equation (4) can be re-written as [1]:

$$\frac{\delta\rho}{\rho} = \frac{1}{\Gamma_{\rm f}} \frac{\delta P}{P} - \left(\frac{\Gamma_\beta}{\Gamma_{\rm f}} - 1 \right) \frac{\delta\chi_{\rm p}}{\rho} \,, \tag{6}$$

where we have defined a new perturbation variable,

$$\delta\chi_{\rm p} \equiv \delta x_{\rm p} / \left(\frac{d x_{\rm p}}{d\rho} \right) \,, \tag{7}$$

and also assumed that $x_{\rm p} = x_{\rm p}(\rho)$. One can show that the difference between Γ_β and $\Gamma_{\rm f}$ is proportional to the composition gradients [1].

In the Cowling approximation, the non-axisymmetric perturbations of rapidly, and uniformly, rotating stratified neutron stars can be described by five dynamical variables, the perturbations δP, $\delta\chi_{\rm p}$ and the flux $\mathbf{f} = \rho\mathbf{v}$. In the frame of the rotating background, the perturbed Euler, mass conservation and frozen composition equations can be written as

$$\partial_t \mathbf{f} = -\nabla\delta P - 2\left(\boldsymbol{\Omega} \times \mathbf{f}\right) + \frac{\nabla \ln P}{\Gamma_{\rm f}} \delta P + \left(1 - \frac{\Gamma_\beta}{\Gamma_{\rm f}}\right) \frac{\nabla P}{\rho} \delta\chi_{\rm p} \,, \tag{8}$$

$$\partial_t \delta P = -\Gamma_{\rm f} \frac{P}{\rho} \nabla \cdot \mathbf{f} + \frac{1}{\rho} \left(\frac{\Gamma_{\rm f}}{\Gamma_\beta} - 1 \right) \mathbf{f} \cdot \nabla P \,, \tag{9}$$

$$\partial_t \delta\chi_{\rm p} = -\frac{\mathbf{f} \cdot \nabla P}{\Gamma_\beta P} \,. \tag{10}$$

Equations (8)-(10) reduce to the barotropic perturbation equations by choosing $\Gamma_\beta = \Gamma_{\rm f}$.

The details of the numerical code are explained in [1], here we report some main properties. The perturbation equations are evolved in time with a Mac-Cormack algorithm, which is a second order accurate numerical method. A key element of the code is the implementation of the fourth-order Kreiss-Oliger numerical dissipation. This prevents the growth of spurious oscillations.

We impose standard boundary conditions for the perturbation functions. At the stellar centre, we guarantee the regularity of the perturbation equations (8)-(10). On the surface, we set the vanishing of the Lagrangian perturbation of the pressure, i.e. $\Delta P = 0$ (see [1] for more details).

4. Results

We study the spectrum of the non-axisymmetric modes by evolving in time the system of equations (8)-(10). We consider generic initial data in order to excite several oscillation modes. The identification of the modes is determined by performing a Fast Fourier Transformation of the perturbation evolution data and using the two-dimensional eigenfunction extraction code developed in [5, 6]. We have been able to properly determine the spectrum of fundamental, pressure modes and inertia-gravity modes in the fast rotating regime. In this context, we discuss the low frequency band of rotating non-barotropic stars, while the high band analysis and further code tests can be found in [1].

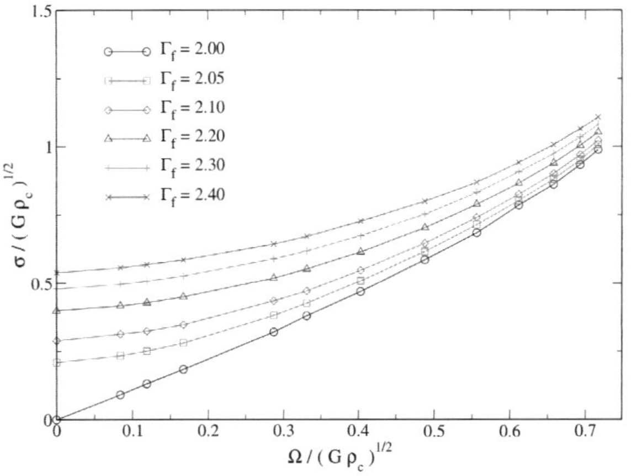

Figure 1. This figure displays the variation of the counter rotating ${}^2\mathrm{g}_1^\mathrm{r}$ mode with the adiabatic index Γ_f for a sequence of rotating stellar models. The polar-led inertial mode ${}^3\mathrm{i}_2$ frequencies of the barotropic models with $\Gamma_\mathrm{f} = 2$ are shown as circle-solid line. This figure suggests that in rapidly rotating stars the counter rotating ${}^2\mathrm{g}_1^\mathrm{r}$ mode behaves similarly to the barotropic ${}^3\mathrm{i}_2$ inertial mode. This is expected when the rotation rate is larger than the Brunt-Väisälä frequency N, namely $2\Omega > N$, and the Coriolis force starts to dominate over the buoyancy force.

From a local analysis of uniformly rotating stars in the Cowling approximation, we can show that the low frequency modes can be approximately described by the following dispersion relation [7]:

$$\sigma^2 \simeq \frac{N^2 k_\perp^2 + (2\boldsymbol{\Omega}\cdot\mathbf{k})^2}{k^2}, \tag{11}$$

where N is the Brunt-Väisälä frequency that estimates the frequency band of the gravity modes [7], $\mathbf{k}$ is the wave vector and $k_\perp$ is its component orthogonal to the apparent gravity. In the barotropic case, $N = 0$, equation (11) describes the inertial waves, while for a non-rotating model it reduces to the usual gravity waves. In stratified and rotating neutron stars, it is thus to be expected that the low-frequency oscillation modes have an hybrid character with gravity dominating at low rotation rates and the Coriolis force taking over at fast stellar spin. It is then natural to call them inertia-gravity waves [7].

In order to study the dependence of the modes on the stellar spin and stratification, we construct a set of equilibrium models by gradually varying the two parameters Ω and Γ_f. We can then follow the change of the mode frequency and eigenfunction shape and correctly distinguish it among the other peaks in the spectrum. We consider, for instance, a sequence of rotating models having increasing composition gradients, i.e. increasing values of the parameter Γ_f. Let us now focus on the counter-rotating $l = m = 2$ g-mode ${}^2\mathrm{g}_1^\mathrm{r}$, and on the $l = 3$ and $m = 2$ barotropic inertial mode ${}^3\mathrm{i}_2$. As shown in Fig. 1, the g-mode frequency approaches the barotropic inertial mode in the limit of weak stratification and large rotational rate. Generally, when $2\Omega > N$, the Coriolis force is dominating and the ${}^2\mathrm{g}_1^\mathrm{r}$ seems to behave as the barotropic ${}^3\mathrm{i}_2$ inertial mode. This property is also noted in the mode eigenfunctions [1].

When we extend this study to the other g-modes, we can determine further associations with the barotropic inertial modes. For barotropic models with $\Gamma_\beta = 2$ and stratified stars with $\Gamma_\mathrm{f} = 2.05$, we select in Fig. 2 some possible associations for the $m = 2$ gravity and inertial modes. Is it evident that the individual non-barotropic g-modes tend toward specific barotropic inertial modes as the rotation rate of the star is increased.

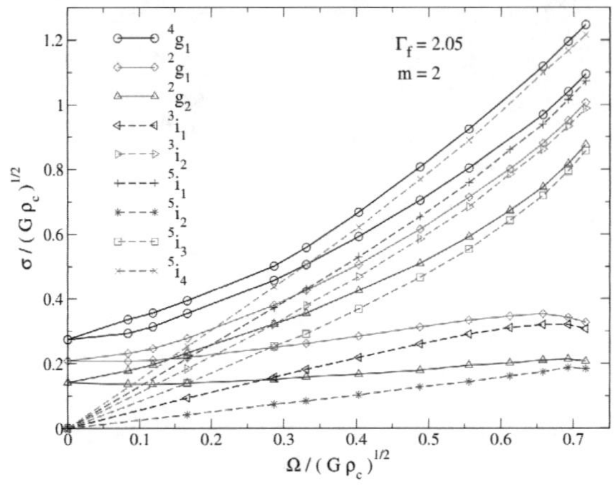

Figure 2. Frequencies of some selected g-modes (solid lines) for $\Gamma_f = 2.05$ models and polarled inertial modes (dashed lines) of the barotropic $\Gamma_\beta = \Gamma_f = 2$ model. All frequencies are measured in the rotating frame. The dominance of the Coriolis force in rapidly rotating stars is evident from this figure. We see that the g-modes are mainly restored by the Coriolis force when $\Omega/\sqrt{G\rho_c} > 0.3$, and thus become similar to the barotropic inertial modes.

The inertia-gravity mode frequencies and their associations with the barotropic inertial modes have been tested with the slow rotation approximation results given in [8]. The authors used a frequency domain approach and studied $\Gamma_\beta = 2$ polytropic models with weak stratification and slow rotation. In this work, we have considerably extended the parameter space of previous analyses by studying stars with large composition gradients and fast rotation.

5. Conclusions

We have studied the pulsations of rapidly rotating stratified neutron stars with the aim of understanding the dependence of the composition g-modes on the rotation rate of the star. We have used the Cowling approximation where the gravitational potential perturbations are neglected. The low frequency band of the spectrum is characterised by the inertia-gravity modes that have a mixed character. They typically behave as g-modes in the slowly rotating limit, while they assume the properties of the inertial modes when the star rotates rapidly. This property make difficult to determine the presence of composition gradients from the inertia-gravity mode spectrum of fast rotating stars. However, as the neutron star ages and spins down, the dependence of the inertia-gravity modes on the star's rotation and their deviation from the inertial barotropic modes can, at least in principle, be used to estimate the Brunt-Väisälä frequency and the degree of stratification.

5.1. Acknowledgments

This work was supported by STFC through grant number PP/E001025/1.

References

[1] Passamonti A, Haskell B, Andersson N, Jones D I and Hawke I 2008 (*Preprint* 0807.3457)
[2] Passamonti A, Haskell B, Andersson N 2008 (*Preprint* 0812.3569)
[3] Hachisu I 1986 *ApJS* **62** 461–499
[4] Reisenegger A and Goldreich P 1992 *ApJ* **395** 240–249
[5] Stergioulas N, Apostolatos T A and Font J A 2004 *MNRAS* **352** 1089 (*Preprint* astro-ph/0312648)
[6] Dimmelmeier H, Stergioulas N and Font J A 2006 *MNRAS* **368** 1609–1630 (*Preprint* astro-ph/0511394)

[7] Unno W, Osaki Y, Ando H, Saio H and Shibahashi H 1989 *Nonradial oscillations of stars* (Tokyo: University of Tokyo Press, 1989, 2nd ed.)
[8] Yoshida S and Lee U 2000 *ApJS* **129** 353–366 (*Preprint* `arXiv:astro-ph/0002300`)

Recent Developments in Gravity (NEB XIII) IOP Publishing
Journal of Physics: Conference Series **189** (2009) 012031 doi:10.1088/1742-6596/189/1/012031

Accelerating universe: recent observations and implications for extended gravity theories

Leandros Perivolaropoulos
Department of Physics, University of Ioannina, Greece
E-mail: leandros@uoi.gr

Abstract. Recent high quality cosmological observations have confirmed and mapped in detail the accelerating expansion of the universe. These observations involve geometrical methods (use of standard candles and standard rulers) and dynamical methods (measured growth rate of cosmological perturbations). The Cosmological Constant (effective equation of state parameter of w=-1) remains consistent with all current data as a driving force of the acceleration and can be generated by quantum fluctuations of the vacuum with a proper cutoff. A signature in the Casimir effect would be expected in that case. An Evolving Dark Energy Density (w=w(t)) is also allowed by the data and a subset of the allowed evolving forms, which corresponds to crossing of the line w=-1, is inconsistent with most models based on General Relativity. On the other hand, scalar tensor extensions of General Relativity are consistent with the full range of allowed expansion histories. Demanding consistency of Scalar-Tensor theories with solar system tests and full range of allowed expansion histories implies constraints on Newtons constant evolution G(t)

1. Introduction

The assumption of large scale homogeneity and isotropy of the universe combined with the assumption that general relativity is the correct theory on cosmological scales leads to the Friedman equation which in a flat universe takes the form

$$H^2(a) = \left(\frac{\dot{a}}{a}\right)^2 = \frac{8\pi G}{3}\rho(a), \tag{1}$$

where $a(t)$ is the scale factor of the universe and ρ its average energy density. Both sides of this equation can be observationally probed directly: The left side using mainly geometrical methods (measuring the luminosity and angular diameter distances $d_L(z)$ [1, 2, 3] and $d_A(z)$ [4, 5, 6, 7] with standard candles and standard rulers) shows an accelerating expansion at recent redshifts and the matter - radiation density part of the right side using dynamical and other methods (cosmic microwave background [7], large scale structure observations [9], lensing [10] etc). These observations have indicated [11] that the two sides of the Friedman equation (44) can not be equal if $\rho(a) = \rho_m(a) \sim a^{-3}$ even if a non-zero curvature is assumed. There are two possible resolutions to this puzzle: Either modify the right side of the Friedman equation (44) introducing a new form of *'dark' energy* ideal fluid component ($\rho(a) = \rho_m(a) + \rho_X(a)$) with suitable evolution in order to restore the equality or modify both sides by changing the way energy density affects geometry thus modifying the Einstein equations.

In the first class of approaches the required gravitational properties of dark energy (see [12, 13, 14, 15, 16, 17] for recent reviews) needed to induce the accelerating expansion are well described by its equation of state $w(z) = \frac{p_X(z)}{\rho_X(z)}$ which enters in the second Friedman equation as

$$\frac{\ddot{a}}{a} = -\frac{4\pi G}{3}(\rho_m + \rho_X(1+3w)), \tag{2}$$

implying that a negative pressure ($w < -1/3$) is necessary in order to induce accelerating expansion. The simplest viable example of dark energy is the cosmological constant [11, 18, 19] ($w = -1$). This example however even though consistent with present data lacks physical motivation. Questions like 'What is the origin of the cosmological constant?' or 'Why is the cosmological constant 10^{120} times smaller than its natural scale so that it starts dominating at recent cosmological times (coincidence problem)?' remain unanswered [20]. Attempts to replace the cosmological constant by a dynamical scalar field (quintessence [21, 22, 23]) have created a new problem regarding the initial conditions of quintessence which even though can be resolved in particular cases (tracker quintessence), can not answer the above questions in a satisfactory way.

The parameter $w(z)$ determines not only the gravitational properties of dark energy but also its evolution. This evolution is easily obtained from the energy momentum conservation

$$d(\rho_X a^3) = -p_X d(a^3), \tag{3}$$

which leads to

$$\rho_X = \rho_{0X} e^{-3\int_1^a \frac{da'}{a'}(1+w(a'))} = \rho_{0X} e^{3\int_0^z \frac{dz'}{1+z'}(1+w(z'))}. \tag{4}$$

Therefore the determination of $w(z)$ is equivalent to that of $\rho_X(z)$ which in turn is equivalent to the observed $H(z)$ from the Friedman equation (44) expressed as

$$H(z) = H_0[\Omega_{0\mathrm{m}}(1+z)^3 + \Omega_{0X} e^{3\int_0^z \frac{dz'}{1+z'}(1+w(z'))}]. \tag{5}$$

Thus, knowledge of $\Omega_{0\mathrm{m}}$ and $H(z)$ suffices to determine $w(z)$ which is obtained from equation (5) as [24]

$$w(z) = \frac{\frac{2}{3}(1+z)\frac{d\ln H}{dz} - 1}{1 - \frac{H_0^2}{H^2}\Omega_{0\mathrm{m}}(1+z)^3}. \tag{6}$$

In the second class of approaches the Einstein equations get modified and the new equations combined with the assumption of homogeneity and isotropy lead to a generalized Friedman equation of the form

$$f(H^2) = g(\rho_m), \tag{7}$$

where f and g are appropriate functions determined by the modified gravity theory [25, 26, 27, 28, 29, 30]. In this class of models, the parameter $w(z)$ can also be defined from equation (6) but it can not be interpreted as $\frac{p_X}{\rho_X}$ of a perfect fluid.

The simplest (but quite general) examples of modified gravity theories are scalar tensor theories [25, 31, 32, 33, 34] where the Newton's constant G is promoted to a function of a field Φ: $8\pi G \to \frac{1}{F(\Phi)}$ whose dynamics at the Lagrangian level is determined by a potential $U(\Phi)$. Assuming homogeneity and isotropy, the modified Friedman equation in these theories takes the form

$$H^2 = \frac{1}{3F}(\rho_m + \frac{1}{2}\dot{\Phi}^2 + U - 3H\dot{F}). \tag{8}$$

This equation reduces to a regular minimally coupled scalar field dark energy (quintessence) in the general relativity limit of a constant $F = \frac{1}{8\pi G}$. The positive nature of the kinetic term

$\dot{\Phi}^2$ however implies that certain types of behaviors of H may not be reproducible without invoking a time-dependent F. These types of behavior of $H(z)$ which include a $w(z)$ crossing the Phantom Divide Line (PDL) $w = -1$ are potential signatures of extended gravity theories and will be discussed in the next section. To identify this type of signatures, a detailed form of the observed $H(z)$ is required which may be obtained by a combination of multiple dark energy probes. Observational probes may be divided in two classes [35] according to the methods used to obtain $H(z)$.

- *Geometric methods* probe the large scale geometry of space-time directly through the redshift dependence of cosmological distances ($d_L(z)$ or $d_A(z)$). They thus determine $H(z)$ independent of the validity of Einstein equations.
- *Dynamical methods* determine $H(z)$ by measuring the evolution of energy density (background or perturbations) and using a gravity theory to relate them with geometry i.e. with $H(z)$. These methods rely on knowledge of the dynamical equations that connect geometry with energy and may therefore be used in combination with geometric methods to test these dynamical equations.

2. Geometrical Probes of Expansion

Examples of geometric probes include

(i) The measured supernova distance redshift relation $d_L(z)$ [1, 2, 3] which for a flat universe, is connected to $H(z)$ as

$$d_L(z) = (1+z)\int_0^z \frac{dz'}{H(z')}, \tag{9}$$

(ii) The measured [7, 8] angular diameter distance $d_A(z_{rec})$ to the sound horizon $r_s(z_{rec})$ at recombination

$$d_A(z_{rec}) = \frac{1}{1+z_{rec}}\int_0^{z_{rec}} \frac{dz'}{H(z')}, \tag{10}$$

(iii) The scale of the sound horizon measured at more recent redshifts (z_{BAO}) through large scale structure redshift survey correlation functions [4]

$$D_V(z) = \left[\left(\int_0^{z_{BAO}} \frac{dz}{H(z)}\right)^2 \frac{z_{BAO}}{H(z_{BAO})}\right]^{1/3}. \tag{11}$$

Early SnIa data put together with more recent such data through the Gold dataset [2, 36] have been used to reconstruct $w(z)$ and have demonstrated a mild preference for a $w(z)$ that crossed the PDL [37, 38]. However, this dataset has been shown to suffer from systematics due to the inhomogeneous origin of the data [39]. Nevertheless, a cosmological constant remained consistent with the Gold dataset but only at the 2σ level. More recent SnIa data (SNLS, ESSENCE, HST) [40] have demonstrated a higher level of consistency with ΛCDM and showed no trend for a redshift dependent equation of state cite.

On the other hand, the use of standard rulers (CMB+BAO) has rarely been studied independent of SnIa due to the small number of datapoints involved (see however [42]). It has been pointed out however that the latest BAO data [43] 'require slightly stronger cosmological acceleration at low redshifts than ΛCDM '. This statement is equivalent to a trend towards a $w(z) < -1$ at low z and therefore a possibility of crossing the PDL $w = -1$. In what follows we quantify this statement in some detail by comparing the best fit form of $w(z)$ obtained from the SnIa data to the corresponding form obtained from the CMB+BAO data. This comparison is done quantitatively by identifying the quality of fit of ΛCDM in the context of each dataset. In particular, we consider the Chevalier-Polarski-Linder (CPL) [44, 45] parametrization

$$w(z) = w_0 + w_1 \frac{z}{1+z}, \tag{12}$$

and assuming flatness, we identify the 'distance' in units of σ (σ-distance) of the parameter space point $(w_0, w_1) = (-1, 0)$ corresponding to ΛCDM from the best fit point $(\bar{w}_0, \bar{w}_1)$ for each dataset (SnIa standard candles or CMB+BAO standard rulers) and for several priors of (Ω_{0m}, Ω_b). We thus identify an interesting systematic difference in trends between the two datasets. We also discuss the implications of this difference in trends on the distance duality relation which measures quantitatively the agreement between luminosity and angular diameter distances. This relation has been shown to be respected when clusters of galaxies are used as standard rulers [46].

Thus, we assume a CPL parametrization for $w(z)$ and apply the maximum likelihood method separately for standard rulers (CMB+BAO) and standard candles (SnIa) assuming flatness. The corresponding late time form of $H(z)$ for the CPL parametrization used in what follows is

$$\begin{aligned} H^2(z) &= H_0^2[\Omega_{0m}(1+z)^3 + \\ &+ (1-\Omega_{0m})(1+z)^{3(1+w_0+w_1)} e^{\frac{-3w_1 z}{(1+z)}}]. \end{aligned} \tag{13}$$

At earlier times this needs to be generalized taking into account radiation i.e.

$$E^2(a) \equiv \frac{H(a)^2}{H_0^2} = \Omega_m (a + a_{eq}) a^{-4} + \Omega_{de} X(a), \tag{14}$$

where $a = 1/(1+z)$, $\Omega_{de} = 1 - \Omega_m - \Omega_{rad}$ and

$$\begin{aligned} X(a) &= \exp\left[-3 \int_1^a \frac{(1+w(a'))}{a'} da'\right] \\ &= a^{-3(1+w_0+w_1)} e^{-3w_1(1-a)}, \end{aligned} \tag{15}$$

with $w(a) = w_0 + w_1(1-a)$

2.1. Standard Rulers

2.1.1. CMB We use the datapoints $(R, l_a, \Omega_b h^2)$ of Ref. [42] where R, l_a are two shift parameters:

- The scaled distance to recombination

$$R = \sqrt{\Omega \frac{H_0^2}{c^2}}\, r(z_{CMB}), \tag{16}$$

where $r(z_{CMB})$ is the comoving distance from the observer to redshift z and is given by

$$r(z) = \frac{c}{H_0} \int_0^z \frac{dz}{E(z)}, \tag{17}$$

with $E(z) = H(z)/H_0$.

- The angular scale of the sound horizon at recombination

$$l_a = \pi \frac{r(a_{CMB})}{r_s(a_{CMB})}, \tag{18}$$

where $r_s(a_{CMB})$ is the comoving sound horizon at recombination given by

$$r_s(a_{CMB}) = \frac{c}{H_0} \int_0^{a_{CMB}} \frac{c_s(a)}{a^2 E(a)} da, \tag{19}$$

with the sound speed being $c_s(a) = 1/\sqrt{3(1+\bar{R}_b a)}$ and $a_{CMB} = \frac{1}{1+z_{CMB}}$, where $z_{CMB} = 1089$. Actually, z_{CMB} has a weak dependence on Ω_m and Ω_b, see Ref. [47] but we have checked that the sound horizon changes only to less than 0.1%. The quantity $\bar{R}_b$, is actually the photon-baryon energy-density ratio, and its value can be calculated using $\bar{R}_b = \frac{3}{4}\frac{\Omega_b h^2}{\Omega_\gamma h^2} = 31500 \Omega_b h^2 (T_{CMB}/2.7K)^{-4}$.

For a flat prior, the 3-year WMAP data (WMAP3) [7] measured best fit values are [42]

$$\mathbf{\bar{V}_{CMB}} = \begin{pmatrix} \bar{R} \\ \bar{l}_a \\ \bar{\Omega}_b h^2 \end{pmatrix} = \begin{pmatrix} 1.70 \pm 0.03 \\ 302.2 \pm 1.2 \\ 0.022 \pm 0.00082 \end{pmatrix}. \tag{20}$$

The corresponding normalized covariance matrix is [42]

$$\mathbf{C^{norm}_{CMB}} = \begin{pmatrix} 1 & -0.09047 & -0.01970 \\ -0.09047 & 1 & -0.6283 \\ -0.01970 & -0.6283 & 1 \end{pmatrix}, \tag{21}$$

from which the covariance matrix can be found to be:

$$(C_{CMB})_{ij} = (C^{norm}_{CMB})_{ij}\, \sigma_{\bar{V}^i_{CMB}} \sigma_{\bar{V}^j_{CMB}}, \tag{22}$$

where $\sigma_{\bar{V}^i_{CMB}}$ are the 1σ errors of the measured best fit values of eq. (20).

We thus use equations (20), (16) and (18) to define

$$\mathbf{X_{CMBs}} = \begin{pmatrix} R - 1.70 \\ l_a - 302.2 \\ \Omega_b h^2 - 0.022 \end{pmatrix}, \tag{23}$$

and construct the contribution of CMB to the χ^2 as

$$\chi^2_{CMB} = \mathbf{X_{CMB}}^{\mathbf{T}} \mathbf{C_{CMB}}^{-1} \mathbf{X_{CMB}}, \tag{24}$$

with

$$\mathbf{C_{CMB}}^{-1} = \begin{pmatrix} 1131.32 & 4.8061 & 5234.42 \\ 4.8061 & 1.1678 & 1077.22 \\ 5234.42 & 1077.22 & 2.48145 \times 10^6 \end{pmatrix}. \tag{25}$$

Notice that χ^2_{CMB} depends on four parameters (Ω, Ω_b, w_0 and w_1). Due to the large number of parameters involved, in what follows we will consider various different priors on the parameters Ω, Ω_b.

2.1.2. BAO As in the case of the CMB, we apply the maximum likelihood method using the datapoints [43]

$$\bar{\mathbf{V}}_{\mathbf{BAO}} = \begin{pmatrix} \frac{r_s(z_{CMB})}{D_V(0.2)} = 0.1980 \pm 0.0058 \\ \frac{r_s(z_{CMB})}{D_V(0.35)} = 0.1094 \pm 0.0033 \end{pmatrix}, \tag{26}$$

where the dilation scale (see also eq. (27))

$$D_V(z_{BAO}) = \left[\left(\int_0^{z_{BAO}} \frac{dz}{H(z)} \right)^2 \frac{z_{BAO}}{H(z_{BAO})} \right]^{1/3}, \tag{27}$$

encodes the visual distortion of a spherical object due to the non-euclidianity of a FRW spacetime, and is equivalent to the geometric mean of the distortion along the line of sight and two orthogonal directions. We thus construct

$$\mathbf{X}_{\mathbf{BAO}} = \begin{pmatrix} \frac{r_s(z_{\rm dec})}{D_V(0.2)} - 0.1980 \\ \frac{r_s(z_{\rm dec})}{D_V(0.35)} - 0.1094 \end{pmatrix}, \tag{28}$$

and using the inverse covariance matrix [43]

$$\mathbf{C}_{\mathbf{BAO}}{}^{-1} = \begin{pmatrix} 35059 & -24031 \\ -24031 & 108300 \end{pmatrix}, \tag{29}$$

we find the contribution of BAO to χ^2 as

$$\chi^2_{BAO} = \mathbf{X}_{\mathbf{BAO}}{}^{\mathbf{T}} \mathbf{C}_{\mathbf{BAO}}{}^{-1} \mathbf{X}_{\mathbf{BAO}}. \tag{30}$$

2.2. Standard Candles

2.2.1. SnIa We use the SnIa dataset of Davis et al. [40] consisting of four subsets: ESSENCE [41] (60 points), SNLS [3] (57 points), nearby [2] (45 points) and HST [36] (30 points).

These observations provide the apparent magnitude $m(z)$ of the supernovae at peak brightness after implementing correction for galactic extinction, K-correction and light curve width-luminosity correction. The resulting apparent magnitude $m(z)$ is related to the luminosity distance $D_L(z)$ through

$$m_{th}(z) = \bar{M}(M, H_0) + 5log_{10}(D_L(z)), \tag{31}$$

where in a flat cosmological model

$$D_L(z) = (1+z) \int_0^z dz' \frac{H_0}{H(z'; \Omega, w_0, w_1)}, \tag{32}$$

is the Hubble free luminosity distance ($H_0 d_L$), and $\bar{M}$ is the magnitude zero point offset and depends on the absolute magnitude M and on the present Hubble parameter H_0 as

$$\begin{aligned} \bar{M} &= M + 5log_{10}(\frac{H_0^{-1}}{Mpc}) + 25 = \\ &= M - 5log_{10}h + 42.38. \end{aligned} \tag{33}$$

The parameter M is the absolute magnitude which is assumed to be constant after the above mentioned corrections have been implemented in $m(z)$.

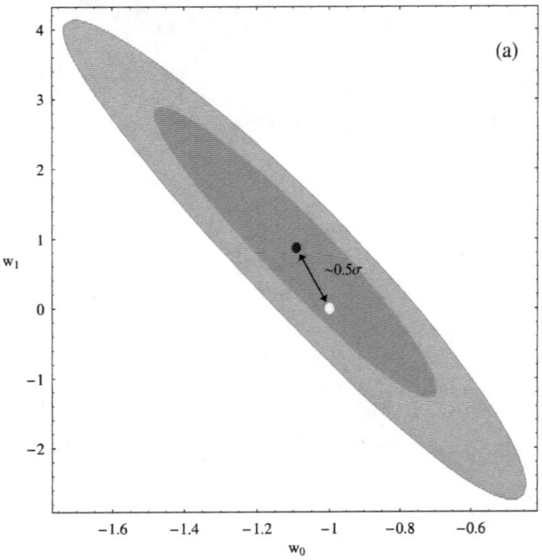

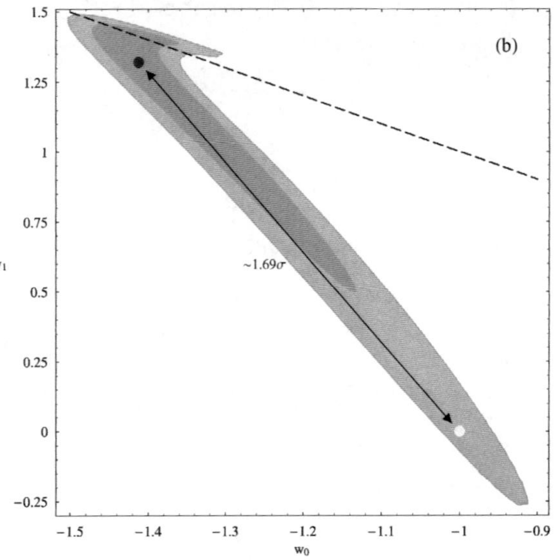

Figure 1. The 68.3% and 95.4% χ^2 confidence contours in the $w_0 - w_1$ parameter space for each dataset category for $\Omega_{0m} = 0.24$, $\Omega_b = 0.042$. The blue dots correspond to the $w_0 - w_1$ best fit while the yellow dots to ΛCDM (-1,0).

The SnIa datapoints are given after the corrections have been implemented, in terms of the distance modulus

$$\mu_{obs}(z_i) \equiv m_{obs}(z_i) - M. \tag{34}$$

The theoretical model parameters are determined by minimizing the quantity

$$\chi^2_{SnIa}(\Omega, w_0, w_1) = \sum_{i=1}^{N} \frac{(\mu_{obs}(z_i) - \mu_{th}(z_i))^2}{\sigma^2_{\mu\, i}}, \tag{35}$$

where $N = 192$ and $\sigma^2_{\mu\, i}$ are the errors due to flux uncertainties, intrinsic dispersion of SnIa absolute magnitude and peculiar velocity dispersion. These errors are assumed to be Gaussian and uncorrelated. The theoretical distance modulus is defined as

$$\mu_{th}(z_i) \equiv m_{th}(z_i) - M = 5log_{10}(D_L(z)) + \mu_0, \tag{36}$$

where

$$\mu_0 = 42.38 - 5log_{10}h, \tag{37}$$

and μ_{obs} is given by (34). The steps we followed for the minimization of (35) are described in detail in Refs [49, 37, 38].

2.3. Results

We consider separately the standard ruler data ($\chi^2_{SR} \equiv \chi^2_{CMB} + \chi^2_{BAO}$) and the standard candle data (χ^2_{SnIa}) and perform minimization of the corresponding χ^2 with respect to the parameters w_0 and w_1 for various priors of Ω and Ω_b in the 2σ range of the the WMAP3 best fit i.e. $0.21 \leq \Omega \leq 0.27$, $0.034 \leq \Omega_b \leq 0.049$.

In Fig 1 we show the 68.3% and 95.4% χ^2 confidence contours in the $w_0 - w_1$ parameter space for the two dataset categories (standard ruler and standard candle data) for $\Omega = 0.24$ (the best fit of the WMAP3 CMB data [7]). Fig. 1a shows the $w_0 - w_1$ contours obtained using SnIa data [40] (standard candles) while Fig. 1b shows the corresponding contours assuming CMB+BAO data [42, 43] (standard rulers). The blue dots correspond to the (w_0, w_1) best fit, the yellow dots to ΛCDM $(w_0, w_1) = (-1, 0)$. The distance in units of σ (σ-distance d_σ) of the best fit to ΛCDM was found by converting $\Delta\chi^2 = \chi^2_{\Lambda CDM} - \chi^2_{min}$ to d_σ i.e. solving [50]

$$1 - \Gamma(1, \Delta\chi^2/2)/\Gamma(1) = Erf(d_\sigma/\sqrt{2}), \tag{38}$$

for d_σ (σ-distance), where $\Delta\chi^2$ is the χ^2 difference between the best-fit and ΛCDM and $Erf()$ is the error function. Notice that ΛCDM is consistent at less than 1σ level according to the SnIa data ($d_\sigma^{SnIa} \simeq 0.5$ in Fig. 1a) while the corresponding consistency level reduces to $d_\sigma^{SR} 1.7\sigma$ for the standard ruler CMB+BAO data. This mild difference in trends between standard candles and standard rulers persists also for all values of Ω in the 2σ range of WMAP3 best fit.

3. Dynamical Tests of LCDM

Even though these tests are presently the most accurate probe of dark energy, the mere determination of the expansion rate $H(z)$ is not able to provide significant insight into the properties of dark energy and distinguish it from models that attribute the accelerating expansion to modifications of general relativity. The additional observational input that is required is the growth function $\delta(z) \equiv \frac{\delta\rho}{\rho}(z)$ of the linear matter density contrast as a function of redshift. The combination of the observed functions $H(z)$ and $\delta(z)$ can provide significant insight into the properties of dark energy (e.g. sound speed, existence of anisotropic stress etc) or even distinguish it from modified gravity theories.

The corresponding parametrization of the linear growth function $\delta(z)$ can be made efficiently by introducing a growth index γ defined by

$$\frac{d\ln\delta(a)}{d\ln a} = \Omega_m(a)^\gamma, \tag{39}$$

where $a = \frac{1}{1+z}$ is the scale factor and

$$\Omega_m(a) \equiv \frac{H_0^2 \Omega_{0\mathrm{m}} a^{-3}}{H^2(a)}. \tag{40}$$

This parametrization was originally introduced by Wang and Steinhardt [51] (see also [52] for more recent discussions) and was shown to provide an excellent fit to $\frac{d\ln\delta(a)}{d\ln a}$ corresponding to various general relativistic cosmological models for specific values of γ. In particular, it was shown that for dark energy with slowly varying $w(z) \simeq w_0$ the parameter γ in a flat universe is

$$\gamma = \frac{3(w_0 - 1)}{6w_0 - 5}, \tag{41}$$

which for the ΛCDM case ($w = -1$) reduces to $\gamma = \frac{6}{11}$. It is therefore clear that the observational determination of the growth index γ can be used to test ΛCDM .

The observational determination of the growth index γ requires knowledge not only of $\delta(z)$ but also of $H(z)$ and $\Omega_{0\mathrm{m}}$ (see equation (40)). It is possible however to construct more direct tests of ΛCDM which require knowledge only of $\delta(z)$ and $H(z)$. Such a null test of ΛCDM was recently proposed in Ref. [53] where it was suggested that the validity of ΛCDM requires the following equality of observables

$$\frac{(H(z)^2)'}{(1+z)^2\delta'(0)^2}\int_0^\infty \frac{\delta(z)\delta'(z)}{(1+z)}dz + 1 = 0, \tag{42}$$

where $' \equiv \frac{d}{dz}$. In fact, as discussed section III, there is an improved version of this null test that does not involve derivative of the observable $H(z)$ and therefore it is less prone to observational errors.

Both of the ΛCDM tests discussed above (the growth index γ and the null test) require observational determination of $\delta(z)$. There are several observational approaches that can lead to the determination of $\delta(z)$. For example, redshift distortions of galaxy power spectra [54], the rms mass fluctuation $\sigma_8(z)$ inferred from galaxy and $Ly-\alpha$ surveys at various redshifts [55]- [56], weak lensing statistics [57], baryon acoustic oscillations [58], Sachs-Wolfe effect etc. Unfortunately, the currently available data are limited in number and accuracy and come mainly from the first two categories. They involve significant error bars and non-trivial assumptions that hinder a reliable determination of $\delta(z)$. In addition, a large part of the available data are at high redshifts ($z > 1$) where ΛCDM is degenerate with most other dark energy models since dark energy is subdominant compared to matter at high redshifts in most models.

Nevertheless, it is still instructive to consider the presently available data to investigate the possible weak constraints that can be imposed on ΛCDM . Such a task serves two purposes

(i) It can be used as a paradigm for the time when more accurate data will be available
(ii) It can provide constraints from a dynamical test which are orthogonal and completely independent from the usual geometrical tests.

Thus, in what follows we use a wide range of presently available data on both redshift distortions and $\sigma_8(z)$ to determine the observed growth index γ and test ΛCDM in two ways

(i) Comparing the measured value of γ with the ΛCDM prediction $\gamma = \frac{6}{11}$
(ii) Implementing a new null test that exploits the consistency between $H(z)$ and $\delta(z)$ in the context of ΛCDM .

3.1. Fitting the Growth Index

According to general relativity, the equations that determine the evolution of the density contrast δ in a flat background consisting of matter with density ρ_m and dark energy with $\rho_{de} = \frac{p_{de}}{w}$ are of the form

$$\ddot{\delta} + 2\frac{\dot{a}}{a}\dot{\delta} = 4\pi G\rho_m\delta, \tag{43}$$

$$\left(\frac{\dot{a}}{a}\right)^2 = \frac{8\pi G}{3}\left(\rho_m + \rho_{de}\right), \tag{44}$$

$$2\frac{\ddot{a}}{a} + \left(\frac{\dot{a}}{a}\right)^2 = -8\pi G w\rho_{de}. \tag{45}$$

It is straightforward to change variables in eq. (43) from t to $\ln a$ ($\frac{d}{dt} = H\frac{d}{d\ln a}$) to obtain [51]

$$(\ln\delta)'' + (\ln\delta)'^2 + (\ln\delta)'\left[\frac{1}{2} - \frac{3}{2}w(1-\Omega_m(a))\right] = \frac{3}{2}\Omega_m(a), \tag{46}$$

where we used (45) and

$$\Omega_m(a) \equiv \frac{\rho_m(a)}{\rho_m(a) + \rho_{de}(a)}, \tag{47}$$

as in (40). A further change of variables from $\ln a$ to $\Omega_m(a)$ can be made by considering the differential of eq. (47) and using energy conservation ($d\rho = -3(\rho + p)d\ln a$) leads to

$$d\Omega_m = 3w\Omega_m(1 + \Omega_m)d\ln a, \tag{48}$$

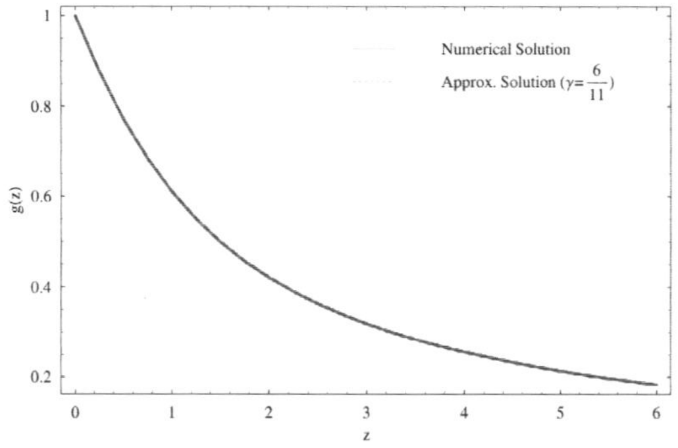

Figure 2. The numerically obtained solution of eq. (46) for the normalized growth of eq. (53) in the case of ΛCDM ($\Omega_{0\mathrm{m}} = 0.3$) along with the corresponding approximate result with $\gamma = \frac{6}{11}$ obtained from eq. (55). The agreement between the two approaches is excellent.

to get [51]

$$3w\Omega_m(1-\Omega_m)\frac{df}{d\Omega_m} + f^2 + f\left[\frac{1}{2} - \frac{3}{2}w(1-\Omega_m)\right] = \frac{3}{2}\Omega_m, \tag{49}$$

where we have set

$$f \equiv \frac{d\ln\delta}{d\ln a}. \tag{50}$$

Using the ansatz

$$f = \Omega_m^{\gamma(\Omega_m)}, \tag{51}$$

in eq. (49) and expanding around $\Omega_m = 1$ (good approximation especially at $z \gtrsim 1$) we find to lowest order

$$\gamma = \frac{3(w-1)}{6w-5}, \tag{52}$$

which reduces to $\gamma = \frac{6}{11}$ for ΛCDM $(w = -1)$.

Equations (51) and (52) provide excellent approximations to the numerically obtained form of $f(z)$. This is demonstrated in Fig. 2 where we plot the numerically obtained solution of eq. (46) for the normalized growth

$$g(z) \equiv \frac{\delta(z)}{\delta(0)}, \tag{53}$$

in the case of ΛCDM ($\Omega_{0\mathrm{m}} = 0.3$) along with the corresponding approximate result

$$g(z) = e^{\int_1^{\frac{1}{1+z}} \Omega_m(a)^\gamma \frac{da}{a}}, \tag{54}$$

with $\gamma = \frac{6}{11}$ obtained from

$$f(\Omega_{0\mathrm{m}}, \gamma, a) = a\frac{d\delta/da}{\delta} = \Omega_m(a)^\gamma. \tag{55}$$

Our goal in this section is to fit the parameter γ using observational data and compare it with the value $\gamma = \frac{6}{11}$ of ΛCDM . The most useful currently available data that can be used to constrain $\delta(z)$ (and γ) involve the redshift distortion parameter β [59] observed through the

Table 1. The currently available data for the parameters β and b at various redshifts along with the inferred growth rates and references.

z	β	b	f_{obs}	**Ref.**
0.15	0.49 ± 0.09	1.04 ± 0.11	0.51 ± 0.11	[54], [9]
0.55	0.45 ± 0.05	1.66 ± 0.35	0.75 ± 0.18	[62]
1.4	$0.60^{+0.14}_{-0.11}$	1.5 ± 0.20	0.90 ± 0.24	[63]
3.0	$-$	$-$	1.46 ± 0.29	[61]

anisotropic pattern of galactic redshifts on cluster scales (for a pedagogical discussion see [38]). The parameter β is related to the growth rate f as

$$\beta = \frac{d\ln\delta/d\ln a}{b} = \frac{f}{b}, \tag{56}$$

where b is the bias factor connecting total matter perturbations δ and galaxy perturbations δ_g ($b = \frac{\delta_g}{\delta}$).

The currently available data for the parameters β and b at various redshifts are shown in Table I along with the inferred growth rates and references. This is an expanded version of the dataset used in Ref. [60] where a similar analysis was performed using a different parametrization suitable for modified gravity models. Notice however that these have assumed ΛCDM (with $\Omega_{0\mathrm{m}} = 0.3$) when converting redshifts to distances for the power spectra and therefore their use to test models different from ΛCDM may not be reliable. In the same Table we also show the growth rate obtained in Ref. [61] directly from the change of power spectrum $Ly-\alpha$ forest data in SDSS at various redshift slices.

Using the data of Table I we can perform a maximum likelihood analysis in order to find γ and check its consistency with the ΛCDM value $\frac{6}{11}$. We thus construct

$$\chi_f^2(\Omega_{0\mathrm{m}}, \gamma) = \sum_i \left[\frac{f_{obs}(z_i) - f_{th}(z_i, \gamma)}{\sigma_{f_{obs}}}\right]^2, \tag{57}$$

where f_{obs} and σ_{fobs} are obtained from Table I while $f_{th}(z_i, \gamma)$ is obtained from eq. (51).

An alternative observational probe of the growth function $\delta(z)$ is the redshift dependence of the rms mass fluctuation $\sigma_8(z)$ defined by

$$\sigma^2(R, z) = \int_0^\infty W^2(kR)\Delta^2(k, z)\frac{dk}{k}, \tag{58}$$

with

$$W(kR) = 3\left(\frac{\sin(kR)}{(kR)^3} - \frac{\cos(kR)}{(kR)^2}\right)r, \tag{59}$$

$$\Delta^2(kz) = 4\pi k^3 P_\delta(k, z), \tag{60}$$

with $R = 8h^{-1}Mpc$ and $P_\delta(k, z)$ the mass power spectrum at redshift z. The function $\sigma_8(z)$ is connected with $\delta(z)$ as

$$\sigma_8(z) = \frac{\delta(z)}{\delta(0)}\sigma_8(z=0), \tag{61}$$

Table 2. The currently available data for the rms fluctuation $\sigma_8(z)$ at various redshifts and references.

z	σ_8	σ_{σ_8}	**Ref.**
2.125	0.95	0.17	[55]
2.72	0.92	0.17	
2.2	0.92	0.16	[56]
2.4	0.89	0.11	
2.6	0.98	0.13	
2.8	1.02	0.09	
3.0	0.94	0.08	
3.2	0.88	0.09	
3.4	0.87	0.12	
3.6	0.95	0.16	
3.8	0.90	0.17	
0.35	0.55	0.10	[64]
0.6	0.62	0.12	
0.8	0.71	0.11	
1.0	0.69	0.14	
1.2	0.75	0.14	
1.65	0.92	0.20	

which implies

$$s_{th}(z_1, z_2) \equiv \frac{\sigma_8(z_1)}{\sigma_8(z_2)} = \frac{\delta(z_1)}{\delta(z_2)} = \frac{e^{\int_1^{\frac{1}{1+z_1}} \Omega_m^\gamma(a) \frac{da}{a}}}{e^{\int_1^{\frac{1}{1+z_2}} \Omega_m^\gamma(a) \frac{da}{a}}}, \tag{62}$$

where we made use of eq. (54). Most of the currently available datapoints $\sigma_8(z_i)$ originate from the observed redshift evolution of the flux power spectrum of $Ly-\alpha$ forest [56]. These datapoints are shown in Table II along with the corresponding reference sources. These data are not as useful as the redshift distortion factors for the determination of γ for two reasons

(i) The rms fluctuation $\sigma_8(z)$ is not connected directly with the growth rate $f(z)$. Instead, it is related with $f(z)$ through the integral of eq. (54).

(ii) Most of the $Ly-\alpha$ σ_8 data appear at high redshifts where ΛCDM is degenerate with most other dark energy models.

Using the data of Table II we construct the corresponding χ_s^2 defined as

$$\chi_s^2(\Omega_{0\mathrm{m}}, \gamma) = \sum_i \left[\frac{s_{obs}(z_i, z_{i+1}) - s_{th}(z_i, z_{i+1})}{\sigma_{obs}}\right]^2, \tag{63}$$

where $\sigma_{s_{obs}}$ is derived by error propagation from the corresponding 1σ errors of $\sigma_8(z_i)$ and $\sigma_8(z_{i+1})$ while $s_{th}(z_i, z_{i+1})$ is defined in eq. (62). We can thus construct the combined $\chi_{tot}^2(\Omega_{0\mathrm{m}}, \gamma)$ as

$$\chi_{tot}^2((\Omega_{0\mathrm{m}}, \gamma) \equiv \chi_f^2((\Omega_{0\mathrm{m}}, \gamma) + \chi_s^2((\Omega_{0\mathrm{m}}, \gamma). \tag{64}$$

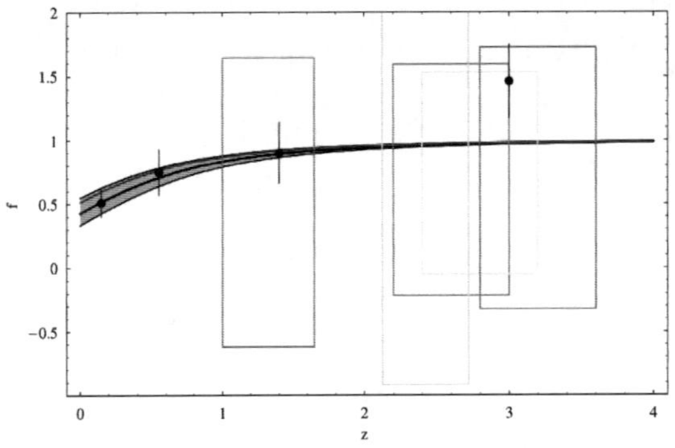

Figure 3. The cosmological data for the growth rate $f(z)$ along with the best theoretical fit $f = \Omega_m(z)^\gamma$ with $\Omega_{0m} = 0.3$ and the corresponding 1σ errors (shaded region). The errorboxes on f are obtained using the ratios at the specific redshifts. Clearly, the best fit shows a minor difference from ΛCDM only at low redshifts.

Setting $\Omega_{0m} = 0.3$ and minimizing χ^2_{tot} with respect to γ we find

$$\gamma = 0.707^{+0.226}_{-0.189}, \tag{65}$$

which differs somewhat from the corresponding result of Ref. [60] because we have used a broader dataset, we have used a different parametrization for f and we have assumed ΛCDM as our fiducial model thus avoiding the marginalization of the parameter w_0. The result (65) indicates that the ΛCDM value of $\gamma = \frac{6}{11} = 0.545$ is well within 1σ from the best fit and is clearly consistent with data. The imposed constraints however are rather weak and even a flat model with matter only (SCDM) predicting $\gamma = 0.6$ (set $w = 0$ in eq. (52)) is consistent with the data.

Ignoring the $\sigma_8(z)$ data of Table II leads to a negligible change in the best fit to $\gamma = 0.693 \pm 0.2$. This is consistent with the above discussion on the usefulness of these data. Alternatively, assuming the ΛCDM value for γ ($\gamma = \frac{6}{11}$) and minimizing with respect to Ω_{0m} we find $\Omega_{0m} = 0.226^{+0.09}_{-0.07}$.

The cosmological data for the growth rate $f(z)$ are shown in Fig.3 along with the best theoretical fit $f = \Omega_m(z)^\gamma$ with $\Omega_{0m} = 0.3$ and the corresponding 1σ errors (shaded region). In the same plot we show the corresponding $f_{\Lambda CDM}(z)$ obtained by solving numerically eq. (46) for $w = -1$ and $\Omega_{0m} = 0.3$. Clearly, the best fit shows a minor difference from ΛCDM only at low redshifts. We therefore conclude that ΛCDM is consistent with current data for the growth factor $\delta(z)$ and the consistency is maximized for $\Omega_{0m} = 0.23 \pm 0.08$.

4. A Possible Origin of the Cosmological Constant

The observed accelerating expansion of the universe indicates that either the universe is dominated by an energy form with repulsive gravitational properties (dark energy) or general relativity needs to be modified on cosmological scales (modified gravity) (see [12]and references therein). Dark energy may appear in the form of a cosmological constant (constant energy density [65]), quintessence (variable energy density due to an evolving scalar field [66]), a perfect fluid filling throughout space [67] etc. As discussed in the previous sections, the simplest among the above possibilities is the cosmological constant which is also favored by most cosmological data [68] compared to alternative more complicated models. It is therefore important to identify the possible origins of such a cosmological constant.

There have been several proposals concerning the origin of the cosmological constant even though none of them is completely satisfactory. Some of them include the zero point energy of the vacuum [69], anthropic considerations [70], brane cosmology [71], degenerate vacua [72] etc. The simplest physical mechanism that could lead to a cosmological constant is the zero-point energy of the vacuum made finite by an ultraviolet cutoff. A natural value of such a cutoff is the Planck scale leading to an energy density of the vacuum $\rho_V = (2.44 \times 10^{27})^4 eV^4$. However, such a value of the energy density leads to a cosmological constant which is 123 orders of magnitude larger than the observed one.

If the accelerating expansion of the universe is due to a cosmological constant emerging due to zero point fluctuations of the vacuum, the required energy density of the vacuum would be $\rho_\Lambda \simeq 10^{-11} eV^4$ corresponding to a cutoff scale (see eq. (66) below) of $\omega_c \simeq 10^{-3} eV$ ($l_c \simeq 0.1mm$). Even though there is no apparent physical motivation for such a cutoff scale, it can not be a priori excluded unless it is shown to be in conflict with specific experimental data.

There are various types of laboratory experiments which are able to probe directly quantities related to the zero point energy of the vacuum. Such experiments include measurements of the Casimir force [73] and measurements of the current noise in Josephson junctions (see [74, 75] for a debate on the effectiveness of Josephson junction experiments on probing the energy of the vacuum). Even though these non-gravitational experiments can only probe changes of the zero-point energy under variations of system parameters or of external couplings, these changes can be modified in the presence of a cutoff of the absolute value of the zero-point energy. Therefore, these experiments can become indirect probes of the absolute value of the zero-point energy under the assumption of the presence of a cutoff.

According to the Casimir effect [76, 77, 78], the presence of macroscopic bodies (like a pair of conducting plates) leads to a modification of the zero point vacuum fluctuations due to the introduction of non-trivial boundary conditions. This modification manifests itself as a force between the macroscopic bodies that distort the vacuum. When the role of the macroscopic bodies is played by a pair of conducting plates, the discreteness imposed on the fluctuation field modes, lowers the vacuum energy and leads to an attractive force between the plates (the *Casimir force*) which has been measured by several experiments [73]. In the presence of a finite cutoff length scale [79] l_c, electromagnetic field modes with wavelengths $\lambda < l_c$ are suppressed and the vacuum energy lowering due to the presence of the plates gets modified. This in turn leads to modifications of the Casimir force between the plates compared to the case where no cutoff has been imposed. In fact it may be shown [80, 75, 81] that when the plate separation d becomes significantly less than the cutoff scale l_c then the force between the plates changes sign and becomes repulsive! However, there is no experimental indication for change of sign of the Casimir force down to separations $d \simeq 100nm$. This imposes a constraint on the vacuum energy cutoff as $l_c < 100nm$ corresponding to a vacuum energy $\rho_V > 1eV^4$. Such a value of ρ_V is clearly inconsistent with the vacuum energy corresponding to the cosmological constant ($\rho_\Lambda \simeq 10^{-11} eV^4$). This inconsistency increases further if the contribution of interactions other that electromagnetism is included since in that case the predicted ρ_V from the Casimir effect would get even larger to include the contribution of other fields. It is therefore clear that either vacuum energy is not responsible for the accelerating expansion of the universe or there is a missing ingredient in the calculation of the Casimir force as a function of the cutoff scale.

One such possible missing ingredient could be the presence of a universal compact extra dimension [83] with compactification scale R. Even though the current experimental experimental bounds on R are quite stringent ($R \leq (300GeV)^{-1} \simeq 10^{-9}nm$ [84]) it is still instructive to consider this possibility in the context of the Casimir effect with a cutoff in the vacuum energy. The propagation of vacuum energy modes along this extra dimension modifies the field spectrum and affects accordingly the Casimir force.

In this section we address the following question: 'What is the effect of an extra compact

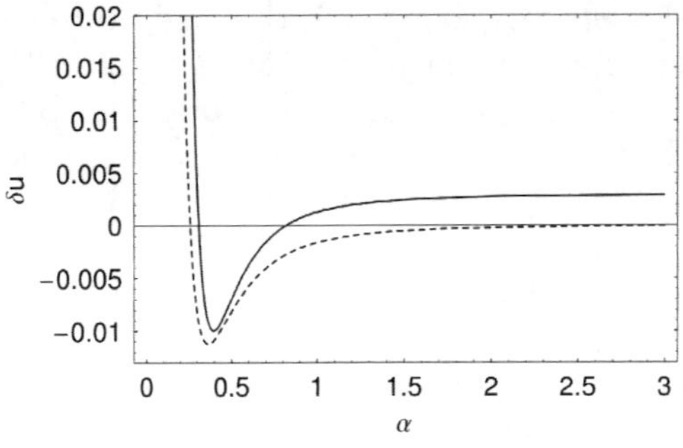

Figure 4. The normalized vacuum energy $\delta u \equiv \frac{\Delta u}{(\frac{3\omega_c^3\hbar}{4\pi c^2})}$ as a function α (dimensionless form of d) for $\beta = 0$ (dashed line) and $\beta = 1$ (continuous line). The curve corresponding to $\beta = 1$ has been shifted to lower values so that the locations of the minima can be compared more easily. Notice that the minimum shifts slightly to larger values as we increase the extra dimension size.

dimension on the Casimir force between two parallel plates and how does this effect change as a function of the vacuum energy cutoff scale'.

We start by deriving the vacuum energy in the region between two parallel plates. Consider a 3 dimensional Euclidean space with an extra compact dimension of scale R. In this space consider a box of square parallel conducting plates of large surface L^2 parallel to the xy plane placed at a small distance d apart. The vacuum energy of the quantized electromagnetic field in the region between the plates [76, 82, 83] is

$$\mathcal{E} = \frac{1}{2}\sum_{\vec{k},\lambda} \hbar\omega_{\vec{k}}\, g(\frac{\omega_{\vec{k}}}{\omega_c}), \tag{66}$$

where $g(x)$ is a UV cutoff regulator ($\lim_{x\to 0} g(x) = 1$, $\lim_{x\to\infty} g(x) = 0$) and λ counts the polarization modes.

Also

$$\omega_{\vec{k}} = c\sqrt{k_x^2 + k_y^2 + k_d^2 + k_R^2}, \tag{67}$$

where

$$k_x = \frac{\pi m_x}{L}\ k_y = \frac{\pi m_y}{L}\ k_d = \frac{\pi n}{d}\ k_R = \frac{N}{R}, \tag{68}$$

with $m_x.m_y, n, N = 0, 1, 2....$ In the large L limit eq. (66) takes the form

$$\frac{\mathcal{E}}{L^2} = \frac{3\hbar c}{4\pi}\sum_{n,N=0}^{\infty}{}' \int_0^\infty dq\, q\, \omega_q\, g(\frac{\omega_{\vec{k}}}{\omega_c}), \tag{69}$$

where $\omega_q^2 = c^2\sqrt{q^2 + (\frac{\pi n}{d})^2 + (\frac{N}{R})^2}$ and the factor 3 is due to the three polarization degrees of freedom in the presence of the extra dimension.

The prime (') on the sum implies that when $N = n = 0$ we should put a factor $\frac{1}{3}$ (only one degree of freedom from polarization) while if only one of the N, n is 0 we should put an extra factor $\frac{2}{3}$ (only two degrees of freedom from polarization). It is straightforward to show that the modification of the vacuum energy due to the presence of the plates is

$$\Delta u(d, R) \equiv u^{vac}(d, R) - u^{vac}_{>>}(d, R) =$$

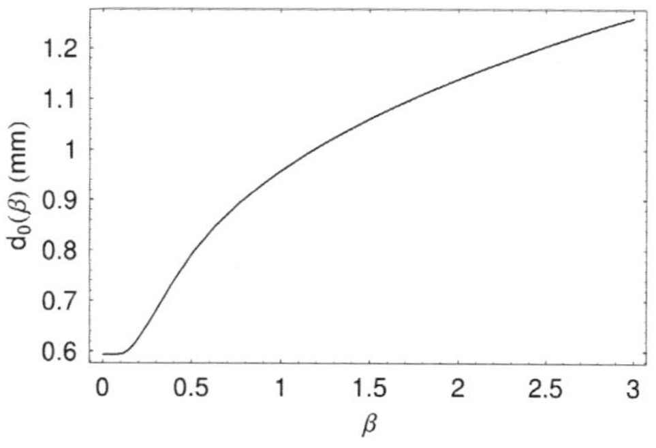

Figure 5. The plate distance $d_0(\beta)$ in mm for which the Casimir force changes sign, as a function of the dimensionless scale of compactification β. In converting from the dimensionless distance $\alpha_0(\beta)$ to the dimensionful distance $d_0(\beta)$ in mm we have assumed that the cutoff is equal to $\omega_0^\Lambda(\beta)$ (eq. (77)) required to reproduce the observed cosmological constant with vacuum energy.

$$= \frac{3\omega_c^3\hbar}{4\pi c^2}\sum_{N=0}^{\infty}\left(\sum_{n=0}^{\infty}F(n,N) - \int_0^\infty dqF(q,N)\right) - \\ -h(n,N), \tag{70}$$

where

$$h(n,N) = \frac{3\omega_c^3\hbar}{4\pi c^2}[\frac{2}{3}F(0,0) + \frac{1}{3}\sum_{n=1}^{\infty}F(n,0) + \\ +\frac{1}{3}\sum_{N=1}^{\infty}F(0,N) - \frac{1}{3}\int_0^\infty dq\; F(q,0)], \tag{71}$$

and

$$F(n,N) \equiv \int_{\sqrt{(\frac{n}{\alpha})^2+(\frac{N}{\beta})^2}}^{\infty} dv\; v^2\; g(v), \tag{72}$$

with

$$\alpha \equiv \frac{\omega_c d}{c\pi} \quad \beta \equiv \frac{\omega_c R}{c}. \tag{73}$$

In the limit $\beta \to 0$ corresponding to $R \to 0$ (no extra dimension), only the $N = 0$ mode contributes and the above expressions reduce to the well known [82, 81] forms of the regularized vacuum energy.

We now specify an exponential form for the UV cutoff function $g(v)$. Other smooth forms of $g(v)$ also lead to similar results. For $g(v) = e^{-v}$ we have $F(n,N) = e^{-s}(2 + s(s+2))$ with $s = \sqrt{(\frac{n}{\alpha})^2 + (\frac{N}{\beta})^2}$. In Fig. 4 we show the normalized vacuum energy $\delta u \equiv \frac{\Delta u}{(\frac{3\omega_c^3\hbar}{4\pi c^2})}$ as a function of α for $\beta = 0$ and $\beta = 1$. Notice that for a small value of R ($R\omega_c \to 0$) we recover the result of ref. [81] where the repulsive nature of the Casimir force was demonstrated for plate separations d much smaller than the cutoff scale $c\omega_c^{-1}$ without the presence of extra dimensions. For $\beta = 0$ (no extra dimension), the Casimir force becomes repulsive for $\alpha < \alpha_0 \simeq 0.36$. For a cutoff leading to the observed value of cosmological constant ($\omega_c \simeq 10^{-3}eV$) we find the critical separation $d_0 \simeq 0.6mm$ such that for separations $d < d_0$ the Casimir force becomes

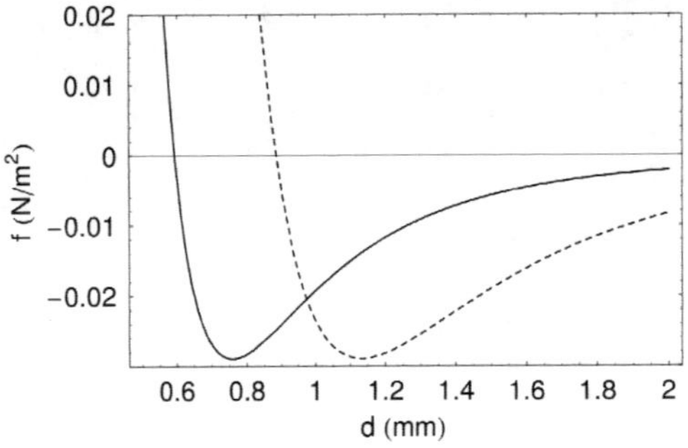

Figure 6. The Casimir force per unit surface predicted for a vacuum cutoff corresponding to the cosmological constant as a function of the plate distance d in mm for $\beta = 0$ (no extra dimension, continuous line) and $\beta = 1$ (dashed line).

repulsive. Since the Casimir force has been experimentally shown to be attractive down to plate separations $d \simeq 100nm$ [73] it becomes clear that the observed cosmological constant can not be due to vacuum energy with appropriate cutoff and no extra dimensions.

The introduction of a compact extra dimension with finite size has two effects:

- The cutoff scale $\omega_c(\beta)$ required to match the observed value of the cosmological constant slowly decreases
- The critical dimensionless separation $\alpha_0(\beta)$ for which the Casimir force changes sign increases (see Fig. 4).

In order to demonstrate the first effect we may evaluate the predicted vacuum energy density as a function of the dimensionless size β of the extra dimension. We have

$$\rho_V(\omega_c, \beta) = \frac{u^{vac}_{>>}(d)}{d} = \frac{3\hbar\omega_c^4}{4\pi^2 c^3} Q(\beta), \tag{74}$$

where

$$Q(\beta) = \sum_{N=0}^{\infty}{}' \int_0^\infty dq' \int_{\sqrt{q'^2+(\frac{N}{\beta})^2}}^\infty v^2 e^{-v}. \tag{75}$$

Demanding

$$\rho_V(\omega_c, \beta) = \rho_\Lambda = 10^{-11} eV^4, \tag{76}$$

we find

$$\omega_c^\Lambda(\beta) = (\frac{4\pi^2}{3} Q(\beta)^{-1} 10^{-11})^{1/4} eV. \tag{77}$$

In Fig. 5 we show a plot of the dimensionful plate distance $d_0(\beta)$ corresponding to a sign change of the Casimir force. This is found by converting the dimensionless minima α_0 of Fig. 4 to the corresponding dimensionful plate distances d_0 assuming a cutoff equal to $\omega_c^\Lambda(\beta)$ in eq. (73) for α. As shown in Fig. 5 $d_0(\beta)$ slowly increases with β as expected by inspection of Fig. 4 which shows the minimum α_0 of $\Delta u(\alpha, \beta)$ shifts to larger values as we increase the dimensionless size β of the extra dimension.

It is easy to convert the dimensionless parameter β to the compactification scale R for a particular value of the cutoff ω_c. For example for the cosmological constant cutoff $\omega_c = \omega_c^\Lambda(\beta)$ we obtain using eqs. (73), (77)

$$R = \frac{3\beta}{(8.2 \times Q(\beta))^{-1/4}}. \tag{78}$$

Finally, it is straightforward to use eqs (70), (73) to evaluate the Casimir force for $\omega_c = \omega_c^\Lambda(\beta)$. The resulting force is shown in Fig. 6 for $\beta = 0$ ($R = 0$) and $\beta = 1$ ($R \simeq 0.7mm$). Clearly, the plate separation where the force changes sign depends weakly on the size of the extra dimension and is always larger than $0.6mm$. We conclude that even in the presence of compact extra dimensions a cosmological constant induced by zero point vacuum fluctuations with appropriate cutoff is in conflict with experimental measurements of the Casimir force which indicate an attractive force down to separations of $d \simeq 100nm$. However, if future Casimir experiments detect a change of sign of the Casimir force at small plate separations, this could be interpreted as an indication of a finite cutoff of the vacuum energy. Even though such a vacuum energy would be too large to be consistent with the observed cosmological constant, the inclusion of other effects (eg fermions which contribute negative vacuum energy) could restore the consistency with cosmology.

5. Observational Signature of Extended Gravity Theories

Two central current questions in cosmology research are the following:

- What theoretical models are consistent with the currently detected form of $H(z)$?
- What are the generic predictions of these models with respect to $H(z)$ so that they can be ruled out or confirmed by more detailed observations of $H(z)$?

In a class of approaches the required gravitational properties of dark energy (see [12, 13, 14, 17] for recent reviews) needed to induce the accelerating expansion are well described by its equation of state $w(z) = \frac{p_X(z)}{\rho_X(z)}$. In this case the simplest model consistent with the currently detected form of $H(z)$ is the flat ΛCDM model. As discussed in section 1, according to this model, the universe is flat and its evolution is determined by general relativity with a cosmological constant through the Friedman equation

$$H(z)^2 = H_0^2(\Omega_{0\mathrm{m}}(1+z)^3 + \Omega_\Lambda), \tag{79}$$

where $\Omega_{0\mathrm{m}} = \frac{\rho_m}{\rho_c}$ is the current matter density normalized on the critical density for flatness ρ_c and $\Omega_\Lambda = 1 - \Omega_{0\mathrm{m}}$ is a constant density due to the cosmological constant. The main advantages of this model are *simplicity* and *predictability*: It has a single free parameter and it can be definitively ruled out by future observations. Its disadvantages are *lack of theoretical motivation* and *fine tuning*: The observed value of the cosmological constant is 120 orders of magnitude smaller than its theoretically expected value [19, 11].

Attempts to replace the cosmological constant by a minimally coupled dynamical scalar field (minimally coupled quintessence (MCQ) [23, 21, 22]) have led to models with a vastly larger number of parameters fueled by the arbitrariness of the scalar field potential. Despite this arbitrariness and vast parameter space, quintessence models are generically constrained to predicting a limited range of functional forms for $H(z)$ [25, 34]. This is a welcome feature which provides ways to either rule out or confirm this class of theories.

The allowed functional space of $H(z)$ can be further increased by considering models based on extensions of general relativity such as braneworlds [27, 85, 86, 87], $f(R)$ theories [88] or scalar-tensor theories [31, 25, 34] (extended quintessence (EXQ) [89]) where the accelerated expansion of the universe is provided by a non-minimally coupled scalar field. This class of theories is strongly motivated theoretically as it is predicted by all theories that attempt to quantize gravity and unify it with the other interactions. On the other hand, its parameter space is even larger than the corresponding space of MCQ since the later is a special case of EXQ. Local (eg solar system) gravitational experiments and cosmological observations constrain the allowed parameter space to be close to general relativity. Despite of these constraints however, the allowed by EXQ functional forms of $H(z)$ are significantly more than those allowed by MCQ.

The detailed identification of the forbidden $H(z)$ functional forms for both MCQ and EXQ is particularly important since it may allow future observations determining $H(z)$ to rule out one or both of these theories.

Previous studies [90] have mainly focused on the $H(z)$ limits of MCQ using a combination of plausibility arguments and numerical simulations of several classes of potentials. The low redshift $H(z)$ parameter space was divided in three sectors: a forbidden sector which could not correspond to any plausible quintessence model, a sector corresponding to the *freezing* quintessence scenario and a sector corresponding to the *thawing* quintessence scenario. In the freezing quintessence models, a field Φ which was already rolling towards its potential minimum prior to the onset of acceleration slows down ($\ddot{\Phi} < 0$) and creeps to a halt (freezes) mimicking a cosmological constant as it comes to dominate the universe. In the thawing quintessence models, the field has been initially halted by Hubble damping at a value displaced from its minimum until recently when it 'thaws' and starts to roll down to the minimum ($\ddot{\Phi} > 0$).

In this section, we extend these studies to the case of EXQ. Instead of using numerical simulations however, applied to specific potential classes, we use generic arguments demanding only the internal consistency of the theory. Thus our 'forbidden' sector when reduced to MCQ is smaller but more generic (applicable to a more general class of models) than that of Ref. [90] (see also Refs. [91, 92]).

The size and location of the sectors of the low z, $H(z)$ parameter space, depends sensitively on the assumed current time derivatives of the Newton's constant $G(t)$ and reduce to well known results in the MCQ limit of $G(t) = G_0 = const.$ Therefore an interesting interplay develops between local gravitational experimental constraints of the current time derivatives of $G(t)$ (eg $\frac{\dot{G}_0}{G_0}$ or $\frac{\ddot{G}_0}{G_0}$) and cosmological observations of $H(z)$ at low redshifts. For example, a constraint on $\frac{\dot{G}_0}{G_0} - \frac{\ddot{G}_0}{G_0}$ from local gravitational experiments defines the forbidden sector in the low z expansion coefficients of $H(z)^2$ in the context of EXQ. If such coefficients are measured to be in the forbidden sector by cosmological observations then EXQ could be ruled out. Alternatively, if such coefficients are measured to be in the forbidden sector for MCQ but in the allowed sector of EXQ (either 'freezing' or 'thawing') then this would rule out MCQ in favor of EXQ. As shown in what follows, current observational constraints on $H(z)$ imply significant overlap with the allowed sectors of both MCQ and EXQ. This however may well change in the near future with more accurate determinations of $H(z)$ and the time derivatives of G_0.

6. The Boundaries of Extended Quintessence

Extended quintessence is based on the simplest but very general (given its simplicity) extension of general relativity: *Scalar-Tensor theories.* In these theories Newton's constant obtains dynamical properties expressed through the potential $F(\Phi)$. The dynamics are determined by the Lagrangian density [25, 31]

$$\mathcal{L} = \frac{F(\Phi)}{2} \, R - \frac{1}{2} \, g^{\mu\nu} \partial_\mu \Phi \partial_\nu \Phi - U(\Phi) + \mathcal{L}_m[\psi_m; g_{\mu\nu}] \; , \tag{80}$$

where $\mathcal{L}_m[\psi_m; g_{\mu\nu}]$ represents matter fields approximated by a pressureless perfect fluid. The function $F(\Phi)$ is observationally constrained as follows:

- $F(\Phi) > 0$ so that gravitons carry positive energy [31].
- $\frac{1}{F}\left(\frac{dF}{d\Phi}\right)^2|_{z=0} < \frac{1}{4}10^{-4}$ from solar system observations [93].

In such a model the effective Newton's constant for the attraction between two test masses is given by

$$G_{eff}(t) = \frac{1}{F(t)} \frac{F(t) + 2\left(\frac{dF}{d\Phi}(t)\right)^2}{F(t) + \frac{3}{2}\left(\frac{dF}{d\Phi}(t)\right)^2} \simeq \frac{1}{F(t)} = G(t), \tag{81}$$

where the approximation of equation (81) is valid at low redshifts. Assuming a homogeneous Φ and varying the action corresponding to (80) in a background of a flat FRW metric, we find the coupled system of generalized Friedman equations

$$3FH^2 = \rho + \frac{1}{2}\dot{\Phi}^2 - 3H\dot{F} + U, \tag{82}$$

$$-2F\dot{H} = \rho + p + \dot{\Phi}^2 + \ddot{F} - H\dot{F}, \tag{83}$$

where we have assumed the presence of a perfect fluid ($\rho = \rho_m, p \simeq 0$) playing the role of matter fields. Expressing in terms of redshift and eliminating the potential U from equations (82), (83) we find [31, 34]

$$\Phi'^2 = -F'' - \left[(lnH)' + \frac{2}{1+z}\right] F' + 2\frac{(lnH)'}{1+z}F - $$
$$-3(1+z)\Omega_{0\mathrm{m}}\left(\frac{H_0}{H}\right)^2 F_0, \tag{84}$$

where $'$ denotes derivative with respect to redshift and F_0 is set to 1 in units of $\frac{1}{8\pi G_0}$ and corresponds to the present value of F. Alternatively, expressing in terms of redshift and eliminating the kinetic term Φ'^2 from equations (82), (83) we find

$$U = \frac{(1+z)^2H^2}{2}[F'' + \left[(\ln H)' - \frac{4}{1+z}\right] F' +$$
$$+ \left[\frac{6}{(1+z)^2} - \frac{2}{1+z}(\ln H)'\right] F -$$
$$- 3(1+z)\left(\frac{H_0}{H}\right)^2 F_0\Omega_{m,0}]. \tag{85}$$

We now wish to explore the observational consequences that emerge from the following generic inequalities anticipated on a purely theoretical level

$$\Phi'(z)^2 > 0, \tag{86}$$

$$U'(z) > 0, \tag{87}$$

$$(\Phi'(z)^2)' > 0, \quad (freezing), \tag{88}$$

$$(\Phi'(z)^2)' < 0, \quad (thawing). \tag{89}$$

The inequality (86) is generic and merely states that the scalar field in scalar-tensor theories is real as it is directly connected to an observable quantity (Newton's constant). The inequality (87) is also very general as it merely states that the scalar field rolls down (not up) its potential. This inequality is not as generic as (86) since it implicitly assumes a monotonic potential. Finally, the inequality (88) ((89)) denotes a scalar field which decelerates (accelerates) as it rolls down its potential thus corresponding to a freezing (thawing) quintessence model.

Since we are interested in the observational consequences of equations (86)-(89) at low redshifts, we consider expansions of equations (84) and (85) around $z = 0$ expanding $F(z)$, $H(z)^2\, U(z)$, $\Phi(z)$ as follows [33]:

$$F(z) = 1 + F_1 z + F_2 z^2 + ..., \tag{90}$$
$$H(z)^2 = 1 + h_1 z + h_2 z^2 + ..., \tag{91}$$
$$U(z) = 1 + U_1 z + U_2 z^2 + ..., \tag{92}$$
$$\Phi(z) = 1 + \Phi_1 z + \Phi_2 z^2 + ..., \tag{93}$$

where we have implicitly normalized over F_0, H_0, U_0 and Φ_0.

It is straightforward to connect the expansion coefficients F_i of equation (90) with the current time derivatives of $G(t)$ using equation (81) and the time-redshift relation

$$\frac{dt}{dz} = -\frac{1}{H(z)(1+z)}. \tag{94}$$

For example for F_1 we have

$$F_1 = \frac{1}{F_0}\frac{dF}{dz}|_{z=0} = \frac{\dot{G}_0}{G_0 H_0} \equiv g_1, \tag{95}$$

where the subscript $_0$ denotes the present time and $H_0 \simeq 10^{-10}\ h\ yrs^{-1}$. Similarly, for F_2 we find

$$F_2 = g_1(g_1 - \frac{h_1+2}{4}) - \frac{g_2}{2}, \tag{96}$$

where we have defined

$$g_n \equiv \frac{G_0^{(n)}}{G_0 H_0^n}, \tag{97}$$

with the superscript $^{(n)}$ denoting the nth time derivative. We may now substitute the expansions (90), (91), (93) in equation (84) replacing the coefficients F_i by the appropriate combination of g_n. Equating terms order by order in z and ignoring terms proportional to g_1 due to solar system constraints which imply [94, 96]

$$|g_1| < 10^{-13} yrs^{-1} H_0^{-1} \simeq 10^{-3} h^{-1} \ll 1, \tag{98}$$

we find for the zeroth and first order in z

$$h_1 - 3\Omega_{0m} + g_2 = \Phi_1^2 > 0, \tag{99}$$
$$-h_1(1+h_1) + 2h_2 - 3\Omega_{0m}(1-h_1) - \\ -g_2(1+h_1) - g_3 = 4\Phi_1\Phi_2 = (\Phi'^2)'(z=0) \quad . \tag{100}$$

The inequality (99) defines a forbidden sector for extended quintessence for each value of g_2. For $g_2 = 0$ this reduces to the well known result that MCQ can not cross the phantom divide line (see equation (116) below). Equation (100) can be used along with (88), (89) to divide the allowed (h_1, h_2) parameter sector into a freezing quintessence sector $(\Phi'(z)^2)' > 0$ and a thawing quintessence sector $(\Phi'(z)^2)' < 0$ for each set of (g_2, g_3).

Unfortunately, solar system gravitational experiments have so far provided constraints for g_1 [94] (equation (98)) but not for g_i with $i \geq 2$. This lack of constraints is not due to lack of observational data quality but simply due to the fact that existing codes have parameterized $G(t)$ in the simplest possible way i.e. as a linear function of t. It is therefore straightforward

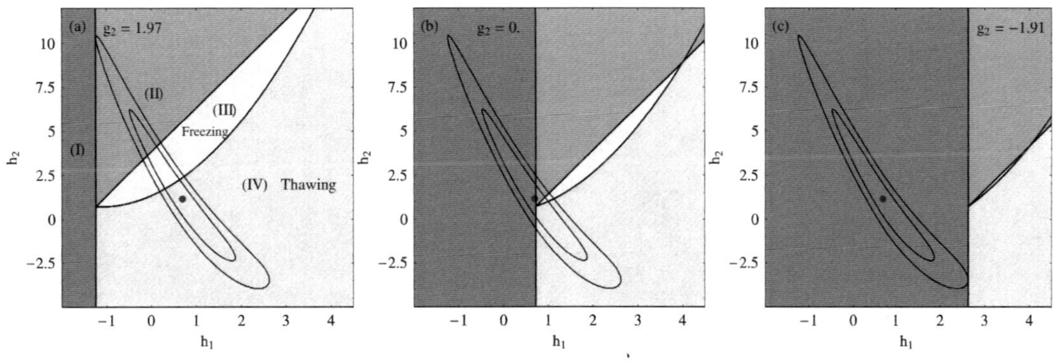

Figure 7. The $h_1 - h_2$ sectors of EXQ: Sector I is the forbidden sector where the scalar field becomes imaginary. Sector II is also forbidden and corresponds to a scalar field that rolls up its potential. Sector III corresponds to 'freezing' EXQ where the field is decelerating down its potential towards freezing. Sector IV corresponds to 'thawing' EXQ where the scalar field is accelerating down its potential. The forbidden sector I shrinks for a Newton's constant G that decreases with time ($g_2 > 0$) while the 'freezing sector' expands (Fig. 7a). The boundaries of the sectors are provided by (99),(100) and (110). The 1σ, 2σ χ^2 contours obtained from the SNLS SnIa dataset for $\Omega_{0m} = 0.24$ using the CPL parameterization, are also shown.

to extend this parameterizations to include more parameters thus obtaining constraints of g_i with $i \geq 2$. Such an analysis is currently in progress [95] but even before the results become available we can use some heuristic arguments to estimate the order of magnitude of the expected constraints on g_i ($i \geq 2$) given the current solar system data.

Current solar system gravity experiments are utilizing lunar laser ranging [96] and high precision planet ephemerides data [94] to compare the trajectories of celestial bodies with those predicted by general relativity. The possible deviation from the general relativity predictions is parameterized [96] using the PPN parameters β and γ, the first current derivative of Newton's constant $\frac{\dot{G}_0}{G_0}$ and a Yukawa coupling correction to Newton's inverse square law. These experiments have been collecting data for a time Δt of several decades [96, 94] i.e. $\Delta t = \mathcal{O}(100yrs)$. The current 1σ constraint [94] of

$$|\frac{\dot{G}_0}{G_0}| = |-0.2 \pm 0.5| \times 10^{-13} yrs^{-1} < 10^{-13} yrs^{-1}, \tag{101}$$

implies an upper bound on the total variation $\frac{\Delta G}{G_0}$ over the time-scale Δt of approximately

$$|\frac{\Delta G}{G_0}| \simeq |\frac{\dot{G}_0}{G_0}|\Delta t < 10^{-11}. \tag{102}$$

The same bound is obtained by considering the relative error in the orbital periods T of the Earth and other planets which at 1σ is [94]

$$\frac{\Delta T}{T} < 10^{-12}. \tag{103}$$

Given that the Keplerian orbital period is

$$T_K = \sqrt{\frac{4\pi^2 r^3}{Gm}} \sim G^{-\frac{1}{2}}, \tag{104}$$

we find

$$|\frac{\Delta T}{T}| = \frac{1}{2}|\frac{\Delta G}{G_0}| < 10^{-12}, \tag{105}$$

in rough agreement with (102).

Using the upper bound (102) and attributing any variation of G to a term quadratic in Δt we get

$$|\frac{\Delta G}{G_0}| \simeq |\frac{\ddot{G}_0}{G_0}|(\Delta t)^2 < 10^{-11}, \tag{106}$$

which implies that

$$|\frac{\ddot{G}_0}{G_0}| < 10^{-15} yrs^{-2} \rightarrow |g_2| < 10^5 h^{-2}, \tag{107}$$

giving a rough order of magnitude estimate of the upcoming constraints on g_2. Preliminary results from the analysis of solar system data indicate that [97]

$$\frac{\ddot{G}_0}{G_0} \simeq (4 \pm 5) \cdot 10^{-15} yrs^{-2}, \tag{108}$$

which is not far off the rough estimate of equation (107). Generalizing the above arguments to arbitrary order in Δt we find

$$|g_n| < 10^{8n-11} h^{-n}, \tag{109}$$

which implies that current solar system tests can not constrain g_n in any cosmologically useful way for $n \geq 2$.

We now return to the constraint equations (99) and (100) and supplement them by the constraint obtained from the inequality (87). Using the expansions (90), (91) and (92) in equation (85) and equating terms of first order in z we find

$$U_1 = U'(z=0) = \frac{1}{2}(5h_1 - 2h_2 - 9\Omega_{0\mathrm{m}} + 5g_2 + g_3) > 0, \tag{110}$$

where as usual we have ignored terms proportional to g_1 due to (98). We now may use (99), (100) and (110) to define the following sectors in the $h_1 - h_2$ parameter space for fixed g_1, g_2:

- A forbidden sector I where the inequality (99) is violated.
- A forbidden sector II where the inequality (110) is violated but not the inequality (99).
- An allowed sector III of freezing EXQ where the inequalities (86), (87) and (88) are respected while (89) is violated.
- An allowed sector IV of thawing EXQ where the inequalities (86), (87) and (89) are respected while (88) is violated.

These sectors are shown in Fig. 7 for $\Omega_{0\mathrm{m}} = 0.24$ [7], $g_3 = 0$ and three values of g_2: $g_2 = 1.97$, $g_2 = 0$ and $g_2 = -1.91$.

It would be interesting to superpose on these sectors of Fig. 7 the parameter χ^2 contours obtained by fitting to the SnIa data. It is not legitimate however to extrapolate the expansions in powers of z out to $z = \mathcal{O}(1)$ where the SnIa data extend. We thus use the CPL ansatz (12), (13). We superpose on the sectors of Fig. 7, the 1σ and 2σ $h_1 - h_2$ χ^2 contours obtained by fitting the Chevalier-Polarski-Linder (CPL) [45, 98] parametrization (12), (13) to the SNLS dataset [3] with the same $\Omega_{0\mathrm{m}}$ prior ($\Omega_{0\mathrm{m}} = 0.24$). Using equation (13) along with the expansion (91) it is easy to show that

$$\begin{aligned} h_1 &= 3(1 + w_0 - \Omega_{0\mathrm{m}} w_0), & (111)\\ h_2 &= \frac{3}{2}(2 + 5w_0(1 - \Omega_{0\mathrm{m}}) + \\ &+ (1 - \Omega_{0\mathrm{m}})(3w_0^2 + w_1)), & (112) \end{aligned}$$

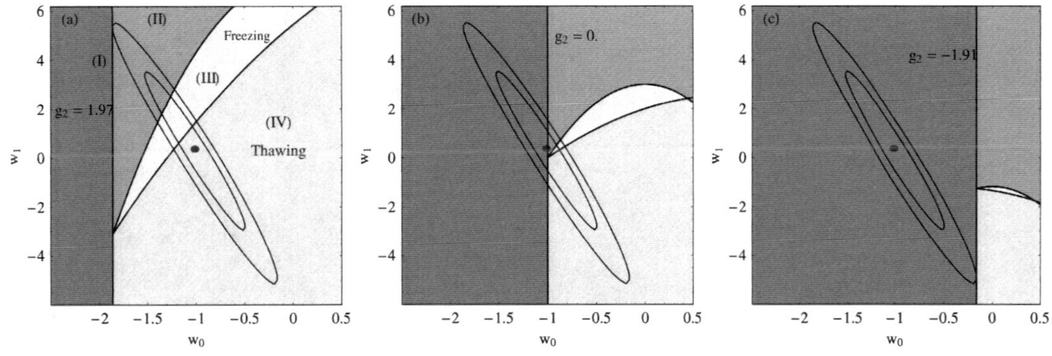

Figure 8. The sectors I-IV of Fig. 7 mapped on the $w_0 - w_1$ parameter space. The allowed sectors extend to $w_0 < -1$ for $g_2 > 0$.

so that the standard χ^2 contours in the $w_0 - w_1$ space (see eg [37]) can be easily translated to the $h_1 - h_2$ space of Fig. 7. The parameter $w(z)$ of eq. (6) is particularly useful and physically relevant in the MCQ limit of $g_i \to 0$. In that limit $w(z)$ becomes the MCQ equation of state parameter i.e.

$$w(z) = \frac{p_{MCQ}}{\rho_{MCQ}} = \frac{\frac{1}{2}\dot{\Phi}^2 - U(\Phi)}{\frac{1}{2}\dot{\Phi}^2 + U(\Phi)}, \tag{113}$$

as may be shown from the MCQ Friedman equations. The values of g_2 used in Figs 7a and 7c were motivated by demanding minimum and maximum overlap of the forbidden sector I with the 2σ $h_1 - h_2$ contour.

The following comments can be made with respect to Fig. 7:

- For values of $g_2 < -1.91$, the 2σ parameter contour obtained from SNLS lie entirely in the forbidden sector. This bound is independent of g_3 which does not enter in the inequality (99). Thus in the context of EXQ the constraint on g_2 obtained by the SNLS data at 2σ level is

$$g_2 = \frac{\ddot{G}_0}{G_0 H_0^2} > -1.91. \tag{114}$$

  Notice the dramatic improvement of this constraint (with respect to the lower bound) compared to the anticipated constraint of $-10^5 < g_2 < 10^5$ anticipated from solar system data (equation (107))!
- The parameter sector III of freezing quintessence is significantly smaller than sector IV of thawing quintessence and the difference is more prominent for smaller g_2.
- For $g_2 > 0$ the allowed parameter space increases significantly compared to MCQ ($g_2 = 0$). Therefore, if future cosmological observations show preference to the forbidden sectors I or II of Fig. 7b (MCQ) this could be interpreted as evidence for EXQ with $g_2 > 0$ (see also comments in Refs. [34, 38, 33])

Even though the plots of Fig. 7 capture the full physical content of our results it is useful to express the sectors I-IV in terms of parameter pairs other than $h_1 - h_2$ which are more common in the literature. Such parameters are the expansion coefficients w_i of $w(z)$

$$w(z) = w_0 + w_1 z + w_2 z^2 + ..., \tag{115}$$

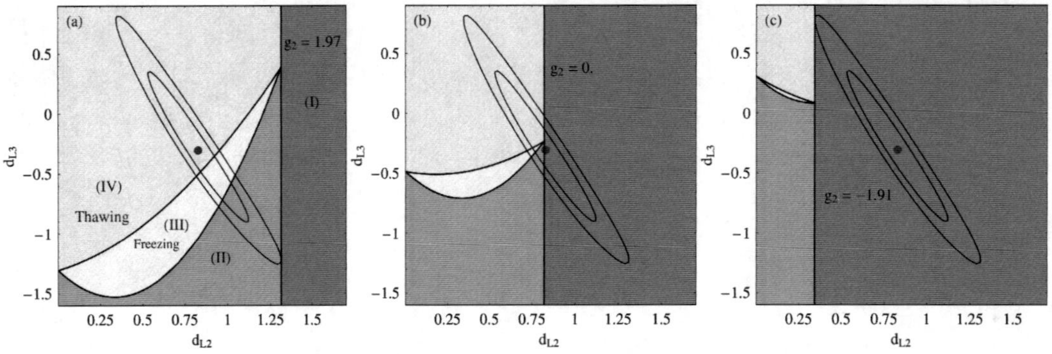

Figure 9. The sectors I-IV of Fig. 7 mapped on the $d_{L2} - d_{L3}$ parameter space. Notice that in this case the forbidden sector I is on the right (large d_{L2}).

which is connected to $H(z)$ via equation (6). By expanding both sides of equation (6) using equations (91) and (115) we may express h_i in terms of w_i thus rederiving equations (111) and (112) for $i = 0$ and $i = 1$. The result is identical since the w_0, w_1 expansion coefficients of equation (115) coincide with the w_0, w_1 coefficients of the CPL parametrization for a $w(z)$ given by equation (12). The advantage of using the parameters $w_0 - w_1$ instead of $h_1 - h_2$ is that they provide better contact with previous studies and can illustrate clearly the fact that a $g_2 > 0$ can provide a phantom behavior $w_0 < -1$ and crossing of the phantom divide line $w = -1$ in EXQ models. Using equations (111) and (112) we may express the constraint equations (99), (100) and (110) that define the sectors I-IV in Fig. 7 in terms of $w_0 - w_1$. The resulting equations are

$$3(1 - \Omega_{0\text{m}})(1 + w_0) + g_2 = \Phi_1^2 > 0, \tag{116}$$

$$3(1 - \Omega_{0\text{m}})((1 + w_0)(3\Omega_{0\text{m}} w_0 - 2) + w_1) - $$
$$-g_2(4 + 3(1 - \Omega_{0\text{m}})w_0) - g_3 = (\Phi'^2)'(z = 0), \tag{117}$$

$$\tfrac{3}{2}(1 - \Omega_{0\text{m}})(3(1 - w_0^2) - w_1) + \tfrac{5}{2} g_2 + \tfrac{g_3}{2} > 0. \tag{118}$$

In the MCQ limit ($g_2 \to 0$, $g_3 \to 0$) equation (118) has also been obtained in Ref. [91] as a generic limit of MCQ. Using now (116)-(118) along with the $w_0 - w_1$ χ^2 contours obtained from the SNLS dataset [3, 37] we construct Fig. 8 which is a mapping of Fig. 7 on the $w_0 - w_1$ parameter space. An interesting point of Fig. 8 is that for $g_2 > 0$ (Fig. 8a) the forbidden sector I shrinks significantly compared to MCQ (Fig. 8b) and allows for a $w_0 < -1$.

A final set of parameters we consider is the set of the expansion coefficients of the luminosity distance $d_L(z)$ which in a flat universe is connected to the Hubble expansion history $H(z)$ as

$$H(z)^{-1} = \frac{d}{dz}\left(\frac{d_L(z)}{1+z}\right). \tag{119}$$

Expanding $d_L(z)$ as

$$d_L(z) = z + d_{L2} z^2 + d_{L3} z^3 + ..., \tag{120}$$

and using the expansion (91) in equation (119) we may express the coefficients $h_1 - h_2$ in terms of $d_{L2} - d_{L3}$ as follows

$$h_1 = 4(1 - d_{L2}), \tag{121}$$
$$h_2 = 6(1 - 3d_{L2} + 2d_{L2}^2 - d_{L3}). \tag{122}$$

Substituting now equations (121), (122) in equations (99), (100) and (110) we obtain the sector equations in $d_{L2} - d_{L3}$ space as

$$4(1 - d_{L2}) - 3\Omega_{0\text{m}} + g_2 = \Phi_1^2 > 0, \tag{123}$$

$$\begin{aligned} &4d_{L2}(2d_{L2} + g_2 - 3\Omega_{0\text{m}}) + 9\Omega_{0\text{m}} - \\ &-8 - 12d_{L3} - 5g_2 - g_3 = (\Phi'^2)'(z = 0), \end{aligned} \tag{124}$$

$$\begin{aligned} &4 + 8d_{L2} - 12d_{L2}^2 + 6d_{L3} - \tfrac{9}{2}\Omega_{0\text{m}} + \\ &+\tfrac{5}{2}g_2 + \tfrac{g_3}{2} > 0. \end{aligned} \tag{125}$$

Using now equations (121), (122) to translate the $h_1 - h_2$ χ^2 contours to the $d_{L2} - d_{L3}$ parameter space and equations (123)-(125) to construct the sectors I-IV in the $d_{L2} - d_{L3}$ space we obtain Fig. 9. The advantage of Fig. 9 compared to Figs. 7 and 8 is that it refers to the parameters $d_{L2} - d_{L3}$ which are directly observable through the luminosity distances of SnIa without the need of any differentiation.

We have used generic theoretically motivated inequalities to investigate the space of observable cosmological expansion parameters that admits viable MCQ and EXQ theoretical models. Our inequalities are generic in the sense that they are independent of the specific features of any scalar field potential (eg scale, tracking behavior etc) and they only require that the models are internally consistent. The derived forbidden sectors which violate the above inequalities already have significant overlap with the parameter space which is consistent with observations at the 2σ level. This overlap which depends on the time derivatives of the Newton's constant G has lead to a useful constraint to the second derivative of G (equation (114)) which is significantly more stringent compared to the corresponding constraint (107) anticipated from solar system gravity experiments .

An important reason that limits the observable parameter space consistent with MCQ and EXQ is the fact that the scalar field potential energy can induce accelerating expansion but not beyond the limit corresponding to the cosmological constant ($w(z) = -1$) obtained when the field's evolution is frozen. Additional acceleration (superacceleration) can only be provided in the context of EXQ through the time variation of Newton's constant G. A decreasing $G(t)$ with time favors accelerating expansion and allows for superacceleration. This physical argument is reflected in our results. We found that given $\dot{G}_0 \simeq 0$ (neglecting g_1 since $|g_1| < 10^{-4}$ from solar system tests) a $\ddot{G}_0 > 0$ ($g_2 > 0$) decreases the forbidden sector and allows superacceleration ($w_0 < -1$). But $\dot{G}_0 \simeq 0$ with $\ddot{G}_0 > 0$ implies that we are currently close to a minimum of $G(t)$ with $G(t)$ being larger in the past i.e.

$$\frac{G(t)}{G_0} \simeq 1 + \frac{1}{2}g_2(H_0(t - t_0))^2. \tag{126}$$

Therefore a decreasing $G(t)$ corresponds to $\ddot{G}_0 > 0$ ($g_2 > 0$) (see Fig. 10) which in turn implies smaller forbidden sectors and allows for superacceleration in agreement with the above physical argument.

An additional interesting point is related to the construction of the χ^2 contours of Figs. 7-9. The SNLS data analysis involved in the construction of these contours did not take into account the possible evolution of SnIa due to the evolving $G(t)$. It is straightforward to take into account the evolution of G in the SnIa data analysis along the lines of Ref. [99]. In order to test the sensitivity of these contours with respect to the evolution of G we have repeated the χ^2 contour construction assuming a varying G according to the ansatz

$$G(z) = G_0(1 + \alpha\frac{z^2}{(1+z)^2}), \tag{127}$$

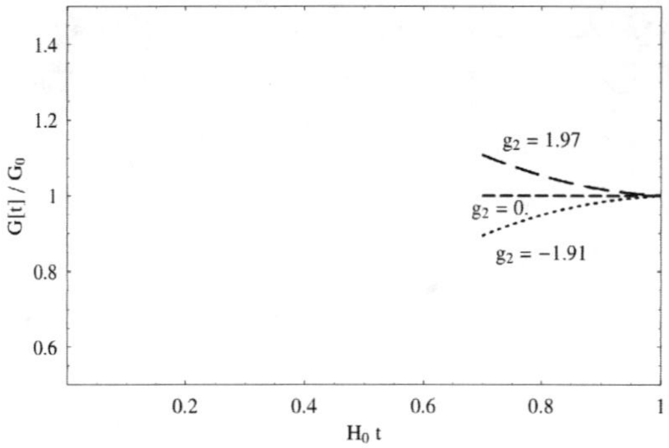

Figure 10. A $g_2 > 0$ implies a decreasing Newton's constant ($\frac{G(t)}{G_0} \simeq 1 + \frac{1}{2}g_2(H_0(t-t_0))^2$ around $t = t_0$).

which smoothly interpolates between the present value of $G = G_0$ and the high redshift value $G = G_0(1+\alpha)$ implying that

$$\alpha = \frac{\Delta G}{G_0}, \tag{128}$$

The parametrization (127) is consistent with both the solar system tests [94] ($\frac{\dot{G}_0}{G_0} \simeq 0$) and the nucleosynthesis constraints [100]

$$|\frac{G_{nuc} - G_0}{G_0}| < 0.2. \tag{129}$$

at 1σ, for $|\alpha| < 0.2$.

It is straightforward to evaluate g_1, g_2, g_3 in terms of α using the parametrization (127) and equation (94). The result is

$$g_1 = 0, \tag{130}$$
$$g_2 = 2\alpha, \tag{131}$$
$$g_3 = 3\alpha\left(-1 + 3w_0(-1+\Omega_{0m})\right). \tag{132}$$

We have considered the value of

$$\alpha = 0.2, \tag{133}$$

and repeated the SNLS data analysis taking into account the evolution of the SnIa absolute magnitude M due to the evolving G [101, 102, 103, 99] as

$$M = M_0 + \frac{15}{4} log\frac{G}{G_0}. \tag{134}$$

The steps involved in this analysis may be summarized as follows:

- Use the following magnitude redshift relation to fit to the SnIa data

$$m_{th}(z) = M_0 + 5 log d_L(z) + \frac{15}{4} log\frac{G(z)}{G_0}, \tag{135}$$

where $G(z)$ is given by (127) and $d_L(z)$ is connected to $H(z)$ in the usual geometrically defined way (119) i.e.

$$d_L(z) = (1+z)\int_0^z dz' \frac{1}{H(z')}, \tag{136}$$

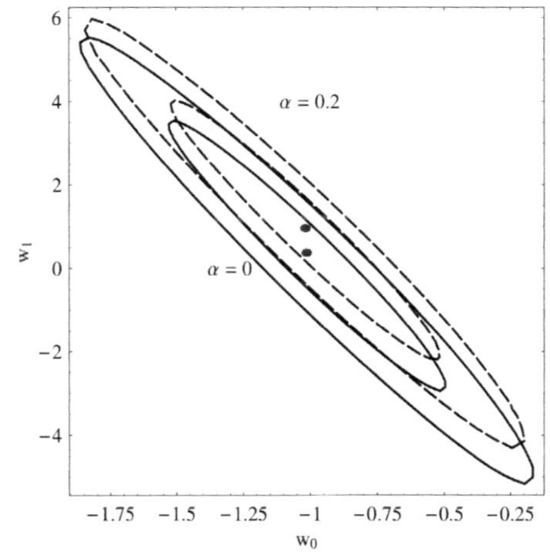

Figure 11. Taking into account a G variation consistent with the nucleosynthesis and solar system bounds introduces modifications to the $w_0 - w_1$ χ^2 contours which are not significant for the currently available SnIa data (for the dashed line $\alpha = 0.2$ and the continuous $\alpha = 0$).

- Use an $H(z)$ parametrization that incorporates the evolution of G (given by (127)) in the context of the CPL parametrization (13) i.e.

$$\begin{aligned} H^2(z) &= \frac{G(z)}{G_0} H_0^2 [\Omega_{0m}(1+z)^3 + \\ &+ (1-\Omega_{0m})(1+z)^{3(1+w_0+w_1)} e^{\frac{-3w_1 z}{(1+z)}}], \end{aligned} \tag{137}$$

- Minimize the χ^2 expression

$$\chi^2(w_0, w_1) = \sum_i \frac{(m^{obs}(z_i) - m^{th}(z_i; w_0, w_1))^2}{\sigma^2_{m^{obs}(z_i)}}, \tag{138}$$

  and find the corresponding 1σ and 2σ χ^2 contours along the lines of Refs. [37, 99]

The resulting χ^2 contours in the $w_0 - w_1$ space are shown in Fig. 11 (dashed lines) superposed with the corresponding contours constructed by neglecting the evolution of G (continuous lines). The change of the best fit $w_0 - w_1$ values and of the corresponding errorbars is minor (especially in the w_0 direction) and such that our main conclusion regarding the limiting values of g_2 remain practically unchanged even after the evolution of G is taken into account in the analysis. This justifies neglecting the evolution of G in the construction of the χ^2 contours for our purposes. However, as the quality of SnIa data improves, it becomes clear from Fig. 11 that the effects of an evolving G *consistent with nucleosynthesis and solar system constraints* on the data analysis can be significant! This possibility is further amplified if there are additional effects of an evolving G on the SnIa data analysis. For example [104, 105] it is possible that the time scale stretch factor s involved in the SnIa data analysis [3, 37] and arising from opacity effects in the stellar atmosphere may have a dependence on the Chandrasekhar mass and therefore on Newton's constant G. We have shown however that even if such effects are included in the SnIa data analysis our results of Fig. 11 (dashed line) do not change more than 10%.

7. Summary-Conclusions

The main points of this review may be summarized as follows:

- Current geometrical and dynamical cosmological observations constraining the Hubble expansion history are consistent with the existence of a cosmological constant corresponding to an equation of state parameter $w = -1$ but they are also consistent with dark energy having an effective evolving $w(z)$ that crosses the $w = -1$ line.
- The cosmological constant is currently the simplest and most probable cause of the accelerating expansion of the universe and may be generated by quantum fluctuations of the vacuum with a cutoff. A change of sign of the Casimir force is predicted in that case.
- An evolving equation of state parameter $w(z)$ that is crossing the phantom divide line $w = -1$ is inconsistent with minimally Coupled quintessence and also with scalar tensor quintessence if the Newton's constant $G(t)$ is increasing with time.
- Demanding consistency of scalar-tensor quintessence with local gravity experiments and with accelerating expansion crossing the $w = -1$ line implies a small present time derivative of the Newton's constant G_0 i.e.

$$|\frac{\dot{G}}{G_0 H_0}| < 10^{-4}, \tag{139}$$

and a positive second derivative of G_0 i.e.

$$\frac{\ddot{G}}{G_0 H_0^2}| > O(1). \tag{140}$$

References

[1] Riess A *et al* , 1998 Astron. J. **116** 1009; Perlmutter S J *et al* , 1999 Astroph. J. **517** 565; Bull.Am.Astron.Soc**29**,1351(1997); Tonry, J L *et al* , 2003 Astroph. J. **594** 1; Barris, B *et al* , 2004 Astroph. J. **602** 571; Knop R *et al* , 2003 Astroph. J. **598** 102;
[2] A. G. Riess *et al.* [Supernova Search Team Collaboration], Astrophys. J. **607**, 665 (2004) [arXiv:astro-ph/0402512].
[3] P. Astier *et al.*, Astron. Astrophys. **447**, 31 (2006) [arXiv:astro-ph/0510447].
[4] D. J. Eisenstein *et al.* [SDSS Collaboration], Astrophys. J. **633**, 560 (2005) [arXiv:astro-ph/0501171].
[5] S. W. Allen, R. W. Schmidt and A. C. Fabian, Mon. Not. Roy. Astron. Soc. **334**, L11 (2002) [arXiv:astro-ph/0205007].
[6] S. W. Allen, R. W. Schmidt, H. Ebeling, A. C. Fabian and L. van Speybroeck, Mon. Not. Roy. Astron. Soc. **353**, 457 (2004) [arXiv:astro-ph/0405340].
[7] D. N. Spergel *et al.*, arXiv:astro-ph/0603449.
[8] Y. Wang and P. Mukherjee, arXiv:astro-ph/0604051.
[9] L. Verde *et al.*, Mon. Not. Roy. Astron. Soc. **335**, 432 (2002) [arXiv:astro-ph/0112161].
[10] H. Hoekstra, H. K. C. Yee and M. D. Gladders, Astrophys. J. **577**, 595 (2002) [arXiv:astro-ph/0204295].
[11] V. Sahni and A. A. Starobinsky, Int. J. Mod. Phys. D **9**, 373 (2000) [arXiv:astro-ph/9904398].
[12] E. J. Copeland, M. Sami and S. Tsujikawa, arXiv:hep-th/0603057.
[13] T. Padmanabhan, arXiv:astro-ph/0603114.
[14] L. Perivolaropoulos, arXiv:astro-ph/0601014.
[15] N. Straumann, Mod. Phys. Lett. A **21**, 1083 (2006) [arXiv:hep-ph/0604231].
[16] J. P. Uzan, arXiv:astro-ph/0605313.
[17] V. Sahni and A. Starobinsky, arXiv:astro-ph/0610026.
[18] P. J. E. Peebles and B. Ratra, Rev. Mod. Phys. **75**, 559 (2003) [arXiv:astro-ph/0207347].
[19] S. M. Carroll, Living Rev. Rel. **4**, 1 (2001) [arXiv:astro-ph/0004075].
[20] S. Weinberg, arXiv:astro-ph/0005265.
[21] R. R. Caldwell, R. Dave and P. J. Steinhardt, Phys. Rev. Lett. **80**, 1582 (1998) [arXiv:astro-ph/9708069].
[22] I. Zlatev, L. M. Wang and P. J. Steinhardt, Phys. Rev. Lett. **82**, 896 (1999) [arXiv:astro-ph/9807002].
[23] B. Ratra and P. J. E. Peebles, Phys. Rev. D **37**, 3406 (1988).
[24] T. D. Saini, S. Raychaudhury, V. Sahni and A. A. Starobinsky, Phys. Rev. Lett. **85**, 1162 (2000) [arXiv:astro-ph/9910231]; D. Huterer and M. S. Turner, Phys. Rev. D **64**, 123527 (2001) [arXiv:astro-ph/0012510].
[25] B. Boisseau, G. Esposito-Farese, D. Polarski and A. A. Starobinsky, Phys. Rev. Lett. **85**, 2236 (2000) [arXiv:gr-qc/0001066].
[26] S. Nojiri, S. D. Odintsov and M. Sami, Phys. Rev. D **74**, 046004 (2006) [arXiv:hep-th/0605039].

[27] V. Sahni and Y. Shtanov, JCAP **0311**, 014 (2003) [arXiv:astro-ph/0202346].
[28] L. P. Chimento, R. Lazkoz, R. Maartens and I. Quiros, arXiv:astro-ph/0605450.
[29] S. M. Carroll, A. De Felice, V. Duvvuri, D. A. Easson, M. Trodden and M. S. Turner, Phys. Rev. D **71**, 063513 (2005) [arXiv:astro-ph/0410031].
[30] C. Deffayet, Phys. Lett. B **502**, 199 (2001) [arXiv:hep-th/0010186].
[31] G. Esposito-Farese and D. Polarski, Phys. Rev. D **63**, 063504 (2001) [arXiv:gr-qc/0009034].
[32] D. F. Torres, Phys. Rev. D **66**, 043522 (2002) [arXiv:astro-ph/0204504].
[33] R. Gannouji, D. Polarski, A. Ranquet and A. A. Starobinsky, arXiv:astro-ph/0606287.
[34] L. Perivolaropoulos, arXiv:astro-ph/0504582.
[35] E. Bertschinger, arXiv:astro-ph/0604485.
[36] A. G. Riess *et al.*, arXiv:astro-ph/0611572.
[37] S. Nesseris and L. Perivolaropoulos, Phys. Rev. D **72**, 123519 (2005) [arXiv:astro-ph/0511040].
[38] S. Nesseris and L. Perivolaropoulos, JCAP **0701**, 018 (2007) [arXiv:astro-ph/0610092].
[39] S. Nesseris and L. Perivolaropoulos, JCAP **0702**, 025 (2007) [arXiv:astro-ph/0612653].
[40] T. M. Davis *et al.*, Astrophys. J. **666**, 716 (2007) [arXiv:astro-ph/0701510].
[41] G. Miknaitis *et al.*, arXiv:astro-ph/0701043;
[42] Y. Wang and P. Mukherjee, arXiv:astro-ph/0703780.
[43] W. J. Percival, S. Cole, D. J. Eisenstein, R. C. Nichol, J. A. Peacock, A. C. Pope and A. S. Szalay, arXiv:0705.3323 [astro-ph].
[44] E. V. Linder, Phys. Rev. D **68**, 083503 (2003) [arXiv:astro-ph/0212301].
[45] M. Chevallier and D. Polarski, Int. J. Mod. Phys. D **10**, 213 (2001) [arXiv:gr-qc/0009008].
[46] J. P. Uzan, N. Aghanim and Y. Mellier, Phys. Rev. D **70**, 083533 (2004) [arXiv:astro-ph/0405620]; F. De Bernardis, E. Giusarma and A. Melchiorri, Int. J. Mod. Phys. D **15**, 759 (2006) [arXiv:gr-qc/0606029].
[47] W. Hu, M. Fukugita, M. Zaldarriaga and M. Tegmark, Astrophys. J. **549**, 669 (2001) [arXiv:astro-ph/0006436]; W. J. Percival *et al.* [The 2dFGRS Team Collaboration], Mon. Not. Roy. Astron. Soc. **337**, 1068 (2002) [arXiv:astro-ph/0206256].
[48] L. P. Chimento and R. Lazkoz, Phys. Lett. B **639**, 591 (2006) [arXiv:astro-ph/0604090].
[49] S. Nesseris and L. Perivolaropoulos, Phys. Rev. D **70**, 043531 (2004) [arXiv:astro-ph/0401556].
[50] W. H. Press *et al.*, 'Numerical Recipes', Cambridge University Press (1994).
[51] L. M. Wang and P. J. Steinhardt, Astrophys. J. **508**, 483 (1998) [arXiv:astro-ph/9804015].
[52] E. V. Linder and R. N. Cahn, arXiv:astro-ph/0701317.
[53] T. Chiba and T. Nakamura, arXiv:0708.3877 [astro-ph].
[54] E. Hawkins *et al.*, Mon. Not. Roy. Astron. Soc. **346**, 78 (2003) [arXiv:astro-ph/0212375].
[55] M. Viel, M. G. Haehnelt and V. Springel, Mon. Not. Roy. Astron. Soc. **354**, 684 (2004) [arXiv:astro-ph/0404600].
[56] M. Viel and M. G. Haehnelt, Mon. Not. Roy. Astron. Soc. **365**, 231 (2006) [arXiv:astro-ph/0508177].
[57] N. Kaiser, Astrophys. J. **498**, 26 (1998) [arXiv:astro-ph/9610120].
[58] H. J. Seo and D. J. Eisenstein, Astrophys. J. **598**, 720 (2003) [arXiv:astro-ph/0307460].
[59] A. J. S. Hamilton, arXiv:astro-ph/9708102; S. Dodelson, 'Modern Cosmology' pp. 270-281, Acad. Press (2003).
[60] C. Di Porto and L. Amendola, arXiv:0707.2686 [astro-ph].
[61] P. McDonald *et al.* [SDSS Collaboration], Astrophys. J. **635**, 761 (2005) [arXiv:astro-ph/0407377].
[62] N. P. Ross *et al.*, arXiv:astro-ph/0612400.
[63] J. da Angela *et al.*, arXiv:astro-ph/0612401.
[64] C. Marinoni *et al.*, A & A **442**, 801825 (2005).
[65] T. Padmanabhan, Phys. Rept. **380**, 235 (2003) [arXiv:hep-th/0212290]; S. M. Carroll, Living Rev. Rel. 4, 1 (2001).
[66] C. Wetterich, Nucl. Phys B. **302**, 668 (1988); B. Ratra and J. Peebles, Phys. Rev D **37**, 321 (1988); R. R. Caldwell, R. Dave and P. J. Steinhardt, Phys. Rev. Lett. **80**, 1582 (1998); I. Zlatev, L. M. Wang and P. J. Steinhardt, Phys. Rev. Lett. **82**, 896 (1999) [arXiv:astro-ph/9807002];
[67] A. Y. Kamenshchik, U. Moschella and V. Pasquier, Phys. Lett. **B 511**, 265 (2001); V. Gorini, A. Kamenshchik and U. Moschella, Phys. Rev. D **67**, 063509 (2003) [arXiv:astro-ph/0209395].
[68] P. Serra, A. Heavens and A. Melchiorri, Mon. Not. Roy. Astron. Soc. **379**, 169 (2007) [arXiv:astro-ph/0701338].
[69] T. Padmanabhan, Class. Quant. Grav. **22**, L107 (2005) [arXiv:hep-th/0406060].
[70] J. Garriga and A. Vilenkin, Prog. Theor. Phys. Suppl. **163**, 245 (2006) [arXiv:hep-th/0508005].
[71] P. Binetruy, J. M. Cline and C. Grojean, Phys. Lett. B **489**, 403 (2000) [arXiv:hep-th/0007029].
[72] J. Yokoyama, Phys. Rev. Lett. **88**, 151302 (2002) [arXiv:hep-th/0110137].
[73] U. Mohideen and A. Roy, Phys. Rev. Lett. **81**, 4549 (1998); S. K. Lamoreaux, Phys. Rev. Lett. **78**, 5 (1997);

G. Bressi et al., Phys. Rev. Lett. **88**, 4 (2002).
[74] C. Beck and M. C. Mackey, Phys. Lett. **B 605**, 295 (2005) [astro-ph/0406504]; Ph. Jetzer and N. Straumann, Phys. Lett. **B 606**, 77 (2005) [astro-ph/0411034]; C. Beck and M. C. Mackey, astro-ph/0603397; P. Jetzer and N. Straumann, Phys. Lett. B **639**, 57 (2006) [arXiv:astro-ph/0604522].
[75] M. Doran and J. Jaeckel, JCAP **0608**, 010 (2006) [arXiv:astro-ph/0605711].
[76] H. B. G. Casimir, *Proc. Kon. Ned. Akad.* **51**, 793 (1948).
[77] K. A. Milton, J. Phys. A **37**, R209 (2004) [arXiv:hep-th/0406024].
[78] K. A. Milton, arXiv:hep-th/0009173.
[79] See e.g. T. Padmanabhan, Phys. Rev. Lett. **78**, 1854 (1997) (arXiv: hep-th/9608182); Phys. Rev. Lett. **60**, 2229 (1988); S. Corley, T. Jacobson, Phys. Rev. **D 54**, 1568 (1996); J. C. Niemeyer, R. Parentani, Phys. Rev. **D 64** 101301 (2001); J. Kowalski-Glikman, Phys. Lett. B **499** 1 (2001); G. Amelino-Camelia, Int. J. Mod. Phys. D **11** 35 (2002). For a review, see L. J. Garay, Int. J. Mod. Phys. A **10**, 145 (1995).
[80] K. Nouicer, J. Phys. A **38**, 10027-10035 (2005) (arXiv:hep-th/0512027); U. Harbach and S. Hossenfelder, Phys. Lett. B **632**, 379-383 (2006).
[81] G. Mahajan, S. Sarkar and T. Padmanabhan, Phys. Lett. B **641**, 6 (2006) [arXiv:astro-ph/0604265].
[82] Ph. A. Martin and P. R. Buenzli, Acta Phys. Polonica B, 37, 2503–2559 (2006) [arXiv:cond-mat/0602559].
[83] K. Poppenhaeger, S. Hossenfelder, S. Hofmann and M. Bleicher, Phys. Lett. B **582**, 1 (2004) [arXiv:hep-th/0309066]; R. Hofmann, P. Kanti and M. Pospelov, Phys. Rev. D **63**, 124020 (2001) [arXiv:hep-ph/0012213]; H. Cheng, Mod. Phys. Lett. A **21**, 1957 (2006) [arXiv:hep-th/0609057].
[84] T. Appelquist, H. C. Cheng and B. A. Dobrescu, Phys. Rev. D **64**, 035002 (2001) [arXiv:hep-ph/0012100].
[85] C. Bogdanos and K. Tamvakis, arXiv:hep-th/0609100.
[86] G. Kofinas, G. Panotopoulos and T. N. Tomaras, JHEP **0601**, 107 (2006) [arXiv:hep-th/0510207].
[87] G. R. Dvali, G. Gabadadze and M. Porrati, Phys. Lett. B **484**, 112 (2000) [arXiv:hep-th/0002190].
[88] S. Nojiri and S. D. Odintsov, Phys. Rev. D **74**, 086005 (2006) [arXiv:hep-th/0608008].
[89] F. Perrotta, C. Baccigalupi and S. Matarrese, Phys. Rev. D **61**, 023507 (2000) [arXiv:astro-ph/9906066].
[90] R. R. Caldwell and E. V. Linder, Phys. Rev. Lett. **95**, 141301 (2005) [arXiv:astro-ph/0505494].
[91] R. J. Scherrer, Phys. Rev. D **73**, 043502 (2006) [arXiv:astro-ph/0509890].
[92] T. Chiba, Phys. Rev. D **73**, 063501 (2006) [arXiv:astro-ph/0510598].
[93] P.D. Scharre, C.M. Will, Phys. Rev. D 65 (2002); E. Poisson, C.M. Will, Phys. Rev. D 52, 848 (1995).
[94] E.V. Pitjeva, Astron. Lett. 31, 340 (2005); Sol. Sys. Res. 39, 176 (2005).
[95] E.V. Pitjeva and J. Muller, private communication.
[96] J. Muller, J. G. Williams and S. G. Turyshev, arXiv:gr-qc/0509114.
[97] J. Muller, private communication.
[98] E. V. Linder, Phys. Rev. Lett. **90**, 091301 (2003) [arXiv:astro-ph/0208512].
[99] S. Nesseris and L. Perivolaropoulos, Phys. Rev. D **73**, 103511 (2006) [arXiv:astro-ph/0602053].
[100] C. J. Copi, A. N. Davis and L. M. Krauss, Phys. Rev. Lett. **92**, 171301 (2004) [arXiv:astro-ph/0311334].
[101] J. R. Bond, W. D. Arnett and B. J. Carr, Astrophys. J. **280**, 825 (1984).
[102] S. A. Colgate and R. H. White, Astrophys. J. **143**, 626 (1966).
[103] L. Amendola, P. S. Corasaniti and F. Occhionero, arXiv:astro-ph/9907222; E. Garcia-Berro, E. Gaztanaga, J. Isern, O. Benvenuto and L. Althaus, arXiv:astro-ph/9907440; E. Gaztanaga, E. Garcia-Berro, J. Isern, E. Bravo and I. Dominguez, Phys. Rev. D **65**, 023506 (2002) [arXiv:astro-ph/0109299].
[104] E. Linder, private communication, http://supernova.lbl.gov/ evlinder/gdotsn.pdf
[105] A. Khokhlov, E. Muller and P. Hoflich, Astron. Astrophys. **270**, 223 (1993).

Alternative high-z cosmic tracers and the dark energy equation of state

M. Plionis[1,2], R. Terlevich[2], S. Basilakos[3], F. Bresolin[4], E. Terlevich[2], J. Melnick[5], I. Georgantopoulos[1]

[1] Institute of Astronomy & Astrophysics, National Observatory of Athens, Palaia Penteli 152 36, Athens, Greece.
[2] Instituto Nacional de Astrofísica Optica y Electrónica, AP 51 y 216, 72000, Puebla, México.
[3]Academy of Athens, Research Center for Astronomy and Applied Mathematics, Soranou Efesiou 4, 11527, Athens, Greece
[4] Institute for Astronomy of the University of Hawaii, 2680 Woodlawn Drive, 96822 Honolulu, HI USA
[5] European Southern Observatory, Alonso de Cordova 3107, Santiago, Chile

E-mail: `mplionis@astro.noa.gr`

Abstract. We propose to use alternative cosmic tracers to measure the dark energy equation of state and the matter content of the Universe [$w(z)$ & Ω_m]. Our proposed method consists of two components: (a) tracing the Hubble relation using HII-like starburst galaxies, as an alternative to supernovae type Ia, which can be detected up to very large redshifts, $z \sim 4$, and (b) measuring the clustering pattern of X-ray selected AGN at a median redshift of ~ 1. Each component of the method can in itself provide interesting constraints on the cosmological parameters, especially under our anticipation that we will reduce the corresponding random and systematic errors significantly. However, by joining their likelihood functions we will be able to put stringent cosmological constraints and break the known degeneracies between the *dark energy* equation of state (whether it is constant or variable) and the matter content of the universe and provide a powerful and alternative route to measure the contribution to the global dynamics and the equation of state of *dark energy*. A preliminary joint analysis of X-ray selected AGN (based on a small XMM survey) and the currently largest SNIa sample (Kowalski et al 2008), provides: $\Omega_{\rm m} = 0.28^{+0.02}_{-0.04}$ and $w = -1.0 \pm 0.1$.

1. Introduction

We live in a very exciting period for our understanding of the Cosmos. Over the past decade the accumulation and detailed analyses of high quality cosmological data (eg., supernovae type Ia, CMB temperature fluctuations, galaxy clustering, high-z clusters of galaxies, etc.) have strongly suggested that we live in a flat and accelerating universe, which contains at least some sort of cold dark matter to explain the clustering of extragalactic sources, and an extra component which acts as having a negative pressure, as for example the energy of the vacuum (or in a more general setting the so called *dark energy*), to explain the observed accelerated cosmic expansion (eg. Riess, et al. 1998; 2004; 2007, Perlmutter et al. 1999; Spergel et al. 2003, 2007, Tonry et al. 2003; Schuecker et al. 2003; Tegmark et al. 2004; Seljak et al. 2004; Allen et al. 2004; Basilakos & Plionis 2005; 2006; Blake et al. 2007; Wood-Vasey et al. 2007, Davies et al. 2007; Kowalski et al. 2008, etc).

Due to the absence of a well-motivated fundamental theory, there have been many theoretical speculations regarding the nature of the exotic *dark energy*, on whether it is a cosmological constant, a scalar or vector fields which provide a time varying dark-energy equation of state, usually parametrized by:

$$p_Q = w(z)\rho_Q \ , \tag{1}$$

with p_Q and ρ_Q the pressure and density of the exotic dark energy fluid and

$$w(z) = w_0 + w_1 f(z) \ , \tag{2}$$

with $w_0 = w(0)$ and $f(z)$ an increasing function of redshift [eg., $f(z) = z/(1+z)$] (see Peebles & Ratra 2003 and references therein, Chevalier & Polarski 2001, Linder 2003, Dicus & Repko 2004; Wang & Mukherjee 2006). Of course, the equation of state could be such that w does not evolve cosmologically. Current measurements do not allow us to put strong constraints on w, with present limits $w \lesssim -0.8$ (eg. Tonry et al. 2003; Riess et al. 2004; Sanchez et al. 2006; Spergel et al. 2006; Wang & Mukherjee 2006; Davies et al. 2007).

Two very extensive recent reports have identified *dark energy* as a top priority for future research: "Report of the Dark Energy Task Force (advising DOE, NASA and NSF) by Albrecht et al. (2006), and "Report of the ESA/ESO Working Group on Fundamental Cosmology", by Peacock et al. (2006). It is clear that one of the most important questions in Cosmology and cosmic structure formation is related to the nature of *dark energy* (as well as whether it is the sole interpretation of the observed accelerated expansion of the Universe) and its interpretation within a fundamental physical theory. To this end a large number of very expensive experiments are planned and are at various stages of development.

Therefore, the paramount importance of the detection and quantification of *dark energy* for our understanding of the cosmos and for fundamental theories implies that the results of the different experiments should not only be scrutinized, but alternative, even higher-risk, methods to measure *dark energy* should be developed and applied as well.

1.1. Methods to estimate the dark energy equation of state.

A large variety of different approaches to determine the cosmological parameters exist. A few of the most important ones are listed below:

- CMB power-spectrum + Hubble relation: By measuring the curvature of the universe (Ω_k) from the CMB angular power spectrum and using SN Ia as standard candles to trace the Hubble relation (eg. Riess et al. 2004) to measure a combination of Ω_m and Ω_Λ, then the values of all the contributing components to the global dynamics can be constrained (using the fact that $\Omega_k + \Omega_m + \Omega_\Lambda = 1$)
- Hubble relation + Clustering of extragalactic sources: If, however, the *dark energy* contribution is not due to a cosmological constant but rather it evolves cosmologically, then the Hubble relation provides a degenerate solution between the contribution to the global dynamics of the total mass and of the *dark energy*, even in the case of an almost flat geometry. In order to break this degeneracy one needs to introduce some other cosmological test (eg., the clustering properties of galaxies, clusters or AGN, which can be compared with the theoretical predictions of an *a priori* selected power-spectrum of density perturbations - say the CDM - to constrain the cosmological parameters such as Ω_m, w, σ_8 and h; eg., Matsubara 2004).
- Baryonic Acoustic Oscillation method, which was identified by the U.S. Dark Energy Task Force as one of the four most promising techniques to measure the properties of the *dark energy* and the one less likely to be limited by systematic uncertainties. BAOs are produced by pressure (acoustic) waves in the photon-baryon plasma in the early universe, generated

by dark matter (DM) overdensities. At the recombination era ($z \sim 1100$), photons decouple from baryons and free stream while the pressure wave stalls. Its frozen scale, which constitutes a standard ruler, is equal to the sound horizon length, $r_s \sim 100\ h^{-1}$ Mpc (e.g. Eisenstein, Hu & Tegmark 1998). This appears as a small, $\sim 10\%$ excess in the galaxy, cluster or AGN power spectrum (and 2-point correlation function) at the scale corresponding to r_s. First evidences of this excess were recently reported in the clustering of luminous SDSS red-galaxies (Eisenstein et al. 2005, Padmanabhan et al. 2007). A large number of photometric surveys are planned in order to measure *dark energy* (eg., the ESO/VST KIDS project, DES: `http://www.darkenergysurvey.org`, Pan-STARRS: `http://pan-starrs.ifa.hawaii.edu`).

- Other important cosmological tests that have been and will be used are based on galaxy clusters. For example, (a) the local cluster mass function and its evolution, $n(M, z)$, which depends on Ω_m, on the linear growth rate of density perturbations and on the normalization of the power-spectrum, σ_8 (eg., Schüecker et al. 2003; Vikhlinin et al. 2003; Newman et al. 2002; Rosati et al., 2002 and references therein), (b) the cluster mass-to-light ratio, M/L, which can be used to estimate Ω_m once the mean luminosity density of the Universe is known, assuming that mass traces light similarly both inside and outside clusters (see Andernach et al. 2005 for a recent application), (c) the baryon fraction in nearby clusters (eg., Fabian, 1991; White et al., 1993). Assuming that it does not evolve, as gasdynamical simulations indicate (eg. Gottlöber & Yepes 2007), then its determination in distant clusters can provide a geometrical constraint on *dark energy* (Allen et al., 2004).

1.2. Objectives of our Approach

We wish to constrain the *dark energy* equation of state using the combination of the Hubble relation and Clustering methods, but utilizing alternative cosmic tracers for both of these components.

From one side we wish to trace the Hubble function using HII-like starburst galaxies, which can be observed at higher redshifts than those sampled by current SNIa surveys and thus at distances where the Hubble function is more sensitive to the cosmological parameters. The HII galaxies can be used as standard candles (Melnick, Terlevich & Terlevich 2000, Melnick 2003; Siegel et al. 2005) due to the correlation between their velocity dispersion, metallicity and H_β luminosity (Melnick 1978, Terlevich & Melnick 1981, Melnick, Terlevich & Moles 1988). Furthermore, the use of such an alternative high-z tracer will enable us to check the SNIa based results and lift any doubts that arise from the fact that they are the only tracers of the Hubble relation used to-date (for possible usage of GRBs see Ghirlanda et al. 2006; Basilakos & Perivolaropoulos 2008)[1]. We therefore plan to improve the $L(H_\beta) - \sigma$ distance estimator by investigating all the parameters that can systematically affect it, like stellar age, metal and dust content, environment etc, in order to also determine with a greater accuracy the zero-point of the relevant distance indicator. The possibility to use effectively HII-like high-z starburst galaxies as cosmological standard candles, relies on our ability to suppress significantly the present distance modulus uncertainties ($\sigma_\mu \simeq 0.52$ mag; Melnick, Terlevich & Terlevich 2000),

[1] GRBs appear to be anything but standard candles, having a very wide range of isotropic equivalent luminosities and energy outputs. Nevertheless, correlations between various properties of the prompt emission and in some cases also the afterglow emission have been used to determine their distances. A serious problem that hampers a straight forward use of GRBs as Cosmological probes is the intrinsic faintness of the nearby events, a fact which introduces a bias towards low (or high) values of GRB observables and therefore the extrapolation of their correlations to low-z events is faced with serious problems. One might also expect a significant evolution of the intrinsic properties of GRBs with redshift (also between intermediate and high redshifts) which can be hard to disentangle from cosmological effects. Finally, even if a reliable scaling relation can be identified and used, the scatter in the resulting luminosity and thus distance modulus is still fairly large.

which are unacceptable large for precision cosmology.

From the other side we wish to determine the clustering pattern of X-ray selected AGN at a median redshift of ~ 1, which is roughly the peak of their redshift distribution (see Basilakos et al. 2004; 2005, Miyaji et al. 2007). To this end we are developing the tools that will enable us to analyse large XMM data-sets, covering up to ~ 300 sq.degrees of sky.

Although each of the previously discussed components of our project (Hubble relation using HII-like starburst galaxies and angular/spatial clustering of X-ray AGNs) will provide interesting and relatively stringent constraints on the cosmological parameters, especially under our anticipation that we will reduce significantly the corresponding random and systematic errors, it is the combined likelihood of these two type of analyses which will enable us to break the known degeneracies between cosmological parameters and determine with great accuracy the *dark energy* equation of state (see preliminary analysis of Basilakos & Plionis 2005; 2006).

Below we present the basic methodology of each of the two main components of our proposal, necessary in order to constrain the *dark energy* equation of state.

2. Cosmological Parameters from the Hubble Relation

It is well known that in the matter dominated epoch the Hubble relation depends on the cosmological parameters via the following equation:

$$H^2(z) = H_0^2 \left[\Omega_m(1+z)^3 + \Omega_w \exp\left(3\int_0^z \frac{1+w(x)}{1+x}\mathrm{d}x\right)\right] , \qquad (3)$$

which is simply derived from Friedman's equation. We remind the reader that Ω_m and Ω_w are the present fractional contributions to the total cosmic mass-energy density of the matter and dark energy source terms, respectively.

Supernovae SNIa are considered standard candles at peak luminosity and therefore they have been used not only to determine the Hubble constant (at relatively low redshifts) but also to trace the curvature of the Hubble relation at high redshifts (see Riess et al. 1998, 2004, 2007; Perlmutter et al. 1998, 1999; Tonry et al. 2003; Astier et al. 2006; Wood-Vasey et al. 2007; Davis et al. 2007; Kowalski et al 2008). Practically one relates the distance modulus of the SNIa to its luminosity distance, through which the cosmological parameters enter:

$$\mu = m - M = 5\log d_L + 25 \quad \text{where} \quad d_L = (1+z)\int_0^z \frac{c}{H(z)}\mathrm{d}z . \qquad (4)$$

The main result of numerous studies using this procedure is that distant SNIa's are dimmer on average by 0.2 mag than what expected in an Einstein-deSitter model, which translates in them being $\sim 10\%$ further away than expected.

The amazing consequence of these results is that they imply that we live in an accelerating phase of the expansion of the Universe, an assertion that needs to be scrutinized on all possible levels, one of which is to verify the accelerated expansion of the Universe using an alternative to SNIa population of extragalactic standard candles. Furthermore, the cause and rate of the acceleration is of paramount importance, ie., the *dark energy* equation of state is the next fundamental item to search for and to these directions we hope to contribute with our current project.

2.1. Theoretical Expectations:

To appreciate the magnitude of the Hubble relation variations due to the different *dark energy* equations of state, we plot in Figure 1 the relative deviations of the distance modulus, $\Delta\mu$, of different *dark-energy* models from a nominal *standard* ($w = -1$) Λ-cosmology (with $\Omega_m = 0.27$ and $\Omega_\Lambda = 0.73$), with the relative deviations defined as:

$$\Delta\mu = \mu_\Lambda - \mu_{\rm model} . \qquad (5)$$

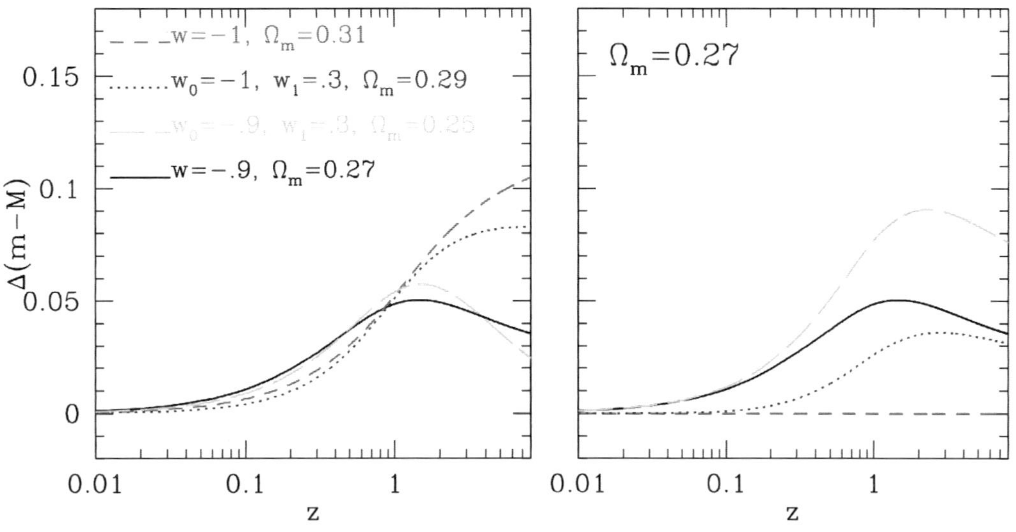

Figure 1. *Left Panel:* The expected distance modulus difference between the *dark-energy* models shown and the reference Λ-model ($w = -1$) with $\Omega_m = 0.27$. *Right Panel:* The expected distance modulus differences once the Ω_m-$w(z)$ degeneracy is broken (imposing the same Ω_m value as the comparison model).

The parameters of the different models used are shown in Figure 1. As far as the *dark-energy* equation of state parameter is concerned, we present the deviations from the *standard* model of two models with a constant w value and of two models with an evolving equation of state parameter, utilizing the form of eq.2. In the left panel of Figure 1 we present results for selected values of Ω_m, while in the right panel we use the same *dark-energy* equations of state parameters but for the same value of $\Omega_m(= 0.27)$ (ie., we eliminate the degeneracy).

Three important observations should be made from Figure 1:

(i) The relative magnitude deviations between the different *dark-energy* models are quite small (typically $\lesssim 0.1$ mag), which puts severe pressure on the necessary photometric accuracy of the relevant observations.

(ii) The largest relative deviations of the distance moduli occur at redshifts $z \gtrsim 1.5$, and thus at quite larger redshifts than those currently traced by SN Ia, and

(iii) There are strong degeneracies between the different cosmological models at redshifts $z \lesssim 1$, but in some occasions even up to much higher redshifts (one such example is shown in Figure 1 between the models with $(\Omega_m, w_0, w_1) = (0.31, -1, 0)$ and $(0.29, -1, 0.2)$.

Luckily, such degeneracies can be broken, as discussed already in the introduction, by using other cosmological tests (eg. the clustering of extragalactic sources, the CMB shift parameter, BAO's, etc). Indeed, current evidence overwhelmingly show that the total matter content of the universe is within the range: $0.2 \lesssim \Omega_m \lesssim 0.3$, a fact that reduces significantly the degeneracies between the cosmological parameters.

2.2. Larger numbers or higher redshifts ?

In order to define an efficient strategy to put stringent constraints on the *dark-energy* equation of state, we have decided to re-analyse two recently compiled SNIa samples, the Davies et al.

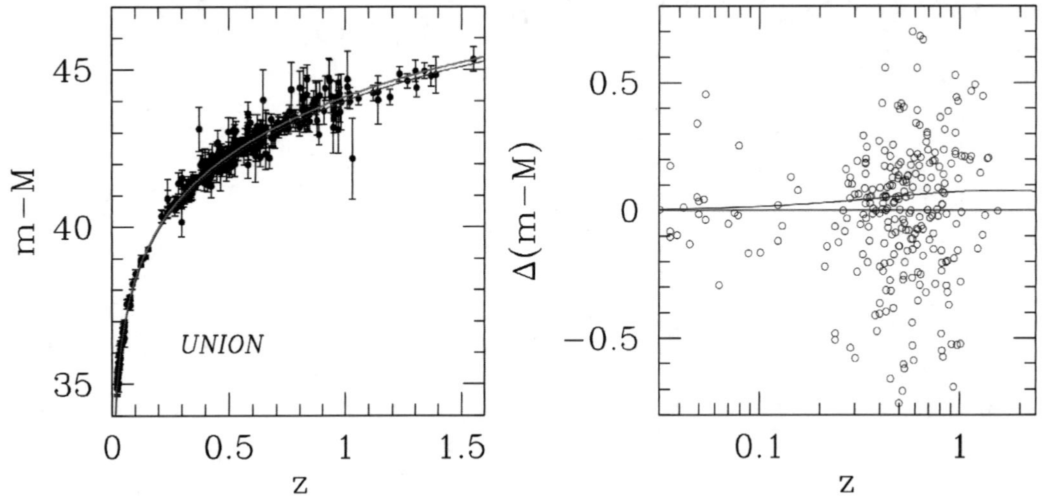

Figure 2. *Left Panel:* SNIa distance moduli as a function of redshift. *Right Panel:* Distance moduli difference between the Λ-model and the SNIa data. The blue line is the corresponding difference between the reference ($w = -1$) and the $w = -0.85$ *dark-energy* models.

(2007) [hereafter *D07*] compilation of 192 SNIa (based on data from Wood-Vasey et al. 2007, Riess et al. 2007 and Astier et al. 2007) and the *UNION* compilation of 307 SNIa (Kowalski et al. 2008). Note that the two samples are not independent since most of the *D07* is included in the *UNION* sample.

Firstly, we present in the left panel of Figure 2 the *UNION* SNIa distance moduli overplotted (red-line) with the theoretical expectation of a flat cosmology with $(\Omega_m, w) = (0.27, -1)$. In the right panel we plot the distance moduli difference between the SNIa data and the previously mentioned model. To appreciate the level of accuracy needed in order to put constraints on the equation of state parameter, we also plot the distance moduli difference between the reference $(\Omega_m, w) = (0.27, -1)$ and the $(\Omega_m, w) = (0.27, -0.85)$ models (thin blue line).

We proceed to analyse the SNIa data by defining the usual likelihood estimator[2] as:

$$\mathcal{L}^{\rm SNIa}(\mathbf{c}) \propto \exp[-\chi^2_{\rm SNIa}(\mathbf{c})/2], \tag{6}$$

where $\mathbf{c}$ is a vector containing the cosmological parameters that we want to fit for, and

$$\chi^2_{\rm SNIa}(\mathbf{c}) = \sum_{i=1}^{N} \left[\frac{\mu^{\rm th}(z_i, \mathbf{c}) - \mu^{\rm obs}(z_i)}{\sigma_i} \right]^2 , \tag{7}$$

where $\mu^{\rm th}$ is given by eq.(3), z_i is the observed redshift and σ_i the observed distance modulus uncertainty. Here we will constrain our analysis within the framework of a flat ($\Omega_{\rm tot} = 1$) cosmology and therefore the corresponding vector $\mathbf{c}$ is: $\mathbf{c} \equiv [\Omega_m, w_0, w_1]$. We will use only SNIa with $z > 0.02$ in order to avoid possible problems related to redshift uncertainties due to the local bulk flow.

- *Larger Numbers?* The first issue that we wish to address is how better have we done in imposing cosmological constraints by increasing the available SNIa sample from 181 to 292

[2] Likelihoods are normalized to their maximum values.

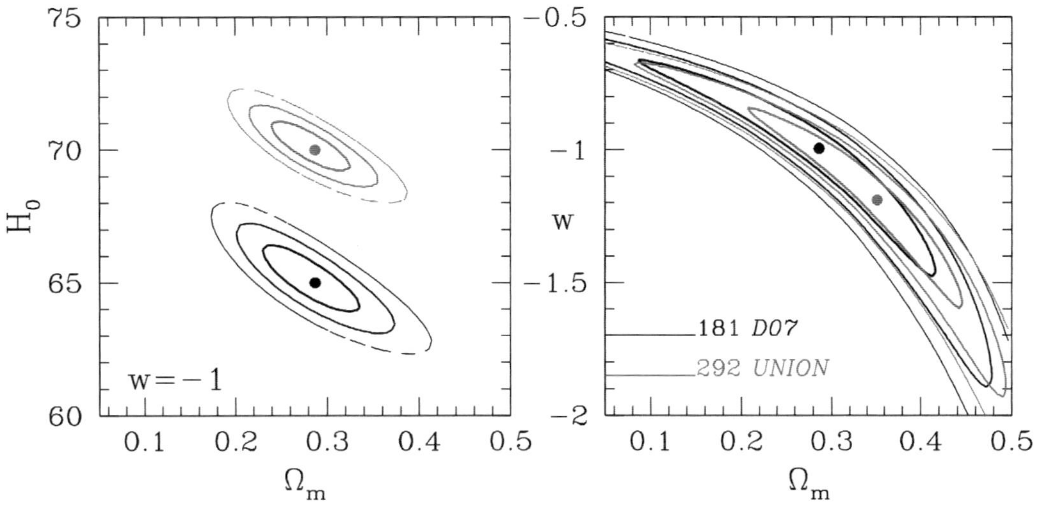

Figure 3. Solution space of the fits to the Cosmological parameters using either of the two SNIa data sets. The contours are plotted where $-2\ln\mathcal{L}/\mathcal{L}_{\max}$ is equal to 2.30, 6.16 and 11.83, corresponding to 1σ, 2σ and 3σ confidence level.

(excluding the $z < 0.02$ SNIa), ie., increasing the sample by more than 60%. In Table 1 we present various solutions using each of the two previously mentioned samples. Note that since only the relative distances of the SNIa are accurate and not their absolute local calibration, we always marginalize with respect to the internally derived Hubble constant (note that fitting procedures exist which do not need to *a priori* marginalize over the internally estimated Hubble constant; eg., Wei 2008).

We present our results in Figure 3. The left panel shows the internally derived Hubble constant for each of the to samples (and for $w = -1$). It is evident that the derived values, which we use to free our analysis of the H_0 dependence of the distance modulus, are well constrained ($h_0 = 0.6515 \pm 0.007$ and 0.7015 ± 0.005 for the *D07* and *UNION* sample, respectively). The right panel shows the main results of interest. The size of the well-known *banana* shape region of the (Ω_m, w) solution space is almost identical for both samples of SNIa.

A first conclusion is therefore that *the increase by* $\sim$ 60% *of the UNION sample has not provided more stringent constraints to the cosmological parameters.* Rather there appears to be an unexpected lateral shift of the contours towards higher values of Ω_m and lower values of w, within, however, the 1 σ contour of the solution space. In order to verify that the larger number of SNIa's in the *UNION* sample are not preferentially located at low-z's - in which case we should have not expected more stringent cosmological constraints using the latter SNIa sample - we plot in Figure 4 the normalized redshift frequency distributions of the two samples (left panel) and the relative increase of SNIa's as a function of redshift in the *UNION* sample with respect to the *D07* sample in the right panel. It is evident that the larger number of *UNION* SNIa's are distributed in all redshifts, except for $z \gtrsim 1$, where there is no appreciable increase of SNIa numbers.

We already have a strong hint, from the previously presented comparison between the *D07* and *UNION* results, that increasing the number of Hubble relation tracers, covering the same redshift range and with the current level of uncertainties as the available SNIa

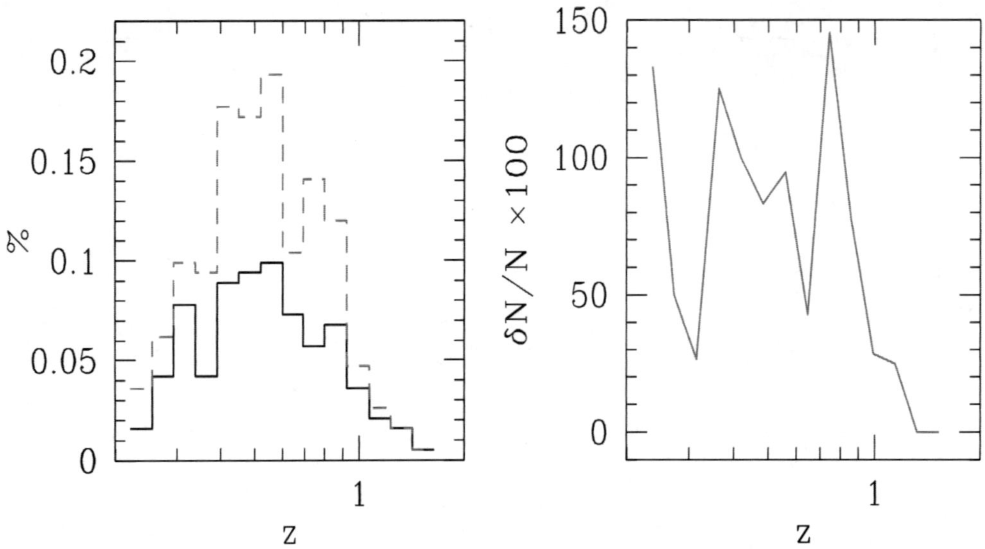

Figure 4. Redshift distribution of the *UNION* and *D07* SNIa's (left panel) and the relative increase of SNIa numbers (in percent) between the two samples.

Table 1. Fitting the SNIa data (with $z > 0.02$ in order to avoid local bulk flow effects) to flat cosmologies. Note that for the case where $\mathbf{c} = (\Omega_m, w)$ (ie., last row), the errors shown are estimated after marginalizing with respect to the other fitted parameter.

	D07			*UNION*		
free params	w	Ω_m	$\chi^2_{\rm min}/{\rm df}$	w	Ω_m	$\chi^2_{\rm min}/{\rm df}$
Ω_m	$\mathbf{-1}$	$0.280^{+0.025}_{-0.015}$	187.03/180	$\mathbf{-1}$	$0.280^{+0.020}_{-0.015}$	301.93/291
Ω_m, w	$-1.025^{+0.060}_{-0.045}$	0.292 ± 0.018	187.02/179	-1.212 ± 0.050	0.355 ± 0.015	301.11/290

samples, appears to be a futile avenue in constraining further the cosmological parameters.

- *Lower uncertainties or higher-z's:* We now resort to a Monte-Carlo procedure which will help us investigate which of the following two directions, which bracket many different possibilities, would provide more stringent cosmological constraints:
  - Reduce significantly the distance modulus uncertainties of SNIa, tracing however the same redshift range as the currently available samples, or
  - use tracers of the Hubble relation located at redshifts where the models show their largest relative differences ($z \gtrsim 2$), with distance modulus uncertainties comparable to that of the highest redshift SNIa's ($\langle \sigma_\mu \rangle \simeq 0.4$)

The Monte-Carlo procedure is based on using the observed high-z SNIa distance modulus uncertainty distribution (σ_μ) and a model to assign random μ-deviations from a reference $H(z)$ function, that reproduces exactly the banana-shaped contours of the (Ω_m, w) solution space of Figure 3 (right panel). Indeed, after a trial and error procedure we have found that by assigning to each *UNION* SNIa (using only their redshift) a distance modulus deviation ($\delta\mu$) from the reference model (in this case we use: $\Omega_m = 0.35, w = -1.21$), using a Gaussian with zero mean and variance given by observed $\langle \sigma_\mu \rangle^2$), and using as the relevant individual distance modulus uncertainty, which enters as a weight in the χ^2 of eq.8, the following: $\sigma_i^2 = \sqrt{(1.1\delta\mu_i)^2 + \phi^2}$, with ϕ a random Poisson deviate within $[-0.01, 0.01]$,

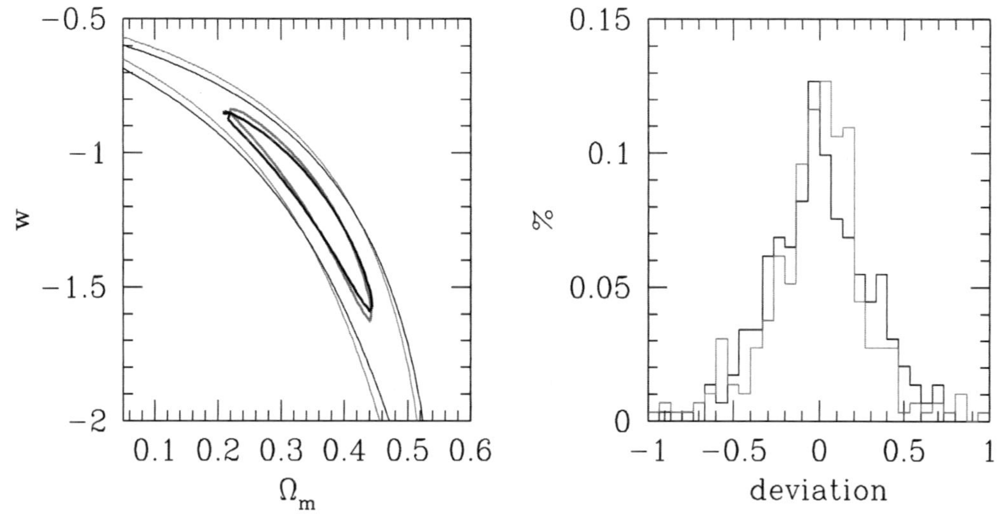

Figure 5. *Left Panel:* Comparison between the *UNION* SNIa constraints (black contours) and those derived by a Monte-Carlo procedure designed to closely reproduce them (for clarity we show only contours corresponding to 1 and 3 σ confidence levels). *Right Panel:* The *UNION* SNIa distance modulus deviations from the best fit model - $(\Omega_m, w) = (0.35, -1.21)$ - and a random realization of the model deviations (see text).

we reproduce exactly the banana-shaped solution range of the reference model (details will appear elsewhere). This can be seen clearly in the upper left panel of Figure 5, where we plot the original *UNION* solution space (black contours) and the model solution space (red contours). In the right panel we show the distribution of the true SNIa deviations from the best fitted model (Table 1) as well as a random realization of the model deviations.
Armed with the above procedure we can now address the questions posed previously. Firstly, we reduce to half the random deviations of the SNIa distance moduli from the reference model (with the corresponding reduction of the relevant uncertainty, σ_i). The results of the likelihood analysis can be seen in the left panel of Figure 6. There is a reduction of the range of the solution space, but indeed quite a small one. Secondly, we have increase artificially the high-z tracers by 88 objects (by additionally using the $z > 0.65$ SNIa and adding a $\delta z = 2$ to their redshift). Note that the new tracers are distributed between $2.65 \lesssim z \lesssim 3.55$, ie., in a range where the largest deviations between the different cosmological models occur (see Figure 1). The deviations from the reference model of these additional SNIa are based on their original μ uncertainty distribution (ie., we have assumed that the new high-z tracers will have similar uncertainties as their $z \gtrsim 0.65$ counterparts, which is $\langle \sigma_\mu \rangle \simeq 0.38$). We now find a significantly reduced solution space (right panel of Figure 6), which shows that indeed by increasing the $H(z)$ tracers by a few tens, at those redshifts where the largest deviations between models occur, can have a significant impact on the recovered cosmological parameter solution space.

The main conclusion of the previous analysis is that the best strategy to decrease the uncertainties of the cosmological parameters based on the Hubble relation is to use standard candles which trace also the redshift range $2 \lesssim z \lesssim 4$. Below we present such a possibility by suggesting an alternative to the SNIa standard candles, namely HII-like starburst galaxies (eg.

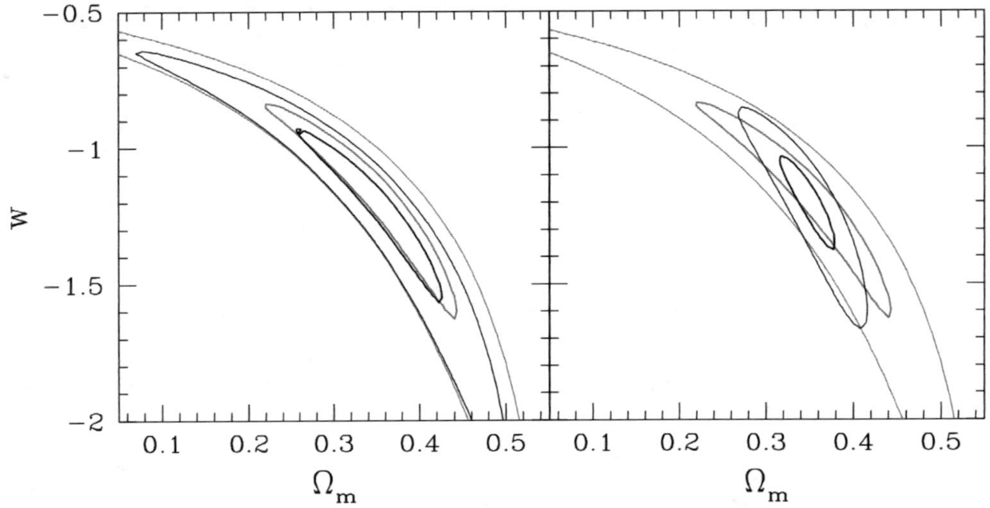

Figure 6. Comparison of the present SNIa constraints (red contours) with *Left Panel*: those derived by reducing to half their uncertainties (black contours) and *Right Panel:* with those derived by adding a sample of 88 high-z tracers ($2.5 \lesssim z \lesssim 3.6$) with a distance modulus mean uncertainty of $\sigma_\mu \simeq 0.38$ (black contours).

Melnick 2003; Siegel et al. 2005).

2.3. Hubble Relation using HII-like starburst galaxies

We now reach our suggestion to use an alternative and potentially very powerful technique to estimate cosmological distances, which is the relation between the luminosity of the H_β line and the stellar velocity dispersion, measured from the line-widths, of HII regions and galaxies (Terlevich & Melnick 1981, Melnick, Terlevich & Moles 1988). The cosmological use of this distance indicator has been tested in Melnick, Terlevich & Terlevich (2000) and Siegel et al (2005) (see also the review by Melnick 2003). It is the presence of O and B-type stars in HII regions that causes the strong Balmer line emission, in both H_α and H_β. Furthermore, the fact that the bolometric luminosities of HII galaxies are dominated by the starburst component implies that their luminosity per unit mass is very large, despite the fact that the galaxies are low-mass. Therefore they can be observed at very large redshifts, and this fact makes them cosmologically very interesting objects. Furthermore, it has been shown that the $L(H_\beta) - \sigma$ correlations holds at large redshifts (Koo et al. 1996, Pettini et al. 2001, Erb et al. 2003) and therefore they can be used to trace the Hubble relation at cosmologically interesting distances. One of the most important prerequisites in using such relations, as distance estimators, is the accurate determination of their zero-point. To this end, Melnick et al (1988) used giant HII regions in nearby late-type galaxies and derived the following empirical relation (using a Hubble constant of $H_0 = 71$ km/sec/Mpc):

$$\log L(H_\beta) = \log M_z + 29.60 \text{ with } M_z = \sigma^5/(O/H), \tag{8}$$

where O/H is the metallicity. Based on the above relation and the work of Melnick, Terlevich & Terlevich (2000), the distance modulus of HII galaxies can be derived from:

$$\mu = 2.5\log(\sigma^5/F_{H\beta}) - 2.5\log(O/H) - A_{H\beta} - 26.44, \tag{9}$$

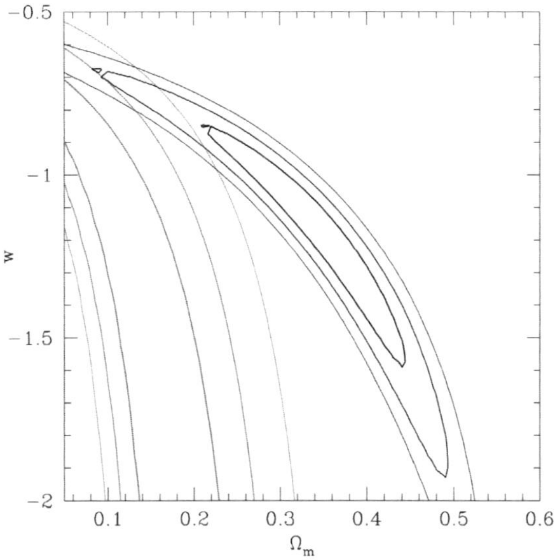

Figure 7. Comparison between the *UNION* SNIa constraints and those derived by using the 15 high-z starburst galaxies of Siegel et al. (2005). Although, the latter constraints are weak and mostly inconsistent with the former, this plot serves to indicate the potential of using high-z starburst galaxies (once of course we have reduced significantly their distance modulus uncertainties).

with $F_{H\beta}$ and $A_{H\beta}$ are the flux and extinction in H_β. The rms dispersion in distance modulus was found to be $\sim$ 0.52 mag. The analysis of Melnick, Terlevich & Terlevich (2000) has shown that most of this dispersion ($\sim$ 0.3 mags) comes from observational errors in the stellar velocity dispersion measurements, from photometric errors and metallicity effects. It is therefore important to understand and correct the sources of random and systematic errors of the $L(H_\beta) - \sigma$ relation, and indeed with the availability of new observing techniques and instruments, we hope to reduce significantly the previously quoted rms scatter.

A few words are also due to the possible systematic effects of the above relation. Such effects may be related to the age of the HII galaxy (this can be dealt with by putting a limit in the equivalent width of the H_β line, eg. $EW(H_\beta) > 25$ Angs; see Melnick 2003), to extinction, to different metallicities and environments. Also the $EW(H_\beta)$ of HII-like galaxies at intermediate and high redshifts are smaller than the local ones, a fact which should be taken into account.

We have commenced an investigation of all these effects by using high-resolution spectroscopy of a relatively large number of SDSS low-z HII galaxies, with a range of H_β equivalent widths, luminosities, metal content and local overdensity, in order to reduce the scatter of the HII-galaxy based distance estimator to about half its present value, ie., our target is $\sim$ 0.25 mag. We will also define a medium and high redshift sample (see Pettini et al. 2001; Erb et al. 2003), consisting of a few hundred objects distributed in the different high-redshift bins, which will finally be used to define the high-z Hubble function.

Summarizing, the use of HII galaxies to trace the Hubble relation, as an alternative to the traditionally used SN Ia, is based on the following facts:

(a) local and high-z HII-like galaxies and HII regions are physically very similar systems (Melnick et al 1987) providing a phenomenological relation between the luminosity of the H_β line, the velocity dispersion and their metallicity as traced by O/H (Melnick, Terlevich & Moles 1988). Therefore HII-like starburst galaxies can be used as alternative standard

candles (Melnick, Terlevich & Terlevich 2000, Melnick 2003; Siegel et al. 2005)

(b) such galaxies can be readily observed at much larger redshifts than those currently probed by SNIa and

(c) it is at such higher redshifts that the differences between the predictions of the different cosmological models appear more vividly.

Already a sample of 15 such high-z starburst galaxies have been used by Siegel et al. (2005) in an attempt to constrain cosmological parameters but the constraints, although in the correct direction, are very weak. Here we perform our own re-analysis of this data-set and the resulting constraints on the Ω_m, w plane (for a flat geometry) can be seen in Figure 7. Note that imposing $w = -1$, our analysis of the Siegel et al (2005) data set provides $\Omega_m = 0.10 \pm 0.05$, which is towards the lower side of the generally accepted values. Comparing these HII-based results to the present constraints of the latest SNIa data (*D07* and *UNION*) clearly indicates the necessity to:

- re-estimate carefully the local zero-point of the $L(H_\beta) - \sigma$ relation,
- suppress the HII-galaxy distance modulus uncertainties,
- increase the high-z starburst sample by a large fraction,
- make sure to select high-z bona-fide HII-galaxies, excluding those that show indications of rotation (Melnick 2003).

3. The Clustering of high-z X-ray AGN

X-ray selected AGNs provide a relatively unbiased census of the AGN phenomenon, since obscured AGNs, largely missed in optical surveys, are included in such surveys. Furthermore, they can be detected out to high redshifts and thus trace the distant density fluctuations providing important constraints on super-massive black hole formation, the relation between AGN activity and Dark Matter (DM) halo hosts, the cosmic evolution of the AGN phenomenon (eg. Mo & White 1996, Sheth et al. 2001), and on cosmological parameters and the dark-energy equation of state (eg. Basilakos & Plionis 2005; 2006).

The earlier ROSAT-based analyses (eg. Boyle & Mo 1993; Vikhlinin & Forman 1995; Carrera et al. 1998; Akylas, Georgantopoulos, Plionis, 2000; Mullis et al. 2004) provided conflicting results on the nature and amplitude of high-z AGN clustering. With the advent of the XMM and *Chandra* X-ray observatories, many groups have attempted to settle this issue, but in vain. Different surveys have provided again a multitude of conflicting results, intensifying the debate (eg. Yang et al. 2003; Manners et al. 2003; Basilakos et al. 2004; Gilli et al. 2005; Basilakos et al 2005; Yang et al. 2006; Puccetti et al. 2006; Miyaji et al. 2007; Gandhi et al. 2006; Carrera et al. 2007). However, the recent indications of a flux-limit dependent clustering appears to remove most of the above inconsistencies (Plionis et al. 2008; see also Figure 8).

Furthermore, there are indications for a quite large high-z AGN clustering length, reaching values $\sim 15 - 18\ h^{-1}$ Mpc at the brightest flux-limits (eg., Basilakos et al 2004; 2005, Puccetti et al. 2006, Plionis et al. 2008), a fact which, if verified, has important consequences for the AGN bias evolution and therefore for the evolution of the AGN phenomenon (eg. Miyaji et al. 2007; Basilakos, Plionis & Ragone-Figueroa 2008). An independent test of these results will be to establish that the environment of high-z AGN is associated with large DM haloes, which being massive should be more clustered.

Below we justify the necessity for a large-area XMM survey in order to unambiguously determine the clustering pattern of high-z ($z \sim 1$) X-ray AGNs. We further show that such measurements can be used to put strong cosmological constraints (see for example Basilakos & Plionis 2005; 2006), and help break the Ω_m, w degeneracies.

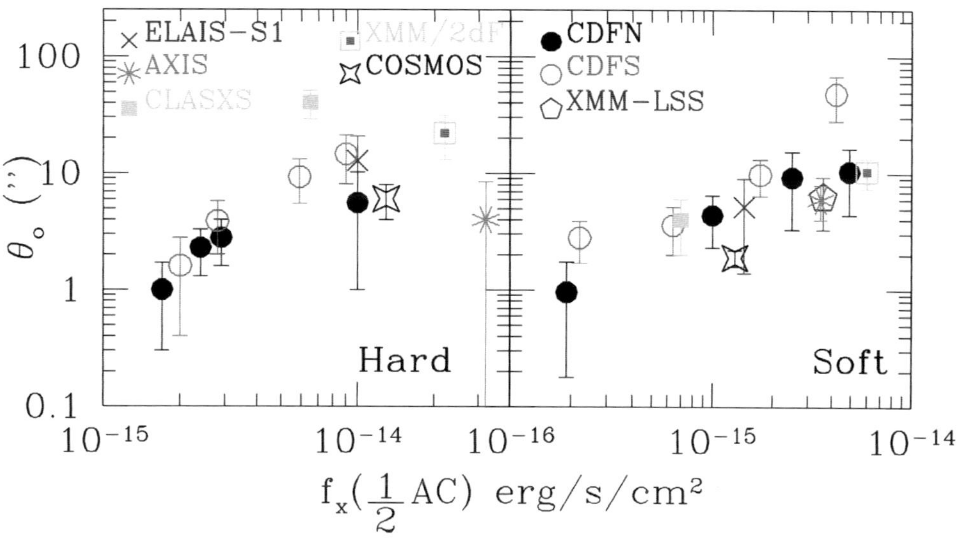

Figure 8. The angular correlation scale, θ_0, as a function of different survey characteristic flux, defined as that corresponding to half the respective survey area-curves. Most results appear to be consistent with the clustering flux-limit dependence, found from the CDF-N and CDF-S (from Plionis et al. 2008).

3.1. Biases and Systematics:

It is important to understand and overcome the shortcomings and problems that one is facing in order to reliably and unambiguously determine the clustering properties of the X-ray selected AGNs. Below we list the source of such problems, some of which can be rectified by considerably increasing the X-ray survey area.

- *Cosmic Variance*: Is the volume surveyed large enough to smooth out inhomogeneities of the large-scale distribution of AGNs? (for example see Stewart et al. 2007). Closely related to this problem is the so-called integral constraint, which practically depends on the unknown true mean density of the cosmic sources under study. If the area is small enough, then the mean density, estimated from the survey itself, is way-out of its true value and thus the usual correlation function analysis will impose the observed mean number density as the true one (an example of this is the CDF-S were a large number of superclusters at $z \sim 0.7$ are found; see Gilli et al. 2003). This usually results into an underestimation of the true correlation amplitude and a shallower zero-crossing of the estimated $\xi(r)$ or $w(\theta)$. A source of the observed scatter between the presently available surveys (see Figure 8) could well be the *cosmic variance*. These problems, however, are rectified with the large-area XMM survey proposed.
- *The amplification bias* which can enhance artificially the clustering signal due to the detector's PSF smoothing of source pairs with intrinsically small angular separations (see Vikhlinin & Forman 1995; Basilakos et al. 2005). This problem can affect clustering results if at the median redshift of the sources under study the XMM PSF angular size corresponds to a rest-frame spatial scale comparable to the typical source pair-wise separations. Furthermore, the possible variability of the PSF size through-out the XMM fields can have an additional effect. This should be modeled and tested with Monte-Carlo simulations in order to establish the extent to which the clustering results are affected. In

large-area surveys it is necessary to take good-care of this effect when using source pairs that belong to different XMM pointings.

- *Reliable production of random source catalogues*: This is an issue which is extremely important and not appreciated at the necessary extent. The random catalogues, with which the observed source-pairs are compared, should be produced to account for all systematic effects from which the observations suffer, among which the different positional sensitivity and edge effects of each XMM pointing. Furthermore, a reliable $\log N - \log S$ distribution (theoretically motivated or observationally determined) should be reproduced in the random "XMM pointings" and the random sources should be observed following the same procedure as in the true observations. Random positioned sources with fluxes lower than that corresponding to the particular position of the sensitivity map of the particular XMM pointing should be removed from each random catalogue.

An optimal approach to unambiguously determine the clustering pattern of X-ray selected AGNs would be to determine both the angular and spatial clustering pattern. The reason being that various systematic effects or uncertainties enter differently in the two types of analyses. On the one side, using $w(\theta)$ and its Limber inversion, one by-passes the effects of redshift-space distortions and uncertainties related to possible misidentification of the optical counter-parts of X-ray sources. On the other side, using spectroscopic or accurate photometric redshifts to measure $\xi(r)$ or $w_p(\theta)$ one by-passes the inherent necessity, in Limber's inversion of $w(\theta)$, of the source redshift-selection function (for the determination of which one uses the integrated X-ray source luminosity function, different models of which exist). For the inversion to work it is also necessary to model the spatial correlation function as a power law, to assume a clustering evolution model, which is taken usually to be that of constant clustering in comoving coordinates (eg. de Zotti et al. 1990; Kundić 1997) and also to assume a cosmological model. Limber's inversion then reads:

$$\theta_{\circ}^{\gamma-1} = H_{\gamma} r_{\circ}^{\gamma} \left(\frac{H_{\circ}}{c}\right)^{\gamma} \int_0^{\infty} \left(\frac{1}{N}\frac{\mathrm{d}N}{\mathrm{d}z}\right)^2 \frac{E(z)}{x^{\gamma-1}(z)} \mathrm{d}z \,, \tag{10}$$

where $\epsilon = \gamma - 3$ for the constant clustering in comoving coordinates model, $x(z)$ is the proper distance, $E(z) = \sqrt{\Omega_{\rm m}(1+z)^3 + \Omega_{\Lambda}}$ and $H_{\gamma} = \Gamma(\frac{1}{2})\Gamma(\frac{\gamma-1}{2})/\Gamma(\frac{\gamma}{2})$. As noted previously, for the inversion to be possible it is necessary to know the X-ray source redshift distribution, $\mathrm{d}N/\mathrm{d}z$, and the total number, N, of the X-ray sources, which can be determined by integrating the X-ray source luminosity function above the minimum luminosity that corresponds to the particular flux-limit used.

3.2. X-ray surveys

In order to reach a flux-limit for which the soft-band clustering appears to converge to its final value (due to the flux-limit-clustering correlation, revealed in Plionis et al. 2008; see also Figure 8) we plan to analyse all >10 ksec XMM pointings, which will allow us to reach a flux-limit of $\sim 2\times10^{-15}$ erg/sec/cm^2 in the soft (0.5-2 keV) and $\sim 10^{-14}$ erg/sec/cm^2 in the hard (2-10 keV) bands, respectively. Such an exposure time will finally provide (taking in to account realistic observational effects) ~250 soft-band and ~100 hard-band X-ray sources per deg^2, according to the Kim et al. (2007) $\log N - \log S$ and since all publicly available XMM pointings, with exposure time more than 10 ksec, add to $\sim$ 300 non-contiguous sq.degrees, implies a resulting sample of $\sim$ 75000 soft and $\sim$ 30000 hard X-ray sources. These numbers will allow us to derive with great accuracy the small-separation angular correlation function, the Limber's inversion of which can provide their spatial correlation function. Furthermore, around 100 non-contiguous sq.degrees of the previous survey (except for a few contiguous regions with areas between 2 and 10 sq.degrees each) are covered also by the SDSS, providing crude photo-zs. This will allow us to derive the angular correlation function in distinct redshift bins in the range $0.5 < z < 2$ and thus quantify the evolution of the bias of X-ray selected AGN, an important ingredient in disentangling the cosmological parameters. Note finally that ~50 sq.degrees are covered also by the publicly available UKIDSS (`http://www.ukidss.org/`) which provide 3 near-IR colours and thus for a subsample of the previous data we will have relatively more accurate photo-zs allowing us to attempt to derive directly the spatial correlation function.

Summarizing, the analysis of such large XXM surveys will allow us to unambiguously determine the soft and hard-band X-ray AGN clustering pattern, minimizing the biases and systematic effects discussed in the previous section, as well as to study the evolution of the AGN correlation function (utilizing photo-z's and dividing the angular sample into 2-3 z-bins).

Finally, together with a large European consortium, we are planning to survey a 2-3 contiguous sky areas adding to $\sim$ 50 deg^2 and covered also by a large number of other multiwavelength surveys (like UKIDSS, NEWFIRM, etc), which will probably allow us to sample the long wavelength regime of the AGN correlation function, and thus measure Baryonic Accoustic Oscillations, which are extremely important for Dark-energy investigations (Peacock et al 2006; Albrecht et al. 2006). To obtain the necessary high photo-z accuracy, one may envision a wide-field multi-filter (say using ~20 broad and narrow-band filters) survey, specially tuned in order to obtain relatively high-accuracy photo-z's of $z \sim 1$ AGN. Thoughts for the acquisition of such an instrument, to be mounted on the 2.3m Hellenic Aristarchos telescope, are already been discussed in the Institute of Astronomy & Astrophysics of the National Observatory of Athens.

3.3. Cosmological Parameter constraints from X-ray AGN Clustering

The unambiguous determination of the correlation function of the $z \sim 1$ X-ray AGNs, even in angular space, will allow us to estimate with good precision (a) their relation to the underlying matter fluctuations at $z \sim 1$ (ie., their bias), (b) the evolution of their bias and therefore the mass of the DM haloes which they inhabit (eg., Miyaji et al. 2007; Basilakos, Plionis & Ragone-Figueroa 2008) and (c) put strong cosmological constraints on the Ω_m, h or Ω_m, σ_8 planes, while with the help of the high-z HII-based Hubble relation on the $\Omega_m, w(z)$ space.

It is well known (Kaiser 1984; Benson et al. 2000) that according to linear biasing the correlation function of the AGN (or any mass-tracer) ($\xi_{\rm AGN}$) and dark-matter one ($\xi_{\rm DM}$), are related by:

$$\xi_{\rm AGN}(r,z) = b^2(z)\xi_{\rm DM}(r,z) \ , \tag{11}$$

where $b(z)$ is the bias evolution function (eg. Mo & White 1996, Matarrese et al. 1997, Basilakos & Plionis 2001; 2003; Basilakos, Plionis & Ragone-Figueroa 2008)

A first outcome of the proposed correlation function analysis will be the accurate determination of the AGN bias at their median redshift (in our case $\bar{z} \simeq 1$), utilizing:

$$b(z) = \left(\frac{r_0}{r_{0,m}}\right)^{\gamma/2} D^{3+\epsilon}(z) \quad \text{with } \gamma = 1.8 \text{ and } \epsilon = -1.2 \ ,$$

where r_0 and $r_{0,m}$ are the measured AGN and dark-matter (from the $P(k)$) clustering lengths, respectively, while $D(z)$ is the perturbation's linear growing mode. Then using a bias evolution model, one will be able to determine the mass of the DM halo within which such AGN live (eg. see Figure 9).

Furthermore, we will compare the observed AGN clustering with the predicted, for different cosmological models, correlation function of the underlying mass, $\xi_{\rm DM}(r,z)$. To this end we can use the Fourier transform of the spatial power spectrum $P(k)$:

$$\xi_{\rm DM}(r,z) = \frac{(1+z)^{-(3+\epsilon)}}{2\pi^2} \int_0^\infty k^2 P(k) \frac{\sin(kr)}{kr} {\rm d}k \ , \tag{12}$$

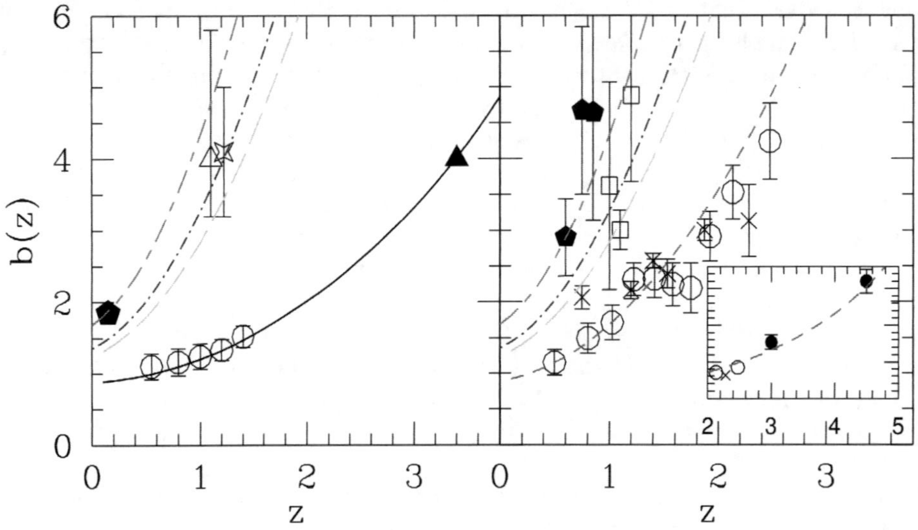

Figure 9. Comparison of the Basilakos et al (2008) $b(z)$ evolution model with different observational data. Different line types represent different halo masses. *Left Panel:* optical galaxies (open points) with solid line corresponding to $M_{\rm DM} \sim 10^{12}\ h^{-1}\ M_{\odot}$, Lyman break galaxies (solid triangle), EROs (star), DRGs (open triangle) and 2dF radio sources (filled pentagon). The dot-dashed line corresponds to $7.7\ \times 10^{13}\ h^{-1}\ M_{\odot}$. *Right Panel:* optically selected quasars (open points and crosses), soft and hard X-ray point sources (open squares and solid diamonds; the large scatter corresponds to the uncertainty of the present day clustering results; see Plionis et al. 2008 and references therein). In the insert we plot, as solid points, the high-z SSRS DR5 QSOs and the same $b(z)$ model that fits their lower redshift counterparts (ie., $M_{\rm DM} \simeq 10^{13}\ h^{-1}\ M_{\odot}$).

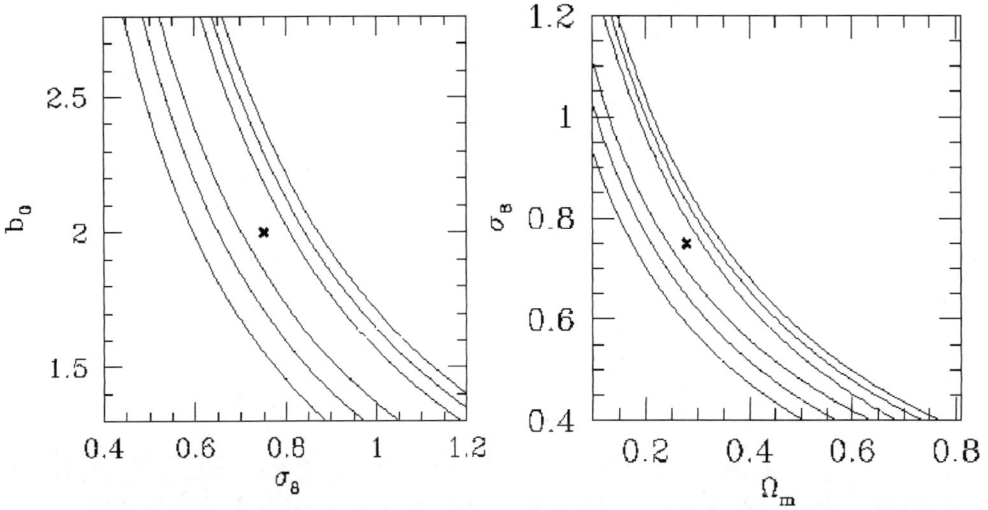

Figure 10. Likelihood contours from the X-ray AGN clustering analysis of Basilakos & Plionis (2005) in the following planes: (Ω_m, σ_8) (right panel), (σ_8, b_0) (left panel). The contours plotted correspond to 1σ, 2σ and 3σ confidence level.

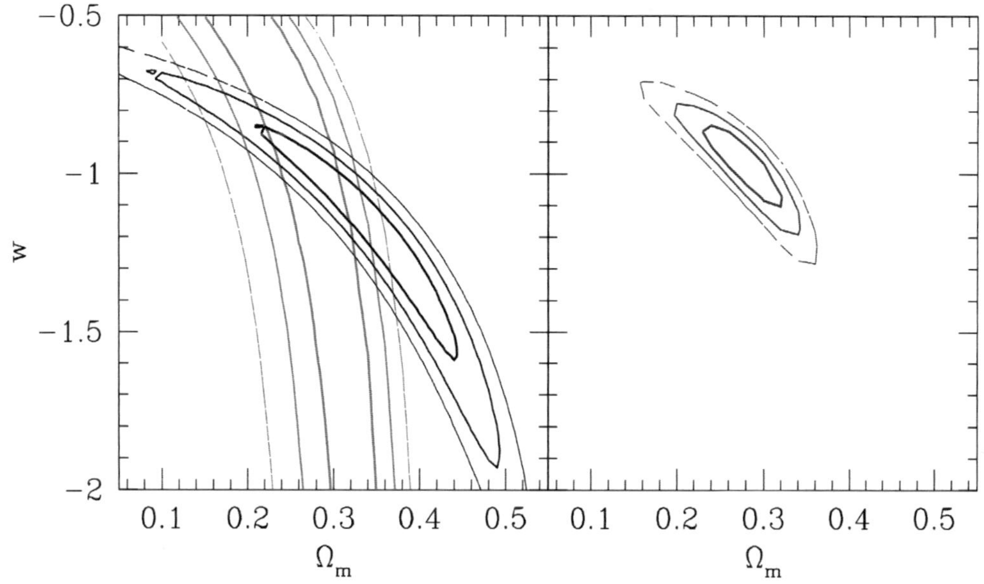

Figure 11. *Left Panel:* Likelihood contours on the Ω_m, w plane from the X-ray AGN clustering analysis of Basilakos & Plionis (2005; 2006) (red contours) and the *UNION* SNIa analysis (black contours). *Right Panel:* The Joint likelihood contours.

where k is the comoving wavenumber, $P(k) = P_0 k^n T^2(k)$ the CDM power-spectrum with scale-invariant ($n = 1$) primeval inflationary fluctuations, while the transfer function parameterization is as in Bardeen et al. (1986), with the corrections given approximately by Sugiyama (1995).

Basilakos & Plionis (2005; 2006) have already used successfully a standard maximum likelihood procedure to compare the measured XMM source angular correlation function from a relatively small (~ 2 sq.degrees survey; Basilakos et al. 2005) with the prediction of different spatially flat cosmological models, and derived interesting cosmological constraints (for flat and constant-w cosmologies).

In Figure 10 we present the constraints provided from the preliminary analysis by Basilakos & Plionis (2006) of a ~ 2 sq.degrees XMM survey, on the present bias-factor (b_0) of the X-ray AGN, on the normalization of the power spectrum (σ_8) and on Ω_m (marginalizing over different parameters). In Figure 11 (left panel) we present the (Ω_m, w) constraints provided by the X-ray AGN clustering analysis (red contours), once we have marginalized over the $\sigma_8(\sim 0.8)$ and the bias factor at the present time (~ 2).

4. Joint Hubble-relation and Clustering analysis

It is evident from Figure 11 (left panel) that w is degenerate with respect to Ω_m and that all the values in the interval $-2 \leq w \leq -0.35$ are acceptable within the 1σ uncertainty. However, we break this degeneracy by adding the constraints provided by the Hubble relation technique, using here the *UNION* SNIa sample. We therefore perform a joint likelihood analysis, assuming that the two data sets are independent (which indeed they are) and thus we can write the joint likelihood as the product of the two individual ones.

Our current joint likelihood analysis, once we impose $h = 0.72$ and $\sigma_8 = 0.8$, provides quite stringent constraints of the Ω_m and w parameters:

$$\Omega_{\rm m} = 0.28^{+0.02}_{-0.04} \quad \text{and} \quad w = -1.0 \pm 0.1 \ .$$

However, the uncertainty on w is still quite large, while the necessity to impose constraints on a more general, time-evolving, *dark-energy* equation of state (eq. 2) implies that there is ample space for great improvement and indeed the aim of the project, detailed in these proceedings, is exactly in this direction.

References

[1] Akylas, A., Georgantopoulos, I., Plionis, M., 2000, MNRAS, 318, 1036
[2] Albrecht, A., et al., 2006, `astro-ph/0609591`
[3] Allen, S. W., et al., 2004, MNRAS, 353, 457
[4] Andernach, H., et al., 2005, ASPC, 329, 289
[5] Antonucci, R., 1993, ARA&A, 31, 473
[6] Astier, P., et al., 2006, A&A, 447, 31
[7] Bardeen, J.M., Bond, J.R., Kaiser, N. & Szalay, A.S., 1986, ApJ, 304, 15
[8] Basilakos, S. & Plionis, M., 2001, ApJ, 550, 522
[9] Basilakos, S. & Plionis, M., 2003, ApJ, 593, L61
[10] Basilakos, S., Georgakakis, A., Plionis, M., Georgantopoulos, I., 2004, ApJ, 607, L79
[11] Basilakos, S., Plionis, M., Georgantopoulos, I., Georgakakis, A., 2005, MNRAS, 356, 183
[12] Basilakos, S. & Plionis, M., 2005, MNRAS, 360, L35
[13] Basilakos, S. & Plionis, M., 2006, ApJ, 650, L1
[14] Basilakos, S., Plionis, M. & Ragone-Figueroa, C., 2008, ApJ, 678, 627
[15] Basilakos, S. & Perivolaropoulos, L., 2008, MNRAS, 391, 411
[16] Benson A. J., Cole S., Frenk S. C., Baugh M. C., Lacey G. C., 2000, MNRAS, 311, 793
[17] Boyle B.J., Mo H.J., 1993, MNRAS, 260, 925
[18] Blake, C., et al., 2007, MNRAS, 374, 1527
[19] Carrera, F.J., Barcons, X., Fabian, A. C., Hasinger, G., Mason, K.O., McMahon, R.G., Mittaz, J.P.D., Page, M.J., 1998, MNRAS, 299, 229
[20] Carrera, F.J., et al. 2007, A&A, 469, 27
[21] Chevallier M., & Polarski D., 2001, Int. J. Mod. Phys. D, 10, 213
[22] de Zotti, G., Persic, M., Franceschini, A., Danese, L., Palumbo, G.G.C., Boldt, E.A., Marshall, F.E., 1990, ApJ, 351, 22
[23] Davis, T.M., et al., 2007, ApJ, 666, 716
[24] Dicus, D.A. & Repko, W.W., 2004, Phys.Rev.D, 70, 3527,
[25] Erb, D.K. et al., 2003, ApJ, 591, 101
[26] Eisenstein, Hu, & Tegmark 1998, ApJ, 504, L57
[27] Eisenstein et al. 2005, ApJ, 633, 560
[28] Fabian, A. C., 1991, MNRAS, 253, 29
[29] Fabian, A., 1999, MNRAS, 308, L39
[30] Gandhi, P., et al., 2006, A&A, 457, 393
[31] Gilli, R., et al. 2003, ApJ, 592, 721
[32] Ghirlanda, G., Ghisellini, G., Firmani, C., 2006, NewJPhys, 8, 123
[33] Göttlober, S. & Yepes, G., 2007, ApJ, 664, 117
[34] Kim, M., Wilkes, B.J., Kim, D-W., Green, P.J., Barkhouse, W.A., Lee, M.G., Silverman, J.D., Tananbaum, H.D., 2007, ApJ, 659, 29
[35] Kaiser N., 1984, ApJ, 284, L9
[36] Kowalski M. et al., 2008, ApJ, 686, 749
[37] Kundić, T., 1997, ApJ, 482, 631
[38] Linder, V. E., 2003, Phys. Rev. Lett., 90, 091301
[39] Matsubara, T., 2004, ApJ, 615, 573
[40] Manners, J.C., et al., 2003, MNRAS, 343, 293
[41] Matarrese, S., Coles, P., Lucchin, F., Moscardini, L., 1997, MNRAS, 286, 115
[42] Melnick, J., 1978, A&A, 70, 157
[43] Melnick J., Terlevich, R., Moles, M., 1988, MNRAS, 235, 313;
[44] Melnick, J., Terlevich, R., Terlevich, E., 2000, MNRAS, 311, 629
[45] Melnick, J., 2003, *Star Formation Through Time'*, ASP Conference Proceedings, Vol. 297, Edited by E.Perez, R.M. Gonzalez Delgado & G. Tenorio-Tagle

[46] Miyaji, T., et al., 2007, ApJS, 172, 396
[47] Mullis C.R., Henry, J. P., Gioia I. M., Böhringer H., Briel, U. G., Voges, W., Huchra, J. P., 2004, ApJ, 617, 192
[48] Mo, H.J, & White, S.D.M 1996, MNRAS, 282, 347
[49] Newman, J. A., et al., 2002, PASP, 114, 29
[50] Padmanabhan et al. 2007, MNRAS, 378, 852
[51] Peacock et al., 2006, `astro-ph/0610906`
[52] Peebles P.J.E., &, Ratra, B., 2003, RvMP, 75, 559
[53] Perlmutter, S., et al., 1998, Nature, 391, 51
[54] Perlmutter, S., et al., 1999, ApJ, 517, 565
[55] Pettini, M., et al., 2001, ApJ, 554, 981
[56] Plionis, M., Rovilos, M., Basilakos, S., Georgantopoulos, I., Bauer, F., 2008, ApJ, 674, L5
[57] Puccetti, S., et al., 2006, A&A, 457, 501
[58] Ragone-Figueroa C. & Plionis M., 2006, MNRAS, 377, 1785
[59] Riess, A. G., et al., 1998, AJ, 116, 1009
[60] Riess, A. G., et al., 2004, ApJ, 607, 665
[61] Riess A.G., 2007, ApJ, 659, 98
[62] Rosati, P., et al., 2002, ARA&A, 40, 539
[63] Sanchez, A. G., et al., 2006, MNRAS, 366, 189
[64] Schuecker, P., et al., 2003, A&A, 402, 53
[65] Seljak, U., et al., 2004, Phys.Rev.D, 71, 3515
[66] Sheth, R.K., Mo, H.J., Tormen, G., 2001, MNRAS, 323, 1
[67] Siegel, E.R., et al., 2005, MNRAS, 356, 1117
[68] Spergel, D. N., et al., 2003, ApJS, 148, 175
[69] Spergel, D.N., et al. 2007, ApJS, 170, 377
[70] Stewart, G.C., 2007, in "X-ray Surveys: Evolution of Accretion, Star-Formation and the Large-Scale Structure", Rodos island, Greece, `http://www.astro.noa.gr/xray07/rodos-talks/`
[71] Sugiyama, N., 1995, ApJS, 100, 281
[72] Tegmark, M., et al., 2004, Phys.Rev.D., 69, 3501
[73] Terlevich, R., Melnick, J., 1981, MNRAS, 195, 839
[74] Tonry, et al. , 2003, ApJ, 594, 1
[75] Vikhlinin, A. & Forman, W., 1995, ApJ, 455, 109
[76] Yang, Y., Mushotzky, R.F., Barger, A.J., Cowie, L.L., Sanders, D.B., Steffen, A.T., 2003, ApJ, 585, L85
[77] Yang, Y., Mushotzky, R.F., Barger, A.J., Cowie, L.L., 2006, ApJ, 645, 68
[78] Vikhlinin, A., et al., 2003, ApJ, 590, 15
[79] Wang, Y. & Mukherjee, P., 2006, ApJ, 650, 1
[80] Wei, , H., 2008, `astro-ph/0809.0057`
[81] White, S.D.M., et al., 1993, Nature, 366, 429
[82] Wood-Vasey W.M., et al., 2007, ApJ, 666, 694

Journal of Physics: Conference Series **189** (2009) 012033 doi:10.1088/1742-6596/189/1/012033

Two approaches to testing general relativity in the strong-field regime

Dimitrios Psaltis

Departments of Astronomy and Physics, University of Arizona, 933 N. Cherry Ave., Tucson, AZ 85721

E-mail: dpsaltis@email.arizona.edu

Abstract. Observations of compact objects in the electromagnetic spectrum and the detection of gravitational waves from them can lead to quantitative tests of the theory of general relativity in the strong-field regime following two very different approaches. In the first approach, the general relativistic field equations are modified at a fundamental level and the magnitudes of the potential deviations are constrained by comparison with observations. In the second approach, the exterior spacetimes of compact objects are parametrized in a phenomenological way, the various parameters are measured observationally, and the results are finally compared against the general relativistic predictions. In this article, I discuss the current status of both approaches, focusing on the lessons learned from a large number of recent investigations.

1. Introduction

Testing the theory of general relativity is the key science objective of many future astrophysical missions as well as of NASA's Beyond Einstein program, as a whole. However, there has been very little consensus, as of now, on the particular direction along which quantitative tests of the theory will be performed.

The parametric post-Newtonian framework, which has been crucial in testing general relativity at solar-system scales, is no longer valid in the strong-field regime. In fact, in the absence of an accepted alternative theory of gravity, there are very few a priori constrains on the potential deviations from general relativistic predictions. However, there are a number of arguments that may guide us in the effort to reach this goal.

As in most other branches of physics, there are two distinct avenues we may follow in testing general relativity in the strong-field regime. In a top-down approach, we may modify the theory at a fundamental level and calculate the consequences of the modifications that can be tested observationally. In a bottom-up approach, we will use a phenomenological description of the observations in order to obtain clues for how to modify the theory at the fundamental level. In this article, I draw a roadmap for testing the theory of general relativity in the strong-field regime along these two avenues, based on general theoretical expectation as well as on a large number of recent investigations.

2. The top-down approach: From Theory to Observations

2.1. The Equivalence Principle and the Einstein Field Equations

The theory of general relativity has two distinct ingredients. The equivalence principle that describes how matter moves in the presence of a gravitational field and the Einstein field equations that describes how the gravitational field is generated in the presence of matter.

The validity of the equivalence principle gives rise to the geometric aspect of the theory (see Will 2006). According to this principle, it is impossible to tell the difference between a reference frame at rest and one free-falling in a gravitational field, by performing local experiments. Moreover, the equivalence principle encompasses the Lorentz symmetry, as well as our beliefs that there is no preferred frame and position anywhere in the universe.

The Einstein equivalence principle has been tested to a very high degree during the last century, mostly in the weak-field regime. The upper bound on possible violations are as low as one part in 10^{12} (Will 2006). It is very unlikely that astrophysical observations of black holes and neutron stars can lead to tests of comparable accuracy. For this reason, most approaches to date have concentrated only on plausible modifications of the field equations although frameworks (such as the Standard Model Extension; see Kostelecky 2003 and references therein) exist that can lead, in principle, to strong-field tests of this principle (see Stairs 2003 for a discussion of testing the strong equivalence principle with double neutron stars; see also Eling & Jacobson 2006 for black-hole solutions in an Einstein-aether theory).

2.2. Modifying the Einstein Field Equations

Contrary to the case of the equivalence principle, there are no arguments one can make that lead uniquely to the Einstein field equations. Indeed, Einstein reached the field equation, more or less, by reverse engineering. For this reason, his approach cannot be easily modified to include additional terms. On the other hand, the "derivation" of the field equations from a Lagrangian action provided by Hilbert allow us to modify the theory, while preserving many of its symmetries and conservation laws.

The Einstein-Hilbert action that leads to the field equations of general relativity is directly proportional to the Ricci scalar, R,

$$S = \frac{c^4}{16\pi G}\int d^4x\,\sqrt{-g}\,(R-2\Lambda)\,, \tag{1}$$

where g is the spacetime metric, c is the speed of light, G is the gravitational constant, and Λ is the cosmological constant.

We can construct a self-consistent theory of gravity from this starting point using any other action that obeys the following four simple requirements (see Misner et al. 1973; p. 410). It has to: *(i)* reproduce the Minkowski spacetime in the absence of matter and cosmological constants, *(ii)* be constructed from only the Riemann curvature tensor and the metric, *(iii)* follow the symmetries and conservation laws of the stress-energy tensor of matter, and *(iv)* reproduce Poisson's equation in the Newtonian limit. Of all the possibilities that meet these requirements, the field equations that are derived from the Einstein-Hilbert action are the only ones that are also linear in the Riemann tensor.

The function form of the Einstein-Hilbert action suggests that we can modify it in one of four ways: In principle, we can *(i)* change the number of spacetime dimensions, *(ii)* change the functional dependence of the Lagrangian density on the Ricci scalar R, *(iii)* allow for the Lagrangian density to depend on other scalars generated from the Riemann curvature tensor, or *(iv)* introduce additional scalar, vector, or tensor fields.

(i) The number of spacetime dimensions.—The possibility that we live in a Universe with more than four spacetime dimensions has been the subject of intense research during the last decade,

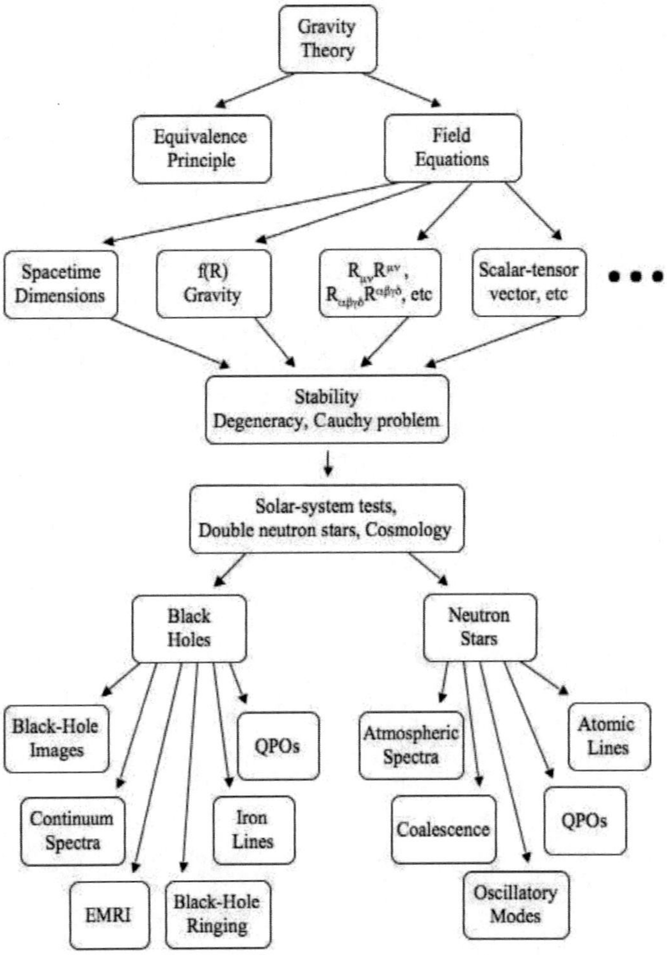

Figure 1. A top-down approach to performing strong-field tests of gravity.

primarily as a solution to the hierarchy problem in physics (Arkani-Hamed et al. 1998a; Randall & Sundrum 1999; Maartens 2005). In these braneworld gravity models, all non-gravitational fields and particles live on a four dimensional spacetime, whereas the gravitational field is allowed to propagate in the extra dimensions. In order for the theory to be consistent with Newton's law of gravity down to the sub-mm level, the effective size of the extra dimensions has to be smaller than $\sim 50\ \mu$m (Kapner et al. 2007). As a result, such a modification to gravity is not expected to have any observable classical consequences even for neutron stars, which are the smallest known astrophysical objects with strong gravitational fields (Arkani-Hamed et al. 1998b).

A plausible consequence of one variant of braneworld gravity (the RS2 scenario of Randall & Sundrum) that can be constrained with astrophysical observations is the rapid evaporation of black holes caused by the emission of gravitons in the extra dimensions (Emparan et al. 2003 and references therein). The rate of rapid evaporation was inferred using the AdS/CFT correspondence and is a matter of debate (Fitzpatrick et al. 2006). If proven accurate, however, it can lead to astrophysical bounds on the asymptotic scale of the large extra dimensions that are comparable to those obtained in table-top experiments (Johannsen et al. 2009 and references therein).

(ii) The functional dependence of the Langrangian density on the Ricci scalar.— A Langrangian action that depends on a non-linear function of the Ricci scalar R, i.e.,

$$S = \frac{c^4}{16\pi G} \int d^4x \sqrt{-g}\, f(R)\,, \tag{2}$$

leads to field equations that obey all the requirements outlined above. Indeed, the action (2) results in a field equation that allows for the Minkowski solution in the absence of matter, is constructed only from the Riemann tensor, obeys the usual symmetries and conservation laws, and can be made to produce negligible corrections at the small curvatures probed by weak-field gravitational experiments. Such non-linear actions for the gravitational field, albeit with negligibly small coefficients, occur naturally in quantum gravity theories and in string theory. This is true because of radiative corrections to the usual Einstein-Hilbert action that induce an infinite series of counterterms that cannot be reabsorbed into the original Lagrangian by adjusting its bare parameters (see discussion in Burgess 2004).

Despite their simplicity and appeal, a number of important limitations to $f(R)$ gravity theories have been recently explored in the cosmological context (see Sotiriou & Faraoni 2009 for an extensive review). Modifications to the Lagrangian action that appear negligible at solar-system scales may lead to order unity deviations of the PPN parameters that are inconsistent with current experiments (Chiba 2003). Moreover, astrophysical objects in non-linear $f(R)$ gravity, including the Universe as a whole, can be violently unstable, depending on the sign of d^2f/dR^2 (Dolgov & Kawasaki 2003). Finally, special attention has to be paid to ensure that a well posed initial value problem can be defined under all physical conditions within the non-linear $f(R)$ theory (Lanahan-Tremblay & Faraoni 2007).

Resolving the above problems requires a carefully constructed $f(R)$ theory with fine-tuned coefficients of the non-linear terms (e.g., Nojiri & Odintsov 2003) or one that leads to the so-called chameleon mechanism (see Hu & Sawicki 2007 and references therein). Alternatively, a simple $f(R)$ function may be considered as an expansion of an unknown, more general theory and its consequences for astrophysical phenomena may be calculated within the context of perturbative constraints (Simon 1990; see also Cooney et al. 2009 and references therein).

In adding terms to the Lagrangian action that are of non-linear order in the curvature, we have the choice of performing the derivation of the parametric Einstein equation either in the metric formalism or in the Palatini formalism. The first corresponds to extremizing the action under variations in the metric only, whereas the second corresponds to extremizing the action under variations in the metric and the connection, as well. For the simple Einstein-Hilbert action, both approaches are equivalent and give rise to the same field equations. However, when the action has non-linear terms in the Ricci scalar, the two approaches diverge. Recent studies of $f(R)$ gravity in the Palatini formalism, however, have revealed a number of serious problems with this approach that are related to the suppression of the new dynamical degrees of freedom and render it a rather unlikely alternative to the usual derivation of the field equations from a Lagrangian action (see Sotiriou & Faraoni 2009 and references therein for a detailed discussion).

(iii) Additional scalar terms constructed from the Riemann curvature tensor.— It is remarkable that the Einstein field equations can be derived from a first-order Lagrangian action that does not involve the Riemann tensor (see Landau & Lifshitz 1992). However, higher-order field equations that obey the Lorenz symmetry have been derived only from Lagrangian actions that

involve different scalar quantities constructed from the Riemann tensor $R_{\alpha\beta\gamma\delta}$, e.g.,

$$\mathcal{S} = \frac{c^4}{16\pi G} \int d^4x\sqrt{-g}\left(R + \alpha R^2 + \beta R_{\sigma\tau}R^{\sigma\tau} + \gamma R^{\alpha\beta\gamma\delta}R_{\alpha\beta\gamma\delta}\right) , \tag{3}$$

where $R_{\sigma\tau}$ is the Ricci tensor and α, β, and γ the three parameters of the theory

Similar to the case of $f(R)$ gravity, such terms arise naturally as high-order corrections in quantum gravity and string theory and their relative importance increases with the curvature of the metric. The resulting field equations suffer, however, from the Ostragardski instability, which can be avoided only by fine tuning of the free parameters of the theory (see Woodard 2006 and references therein).

(iv) The presence of additional gravitational fields.—The single, rank-2 tensor field $g_{\mu\nu}$ (i.e., the metric) of the Einstein-Hilbert action may also not be adequate to describe completely the gravitational force (although, if additional fields are introduced, then the strong equivalence principle is violated, with important implications for the frame- and time- dependence of gravitational experiments).

The simplest case of a theory with additional gravitational fields involves a single additional scalar and has been studied for more than 40 years in the form of Brans-Dicke gravity (e.g., Will 2006). Because of its particular properties, the fractional deviation of the Brans-Dicke predictions from general relativity are comparable in both the weak and strong-field regimes. In a more general case, however, depending on the coupling between the metric, the scalar field, and matter, the relative contribution of such additional fields may become significant only at the high curvatures found in the vicinity of compact objects.

The general form of the Lagrangian of a scalar-tensor theory is given, in the Einstein frame, by the Bregmann-Wagoner action (see Will 2006 for details)

$$S = \frac{c^4}{16\pi G} \int d^4x\sqrt{-g_*}\left[R_* \pm g_*^{\mu\nu}\partial_\mu\phi\partial_\nu\phi + 2\lambda(\phi)\right] + S_{\rm m}[\phi_m, A^2(\phi)g_{\mu\nu}] , \tag{4}$$

where $A(\phi)$ and $\lambda(\phi)$ are two arbitrary functions, and S_m is the action for the matter field ϕ_m. In the strong-field regime, the potential term $\lambda(\phi)$ in the action (4) is typically negligible. The single coupling function $A(\phi)$ encodes the properties of the theory and may be parametrized as $A(\phi) = exp[\phi_0 + \alpha_\phi\phi + (1/2)\beta_\phi\phi^2]$, with $\phi_0 \to 0$ being the cosmological value of the field (Damour & Esposito-Farese 1993).

2.3. Studying the Properties of a Parametric Lagrangian Theory

The parametric addition of terms in the Einstein-Hilbert action offers a systematic and self-consistent way of characterizing potential deviations from general relativistic predictions that can be constructed to become arbitrarily large in the strong-field regime. It may also lead, however, to a number of pitfalls that need to be studied and understood before embarking onto a detailed comparison to observational data.

First, a parametric extension to the Einstein-Hilbert action is guided only by the mathematical properties of the form that gave rise to the field equations of general relativity and not by a new requirement of the physics. As such, the extensions discussed above will not cover the entire range of possibilities, let alone a theory that is fundamentally different from general relativity in the strong-field regime. In some sense, using the parametric extension of the Einstein-Hilbert action will allow us to look for deviations from general relativity albeit not necessarily the ones that are described by the particular parameters measured to differ from their general relativistic values. If future experiments do measure such a deviation, then a more fundamental study will be required to account for it. There is a close similarity between this situation and the current measurements of the cosmological constant in the very weak field regime. Indeed, the non-zero measured value of the cosmological constant may just be an indication for the presence of a dynamical scalar field (such as quintessence) as opposed to a constant in the Einstein field equations. However, until a non-zero cosmological constant was measured, it made little sense to introduce the additional complication of a dynamical equation for the scalar field in the Friedman equation.

Second, taken at face value, the predictions of a parametric Lagrangian theory may describe unstable physical systems, as discussed above. Albeit challenging to deal with, such theories need not be *a priori* excluded from quantitative comparison with astrophysical data. There have been, indeed, many cases in physics where unstable solutions have been tested against data (e.g., the cosmological solution with $\Omega \neq 1$ is highly unstable but is still being tested against WMAP data) and even some situations in which the unstable solution was preferred over the stable one (e.g., the Rutherford model of the atom). When the experimental data prefer a steady-state solution that is unstable, it is almost always an indication of new physics, which would have been missed had such a solution been excluded.

The same argument can also be made for predictions that appear to be unphysical in the context of the parametric theory but are not at a more fundamental level. This is similar to some predictions of the PPN formalism related, e.g., to non-zero values of the parameters α_3, and ζ_1 through ζ_4 that describe violations of conservation of total momentum. Measuring any of these "unphysical" parameters to be anything other than zero would be interpreted simply as a smoking gun for the presence of an additional field that has not been incorporated in the PPN formalism but carries some momentum away from the system under study.

Finally, it is also worth emphasizing that there is significant degeneracy between theories that arise from different Lagrangian actions and, therefore, not every astrophysical observation can distinguish between them. For example, it is well known that the field equations of gravity theories with higher-order terms in the Ricci scalar are completely equivalent to those of a scalar-tensor theory (see, e.g., Magnano & Sokolowski 1994) although this is no longer true when additional terms are introduced to the action that involve the Riemann and Ricci tensors, i.e., $R^{\mu\nu}R_{\mu\nu}$ and $R^{\alpha\beta\mu\nu}R_{\alpha\beta\mu\nu}$. Even in the latter case, however, the particular predictions of the theory for specific astrophysical systems is not always different from their general relativistic counterparts.

Consider, for example, the theory given by the Lagrangian action (3). Because of the Gauss-Bonnet identity,

$$\frac{\delta}{\delta g_{\mu\nu}} \int d^4x \left(R^2 - 4R_{\sigma\tau}R^{\sigma\tau} + R^{\alpha\beta\gamma\delta}R_{\alpha\beta\gamma\delta}\right) = 0\,, \tag{5}$$

variations of the term proportional to γ in equation (3), with respect to the metric, can be expressed as variations of the terms proportional to α and β. Therefore, for all classical tests, the predictions of the theory described by the Lagrangian action (3) are identical to those of the Lagrangian

$$\mathcal{S} = \frac{c^4}{16\pi G} \int d^4x\sqrt{-g}\left[R + (\alpha - \gamma)R^2 + (\beta + 4\gamma)R_{\sigma\tau}R^{\sigma\tau}\right]\,. \tag{6}$$

As a result, classical astrophysical tests can only constrain a particular combination of the parameters, i.e., $\alpha - \gamma$ and $\beta + 4\gamma$. It is only through phenomena related to quantum gravity, such as the evaporation timescale of black holes, that the parameter γ may be constrained.

When the spacetime is isotropic and homogeneous, as in the case of cosmological tests, an additional identity is satisfied, i.e.,

$$\frac{\delta}{\delta g_{\mu\nu}} \int d^4x \left(R^2 - 3R_{\sigma\tau}R^{\sigma\tau}\right) = 0\,. \tag{7}$$

This implies that, for cosmological tests, the predictions of the theory described by the Lagrangian action (3) are identical to those of the Lagrangian

$$\mathcal{S} = \frac{c^4}{16\pi G}\int d^4x\sqrt{-g}\left[R + (\alpha + \frac{1}{3}\beta + \frac{1}{3}\gamma)R^2\right] . \quad (8)$$

As a result, cosmological tests of gravity can only constrain a particular combination of the parameters, i.e., $\alpha + 1/3\beta + 1/3\gamma$.

2.4. Reducing the Size of the Parameter Space of Alternative Theories

The general relativistic field equations in the presence of matter are highly non-linear and the same is true for all alternatives explored so far. Because of this non-linearity, modifications to the theory that are apparently significant, e.g., only in the strong-field regime may have observable consequences at very different astrophysical or cosmological scales and vice versa. Recent studies of $f(R) = R \pm \mu^4/R$ gravity theories have actually provided one of the most straightforward demonstrations of this effect. Although the term μ^4/R in the action appears to become relevant only when the curvature of the field is $R \leq \mu^2$, the PPN parameter γ for such a theory is equal to 1/2, is independent of the value of the small parameter μ^2, and is demonstratively different that the general relativistic prediction of $\gamma = 1$ (Chiba 2003).

Because of this non-linearity in the field equations, the vast parameter space of alternatives to general relativity can be reduced by examining their behavior in the weak-field regime or even at cosmological scales. In particular, every modification to general relativity has to account for

(i) the observed constraints on the PPN parameters.— In particular, the PPN parameters β and γ have been constrained to lie within $\simeq 10^{-5}$ of their general relativistic values using tests involving bodies in the solar-system as well as man-made spacecraft (Will 2006 and references therein). Calculations of the PPN parameters in scalar-tensor theories specifically designed to perform strong-field tests have been performed by Damour & Esposito-Farese (1993), whereas a large number of similar studies have been carried out for $f(R)$ gravity theories (Barrause & Sotiriou 2009).

(ii) the orbital period decay of double neutron stars.— This indirect evidence for the emission of gravitational waves offers a unique setting in which the predictions of a gravity theory for the time dependence of a dynamical spacetime can be compared against highly accurate observations (Stairs 2003). Damour & Esposito-Farese (1993) have exploited these observations to place strong constraints on strong-field modifications to general relativity that involve dynamical scalar fields.

(iii) the evolution of the Universe predicted by the modified Friedman equation. Reversing the argument given earlier for cosmologically motivated theories, it is important that a modification to gravity constructed for use in strong-field tests does not contradict the constraints imposed by the observations of the high-z supernovae (which probe the nearby universe, up to a redshift of ~ 1), of the cosmic microwave background (which is sensitive to the expansion of the universe up to a redshift of ~ 1000), and of big-bang nucleosynthesis (which is highly sensitive to the rate of the expansion of the universe at a much earlier time).

2.5. Black Holes and Neutron Stars in Alternative Gravity Theories

The final step in setting up tests of strong-field gravity in the top-bottom approach discussed here involves calculating the properties of black holes and neutron stars in the parametric Lagrangian theories.

In the case of neutron stars, the presence of matter in the domain of solution guarantees that the addition of terms to the Einstein field equations will lead to modifications in the structure of the stars and their exterior spacetimes. Indeed, studies of neutron stars in alternative gravity theories have invariably resulted in larger and more massive neutron stars compared to their general relativistic counterparts. Zaglauer (1992) and Damour & Esposito-Farese (1993) studied the structure of neutron stars in gravity theories that involve a parametric addition of a scalar field to the Einstein-Hilbert action. More recently, a number of studies investigated compact objects in $f(R)$ gravity and in particular the role of the chameleon mechanism in allowing neutron stars to exist in a large class of theories (see Upadhye & Hu 2009 and references therein).

Contrary to the case of neutron stars, most parametric extensions to the Einstein-Hilbert action lead to field equations that admit as a solution the Kerr spacetime of a black hole (Psaltis et al. 2008). This is formally true only for steady-state black holes, as the time-dependent spacetime of a perturbed black hole that can be probed with observations of gravitational waves is different in different theories (Barausse & Sotiriou 2008). The reason behind the similarity of black hole solutions among widely different theories is the absence of matter everywhere in the domain of solution. Indeed, the entire spacetime of a black hole as well as that of vacuum are characterized by a constant Ricci scalar curvature R. As a result, calculating the spacetime of a black hole requires simply solving the equation R =constant, independent of the underlying field equations of the theory.

Recently, Yunes & Pretorius (2009) as well as Konno et al. (2009) showed that, in Chern-Simons gravity, rotating black-hole solutions are different from the Kerr spacetimes even though the theory admits the normal vacuum solution. They key difference between Chern-Simons gravity and general relativity is the addition of a term in the Einstein-Hilbert action that allows for solutions that violate parity. The spacetime for the vacuum and for a Schwarzschild black hole have zero parity and are, therefore, the same in both theories. On the other hand, the spacetime of a rotating black hole has non-zero parity and, in Chern-Simons gravity, is parametrically different from the Kerr solution.

Having obtained the structure of neutron stars and exterior spacetimes of black holes, the theory can now be put to test against observations that probe strong gravitational fields. In the electromagnetic spectrum, such tests can be provided by observations of the images of black-hole accretion flows, of the continuum and line spectra of neutron stars and black holes, as well as of their rapid variability properties (see Psaltis 2008 for a review). In the near future, gravitational wave observations of coalescing compact objects, of extreme mass-ratio inspirals, as well as of the ringing of perturbed black holes and of the global oscillatory modes of neutron stars will open a new window to the study of the strong-field properties of gravity (see Flanagan & Hughes 2005 for a review).

3. The bottom-up approach: From Phenomenology to Observations

The top-down approach to testing the strong-field regime of gravity discussed above is appealing for a number of reasons. First, the terms added to the Einstein-Hilbert action are often motivated by quantum gravity and string theory, offering therefore the possibility of testing predictions of a fundamental modification to gravity. Second, starting from a Lagrangian action guarantees that the theory will encompass important symmetries and conservation laws. Finally, the parameters of the same theory can be constrained by a wide range of observations of neutron stars and black holes, in the electromagnetic spectrum and with gravitational waves, or even with solar-system experiments and in cosmological settings. As a result, very tight constraints can be obtained and degeneracies between parameters can be broken.

The key drawback of the top-down approach, however, is the fact that observational data are interpreted within the narrow confines of a particular theoretical framework. The case of black-hole spacetimes discussed earlier is the most striking example of the limitations of this

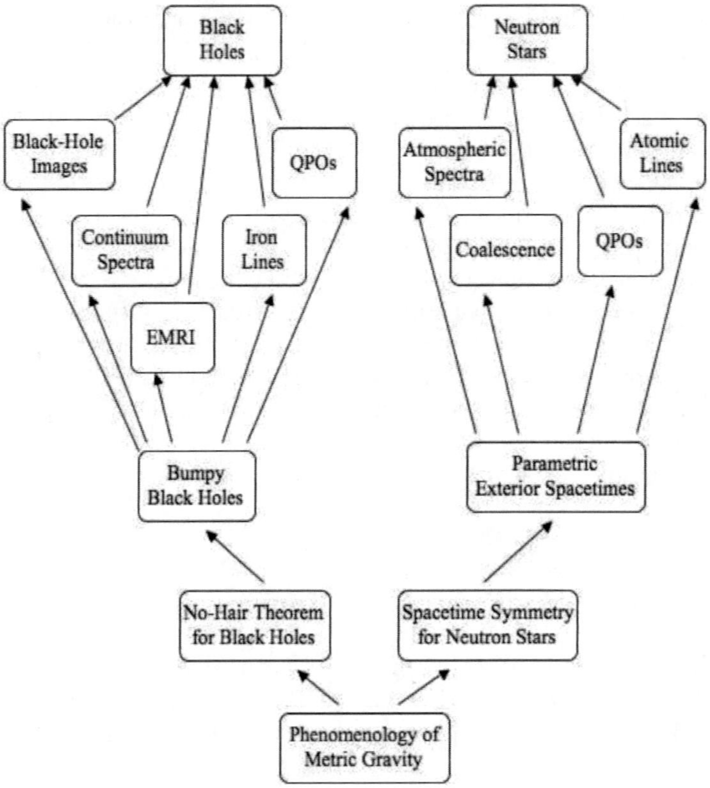

Figure 2. A bottom-up approach to performing strong-field tests of gravity.

approach: a very wide range of gravity theories lead to only one black-hole solution potentially modified by a parity violating term, thereby allowing only for a very limited range of possibilities to be tested against observations.

As an alternative to this approach, we might follow a different, phenomenologically driven path. We will start again by assuming the validity of the equivalence principle and hence of the fact that test particles and photons follow geodesics of the spacetime. In this approach, however, we will not aim to obtain the spacetime as a solution to a field equation. Instead, we will prescribe the external spacetime of a black hole or a neutron star in a way that is parametrically different from their general relativistic counterparts. Using these spacetimes we can then predict observational signatures of deviations from general relativistic predictions to be tested observationally. This approach has been followed only recently and for a limited number of cases, which I will discuss below.

3.1. Parametric Spacetimes of Black Holes and Neutron Stars

The exterior spacetime of an astrophysical black hole in general relativity is completely described by only two parameters: its mass and its spin (any charge on the black hole will be very quickly neutralized by attracting opposite charges from the surrounding medium). This is the so-called no-hair theorem that provides the framework on which to build a phenomenological model of black-hole spacetimes (this approach was pioneered by Ryan 1995 and was further developed more recently by Collins & Hughes 2004 and Glampedakis & Babak 2006).

Because of the no-hair theorem, if we expand the exterior spacetime of a black hole in multipole moments, the coefficients of only two of them are independent quantities: the monopole is related to the black-hole mass M and the dipole to its spin a. All higher-order moments depend in a particular way on these two. For example, the dimensionless quadrupole moment of a black hole in general relativity is equal to $q = -a^2$. We can, therefore, write the most general spacetime with arbitrary coefficients of its multipole moments and use it to predict observables such as the spectra of accretion flows around black holes or the waveforms of gravitational waves during extreme mass-ratio inspirals. Measuring the coefficients of at least three multipole moments of a black hole spacetime by comparing these predictions to observations we can, therefore, test this fundamental property of general relativistic black-hole spacetimes and search for potential deviations.

This approach is very powerful for performing tests of strong-field gravity with black holes and perhaps the one that is the least model dependent. This is especially true for tests that will use observations in the electromagnetic spectrum, because the calculation of the propagation of photons from the strong gravitational fields to the distant observers does not require the knowledge of the underlying field equations that gave rise to the parametric spacetimes. In order, however, to predict the waveforms of gravitational waves emitted, for example, from an extreme mass-ratio inspiral into such a spacetime one needs to use a field equation to describe the time-dependence of the parametric spacetime. As a result, gravitational wave observations can be used within this parametric framework only as null hypothesis tests of general relativistic predictions (see discussion in Hughes 2006).

Contrary to the case of black holes, the exterior spacetimes of neutron stars depend, in general, on the mass distribution in their interiors and are not uniquely described by only two coefficients of their multiple moments. On the other hand, a number of phenomena have been observed from neutron stars that take place in their atmospheres, which is a very thin shell on their surfaces. In order to predict observable quantities related to these phenomena we need to parametrize only the values of the metric elements on their surfaces and not throughout their exterior spacetimes. We can, therefore, use observations to read directly different combinations of the metric elements on the surfaces of the stars and compare them against general relativistic predictions.

As an example, the atmospheric spectrum from a slowly spinning bursting neutron star is determined by various metric elements that depend on three parameters: the coordinate radius of the neutron-star surface, the redshift from it, and the effective gravitational acceleration at the same place. In general relativity all three parameters depend only on the mass and radius of the neutron star. Using, therefore, at least three atmospheric phenomena to measure these three parameters independently can lead to a quantitative test of this general relativistic prediction (Psaltis 2008).

Conclusions

Testing general relativity in the strong-field regime may follow one of two main approaches that are familiar from other areas of physics or astrophysics. We can again draw a parallel between the directions described in this article with current investigations of gravity at cosmological scales. In one set of studies that aim to characterize the accelerated expansion of the universe, the Friedman equation is modified at a fundamental level by postulating, e.g., a scalar field such as quintessence, and the parameters of the theory are constrained against observations. In a different set of studies, the field that provides the acceleration of the universe is described phenomenologically by an equation of state, with parameters that can be measured observationally. The results of these two sets of studies offer a complimentary view of the observational data. At the same time, the lessons learned and the interplay between the two approaches allows for both to be further developed. The quality of high-energy and of gravitational-wave data that are expected from future missions and observatories require that both approaches be exploited as well in the case of strong-field tests of general relativity.

References

Arkani-Hamed N. Dimopoulos S. and Dvali G. 1998a *Phys. Lett. B* **429** 263
Arkani-Hamed N. Dimopoulos S. and Dvali G. 1998b *Phys. Rev. D.* **59** 086044
Barausse E. and Sotiriou T. P. 2008 *Phys. Rev. Let.* **101** 9001
Burgess C. P. 2004 *Living Rev. Relativity* **7** lrr-2004-5
Chiba T. 2003 *Phys. Lett. B.* **575** 1
Collins N. A. and Hughes S. A. 2004 *Phys. Rev. D.* **69** 124022
Cooney A. DeDeo S. and Psaltis D. 2009 *Phys. Rev. D.* **79** 4033
Damour T. and Esposito-Farese G. 1993 *Phys. Rev. Let.* **70** 2220
Dolgov A. D. and Kawasaki M. 2003 *Phys. Lett. B.* **573** 1
Eling C. and Jacobson T. 2006 *Class. Quantum Grav.* **23** 5642
Emparan R. Garcia-Bellido J. and Kaloper N. 2003 *High Energy Phys.* **1** 79
Lanahan-Tremblay N. and Faraoni V. 2007 *Class. Quantum Grav.* **24** 5667
Fitzpatrick A. L. Randall L. and Wiseman T. 2006 *JHEP* **0611** 033
Flanagan E. E. and Hughes S. A. 2005 *New J. Phys.* **7** 204
Glampedakis K. and Babak S. 2006 *Class. Quantum Grav.* **23** 4167
Hu W. and Sawicki I. 2007 *Phys. Rev. D.* **76** 4004
Hughes S. A. 2006 *AIP Conference Series* **873** 233
Johannsen T. Psaltis D. and McClintock J. 2009 *Astrophys. J.* **691** 997
Kapner D. J. et al. 2007 *Phys. Rev. Let.* **98** 021101
Konno K. Matsuyama T. and Tanda S. 2009 *Preprint* arXiv:0902.4767
Kostelecky A. 2003 *Phys. Rev. D.* **69** 105009
Landau L. D. and Lifshitz E. M. 1980 *The Classical Theory of Fields* (Butterworth-Heinemann; 5th edition)
Maartens R. 2005 *Living Rev. Relativity* **7**, lrr-2004-7
Magnano G. and Sokolowski 1994 *Phys. Rev. D* **50** 5039
Misner C. W. Thorne K. S. and Wheeler J. A. 1973 *Gravitation* (W.H. Freeman: San Fransisco)
Nojiri S. and Odintsov S. D. 2003 *Phys. Lett. B* **576** 5
Psaltis D. 2009 *Living Rev. Relativity* **11**, lrr-2009-9
Psaltis D. Perrodin D. Dienes K. and Mocioiu I. 2008 *Phys. Rev. Let.* **100** 1101
Randall L. and Sundrum R. 1999 *Phys. Rev. Lett.* **83** 4690
Ryan F. D. 1995 *Phys. Rev. D.* **52** 5707
Simon J. Z. 1990 *Phys. Rev. D.* **41** 3720
Sotiriou T. P. and Faraoni V. 2009 *Rev. Mod. Phys. Preprint* arXiv:0805.1726
Stairs I. H. 2003 *Living Rev. Relativity* **6**, lrr-2003-5
Upadhye A. and Hu W. 2009 *Preprint* arXiv:0905.4055
Yunes N. and Pretorious F. 2009 *Phys. Rev. D. Preprint* arXiv:0902.4669
Will C. M. 2006 *Living Rev. Relativity* **9**, lrr-2006-3
Woodard R. P. 2006 *Preprint* astro-ph/0601672
Zaglauer H. 1992 *Astrophys. J.* **393** 685

Journal of Physics: Conference Series **189** (2009) 012034 doi:10.1088/1742-6596/189/1/012034

Black holes in loop quantum gravity

Rodolfo Gambini[1], **Jorge Pullin**[2]

1. Instituto de Física, Facultad de Ciencias, Iguá 4225, esq. Mataojo, Montevideo, Uruguay.
2. Department of Physics and Astronomy, Louisiana State University, Baton Rouge, LA 70803-4001

Abstract. We summarize recent results concerning the quantization of the complete extension of the Schwarzschild space-time using spherically symmetric loop quantum gravity. We find an exact solution of the polymerized theory, that is expected to capture features of the semi-classical limit. The singularity is eliminated but the space-time still contains a horizon. Although the solution is known partially numerically and therefore a proper global analysis is not possible, a global structure akin to a singularity-free Reissner–Nordström space-time including a Cauchy horizon is suggested.

Loop quantum gravity inspired techniques have been applied to mini-superspace models (loop quantum cosmology) in the last few years. Among the results one generically finds in these models that the big bang singularity is replaced by a bounce (see for instance [1] and references therein). Since the interior of a black hole is classically isometric to a Kantowski–Sachs cosmology (that also sees its singularity eliminated in certain treatments as a loop quantum cosmology), it is natural to expect that the black hole singularity may also disappear in a similar way [2, 3]. A complete treatment of the space-time of a black hole in loop quantum gravity was lacking, even within a midi-superspace type of quantization, and we made progress on this issue recently [4]. What we did was to consider space-times with spherical symmetry and set up their canonical theory. Using a further gauge fixing to avoid the hard problem of having structure functions in the constraint algebra (see [5] for a good discussion) we obtain a model that has only Abelian constraints and a true Hamiltonian for evolution. We proceed to study classically the "polymerized" theory of such a model that can be straightforwardly quantized in the loop representation. It is known that such polymerized theories can capture many effects that one would find in a more systematic quantization followed by a semi-classical approximation. We found that the complete space-time can be covered and a solution can be constructed that replaces the singularities (black and white hole) of the usual Kruskal diagram by regular surfaces. Such surfaces can be smoothly matched so where one expected a "black hole" one tunnels into a "white hole" region of another universe and this can be continued indefinitely. The resulting solution therefore has a Cauchy horizon and can be characterized as the analog in semiclassical loop quantum gravity of an eternal black hole.

We used the Ashtekar new variables to describe the spherically symmetric space-times. Previous work on this subject was done in modern language by Bojowald and Swiderski [6] so we refer the reader to them for details. There is only one non-trivial spatial direction (the radial) which we call x since it is not necessarily parameterized by the usual radial coordinate. We will elaborate more on the range of x later. The canonical variables usual in loop quantum gravity are a set of triads E^a_i and $SO(3)$ connections A^i_a; after the imposition of spherical symmetry one is left with three pairs of canonical variables $(\eta, P^\eta, A_\varphi, E^\varphi, A_x, E^x)$. Instead of

Journal of Physics: Conference Series **189** (2009) 012034 doi:10.1088/1742-6596/189/1/012034

using triads in the directions transverse to the radial one, a "polar" set of variables E^φ, η and their canonical momenta is chosen. It is convenient to introduce the gauge invariant variable K_x defined by $2\gamma K_x = A_x + \eta'$ and also K_φ defined as $A_\varphi = 2\gamma K_\varphi$, where γ is the Immirzi parameter of loop quantum gravity. The canonically conjugate pairs are now E^x, K_x and E^φ, K_φ. The relationship to more traditional metric variables is,

$$\begin{aligned} g_{xx} &= \frac{(E^\varphi)^2}{|E^x|}, \qquad g_{\theta\theta} = |E^x|, \\ K_{xx} &= -K_x \text{sign}(E^x)\frac{(E^\varphi)^2}{\sqrt{|E^x|}}, \quad K_{\theta\theta} = -\sqrt{|E^x|}K_\varphi, \end{aligned} \tag{1}$$

and the latter two are the components of the extrinsic curvature. The diffeomorphism and Hamiltonian constraints can be seen in detail in ref. [7]. These constraints have the usual constraint algebra for gravity in $1+1$ dimensions, which includes structure functions. This implies the usual "problem of dynamics" of canonical quantum gravity. Our strategy to treat this model will be to further fix the gauge so we are left with a model with Abelian constraints and a true Hamiltonian. That way it can be treated using the standard Dirac procedure and it can be quantized with loop quantum gravity techniques. If one were to fix the gauge further before quantization, one would be led back to the standard quantization of Kuchař [8], which cannot add insights on the question of singularities. We eliminate the diffeomorphism constraint by choosing a gauge that determines the functional form for $E^x = f(x,t)$. Imposing the constraint strongly determines K_x. This also fixes the corresponding Lagrange multiplier (the shift) $N^r = -\dot{f}(x,t)/f'(x,t)$ and also breaks reparametrization invariance. One is left with a theory with Abelian constraints and with a true Hamiltonian, the dynamical variables are E^φ and K_φ.

The quantization of the Abelian constraints is straightforward and can be carried out in the same Hilbert space that was considered in the exterior case [7]. In brief, one discretizes the radial direction and the Hilbert space is a tensor product of Hilbert spaces of loop quantum cosmology, one per spatial point. In such a space the constraint of the model is not well defined, but one can work with an expression where K_φ is replaced by $\sin(\mu K_\varphi)/\mu$. The latter is immediately expressible in terms of holonomies and therefore naturally exists in the loop representation. The resulting theory agrees with general relativity in the limit $\mu \to 0$. In loop quantum gravity it is natural to consider a finite value of μ, usually associated with the elementary quantum of area [1].

Instead of quantizing the theory and then studying the semiclassical limit, we will follow a procedure that is known [3] to capture some of the semiclassical behaviors, in particular the elimination of the singularity, at least in simple examples with a constant value of μ as the one we are considering. We analyze the resulting classical "polymerized" theory with finite μ. One is then considering a classical theory of gravitation, different from general relativity that contains some of the ingredients of the quantum theory, akin to when one works out in an effective theory.

We wish to choose the function $f(t,x)$ in such a way that in the limit $\mu \to 0$ one recovers the standard Schwarzschild metric in Kruskal-like coordinates. That is, a metric with a singularity at $x^2 - t^2 = -1$. On the other hand, in the case of finite μ we will make a gauge choice such that no singularities are present on the surface $x^2 - t^2 = -1$ (one could choose gauges with coordinate singularities there). To be more specific we will choose $E^x = f(u,t,\delta)$, where $u = x^2 - t^2 + 1$ and $\delta(\mu)$ a positive parameter such that when $\mu \to 0$, $\delta \to 0$ and we recover the standard Kruskal form of the Schwarzschild space-time. To completely fix the gauge and obtain an explicit solution we set $K_\varphi = g(u,t,\delta)$ after polymerization. In the quantum theory such a gauge fixing would be equivalent to the study of an evolving constant [9, 10] E^φ in terms of c-number variable K_φ.

We require the following conditions on the gauge fixing. We choose the u in the range $[0, \infty]$ and is such that the radial variable has a logarithmic dependence on u for $u \to \infty$ just like in ordinary Kruskal coordinates. Moreover, we want asymptotically $E^\varphi \sim r + M$ in ordinary Schwarzschild coordinates, which translates conditions on the falloff in u. The conjugate variables are exponentially small in the radial coordinate $K_x \sim K_\varphi \sim 1/\sqrt{u}$. These boundary conditions are very similar to those in Kruskal coordinates [11]. At $u = 0$ we will require that all variables be t-independent and we will choose their derivatives to vanish. This ensures that one can easily continue the manifold without shells of matter present at $u = 0$. There might be other possibilities for this boundary condition but we have not explored them. Finally we would like that in the limit $\delta \to 0$ we get a gauge choice that covers the entire extension of the Schwarzschild space-time. Although the choice of coordinates we are making is not unique, it is computationally laborious to actually find a coordinate system that satisfies all the conditions we listed and that involves variables that do not turn complex in certain regions and that has the variable K_φ taking correct values in the Bohr compactification.

The specific choice we make for E^x is,

$$E^x = \left\{ \frac{\left[\delta(1+u) + \left(10u^2 + u^{7/2}\right)\left(\delta\left(t^2-1\right)+1\right)\right]}{u^{7/2} + (t^2-1)\left(\delta u^{7/2} + \delta^2\right) + \frac{1}{2}\delta^2 u} \times [\ln(1+u)]^2 + \delta^8 \right\} M^2. \tag{2}$$

This choice has the property that for $u \to 0$ $E^x = M^2\delta^8$ independent of t and in the limit $\delta \to 0$ we have that $E^x = M^2(10u^{3/2} + u^3)\ln^2(1+u)/u^3$ tends to 0 when $u = 0$, as in the Kruskal coordinates, giving rise to the singularity. It can be checked that the first derivative with respect to x of E^x vanishes for $u = 0$ for any finite value of δ. This choice for E^x is not unique, in the sense that other choices may satisfy the above conditions. It might be possible to find simpler choices.

For K_φ we choose,

$$K_\varphi = \frac{1}{2}\frac{\delta^{5/2}\pi\left(1+\ln\left(1+u^2\right)\right)}{\mu\left(\delta^{5/2} + \ln(1+u)^2\right)} + \frac{|t|\ln\left(1+u^3\right)}{u^{3/8}} \times \frac{\left(-1 + \frac{u}{(10+u\ln(1+u))} + \frac{9u}{\left(100+u\ln(1+u)^2\right)}\right)}{\left(\delta^2 t + \ln\left(1+u^3\right)\right)\left(1+u^{1/8}\right)}. \tag{3}$$

This choice has the property that for $u \to 0$ $K_\varphi = \pi/(2\mu)$ independent of t, so the term that appears in the Hamiltonian goes as $\sin(\mu K_\varphi) \sim 1$. This means that the departure of the polymerized theory from classical general relativity is maximum at the point where the singularity would have occurred in the continuum theory. Therefore loop quantum gravity could remove the classical Schwarzschild singularity. In the limit $\delta \to 0$ we have that K_φ blows up when $u = 0$, as in the Kruskal coordinates, also compatible with the presence of the singularity in the continuum theory. It can be checked that the first derivative with respect to x of $\sin(\mu K_\varphi)$ vanishes for $u = 0$. As in the case of E^x, the choice is not unique. It should also be noted that the choice is only valid in $|t| > 1$. We have extended the solution beyond that domain. The extension is symmetric under $t \to -t, x \to -x$, but it makes the expressions too lengthy, so for reasons of space here we concentrate in the region $|t| > 1$ since it includes the singularity.

We then solve the diffeomorphism constraint for K_x and the remaining constraint for E^φ. The preservation of the gauge conditions in time determine the lapse and shift. The consistency of the system, that is, the preservation of the constraints upon the Hamiltonian evolution guarantees that the evolution equations for the canonical variables are automatically satisfied. All these calculations can be carried out explicitly, except for the determination of the lapse that requires a difficult integral that can be handled numerically. Studying the components of the metric one notices that they behave very closely to the usual Schwarzschild solution except that when one approaches the singularity all relevant quantities remain finite. The singularity is avoided and a picture is suggested in which the spacetime of a (highly idealized) eternal black hole is continued into another region containing a Cauchy horizon, similar to a Reissner–Nordström space-time but without the singularity. In spite of the lack of singularity, there still is a horizon and a causal behavior far away from the singularity similar to that of the usual Schwarzschild solution.

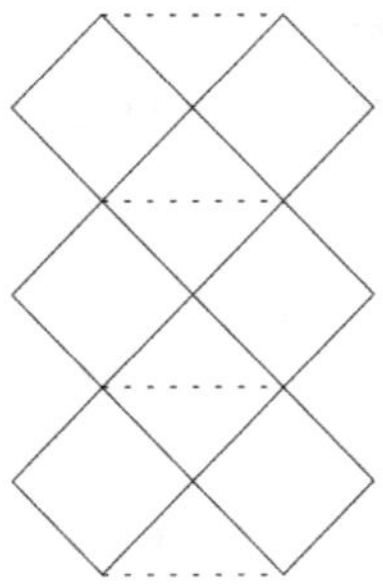

Figure 1. The conjectured global structure of the solution. The singularity is replaced by a regular region indicated with a dashed line. The space-time is continued through into another copy of the same solution. The solution would have a Cauchy horizon similar to that in a Reissner–Nordstöm solution, presumably unstable.

Is the solution unique? At this point we cannot say. There clearly are parameters that can be changed, and choices that were made, but it is not clear if they just correspond to diffeomorphisms. In particular we do not know if all possible choices will lead to non-singular solutions. During our work towards constructing the solution we display here we encountered solutions with singularities, but they ended up being coordinate singularities. Although the treatment of the exterior carried out previously [7] yields a single solution up to diffeomorphisms, it is known that in the treatments of the interior the "polymerization" breaks Birkhoff's theorem [2, 3] suggesting it may not hold in the complete case either. In the interior treatment there appears an additional parameter in the solution which, for instance, controls if the "bounce" is symmetric or not and the extent of the region where the polymerized theory departs from general relativity. Our solution appears to have several free parameters, even though we have imposed by hand that the bounce be symmetric. Clarifying the uniqueness point may shed light on the degrees of freedom that are remnant of the elimination of the singularity in loop quantum gravity and may yield a picture with elements in common with the "fuzzballs" [12] of string theory, although our solutions do not exhibit significant departures from general relativity at the position of the horizon.

Since we completed our manuscript other approaches to the Schwarzschild space-time have appeared. Modesto [13] considered the interior solution to the polymerized theory and analytically continued by exchanging t and r to produce an "exterior" solution. This has the advantage that the solution is completely known analytically. The disadvantage is that the solution solves the analytically continued polymerized equations of the interior. It is not clear if such equations would result from a plausible polymerization of the exterior. Peltola and Kunstatter [14] have considered polymerizing in the traditional variables and noted that perhaps only some of the variables should be polymerized. The result is a spacetime without a Cauchy horizon. This highlights that choices in polymerization may alter importantly the picture of the space-time one obtains. This clearly requires further investigation. Another point of interest is that when one finds solutions in ordinary general relativity one usually works in a given gauge noting that the solution can be later recast in any gauge one wants. In the polymerized theory there are also gauge transformations, but they are somewhat more limited than the full diffeomorphism invariance of ordinary general relativity, as we discussed in [15]. This is another issue that will require further investigation.

We wish to thank Abhay Ashtekar and Rafael Porto for comments and Miguel Campiglia for help with early calculations. This work was supported in part by grants NSF-PHY0650715, and funds of the Hearne Institute for Theoretical Physics, FQXi, PEDECIBA and PDT #63/076 (Uruguay) and CCT-LSU.

References

[1] Ashtekar A, Pawlowski T and Singh P 2006 *Phys. Rev.* D **73** 124038
[2] Ashtekar A and Bojowald M 2006 *Class. Quantum Grav.* **23** 391; Boehmer C G and Vandersloot K 2007 *Phys. Rev.* D **76** 104030
[3] Campiglia M, Gambini R and Pullin J 2008 *AIP Conf. Proc.* **977** 52
[4] Gambini R and Pullin J 2008 *Phys. Rev. Lett.* **101** 161301
[5] Giesel K and Thiemann T 2007 *Class. Quantum Grav.* **24** 2465
[6] Bojowald M and Swiderski R 2006 *Class. Quantum Grav.* **23** 2129
[7] Campiglia M, Gambini R and Pullin J 2007 *Class. Quantum Grav.* **24** 3649
[8] Kuchar K V 1994 *Phys. Rev.* D **50** 3961
[9] Rovelli C 1991 *Phys. Rev.* D **43** 4
[10] Gambini R and Porto R A 2001 *Phys. Rev.* D **63** 105014
[11] Varadarajan M 2001 *Phys. Rev.* D **63** 084007
[12] Mathur S 1998 *Nucl. Phys.* B **529** 295
[13] Modesto L 2008 Space-Time Structure of Loop Quantum Black Hole *Preprint* arXiv:0811.2196
[14] Peltola A and Kunstatter G 2008 A Complete, Single-Horizon Quantum Corrected Black Hole Spacetime *Preprint* arXiv:0811.3240
[15] Gambini R and Pullin J 2008 Diffeomorphism invariance in spherically symmetric loop quantum gravity *Preprint* arXiv:0807.4748

Lattice refinement in loop quantum cosmology

Mairi Sakellariadou

Department of Physics, King's College London, University of London, Strand WC2R 2LS, London, U.K.

E-mail: mairi.sakellariadou@kcl.ac.uk

Abstract. Lattice refinement in LQC, its meaning and its necessity are discussed. The rôle of lattice refinement for the realisation of a successful inflationary model is explicitly shown. A simple and effective numerical technique to solve the constraint equation for any choice of lattice refinement model is briefly illustrated. Phenomenological and consistency requirements leading to a particular choice of lattice refinement model are presented, while it is subsequently proved that only this choice of lattice refinement leads to a unique factor ordering in the Wheeler-De Witt equation, which is the continuum limit of LQC.

1. Introduction

Loop Quantum Gravity (LQG) [1, 2] is one of the most promising candidate theories for describing the quantum degrees of freedom of the gravitational field. Quantum gravity, combining consistently quantum mechanics and general relativity, is essential when curvature becomes large, as for example in the early stages of the evolution of our universe. LQG is a non-perturbative and background independent[1] canonical quantisation of General Relativity (GR) in four space-time dimensions. Considering a fundamental theory of quantum gravity and following the main concept of GR, namely that gravity is indeed geometry, it follows that there is no background metric, there is only a manifold, while geometry and matter should both have a quantum mechanical origin. This indeed differentiates LQG from other approaches of quantising gravity, which have been developed in the framework of particle physics[2]. While the full theory is not yet complete, LQG already has a number of successes, such as the construction of quantum geometry, the prediction of Planck discreteness in geometric operators, as well as a quantum accounting for black hole entropy on the horizon.

The application of ideas and mathematical methods of the full LQG theory to the cosmological sector — the dynamical variables are reduced to first homogeneity and then possibly also to isotropic models — led to Loop Quantum Cosmology (LQC) [3], which is not a field theory[3]. LQC, which gains a constantly increasing interest from the scientific community and has recently made significant progress, differs from other attempts of quantising gravity in the sense that it

[1] Background independence means that quantisation is achieved in the absence of a metric other that the physical one determined by the densitised triad. At the classical level, the fact that the laws of physics are background independent is mathematically expressed by the Einstein equations being four-diffeomorphism covariant.

[2] The only known consistent perturbative approach to quantum gravity is string theory, a theory aiming at unifying all interactions.

[3] The quantised Hamiltonian is given in terms of symmetry reduced variables, so there is only a finite number of degrees of freedom.

gets its input from the full theory of LQG. In short, it is a cosmological mini-superspace model quantised with methods of the full LQG theory. Considering a flat isotropic model within LQC, the extrinsic curvature scale $k = \dot{a} = \sqrt{8\pi G a^2 \Lambda/3}$ (a stands for the scale factor and Λ denotes a positive cosmological constant) appears in the holonomies in such a way that only $e^{i\alpha k}$, with $\alpha \in \mathbf{R}$, can be represented as operators, k itself cannot [4]. The parameter α is related to the edge length used in holonomies, which are playing the rôle of the basic operators and they imply that the Hamiltonian constraint — a Hamiltonian density which is constrained to vanish by the equations of motion — is quantised to a second order difference equation, instead of the second order differential equation of the Wheeler-De Witt (WDW) approach to quantum cosmology.

Following the approach of LQC is useful in two ways. Firstly, it allows us to get some useful insight about open issues of the full LQG theory, and secondly, by using symmetry reduction of the infinite dimensional phase space of LQG, the theory becomes often tractable leading to satisfactory answers about various interesting physical questions. The discreteness of spatial geometry, a key element of the full theory, leads to successes in LQC which do not hold on the WDW approach. In particular, it has been shown that classical big bang [5] and black hole singularities [6] are removed in LQC, in a well defined manner.

In LQC, the evolution of the universe is divided into three distinct phases depending on the value of the scales probed by the universe itself. They are the following:

- *The discrete quantum phase*: Very close to the Planck scale, the concept of space-time has no meaning and one should consider the full quantum gravity theory. Applying LQC during this phase, leads to a finite bounded spectrum for eigenvalues of inverse powers of the three-volume density, called the geometrical density.
- *The intermediate phase*: As the volume of the universe increases with time, the universe enters a semi-classical phase. More precisely, for lengths above $L_{\rm Pl} \equiv \sqrt{\gamma} l_{\rm Pl}$ ($\gamma \approx 0.2375$ is the Barbero-Immirzi parameter — a constant ambiguity parameter, whose value is fixed by black hole entropy calculations — and $l_{\rm Pl}$ is the Planck length with $l_{\rm Pl}^2 = (8\pi G)^{-1}$) the space-time can be approximated by a continuous manifold and the equations of motion take a continuous form. This intermediate phase is characterised by a second scale $L_\star$ with $L_\star \equiv \sqrt{(\gamma J \mu_0)/3} l_{\rm Pl}$, below which the geometrical density is significantly different from its classical form[4]. For scales below $L_\star$ quantum corrections cannot be neglected. This phase is the most relevant one regarding phenomenological consequences of LQC.
- *The classical phase*: At later times, and therefore large scales, the universe enters the full classical phase and standard cosmology becomes valid.

LQG/LQC is formulated in terms of SU(2) holonomies of the connection and triads[5] [7]. To obtain a quantum constraint we introduce an operator representing the curvature of the gravitational connection. A feature of LQC, which is a direct consequence of LQG, is that while there are well-defined analogs of holonomies, there is no operator corresponding to the connection; one defines a curvature operator in terms of holonomies. In the classical level, curvature can be expressed as a limit of the holonomies around a loop as the area enclosed by the loop shrinks to zero. However, in quantum geometry one cannot continuously shrink a loop to zero area, since the eigenvalues of the area operator are discrete, implying that there is a smallest non-zero area eigenvalue, called the *area gap* [8, 9]. In a canonical quantisation scheme, as the one followed in this approach, one first writes the action of GR into a Hamiltonian formulation and

[4] The half-integer J labels the ambiguity in choosing the representation in which the matter part of the Hamiltonian constraint for a scalar field is quantised. The length parameter μ_0 is related to the underlying discrete structure.

[5] The connection in LQG determines the parallel transport of chiral fermions, mathematically represented by spinors, while the conjugate momenta can be interpreted as spatial triads (i.e., *square-roots* of the metric of the 3-dimensional space).

then one quantises this classical Hamiltonian. In the *old* quantisation, the quantised holonomies were taken to be shift operators with a fixed magnitude. However, it was shown [10, 11] that this approach leads to unavoidable instabilities in the continuum semi-classical limit, where the WDW wave-function becomes a good approximation to the difference equation of LQC. For a large semi-classical universe, the WDW wave-function would be oscillating on scales of the order $(a\sqrt{\Lambda})^{-1}$. As the universe expands, this scale becomes eventually smaller than the discreteness scale of the difference equation of LQC, implying that discreteness of spatial geometry would become apparent in the behaviour of wave-functions describing a classical universe. In the underlying LQG theory, the contributions to the discrete Hamiltonian operator depend on the state which describes the universe. As the universe expands, the full Hamiltonian constraint operator creates new vertices of a lattice state, in addition to changing their edge labels. As the extrinsic curvature scale k increases with increasing volume, the corresponding α decreases since the lattice is being continuously refined. Thus, in the context of LQC one has a refinement of the discrete lattice[6]. One can choose such a lattice refinement model, so that the increase in extrinsic curvature scale k can be balanced by the decrease of α such that αk remains small and semi-classical behaviour is achieved for any macroscopic volume even in the presence of a positive cosmological constant.

Lattice refinement is also required from phenomenological reasons [11, 12]. For example, as we will discuss, lattice refinement renders a successful inflationary era[7] more natural [12]. The effect of lattice refinement has been modelled and the elimination of the instabilities in the continuum era has been explicitly shown.

The *correct* lattice refinement model should be obtained from the full LQG theory. In principle, one should use the full Hamiltonian constraint and find the way that its action balances the creation of new vertices as the volume increases. Instead, phenomenological arguments have been used, where the choice of the lattice refinement model is constrained by the form of the matter Hamiltonian [17]. In particular, we have shown [17] that for LQC to generically support inflation, and other matter fields, without the onset of large scale quantum gravity corrections, one should adopt a particular model of lattice refinement. This choice has been then found [18] to be the only one, for which physical quantities are independent of the choice of the elementary cell used to regulate the spatial integrations. Amazingly enough, this is exactly the choice required for the uniqueness of the factor ordering of the Wheeler-De Witt equation [19].

Lattice refinement leads to new dynamical difference equations which, in general, do not have a uniform step-size making their study quite involved, particularly for anisotropic cases, as for example for Bianchi models or black hole interiors. Numerical techniques have been developed [20, 21] to address this issue.

Our primary interest here concerns cosmological predictions of quantum gravity, and therefore we only focus on questions dealing with the early universe and the initial conditions. A quantum theory of gravity is expected, on the one hand, to cure the classical singularities of GR and, on the other hand, either to provide the conditions suitable for the onset of inflation, or to suggest an alternative scenario for alleviating the strong fine-tuning of the standard cosmological model.

[6] The parameter, denoted earlier by μ_0, which appears in the regularisation of the Hamiltonian constraint will no longer be constant within the lattice refinement context.

[7] It was hoped that LQC would help to overcome the extreme fine tuning necessary to achieve successful inflation in GR [13, 14, 15]. However, it has been shown [16] that semi-classical corrections are insufficient to alleviate this difficulty. Certainly, inflation could be generic in the deep quantum regime.

Journal of Physics: Conference Series **189** (2009) 012035 doi:10.1088/1742-6596/189/1/012035

2. Elements of LQG/LQC

LQG, as well as LQC, are both based on a Hamiltonian formulation of GR with basic variables an SU(2) valued connection and the conjugate momentum variable which is a densitised triad[8], a derivative operator quantised in the full LQG theory in the form of fluxes. By using connection-triad variables, arising from a canonical transformation of Arnowitt-Desner-Misner (ADM) variables, we make an analogy with gauge theories; this will be helpful when dealing with quantisation issues.

The densitised triad carries information about the spatial geometry, encoded in the three-metric; the connection carries information about the spatial curvature, in the form of the spin-connection, and the extrinsic curvature. More precisely, the densitised triad, E_i^a, is related to the three-metric, q_{ab}, by

$$E_i^a = \sqrt{|q|}e_i^a \ , \tag{1}$$

where e_i^a is a physical triad, dual ($e_i^a e_a^j = \delta_i^j$) to the co-triad, e_a^j, and satisfying

$$q_{ab} = e_a^i e_b^j \delta_{ij} \ . \tag{2}$$

Note that i refers to the Lie algebra index and a is a spatial index with $a, i = 1, 2, 3$.

The connection, A_a^i, can be related to the ADM variables as

$$A_a^i = \Gamma_a^i + \gamma K_a^i \ , \tag{3}$$

where $\Gamma_a^i = -(1/2)\epsilon^{ijk}e_j^b(2\partial_{[a}e_{b]}^k + e_k^c e_a^l \partial_c e_b^l)$ stands for the spin-connection compatible with the co-triad, γ is the Barbero-Immirzi parameter — classically it has no physical consequence, while at the quantum level it plays a rôle in the level spacing of discrete geometric eigenvalues — and K_a^i stands for the extrinsic curvature one-form $K_a^i = e^{bi}K_{ab}$ (with K_{ab} the extrinsic curvature), it is the Lie derivative of q with respect to the normal vector to the spatial slice, $K_a^i = (\mathcal{L}_{\vec{n}}q_{ab})\delta^{ij}e_j^b$.

The Poisson bracket of the densitised triad and the connection reads

$$\{A_a^i(x), E_j^b(y)\} = \kappa\gamma\delta_a^b\delta_j^i\delta^3(x,y) \ , \tag{4}$$

where $\kappa \equiv 8\pi G$.

The Hamiltonian for GR is given by the sum of constraints, with the scalar constraint

$$C_{\rm GR} = \frac{1}{2\kappa}\int_\Sigma {\rm d}^3x N(x)\left[\frac{E_i^a E_j^b}{\sqrt{|{\rm det}E|}}\epsilon_k^{ij}F_{[ab]}^k - 2(1+\gamma^2)\frac{E_i^a E_j^b}{\sqrt{|{\rm det}E|}}K_{[a}^i K_{b]}^j\right] \ , \tag{5}$$

being the most important one. Note that $F_{[ab]}^k = 2\partial_{[a}A_{b]}^k + \epsilon^k{}_{ij}A_{[a}^i A_{b]}^j$ denotes the curvature two-form of the connection, and N stands for the lapse function.

Before proceeding, let us note that for any quantisation scheme based on a Hamiltonian framework, as for example LQC, or an action principle, as for example path integral approaches, for the homogeneous flat ($k = 0$) model, one should regularise the divergences which appear due to the homogeneity as the action and Hamiltonian are integrated over spatial hyper-surfaces. To do so, the spatial homogeneity and Hamiltonian are restricted to a fiducial cell[9], with finite volume $V_0 = \int {\rm d}^3x\sqrt{|^0q|}$ [4].

[8] The variables E_i^a and A_a^i were introduced by Barbero [22] as an alternative to the complex Ashtekar [23] variables. Both real and complex connections have been used for canonical gravity. The complex connection has SL(2,C) as gauge group, while the real connection has SU(2) as gauge group. Mathematical techniques can only cope with a quantum theory based on SU(2).

[9] Note that μ_0, which we will discuss later is the scale of the finite fiducial cell that spatial integration is restricted to, so as to remove the divergences that occur in non-compact topologies.

The expressions can be simplified a lot by restricting the analysis to homogeneous and isotropic geometries:

$$q = [a(t)]^2 \delta_{ij}\ {}^0\omega^i_a\ {}^0\omega^j_b \mathrm{d}x^a \mathrm{d}x^b \ , \tag{6}$$

where the one-forms satisfy $\partial_{[a}\ {}^0\omega^i_{b]} = 0$, leading to

$$q = [a(t)]^2[(1-kr^2)^{-1}\mathrm{d}r^2 + r^2\mathrm{d}\Omega^2] \ . \tag{7}$$

The symmetry reduced variables E, A are given by

$$A^i_a = V_0^{-1/3} c\ {}^0\omega^i_a \quad , \quad E^a_i = pV_0^{-2/3}|\det {}^0\omega|\ {}^0e^a_i \ , \tag{8}$$

where the vector fields ${}^0e^a_i$ are dual to the 1-forms: ${}^0\omega^i_a\ {}^0e^a_j = \delta^i_j$.

Concentrating on isotropic cosmological models with constant spatial curvature, we will first consider symmetric connections and triads, and we will then insert them into the full action leading to a symmetry reduced action. The symmetric connections and triads can be decomposed using basis one-forms and vector fields obtained by Bianchi models. One can show that the actions lead to the correct equations of motion of GR, implying that they are equivalent to the Einstein-Hilbert action on metric variables.

The loop quantisation of the flat FLRW scalar constraint changes the curvature 2-form F, its ab component reads [24]:

$$F^k_{ab} = -2 \lim_{Ar_\Box \to 0} \mathrm{Tr}\left(\frac{h^{(\bar\mu)}_{\Box_{ij}} - 1}{\bar\mu^2 V_0^{2/3}}\right)\tau^k\ {}^0\omega^i_a\ {}^0\omega^j_b \ ; \tag{9}$$

$Ar_\Box$ is the area of the square $\Box_{ij}$ in the (i,j)-plane swept by a face of the elementary cell, the holonomy $h^{(\mu_0)}_{\Box_{ij}}$ around the square $\Box_{ij}$ is the product of holonomies along the four edges of $\Box_{ij}$, and $\bar\mu$ is $\bar\mu = \sqrt{\Delta}/|p|^{1/2}$, with Δ the eigenvalue of the area operator.

In the case of a spatially flat background, derived from the Bianchi I model, the isotropic connection can be expressed in terms of the dynamical component of the connection $\tilde c(t)$ as

$$A^i_a = \tilde c(t)\omega^i_a \ , \tag{10}$$

with ω^i_a a basis of left-invariant one-forms $\omega^i_a = \mathrm{d}x^i$. The densitised triad can be decomposed using the Bianchi I basis vector fields $X^a_i = \delta^a_i$ as

$$E^a_i = \sqrt{{}^0q}\,\tilde p(t) X^a_i \ , \tag{11}$$

where ${}^0q$ stands for the determinant of the fiducial background metric,

$${}^0q_{ab} = \omega^i_a \omega_{bi} \ , \tag{12}$$

and $\tilde p(t)$ denotes the remaining dynamical quantity after symmetry reduction.

In terms of the metric variables with three-metric $q_{ab} = a^2\omega^i_a\omega_{bi}$, the dynamical quantity is just the scale factor $a(t)$. Given that the Bianchi I basis vectors are $X^a_i = \delta^a_i$,

$$|\tilde p| = a^2 \ , \tag{13}$$

where the absolute value is taken because the triad has an orientation. Since the basis vector fields are spatially constant in the spatially flat model, the connection component is

$$\tilde c = \mathrm{sgn}(\tilde p)\gamma\frac{\dot a}{N} \ . \tag{14}$$

Note that in what follows, the lapse function which is a constant due to spatial homogeneity, will be set equal to 1. Thus, GR can be formulated as a gauge theory in Ashtekar variables.

The canonical variables $\tilde{c}, \tilde{p}$ are related through

$$\{\tilde{c}, \tilde{p}\} = \frac{\kappa\gamma}{3} V_0 , \tag{15}$$

where V_0 the volume of the elementary cell adapted to the fiducial triad.

Defining the triad component p, determining the physical volume of the fiducial cell, and the connection component c, determining the rate of change of the physical edge length of the fiducial cell, as

$$p = V_0^{2/3}\tilde{p} \quad , \quad c = V_0^{1/3}\tilde{c} , \tag{16}$$

respectively, we obtain

$$\{c, p\} = \frac{\kappa\gamma}{3} , \tag{17}$$

independent of the volume V_0 of the fiducial cell.

To quantise, we follow the approach used in the full LQG theory. The metric itself is a physical field which must be quantised; it cannot be considered as a fixed background. Thus, to quantise gravity we use gauge theory variables to define holonomies of the connection along a given edge

$$h_e(A) = \mathcal{P}\exp \int \mathrm{d}s\dot{\gamma}^\mu(s)A^i_\mu(\gamma(s))\tau_i , \tag{18}$$

where $\mathcal{P}$ indicates a path ordering of the exponential, γ^μ is a vector tangent to the edge and $\tau_i = -i\sigma_i/2$, with σ_i the Pauli spin matrices, and fluxes of a triad along an S surface

$$E(S, f) = \int_S \epsilon_{abc}E^{ci}f_i \mathrm{d}x^a \mathrm{d}x^b , \tag{19}$$

with f_i an SU(2) valued test function. Note that even though these variables appear rather artificial, as presented here in the context of LQC, they nevertheless arise naturally within the full LQG theory.

Thus, the basic configuration variables in LQC are holonomies of the connection

$$h_i^{(\mu_0)}(A) = \cos\left(\frac{\mu_0 c}{2}\right)\mathbf{1} + 2\sin\left(\frac{\mu_0 c}{2}\right)\tau_i , \tag{20}$$

along a line segment $\mu_0 \, {}^0e^a_i$ and the flux of the triad

$$F_S(E, f) \propto p ;$$

the basic momentum variable is the triad component p. Note that $\mathbf{1}$ is the identity 2×2 matrix and $\tau_i = -i\sigma_i/2$ is a basis in the Lie algebra SU(2) satisfying the relation

$$\tau_i\tau_j = (1/2)\epsilon_{ijk}\tau^k - (1/4)\delta_{ij} .$$

Let us first review the *old quantisation* procedure. We follow the same approach as in LQG. We thus take $e^{i\mu_0 c/2}$ (with μ_0 an arbitrary real number) and p, as the elementary classical variables, which have well-defined analogues [4]. Using the Dirac bra-ket notation and setting $e^{i\mu_0 c/2} = \langle c|\mu\rangle$, the action of the operator $\hat{p}$ acting on the basis states $|\mu\rangle$ is

$$\hat{p}|\mu\rangle = \frac{\kappa\gamma\hbar|\mu|}{6}|\mu\rangle , \tag{21}$$

where μ (a real number) stands for the eigenstates of $\hat{p}$, satisfying the orthonormality relation

$$\langle\mu_1|\mu_2\rangle = \delta_{\mu_1,\mu_2} . \tag{22}$$

The action of the $\widehat{\exp\left[\frac{i\mu_0}{2}c\right]}$ operator acting on basis states $|\mu\rangle$ is

$$\widehat{\exp\left[\frac{i\mu_0}{2}c\right]}|\mu\rangle = \exp\left[\mu_0\frac{\mathrm{d}}{\mathrm{d}\mu}\right]|\mu\rangle = |\mu+\mu_0\rangle , \tag{23}$$

where μ_0 is any real number. Thus, in the old quantisation, the operator $e^{i\mu_0 c/2}$ acts as a simple shift operator.

Using the volume operator $\hat{V} = |\hat{p}|^{3/2}$, representing the volume of the elementary cell with eigenvalues $V_\mu = (\kappa\gamma\hbar|\mu|/6)^{3/2}$, we get[10] [24]

$$\hat{V}|\mu\rangle = \left(\frac{\kappa\gamma\hbar|\mu|}{6}\right)^{3/2}|\mu\rangle . \tag{24}$$

To define the inverse volume operator one has to trace over SU(2) valued holonomies. Since there is a freedom in choosing the irreducible representation to perform the trace, an ambiguity J arises[11]. Let us use the $J = 1/2$ irreducible representation of SU(2). The inverse volume operator is diagonal in the $|\mu\rangle$ basis and is given by [4]

$$\widehat{V^{-1}}|\mu\rangle = \left|\frac{6}{\kappa\gamma\hbar\mu_0}\left(\{V(\mu+\mu_0)\}^{1/3} - \{V(\mu-\mu_0)\}^{1/3}\right)\right|^3|\mu\rangle , \tag{25}$$

where μ_0 is proportional to the length of the holonomy. Note that the regulating length μ_0 is the crucial parameter in the quantum corrections; μ_0 determines the step-size of the resulting difference equation. In the above equation, Eq. (25), the eigenvalues are bounded and approach zero near the classical singularity; in the classical case the eigenvalues diverge at the singularity $\mu = 0$. The eigenvalues reach their maximum at a characteristic scale μ_0, at larger scales they approach the classical values and at smaller scales they are suppressed [25].

As in the full LQG theory, there is no operator corresponding to the connection. Nevertheless, the action of its holonomy is well-defined. Let us denote by $\hat{h}_i^{(\mu_0)}$ the holonomy along the edge parallel to the $i^{\rm th}$ basis vector of length $\mu_0 V_0^{1/3}$ with respect to the fiducial metric. Its action on the basis states is given by [26]

$$\hat{h}_i^{(\mu_0)}|\mu\rangle = (\widehat{\rm cs}\mathbf{1} + 2\widehat{\rm sn}\tau_i)|\mu\rangle , \tag{26}$$

where,

$$\begin{aligned}\widehat{\rm cs}|\mu\rangle \equiv \widehat{\cos(\mu_0 c/2)}|\mu\rangle &= [\,|\mu+\mu_0\rangle + |\mu-\mu_0\rangle\,]/2 ,\\ \widehat{\rm sn}|\mu\rangle \equiv \widehat{\sin(\mu_0 c/2)}|\mu\rangle &= [\,|\mu+\mu_0\rangle - |\mu-\mu_0\rangle\,]/(2i) .\end{aligned} \tag{27}$$

Thus,

$$\begin{aligned}&\hat{h}_i^{(\mu_0)}\hat{h}_j^{(\mu_0)}\hat{h}_i^{(\mu_0)-1}\hat{h}_j^{(\mu_0)-1}|\mu\rangle\\ &= \left[\left(\widehat{\rm cs}^4 - \widehat{\rm sn}^4\right)\mathbf{1} + 2\left(\mathbf{1} - 4\tau_j\tau_i\right)\widehat{\rm cs}^2\widehat{\rm sn}^2 + 4\left(\tau_i - \tau_i\right)\mathbf{1}\widehat{\rm cs}\;\widehat{\rm sn}^3\right]|\mu\rangle ,\end{aligned} \tag{28}$$

[10] Being concerned with the large scale behaviour of the LQC equations, we neglect the sign ambiguity.

[11] Note that $J/2$ stands for the spin of the representation. Usually one quantises the gravitational part of the Hamiltonian constraint using the fundamental $J = 1/2$ representation, and the ambiguity is only investigated for the matter part.

and

$$\hat{h}_i^{(\mu_0)}\left[\hat{h}_i^{(\mu_0)-1},\hat{V}\right]|\mu\rangle$$
$$= \left(\hat{V}-\widehat{\mathrm{cs}}\hat{V}\widehat{\mathrm{cs}}-\widehat{\mathrm{sn}}\hat{V}\widehat{\mathrm{sn}}\right)\mathbf{1}|\mu\rangle + 2\tau_i\left(\widehat{\mathrm{cs}}\hat{V}\widehat{\mathrm{sn}}-\widehat{\mathrm{sn}}\hat{V}\widehat{\mathrm{cs}}\right)|\mu\rangle \ . \tag{29}$$

The gravitational part of the Hamiltonian operator in terms of SU(2) holonomies and the triad component, in the irreducible $J = 1/2$ representation[12] , reads [25, 26]

$$\hat{\mathcal{C}}_{\rm grav} = \frac{2i}{\kappa^2\hbar\gamma^3\mu_0^3}{\rm tr}\sum_{ijk}\epsilon^{ijk}\left(\hat{h}_i^{(\mu_0)}\hat{h}_j^{(\mu_0)}\hat{h}i^{(\mu_0)-1}\hat{h}_j^{(\mu_0)-1}\hat{h}_k^{(\mu_0)}\left[\hat{h}_k^{(\mu_0)-1},\hat{V}\right]\right) \ . \tag{30}$$

As in the full LQG theory, curvature is defined in LQC in terms of holonomies around closed loops. This implies that the limit $\mu_0 \to 0$ of the above operator does not exist, since it would mean that the area enclosed by loops should be shrunk to zero. In the underlying quantum geometry, the eigenvalues of the area operator are discrete, implying that there is a smallest nonzero eigenvalue, an *area gap* Δ [9]. This is indeed the reason for which the WDW differential equation gets replaced by a difference equation whose step size is controlled by Δ. Since μ_0 enters through the holonomies, its value in the fixed lattice case was fixed by demanding that the eigenvalue of the area operator be the area gap: $\Delta = 2\sqrt{3}\pi\gamma l_{\rm Pl}^2$, implying $\mu_0 = 3\sqrt{3}/2$.

The action of the self-adjoint Hamiltonian constraint operator, $\hat{\mathcal{H}}_{\rm grav} = (\hat{\mathcal{C}}_{\rm grav} + \hat{\mathcal{C}}_{\rm grav}^{\dagger})/2$, on the basis states, $|\mu\rangle$, is

$$\hat{\mathcal{H}}_{\rm grav}|\mu\rangle = \frac{3}{4\kappa^2\gamma^3\hbar\mu_0^3}\Big\{[R(\mu)+R(\mu+4\mu_0)]|\mu+4\mu_0\rangle - 4R(\mu)|\mu\rangle + [R(\mu)+R(\mu-4\mu_0)]|\mu-4\mu_0\rangle\Big\} \ , \tag{31}$$

where

$$R(\mu) = (\kappa\gamma\hbar/6)^{3/2}\left||\mu+\mu_0|^{3/2} - |\mu-\mu_0|^{3/2}\right| \ . \tag{32}$$

Having the Hamiltonian operator, dynamics are determined by the Hamiltonian constraint[13]

$$\left(\hat{\mathcal{H}}_{\rm grav} + \hat{\mathcal{H}}_\phi\right)|\Psi\rangle = 0 \ . \tag{33}$$

Note that in the full LQG theory, there is an infinite number of constraints, whereas in the reduced homogeneous and isotropic case there is only one integrated Hamiltonian constraint. Matter is then introduced by just adding the actions of matter components to the gravitational action. One finally obtains difference equations analogous to the differential WDW equations.[14]

Let us impose the constraint equation on the physical wave-functions $|\Psi\rangle$, which can be expanded using the basis states as $|\Psi\rangle = \sum_\mu \Psi_\mu(\phi)|\mu\rangle$, with summation over values of μ and where the dependence of the coefficients on ϕ represents the matter degrees of freedom. Since the states $|\mu\rangle$ are eigenstates of the triad operator, the coefficients $\Psi_\mu(\phi)$ represent the state in the triad representation. Thus, quantising the Friedmann equation along the lines of the constraint in the full LQG theory, one gets the following difference equation [27]

$$\left[\left|V_{\mu+5\mu_0} - V_{\mu+3\mu_0}\right| + \left|V_{\mu+\mu_0} - V_{\mu-\mu_0}\right|\right]\Psi_{\mu+4\mu_0}(\phi) - 4\left|V_{\mu+\mu_0}V_{\mu-\mu_0}\right|\Psi_\mu(\phi)$$
$$+\left[\left|V_{\mu-3\mu_0} - V_{\mu-5\mu_0}\right| + \left|V_{\mu+\mu_0} - V_{\mu-\mu_0}\right|\right]\Psi_{\mu-4\mu_0}(\phi) = -\frac{4\kappa^2\gamma^3\hbar\mu_0^3}{3}\mathcal{H}_\phi(\mu)\Psi_\mu(\phi) \ , \tag{34}$$

[12] With this choice, the Hamiltonian constraint is free of the ill-behaving spurious solutions.

[13] Note that the Gauss and the diffeomorphism constraints are automatically satisfied by appropriate gauge fixing.

[14] The reader should note that in LQC the physical fundamental object is the discrete difference equation, while the differential equation is just the approximation in the continuum limit.

where the matter Hamiltonian $\hat{\mathcal{H}}_\phi$ is assumed to act diagonally on the basis states with eigenvalue $\mathcal{H}_\phi(\mu)$. Equation (34) is indeed the quantum evolution (in internal time μ) equation. There is no continuous variable (the scale factor in classical cosmology), but a label μ with discrete steps. The wave-function $\Psi_\mu(\phi)$, depending on internal time μ and matter fields ϕ, determines the dependence of matter fields on the evolution of the universe. A massless scalar field plays the rôle of the *emergent time*. Thus, in LQC the quantum evolution is governed by a second order difference equation, rather than the second order differential equation of the WDW quantum cosmology. As the universe becomes large and enters the semi-classical regime, the WDW differential equation becomes a very good approximation to the difference equation of LQC.

3. Lattice refinement

Consider the continuum limit (namely that $\mu \gg \mu_0$) of the Hamiltonian constraint operator acting on the physical states. In the small regulating length μ_0 limit, one obtains [17] a second order difference equation which distinguishes the components of the wave-functions in different lattices of spacing $4\mu_0$. Assuming that Ψ does not vary much on scales of the order of $4\mu_0$, known as *pre-classicality* [28], one can smoothly interpolate between the points on the discrete function $\Psi_\mu(\phi)$ and approximate them by the continuous function $\Psi(\mu,\phi)$. In this way, one approximates the difference equation by a differential equation for a continuous wave-function.

The form of the wave-functions indicates that the period of oscillations can decrease as the scale increases, which implies that at sufficiently large scales the assumption of *pre-classicality* breaks down[15]. This would then lead to quantum gravity corrections at large scales (i.e., classical) physics. To avoid this undesired event, was one of the motivations behind lattice refinement [27, 29]. Allowing the length scale of the holonomies to vary, the form of the difference equation changes. Assuming that the lattice size is growing, the step-size of the difference equation is not constant in the original triad variables. The exact form of the difference equation depends on the lattice refinement used.

Consider the particular model

$$\mu_0 \to \tilde{\mu}(\mu) = \mu_0 \mu^{-1/2} ; \tag{35}$$

we will come back to the issue of the lattice refinement choice in a subsequent section.

The basic operators are given by replacing μ_0 with $\tilde{\mu}$. Upon quantisation [26]

$$\widehat{e^{i\tilde{\mu}c/2}}|\mu\rangle = e^{-i\tilde{\mu}\frac{\mathrm{d}}{\mathrm{d}\mu}}|\mu\rangle , \tag{36}$$

which is no longer a simple shift operator since $\tilde{\mu}$ is a function of μ. Changing the basis to

$$\nu = \mu_0 \int \frac{\mathrm{d}\mu}{\tilde{\mu}} = \frac{2}{3}\mu^{3/2} , \tag{37}$$

one gets

$$e^{-i\tilde{\mu}\frac{\mathrm{d}}{\mathrm{d}\mu}}|\nu\rangle = e^{-i\mu_0\frac{\mathrm{d}}{\mathrm{d}\nu}}|\nu\rangle = |\nu+\mu_0\rangle . \tag{38}$$

The volume operator acts on these basis states as

$$\hat{V}|\nu\rangle = \frac{3\nu}{2}\left(\frac{\kappa\gamma\hbar}{6}\right)^{3/2}|\nu\rangle , \tag{39}$$

[15] The constant lattice, in the *old quantisation* approach, does not take into account the expansion of the fiducial cell in a FLRW background.

and the self-adjoint Hamiltonian constraint operator acts as [25]

$$\hat{\mathcal{H}}_{\rm g}|\nu\rangle = \frac{9|\nu|}{16\mu_0^3}\left(\frac{\hbar}{6\kappa\gamma^3}\right)^{1/2}$$
$$\times\left[\frac{1}{2}\Big\{U(\nu)+U(\nu+4\mu_0)\Big\}|\nu+4\mu_0\rangle - 2U(\nu)|\nu\rangle + \frac{1}{2}\Big\{U(\nu)+U(\nu-4\mu_0)\Big\}|\nu-4\mu_0\rangle\right] \quad (40)$$

where

$$U(\nu) = |\nu+\mu_0| - |\nu-\mu_0| \ . \quad (41)$$

Expanding $|\Psi\rangle = \sum_\nu \Psi_\nu(\phi)|\nu\rangle$ the Hamiltonian constraint reads [17]

$$\frac{1}{2}\Big|\nu+4\mu_0\Big|\Big[U(\nu+4\mu_0)+U(\nu)\Big]\Psi_{\nu+4\mu_0}(\phi) + 2|\nu|U(\nu)\Psi_\nu(\nu)$$
$$+\frac{1}{2}\Big|\nu-4\mu_0\Big|\Big[U(\nu-4\mu_0)+U(\nu)\Big]\Psi_{\nu-4\mu_0}(\phi)$$
$$= -\frac{16\mu_0^3}{9}\left(\frac{6\kappa\gamma^3}{\hbar}\right)^{1/2}\mathcal{H}_\phi(\nu)\Psi_\nu(\phi) \ ; \quad (42)$$

$\mathcal{H}_\phi$ stands for the matter part of the Hamiltonian, which for a massive scalar field is given by

$$\mathcal{H}_\phi = \kappa\left[\frac{P_\phi^2}{2a^3} + a^3V(\phi)\right] \ , \quad (43)$$

with P_ϕ the momentum and $V(\phi)$ the potential of the scalar field ϕ. Quantising the Hamiltonian constraint we obtain, in terms of $p = \kappa\gamma\hbar\mu/6$,

$$\sqrt{p}\frac{\partial^2\Psi(p,\phi)}{\partial p^2} + \frac{\partial^2}{\partial p^2}\left(\sqrt{p}\Psi(p,\phi)\right) - 3p^{-3/2}\frac{\partial^2\Psi(p,\phi)}{\partial\phi^2} + \frac{6}{\kappa\hbar^2}p^{3/2}V(\phi)\Psi(p,\phi) + \mathcal{O}(\mu_0) + \cdots = 0 \ , \quad (44)$$

which is a particular factor ordering of the WDW equation for a massive scalar field.

3.1. Implications for inflation

We will show that lattice refinement is essential in order to achieve a successful inflationary era. Let us consider a fixed and a dynamically varying lattice and solve in the continuum limit the second order difference equation which governs the quantum evolution. By doing so we constrain the potential of a scalar field (the inflaton) so that the continuum approximation is valid. A second constraint is imposed on the inflationary potential so that there is consistency with measurements of the cosmic microwave background temperature anisotropies on large angular scales. Combining the two constraints in a particular inflationary model, for either a fixed lattice or a given lattice refinement model, we deduce the conditions for natural and successful inflation within LQC in each of the two cases.

More precisely, let us separate the wave-function $\Psi(p,\phi)$ into $\Psi(p,\phi) = \Upsilon(p)\Phi(\phi)$ and approximate the dynamics of the inflaton field, ϕ, by setting $V(\phi) = V_\phi p^{\delta-3/2}$, where V_ϕ is a constant and $\delta = 3/2$ in the case of slow-roll, to get [17]

$$p^{-1/2}\frac{\rm d}{{\rm d}p}\left[p^{-1/2}\frac{\rm d}{{\rm d}p}\left(p^{3/2}\Upsilon(p)\right)\right] + \beta V_\phi p^\delta\Upsilon(p) = 0 \ , \quad (45)$$

with solutions [17]

$$\Upsilon(p) \approx p^{-(9+2\delta)/8}\sqrt{\frac{2\delta+3}{2\sqrt{\beta V_\phi}\pi}}\left[C_1\cos\left(x-\frac{3\pi}{2(2\delta+3)}-\frac{\pi}{4}\right)+C_2\sin\left(x-\frac{3\pi}{2(2\delta+3)}-\frac{\pi}{4}\right)\right], \tag{46}$$

where

$$x = 4\sqrt{\beta V_\phi}(2\delta+3)^{-1}p^{(2\delta+3)/4}, \tag{47}$$

and $\beta = 96/(\kappa\hbar^2)$.

Without lattice refinement, the discrete nature of the underlying lattice would eventually be unable to support the oscillations and the assumption of pre-classicality will break down, implying that the discrete nature of space-time becomes significant on very large scales. For the end of inflation to be describable using classical GR, it must end before a scale, at which the assumption of pre-classicality breaks down and the semi-classical description is no longer valid, is reached. We will quantify this constraint.

The separation between two successive zeros of $\Upsilon(p)$ is

$$\Delta p = \frac{\pi}{\sqrt{\beta V_\phi}}p^{(1-2\delta)/4}. \tag{48}$$

For the continuum limit to be valid, the wave-function must vary slowly on scales of the order of $\mu_c = 4\tilde{\mu}$. Thus, we impose the constraint [17]

$$\Delta p > 4\mu_0\left(\frac{\kappa\gamma\hbar}{6}\right)^{3/2}p^{-1/2}, \tag{49}$$

which implies the following constraint on V_ϕ [17]

$$V_\phi < \frac{27\pi^2}{192\mu_0^2\gamma^3\kappa^2\hbar}p^{(3-2\delta)/2}. \tag{50}$$

For slow-roll inflation, $V(\phi)$ must be approximately constant, implying $\delta \approx 3/2$. For $\mu_0 = 3\sqrt{3}/2$ and $\gamma \approx 0.24$, the constraint on the inflationary potential in units of $\hbar = 1$ reads

$$V(\phi) \lesssim 2.35 \times 10^{-2} l_{\rm Pl}^{-4}. \tag{51}$$

This is a softer constraint than the one imposed for fixed lattices, namely [17]

$$V_\phi \ll 10^{-28} l_{\rm pl}^{-4}, \tag{52}$$

assuming that half of inflation takes place during the classical era.

Selecting a successful and simple inflationary model, for example $V(\phi) = m^2\phi^2/2$, an additional constraint can be imposed on the potential so that the fractional over-density in Fourier space at horizon crossing is consistent with the COBE-DMR measurements. Combining the two constraints we obtain [17] for the fixed and varying lattices

$$m \lesssim 70(e^{-2N_{\rm cl}})\, M_{\rm Pl} \tag{53}$$

$$\text{and } m \lesssim 10\, M_{\rm Pl}, \tag{54}$$

respectively.

Thus, for any significant proportion of inflation to take place during the classical era, the constraint imposed on the inflaton mass is very strong, while it becomes natural once lattice refinement is taken into account.

3.2. Relation between lattice refinement model and matter Hamiltonian
A particular lattice refinement model can support only certain types of matter [12]. To prove it, let us parametrise the lattice refinement by a parameter A and the matter Hamiltonian by a parameter δ and solve the Hamiltonian constraint. The restrictions on the two-dimensional parameter space will become apparent once we impose some physical restrictions to the solutions of the wave-functions [12].

More precisely, assuming

$$\tilde{\mu} = \mu_0 \mu^A \ , \tag{55}$$

we obtain

$$\nu = \frac{\tilde{\mu}_0 \mu^{1-A}}{\mu_0(1-A)} \ . \tag{56}$$

The WDW constraint equation reads

$$\left(\hat{\mathcal{H}}_{\rm grav} + \hat{\mathcal{H}}_\phi\right)\Psi = 0 \ . \tag{57}$$

Being interested in the large scale limit, we can approximate [12] the matter Hamiltonian with $\hat{\mathcal{H}}_\phi = \hat{\nu}^\delta \hat{\epsilon}\,(\phi)$, leading to

$$\hat{\epsilon}\,(\phi)\,\Psi \equiv \epsilon\,(\phi)\,\Psi = -\nu^{-\delta}\hat{\mathcal{H}}_{\rm grav}\Psi \ . \tag{58}$$

A necessary condition for the wave-functions to be physical is that the finite norm of the physical wave-functions, which is defined by $\int_{\phi=\phi_0} d\nu |\nu|^\delta \overline{\Psi}_1 \Psi_2$, must be independent of the choice of $\phi = \phi_0$. The solutions of the constraint are renormalisable provided they decay, on large scales, faster than $\nu^{-1/(2\delta)}$.

To solve the constraint equation, we need to specify the from of $\mathcal{H}_\phi$, which has in general two terms with different scale dependence. Being interested in the large scale limit, one will be the dominant term, allowing to write [12]

$$\beta \mathcal{H}_\phi = \epsilon_\nu(\phi)\nu^{\delta_\nu} \ , \tag{59}$$

where the function ϵ_ν is constant with respect to ν.

We then solve [12] the constraint equation and consider only those solutions which are the physical ones. The large scale behaviour of the wave-functions must be normalisable — a necessary condition for having physical wave-functions — and the wave-functions should preserve pre-classicality at large scales — a necessary condition for the validity of the continuum limit [12]. This procedure leads to constraints to the two-dimensional parameter space (A, δ), which we illustrate below [12] in Fig. 1 for A in the range $0 < A < -1/2$, imposed from full LQG theory considerations [11].

We thus conclude that the continuum limit of the Hamiltonian constraint equation is sensitive to the choice of model and only a limited range of matter components can be supported within a particular choice [12].

3.3. Numerical methods for solving the constraint equation for any lattice refinement model
The lattice refinement implies new dynamical difference equations, which are not expected to have a uniform step-size, leading to technical complications. This becomes apparent in the case of two-dimensional wave-functions, such as those necessary to study Bianchi models of black hole interiors. More precisely, the information needed to calculate the wave-function at a given lattice point is not provided by previous iterations. We prescribe below a method [21] based on Taylor expansion that can be used to perform this desired interpolations with a well-defined and predictable accuracy[16].

[16] A simple local interpolation scheme to approximate the necessary data points, allowing direct numerical evolution of two-dimensional systems has been also proposed in Ref. [20].

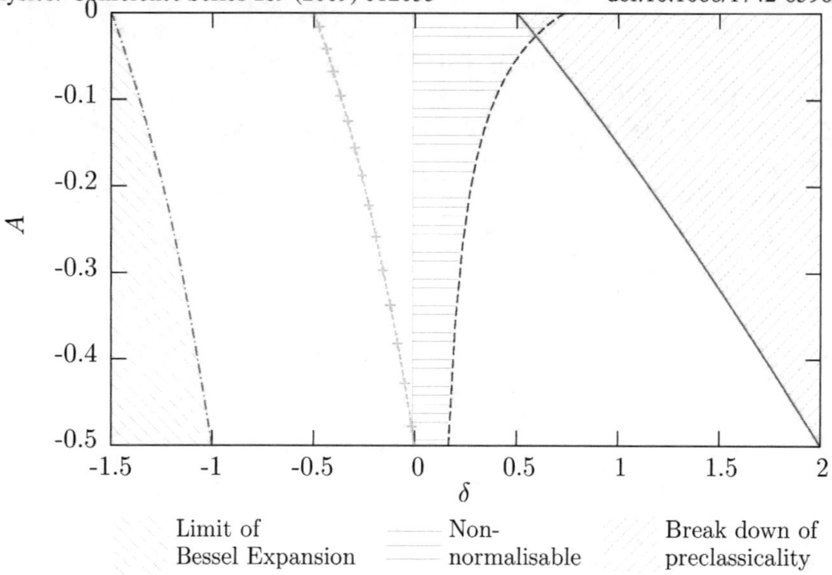

Figure 1. The allowed types of matter content are significantly restricted. Note that in the case of a varying lattice ($A \neq 0$) it is not always possible to treat the large scale behaviour of the wave-functions perturbatively (dashed line with crosses) [12].

Let us first note that the Hamiltonian constraint for a one-dimensional difference equation defined on a varying lattice, can be mapped onto a fixed lattice simply by a change of basis [21][17]. However this method is of no help for the two-dimensional case, where the Hamiltonian constraint is a difference equation on a varying lattice [11],

$$\begin{aligned}
&C_+(\mu,\tau)\left[\Psi_{\mu+2\delta_\mu,\tau+2\delta_\tau}-\Psi_{\mu-2\delta_\mu,\tau+2\delta_\tau}\right]\\
&+C_0(\mu,\tau)\left[(\mu+2\delta_\mu)\Psi_{\mu+4\delta_\mu,\tau}-2\left(1+2\gamma^2\delta_\mu^2\right)\mu\Psi_{\mu,\tau}+(\mu-2\delta_\mu)\Psi_{\mu-4\delta_\mu,\tau}\right]\\
&+C_-(\mu,\tau)\left[\Psi_{\mu-2\delta_\mu,\tau-2\delta_\tau}-\Psi_{\mu+2\delta_\mu,\tau-2\delta_\tau}\right]=\frac{\delta_\tau\delta_\mu^2}{\delta^3}\mathcal{H}_\phi\Psi_{\mu,\tau}\ ,
\end{aligned} \tag{60}$$

with

$$C_\pm \equiv 2\delta_\mu\left(\sqrt{|\tau\pm 2\delta_\tau|}+\sqrt{|\tau|}\right)\ , \tag{61}$$

$$C_0 \equiv \sqrt{|\tau+\delta_\tau|}-\sqrt{|\tau-\delta_\tau|}\ , \tag{62}$$

where we have defined δ_μ and δ_τ as the step-sizes along the μ and τ directions, respectively. The parameter δ, with $0<\delta<1$, gives the fraction of a lattice edge that the underlying graph changing Hamiltonian uses [11].

In the case of lattice refining, δ_μ and δ_τ are decreasing functions of μ and τ, respectively, and the data needed to calculate the value of the wave-function at a particular lattice site are not given by previous iterations. One can use Taylor expansions to calculate the necessary data

[17] We have checked [21] the validity of our Taylor expansion numerical method in calculating the wave-function by comparing our results with those obtained by mapping the one-dimensional difference equation defined on a varying lattice, onto a fixed lattice by performing a change of basis.

points [21]. More precisely, let us assume that the matter Hamiltonian acts diagonally on the basis states of the wave-function, namely

$$\hat{\mathcal{H}}_\phi|\Psi\rangle \equiv \hat{\mathcal{H}}_\phi \sum_{\mu,\tau} \Psi_{\mu,\tau}|\mu,\tau\rangle = \sum_{\mu,\tau} \mathcal{H}_\phi \Psi_{\mu,\tau}|\mu,\tau\rangle \ . \tag{63}$$

Given a function evaluated at three (noncolinear) coordinates, the Taylor approximation to the value at a fourth position is

$$\begin{aligned} f(x_4,y_4) &= f(x_2,y_2) + \delta^x_{42}\frac{\partial f}{\partial x}\Big|_{x_2,y_2} + \delta^y_{42}\frac{\partial f}{\partial y}\Big|_{x_2,y_2} \\ &+\mathcal{O}\left((\delta^x_{42})^2\frac{\partial^2 f}{\partial x^2}\Big|_{x_2,y_2}\right) + \mathcal{O}\left((\delta^y_{42})^2\frac{\partial^2 f}{\partial y^2}\Big|_{x_2,y_2}\right) , \end{aligned} \tag{64}$$

where the Taylor expansion is taken about the position (x_2, y_2), we have defined $\delta^x_{ij} \equiv x_i - x_j$ and $\delta^y_{ij} \equiv y_i - y_j$, and the differentials can be approximated using

$$f(x_1,y_1) = f(x_2,y_2) + \delta^x_{12}\frac{\partial f}{\partial x}\Big|_{x_2,y_2} + \delta^y_{12}\frac{\partial f}{\partial y}\Big|_{x_2,y_2} + \cdots , \tag{65}$$

$$f(x_3,y_3) = f(x_2,y_2) + \delta^x_{32}\frac{\partial f}{\partial x}\Big|_{x_2,y_2} + \delta^y_{32}\frac{\partial f}{\partial y}\Big|_{x_2,y_2} + \cdots , \tag{66}$$

where the dots indicate higher order terms.

For slowly varying wave-functions, it has been shown [21] that linear approximation is very accurate and higher order terms in Taylor expansion can only improve the accuracy by 10^{-2}%. This method can be applied in any lattice refinement model, while its accuracy can be estimated. Even though we have illustrated it in the case of black hole interiors, this method can be applied to anisotropic Bianchi models and in general to systems with anisotropic symmetries.

By using this Taylor expansions method, we were able to confirm [21] numerically the stability criterion of the Schwarzchild interior, which was earlier found [11] using a von Neumann analysis, and investigate [21] how lattice refinement can change the stability properties of the system. Finally, the underlying discreteness of space-time leads to a twist [21] in the wave-functions, for both a constant lattice, as well as lattice refinement models.

3.4. Uniqueness of WDW factor ordering and the lattice refinement choice

The *correct* lattice refinement model should in principle be given by the full LQG theory. In this sense, one should consider the full Hamiltonian constraint and find the way that its action balances the creation of new vertices while the volume increases. Instead of doing so, as we have already discussed earlier, phenomenological arguments [12, 17] have been used to constrain the choice of the lattice refinement model by the form of the matter Hamiltonian. Later on, it has been argued [18] that only the lattice refinement model $\tilde{\mu} = \mu_0\mu^{-1/2}$ [19], can be achieved by physical considerations of large scale physics and consistency of the quantisation structure. We will show below that this choice is also the only one which makes the factor order ambiguities of LQC to disappear in the continuum limit [19].

Indeed, there are many ways of writing the gravitational part of the Hamiltonian constraint in terms of the triad and the holonomies of the connection, our quantisable variables. Writing [26] for example,

$$\hat{C}_{\rm grav} = \frac{2i}{\kappa^2\hbar\gamma^3 k^3}{\rm tr}\sum_{\rm ijk}\epsilon^{\rm ijk}\left(\hat{h}_{\rm i}\hat{h}_{\rm j}\hat{h}_{\rm i}^{-1}\hat{h}_{\rm j}^{-1}\hat{h}_{\rm k}\left[\hat{h}_{\rm k}^{-1},\hat{V}\right]\right) , \tag{67}$$

one immediately realises that there are many possible choices of factor ordering that could have been made at this point, since classically the actions of the holonomies commute. However, each of these factor ordering choices leads to a different factor ordering of the WDW equation in the continuum limit.

The action of the factor ordering chosen in Eq. (67) above gives [19]

$$\epsilon_{\rm ijk}{\rm tr}\left(\hat{h}_{\rm i}\hat{h}_{\rm j}\hat{h}_{\rm i}^{-1}\hat{h}_{\rm j}^{-1}\hat{h}_{\rm k}\left[\hat{h}_{\rm k}^{-1},\hat{V}\right]\right) = -24\widehat{\rm sn}^2\widehat{\rm cs}^2\left(\widehat{\rm cs}\hat{V}\widehat{\rm sn}-\widehat{\rm sn}\hat{V}\widehat{\rm cs}\right)~, \tag{68}$$

while other choices have different action, as it has been explicitly found in Ref. [19].

Defining $\hat{V}|\nu\rangle = V_\nu|\nu\rangle$, the action of the above factor ordering on a general state in the Hilbert space given by $|\Psi\rangle = \sum_\nu \psi_\nu|\nu\rangle$ reads [19]

$$\epsilon_{\rm ijk}{\rm tr}\left(\hat{h}_{\rm i}\hat{h}_{\rm j}\hat{h}_{\rm i}^{-1}\hat{h}_{\rm j}^{-1}\hat{h}_{\rm k}\left[\hat{h}_{\rm k}^{-1},\hat{V}\right]\right)|\Psi\rangle = \frac{-3i}{4}\sum_\nu\Big[\Big(V_{\nu-3k}-V_{\nu-5k}\Big)\psi_{\nu-4k}-2\Big(V_{\nu+k}-V_{\nu-k}\Big)\psi_\nu + \Big(V_{\nu+5k}-V_{\nu+3k}\Big)\psi_{\nu+4k}\Big]|\nu\rangle~. \tag{69}$$

Similarly, one can find [19] the action of any other factor ordering choice.

Then one can take the continuum limit of these expressions by expanding $\psi_\nu \approx \psi(\nu)$ as a Taylor expansion in small k/ν. For the factor ordering chosen for this illustration here, the large scale continuum limit of the Hamiltonian constraint reads [19]:

$$\lim_{k/\nu\to 0}\epsilon_{\rm ijk}{\rm tr}\left(\hat{h}_{\rm i}\hat{h}_{\rm j}\hat{h}_{\rm i}^{-1}\hat{h}_{\rm j}^{-1}\hat{h}_{\rm k}\left[\hat{h}_{\rm k}^{-1},\hat{V}\right]\right)|\Psi\rangle \sim \frac{-36i}{1-A}\alpha^{3/[2(1-A)]}k^3\sum_\nu \nu^{(1+2A)/[2(1-A)]}\left[\frac{{\rm d}^2\psi}{{\rm d}\nu^2}+\frac{1+2A}{1-A}\frac{1}{\nu}\frac{{\rm d}\psi}{{\rm d}\nu}+\frac{(1+2A)(4A-1)}{(1-A)^2}\frac{1}{4\nu^2}\psi(\nu)\right]|\nu\rangle~. \tag{70}$$

Setting $\alpha = 3\mu_0/(2k)$ and $\mu_0 = k$, all lattice refinement models will lead to the same continuum limit for the WDW equation, only for $A = -1/2$ [19], in which case the WDW equation reads [19]

$$\lim_{k/\nu\to 0} C_{\rm grav}|\Psi\rangle = \frac{72}{\kappa^2\hbar\gamma^3}\left(\frac{\kappa\gamma\hbar}{6}\right)^{3/2}\sum_\nu\frac{{\rm d}^2\psi}{{\rm d}\nu^2}|\nu\rangle~. \tag{71}$$

Thus, there is only one lattice refinement model, namely $\tilde{\mu} = \mu_0\mu^{-1/2}$, with a non ambiguous continuum limit[18].

In conclusion, phenomenological and consistency requirements lead to a particular lattice refinement model, implying that LQC predicts a unique factor ordering of the WDW equation in its continuum limit. Alternatively, demanding that factor ordering ambiguities disappear in LQC at the level of WDW equation leads to a unique choice for the lattice refinement model.

4. Conclusions

LQG canonically quantises space-time via triad and holonomies of the connection. Full understanding of the theory has not yet been reached, nevertheless symmetry reduction versions akin to WDW mini-superspace model have been successfully developed.

[18] The $A = -1/2$ choice can be easily understood with the following simple argument [30]. In LQC, the basis sates are $|\mu\rangle$ and the physical area of the fiducial cell is $\mu V_0^{2/3}$. Consider, in the full LQG theory, N fluxes passing through a side of the fiducial cell and divide its surface in Δ elementary surfaces. Then $N\Delta = \mu V_0^{2/3}$, implying $N = \mu V_0^{2/3}/\Delta$. The holonomies of the connection are $e^{i\lambda c/2} = e^{i\lambda\bar{c}V_0^{1/3}/2}$ and the fiducial area is $\lambda^2 = V_0^{2/3}/N$. Thus, $\lambda \propto \mu^{-1/2}$.

As a first approximation the quantised holonomies were taken to be shift operators with a fixed magnitude. This results in the quantised Hamiltonian constraint being a difference equation with a constant interval between points on the lattice. These models lead to serious instabilities in the continuum semi-classical limit.

In the underlying LQG theory, the contributions to the discrete Hamiltonian operator depend on the state which describes the universe. As the universe expands, the number of contributions increases, so that the Hamiltonian constraint operator is creating new vertices of a lattice state, leading to a refinement of the discrete lattice in LQC.

The lattice refinement can be modelled and the instabilities in the continuum era get eliminated. We have discussed here why lattice refinement seems to be necessary to achieve a natural inflationary era, and we have illustrated that only a limited range of matter components can be supported within a particular lattice refinement choice. We have then shown that factor ordering ambiguities in the continuum limit of the gravitational part of the Hamiltonian constraint disappear only for a particular choice of lattice refinement.

Whilst the continuum limit of the lattice refinement models can be taken, there is a complication in directly evolving two-dimensional wave-functions, such as those necessary to study Bianchi models or black hole interiors. The information needed to calculate the wave-function at a given lattice point is not provided by previous iterations. We have shown that Taylor expansions can be used to perform this interpolation with a well-defined and predictable accuracy. We have then discussed how lattice refinement can alter stability conditions of the system.

We have only focused here in a few aspects of LQC concerning lattice refinement. There is certainly a much vaster arena of themes within LQC, an area which is gaining a lot of interest from the scientific community. Its basic advantage is that it allows us to successfully address some physical features of our universe, while it gives us some valuable insight for the full LQG theory.

Acknowledgments

It is a pleasure to thank the organisers of the NEBXIII meeting[19] *Recent Developments in Gravity* — $N\epsilon\omega\tau\epsilon\rho\epsilon s$ $E\xi\epsilon\lambda\iota\xi\epsilon\iota s$ $\sigma\tau\eta$ $B\alpha\rho\upsilon\tau\eta\tau\alpha$ — for inviting me to give this talk, in the beautiful city of Thessaloniki in Greece. This work is partially supported by the European Union through the Marie Curie Research and Training Network *UniverseNet* (MRTN-CT-2006-035863).

References

[1] Rovelli C 2004 *Quantum Gravity* (Cambridge: Cambridge University Press)
[2] Thiemann T 2007 *Modern Canonical Quantum General Relativity* (Cambridge: Cambridge University Press)
[3] Bojowald M 2005 *Living Rev. Rel.* **8** 11
[4] Ashtekar A, Bojowald M and Lewandowski J 2003 *Adv. Theor. Math. Phys.* **7** 233
[5] Bojowad M 2001 *Phys. Rev. Lett.* **86** 5227
[6] Ashtekar A, Baez J C and Knasnov K 2000 *Adv. Theor. Math. Phys.* **4** 1
[7] Daniel M and Viallet C-M 1980 *Rev. Mod. Phys.* **52** 175
[8] Ashtekar A, Lewandowski J, Marolf D, Mourao J and Thiemann T 1995 *J. Math. Phys.* **36** 6456.
[9] Ashtekar A and Lewandowski J 1997 *Class. & Quant. Grav.* **14** A55
[10] Rosen J, Jung J H and Khanna G 2006 *Class. Quant. Grav.* **23** 7075
[11] Bojowald M, Cartin D and Khanna G 2007 *Phys. Rev. D* **76** 064018
[12] Nelson W and Sakellariadou M 2007 *Phys. Rev. D* **76** 104003
[13] Calzetta E and Sakellariadou M 1992 *Phys. Rev. D* **45** 2802
[14] Calzetta E and Sakellariadou M 1993 *Phys. Rev. D* **47** 3184
[15] Gibbons G W and Turok N 2006 arXiv:hep-th/0609095

[19] http://www.astro.auth.gr/~neb-13/

[16] Germani C, Nelson W and Sakellariadou M 2007 *Phys. Rev. D* **76** 043529
[17] Nelson W and Sakellariadou M 2007 *Phys. Rev. D* **76** 044015
[18] Corichi A and Singh P 2008 *Phys. Rev. D* **78** 024034
[19] Nelson W and Sakellariadou M 2008 *Phys. Rev. D* **78** 024006
[20] Sabhararwal S and Khanna G 2008 *Class. & Quant. Grav.* **25** 085009
[21] Nelson W and Sakellariadou M 2008 *Phys. Rev. D* **78** 024030
[22] Barbero J 2005 *Phys. Rev. D* **51** 5507
[23] Ashtekar A 1986 *Phys. Rev. Lett.* **57** 2244
[24] Ashtekar A, Pawlowski T and Singh P 2006 *Phys. Rev. D* **73** 124038
[25] Vandersloot K 2006 *Loop quantum cosmology* (Ph.D. Thesis Pennsylvania State University)
[26] Ashtekar A, Pawlowski T and Singh P 2006 *Phys. Rev. D* **74** 084003
[27] Vandersloot K 2005 *Phys. Rev. D* **71** 103506
[28] Bojowald M 2002 *Clas. & Quant. Grav.* **19** 2717
[29] Bojowald M and Hinterleitner F 2002 *Phys. Rev. D* **66** 104003
[30] Ashtekar A 2008 *Private communication*

Black holes and the quark-gluon plasma

George Siopsis
Department of Physics and Astronomy, The University of Tennessee, Knoxville, TN 37996 - 1200, USA
E-mail: siopsis@tennessee.edu

Abstract. I discuss the possibility that the quark-gluon plasma at strong coupling admits a description in terms of a black hole in asymptotically anti-de Sitter space.

1. Introduction

"A second unexpected connection comes from studies carried out using the Relativistic Heavy Ion Collider, a particle accelerator at Brookhaven National Laboratory. This machine smashes together nuclei at high energy to produce a hot, strongly interacting plasma. Physicists have found that some of the properties of this plasma are better modeled (via duality) as a tiny black hole in a space with extra dimensions than as the expected clump of elementary particles in the usual four dimensions of spacetime. The prediction here is again not a sharp one, as the string model works much better than expected. String-theory skeptics could take the point of view that it is just a mathematical spinoff. However, one of the repeated lessons of physics is unity - nature uses a small number of principles in diverse ways. And so the quantum gravity that is manifesting itself in dual form at Brookhaven is likely to be the same one that operates everywhere else in the universe."

– Joe Polchinski

The AdS/CFT correspondence [1, 2, 3, 4] which has emerged from string theory has provided a path for understanding a conformally invariant gauge theory (CFT) at strong coupling in terms of a dual gravitational description in an asymptotically anti-de Sitter (AdS) space in one higher dimension. The quark-gluon plasma observed at the Relativistic Heavy Ion Collider (RHIC) is thought to be strongly interacting [5, 6, 7]. Even though it is described by a different gauge theory, namely Quantum Chromodynamics (QCD), one hopes that there are enough similarities between the two theories allowing us to acquire a good qualitative (if not quantitative) understanding of its behavior. This has cultivated considerable interest in the connection between string theory and the experimental results at RHIC [8, 9, 10, 11, 12, 13].

Via the AdS/CFT correspondence, information on the strongly interacting plasma is gained by studying the supergravity solution in AdS space. From the AdS metric one may then calculate the stress-energy tensor of the dual gauge theory using holographic renormalization [14, 15]. The simplest supergravity solution to study is the AdS Schwarzschild metric which has been shown to be dual to a static fluid on the AdS boundary. This analysis has been extended to include small deformations of the Schwarzschild metric governed by quasinormal modes that dictate the late-time behavior of the black hole perturbations [16]. These modes have been studied in great detail (see [17, 18, 19, 20] and references therein). The lowest frequency modes govern the hydrodynamic behavior of the conformal field theory on the boundary [21, 22, 23].

The lowest lying gravitational quasinormal modes for an AdS Schwarzschild black hole with a spherical horizon were calculated numerically in four and five dimensions in [16, 24] and analytically in arbitrary dimension in [23]. They were shown to be in agreement with appropriate sinusoidal perturbations of the gauge theory plasma on the AdS boundary. To make contact with experimental results, the boundary is defined through a foliation of flat surfaces turning it into flat Minkowski space. This results in a finite "conformal soliton flow" once the spherical AdS boundary one obtains in global coordinates is conformally mapped to the physically relevant flat Minkowski spacetime [16]. The perturbations also allow for calculations of the elliptic flow of the plasma and its thermalization time – both observable at RHIC. While there is still work to be done, these calculations compare well with what has been found experimentally.

An alternative to a spherical AdS black hole is one with a hyperbolic horizon [25, 26, 27, 28, 29, 30, 31, 32]. They are usually referred to as topological AdS black holes because they possess topologically non-trivial horizons. Their effect on the gauge theory plasma on the AdS boundary was elucidated in [33]. Gravitational perturbations were shown to possess quasinormal modes whose lifetime was comparable to or longer than their counterparts in the case of spherical horizons. These results were in agreement with those obtained by studying the hydrodynamics of the gauge theory plasma on the boundary. Therefore, topological AdS black holes might have a significant effect on the behavior of the quark-gluon plasma in heavy ion collisions at RHIC and the LHC.

The AdS/CFT correspondence has been applied to the description of the quark-gluon plasma in several different scenarios; see [34, 35] for nice reviews. Janik and Peschanski showed that a model for relativistic heavy ion collisions, suggested two decades previously by Bjorken [36], can be seen to be a consequence of the AdS/CFT correspondence [37]. They obtained an asymptotic solution to the Einstein equations in the five-dimensional bulk in the large longitudinal proper time (τ) limit and showed that it gave rise to hydrodynamics on the four-dimensional boundary matching the Bjorken flow of an ideal fluid [36]. The work was subsequently continued to include subleading terms in the large τ expansion which could be understood as dissipative effects in the hydrodynamic behavior of the gauge theory plasma [38, 39, 40, 41].

Kajantie, *et al.,* [42] showed that the two-dimensional Bjorken flow could be derived from a static black hole in the three-dimensional bulk spacetime by an appropriate time-dependent slicing near its boundary. This result was then extended to arbitrary dimensions [43] by showing that to leading order in τ there exists a foliation near the boundary of an AdS Schwarzschild black hole which corresponds to Bjorken flow in the dual gauge theory plasma. The Schwarzschild metric to leading order in τ agreed with the time-dependent asymptotic solution of Janik and Peschanski [37] in five dimensions. In three dimensions it reduced to the result of Kajantie, *et al.,* [42].

Subleading corrections in the large τ expansion were discussed in [44]. It was shown that next-to-leading-order corrections correspond to viscosity in the gauge theory plasma. At this level the coefficient of viscosity η is arbitrary, in agreement with results in five dimensions based on an asymptotic time-dependent solution to the Einstein equations [38]. At next-to-next-to-leading order the Schwarzschild metric yields a flow which is not boost-invariant no matter how one chooses the foliation near the AdS boundary. Boost invariance is recovered after the Schwarzschild metric is perturbed by a power law τ-dependent perturbation. The perturbed metric was shown to be non-singular in the bulk, provided

$$\frac{\eta}{s} = \frac{1}{4\pi}, \tag{1}$$

where s is the entropy density, in agreement with asymptotic time-dependent solutions in five dimensions [39]. This special value of the ratio η/s is also in agreement with the case of sinusoidal perturbations of an AdS Schwarzschild black hole [11, 12, 13].

The outline of my presentation is as follows. In section 2, I discuss the various types of AdS Schwarzschild black holes and the dual static gauge theory fluid on the AdS boundary. In section 3 I discuss perturbations and the calculation of low frequency quasinormal modes. These modes are then shown to agree with the hydrodynamic behavior of the gauge theory fluid in section 4. In section 5 a different foliation of the AdS Schwarzschild spacetime near the AdS boundary is shown to lead to Bjorken flow for the gauge theory fluid. Finally section 6 contains my concluding remarks.

2. AdS Schwarzschild black holes

The metric of an AdS Schwarzschild black hole in d dimensions can be written in static coordinates as

$$ds^2_{\text{b.h.}} = -\left(\frac{r^2}{R^2} + K - \frac{2\mu}{r^{d-3}}\right) dt^2 + \frac{dr^2}{\frac{r^2}{R^2} + K - \frac{2\mu}{r^{d-3}}} + r^2 d\Sigma^2_{K,d-2}. \tag{2}$$

For simplicity I shall choose units so that the AdS radius $R = 1$. The horizon radius and Hawking temperature are, respectively,

$$2\mu = r_+^{d-1}\left(1 + \frac{K}{r_+^2}\right), \quad T_H = \frac{(d-1)r_+^2 + K(d-3)}{4\pi r_+}. \tag{3}$$

The mass and entropy of the hole are, respectively,

$$M = (d-2)(K + r_+^2)\frac{r_+^{d-3}}{16\pi G} Vol(\Sigma_{K,d-2}), \quad S = \frac{r_+^{d-2}}{4G} Vol(\Sigma_{K,d-2}), \tag{4}$$

$K = 0$ corresponds to a flat horizon ($\mathbb{R}^{d-2}$), $K = +1$ to a spherical horizon ($\mathbb{S}^{d-2}$) and $K = -1$ to a hyperbolic horizon ($\mathbb{H}^{d-2}/\Gamma$, where Γ is a discrete group of isometries of $\mathbb{H}^{d-2}$ (topological black hole)).

The harmonics on $\Sigma_{K,d-2}$ obey the equation

$$\left(\nabla^2 + k^2\right)\mathbb{T} = 0. \tag{5}$$

For $K = 0$, k represents momentum (continuous parameter), for $K = +1$,

$$k^2 = l(l + d - 3) - \delta, \tag{6}$$

where l is the (discrete) angular momentum quantum number and for $K = -1$,

$$k^2 = \xi^2 + \left(\frac{d-3}{2}\right)^2 + \delta, \tag{7}$$

where ξ is a discrete parameter for non-trivial Γ [45, 46, 47, 48, 49, 50, 51]. $\delta = 0, 1, 2$ for scalar, vector, or tensor harmonics, respectively.

To study the consequencies of the AdS/CFT correspondence, it is convenient to switch coordinates to

$$z = \frac{1}{r}, \tag{8}$$

placing the boundary of AdS space at $z = 0$. For the flat (large) Schwarzschild black hole ($K = 0$), the metric (2) reads

$$ds^2_{\text{b.h.}} = \frac{1}{z^2}\left(-(1-2\mu z^{d-1})dt^2 + d\vec{x}^2 + \frac{dz^2}{1-2\mu z^{d-1}}\right), \tag{9}$$

where $\vec{x} \in \mathbb{R}^{d-2}$. The horizon is at

$$z_+ = (2\mu)^{-\frac{1}{d-1}}, \tag{10}$$

and the Hawking temperature (3) reads

$$T_H = \frac{d-1}{4\pi z_+}. \tag{11}$$

The above expressions are related to a gauge theory fluid on the boundary via holographic renormalization [14, 15]. To see this, write the metric (9) in Fefferman-Graham coordinates

$$ds^2 = \frac{g_{\mu\nu}dx^\mu dx^\nu + dz_{FG}^2}{z_{FG}^2}. \tag{12}$$

Then near the boundary at $z_{FG} = 0$ expand

$$g_{\mu\nu} = g^{(0)}_{\mu\nu} + z^2_{FG}g^{(2)}_{\mu\nu} + \cdots + z^{d-1}_{FG}g^{(d-1)}_{\mu\nu} + h^{(d-1)}_{\mu\nu}z^{d-1}_{FG}\ln z^2_{FG} + \mathcal{O}(z^d_{FG}), \tag{13}$$

where $g^{(0)}_{\mu\nu} = \eta_{\mu\nu}$.

One obtains the stress-energy tensor of the plasma

$$\langle T_{\mu\nu}\rangle = \frac{d-1}{16\pi G}g^{(d-1)}_{\mu\nu}, \tag{14}$$

leading to the energy density and pressure, respectively,

$$\epsilon = \langle T^{tt}\rangle = (d-2)\frac{\mu}{8\pi G} \quad , \quad p = \langle T^{ii}\rangle = \frac{\mu}{8\pi G}, \tag{15}$$

obeying $p = \frac{1}{d-2}\epsilon$, as expected of a conformal fluid in $d-2$ spatial dimensions.

The equation of state of this fluid is

$$p = \frac{1}{16\pi G_d}\left(\frac{4\pi T_H}{d-1}\right)^{d-1}. \tag{16}$$

We may also write the energy and entropy densities, respectively, as functions of the temperature (11)

$$\epsilon = \frac{d-2}{16\pi G}\left(\frac{4\pi T_H}{d-1}\right)^{d-1} \quad , \quad s = \frac{dp}{dT} = \frac{1}{4G}\left(\frac{4\pi T_H}{d-1}\right)^{d-2}. \tag{17}$$

It should be noted that we obtained a static fluid.

The above results can be easily generalized to the case of a curved horizon (boundary), i.e., $K = \pm 1$.

3. Perturbations

As was shown by Ishibashi and Kodama [52], any perturbation of the black hole background (2) obeys a radial wave equation of the Schrödinger-like form (master equation)

$$-\frac{d^2\phi}{dr_*^2} + V[r(r_*)]\phi = \omega^2\phi, \tag{18}$$

where $V(r)$ is a potential whose form depends on the perturbation and the "tortoise coordinate" r_* is defined by

$$\frac{dr_*}{dr} = \frac{1}{f(r)} \quad , \quad f(r) = r^2 + K - \frac{2\mu}{r^{d-3}}. \tag{19}$$

This wave equation is to be solved for fluctuations subject to the conditions that the flux be ingoing at the horizon and outgoing at asymptotic infinity. One thus obtains the quasinormal (QNM) modes of the black hole possessing in general a discrete spectrum of complex frequencies whose imaginary part determines the decay time of the small fluctuations,

$$\Im\omega = \frac{1}{\tau}. \tag{20}$$

The AdS/CFT correspondence maps the QNMs onto physical properties of the gauge theory fluid on the boundary. In particular, low frequency modes determine the hydrodynamic behavior of the fluid [11], because the long-distance, low-frequency behavior of any interacting theory at finite temperature must be described by fluid mechanics (hydrodynamics). This leads to a universality as hydrodynamics implies very precise constraints on correlation functions of conserved currents and the stress-energy tensor. In the hydrodynamic limit, gauge theory correlators are fixed once a few transport coefficients are known. The latter may be determined by the low frequency QNMs of the dual black hole.

I shall consider various types of gravitational perturbations and calculate the corresponding low frequency QNMs following the steps in [23, 33].

3.1. Vector Perturbations

The vector gravitational potential is

$$V_V(r) = \frac{f(r)}{r^2}\left(k_V^2 + K + \frac{(d-2)(d-4)}{4}(K+r^2) - \frac{3(d-2)^2\mu}{r^{d-3}}\right), \tag{21}$$

Introducing the variable

$$u = \left(\frac{r_+}{r}\right)^{d-3}, \tag{22}$$

the wave equation (18) takes the form

$$-(d-3)^2 u^{\frac{d-4}{d-3}}\hat{f}(u)\partial_u\left(u^{\frac{d-4}{d-3}}\hat{f}(u)\partial_u\phi\right) + \hat{V}_V(u)\phi = \hat{\omega}^2\phi, \tag{23}$$

where

$$\begin{aligned}
\hat{V}_V(u) &= \hat{f}(u)\left[\hat{k}_V^2 + \frac{(d-2)(d-4)}{4}u^{\frac{2}{3-d}} - \frac{3(d-2)^2}{4}u \right.\\
&\left. + \frac{K}{r_+^2}\left(1 + \frac{(d-2)(d-4)}{4} - \frac{3(d-2)^2}{4}u\right)\right]\\
\hat{f}(u) &= \frac{f(r)}{r^2} = 1 - u^{\frac{2}{d-3}}\left(u - \frac{K(1-u)}{r_+^2}\right), \quad \hat{\omega} = \frac{\omega}{r_+}, \quad \hat{k}_V^2 = \frac{k_V^2}{r_+^2}.
\end{aligned} \tag{24}$$

Factoring out the behavior of the wavefunction ϕ at the horizon ($u = 1$),

$$\phi(u) = (1-u)^{-i\frac{\hat{\omega}}{d-1}} F(u), \tag{25}$$

the wave equation in the large r_+ limit (up to $\mathcal{O}(1/r_+^2)$) can be written as

$$\mathcal{H}F \equiv \mathcal{A}F'' + \mathcal{B}F' + \mathcal{C}F = 0, \tag{26}$$

where

$$\begin{aligned}
\mathcal{A} &= -(d-3)^2 u^{\frac{2d-8}{d-3}}(1-u^{\frac{d-1}{d-3}}) - K\frac{2(d-3)^2}{r_+^2}u^2(1-u), \\
\mathcal{B} &= -(d-3)[d-4-(2d-5)u^{\frac{d-1}{d-3}}]u^{\frac{d-5}{d-3}} - 2(d-3)^2\frac{i\hat{\omega}}{d-1}\frac{u^{\frac{2d-8}{d-3}}(1-u^{\frac{d-1}{d-3}})}{1-u} \\
&\quad -K\frac{d-3}{r_+^2}u\left[(d-3)(2-3u)-(d-1)\frac{1-u}{1-u^{\frac{d-1}{d-3}}}u^{\frac{d-1}{d-3}}\right], \\
\mathcal{C} &= \hat{k}_V^2 + \frac{(d-2)[d-4-3(d-2)u^{\frac{d-1}{d-3}}]}{4}u^{-\frac{2}{d-3}}, \\
&\quad -(d-3)\frac{i\hat{\omega}}{d-1}\frac{[d-4-(2d-5)u^{\frac{d-1}{d-3}}]u^{\frac{d-5}{d-3}}}{1-u} - (d-3)^2\frac{i\hat{\omega}}{d-1}\frac{u^{\frac{2d-8}{d-3}}(1-u^{\frac{d-1}{d-3}})}{(1-u)^2} \\
&\quad +K\frac{d-2}{2r_+^2}\left[d-4-(2d-5)u-(d-1)\frac{1-u}{1-u^{\frac{d-1}{d-3}}}u^{\frac{d-1}{d-3}}\right].
\end{aligned} \tag{27}$$

Expanding the wavefunction,

$$F = F_0 + F_1 + \ldots. \tag{28}$$

I shall solve the wave equation perturbatively.

The zeroth order wave equation,

$$\mathcal{H}_0 F_0 = 0, \tag{29}$$

is obtained in the limit $\hat{\omega}, \hat{k}_V, 1/r_+^2 \to 0$. Explicitly,

$$\mathcal{H}_0 F_0 = \mathcal{A}_0 F_0'' + \mathcal{B}_0 F_0' + \mathcal{C}_0 F_0, \tag{30}$$

where

$$\begin{aligned}
\mathcal{A}_0 &= -(d-3)^2 u^{\frac{2d-8}{d-3}}(1-u^{\frac{d-1}{d-3}}), \\
\mathcal{B}_0 &= -(d-3)[d-4-(2d-5)u^{\frac{d-1}{d-3}}]u^{\frac{d-5}{d-3}}, \\
\mathcal{C}_0 &= \frac{(d-2)[d-4-3(d-2)u^{\frac{d-1}{d-3}}]}{4}u^{-\frac{2}{d-3}}.
\end{aligned} \tag{31}$$

Two exact linearly independent solutions are

$$F_0 = u^{\frac{d-2}{2(d-3)}}, \quad \check{F}_0 = u^{-\frac{d-4}{2(d-3)}}\,{}_2F_1\left(1, -\frac{d-3}{d-1}, \frac{2}{d-1}; u^{\frac{d-1}{d-3}}\right). \tag{32}$$

F_0 is well behaved at both the horizon ($u \to 1$) and the boundary ($u \to 0$), whereas $\check{F}_0$ diverges at both ends making it unacceptable.

The first order equation

$$\mathcal{H}_0 F_1 + \mathcal{H}_1 F_0 = 0, \tag{33}$$

is solved by

$$F_1 = -F_0 \int \frac{\check{F}_0 \mathcal{H}_1 F_0}{\mathcal{A}_0 \mathcal{W}_0} + \check{F}_0 \int \frac{F_0 \mathcal{H}_1 F_0}{\mathcal{A}_0 \mathcal{W}_0}, \tag{34}$$

where $\mathcal{W}_0$ is the zeroth order Wronskian

$$\mathcal{W}_0 = \frac{1}{u^{\frac{d-4}{d-3}} \left(1 - u^{\frac{d-1}{d-3}}\right)}. \tag{35}$$

One arrives at the constraint

$$\int_0^1 \frac{F_0 \mathcal{H}_1 F_0}{\mathcal{A}_0 \mathcal{W}_0} = 0, \tag{36}$$

which is a linear equation in $\hat{\omega}$ whose solution is

$$\hat{\omega} = -i \frac{\hat{k}_V^2 - K \frac{d-3}{r_+^2}}{d-1}, \tag{37}$$

which yields the frequencies of the lowest-lying vector quasinormal modes.

For $K = +1$, we obtain the frequency

$$\omega = -i \frac{(l+d-2)(l-1)}{(d-1) r_+}, \tag{38}$$

and the maximum lifetime

$$\tau_{\max} = \frac{4\pi}{d} T_H. \tag{39}$$

For $K = 0$, the frequency is

$$\omega = -i \frac{k^2}{(d-1) r_+}, \tag{40}$$

leading to the diffusion constant $D = \frac{1}{4\pi T_H}$.

Finally for $K = -1$,

$$\omega = -i \frac{\xi^2 + \frac{(d-1)^2}{4}}{(d-1) r_+}, \quad \tau = \frac{1}{|\omega|} < \frac{16\pi}{(d-1)^2} T_H. \tag{41}$$

It should be noted that in the physically relevant case $d = 5$, the $K = -1$ modes live the longest, which is important for the behavior of the plasma.

Journal of Physics: Conference Series **189** (2009) 012036 doi:10.1088/1742-6596/189/1/012036

3.2. Scalar Perturbations

In the case of scalar gravitational perturbations the potential is

$$\begin{aligned}\hat{V}_S(u) &= \frac{u^{-\frac{2}{d-3}} - u + \frac{K}{r_+^2}(1-u)}{4(\hat{m}+u)^2} \\ &\times \Bigg\{(-6+d)(-4+d)\hat{m}^2 - 6(-4+d)(-2+d)\hat{m}u + (-2+d)du^2 \\ &- 3(-6+d)(-2+d)\hat{m}^2 u^{\frac{d-1}{d-3}} \\ &+ 2(18+d(-11+2d))\hat{m}u^{\frac{2(d-2)}{d-3}} + (-2+d)^2 u^{\frac{3d-7}{d-3}} + 2(-2+d)(-1+d)\hat{m}^3 u^{\frac{2}{-3+d}} \\ &+ \frac{Ku^{\frac{2}{d-3}}}{r_+^2}[(-2+d)\hat{m}^2(d+2(-1+d)\hat{m}) - 3(-2+d)\hat{m}(-8-6\hat{m}+d(2+\hat{m}))u \\ &+ (24+36\hat{m}+d(-10+d-22\hat{m}+4d\hat{m}))u^2 + (-2+d)^2u^3]\Bigg\}, \end{aligned} \tag{42}$$

where

$$\hat{m} = 2\frac{k_S^2 - K(d-2)}{(d-1)(d-2)(r_+^2+K)}, \tag{43}$$

Evidently, this potential possesses a new singularity at $u = -\hat{m}$. It is convenient to factor out the behavior at the singularity as well as the two ends ($u = 0, 1$),

$$\phi(u) = (1-u)^{-i\frac{\hat{\omega}}{d-1}} \frac{u^{\frac{d-4}{2(d-3)}}}{\hat{m}+u} F(u). \tag{44}$$

Proceeding as with vector perturbations, after some tedious algebra [23, 33], one arrives at the constraint

$$\frac{d-1}{2}\frac{1+(d-2)\hat{m}}{(1+\hat{m})^2} + \frac{K}{r_+^2}\left(\frac{1}{\hat{m}} + \mathcal{O}(1)\right) - i\hat{\omega}\frac{d-3}{(1+\hat{m})^2} - \hat{\omega}^2\left(\frac{1}{\hat{m}} + \mathcal{O}(1)\right) = 0, \tag{45}$$

whose solution for small $\hat{m}$ is

$$\hat{\omega} = \pm\sqrt{\frac{d-1}{2}\hat{m} + \frac{K}{r_+^2}} - i\frac{d-3}{2}\hat{m}, \tag{46}$$

yielding the spectrum

$$\omega = \pm\frac{k_S}{\sqrt{d-2}} - i\frac{d-3}{(d-1)(d-2)r_+}\left[k_S^2 - K(d-2)\right]. \tag{47}$$

For $K = +1$, the frequencies are

$$\omega = \frac{l(l+d-3)}{\sqrt{d-2}} - i\frac{(d-3)(l+d-2)(l-1)}{(d-1)(d-2)r_+}, \tag{48}$$

and the maximum lifetime is

$$\tau_{\max} = \frac{d-2}{(d-3)d}4\pi T_H, \tag{49}$$

For $K = 0$,

$$\omega = \pm \frac{k}{\sqrt{d-2}} - i \frac{d-3}{(d-1)(d-2)r_+} k^2, \tag{50}$$

leading to the speed of sound $v = \frac{1}{\sqrt{d-2}}$ (which is the correct value for a CFT) and the diffusion constant $D = \frac{d-3}{d-2} \frac{1}{4\pi T_H}$.

For $K = -1$,

$$\omega = \pm \sqrt{\frac{\xi^2 + (\frac{d-3}{2})^2}{d-2}} - i \frac{(d-3)[\xi^2 + \frac{(d-1)^2}{4}]}{(d-1)(d-2)r_+} \quad , \quad \tau < \frac{4(d-2)}{(d-3)(d-1)^2} 4\pi T_H. \tag{51}$$

Notice that for $d = 5$, $K = -1$ scalar modes live longer than any other modes.

3.3. Tensor Perturbations

For tensor gravitational perturbations, the potential in the large r_+ limit is

$$\hat{V}_T(u) = \frac{d-2}{4} \left(du^{-\frac{2}{d-3}} - (d-2)u^{\frac{2(d-2)}{d-3}} - 2u \right) + \hat{k}_T^2 \left(1 - u^{\frac{d-1}{d-3}} \right). \tag{52}$$

The zeroth order wave equation can be solved to find two independent solutions

$$\phi_0 = u^{-\frac{d-2}{2(d-3)}} \quad , \quad \check{\phi}_0 = u^{-\frac{d-2}{2(d-3)}} \ln \left(1 - u^{\frac{d-1}{d-3}} \right). \tag{53}$$

They are both unacceptable as they diverge at the boundary ($u \to 0$) and the horizon ($u \to 1$). Therefore, no low lying tensor modes affecting the hydrodynamic behavior of the plasma exist.

4. Hydrodynamics on the AdS boundary

The QNMs calculated above affect the hydrodynamic behavior of the dual plasma on the AdS boundary. This can be seen by calculating the hydrodynamics in the linearized regime of a $d-1$ dimensional fluid with dissipative effects.

The metric on the AdS boundary is

$$ds_\partial^2 = -dt^2 + d\Sigma_{K,d-2}^2.$$

The hydrodynamic equations amount to the conservation law of the stress-energy tensor,

$$\nabla_\mu T^{\mu\nu} = 0. \tag{54}$$

The stress-energy tensor can be cast in the general form

$$T^{\mu\nu} = (\epsilon + p)u^\mu u^\nu + pg^{\mu\nu} - \eta \left(\Delta^{\mu\lambda} \nabla_\lambda u^\nu + \Delta^{\nu\lambda} \nabla_\lambda u^\mu - \frac{2}{d-2} \Delta^{\mu\nu} \nabla_\lambda u^\lambda \right) - \zeta \Delta^{\mu\nu} \nabla_\lambda u^\lambda, \tag{55}$$

where $\Delta_{\mu\nu} = g_{\mu\nu} + u_\mu u_\nu$ and ϵ, p, η and ζ represent the energy density, pressure, shear viscosity and bulk viscosity, respectively, of the gauge theory fluid. Two constraints on the parameters immediately follow from the requirement of conformal invariance,

$$\epsilon = (d-2)p \quad , \quad \zeta = 0, \tag{56}$$

u^μ is the velocity field of the fluid. In its rest frame, $u^\mu = (1, 0, 0, 0)$ and the pressure is constant, $p = p_0$. Perturbations introduce small disturbances,

$$u^\mu = (1, u^i), \tag{57}$$

where u^i is small and also allow for small corrections to the pressure so that

$$p = p_0 + \delta p. \tag{58}$$

The hydrodynamic equations imply

$$\begin{aligned}(d-2)\partial_t \delta p + (d-1)p_0 \nabla_i u^i &= 0,\\ (d-1)p_0\partial_t u^i + \partial^i \delta p - \eta\left[\nabla^j\nabla_j u^i + K(d-3)u^i + \frac{d-4}{d-2}\partial^i(\nabla_j u^j)\right] &= 0,\end{aligned} \tag{59}$$

where I used $R_{ij} = K(d-3)g_{ij}$.

For vector perturbations, use the *ansatz*

$$\delta p = 0\ , \quad u^i = \mathcal{C}_V e^{-i\omega t}\mathbb{V}^i, \tag{60}$$

where $\mathbb{V}^i$ is a vector harmonic.

The hydrodynamic equations become

$$-i\omega(d-1)p_0 + \eta\left[k_V^2 - K(d-3)\right] = 0. \tag{61}$$

Using

$$\frac{\eta}{p_0} = (d-2)\frac{\eta}{s}\frac{S}{M} = \frac{4\pi\eta}{s}\frac{r_+}{K+r_+^2}, \tag{62}$$

with ω from the gravity dual (eq. (37)), we obtain for large r_+,

$$\frac{\eta}{s} = \frac{1}{4\pi}, \tag{63}$$

as expected [11].

For scalar perturbations, use instead the *ansatz*

$$u^i = \mathcal{A}_S e^{-i\omega t}\partial^i\mathbb{S}\ , \quad \delta p = \mathcal{B}_S e^{-i\omega t}\mathbb{S}, \tag{64}$$

where $\mathbb{S}$ is a scalar harmonic.

The hydrodynamic equations in this case read

$$\begin{aligned}&(d-2)i\omega\mathcal{B}_S + (d-1)p_0 k_S^2\mathcal{A}_S = 0,\\ &\mathcal{B}_S + \mathcal{A}_S\left[-i\omega(d-1)p_0 - 2(d-3)K\eta + 2\eta k_S^2\frac{d-3}{d-2}\right] = 0.\end{aligned} \tag{65}$$

This is a linear system whose determinant must vanish,

$$\begin{vmatrix}(d-2)i\omega & (d-1)p_0k_S^2\\ 1 & -i\omega(d-1)p_0 - 2(d-3)K\eta + 2\eta k_S^2\frac{d-3}{d-2}\end{vmatrix} = 0. \tag{66}$$

Using this constraint and arguing along the same lines as with vector perturbations (eq. (62)), one arrives at the same special value of the ratio η/s (63), as before.

The fluid described above is static with small sinusoidal perturbations. This is hardly the behavior of the quark-gluon plasma observed at RHIC. However, it was argued in [16] that a flowing behavior can be obtained in the spherical case ($K = +1$) by choosing a foliation near the AdS boundary which consists of flat Minkowksi hypersurfaces instead of ones consisting of spatial spheres ($S^{d-2}\times\mathbb{R}$). The latter are obtained by fixing the radial distance in the metric

(r = const. hypersurfaces). The former (which are physically relevant) are obtained from the latter via a conformal transformation. The static fluid is then mapped onto a "conformal soliton flow" whilst the perturbations induce an elliptic flow similar to the one observed at RHIC. One may then calculate physical quantities and compare with experimental results.

Let $v_2 = \langle \cos 2\phi \rangle$ at $\theta = \frac{\pi}{2}$ (mid-rapidity) averaged with respect to the energy density at late times and $\delta = \frac{\langle y^2 - x^2 \rangle}{\langle y^2 + x^2 \rangle}$ (eccentricity at time $t = 0$). The ratio of these two quantities is found to be [16]

$$\frac{v_2}{\delta} = \frac{1}{6\pi} \Re \frac{\omega^4 - 40\omega^2 + 72}{\omega^3 - 4\omega} \sin \frac{\pi\omega}{2}, \tag{67}$$

where ω is the frequency of the longest lived (scalar) QNM. Numerically, $\frac{v_2}{\delta} = 0.37$, which compares well with the result from RHIC data, $\frac{v_2}{\delta} \approx 0.323$ [53].

Another quantity that can be predicted is the thermalization time,

$$\tau = \frac{1}{2|\Im\omega|} \approx \frac{1}{8.6 T_{\text{peak}}} \approx 0.08 \text{ fm/c} \quad , \qquad T_{\text{peak}} = 300 \text{ MeV}, \tag{68}$$

to be compared with the RHIC result $\tau \sim 0.6$ fm/c [54]. While not in good agreement, it is encouragingly small. For comparison, perturbative QCD yields $\tau \gtrsim 2.5$ fm/c [55, 56].

A similar comparison with experimental results can also be done in the case of a hyperbolic horizon (topological black hole, $K = -1$). Recall that this is the case of longest lived QNMs, so these modes may play an important role in the late time behavior of the quark-gluon plasma. However, it is much harder than with spherical horizons to extract predictions because of the complexity of the conformal map $\mathbb{H}^{d-2}/\Gamma \times \mathbb{R} \mapsto (d-1)$−dim Minkowski space [33].

5. Bjorken flow

As was first suggested by Bjorken [36], in high energy collisions of heavy nuclei, there is a "plateau" in particle production in the central rapidity region since the interactions are independent of the Lorentz frame (boost-invariant) and the emergent nuclei are highly Lorentz-contracted pancakes receding almost at the speed of light.

The initial conditions are shown in table 1 for RHIC and the LHC. τ is the longitudinal proper time, ϵ is the energy density, T is the temperature, and $\sqrt{s}$ is the center of mass energy of the beams.

Table 1. Initial conditions.

	τ_0 (fm/c)	ϵ_0 (GeV/fm^3)	T (GeV)	$\sqrt{s}$ (GeV)
RHIC	0.2	10	0.5	200
LHC	0.1	10	1	5,500

The hydrodynamic equations respect the symmetry of the initial conditions (boost invariance) and lead to simple solutions. For a conformal fluid,

$$\epsilon \sim \tau^{-4/3} \quad , \quad T \sim \tau^{-1/3} \quad , \quad s \sim \tau^{-1}. \tag{69}$$

To understand Bjorken Hydrodynamics, consider a plasma on a $(d-1)$-dimensional Minkowski space spanned by coordinates $\tilde{x}^\mu$ ($\mu = 0, 1, \ldots, d-2$) with the colliding beams along the $\tilde{x}^1$ direction.

Define coordinates τ, y (proper time and rapidity y in the longitudinal plane, respectively) by

$$\tilde{x}^0 = \tau \cosh y \ , \quad \tilde{x}^1 = \tau \sinh y. \tag{70}$$

The Minkowski metric reads

$$d\tilde{s}^2 = d\tilde{x}_\mu d\tilde{x}^\mu = -d\tau^2 + \tau^2 dy^2 + (d\tilde{x}^\perp)^2, \tag{71}$$

where $\tilde{x}^\perp = (\tilde{x}^2, \dots, \tilde{x}^{d-2})$ are the transverse coordinates.

Assuming no viscosity, the stress-energy tensor is

$$T^{\mu\nu} = \begin{pmatrix} \epsilon(\tau) & 0 & \dots & 0 \\ 0 & p(\tau)/\tau^2 & \dots & 0 \\ & & \ddots & \\ 0 & 0 & \dots & p(\tau) \end{pmatrix}, \tag{72}$$

where ϵ and p are the energy density and pressure of the fluid, respectively. The local conservation law of the stress-energy tensor implies

$$\partial_\tau \epsilon + \frac{1}{\tau}(\epsilon + p) = 0. \tag{73}$$

Conformal invariance implies tracelessness, $-\epsilon + (d-2)p = 0$, therefore

$$\epsilon = (d-2)p = \frac{\epsilon_0}{\tau^{\frac{d-1}{d-2}}}. \tag{74}$$

Then conservation of entropy in a perfect fluid implies

$$T = \frac{T_0}{\tau^{1/(d-2)}} \ , \quad s = \frac{\dot{p}}{\dot{T}} = \frac{s_0}{\tau} \ , \quad s_0 = \frac{d-1}{d-2}\frac{\epsilon_0}{T_0}. \tag{75}$$

Note that the energy and entropy densities have the same dependence on the temperature as in static case (discussed in section 2 in terms of a gravity dual).

A gravity dual of the above perfect fluid flow was obtained by Janik and Peschanski [37]. They obtained an asymptotic solution of the Einstein equations which was valid for large longitudinal proper time τ. Their metric reads

$$ds^2_{\text{Bjorken}} = \frac{1}{\tilde{z}^2}\left[-\left(1 - 2\mu \frac{\tilde{z}^{d-1}}{\tau^{\frac{d-1}{d-2}}}\right) d\tau^2 + \tau^2 dy^2 + (d\tilde{x}^\perp)^2 + \frac{d\tilde{z}^2}{1 - 2\mu \frac{\tilde{z}^{d-1}}{\tau^{\frac{d-1}{d-2}}}} \right], \tag{76}$$

and leads to Bjorken hydrodynamics via holographic renormalization.

In [43] it was shown that Bjorken hydrodynamics can be obtained from an AdS Schwarzschild black hole. To see this, instead of approximating the boundary of the flat AdS black hole (9) with $z =$ const. hypersurfaces (as $z \to 0$), slice with $\tilde{z} =$ const. hypersurfaces, where

$$t = \frac{d-2}{d-3}\tau^{\frac{d-3}{d-2}} \ , \quad x^1 = \tau^{\frac{d-3}{d-2}} y \ , \quad x^\perp = \frac{\tilde{x}^\perp}{\tau^{1/(d-2)}} \ , \quad z = \frac{\tilde{z}}{\tau^{1/(d-2)}}. \tag{77}$$

The two different foliations coincide "initially" (at $\tau = 1$ in arbitrary units; see table 1 for conventional units). Let us identify initial data with their corresponding values in the static case (eqs. (11) and (15)),

$$T_0 = T_H \ , \quad \epsilon_0 = \frac{(d-2)\mu}{8\pi G}. \tag{78}$$

We obtain a flowing fluid on the boundary because the new coordinates (77) of the black hole metric (9) are τ-dependent.

Applying the transformation (77) to the exact black hole metric (9), or more precisely, to a patch which includes the boundary $z \to 0$, the black hole metric becomes

$$ds^2_{\text{b.h.}} = \frac{1}{\tilde{z}^2}\left[-\left(1 - 2\mu\frac{\tilde{z}^{d-1}}{\tau^{\frac{d-1}{d-2}}}\right)d\tau^2 + \tau^2 dy^2 + (d\tilde{x}^\perp)^2 + \frac{d\tilde{z}^2}{1 - 2\mu\frac{\tilde{z}^{d-1}}{\tau^{(d-1)/(d-2)}}}\right] + \dots, \tag{79}$$

where the dots represent higher-order contributions in the large τ expansion. This form matches the bulk metric of Bjorken flow (76) to leading order in $1/\tau$.

Having a dual black hole also allows us to understand the origin of the temperature of the quark-gluon plasma. Recall that the Hawking temperature of the hole T_H (11) is the temperature of the *static* fluid on the $z = 0$ hypersurface whose metric is

$$ds^2_{z\to 0} = -dt^2 + d\vec{x}^2. \tag{80}$$

On the other hand, the $\tilde{z} \to 0$ hypersurface has metric

$$ds^2_{\tilde{z}\to 0} = -d\tau^2 + \tau^2 dy^2 + (d\tilde{x}^\perp)^2. \tag{81}$$

The two metrics are related through the transformation

$$t = \frac{d-2}{d-3}\tau^{\frac{d-3}{d-2}} \quad , \quad x^1 = \tau^{\frac{d-3}{d-2}} y \quad , \quad x^\perp = \frac{\tilde{x}^\perp}{\tau^{1/(d-2)}}, \tag{82}$$

by a leading-order conformal factor,

$$ds^2_{z\to 0} = \tau^{-\frac{2}{d-2}}\left[ds^2_{\tilde{z}\to 0} + \dots\right]. \tag{83}$$

It follows that the period of thermal Green functions on the Bjorken boundary scales as $\tau^{1/(d-2)}$. Since the period is inversely proportional to the temperature, the latter scales as $\tau^{-1/(d-2)}$, in agreement with Bjorken hydrodynamics. The two hypersurfaces coincide at $\tau = 1$ at which time $T = T_H$.

The above asymptotic expressions are exact (no higher order corrections) in three dimensions ($d = 3$) as was discovered in [42]. Indeed, letting $d \to 3$, the coordinate transformation (77) becomes

$$t = \ln \tau \quad , \quad x^1 = y \quad , \quad z = \frac{\tilde{z}}{\tau}, \tag{84}$$

and the transformation to Fefferman-Graham coordinates can be found exactly in this case,

$$z = \frac{\tilde{z}_{FG}}{\tau}\left(1 + \frac{\mu}{2}\frac{\tilde{z}^2_{FG}}{\tau^2}\right)^{-1}, \tag{85}$$

leading to explicit expressions (in closed form) for the physical properties of the plasma.

Higher order corrections to the Bjorken flow dictated by the black hole may be found by refining the transformation (77). At next-to-leading order, viscosity is introduced whereas at higher orders it is no longer possible to maintain a metric which is independent of the rapidity and the transverse coordinates. For Bjorken flow, it is necessary to introduce a perturbation. The resulting metric leads to a fluid on which the ratio η/s takes on the special value (1) [44] in agreement with sinusoidal perturbations [11].

6. Conclusion

Quasi-normal modes are a powerful tool in understanding the hydrodynamic behavior of the gauge theory fluid at strong coupling via the AdS/CFT correspondence. I analytically calculated the low-lying quasinormal modes of AdS black holes in arbitrary dimension in the high temperature limit. I considered all three different types of perturbations (scalar, vector and tensor) and solved the wave equation [52] in each case by applying the methods in [23, 33]. I obtained quasinormal frequencies which were in agreement with the frequencies obtained by considering perturbations of the gauge theory fluid on the boundary [16, 24]. These modes may contribute to the properties of the quark-gluon plasma produced in heavy ion collisions.

I also discussed the possibility of obtaining viscous Bjorken hydrodynamics on a $(d-1)$-dimensional Minkowski space from a large AdS_d Schwarzschild black hole (of flat horizon). The latter is normally considered dual to a static gauge theory fluid on the boudary whose temperature coincides with the Hawking temperature. By appropriately modifying the boundary conditions, I obtained Bjorken hydrodynamics on the boundary in the limit of large longitudinal proper time [43, 44]. These results are in agreement with those obtained by considering time-dependent asymptotic solutions of the Einstein equations in five dimensions [37, 38, 40]. Moreover, since the bulk space consists of a Schwarzschild black hole, I could determine the temperature of the conformal fluid on the boundary in terms of the Hawking temperature of the hole. It turns out that at next-to-next-to-leading order, no choice of boundary conditions leads to a boost-invariant flow [44]. In order to obtain a dual Bjorken flow at that order, the black hole metric needs to be perturbed. This leads to a constraint on the viscosity coefficient and the viscosity to entropy density ratio is then fixed to the value $1/(4\pi)$ as in the case of sinusoidal perturbations [11, 12, 13]. This is also in agreement with a next-to-next-to-leading order calculation of a time-dependent asymptotic solution [39].

It should be pointed out that a constraint on the viscosity is not necessary if one does not perturb the black hole metric. In this case, one obtains deviations from Bjorken flow which are a subleading effect at late times. It might be worth exploring the connection of such deviations (coming from a dual Schwarzschild black hole) to experimental data (RHIC and the LHC).

In conclusion, it is an exciting possibility that heavy ion collisions at RHIC and the LHC will probe black holes and provide information on string theory as well as non-perturbative QCD effects.

Acknowledgments

Work supported in part by the US Department of Energy under grant DE-FG05-91ER40627.

References

[1] Maldacena J M 1998 The large-N limit of superconformal field theories and supergravity *Adv. Theor. Math. Phys.* **2** 231 (*Preprint* arXiv:hep-th/9711200)

[2] Gubser S S, Klebanov I R and Polyakov A M 1998 Gauge theory correlators from non-critical string theory *Phys. Lett.* B **428** 105 (*Preprint* arXiv:hep-th/9802109)

[3] Witten E 1998 Anti-de Sitter space and holography *Adv. Theor. Math. Phys.* **2** 253 (*Preprint* arXiv:hep-th/9802150)

[4] Aharony O, Gubser S S, Maldacena J M, Ooguri H and Oz Y 2000 Large-N field theories, string theory and gravity *Phys. Rept.* **323** 183 (*Preprint* arXiv:hep-th/9905111)

[5] Muller B and Nagle J L 2006 Results from the relativistic heavy ion collider *Ann. Rev. Nucl. Part. Sci.* **56** 93 (*Preprint* arXiv:nucl-th/0602029)

[6] Kolb P F and Heinz U W 2003 Hydrodynamic description of ultrarelativistic heavy-ion collisions *Preprint* arXiv:nucl-th/0305084

[7] Liu H, Rajagopal K and Shi Y 2008 Robustness and infrared sensitivity of various observables in the application of AdS/CFT to heavy ion collisions *J. High Energy Phys.* JHEP08(2008)048 (*Preprint* arXiv:0803.3214)

[8] Sin S J and Zahed I 2005 Holography of radiation and jet quenching *Phys. Lett.* B **608** 265 (*Preprint* arXiv:hep-th/0407215)
[9] Son D T and Starinets A O 2006 Hydrodynamics of R-charged black holes *J. High Energy Phys.* JHEP06(2003)052 (*Preprint* arXiv:hep-th/0601157)
[10] Shuryak E, Sin S J and Zahed I 2005 A gravity dual of RHIC collisions *Preprint* arXiv:hep-th/0511199
[11] Policastro G, Son D T and Starinets A O 2001 The shear viscosity of strongly coupled $N = 4$ supersymmetric Yang-Mills plasma *Phys. Rev. Lett.* **87** 081601 (*Preprint* arXiv:hep-th/0104066)
[12] Kovtun P, Son D T and Starinets A O 2005 Viscosity in strongly interacting quantum field theories from black hole physics *Phys. Rev. Lett.* **94** 111601 (*Preprint* arXiv:hep-th/0405231)
[13] Buchel A and Liu J T 2004 Universality of the shear viscosity in supergravity *Phys. Rev. Lett.* **93** 090602 (*Preprint* arXiv:hep-th/0311175)
[14] de Haro S, Solodukhin S N and Skenderis K 2001 Holographic reconstruction of spacetime and renormalization in the AdS/CFT correspondence *Commun. Math. Phys.* **217** 595 (*Preprint* arXiv:hep-th/0002230)
[15] Skenderis K 2002 Lecture notes on holographic renormalization *Class. Quant. Grav.* **19** 5849 (*Preprint* arXiv:hep-th/0209067)
[16] Friess J J, Gubser S S, Michalogiorgakis G and Pufu S S 2007 Expanding plasmas and quasinormal modes of Anti-de Sitter black holes *J. High Energy Phys.* JHEP04(2007)080 (*Preprint* arXiv:hep-th/0611005)
[17] Horowitz G T and Hubeny V E 2000 Quasinormal modes of AdS black holes and the approach to thermal equilibrium *Phys. Rev.* D **62** 024027 (*Preprint* arXiv:hep-th/9909056)
[18] Cardoso V and Lemos J P S 2001 Quasi-normal modes of Schwarzschild Anti-de Sitter black holes: electromagnetic and gravitational perturbations *Phys. Rev.* D **64** 084017 (*Preprint* arXiv:gr-qc/0105103)
[19] Siopsis G 2004 Large mass expansion of quasinormal modes in AdS_5 *Phys. Lett.* B **590** 105 (*Preprint* arXiv:hep-th/0402083)
[20] Musiri S, Ness S and Siopsis G 2006 Perturbative calculation of quasinormal modes of AdS Schwarzschild black holes *Phys. Rev.* D **73** 064001 (*Preprint* arXiv:hep-th/0511113)
[21] Son D T and Starinets A O 2002 Minkowski-space correlators in AdS/CFT correspondence: recipe and applications *J. High Energy Phys.* JHEP09(2002)042 (*Preprint* arXiv:hep-th/0205051)
[22] Policastro G, Son D T and Starinets A O 2002 From AdS/CFT correspondence to hydrodynamics *J. High Energy Phys.* JHEP09(2002)043 (*Preprint* arXiv:hep-th/0205052)
[23] Siopsis G 2007 Low frequency quasinormal modes of AdS black holes *J. High Energy Phys.* JHEP05(2007)042 (*Preprint* arXiv:hep-th/0702079)
[24] Michalogiorgakis G and Pufu S S 2007 Low-lying gravitational modes in the scalar sector of the global AdS_4 black hole *J. High Energy Phys.* JHEP02(2007)023 (*Preprint* arXiv:hep-th/0612065)
[25] Mann R B 1997 Pair production of topological anti-de Sitter black holes *Class. Quant. Grav.* **14** L109 (*Preprint* arXiv:gr-qc/9607071)
[26] Mann R B 1998 Charged topological black hole pair creation *Nucl. Phys.* B **516** 357 (*Preprint* arXiv:hep-th/9705223)
[27] Vanzo L 1997 Black holes with unusual topology *Phys. Rev.* D **56** 6475 (*Preprint* arXiv:gr-qc/9705004)
[28] Brill D R, Louko J and Peldan P 1997 Thermodynamics of (3+1)-dimensional black holes with toroidal or higher genus horizons *Phys. Rev.* D **56** 3600 (*Preprint* arXiv:gr-qc/9705012)
[29] Birmingham D 1999 Topological black holes in anti-de Sitter space *Class. Quant. Grav.* **16** 1197 (*Preprint* arXiv:hep-th/9808032)
[30] Emparan R 1999 AdS/CFT duals of topological black holes and the entropy of zero-energy states *J. High Energy Phys.* JHEP06(1999)036 (*Preprint* arXiv:hep-th/9906040)
[31] Birmingham D and Mokhtari S 2006 Exact gravitational quasinormal frequencies of topological black holes *Phys. Rev.* D **74** 084026 (*Preprint* arXiv:hep-th/0609028)
[32] Aros R, Martinez C, Troncoso R and Zanelli J 2003 Quasinormal modes for massless topological black holes *Phys. Rev.* D **67** 044014 (*Preprint* arXiv:hep-th/0211024)
[33] Alsup J and Siopsis G 2008 Low-lying quasinormal modes of topological AdS black holes and hydrodynamics *Phys. Rev.* D **78** 086001 (*Preprint* arXiv:0805.0287)
[34] Myers R C and Vazquez S E 2008 Quark soup al dente: applied superstring theory *Class. Quant. Grav.* **25** 114008 (*Preprint* arXiv:0804.2423)
[35] Lin S and Shuryak E 2008 Toward the AdS/CFT gravity dual for High Energy Collisions: 1. Falling into the AdS *Phys. Rev.* D **77** 085013 (*Preprint* arXiv:hep-ph/0610168)
Toward the AdS/CFT gravity dual for High Energy Collisions: 2. The stress tensor on the boundary *Phys. Rev.* D **77** 085014 (*Preprint* arXiv:0711.0736)
Toward the AdS/CFT gravity dual for High Energy Collisions: 3. Gravitationally Collapsing Shell and Quasiequilibrium *Phys.Rev.* D **78** 125018 (*Preprint* arXiv:0808.0910)
[36] Bjorken J D 1983 Highly Relativistic Nucleus-Nucleus Collisions: The Central Rapidity Region *Phys. Rev.*

D **27** 140

[37] Janik R A and Peschanski R 2006 Asymptotic perfect fluid dynamics as a consequence of AdS/CFT *Phys. Rev.* D **73** 045013 (*Preprint* arXiv:hep-th/0512162)

[38] Nakamura S and Sin S-J 2006 A Holographic Dual of Hydrodynamics *J. High Energy Phys.* JHEP09(2006)020 (*Preprint* arXiv:hep-th/0607123)

[39] Janik R A 2007 Viscous plasma evolution from gravity using AdS/CFT *Phys. Rev. Lett.* **98** 022302 (*Preprint* arXiv:hep-th/0610144)

[40] Sin S-J, Nakamura S and Kim S P 2006 Elliptic flow, Kasner universe and holographic dual of RHIC fireball *J. High Energy Phys.* JHEP12(2006)075 (*Preprint* arXiv:hep-th/0610113)

[41] Alsup J, Siopsis G and Middleton C 2007 AdS/CFT Correspondence with Heat Conduction *Phys. Lett.* B **654** 35 (*Preprint* arXiv:hep-th/0607139)

[42] Kajantie K, Louko J and Tahkokallio T 2007 The gravity dual of 1+1 dimensional Bjorken expansion *Phys. Rev.* D **76** 106006 (*Preprint* arXiv:0705.1791)

[43] Alsup J and Siopsis G 2008 Bjorken flow from an AdS Schwarzschild black hole *Phys. Rev. Lett.* **101** 031602 (*Preprint* arXiv:0712.2164)

[44] Alsup J and Siopsis G 2008 Dissipative Bjorken hydrodynamics from an AdS Schwarzschild black hole *Preprint* arXiv:0812.1818

[45] Weeks J "SnapPea": A computer program for creating and studying hyperbolic 3-manifolds `http://www.geometrygames.org/SnapPea/index.html`

[46] Cornish N J and Spergel D N 1999 On the eigenmodes of compact hyperbolic 3-manifolds *Preprint* arXiv:math/9906017

[47] Inoue K 2000 Numerical study of length spectra and low-lying eigenvalue spectra of compact hyperbolic 3-manifolds *YITP* **00-59** 19 (*Preprint* arXiv:math-ph/0011012)

[48] Inoue K 1999 Computation of eigenmodes on a compact hyperbolic 3-space *Class. Quant. Grav.* **16** 3071 (*Preprint* arXiv:astro-ph/9810034)

[49] Muller D, Fagundes H V and Opher R 2002 Casimir energy in multiply connected static hyperbolic universes *Phys. Rev.* D **66** 083507 (*Preprint* arXiv:gr-qc/0209103)

[50] Balazs N L and Voros A 1986 Chaos on the pseudosphere *Phys. Reports* **143(3)** 109

[51] Sausset F and Tarjus G 2007 Periodic boundary conditions on the pseudosphere *J. Phys.* A **40** 12873 (*Preprint* arXiv:cond-mat/0703326)

[52] Ishibashi A and Kodama H 2003 A Master Equation for Gravitational Perturbations of Maximally Symmetric Black Holes in Higher Dimensions *Prog. Theor. Phys.* **110** 701 (*Preprint* arXiv:hep-th/0305147)

[53] PHENIX Collaboration 2007 Scaling properties of azimuthal anisotropy in Au+Au and Cu+Cu collisions at $\sqrt{s_{NN}} = 200$ GeV *Phys. Rev. Lett.* **98** 162301 (*Preprint* arXiv:nucl-ex/0608033)

[54] Arnold P, Lenaghan J, Moore G D and Yaffe L G 2005 Apparent Thermalization due to Plasma Instabilities in Quark-Gluon Plasma *Phys. Rev. Lett.* **94** 072302 (*Preprint* arXiv:nucl-th/0409068)

[55] Baier R, Mueller A H, Schiff D and Son D T 2001 Bottom-up thermalization in heavy ion collisions *Phys. Lett.* B **502** 51 (*Preprint* arXiv:hep-ph/0009237)

[56] Molnar D and Gyulassy M 2002 Saturation of Elliptic Flow and the Transport Opacity of the Gluon Plasma at RHIC *Nucl. Phys.* A **697** 495 Erratum *ibid.* A **703** 893 (*Preprint* arXiv:nucl-th/0104073)

Recent Developments in Gravity (NEB XIII) IOP Publishing
Journal of Physics: Conference Series **189** (2009) 012037 doi:10.1088/1742-6596/189/1/012037

Birefringence in pseudo–Finsler spacetimes

Jozef Skákala and Matt Visser

School of Mathematics, Statistics, and Computer Science,
Victoria University of Wellington, Wellington, New Zealand

E-mail: {jozef.skakala,matt.visser}@mcs.vuw.ac.nz

Abstract. Based on the analogue spacetime programme, and many other ideas currently mooted in "quantum gravity", there is considerable ongoing speculation that the usual pseudo-Riemannian (Lorentzian) manifolds of general relativity might eventually be modified at short distances. Two specific modifications that are often advocated are the adoption of Finsler geometries (or more specifically, pseudo-Finsler spacetimes) and the possibility of birefringence (or more generally, multi-refringence). We have investigated the possibility of whether it is possible to usefully and cleanly deal with these two possibilities *simultaneously*. That is, given two (or more) "signal cones": Is it possible to naturally and intuitively construct a "unified" pseudo-Finsler spacetime such that the pseudo-Finsler metric is null on these "signal cones", *but has no other zeros or singularities*? Our results are much less encouraging than we had originally hoped, and suggest that while pseudo-Finsler spacetimes are certainly useful constructs, it is physically more appropriate to think of physics as taking place in a single topological manifold that carries several distinct pseudo-Finsler metrics, one for each polarization mode.

1. Introduction

Motivation for our work [1] comes from various speculations that came up mainly from:

- the analogue spacetime programme [2];
- various different approaches to "quantum gravity",
  (loop quantum gravity [3], string models [4], causal dynamical triangulations [5]);
- various different approaches to "quantum gravity phenomenology",
  (DSR [6], VSR [7], noncommutative geometry [8]);
- various different approaches to "exotic" cosmologies [9, 10].

Many of these approaches strongly suggest the possibility of high energy Lorentz symmetry breaking [11, 12] and/or birefringence (or multi-refringence) [13].

If we wish to create a geometric theory encoding causal relations more complicated than those of special relativity, then we must unfortunately depart from the standard concept of pseudo-Riemannian geometry, and adopt something more general. This more general geometry is almost inevitably a non-positive definite analogue of Finsler geometry [14, 15, 16], that is, a geometry which can most descriptively be called a pseudo-Finsler geometry [17]. (One might also try to generalize the spacetime in a slightly different direction by adopting the notion of an "area metric" [18], but we shall not take that particular route for now.)

One particularly attractive route to constructing Lorentz symmetry breaking geometries is to consider analogue spacetimes. These are classical systems providing us, at the macroscopic level, with emergent phenomena similar to gravity, while being radically different at the microscopic level — and at the same time are physically well understood [2, 12]. They provide us with analogues of many aspects of quantum gravity phenomenology, and are also possible sources of experimental tests for quantum gravity phenomenology. (We can even hope for the detection of Hawking radiation [2].)

In our specific case [1] we chose a particular analogue spacetime model — that of a birefringent crystal [19] — providing us with explicit Lorentz symmetry breaking relations, which at the same time has a simple enough structure (the dispersion relations factorize, in general, to give a two-cone structure) so as to give us a chance to create in an intuitive way a full spacetime pseudo-Finsler geometry (with both metric and co-metric structures).

Birefringent crystals have in the past traditionally been analysed in terms of 3-dimensional *spatial* Finsler geometry. (See, for instance, Antonelli, Ingarden, and Matsumoto [15], and Duval [16].) But these analyses are purely *spatial* — they effectively pick a fixed inertial reference frame, the rest frame of the crystal, and hence pick a preferred $3d$ spacelike slices, so these approaches do not cross the boundaries of classical nonrelativistic mechanics. In contrast, we wish to construct a full $4d$ *spacetime* pseudo-geometry based on birefringent crystals — a fully $(3+1)$-dimensional construction.

2. Basics of pseudo-Finsler geometry

How should the theory of crystal optics provide us with a Lorentz symmetry breaking geometry? The theory gives us a clear ideas how the waves propagate, specified by the possible four-wavevectors $k = (\omega; \mathbf{k})$ with their corresponding phase velocities $\mathbf{v}_p = \omega\, \mathbf{k}/||\mathbf{k}||^2 = \omega\, \hat{\mathbf{k}}/||\mathbf{k}||$, and a clear idea about how the energy is transferred, specified by the possible group three-velocities $\mathbf{v}_g$. The group three-velocity can most naively be extended to a group four-velocity by the prescription $v = (1; \mathbf{v}_g)$. More generally, if we choose to parameterize the integral curves of the group velocity by some arbitrary parameter s, then the possible group four-velocities become $v = (1;\, \mathbf{v}_g) \times [\mathrm{d}t/\mathrm{d}s] = (v^0;\, \mathbf{v}) \propto (1;\, \mathbf{v}_g)$. These objects can be explicitly interpreted as describing how quanta of electromagnetic energy (hence photons) propagate. Note that any multiple of a possible four-wavevector is again a possible four-wavevector, and similarly by construction any multiple of a possible group four-velocity is again a possible group four-velocity.

So the spacetime geometry we are seeking should provide us with a pair of pseudo-norms $N(v)$ and $N'(k)$ specifying as null four-vectors exactly the possible group four-velocities and four-wavevectors respectively. That is, the pseudo-norm should give us the relation $N(v) = 0$ or $N'(k) = 0$ for the possible group four-velocities/four-wavevectors, and *only* for the possible group four-velocities/four-wavevectors.

Let us now precisely state the conditions we should impose on a pseudo-norm in order for it to be physically interesting: A pseudo-norm $N(v)$ on four-vectors is a map from $V \to i\mathbb{R}_+ \cup \mathbb{R}_+ \cup 0$, which is [1]:

a) 1-homogeneous,
b) the square $N^2(v)$ is smooth on $V/\{0\}$, and
c) the gradient $\partial_{v_i}[N^2(v)]$ is a bijection from $V \to V^*$, the vector-space dual to V.

This concept of pseudo-norm now specifies what we mean by pseudo-Finsler geometry [1].

To see that we can build up a full spacetime geometry out of this concept, we shall define a few other geometric concepts in complete analogy to the traditional positive definite Finsler case. The metric tensor is defined, as usual, by

$$g_{ij}(v) = \frac{1}{2}\, \frac{\partial^2[N^2(v)]}{\partial v^i\, \partial v^j}. \tag{2.1}$$

We see that this definition reduces to standard pseudo-Riemann case for

$$N(v) = \sqrt{g_{ij}\ v^i\ v^j}, \tag{2.2}$$

where (in this case) g_{ij} does not depend on v. By using Euler's theorem, which states that for n-homogeneous functions

$$\frac{\partial f(v)}{\partial v^i}\ v^i = n\ f(v), \tag{2.3}$$

we can easily define a mapping from vectors (four-velocities) to co-vectors (four-wavevectors) as

$$k_i(v) = \frac{\partial[N^2(v)]}{\partial v^i} = g_{ij}(v)\ v^j. \tag{2.4}$$

Then by using a Legendre transform, we can easily define the co-Finsler pseudo-norm $N'(k)$ as

$$N'(k(v)) = N(v), \tag{2.5}$$

and co-Finsler metric tensor as

$$g^{ij}(k(v)) = \frac{1}{2}\frac{\partial^2[N'^2(k(v))]}{\partial k^i(v)\ \partial k^j(v)}. \tag{2.6}$$

Then it can be proven that

$$v^i(k) = \frac{\partial[N'^2(k)]}{\partial k_i} = g^{ij}(k)\ k_j, \tag{2.7}$$

where $v(k)$ and $k(v)$ are inverse mappings: $v(k(v)) = v$. This shows that we have a meaningful generalization of pseudo-Riemann geometry. Note that we can easily define a pseudo-Finsler connection, and hence curvature, in this formalism simply by extremalizing the Finsler distance ($v = \mathrm{d}x/\mathrm{d}s$):

$$\ell(a,b) = \int_a^b \sqrt{g_{ij}(v)\ v^i\ v^j}\ \mathrm{d}s. \tag{2.8}$$

3. Birefringent crystal

3.1. General constraints

What are the specific physical constraints given by birefringent crystals on our geometry? We consider it meaningful to impose the following constraints:

a) The pseudo-norm should, for vectors approaching the purely timelike $(a, \mathbf{0})$ vector (in the fixed coordinates "at rest" with respect to the crystal), approach the pseudo-Riemann flat-space pseudo-norm.

b) The pseudo-norm should break Lorentz invariance as in a birefringent crystal.

c) Each "constant proper time" hypersurface, specified by $N^2(v) = -\tau_0^2$, should be connected, and should contain the point $(a, \mathbf{0})$.

Let us discuss these constraints a little bit: The first is just requirement that we want to recover low-energy physics as we know it (in a pseudo-Riemann form). The second is trivial: it is the specific contribution of using birefringent crystal as an analogue model (unlike the first and the third condition, which can be considered as generic). The third condition states that if we have a massive particle, it should not be able to have two distinct sets of four-velocities between which it is not able to undergo a smooth transition. It also means that any massive particle should be able to move arbitrarily slowly in the rest frame of the crystal.

3.2. Classical theory of birefringent crystals giving rise to co-metric structure

Now focus on the classical theory of birefringent crystals (see [19] for a detailed exposition). It is typically formulated in a preferred coordinate frame, which is the inertial frame of the crystal, with spatial coordinate axes aligned with the principal axes of the crystal. In fact in the rest fame the crystal is described by a 3×3 dielectric tensor ϵ_{ij}, where the principal axes are defined as axes where the dielectric tensor is diagonalized.

The relation between the components of the possible four-wavevectors can be extracted from the Fresnel equation for wave normals [1, 19]:

$$1 = \sum_{i=1}^{3} \frac{k_i^2}{\mathbf{k}^2 - \omega^2 \mu \epsilon_i}. \tag{3.9}$$

This factorizes into

$$F'(k) = F'_1\, F'_2 = \left(\omega^2 + b(\mathbf{k}) - \sqrt{b^2(\mathbf{k}) - c(\mathbf{k})}\right)\, \left(\omega^2 + b(\mathbf{k}) + \sqrt{b^2(\mathbf{k}) - c(\mathbf{k})}\right) = 0, \tag{3.10}$$

where

$$b(\mathbf{k}) = -\frac{k_1^2\, \epsilon_1(\epsilon_2 + \epsilon_3) + k_2^2\, \epsilon_2(\epsilon_1 + \epsilon_3) + k_3^2\, \epsilon_3(\epsilon_1 + \epsilon_2)}{2c^2 \mu \epsilon_1 \epsilon_2 \epsilon_3}, \tag{3.11}$$

and

$$c(\mathbf{k}) = \frac{\sum_{i=1}^{3} k_i^4\, \epsilon_i + k_1^2 k_2^2\, (\epsilon_1 + \epsilon_2) + k_1^2 k_3^2\, (\epsilon_1 + \epsilon_3) + k_2^2 k_3^2\, (\epsilon_2 + \epsilon_3)}{c^4 \mu^2 \epsilon_1 \epsilon_2 \epsilon_3}. \tag{3.12}$$

As we see, in the general case this relation encodes a two-cone structure, one for each polarization, each taking in general a distinct phase velocity $\mathbf{v}_p = \omega\, \mathbf{k}/||\mathbf{k}||^2 = \omega\, \hat{\mathbf{k}}/||\mathbf{k}||$. So we see that here the space-time causal structure is much more complicated than in special relativity.

3.3. Co-metric structure construction

Since the four-wavevector is in fact a covector, these relations give rise to a co-Finsler structure rather then to a Finsler structure. Hence they give us a generic candidate for the co-pseudo-Finsler norm, which can be written as $N'(k)^2 \propto F'(k)$ or more precisely $N'(k)^2 = M'(k)\, F'(k)$, where $M'(k)$ is an otherwise arbitrary function fulfilling following constraints:

a) $\dfrac{\partial [M'(k) F'(k)]}{\partial k^i}$ should be a bijection from $V^* \to V$;
b) $M'(k)$ is a -2 homogeneous map to $\mathbb{R}$;
c) smooth on $V^*/\{0\}$;
d) inside both signal cones negative, outside both positive, between them nonzero;
e) $M'(k)\, F'(k)$ should approximate the flat-space pseudo-Riemann norm (squared) in the given coordinates for k close to $(a, \mathbf{0})$ and such vectors should lie on some hypersurface (mass-shell) given by: $M'(k)\, F'(k) = -m^2$, which should be always connected.

These conditions follow trivially from the conditions we imposed on any physically meaningful Finslerian geometry.

3.4. Classical theory of birefringent crystals giving rise to metric structure

We should also consider the Finsler pseudo-norm; this should be extracted from the equations describing how energy propagates, which is the equation determining how photons propagate (their group velocities). This should be put into the form $g_{\mu\nu}(v)\; dx^\mu\; dx^\nu = 0$, so that it

Journal of Physics: Conference Series **189** (2009) 012037 doi:10.1088/1742-6596/189/1/012037

encodes the pseudo-Finsler metric tensor as opposed to co-Finsler metric tensor. The information regarding energy propagation is given by the Fresnel ray equation, which is:

$$\frac{t_1^2}{\frac{1}{v_r^2}-\frac{1}{v_2^2}}+\frac{t_3^2}{\frac{1}{v_r^2}-\frac{1}{v_1^2}}+\frac{t_3^2}{\frac{1}{v_r^2}-\frac{1}{v_3^2}}=0. \tag{3.13}$$

Here

$$v_i^2=\frac{1}{\mu\epsilon_i}. \tag{3.14}$$

Using the four vector $x=(t;\mathbf{x})$, the Fresnel ray equation can be transformed into

$$F(x)=F_1(x)\,F_2(x)=\left(t^2+b(\mathbf{x})-\sqrt{b^2(\mathbf{x})-c(\mathbf{x})}\right)\,\left(t^2+b(\mathbf{x})+\sqrt{b^2(\mathbf{x})-c(\mathbf{x})}\right)=0, \tag{3.15}$$

where

$$b(\mathbf{x})=-c^2\frac{v_1^2(v_z^2-v_2^2)\,x_1^2+v_2^2(v_1^2+v_3^2)\,x_2^2+v_3^2(v_1^2+v_2^2)\,x_3^2}{2v_1^2v_2^2v_3^2}, \tag{3.16}$$

and

$$c(\mathbf{x})=\frac{c^4\mathbf{x}^2\,\left(\sum_{i=1}^3 v_i^2 x_i^2\right)}{v_1^2v_2^2v_3^2}. \tag{3.17}$$

Obviously the the ray equation also encodes a two-cone structure.

3.5. Metric structure construction and interconnecting the two structures

The generic Finsler pseudo-norm is in this case expressed exactly in the same way as was the co-Finsler pseudo-norm in the previous case, by the function $N(v)^2=M(v)\ F(v)$, where $M(v)$ is again an arbitrary function fulfilling exactly the same conditions as the $M'(k)$ function, we just have to exchange V^* for V.

Now the last step in putting constraints on geometric construction is to relate the Finsler and co-Finsler structures. The items which are as yet undetermined are the functions $M(v)$, and $M'(k)$. So they have to fulfill the last condition, which is the condition of forming the full united geometry. This condition is given by the Legendre transform relation:

$$\left[N'\left(\frac{\partial[N(v)^2]}{\partial v^i}\right)\right]^2=N(v)^2. \tag{3.18}$$

In our language, the function $M(v)$ must be connected to $M'(k)$ by:

$$M'\left(\frac{\partial[M(v)F(v)]}{\partial v^i}\right)\ F'\left(\frac{\partial[M(v)F(v)]}{\partial v^i}\right)=M(v)\ F(v). \tag{3.19}$$

Thus we see that if we have somehow found an appropriate $M(v)$, then $M'(k)$ will be uniquely determined.

To find solutions of such an abstract (although precisely formulated) mathematical exercise is not an easy task. At this early stage it is quite doubtful whether general solutions exist. We shall look at particular cases, and try to check some intuitive solutions, since our original objective was to build up geometry in an intuitive way.

4. Attempts to solve our problem and conclusions

The simplest case of crystal optics is an isotropic medium $v_1 = v_2 = v_3 = v_o$. The obvious interpretation in this case is as a pseudo-Riemann metric, but with v_o playing the role of the speed of light [1, 19]. (Strictly speaking, even this obvious solution does not fulfill our condition (e), unless $v_o = c$, which implies there is no crystal at all.)

In a uniaxial crystal ($v_1 = v_2 \neq v_3$), the $F(v)$ and $F'(k)$ functions decompose into the product of two Riemannian structures [1], but simple and intuitive ansatzë for the $N(k)$ and $N'(k)$ functions lead to serious difficulties. For instance a suggestion where either

$$N(v)^2 = \sqrt{F(v)} = \sqrt{F_1(v) \cdot F_2(v)}, \qquad \text{or} \qquad N'(k)^2 = \sqrt{F'(k)} = \sqrt{F'_1(k) \cdot F'_2(k)}, \tag{4.20}$$

(with the two relations being related by Legendre transformation) encounters problems with condition (e), the putative Finsler norms are nonreal between the two signal cones, and not smooth at the signal cones. (Even second derivatives do not exist there, hence the pseudo-(co)-Finsler metric tensor is undefined there [1]).

In the case of biaxial crystal ($v_1 \neq v_2 \neq v_3 \neq v_1$) there are even bigger problems with this "solution" — even the individual $F_1(v)$, $F_2(v)$ and $F'_1(k)$, $F'_2(k)$ are not themselves pseudo-Riemannian structures; they are nonsmooth at the optical axes [1]. Obviously, the previous problems with the "unified" $\sqrt{F_1 \cdot F_2}$ or Legendre-equivalent $\sqrt{F'_1 \cdot F'_2}$ structures remain.

We could try to work out more sophisticated solutions to our "compatibility problem" (3.19), and at least prove existence. We have found a partial solution for this problem, except for proving the bijection condition, which is quite tricky. But the fact that this construction lacks any intuitive basis is quite disturbing. In these circumstances we have to accept the unwelcome conclusion that although birefringent crystal have given us many ideas about the construction of pseudo-Finsler geometries, they have not provided us with a clear and intuitive example of a "unified" geometry capable of handling both polarization modes simultaneously. So the original aim we posed at the beginning of this article has not been accomplished.

In conclusion: Our results are much less encouraging than we had originally hoped, and suggest that while pseudo-Finsler spacetimes are certainly useful constructs, in birefringent situations it does not appear possible to naturally and intuitively construct a "unified" pseudo-Finsler spacetime such that the pseudo-Finsler metric is null on both "signal cones", but has no other zeros or singularities — it seems physically more appropriate to think of physics as taking place in a single topological manifold that carries distinct pseudo-Finsler metrics, one for each polarization mode.

References

[1] J. Skakala and M. Visser, "The pseudo-Finslerian spacetimes of birefringent optics," arXiv:0806.0950 [gr-qc].
[2] M. Visser, "Acoustic propagation in fluids: An Unexpected example of Lorentzian geometry," arXiv:gr-qc/9311028.
M. Visser, "Acoustic black holes: Horizons, ergospheres, and Hawking radiation," Class. Quant. Grav. **15** (1998) 1767 [arXiv:gr-qc/9712010].
C. Barcelo, S. Liberati and M. Visser, "Analogue gravity," Living Rev. Rel. **8** (2005) 12 [arXiv:gr-qc/0505065].
[3] R. Gambini and J. Pullin, "Nonstandard optics from quantum spacetime," Phys. Rev. D **59** (1999) 124021 [arXiv:gr-qc/9809038].
M. Bojowald, H. A. Morales-Tecotl and H. Sahlmann, "On loop quantum gravity phenomenology and the issue of Lorentz invariance," Phys. Rev. D **71** (2005) 084012 [arXiv:gr-qc/0411101].
[4] N. E. Mavromatos, "Lorentz Invariance Violation from String Theory," arXiv:0708.2250 [hep-th].
N. E. Mavromatos, "Quantum gravity, cosmology, (Liouville) strings and Lorentz invariance," arXiv:hep-th/0210079.
V. A. Kostelecky and R. Potting, Phys. Lett. B **381**, 89 (1996) [arXiv:hep-th/9605088].
S. I. Vacaru, "Finsler and Lagrange Geometries in Einstein and String Gravity", arXiv:0801.4958 [gr-qc].

[5] J. Ambjorn, J. Jurkiewicz and R. Loll, "Emergence of a 4D world from causal quantum gravity," Phys. Rev. Lett. **93** (2004) 131301 [arXiv:hep-th/0404156].
J. Ambjorn, R. Loll, Y. Watabiki, W. Westra and S. Zohren, "A String Field Theory based on Causal Dynamical Triangulations," JHEP **0805** (2008) 032 [arXiv:0802.0719 [hep-th]].
[6] S. Judes and M. Visser, "Conservation Laws in Doubly Special Relativity," Phys. Rev. D **68** (2003) 045001 [arXiv:gr-qc/0205067].
S. Liberati, S. Sonego and M. Visser, "Interpreting doubly special relativity as a modified theory of measurement," Phys. Rev. D **71** (2005) 045001 [arXiv:gr-qc/0410113].
[7] A. G. Cohen and S. L. Glashow, "Very special relativity," Phys. Rev. Lett. **97** (2006) 021601 [arXiv:hep-ph/0601236].
G. W. Gibbons, J. Gomis and C. N. Pope, "General Very Special Relativity is Finsler Geometry," Phys. Rev. D **76** (2007) 081701 [arXiv:0707.2174 [hep-th]].
A. P. Kouretsis, M. Stathakopoulos and P. C. Stavrinos, "The General Very Special Relativity in Finsler Cosmology," arXiv:0810.3267 [gr-qc].
[8] F. Girelli and E. R. Livine, "Special Relativity as a non commutative geometry: Lessons for Deformed Special Relativity," arXiv:gr-qc/0407098.
A. Connes, M. R. Douglas and A. S. Schwarz, "Noncommutative geometry and matrix theory: Compactification on tori," JHEP **9802** (1998) 003 [arXiv:hep-th/9711162].
[9] G. F. R. Ellis and J. P. R. Uzan, "'c' is the speed of light, isn't it?," Am. J. Phys. **73** (2005) 240 [arXiv:gr-qc/0305099].
G. F. R. Ellis, "Note on varying speed of light cosmologies," Gen. Rel. Grav. **39** (2007) 511 [arXiv:astro-ph/0703751].
B. A. Bassett, S. Liberati, C. Molina-Paris and M. Visser, "Geometrodynamics of variable speed of light cosmologies," Phys. Rev. D **62** (2000) 103518 [arXiv:astro-ph/0001441].
S. Liberati, B. A. Bassett, C. Molina-Paris and M. Visser, "Chi Variable-Speed-Of-Light Cosmologies," Nucl. Phys. Proc. Suppl. **88** (2000) 259 [arXiv:astro-ph/0001481].
[10] G. Ellis, R. Maartens and M. A. H. MacCallum, "Causality and the speed of sound," Gen. Rel. Grav. **39** (2007) 1651 [arXiv:gr-qc/0703121].
[11] T. Jacobson, S. Liberati and D. Mattingly, "Lorentz violation at high energy: Concepts, phenomena and astrophysical constraints," Annals Phys. **321** (2006) 150 [arXiv:astro-ph/0505267].
T. A. Jacobson, S. Liberati, D. Mattingly and F. W. Stecker, "New limits on Planck scale Lorentz violation in QED," Phys. Rev. Lett. **93** (2004) 021101 [arXiv:astro-ph/0309681].
T. Jacobson, S. Liberati and D. Mattingly, "Lorentz violation and Crab synchrotron emission: A new constraint far beyond the Planck scale," Nature **424** (2003) 1019 [arXiv:astro-ph/0212190].
T. Jacobson, S. Liberati and D. Mattingly, "Threshold effects and Planck scale Lorentz violation: Combined constraints from high energy astrophysics," Phys. Rev. D **67** (2003) 124011 [arXiv:hep-ph/0209264].
T. Jacobson, S. Liberati and D. Mattingly, "TeV astrophysics constraints on Planck scale Lorentz violation," Phys. Rev. D **66** (2002) 081302 [arXiv:hep-ph/0112207].
[12] M. Visser, C. Barcelo and S. Liberati, "Acoustics in Bose-Einstein condensates as an example of broken Lorentz symmetry," arXiv:hep-th/0109033.
S. Weinfurtner, S. Liberati and M. Visser, "Modelling Planck-scale Lorentz violation via analogue models," J. Phys. Conf. Ser. **33** (2006) 373 [arXiv:gr-qc/0512127].
S. Liberati, M. Visser and S. Weinfurtner, "Naturalness in emergent spacetime," Phys. Rev. Lett. **96** (2006) 151301 [arXiv:gr-qc/0512139].
S. Weinfurtner, S. Liberati and M. Visser, "Analogue spacetime based on 2-component Bose-Einstein condensates," Lect. Notes Phys. **718** (2007) 115 [arXiv:gr-qc/0605121].
S. Weinfurtner, S. Liberati and M. Visser, "Analogue model for quantum gravity phenomenology," J. Phys. A **39** (2006) 6807 [arXiv:gr-qc/0511105].
P. Jain, S. Weinfurtner, M. Visser and C. W. Gardiner, "Analogue model of a FRW universe in Bose-Einstein condensates: Application of the classical field method," Phys. Rev. **A 76** (2007) 033616 [arXiv:0705.2077 [cond-mat.other]].
S. Weinfurtner, P. Jain, M. Visser and C. W. Gardiner, "Cosmological particle production in emergent rainbow spacetimes," arXiv:0801.2673 [gr-qc].
M. Visser and S. Weinfurtner, "Analogue spacetimes: Toy models for 'quantum gravity'," PoS **QG-Ph** (2008) 042 [arXiv:0712.0427 [gr-qc]].
S. Weinfurtner, M. Visser, P. Jain and C. W. Gardiner, "On the phenomenon of emergent spacetimes: An instruction guide for experimental cosmology," PoS **QG-Ph** (2008) 044 [arXiv:0804.1346 [gr-qc]].
M. Visser, "Emergent rainbow spacetimes: Two pedagogical examples," arXiv:0712.0810 [gr-qc].

[13] C. Barceló, S. Liberati and M. Visser, "Analog gravity from field theory normal modes?", Class. Quant. Grav. **18** (2001) 3595 [arXiv:gr-qc/0104001].
C. Barceló, S. Liberati and M. Visser, "Refringence, field theory, and normal modes", Class. Quant. Grav. **19** (2002) 2961 [arXiv:gr-qc/0111059].
M. Visser, C. Barceló and S. Liberati, "Bi-refringence versus bi-metricity", arXiv:gr-qc/0204017.

[14] P. Finsler, "Über Kurven und Flachen in allgemeinen Raumen", Dissertation, Gottingen, 1918, published by Verlag Birkhauser Basel, 1951.
H. Buseman, "The geometry of Finsler space", Bull. Amer. Math. Soc. **56**, Number 1, Part 1 (1950), 5-16.
H. Rund, *The differential geometry of Finsler spaces*, (Springer, Berlin, 1959).
D. Bao, S.S. Chern, and Z. Shen, *An introduction to Riemann–Finsler geometry*, (Springer, New York, 2000).
D. Bao, S.S. Chern, and Z. Shen (editors), *Finsler Geometry*, Proceedings of the Joint Summer Research Conference on Finsler Geometry, July 1995, Seattle, Washington, (American Mathematical Society, Providence, Rhode Island, 1996).
S.S. Chern, "Finsler geometry is just Riemannian geometry without the quadratic restriction", Not. Amer. Math. Soc. **43** (1996) 959–963.
Z. Shen, *Differential Geometry of Spray and Finsler Spaces*, (Kluwer, Dordrecht, 2001).
Z. Shen, *Lectures on Finsler Geometry*, (World Scientific, Singapore, 2001).
P.L. Antonelli and B.C. Lackey (editors), *The theory of Finslerian Laplacians and applications*, (Kluwer, Dordrecht, 1998).
A. Bejancu, *Finsler geometry and applications*, (Ellis Horwood, Chichester, England, 1990).

[15] P.L. Antonelli, R.S. Ingarden, M. Matsumoto, *The theory of sprays and Finsler spaces with applications in physics and biology*, (Springer [Kluwer], Berlin, 1993).

[16] C. Duval, "Finsler Spinoptics," Commun. Math. Phys. **283**, 701 (2008) [arXiv:0707.0200 [math-ph]].

[17] S. Liberati, "Quantum gravity phenomenology via Lorentz violations", PoS **P2GC** (2007) 018 [arXiv:0706.0142 [gr-qc]].
F. Girelli, S. Liberati and L. Sindoni, "Phenomenology of quantum gravity and Finsler geometry", Phys. Rev. D **75** (2007) 064015 [arXiv:gr-qc/0611024].
L. Sindoni, "The Higgs mechanism in Finsler spacetimes", arXiv:0712.3518 [gr-qc].
V. Perlick, "Fermat Principle in Finsler Spacetimes", Gen. Rel. Grav. **38** (2006) 365 [arXiv:gr-qc/0508029].
S. Siparov, Introduction to the problem of anisotropy in geometrodynamics, arXiv 0809.1817 [gr-qc].

[18] R. Punzi, M. N. R. Wohlfarth and F. P. Schuller, "Propagation of light and fluids in string backgrounds," arXiv:0711.3771 [hep-th].
F. P. Schuller and M. N. R. Wohlfarth, "Radiation-dominated area metric cosmology," JCAP **0712** (2007) 013 [arXiv:0705.4656 [hep-th]].
R. Punzi, F. P. Schuller and M. N. R. Wohlfarth, "Area metric gravity and accelerating cosmology," JHEP **0702** (2007) 030 [arXiv:hep-th/0612141].
F. P. Schuller and M. N. R. Wohlfarth, "Geometry of manifolds with area metric," Nucl. Phys. B **747** (2006) 398 [arXiv:hep-th/0508170].

[19] M. Born and E. Wolf, *Principles of Optics*, 7th edition, (Cambridge University Press, England, 1999).

Alfvén QPOs in Magnetars

Hajime Sotani[1], Kostas D. Kokkotas[1,2], and Nikolaos Stergioulas[2]
[1]Theoretical Astrophysics, University of Tübingen, 72076, Tübingen, Germany
[2]Department of Physics, Aristotle University of Thessaloniki, Thessaloniki 54124, Greece
E-mail: sotani@astro.auth.gr

Abstract. We investigate torsional Alfvén oscillations of relativistic stars with a global dipole magnetic field, via 2D numerical simulations. We find that a) there exist two families of quasi-periodic oscillations (QPOs) with harmonics at integer multiples of the fundamental frequency, b) the lower-frequency QPO is related to the region of closed field lines, near the equator, while the higher-frequency QPO is generated near the magnetic axis, c) the QPOs are long-lived, d) for the chosen form of dipolar magnetic field, the frequency ratio of the lower to upper fundamental QPOs is about 0.6, independent of the equilibrium model or of the strength of the magnetic field, and e) within a representative sample of EOSs and of various magnetar masses, the Alfvén QPO frequencies are given by accurate empirical relations that depend only on the compactness of the star and on the magnetic field strength. Compared to the observational frequencies, we also obtain an upper limit on the strength of magnetic field of SGR 1806-20 (if it is dominated by a dipolar component) between ~ 3 and 7×10^{15} Gauss.

1. Introduction

The phenomenon of Soft Gamma Repeaters (SGRs) may allow us in the near future to determine fundamental properties of strongly magnetized, compact stars. SGRs produce giant flares with peak luminosities of $10^{44} - 10^{46}$ erg/s, which display a decaying tail for several hundred seconds. Up to now, three giant flares have been detected, SGR 0526-66 in 1979, SGR 1900+14 in 1998, and SGR 1806-20 in 2004. The timing analysis of the latter two events revealed several quasi-periodic oscillations (QPOs) in the decaying tail, whose frequencies are approximately 18, 26, 30, 92, 150, 625, and 1840 Hz for SGR 1806-20, and 28, 53, 84, and 155 Hz for SGR 1900+14 [1]. Some of these oscillation frequencies are similar to those of torsional modes of the solid crust of a compact star. This observation is in support of the proposal that SGRs are magnetars (compact objects with very strong magnetic fields) [2]. During an SGR event, torsional oscillations in the solid crust of the star could be excited [3], leading to the observed frequencies in the X-ray tail. Spherical stars have generally two types of oscillations, *spheroidal* with polar parity and *toroidal* with axial parity. The observed QPOs in SGR X-ray tails may originate from toroidal oscillations, since these could be excited more easily than poloidal oscillations, because they do not involve density variations. In Newtonian theory, there have been several investigations of torsional oscillations in the crust region of neutron stars (see e.g., [4] for references), while only few studies have taken general relativity into account [5, 6, 7, 8, 9, 10]. On the other hand, as an alternative possibility for explaining the QPO frequencies, one can also consider global Alfvén oscillations. Levin stressed the importance of crust-core coupling by a global magnetic field and of the existence of an Alfvén continuum [11], while Glampedakis et al. considered a model with simplified geometry, in which Alfvén oscillations form a discrete spectrum of normal modes, that could be associated with the observed low-frequency QPOs [12]. In [13], the existence of a continuum was stressed further and it was shown that the edges or turning points of the continuum can yield long-lived QPOs. In addition, numerical simulations showed that drifting QPOs within the continuum become amplified near the frequencies of the crustal normal modes. Within this model, Levin suggested a likely identification of the 18Hz QPO in SGR 1806-20 with the lowest frequency of the MHD continuum or its first overtone. The above results were obtained in toy models with simplified geometry and Newtonian gravity, i.e., still we do not know how the Alfvén oscillations behave in a realistic magnetar models. Thus in this article, we perform two-dimensional numerical simulations of linearized Alfvén oscillations in magnetars. Our model improves on the previously considered toy models in various ways: relativistic gravity is assumed, various realistic equations of state (EOSs) are considered and a dipolar magnetic field is constructed. We do not consider the presence of a solid crust, but only examine the response of the ideal magnetofluid to a chosen initial perturbation. Unless otherwise noted, we adopt units of $c = G = 1$, where c and G denote the speed of light and the gravitational constant, respectively, while the metric signature is $(-,+,+,+)$. Extensive details of our work were presented in [14].

2. Magnetar Models

We adopt the ideal MHD approximation, for which the electric fields are zero for a comoving observer. And the stellar deformation due to the magnetic field is also neglected, because the magnetic energy is much smaller than the gravitational binding energy even for the magnetar models considered here. Therefore, the equilibrium configuration is static and spherically symmetric. To a non-magnetized TOV model, we superimpose an axisymmetric, dipolar magnetic field.

The perturbation equations can be derived by linearizing the equations of motion and Maxwell's equations. We focus only on the axisymmetric axial perturbations, i.e., $m = 0$, which are independent from the polar perturbations, under our assumptions. Additionally, we adopt the Cowling approximation by neglecting the metric perturbations, i.e. we assume $\delta g_{\mu\nu} = 0$.

3. Torsional Oscillations of the Crust

Before showing the results for the Alfvén oscillations, we briefly review the properties of torsional oscillations of the crust, ignoring the coupling with the core region, for which details can be found in [7]. The effect of magnetic fields on crustal oscillations becomes important for $B > B_\mu$, where $B_\mu = 4 \times 10^{15}$ Gauss. The frequencies for fundamental modes depend strongly on the stellar properties, differing by about $30 \sim 50$ %. On the other hand, frequencies of overtones are nearly independent of the harmonics index ℓ, but depend strongly on both the EOS for crust and for the core region. Furthermore, using our numerical results for a large number of stellar models with different EOSs and different stellar masses, we derive the following empirical formula for the frequencies of torsional oscillations of the crust

$$ {}_\ell t_n \simeq \left[1 + {}_\ell\alpha_n \left(\frac{B}{B_\mu}\right)^2\right]^{1/2} {}_\ell t_n^{(0)}, \tag{1} $$

where ${}_\ell t_n^{(0)}$ is the corresponding frequency in the limit of no magnetic field and ${}_\ell\alpha_n$ and is a coefficient that depends on the properties of the stellar model. From these calculations we find that the attempt to explain the observed frequencies of QPOs is partially successful, but also notice that it may not be possible to explain all data by using only torsional oscillations of the crust. Especially the observed frequencies of 18, 26 and 30 Hz in SGR 1806-20 are difficult to be made compatible with only pure crustal frequencies, because the differences between these observed frequencies are smaller than the frequency spacing of the crustal torsional oscillations with different harmonics index ℓ. This is one of the motivations to examine global Alfvén oscillation.

4. Alfvén Oscillations

In order to examine the global Alfvén oscillations, we consider the magnetar models with axisymmetric poloidal magnetic field and without crust regions. For out numerical calculations we add a Kreiss-Oliger numerical dissipation [15]. Then, defining our main perturbation variable $\mathcal{Y}$ trough $\delta u^\phi \sim \partial_t \mathcal{Y}(t,r,\theta)$, (where u^ϕ is the ϕ-component of the 4-velocity) we are led to a single, two-dimensional, wave-like evolutionary equation

$$\mathcal{A}_{tt}\frac{\partial^2\mathcal{Y}}{\partial t^2} = \mathcal{A}_{20}\frac{\partial^2\mathcal{Y}}{\partial r^2} + \mathcal{A}_{11}\frac{\partial^2\mathcal{Y}}{\partial r\partial\theta} + \mathcal{A}_{02}\frac{\partial^2\mathcal{Y}}{\partial\theta^2} + \mathcal{A}_{10}\frac{\partial\mathcal{Y}}{\partial r} + \mathcal{A}_{01}\frac{\partial\mathcal{Y}}{\partial\theta} + \epsilon_D\mathcal{D}_4\mathcal{Y}, \tag{2}$$

where the coefficients $\mathcal{A}_{tt}$, $\mathcal{A}_{20}$, $\mathcal{A}_{11}$, $\mathcal{A}_{02}$, $\mathcal{A}_{10}$, and $\mathcal{A}_{01}$ are functions of the equilibrium properties and $\epsilon_D\mathcal{D}_4\mathcal{Y}$ corresponds to the 4th-order Kreiss-Oliger dissipation (see [14] for the detail of this equation). With appropriate boundary conditions we can thus obtain the time-evolution of oscillations throughout the magnetar interior. Fourier Transforms of this time-evolution at various points inside the star are shown in figure 1. At a first glance we can see the same peaks on each figure, whose fundamental frequency ($n = 0$) is around 41 Hz, with additional peaks being integer multiples of the fundamental. But for the panel at $\theta \sim \pi/2$, i.e., near the equatorial plane, we can see the additional peaks, for which the fundamental frequency is around 25 Hz. Using Levin's toy model for the Alfvén continuum in magnetars [13], these results can be interpreted as the frequency peaks of two families of QPOs generated by the edges or turning points of an Alfvén continuum with the addition of overtones. Hereafter, we call these two families lower and upper QPOs and label n-th overtones as L_n and U_n correspondingly, with L_0 and U_0 being the fundamental QPOs.

In the left panel of Figure 2 we plot the phase of these Alfvén oscillations, for a representative case. From this figure one can see that the oscillations near the axis ($\theta = 0$) are practically at the same phase, while at other points inside the star the phase differs. This means that the Alfvén oscillations are not discrete normal modes, but form a continuum. The continuum can be seen more strongly near the equatorial plane. In the right panel of Figure 2 we show the evolution of $\partial_t\mathcal{Y}$ at the location inside the star where the effective amplitude of the fundamental upper QPO, U_0, attains its maximum value (see [14] for our definition of the effective amplitude). It is evident that, even though the oscillations are not discrete, the fundamental upper QPO is long-lived, since the amplitude of the oscillations barely diminishes with time.

5. Realistic EOS and Empirical Formula

We have obtained the lower and upper Alfvén QPO frequencies for a representative sample of magnetar models with realistic EOSs and various masses. These results are summarized in Table 1. In this Table we also show the ratio of the lower to upper QPO frequencies and we can see that these ratios are almost 0.6, independently of the stellar models. Additionally, we can observe that the overtones are nearly integer multiples of the fundamental frequencies such that

$$f_{L_n} \simeq (n+1)f_{L_0} \quad \text{and} \quad f_{U_n} \simeq (n+1)f_{U_0}. \tag{3}$$

Furthermore, we found that the frequencies of the Alfvén QPOs scale linearly with the strength of the magnetic field, at least for magnetic field strengths of order B_μ. Through these numerical results, the frequencies of all lower and upper Alfvén QPOs are given by the following two

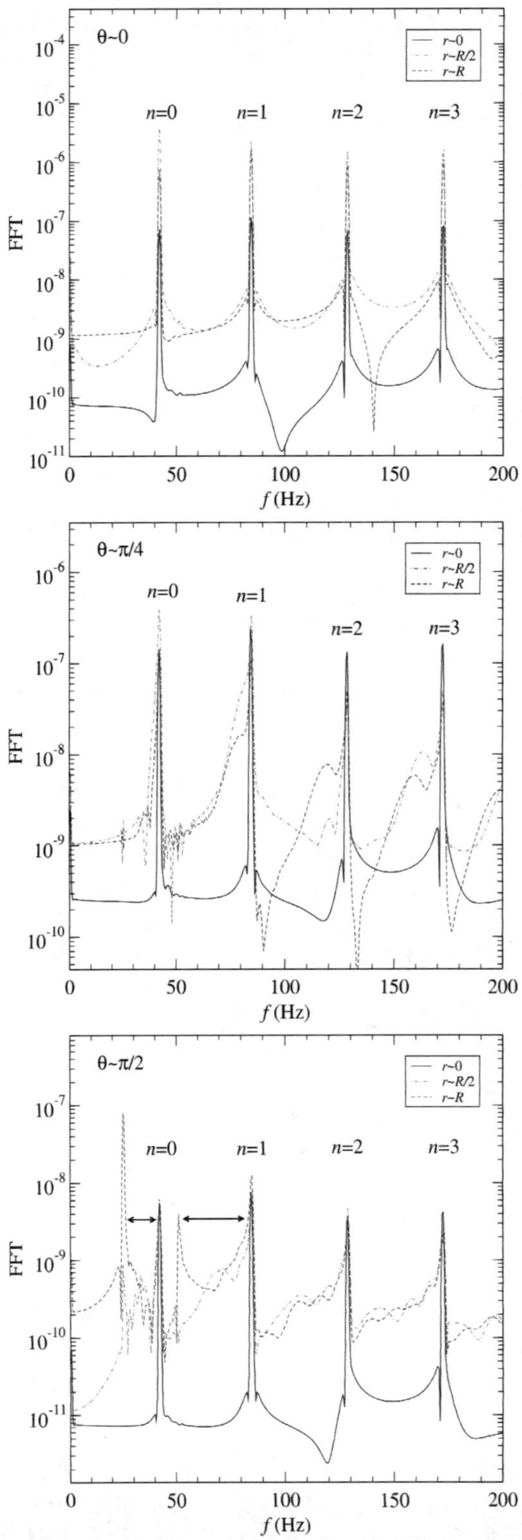

Figure 1. Fourier transform obtained from the 2D simulation with $B = B_{\mu}$, whose evolution time is 2 seconds. In the figures the different lines correspond to the different radial observers of inside star, at $r \sim 0$, $r \sim R/2$, and $r \sim R$ and the different panels correspond to different angular observers inside the star, at $\theta \sim 0$, $\theta \sim \pi/4$, and $\theta \sim \pi/2$.

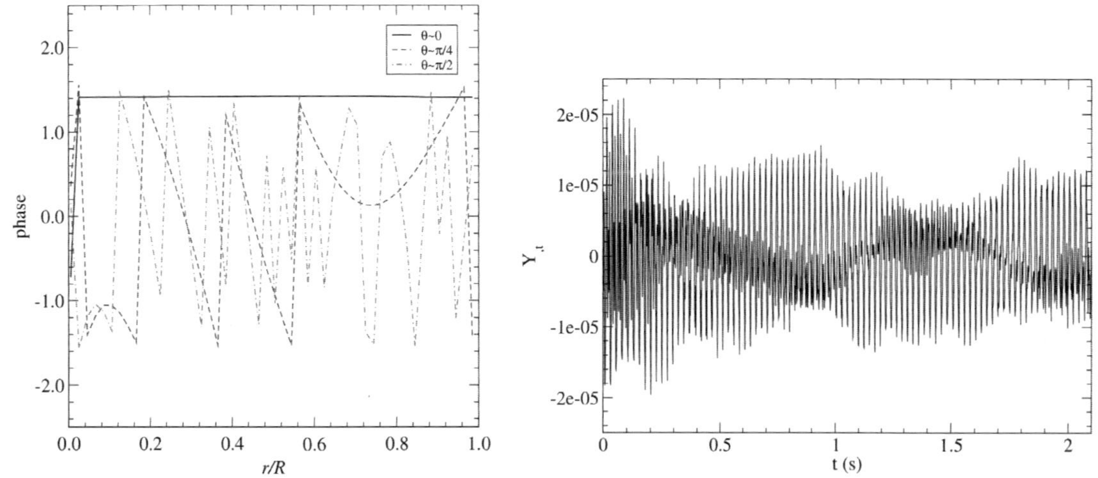

Figure 2. The left panel shows the phase of the Alfvén oscillations at different angular positions, while the right panel shows the time evolution of $\partial_t \mathcal{Y}$ at a point where at the location inside the star where the effective amplitude of the fundamental upper QPO, U_0, attains its maximum value.

Table 1. Frequencies of lower and upper Alfvén QPOs and their ratios, for a representative sample of equilibrium models, constructed with various EOSs and masses and for $B = B_\mu$.

Model	M/R	f_{L_0} (Hz)	f_{U_0} (Hz)	ratio	f_{L_1} (Hz)	f_{U_1} (Hz)	ratio	f_{U_2} (Hz)
A+DH_{14}	0.218	15.4	25.0	0.616	30.7	49.4	0.621	74.4
A+DH_{16}	0.264	11.7	18.3	0.639	23.5	35.7	0.658	54.0
WFF3+DH_{14}	0.191	17.9	29.8	0.601	36.2	59.2	0.611	89.8
WFF3+DH_{18}	0.265	11.7	18.0	0.650	23.5	35.5	0.662	53.3
APR+DH_{14}	0.171	20.4	34.1	0.598	41.3	68.6	0.602	104.6
APR+DH_{20}	0.248	12.8	20.6	0.621	26.0	40.3	0.645	61.0
L+DH_{14}	0.141	23.7	40.8	0.581	47.5	81.6	0.582	123.8
L+DH_{20}	0.199	16.4	27.8	0.590	33.1	54.7	0.605	82.6

empirical relations as

$$f_{L_n}(\mathrm{Hz}) \simeq 48.9(n+1)\left[1 - 4.51\left(\frac{M}{R}\right) + 6.18\left(\frac{M}{R}\right)^2\right]\left(\frac{B}{B_\mu}\right), \tag{4}$$

$$f_{U_n}(\mathrm{Hz}) \simeq 86.1(n+1)\left[1 - 4.58\left(\frac{M}{R}\right) + 6.06\left(\frac{M}{R}\right)^2\right]\left(\frac{B}{B_\mu}\right). \tag{5}$$

Updated empirical relations for improved boundary conditions can be found in [16, 17].

6. Constraints on Magnetic Field Strength

Using three empirical formulas, i.e., one for the frequencies of crustal torsional oscillations (1) and two for the Alfvén QPO frequencies, (4) and (5) we can attempt to identify several of the low-frequency QPOs observed in the X-ray tail of SGR 1806-20. In particular, one could identify the 18Hz and 30Hz observed frequencies with the fundamental lower and upper QPOs, correspondingly, while the observed frequencies of 92Hz and 150Hz would then be integer multiples of the fundamental upper QPO frequency (three times and five times, correspondingly). With such an identification, our empirical relations (4) and (5) constrain the magnetic field strength of SGR 1806-20 (if it is dominated by a dipolar component) to be between 3×10^{15}Gauss and 7×10^{15}Gauss. Furthermore, an identification of the observed frequency of 26Hz with the frequency of the fundamental torsional $\ell=2$ oscillation of the magnetar's crust implies a very stiff equation of state and a mass of about 1.4 to $1.6M_\odot$. For example, for the $1.4M_\odot$ model constructed with EOS L+DH, one obtains the following frequencies: ${}_2t_0 = 25.8$Hz, $f_{L_0} = 17.5$Hz, $f_{U_0} = 30.0$Hz, $f_{U_3} = 90.1$Hz and $f_{U_5} = 150.2$Hz, for $B = 2.94\times10^{15}$ Gauss. Alternatively, one could also identify the 18Hz and 30Hz observed frequencies with overtones (which are also at a near 0.6 ratio). In this case, the strength of the magnetic field derived above is only an upper limit and the actual magnetic field may be weaker. Then, if one assumes that the observed frequency of 26Hz is due to the fundamental $\ell=2$ crust mode for a weak magnetic field, our numerical data agree best with a $1.4M_\odot$ model constructed with an EOS of moderate stiffness. For example, for the $1.4M_\odot$ model constructed with the APR+DH EOS one obtains ${}_2t_0 = 25.9$Hz, $f_{L_1} = 17.7$Hz, $f_{U_1} = 30.0$Hz, $f_{U_5} = 90.1$Hz and $f_{U_9} = 150.1$Hz, for $B = 1.77\times10^{15}$ Gauss.

7. Discussion

Using torsional oscillations of the magnetar models, we try to explain observed frequencies of QPOs in SGRs. This attempt is partially successful, when one considers only crustal torsional oscillations. To arrive at a more consistent model, we also consider global Alfvén oscillations, finding a continuum that allows for two families of QPOs. The QPOs produced near the magnetic axis are long-lived. Furthermore, we examined the possibility of explaining the observed data, using empirical formulas for the frequencies of crustal torsional oscillations and for global the Alfvén QPOs. Such a fit can constrain the strength of the magnetic field. In [16, 17], this investigation is continued, using improved boundary conditions and arriving at a simplified description of Alfvén oscillations along individual magnetic field lines.

References

[1] Watts A L and Strohmayer T E 2006 *Proceedings of the 2006 Beijing COSPAR assembly* in press
[2] Duncan R C and Thompson C 1992 *ApJ* **392** L9
[3] Duncan R C 1998 *ApJ* **498** L45
[4] Lee U 2007 *Mon. Not. R. Astron. Soc.* **374** 1015
[5] Messios N, Papadopoulos D B, and Stergioulas N 2001 *Mon. Not. R. Astron. Soc.* **328** 1161
[6] Samuelsson L and Andersson N 2007 *Mon. Not. R. Astron. Soc.* **374** 256
[7] Sotani H, Kokkotas K D, and Stergioulas N 2007 *Mon. Not. R. Astron. Soc.* **375** 261
[8] Sotani H, Kokkotas K D, Stergioulas N, and Vavoulidis M 2006 (*Preprint* astro-ph/0611666)
[9] Vavoulidis M, Stavridis A, Kokkotas K D, and Beyer H 2007 *Mon. Not. R. Astron. Soc.* **377** 1553
[10] Sotani H, Colaiuda A, and Kokkotas K D 2008 *Mon. Not. R. Astron. Soc.* **385** 2161
[11] Levin Y 2006 *Mon. Not. R. Astron. Soc.* **368** L35
[12] Glampedakis K, Samuelsson L, and Andersson N 2006 *Mon. Not. R. Astron. Soc.* **371** L74
[13] Levin Y 2007 *Mon. Not. R. Astron. Soc.* **377** 159
[14] Sotani H, Kokkotas K D, and Stergioulas N 2008 *Mon. Not. R. Astron. Soc.* **385** L5
[15] Kreiss H -O and Oliger J 1973 *Methods for the approximate solution of time dependent problems* Series 10 (GARP Publications)
[16] Colaiuda A, Beyer H, and Kokkotas K D 2009 *Mon. Not. R. Astron. Soc.* **396** 1441
[17] Cerdá-Durán P, Stergioulas N, and Font, J A 2009 *Mon. Not. R. Astron. Soc.* **397** 1607

Journal of Physics: Conference Series **189** (2009) 012039 doi:10.1088/1742-6596/189/1/012039

6+1 lessons from $f(R)$ gravity

Thomas P. Sotiriou

Center for Fundamental Physics, University of Maryland, College Park, MD 20742-4111, USA

E-mail: sotiriou@umd.edu

Abstract. There has been a recent stimulus in the study of alternative theories of gravity lately, mostly triggered from combined motivation coming from cosmology/astrophysics and high energy physics. Among the proposed theories, one that has attracted much attention is $f(R)$ gravity. It is certainly debatable whether such a simplistic modification of General Relativity can constitute a viable alternative theory of gravitation. However, it is quite straightforward to see the merits of such a theory when viewed as a toy theory whose role is to help us understand the implications and difficulties of beyond-Einstein gravity. Under this perspective, I review some of the main lessons we seem to have learned from the study of $f(R)$ gravity in the recent past.

1. Introduction

Modifications of General Relativity (GR) by including higher order curvature invariants in the gravitational action have a long history. Indeed, it was just 1919 when Weyl, and 1922 when Eddington (the very man that three years earlier had provided the first experimental verification of GR by measuring light bending during a solar eclipse) started considering such terms in the action [1, 2]. These early attempts were mostly driven by scientific curiosity and a need to question and, therefore, understand the then newly proposed theory. Thus, they were quickly abandoned. However, by the 1960's and 1970's new motivation coming from theoretical physics revived the field of higher-order gravity theories: Utiyama and De Witt showed that renormalization at one-loop demands that the Einstein–Hilbert action be supplemented by higher order curvature terms [3]. Later on, Stelle showed that higher order actions are renormalizable (but not unitary) [4] (see [5] for a historical review and a list of references to early work). More recent results show that when quantum corrections or string theory are taken into account, the effective low energy gravitational action admits higher order curvature invariants [6, 7, 8]. However, the relevance of such terms in the action was considered to be restricted to very strong gravity regimes and they were expected to be strongly suppressed by small couplings, as one would expect when simple effective field theory considerations are taken into account.

Recently, a new stimulus appeared in higher-order theories of gravity. This time the motivation came from cosmological and astrophysical observations. The latest datasets coming from different sources, such as the Cosmic Microwave Background Radiation (CMBR) and supernovae surveys, seem to indicate that the energy budget of the universe is 4% ordinary baryonic matter, 20% *dark matter* and 76% *dark energy* [9, 10, 11, 12]. The term dark matter refers to an unknown form of matter, which has the clustering properties of ordinary matter but has not yet been detected in the laboratory. The term dark energy is reserved for an unknown form of energy which not only has not been detected directly, but also does not cluster as ordinary

Journal of Physics: Conference Series **189** (2009) 012039 doi:10.1088/1742-6596/189/1/012039

matter does. The simplest model that fits the data adequately, is the so called concordance model or ΛCDM model (supplemented with some inflationary scenario), where a cosmological constant plays the role of dark energy and the nature of dark matter remains undetermined apart from the fact that it does not interact but gravitationally. However, even in the most optimistic scenario in which the nature of dark matter will soon be determined in some accelerator, the known cosmological constant problems [13, 14] would still plague the concordance model. One of the alternatives is to consider the unexpected observation as shortcoming of our macroscopic description of gravity itself: gravity should be by far the dominant interaction at cosmological scales so, could it be that a modification of GR could account for the unexplained observations?

It is tempting to combine the earlier motivation for corrections to GR coming from high energy physics with the more recent cosmological/astrophysical stimulus. This is indeed often done in the literature by proposing that some theory of gravity, which comes about as a low energy limit of a more fundamental theory and includes both ultraviolet and infrared corrections, could maybe account for all or some of the unexplained observations. However, there are two loopholes in such a suggestion: a) It is not clear exactly which one the fundamental theory should be or what is its low energy limit, and b) the length scales at which we expect high-energy inspired corrections to GR to be important are orders of magnitude smaller than the length scales at which our puzzling observations have been made.

These two issues can be seen as caveats against modified gravity; or, in a more optimistic approach, they can be seen as opportunities: Problems, such as the "unnatural" value of the cosmological constant have not found a solution within the realms of conventional (or even slightly unconventional) thinking. Therefore, they might as well be indications that something is wrong with the very core of this way of thinking. The large ambiguities on how gravity works at small scales and which is the fundamental theory describing it seem to leave enough room for such speculations (indeed there are results already in the literature claiming that terms responsible for late time gravitational phenomenology might be predicted by some more fundamental theory, such as string theory [15]).

What seems to be the best way to proceed is probably the usual way when we are diving into the unknown in physics: consider a toy theory, probably of a very simple form, and use it as a tool to explore the limitations of the theory being questioned, in our case GR. This would allow us to test whether the very idea of modified gravity accounting for unexplained cosmological observations can be correct in principle, or what is the relation between the modifications required for that and those predicted by high energy physics.

$f(R)$ theories of gravity seem to be very good candidates for such toy theories. They come about by straightforwardly generalizing the Lagrangian in the Einstein–Hilbert Lagrangian to be a general function f of the Ricci scalar R, so that the action becomes

$$S = \frac{1}{2\kappa}\int d^4x\sqrt{-g}\,f(R), \tag{1}$$

where $\kappa \equiv 8\pi G$, G is the gravitational constant, g is the determinant of the metric and R is the Ricci scalar ($c = \hbar = 1$). $f(R)$ theories of gravity make good toy theories for two reasons: a) They are sufficiently general to encapsulate some of the basic characteristics of higher-order gravity, but at the same time they are simple enough to be easy to handle; b) there are unique among higher-order gravity theories, in the sense that they seem to be the only ones which can avoid the long known and fatal Ostrogradski instability [16].

Now, if we are to consider $f(R)$ gravity as a toy theory whose main scope was to help us learn something about beyond GR gravity, the obvious question to ask is what did we really learn so far? I attempt to answer this question here by going through some "lessons" coming from $f(R)$ gravity (hence, the title). This talk is based on [17], where a longer discussion on the motivation of $f(R)$ gravity, as well as all on the other issues that are discussed here, is given (see also [18]).

2. What have we learned from $f(R)$ gravity?

2.1. Lesson 1: Choice of fundamental variables is crucial

There are actually three versions of $f(R)$ gravity: Metric $f(R)$ gravity, Palatini $f(R)$ gravity, and metric-affine $f(R)$ gravity. In metric $f(R)$ gravity, the action

$$S_{met} = \frac{1}{2\kappa}\int d^4x\sqrt{-g}\, f(R) + S_M(g_{\mu\nu},\psi), \tag{2}$$

where a matter action S_M has been included and ψ collectively denotes the matter fields, is varied with respect to the metric as usual. This yields the field equations

$$f'(R)R_{\mu\nu} - \frac{1}{2}f(R)g_{\mu\nu} - \left[\nabla_\mu\nabla_\nu - g_{\mu\nu}\Box\right] f'(R) = \kappa\, T_{\mu\nu}, \tag{3}$$

where

$$T_{\mu\nu} = \frac{-2}{\sqrt{-g}}\frac{\delta S_M}{\delta g^{\mu\nu}}, \tag{4}$$

a prime denotes differentiation with respect to the argument, ∇_μ is the covariant derivative associated with the Levi-Civita connection of the metric, and $\Box \equiv \nabla^\mu\nabla_\mu$. Clearly, these equation are fourth order differential equations in the metric.

Palatini $f(R)$ gravity comes about from the same action if we decide to treat the connection as an independent quantity. In this case, for the sake of clarity, we denote as $\mathcal{R}_{\mu\nu}$ the Ricci tensor constructed with this independent connection. $\mathcal{R} = g^{\mu\nu}\mathcal{R}_{\mu\nu}$ is the corresponding Ricci scalar and the action takes the form:

$$S_{pal} = \frac{1}{2\kappa}\int d^4x\sqrt{-g}\, f(\mathcal{R}) + S_M(g_{\mu\nu},\psi). \tag{5}$$

Independent variation with respect to the metric and the connection gives, after some manipulations, the following field equations

$$f'(\mathcal{R})\mathcal{R}_{(\mu\nu)} - \frac{1}{2}f(\mathcal{R})g_{\mu\nu} = \kappa G\, T_{\mu\nu}, \tag{6}$$

$$\bar{\nabla}_\lambda\left(\sqrt{-g}f'(\mathcal{R})g^{\mu\nu}\right) = 0, \tag{7}$$

where $T_{\mu\nu}$ is defined in the usual way as in eq. (4) and $\bar{\nabla}_\mu$ denotes the covariant derivative defined with the independent connection. Notice that when $f(\mathcal{R}) = \mathcal{R}$, eq. (7) becomes the definition of the Levi-Civita connection and eq. (6) becomes Einstein's equation. So Palatini and metric variations both lead to GR for an action linear in the Ricci scalar. However, they lead to different theories for a more general action. In this sense, it is not straightforward which variation on should choose when generalizing the Lagrangian to be a general function of the Ricci scalar. Both metric and Palatini $f(R)$ gravity was first rigorously studied in [19].

In Palatini $f(R)$ gravity the connection, although independent, does not enter the matter action (this means that Palatini $f(R)$ gravity is a metric theory, according to the definition of [20]). This implies that the covariant derivative of the matter fields which are usually present in the matter action are defined with the Levi-Civita connection (alternatively, the theory would include only very specific matter which does not couple to the connection, such as scalar fields, but this is not really an option for a gravity theory) [21, 22, 23]. Additionally, the independent connection is really an auxiliary field: eq. (7) can be solved algebraically for the connection to give

$$\Gamma^\lambda{}_{\mu\nu} = \frac{1}{f'(\mathcal{R})}g^{\lambda\sigma}\Big[\partial_\mu\left(f'(\mathcal{R})g_{\nu\sigma}\right) + \partial_\nu\left(f'(\mathcal{R})g_{\mu\sigma}\right) - \partial_\sigma\left(f'(\mathcal{R})g_{\mu\nu}\right)\Big]. \tag{8}$$

At the same time, a contraction of eq. (6) with the metric gives

$$f'(\mathcal{R})\mathcal{R} - 2f(\mathcal{R}) = \kappa T, \tag{9}$$

where $T = g^{\mu\nu}T_{\mu\nu}$. for a given f this is an algebraic equation which can be solved to express $\mathcal{R}$ in terms of T.[1] Thus, using eqs. (8) and (9) one can algebraically express the independent connection in terms of the metric and the matter fields and eliminate it from the field equations.

The way to give more substance to the connection is to restore its geometrical role and use it to define the covariant derivatives in the matter action, as proposed in [22]. In this case the matter action would be of the form $S_M(g_{\mu\nu}, \Gamma^{\lambda}{}_{\mu\nu}, \psi)$. Such theories are called metric-affine $f(R)$ theories. We refrain from giving their field equations here as we will not use them in what comes next. We refer the reader to [22] for more details.

What should be clear after this discussion is that it is crucial to pick, not only the action, but also the fundamental fields which will be varied and assign to them a precise meaning. Remarkably, the distinction between Palatini and metric variation, which (at least at a classical level) was of little importance for the Einstein–Hilbert action, is critical for more general actions.

2.2. Lesson 2: Equivalence with Brans–Dicke theory

Metric $f(R)$ gravity appears as an extension of GR that includes no extra fields mediating gravity. Palatini $f(R)$ gravity appears to include a independent connection, so one could be trick to think that the underlying geometry is not pseudo-Riemannian. We already argued that the independent connection is not really related to the geometry (at least as the latter is felt by the matter field which act as our probe). So appearances can deceive; and they do: both metric and Palatini $f(R)$ can be written as specific versions of Brans–Dicke theory [25, 26, 27, 28, 29, 30, 26, 31, 32, 23]. Let us see how this is possible.

Starting from the action (2) for metric $f(R)$ gravity, one can introduce a new field χ and write the dynamically equivalent action

$$S_{met} = \frac{1}{2\kappa}\int d^4x\sqrt{-g}\left[f(\chi) + f'(\chi)(R-\chi)\right] S_M(g_{\mu\nu}, \psi). \tag{10}$$

Variation with respect to χ leads to the equation

$$f''(\chi)(R-\chi) = 0. \tag{11}$$

Therefore, $\chi = R$ if $f''(\chi) \neq 0$, which reproduces the action (2). Redefining the field χ by $\phi = f'(\chi)$ and setting

$$V(\phi) = \chi(\phi)\phi - f(\chi(\phi)), \tag{12}$$

the action takes the form

$$S_{met} = \frac{1}{2\kappa}\int d^4x\sqrt{-g}\left[\phi R - V(\phi)\right] + S_M(g_{\mu\nu}, \psi). \tag{13}$$

This is the Jordan frame representation of the action of a Brans–Dicke theory with Brans–Dicke parameter $\omega_0 = 0$ (also called "O'Hanlon action" as it was originally proposed by O'Hanlon in [33]).

We could follow exactly the same steps starting from action (5) for Palatini $f(R)$ gravity. We would then arrive to the action

$$S_{pal} = \frac{1}{2\kappa}\int d^4x\sqrt{-g}\left[\phi\mathcal{R} - V(\phi)\right] + S_M(g_{\mu\nu}, \psi). \tag{14}$$

[1] See [24, 17] for a discussion on the exceptional case where this equation has no solutions or when $f(\mathcal{R}) \propto \mathcal{R}^2$.

This action is identical to (13) apart from the fact that R has been now replaced by $\mathcal{R}$. This is critical, as $\mathcal{R}$ is not the Ricci scalar of the metric $g_{\mu\nu}$ and, therefore, action (14) is not that of a Brans–Dicke theory with $\omega_0 = 0$. However, taking into account also the definition of ϕ, we known that we can express the independent connection algebraically with respect to the metric and ϕ using eq. 8. Using this expression action (14) takes the form

$$S_{pal} = \frac{1}{2\kappa}\int d^4x\sqrt{-g}\left(\phi R + \frac{3}{2\phi}\partial_\mu\phi\partial^\mu\phi - V(\phi)\right) + S_M(g_{\mu\nu}, \psi). \tag{15}$$

This is the action of a Brans–Dicke theory with Brans–Dicke parameter $\omega_0 = -3/2$.

I will avoid to present the complete set of field equations for Brans–Dicke theory for the sake of brevity and give only the equation governing the dynamics of the scalar:

$$(2\omega_0 + 3)\Box\phi + 2V - \phi V' = \kappa T. \tag{16}$$

This equation is enough to reveal that, it is not the $\omega_0 = 0$ case the one for which ϕ carries no dynamics as one might have expected judging from the absence of a kinetic term in the action, but actually the $\omega_0 = -3/2$, which corresponds to Palatini $f(R)$ gravity.

The outcome of this discussion is twofold: Firstly, both metric and Palatini $f(R)$ gravity are equivalent to some version of Brans–Dicke theory (with all the implication this carries: for instance, this highlights the fact that Palatini $f(R)$ gravity is really a metric theory). Secondly, in metric $f(R)$ gravity there is only one extra scalar degree of freedom, where as in Palatini, the Brans–Dicke scalar is non-dynamical (so it is an auxiliary field, much like the independent connection). The moral is that not all representations of a theory are equally suitable for straightforwardly exhibiting some of the features of the theory (for a longer discussion see also [34]).

2.3. Lesson 3: Devil hides in the details

During the last few years it has been understood that it is not that difficult to construct models within the framework of $f(R)$ gravity that lead to a desired cosmological background evolution (see [35, 36, 37, 38, 39, 40, 41, 42, 43, 44, 45] for some examples). Without getting into the details, the procedure is quite simple: Under the usual cosmological assumptions regarding symmetry and matter description, $f(R)$ gravity leads to modified Friedmann equations governing the dynamics of the universe. Choosing the function f properly can lead to the desirable modification of the dynamics (*e.g.* late time expansion). In fact, it has even been shown that, starting from a given scale factor $a(t)$ as a function of cosmic time, one could "reconstruct" the function f (or at least a family of functions f) that leads to this evolution [46, 47, 48].

However, the difficulties arise when one goes beyond background evolution and requires that some $f(R)$ can account also for the fine details in cosmological observations. For instance, the requirement that de Sitter spacetime (in models which do not admit Minkowski space as a maximally symmetric solution) be a stable solution gives important constraints [49, 50, 51]. Big Bang Nucleosynthesis is a delicate process that can lead to severe constraints as well (*e.g.* [52]). The growth of cosmological perturbation and structure formation are other examples of processes that can rule out models and lead to strong viability constraints [53, 54, 55]. See [17] for a more detailed discussion on cosmological constraints and a more complete list of references.

The lesson learned here is that the devil always hides in the details and proposing cosmologically viable $f(R)$ gravity models can be significantly trickier than it seems. The subtle details of the cosmological evolution are quite sensitive to the underlying gravity theory and as we are entering the era of precision cosmology it should become more and more difficult to deviate from GR without affecting the delicate picture of the standard model of cosmology.[2]

[2] Besides cosmology, another example where perturbations around the spacetime are much more revealing than

2.4. Lesson 4: Extra degrees of freedom are troublesome

As mentioned in the Introduction, metric $f(R)$ gravity can avoid the Ostrogradski instability, unlike higher order gravity theories in general [16]. At the same time $f(R)$ theories have no ghosts (massive states of negative norm that cause apparent lack of unitarity), which is also not true for generic higher order theories of gravity [8, 6, 58, 3, 4, 59, 60]. Therefore, one could say that metric $f(R)$ gravity has been carefully chosen so that the extra degrees of freedom with respect to those of GR will not evidently lead to serious viability issues.

Nevertheless, these criteria have proven inadequate and problems related with the presence of extra degrees of freedom are still plaguing metric $f(R)$ gravity. The most prominent is the instability initially found by Dolgov and Kawasaki [61] for the model proposed in [36] but can appear for more general functions f [62].[3] To see how this comes about let us parametrize the deviations from Einstein gravity as

$$f(R) = R + \epsilon\,\varphi(R), \tag{17}$$

where ϵ is a small parameter with the dimensions of a mass squared and φ is arranged to be dimensionless. The trace of eq. (3) yields

$$3\Box f'(R) + f'(R)R - 2f(R) = \kappa T, \tag{18}$$

and substituting eq. (17) gives

$$\Box R + \frac{\varphi'''}{\varphi''}\nabla^\alpha R\,\nabla_\alpha R + \frac{(\epsilon\varphi' - 1)}{3\epsilon\,\varphi''}R = \frac{\kappa\,T}{3\epsilon\,\varphi''} + \frac{2\varphi}{3\varphi''}. \tag{19}$$

where it is assumed that $\varphi'' \neq 0$. Consider now a small region of spacetime in the weak-field regime and approximate *locally* the metric and the curvature by

$$g_{\mu\nu} = \eta_{\mu\nu} + h_{\mu\nu}, \qquad R = -\kappa\,T + R_1, \tag{20}$$

where $\eta_{\mu\nu}$ is the Minkowski metric and $|R_1/\kappa\,T| \ll 1$. Equation (20) yields, to first order in R_1,

$$\begin{aligned}\ddot{R}_1 - \nabla^2 R_1 - \frac{2\kappa\,\varphi'''}{\varphi''}\dot{T}\dot{R}_1 + \frac{2\kappa\,\varphi'''}{\varphi''}\vec{\nabla}T\cdot\vec{\nabla}R_1 \\ +\frac{1}{3\varphi''}\left(\frac{1}{\epsilon} - \varphi'\right)R_1 = \kappa\,\ddot{T} - \kappa\nabla^2 T - \frac{(\kappa\,T\varphi' + 2\varphi)}{3\varphi''},\end{aligned} \tag{21}$$

where $\vec{\nabla}$ and ∇^2 are the gradient and Laplacian in Euclidean three-dimensional space, respectively, and an overdot denotes differentiation with respect to time. The function φ and its derivatives are now evaluated at $R = -\kappa\,T$. The coefficient of R_1 in the fifth term on the left hand side is the square of an effective mass and is dominated by the term $(3\epsilon\,\varphi'')^{-1}$ due to the extremely small value of ϵ needed for these theories to reproduce the correct cosmological dynamics. Then, the scalar mode R_1 of the $f(R)$ theory is stable if $\varphi'' = f'' > 0$, and unstable if this effective mass is negative, *i.e.*, if $\varphi'' = f'' < 0$. The time scale for this instability to manifest is estimated to be of the order of the inverse effective mass $\sim 10^{-26}$ s in the example

the background itself, is black holes: even though the black hole solutions of GR are also solutions of $f(R)$ gravity model and one cannot distinguish between theories by mapping the spacetime around a black hole [56], the perturbations around a black hole solution behave differently in different theories and, therefore, they allow for distinctions [57].

[3] This instability does not appear in Palatini $f(R)$ gravity [63].

$\epsilon\varphi(R) = -\mu^4/R$ [61]. The small value of φ'' gives a large effective mass and is responsible for the small time scale over which the instability develops.

This fatal instability for models with $f'' < 0$ manifests itself in the linearized version of equation (18). There are also recent claims that R can be driven to infinity due to strong non-linear effects related to the same equation [64, 65, 66]. Both of these problems are strictly related to the extra scalar degree of freedom in metric $f(R)$ gravity and hint towards the following conclusion: no matter how carefully one constructs a theory with extra degrees of freedom, complications related to them can arise as extra degrees of freedom are very difficult to control.

2.5. *Lesson 5: Parametrized post-Newtonian expansion is difficult but fruitful*

Even though one might think that the parametrized post-Newtonian (PPN) expansion is more or less a standardized procedure, it actually turned out to hold a lot of surprises in $f(R)$ gravity. It took several papers and lengthy debates for this issue to be settled (*e.g.* [67, 68, 69, 31, 70, 71, 72, 73], see also [17] for a more detailed list of references). The main problems which were encountered and led to debates were: a) the equivalence with Brans–Dicke theory and whether this can be used for the PPN expansion, b) confusion related to the absence of flat spacetime as a solution in many models, c) misunderstandings regarding the difference between the existence of the Schwarzschild(-de Sitter) metric as a vacuum solution and a correct (post-)Newtonian limit, d) difficulty with non-analytic functions $f(R)$ and their expansion. Most of these problems were not specific to $f(R)$ gravity and, therefore, solving them has led to a better understanding of the post-Newtonian expansion for alternative theories of gravity in general.

In spite of its difficulties, the post-Newtonian limit of $f(R)$ gravity has proved to be very fruitful after all: it led to strong viability constraints. Briefly summarized, the outcome of the relevant studies is the following: For metric $f(R)$ gravity, one can indeed use the equivalence with Brans–Dicke theory, but with some care (*e.g.* the scalar is algebraically related to the curvature so it does not tend to settle to the minimum of the potential). This implies that, since the theory corresponds to a Brans–Dicke parameter $\omega_0 = 0$, it is ruled out (experiment require $\omega_0 > 40000$) unless the effective mass of the scalar is large. The effective mass is essentially the second derivative of the potential whose functional form depends on the form of f. It has also been shown that certain models of metric $f(R)$ gravity can exhibit "chameleon" behavior: the effective mass of the scalar depends on the curvature and, therefore, it can appear as large at small scales, making the mediated force short ranged and helping the theory avoid solar system constraints; however, the effective mass can become small at large scales, making the force long range and allowing the theory to have cosmological effects [74, 75, 76, 77]. In any case, the models have to be very carefully selected in order to be viable and at the same time lead to some interesting late time phenomenology (some examples do exist, *e.g.* [77, 78])

The results for Palatini $f(R)$ gravity are much more intriguing. In this case, the known results from Brans–Dicke theory cannot be used because the $\omega_0 = -3/2$ case is exceptional in the sense that the scalar is non-dynamical (the standard treatment assumes $\omega_0 \neq -3/2$). In [70, 71] the PPN expansion for $\omega_0 = -3/2$ Brans–Dicke theory and, consequently, Palatini $f(R)$ gravity was rigorously performed. Remarkably, as is pointed out there, and also in [72] without the use of the equivalence between theories, whether a model will lead to the correct post-Newtonian behavior depends on the density. This had not been encountered before in alternative theories of gravity. Just to give a characteristic example, the zero-zero component of the PPN metric is given by

$$h_{00}^{(1)}(t,\vec{x}) = \frac{2G_{\rm eff}M_\odot}{r} + \frac{V_0}{6\phi_0}r^2 + \Omega(T), \tag{22}$$

where V_0 and ϕ_0 are the background values of the potential and the scalar respectively and $M_\odot \equiv \phi_0 \int d^3\vec{x}'\rho(t,\vec{x}')/\phi$. The effective Newton constant $G_{\rm eff}$ and the post-Newtonian

Journal of Physics: Conference Series **189** (2009) 012039 doi:10.1088/1742-6596/189/1/012039

parameter γ are defined as

$$G_{\rm eff} \equiv \frac{G}{\phi_0}\left(1+\frac{M_V}{M_\odot}\right), \quad \gamma \equiv \frac{M_\odot - M_V}{M_\odot + M_V}, \tag{23}$$

where $M_V \equiv \kappa^{-1}\phi_0 \int d^3\vec{x}' \, [V_0/\phi_0 - V(\phi)/\phi]$. The striking term in eq. (22) is the last one: $\Omega(T)$ is some algebraic function of the trace of the stress-energy tensor T whose functional form depends on the functional form of f. T is in turn algebraically related to the density ρ and so the PPN metric picks up an algebraic dependence on the density! This signals a serious shortcoming of Palatini $f(R)$ gravity, non-cumulativity, which will be discussed in the following section. Therefore, I refrain from saying more now.

To summarize, the PPN expansion for $f(R)$ gravity, even though it was a tedious procedure, deepened our understanding of post-Newtonian gravity in general and at the same time provided important viability constraints.

2.6. Lesson 6: Non-cumulativity as the root of all evils!

We just saw that in Palatini $f(R)$ gravity the PPN metric depends algebraically on the density. This issue turns out to be related with two other problems: a conflict between Palatini $f(R)$ gravity and the standard model of particle physics [79, 80] and the appearance of singularities on the surface of spherically symmetric matter configurations [81, 82, 83].

More precisely, it was shown in [79] assuming that matter is a Dirac field and in [80] assuming that matter is a scalar field that, when one goes to the local frame where spacetime is flat plus second order corrections, non-perturbative corrections and strong couplings appear in the matter action even at low energies. Since non-gravitational experiments are held at a local frame, these deviations from the standard matter actions contradict the results of such experiments. In [81, 82, 83] on the other hand, spherically symmetric matter configuration where considered. It was shown that if one assumes a polytropic equation of state and attempts to match any interior solution to the unique Schwarzschild-de Sitter exterior, then singularities appear at the surface of the matter configuration when the polytropic index Γ is in the range $3/2 < \Gamma < 2$. The presence of the singularities is essentially insensitive to the choice of the function $f(R)$ [81, 82]. Even though a polytrope is a very idealized form of matter, the importance of this results lies on two factors: a) There are physical matter configurations, such as a degenerate non-relativistic electron gas or an isentropic monoatomic gas, which can be described accurately by a $\Gamma = 5/3$ polytrope; such configurations do not have a description in Palatini $f(R)$ gravity, rendering the theory incomplete. b) The polytrope is just used as an idealized approximation so that analytic results can be easily obtained, and the root of the problem lies deep within the structure of the theory.

Lets us elaborate on this second point. This will also allow us to make the connection with the other problems discussed here. As mentioned earlier, the independent connection in Palatini $f(R)$ gravity is an auxiliary field, and one can express it algebraically in terms of the metric and the matter fields using eqs. (8) and (9) and eliminate it from the field equations. Indeed, if this is done, eqs. (6) and (7) reduce to a single equation:

$$\begin{aligned} G_{\mu\nu} &= \frac{\kappa}{f'}T_{\mu\nu} - \frac{1}{2}g_{\mu\nu}\left(\mathcal{R}-\frac{f}{f'}\right) + \frac{1}{f'}\left(\nabla_\mu\nabla_\nu - g_{\mu\nu}\Box\right)f' - \\ &\quad -\frac{3}{2}\frac{1}{f'^2}\left[(\nabla_\mu f')(\nabla_\nu f') - \frac{1}{2}g_{\mu\nu}(\nabla f')^2\right], \end{aligned}$$

where $G_{\mu\nu}$ is the Einstein tensor of the metric $g_{\mu\nu}$. Note that, since $\mathcal{R}$ is algebraically related to T, $f(\mathcal{R})$ and $f'(\mathcal{R})$ are essentially function of T. T contains usually first derivatives of the matter fields (or is algebraically related to the density for a perfect fluid). These imply that the right hand side (rhs) of eq. (24) contains third derivatives of the matter fields (or second derivatives of the density). However, this equation is only second order in the metric. Therefore, the metric will have some algebraic dependence on the matter fields or the density! Gravity will not be cumulative, as in GR or most theories, and the characteristics of the gravitational field will depend on the local matter content.

This explains the behavior of the PPN limit. It also explains the conflict with particle physics (the terms on the rhs of eq. (24) can be considered as coming from strong interactions). An easier way to think about this is to recall that in the equivalent Brans–Dicke theory the scalar is non-dynamical and algebraically related to the matter but non-minimally coupled to gravity. This leads to the strong couplings: when going to the local frame ϕ cannot be treated perturbatively. One has to solve for it in terms of the matter and replace back, introducing the new terms. *I.e.* even though ϕ appears not to couple to matter, it enters the matter action though the coupling with the metric.

The singularities found in [81, 82, 83] are a manifestation of the non-cumulativity as well, in the sense that, in Palatini $f(R)$ gravity, the metric is bound to inherit any discontinuities in the matter and this would lead to a singularity. The simple example of polytropic matter configurations is a good demonstration of this, as the range $3/2 < \Gamma < 2$ for the polytropic index is exactly the range for which the density on the surface is finite (and so there is a discontinuity since the density is zero in the exterior). However, matter is classically allowed to exhibit discontinuities. Remarkably, as it was mentioned, the singularities appear mathematically for essentially all functions f (besides GR). There is only one case when one can question their physical significance: if they manifest themselves at scales for which the microstructure of matter has to be taken into account and the averaging procedure leading to a perfect fluid or a polytropic description is questionable. In this sense requiring the absence of such singularities provides a very strong constraint, ruling out all models with late time cosmological phenomenology, but even models with ultraviolet corrections [81, 82]. Only models with Planck suppressed correction seems to survive [84].

3. Conclusions

As is probably obvious by now, proposing and studying gravity theories other than GR is not an easy task. One can easily end up causing more problems that those the proposed modification of gravity is attempting to solve. Many of the shortcomings of $f(R)$ gravity become much more obvious after field redefinition and representation changes. The fact that it took some time to discover them seems to indicate that habit and affinity to specific formalisms and representation of a theory seem to have played a crucial role in the misunderstandings and the difficulties of $f(R)$ gravity. In any case, $f(R)$ gravity has been a very successful toy theory. It seems that we may have to go beyond simple $f(R)$ models if we want to have a chance of explaining some of the current riddles of physics with a deviation from GR. However, our experience with $f(R)$ gravity indicates that pursuing such an idea might be fruitful: possibly in providing a solution to some specific open problem, but most certainly in providing insight into the most puzzling interaction, gravity. And this is an important lesson on it own.

Acknowledgments

I am grateful to Valerio Faraoni, which was the coauthor of [17] on which this talk was based, for numerous discussions on $f(R)$ gravity. This work was supported by the National Science Foundation under grant PHYS-0601800.

References

[1] Weyl H 1919 *Ann. Phys.* **59** 101

[2] Eddington A S 1923 *The Mathematical Theory of Relativity* (Cambridge University Press, Cambridge)
[3] Utiyama R and DeWitt B S 1962 *J. Math. Phys.* **3** 608–618
[4] Stelle K S 1977 *Phys. Rev.* **D16** 953–969
[5] Schmidt H J 2007 *Int. J. Geom. Meth. Phys.* **4** 209–248 (*Preprint* gr-qc/0602017)
[6] Buchbinder I L, Odintsov S D and Shapiro I L 1992 *Effective Actions in Quantum Gravity* (IOP Publishing, Bristol)
[7] Birrell N D and Davies P C W 1982 *Quantum Fields in Curved Spacetime* (Cambridge University Press, Cambridge)
[8] Vilkovisky G A 1992 *Class. Quant. Grav.* **9** 895
[9] Spergel D N *et al.* (WMAP) 2007 *Astrophys. J. Suppl.* **170** 377 (*Preprint* astro-ph/0603449)
[10] Riess A G *et al.* (Supernova Search Team) 2004 *Astrophys. J.* **607** 665–687 (*Preprint* astro-ph/0402512)
[11] Astier P *et al.* (The SNLS) 2006 *Astron. Astrophys.* **447** 31–48 (*Preprint* astro-ph/0510447)
[12] Eisenstein D J *et al.* (SDSS) 2005 *Astrophys. J.* **633** 560–574 (*Preprint* astro-ph/0501171)
[13] Weinberg S 1989 *Rev. Mod. Phys.* **61** 1–23
[14] Carroll S M 2001 *Living Rev. Rel.* **4** 1 (*Preprint* astro-ph/0004075)
[15] Nojiri S and Odintsov S D 2003 *Phys. Lett.* **B576** 5–11 (*Preprint* hep-th/0307071)
[16] Woodard R P 2007 *Lect. Notes Phys.* **720** 403–433 (*Preprint* astro-ph/0601672)
[17] Sotiriou T P and Faraoni V 2008 (*Preprint* 0805.1726)
[18] Sotiriou T P 2007 *Modified Actions for Gravity: Theory and Phenomenology* Ph.D. thesis SISSA (*Preprint* 0710.4438)
[19] Buchdahl H A 1970 *Mon. Not. Roy. Astron. Soc.* **150** 1
[20] Will C M 1981 *Theory and Experiment in Gravitational Physics* (Cambridge University Press, New York)
[21] Sotiriou T P 2006 (*Preprint* gr-qc/0611158)
[22] Sotiriou T P and Liberati S 2007 *Annals Phys.* **322** 935–966 (*Preprint* gr-qc/0604006)
[23] Sotiriou T P 2006 *Class. Quant. Grav.* **23** 5117–5128 (*Preprint* gr-qc/0604028)
[24] Ferraris M, Francaviglia M and Volovich I 1992 (*Preprint* gr-qc/9303007)
[25] Teyssandier P and Tourrenc P 1983 *J. Math. Phys.* **24** 2793
[26] Wands D 1994 *Class. Quant. Grav.* **11** 269–280 (*Preprint* gr-qc/9307034)
[27] Barrow J D and Cotsakis S 1988 *Phys. Lett.* **B214** 515–518
[28] Barrow J D 1988 *Nucl. Phys.* **B296** 697–709
[29] Flanagan E E 2003 *Class. Quant. Grav.* **21** 417–426 (*Preprint* gr-qc/0309015)
[30] Cecotti S 1987 *Phys. Lett.* **B190** 86
[31] Chiba T 2003 *Phys. Lett.* **B575** 1–3 (*Preprint* astro-ph/0307338)
[32] Flanagan E E 2004 *Class. Quant. Grav.* **21** 3817 (*Preprint* gr-qc/0403063)
[33] O' Hanlon J 1972 *Phys. Rev. Lett.* **29** 137–138
[34] Sotiriou T P, Faraoni V and Liberati S 2007 (*Preprint* 0707.2748)
[35] Capozziello S, Carloni S and Troisi A 2003 *Recent Res. Dev. Astron. Astrophys.* **1** 625 (*Preprint* astro-ph/0303041)
[36] Carroll S M, Duvvuri V, Trodden M and Turner M S 2004 *Phys. Rev.* **D70** 043528 (*Preprint* astro-ph/0306438)
[37] Nojiri S and Odintsov S D 2003 *Phys. Rev.* **D68** 123512 (*Preprint* hep-th/0307288)
[38] Vollick D N 2003 *Phys. Rev.* **D68** 063510 (*Preprint* astro-ph/0306630)
[39] Meng X H and Wang P 2003 (*Preprint* astro-ph/0308284)
[40] Meng X H and Wang P 2004 *Class. Quant. Grav.* **21** 2029–2036 (*Preprint* gr-qc/0402011)
[41] Meng X H and Wang P 2005 *Class. Quant. Grav.* **22** 23–32 (*Preprint* gr-qc/0411007)
[42] Allemandi G, Borowiec A and Francaviglia M 2004 *Phys. Rev.* **D70** 043524 (*Preprint* hep-th/0403264)
[43] Allemandi G, Borowiec A, Francaviglia M and Odintsov S D 2005 *Phys. Rev.* **D72** 063505 (*Preprint* gr-qc/0504057)
[44] Sotiriou T P 2006 *Phys. Rev.* **D73** 063515 (*Preprint* gr-qc/0509029)
[45] Sotiriou T P 2006 *Class. Quant. Grav.* **23** 1253–1267 (*Preprint* gr-qc/0512017)
[46] Capozziello S, Cardone V F and Troisi A 2005 *Phys. Rev.* **D71** 043503 (*Preprint* astro-ph/0501426)
[47] Capozziello S, Nojiri S, Odintsov S D and Troisi A 2006 *Phys. Lett.* **B639** 135–143 (*Preprint* astro-ph/0604431)
[48] Nojiri S and Odintsov S D 2006 *Phys. Rev.* **D74** 086005 (*Preprint* hep-th/0608008)
[49] Faraoni V 2004 *Phys. Rev.* **D69** 123520 (*Preprint* gr-qc/0404078)
[50] Faraoni V 2005 *Phys. Rev.* **D72** 061501 (*Preprint* gr-qc/0509008)
[51] Faraoni V and Nadeau S 2005 *Phys. Rev.* **D72** 124005 (*Preprint* gr-qc/0511094)
[52] Brookfield A W, van de Bruck C and Hall L M H 2006 *Phys. Rev.* **D74** 064028 (*Preprint* hep-th/0608015)
[53] Bean R, Bernat D, Pogosian L, Silvestri A and Trodden M 2007 *Phys. Rev.* **D75** 064020 (*Preprint*

astro-ph/0611321)
[54] Koivisto T and Kurki-Suonio H 2006 *Class. Quant. Grav.* **23** 2355–2369 (*Preprint* astro-ph/0509422)
[55] Hwang J C and Noh H 2002 *Phys. Rev.* **D65** 023512 (*Preprint* astro-ph/0102005)
[56] Psaltis D, Perrodin D, Dienes K R and Mocioiu I 2008 *Phys. Rev. Lett.* **100** 091101 (*Preprint* 0710.4564)
[57] Barausse E and Sotiriou T P 2008 *Phys. Rev. Lett.* **101** 099001 (*Preprint* 0803.3433)
[58] Strominger A 1984 *Phys. Rev.* **D30** 2257
[59] Stelle K 1978 *Gen. Rel. Grav.* **9** 353
[60] Ferraris M, Francaviglia M and Magnano G 1988 *Class. Quant. Grav.* **5** L95
[61] Dolgov A D and Kawasaki M 2003 *Phys. Lett.* **B573** 1–4 (*Preprint* astro-ph/0307285)
[62] Faraoni V 2006 *Phys. Rev.* **D74** 104017 (*Preprint* astro-ph/0610734)
[63] Sotiriou T P 2007 *Phys. Lett.* **B645** 389–392 (*Preprint* gr-qc/0611107)
[64] Tsujikawa S 2008 *Phys. Rev.* **D77** 023507 (*Preprint* arXiv:0709.1391 [astro-ph])
[65] Frolov A V 2008 (*Preprint* 0803.2500)
[66] Appleby S A and Battye R A 2008 (*Preprint* 0803.1081)
[67] Barraco D E and Hamity V H 2000 *Phys. Rev.* **D62** 044027
[68] Dick R 2004 *Gen. Rel. Grav.* **36** 217–224 (*Preprint* gr-qc/0307052)
[69] Meng X H and Wang P 2004 *Gen. Rel. Grav.* **36** 1947–1954 (*Preprint* gr-qc/0311019)
[70] Olmo G J 2005 *Phys. Rev. Lett.* **95** 261102 (*Preprint* gr-qc/0505101)
[71] Olmo G J 2005 *Phys. Rev.* **D72** 083505
[72] Sotiriou T P 2006 *Gen. Rel. Grav.* **38** 1407–1417 (*Preprint* gr-qc/0507027)
[73] Chiba T, Smith T L and Erickcek A L 2007 *Phys. Rev.* **D75** 124014 (*Preprint* astro-ph/0611867)
[74] Cembranos J A R 2006 *Phys. Rev.* **D73** 064029 (*Preprint* gr-qc/0507039)
[75] Navarro I and Van Acoleyen K 2007 *JCAP* **0702** 022 (*Preprint* gr-qc/0611127)
[76] Faulkner T, Tegmark M, Bunn E F and Mao Y 2007 *Phys. Rev.* **D76** 063505 (*Preprint* astro-ph/0612569)
[77] Starobinsky A A 2007 *JETP Lett.* **86** 157–163 (*Preprint* arXiv:0706.2041 [astro-ph])
[78] Nojiri S and Odintsov S D 2007 *Phys. Lett.* **B657** 238–245 (*Preprint* 0707.1941)
[79] Flanagan E E 2004 *Phys. Rev. Lett.* **92** 071101 (*Preprint* astro-ph/0308111)
[80] Iglesias A, Kaloper N, Padilla A and Park M 2007 *Phys. Rev.* **D76** 104001 (*Preprint* 0708.1163)
[81] Barausse E, Sotiriou T P and Miller J C 2008 *Class. Quant. Grav.* **25** 062001 (*Preprint* gr-qc/0703132)
[82] Barausse E, Sotiriou T P and Miller J C 2008 *Class. Quant. Grav.* **25** 105008 (*Preprint* 0712.1141)
[83] Barausse E, Sotiriou T P and Miller J C 2008 (*Preprint* 0801.4852)
[84] Olmo G J 2008 (*Preprint* 0810.3593)

Negative mass and repulsive gravity in Newtonian theory, and consequences

N. K. Spyrou

Astronomy Department, Aristoteleion University of Thessaloniki,
541.24 Thessaloniki, Macedonia, Hellas (Greece)

Email: spyrou@astro.auth.gr

Abstract. In the context of the Newtonian theory of gravity, the dynamical equivalence of hydrodynamic flows with geodesic lines, in the interior of a bounded, gravitating perfect-fluid source, results in the possibility of negative mass and, hence, of repulsive gravity. The consequences are outlined for the overall picture of the Solar System and the large-scale cosmological, structures, and some predictions are attempted based on some current and mostly unexplained so far observational data.

1. Introduction

According to many current observational data, the realistic picture and morphology of an astrophysical-cosmological structure differs greatly from its corresponding optical picture. This is true for the Solar System, whose linear dimensions are of the order of 10^5 AU, namely, approximately half the distance to the nearest star to the Sun, and also for the Milky Way, whose linear dimensions are at least 200 kpc, almost ten times larger than its optical linear dimensions (~30 kpc). Similar arguments hold for other types of galaxies and for even larger cosmological structures, namely, clusters of galaxies and super-clusters of galaxies. Therefore, it appears appropriate that the large-scale cosmological structures be treated as almost spherically symmetric, very complex, practically continuous, and of much larger linear dimensions cosmological structures than previously assumed. Consequently, the *constituent elements of the Universe and the Universe as a whole, can quite satisfactorily be treated as continuous gravitational systems and, more specifically, bounded, gravitating perfect-fluid sources, the physical-dynamical description of which is very well established at both the Newtonian and the general-relativistic levels.* So, we arrive at the very crucial result, that the motions of and in these constituents should be considered as hydrodynamic flows rather than geodesic motions. Now, in view of the wealth of such observational data and strong indications, it has been suggested [1,14] that, in both the Newtonian and the general-relativistic theories of gravity, and at all levels, namely, cosmological [1,7,8,9,10,13], galactic [1,2,3,5,6,7,8,10,11], and stellar [1,2,4,6], it is possible to give to the equations of hydrodynamic (and hydromagnetic) flow motions in the interior of a bounded gravitating perfect-fluid source the form of the equations of the equations of geodesic motion in it. This approach is usually referred to as the *dynamical-equivalence approach.*

It is exactly this *dynamical-equivalence approach,* that enables us to recast the geodesic motions, mostly applied in the observational determination of masses, now taking into account, as sources of geodesic motions, not simply the mass density but all of the source's internal physical characteristics (like e.g. mass density, pressure, internal thermodynamic energy, velocities of both ordered and statistical thermodynamic motions). Similarly, this approach enables us to determine the contribution to the observationally determined mass of the fluid source of all of the source's internal physical characteristics.

In the context of the dynamical-equivalence approach, especially it has been proved [1,11,12,14] that, in the Newtonian theory of gravity, the flows take on the functional form of the geodesic motions in an explicitly known generalized potential containing, beyond the usual gravitational potential, also a part depending on the internal physical characteristics of the source. Quite interestingly, the generalized density producing the generalized potential can be positive, vanishing, or negative, defining, respectively, positive, vanishing, or negative mass, and so causing, respectively, decelerated geodesic motion (attractive gravity), unchanged acceleration, and accelerated geodesic motion (repulsive gravity).

Finally we recall, that according to the observations of the cosmic microwave background by the *Wilkinson Microwave Anisotropy Probe* (WMAP), the Cosmos is composed of heavy elements (0.03 %), ghostly neutrinos (0.3 %), stars (0.5 %), free hydrogen and helium (4%), dark matter (22 %), and dark energy (73 %).

In the following we shall briefly expose some the consequences of this equivalence on problems like dark-matter problem, the flat-rotation curves problem, the problem of the true linear dimensions of the cosmological structures, and, finally, the problem related to the so-called *Pioneer Anomaly Effect* in the Solar System .

2. Applications

In this section we shall give only a brief account of the planetary, stellar, astrophysical, and cosmological consequences, explicit results on which can be found in the literature at the end.

According the dynamical equivalence of the hydrodynamic flows and geodesic motions, the observationally determined mass is not the baryonic mass, m, but the *generalized mass* , $m_v = m + m_i$ where the internal mass *internal mass*, m_i, is the contribution to the observationally determined mass of the internal physical characteristics of the source considered. I most of the cases of isentropic and isothermal flows, m_i (and the corresponding *internal mass density)* is a negative quantity. Therefore, the baryonic mass is larger than the observationally determined mass, and there is a plentitude not luck of baryonic matter, a conclusion of particular interest to the dark-matter problem.

Furthermore, it has been proved that in the case of a spherically-symmetric, gravitating, isentropic perfect-fluid source, the total velocity of an equatorial, circular geodesic orbit at a distance from the centre depends, through an isothermal equation of state, also on the temperature and chemical composition of the source. It is straightforward to verify that, for an overall temperature of the order of 10^4 K, the (circular) velocity curve in the Milky Way remains flat for practically the whole optical part of the galaxy, and continues so up to a radial distance of at least 200 kpc retaining a constant value in the range ~100-200 km/s, in agreement with standard observational data.

Moreover, in the case of a super cluster of galaxies, additionally, the requirement for a vanishing total acceleration (or, independently, the requirement that the thermal velocities are not larger than the escape velocity at the free surface of the super cluster) specifies the true linear dimensions of the super cluster, which are of the order of those of the observable Universe. This prediction is in complete agreement with the fact that no third-order clusters of galaxies have been observed so far (simply because their existence is not necessary!!!).

Furthermore, in the stellar and galactic levels, the notion of the repulsive gravity, predicted in the dynamical-equivalence approach, can be used to tackle problems like the formation of the (stellar and galactic) winds, and, in the case of magnetohydrodynamic flows, of the (stellar and galactic) jets [15] .

Finally, in the case of the (spherical) Solar System, from the functional form of the generalized potential, on the one hand, one can determine the true linear dimensions of the Solar System (approximately, 10^5 AU, in agreement with observational data), and on the other hand, the anomalous behaviour of the acceleration of various space probes can be explained, like the famous "Pioneer-Anomaly Effect", and other similar anomalies for other space probes.

It still remains to be examined whether the dynamical-equivalence approach could, in principle, be of some importance to the celebrated dark-energy problem [16], beyond the one to the dark-matter problem [17].

3. Concluding Remarks

From all the above we conclude that, if we insist in using the notion of the geodesic motions (inside and outside) of the cosmological structures for the description of their physical motions and for the observational determination of masses, in conjunction with the results of the dynamical equivalence, then *the internal mass* m_i, *due to the internal physical characteristics of the source, is revealed to be there*, it is negative and, in many cases, absolutely it can largely exceed the rest (baryonic) mass of the gas of the (corresponding region of the) large-scale cosmological structure under consideration. It is interesting that this negative extra mass "shows up", in planetary, stellar, astrophysical and cosmological levels, when, in the context of the dynamical-equivalence approach, the geodesic motion in the generalized potential is used, and *not* when the standard geodesic motion of a test particle in the original gravitational potential is used. Moreover, since the internal mass m_i is negative, the observationally determined mass of the cosmological structure is smaller than its baryonic mass. These two general results can give a very simple, classical solution to the dark-matter and the flat-velocity-curves problems. Finally, it is stressed that the derivation of the these results is valid also in the exact general-relativistic theory of gravity, and that in this derivation nor any modifications of Newton's law of gravitational attraction or of the Einstein equations are required, neither any other theories, beyond classical Newtonian and relativity theory, are introduced and used. Detailed results can be found in the literature cited below.

Acknowledgements

It is a pleasure to express, from this position also, my sincere thanks to the LOC for their nice organizing and warm hospitality during the Conference.

Appendix (Address during the opening ceremony):

Dear Guests and Participants,

On behalf of the Scientific Committee of the 13th Conference "Recent Developments in Gravity", I wish to welcome all of you, participants and guests, here in Thessaloniki, and to wish you a pleasant and fruitful stay in our city, the capital of Macedonia.

It is well-known that, in their initial form, these conferences intended to bring together, for a week or so every second year, all the relativists of Greece, for communicating their scientific results and for exchanging ideas, views and suggestions.

But now, after 24 whole years since the first conference in 1984, these conferences have developed into international reputable conferences on gravitational theories, both Newtonian and relativistic, and their applications, with, however, the responsibilities and demands of the eventual organizers continuously and naturally increasing.

So, we, at the Astronomy Department of our University, are rally happy organizing and hosting the 13th Conference of the series, now in Greek, «Νεώτερες Εξελίξεις στη Βαρύτητα» .

In the name of its Chairman, our colleague Dr Nikos Stergioulas, I wish to congratulate all the members of the Local Organizing Committee for their continuous and successful efforts with this superb result.

As the National Delegate of Greece to the European Space Agency, I grasp this opportunity to point out, for the future, a new reality, of interest to these conferences. And I wish to do it, additionally, because of some recent, happy and interesting coincidences.

It is quite probable that some of you are not aware of three recent events, during the last few months, of importance to the space programme of Europe.

The first one is the launch of the European Space Laboratory "Columbus" of the European Space Agency, which is already docked to the International Space Station orbiting the Earth at a height of approximately 400 Km. This laboratory is Europe's "house in space".

The second is the launch of the European Automated Transfer Vehicle (ATV) "Jules Verne", the European cargo ship, already docked to the International Space Station.

The third, as a consequence also of the above two events, is that this year, 2008, is the year for the recruitment of the new European astronauts. Recently, an informing press conference was co-organized at the University of Thessaloniki by the European Space Agency, the General Secretariat for Research and Technology of the Greek Ministry of Development and, of course, by the University of Thessaloniki.

To the above three important events one must add the new, and, this time, quite clear in its details, Announcement of Opportunity for the research programmes of the European Space Agency. This development is of particular interest on the one hand to Universities and Research Centers and on the other hand to the imperative involvement of the Greek Industry in the activities of Greece in the context of the European Space Agency, for the sake, of course, of the success of Greece joining the European Space Agency. This Announcement of Opportunity is expected to be released by the end of June 2008.

Also, to all these one must add the Announcement of Opportunity for a number of post-graduate scholarships for young Greek scientists to go to various Centers of the European Space Agency and be trained there on matters of Space Physics and technology of interest to the European Space Agency. These scholarships will be funded by the Greek Foundation of Scholarships (IKY, in Greek) and the Task Force for Greece and the European Space Agency, and the Announcement of Opportunity is expected to be released soon, perhaps, during June 2008.

All the above are of extreme importance to Greece and to us as scientists and researchers. Especially, General-Relativistic Effects in Space, Fundamental Interactions, and Quantum Physics in Space Time are of central importance to e.g. the so-called E.L.I.P.S. programme, namely, the European Life and Physical Sciences programme.

In relation to this, just recall that the International Space Station, at a height of approximately 400 km, moves as fast as about 7.5 km/sec.

So we all can understand that, for example, time measurement from space, using various types of clocks, like optical clocks in space and hydrogen maser clocks, is very important for a large number of problems, like e.g., just to mention a couple of them, the massive particle interferometers, and the Very Long-Baseline Interferometry of great importance to Cosmology and, more generally, to Astronomy.

I am in the happy position to tell you that the national priorities of Greece for Human Spaceflight, Microgravity, and Exploration, to be finalized in the coming Ministerial Council of Europe next November, are

1. The full use and exploitation of the International Space Station and Columbus,
2. The promotion and use of the E.L.I.P.S. programme, and
3. The promotion and use of the really rich educational material of ESA.

Finally, let me point it out, of particular interest is the fact that, with all these developments occurring "over our heads", on which every European citizen-taxpayer must be informed, the notion "environment" is changing, because, now, in a decisive way, the Near Earth Space Environment enters the environmental game, beyond the standard forms of environment, namely, the atmospheric and sub-atmospheric environments with all their serious problems.

Also, the notion "astronaut" has changed, because, now, it is possible for a young European scientist, under certain circumstances, of course, to fly to the International Space Station and Columbus and perform his/her research **up there** .

In conclusion, in view of all the above developments, new opportunities emerge for the Greek science, research, technology and industry, and new positions of work are created especially for our younger colleagues

This is something that we must soon realize and do our best for contributing to the success of the joining of Greece with the large family called Europe and, of course, the European Space Agency.

So, I look forward to the active, close, and intense collaboration, involvement and interaction of our scientific community as well, with the European Space Agency, and, of course, finally, I thank you very much for your patience and attention.

References

[1] Kleidis K and Spyrou N K 2000 *Class. Quantum Grav. 17 2965-2982*

[2] Spyrou N K 1997 in The Physics of Ionized Gases, Proceedings of the 18th Summer School and International Symposium on the Physics of Ionizes Gases (Kotor, Jugoslavia, 2-6 September 1996), eds. B.Vujicic, S. Djurovic and I. Puric, Institute of Physics, Novi Sad University, Yugoslavia, 417-446

[3] Spyrou N K 1997 in The Earth and the Universe, Volume in Honorem L. Mavrides, eds. G. Asteriadis, A. Bandellas, M. Contadakis, K. Katsambalos, A. Papademetriou and I. Tziavos, Thessaloniki, Greece, 277-291

[4] Spyrou N K 1997 *Facta Universitatis* **4** 7-14

[5] Spyrou N K and Kleidis K 1999 On the Nature of Nuclear Galactic Masses, Proceedings of JENAM 1999, Samos

[6] Spyrou N K 1999 in Current Issues of Astronomical and Planetary Environmental Concern, Proceedings of International Seminar (Thessaloniki, 6-7 April 1999) ed. N,K.Spyrou, Astronomy Department, Aristoteleion University of Thessaloniki, Thessaloniki, Greece, 23-35

[7] Spyrou N K 2001 Conformal Invariance and the Nature of Cosmological Structures, Invited Talk, Proceedings of The Conference on Applied Differential Geometry-General Relativity, and The Workshop on Global Analysis, Differential Geometry, Lie Groups, (Thessaloniki, 27 June-1July 2001) eds. G. Tsagas, C. Udriste, and D. Papadopoulos, 101-107

[8] Spyrou N K 2002, in Modern Theoretical and Observational Cosmology, Proceedings of the 2nd Hellenic Cosmology Meeting (Athens, 19-20 April 2001), eds. M.Plionis and S.Cotsakis, Kluwer Academic Publishers, Dordrecht, **276** 35-43

[9] Spyrou N K 2003 in Proceedings of the 10th Hellenic Relativity Conference Recent Developments in Gravity, eds. K. Kokkotas and N. Stergioulas, World Scientific, 90-96.

[10] Spyrou N K 2004 in Laskarides P. G., ed., Proc. 6th Helenic Astronomical Conference, Editing Office of the Univ. of Athens, Athens, 229

[11] Spyrou N K, Tsagas C G 2004 *Class. Quantum Grav.* **21** 2435

[12] Spyrou N K 2005 *J. Phys. Conf. Ser.* **8** 122

[13] Spyrou N K 2006 in Proceedings of the International Scientific Workshop Cosmology and Gravitational Physics, eds. N.K. Spyrou, N. Stergioulas,and C.Tsagas, Publishing ZITI, Perea-Thessaloniki, 57-66

[14] Spyrou N K and Tsagas C G 2008 *Mon. Not. R. Astron. Soc.* **388** 187–196

[15] Spyrou N K and Tsagas C G 2009 in preparation

[16] Kleidis K and Spyrou N K 2009 in preparation

[17] Plionis M, Vasilakos S and Spyrou N K 2009 in preparation

The generic model of General Relativity

Michael Tsamparlis

Department of Physics, Section Astrophysics Astronomy Mechanics, University of Athens, University of Athens, Zografos 15783, Athens, Greece

E-mail: mtsampa@phys.uoa.gr

Abstract. We develop a generic spacetime model in General Relativity from which all existing model results are produced under specific assumptions, depending on the case. We classify each type of possible assumption, especially the role of observers and that of symmetries, and discuss their role in the development of a model. We apply the results in a step by step approach to the case of a Bianchi I spacetime and a string fluid.

1. Introduction

General Relativity is the first theory of Physics which uses geometry to a great extend. Newtonian Physics is also a geometric theory but the difference is that the 3-space geometry which is used has direct sensory representation therefore 'couples' much easier and direct with the description of physical phenomena. In General Relativity the space used (spacetime) and the geometry employed (Riemannian Geometry) do not allow for a direct comparison of physical phenomena with the mathematical description. Most frequently this results in a confusion as to where geometry starts and stops, how much Physics is used and done, what is the physical interpretation of the geometrically derived results and how it will be done etc. Even most important it does not seem to exist a concise exposition of the structure of General Relativity and a clear description of the various hypotheses made both at geometric and physical level.

In this article we shall develop systematically a generic model for GR. These model will have all necessary parameters which can be used to classify all existing spacetime models. Furthermore this approach can be used to employ physical and geometric assumptions consistently so that one will not conflict with the other.

2. Universal model of GR

Every model-theory of Physics has three constituents:

a. Fundamental variables and equations between these variables

b. Constraints expressed by relations among these variables

c. Interpretation of the mathematical results and assumptions in terms of (directly or indirectly) observational phenomena.

An additional assumption/ constraint on the theory is the economy one, that is, we demand that both the number of fundamental variables and the equations which relate them shall be of a minimum number and will satisfy an aesthetic 'simplicity principle', that is, they will have the simplest possible structure. This requirement is vague but it summarizes the feeling one attains with the years for GR.

Journal of Physics: Conference Series **189** (2009) 012041 doi:10.1088/1742-6596/189/1/012041

The basic objects which are required to build up a theory of Physics are:

1. A background mathematical space into which the 'motion' or 'evolution' will be described.

2. Background symmetries which will simplify the fundamental geometric objects of the theory.

b. A class of observers who will observe and describe motion.

c. Matter which will make up the environment of the observers and modulate motion and evolution.

d. Other physical assumptions.

These objects are related by the Relativity Principle of the Theory, which states that:

The physical quantities will be modeled mathematically with the geometric objects defined by the transformation group of the theory.

In the following we comment briefly on each of the above 'components' of a physical theory.

3. The background space

Physics does not build its models in the 'real' space i.e. the space we live in. In Newtonian Physics we have a direct comparison of the two concepts of space due to our direct sensory conception of the world. In more recent models of Physics we have an 'indirect sensory' conception of the world by means of our measuring instruments. The first theory which did that has been Special Relativity and this is the reason people were unable to understand and anxious to disprove it by means of the paradoxes. It violated their 'common sense'.

The trouble is the use of the word 'space' to describe both a mathematical entity (software) and a physical entity (hardware). The physical entity is what it is. The mathematical entity is a set with certain mathematical structures. Our intention here is not to enter into details about these delicate questions, which in any case are not first priority questions to a general physicist.

The mathematical space used by the theories of Physics is a manifold. A manifold is a structure on a set which locally is diffeomorphic to an open set in a flat space of some dimension (called the dimension of the manifold) therefore it can be covered by a set of coordinate patches form the flat space so that whenever two patches coincide there is a differentiable map which relates the patches either way. In General Relativity the set of points are the events which correspond to various facts in the real world, the flat space is the space R^4 and the patches which cover the manifold are called coordinate systems. The differentiable maps which relate two coordinate systems are called coordinate transformations. This manifold is called space time[1]. The coordinate transformations form a group under the action of composition of maps. This group is an infinite dimensional Lie group called the Manifold Mapping Group (MMG) ([?]). The mathematical quantities which if given in one coordinate patch can be computed in any other overlapping this patch by means of the appropriate coordinate transformation we call geometric objects. We say that the geometric objects form a representation of the MMG. The geometric objects which are related in a linear and homogeneous manner we call general tensors. This is as far as one can go with the assumption of structure of the mathematical space.

In such a space there is not much one can do because there is no structure to be associated with a physical entity. The main interest of General Relativity is to develop a theory for the gravitational field. Each structure in a space is realized by a geometric object (not necessarily a tensor e.g. the connection). Therefore in General Relativity one introduces a new structure in the background space, the metric which is a $(0,2)$ tensor field and it is assumed that it geometrizes the gravitational field. Furthermore because at each point we want the theory to reduce to Special Relativity, it is postulated that the signature of the metric will be that of the

[1] In addition to the local structure spacetime has a global structure described by its topology. However these topological properties do not interest us here.

Journal of Physics: Conference Series **189** (2009) 012041 doi:10.1088/1742-6596/189/1/012041

Lorentz metric, that is -2. It is well known that having a metric on a manifold one can define a covariant derivative which is symmetric (torsion vanishes) and also the covariant derivative of the metric vanishes (zero metricity field). This covariant derivative is called the Riemannian derivative and it is defined by means of the geometric object (not a tensor!) which is the well known Christoffel symbols. The manifold structure together endowed with the fields of the metric and the Riemannian connection we call spacetime. Spacetime is the background space of General Relativity.

In spacetime all tensor fields must satisfy certain geometric identities. These identities together with other constraints which will follow will make up the framework within which General Relativity operates. In the following we classify these relations - constraints in various levels according to the wideness of applicability of their action.

Level G1

This level contains the geometric identities which result directly from the Riemannian structure of spacetime. These identities must be satisfied by whatever new fields will be introduced at later stages and are the following:

a. Ricci identity. For a tensor of type (r, s) this identity is:

$$T^{j_1\ldots j_r}{}_{l_1\ldots l_s;h;k} - T^{j_1\ldots j_r}{}_{l_1\ldots l_s;k;h} = \sum_{a=1}^{r} R_m{}^{j_a}{}_{hk} T^{j_1..j_{a-1}m.j_{a+1}....j_r}{}_{l_1\ldots l_s} - \sum_{\beta=1}^{r} R_m{}^{j_a}{}_{hk} T^{j_1\ldots j_r}{}_{l_1..l_{\beta-1}ml_{b+1}\ldots l_s}. \tag{1}$$

In particular for a vector field X^a the Ricci identity reads:

$$X_{a;bc} - X_{a;cb} = R_{dabc}X^d. \tag{2}$$

b. Bianchi identities.

The Bianchi identities in Riemannian geometry are the following:

$$R_{ab[cd;e]} = 0, \tag{3}$$

$$C^{abcd}{}_{..;d} = R^{c[a;b]} - \frac{1}{6}g^{c[a}R^{;b]}. \tag{4}$$

In a four dimensional Riemannian space these two identities are equivalent[2], therefore in spacetime they reduce to one identity. These identities imply the contracted Bianchi identity, which reads:

$$R^{ab}{}_{....;b} - \frac{1}{2}R^{;a} = 0 \Leftrightarrow G^{ab}{}_{...;b} = 0, \tag{5}$$

where:

$$G_{ab} = R_{ab} - \frac{1}{2}Rg_{ab}, \tag{6}$$

is the Einstein tensor.

Level G2

This level contains the symmetries (collineations) of the background spacetime. Symmetries are perhaps the most important aspect of physical theories. A symmetry is an additional constraint which we introduce mainly on some aesthetic or mathematical ground for simplifying the field equations or the geometry. Symmetries have various levels of effects.

a. Primary effects.

They are expressions of the form:

$$L_\xi T = A, \tag{7}$$

[2] This equivalence holds for $n = 4$ only. See Kundt and Trumper Abh. Akad.Wiss. and Lit. Mains., Mat. Nat.Kl.,No **12** (1962).

where T is a geometric object in spacetime and A is a tensor field and L_ξ stands for the Lie derivative wrt the vector ξ^a. In spacetime we can define many geometric objects by considering combinations of the partial derivatives of the metric g_{ab} which do not vanish. The main such object is the connection coefficients (Christoffel symbols), the curvature tensor and its irreducible parts (Weyl tensor, Ricci tensor, curvature scalar). In general all geometric objects which are defined in terms of the metric and its derivatives we shall call metric geometric objects. A collineation is a symmetry of a metric geometric object, that is, a relation of the form (7) where T is a metric geometric object. The following are examples of collineations:

$\mathcal{L}_\xi g_{ab} = 0,$	Killing Vector
$\mathcal{L}_\xi g_{ab} = 2cg_{ab},\ c = const.$	Homothetic Killing Vector
$\mathcal{L}_\xi g_{ab} = 2\psi g_{ab},\ \psi \neq const.$	Conformal Killing Vector
$\mathcal{L}_\xi \Gamma^i_{jk} = 0,$	Affine collineation Vector
$\mathcal{L}_\xi R_{ab} = 0,$	Ricci collineation
$\mathcal{L}_\xi R^i_{jkl} = 0,$	Curvature collineation
$g^{ab}\mathcal{L}_\xi R_{ab} = 0,$	Contracted Ricci collineation.
$\mathcal{L}_\xi W^i_{jkl} = 0,$	Contracted Ricci collineation.

b. Secondary effects.

These are identities satisfied by the metric geometric objects and are due to primary collineations. For instance if ξ^a is a conformal Killing vector (CKV) the a secondary effect is(see [22] Eqn.(5.1)):

$$\left(R^{ab}\xi_b\right)_{;a} = -3\Box\psi, \tag{8}$$

where $\Box\psi = g^{ab}\psi_{;ab}$.

c. Mathematical constraints

These are mathematical requirements introduced 'by hand' after the previous two sublevels of simplifying assumptions have been exhausted. The purpose of their introduction is to simplify further the equations obtained by the previous simplifying assumptions. Their form is indicated by the form of the equations they have to simplify. For example in the case of a CKV such a requirement is $\psi_{;ab} = 0$ because this condition removes $\psi_{;ab}$ from the field equations. A CKV which satisfies this condition is called a special CKV.

It is to be noted that up to now we discuss the background space only, therefore the results apply to all physical fields introduced and to be more specific to both the kinematics and the dynamics of the theory we shall develop on this space.

3.1. Some useful material about collineations

Each collineation has a different constraint 'power'. The strongest is the $\mathcal{L}_\xi g_{ab} = 0$ because if ξ^a satisfies this (i.e. it is a Killing vector) then satisfies all the rest. A collineation is called proper if it cannot be reduced to a 'simpler' one. For example conformal Killing vectors (CKV) contain the homothetic Killing vectors (HKV) when the conformal function ψ =constant. A proper CKV is one for which ψ is not constant.

The vectors which satisfy $\mathcal{L}_\xi g_{ab} = 0$ are called Killing vectors and the equation$\mathcal{L}_\xi g_{ab} = 0$ Killing equation. In an n-dimensional space with a non-degenerate metric there exist at most $\frac{n(n+1)}{2}$ KVs. If this is the case then the space is called a space of constant curvature[3]. The KVs serve all standard geometric symmetry concepts e.g. spherical symmetry, cylindrical symmetry etc.

The KVs form a finite dimensional Lie algebra. This algebra can be used to classify the metrics in various classes in the sense that one is able to write the metric in a form that takes

[3] It should be called metric of constant curvature.

into account the symmetry and it is written in terms of a small number of parameters. An well known example is the Freedman - Robertson - Walker (FRW) spacetime which is used in the standard cosmology. The metric of this spacetime has the form:

$$ds_{RW}^2 = -dt^2 + R(t)^2(dx^2 + dy^2 + dz^2), \tag{9}$$

where (t, x, y, z) are Cartesian coordinates and $R(t)$ is the only unknown function of (the coordinate) time t and can be defined by means of the following symmetry assumptions [2]:

1. Spacetime admits a timelike gradient Conformal Killing Vector (CKV) such that there exists another reduced metric for which the CKV is a KV
2. The 3-D hypersurfaces orthogonal to the CKV are spaces of constant curvature.

The collineations other than Killing vectors fix the metric to a lesser degree, however they do act as constraints and can be used in any case as such.

3.2. The generic collineation

One can prove that the Lie derivative of every metric geometric object can be expressed in terms of the quantity $\mathcal{L}_\xi g_{ab}$ and its derivatives. For example we have the relations:

$$\begin{aligned}
L_\xi \Gamma^c{}_{ab} &= \frac{1}{2} g^{cd} \left[(L_\xi g_{ad})_{;b} + (L_\xi g_{bd})_{;a} - (L_\xi g_{ab})_{;d} \right], && (10) \\
L_\xi R^a_{\ bcd} &= (L_\xi \Gamma^a{}_{bd})_{;c} - (L_\xi \Gamma^a{}_{bc})_{;d}, && (11) \\
L_\xi R_{\ ab} &= (L_X \Gamma^c{}_{ab})_{;c} - (L_X \Gamma^c{}_{ca})_{;b}. && (12)
\end{aligned}$$

This result leads us to introduce the concept of **generic collineation** by the identity:

$$L_\xi g_{ab} = 2\psi g_{ab} + 2H_{ab}. \tag{13}$$

where H_{ab} is a symmetric traceless tensor. Then it is possible to express every collineation in terms of the symmetry variables ψ, H_{ab} and their derivatives. This approach greatly unifies and simplifies the study of the effects of each collineation and reveals its relative significance.

As an example we refer the following general result:

$$L_\xi R_{ab} = -(n-2)\,\psi_{;ab} - g_{ab}\Box\psi + 2H^d_{.(a;b)d} - \Box H_{ab}. \tag{14}$$

In terms of trace and traceless parts $L_\xi R_{ab}$ is written as follows:

$$L_\xi R_{ab} = -(n-2)\,A_{ab} + 2K_{ab} - \Box H_{ab} + \frac{1}{n} g_{ab} \left[-2(n-1)\Box\psi + 2H^{ab}_{.;ab} \right], \tag{15}$$

where the traceless tensors:

$$A_{ab} = \psi_{;ab} - \frac{1}{n} g_{ab}\Box\psi \text{ and } K_{ab} = H^d_{.(a;b)d} - \frac{1}{n} g_{ab} H^{cd}_{.;cd}. \tag{16}$$

In spacetime $(n = 4)$ the above formulae read:

$$L_\xi R_{ab} = -2\psi_{;ab} - g_{ab}\Box\psi + 2H^d_{.(a;b)d} - \Box H_{ab}, \tag{17}$$

$$L_\xi R_{ab} = -2A_{ab} + 2K_{ab} - \Box H_{ab} + \frac{1}{4} g_{ab} \left(-6\Box\psi + 2H^{ab}_{.;ab} \right), \tag{18}$$

$$A_{ab} = \psi_{;ab} - \frac{1}{4} g_{ab}\Box\psi \text{ and } K_{ab} = H^d_{.(a;b)d} - \frac{1}{4} g_{ab} H^{cd}_{.;cd}. \tag{19}$$

This is as far as one can go with the spacetime structure (i.e. the metric tensor). It is possible to continue making Geometry in spacetime but no Physics. To do the latter one has to introduce more tensor fields which will describe / correspond to physical quantities.

Journal of Physics: Conference Series **189** (2009) 012041 doi:10.1088/1742-6596/189/1/012041

4. The observers - Kinematics

To begin making Physics in spacetime one has to introduce observers. In accordance with Special Relativity the observers are introduced by the four-velocity i.e. a unit timelike vector field or, equivalently, by a timelike congruence of differentiable curves, which are the integral curves of the vector field. The Physics of the new timelike vector field we call the kinematics in spacetime. Obviously the Kinematics of a *certain spacetime* is different for different observers. The standard approach is to consider the observers as a 'fluid' in spacetime and express the kinematics of these observers in terms of fluid flow quantities.

One point which must be emphasized is the following. In Special Relativity the curvature tensor vanishes, the geometry is absolute (that of flat space) and does not give conditions/constraints on four-velocity. In General Relativity the situation is different and the four-velocity one considers *must* satisfy Ricci's identity. Concerning Bianchi's second identity we note that it involves only the metric (geometry) therefore it does not restrict directly the four-velocity. However it does impose restrictions on the metric and consequently in the symmetries of the metric, hence possibly on the four-velocity. The conclusion is that when we introduce a four-velocity to describe the mater in a given spacetime and compute the kinematic variables (as we shall do below) we are led to a set of identities / constraints which the kinematic variables must satisfy.

The introduction of a non-null vector field in spacetime allows one to decompose a tensor equation and an individual tensor parallel and normal to the vector field. The vector field can be either timelike or spacelike. This decomposition, called **1+3 decomposition** for obvious reasons, is of primary importance to Physics because it is covariant, in the sense that each irreducible part transforms independently of the other, therefore one can break the study the Physics (Kinematics or Dynamics) of a tensor field / equation by studying the simpler Physics of its irreducible parts. In the following we develop the basic mathematics of the 1+3 decomposition[4] wrt P^a.

4.1. 1+3 decomposition wrt a non-null vector field

Consider a non-null ($|P^iP_i| \neq 0$) vector field P^i with signature $\varepsilon(P) = P^iP_i/|P^iP_i| = \pm 1$, where the + sign applies to a spacelike 4-vector and the − sign to a timelike 4-vector in a metric space with metric g_{ij}. The projection tensor $h_{ij}(P)$ associated with P^i is defined by the equation:

$$h_{ij}(P) = g_{ij} - \frac{\varepsilon(P)}{P^2} P_i P_j, \tag{20}$$

where $P^2 = |P^iP_i| = \varepsilon(P)P^iP_i > 0$. It is easy to prove the properties:

$$h_{ij}(P)P^j = 0, h^{ij}(P)h_{jk}(P) = h^i_k(P), h^i_i(P) = 3.$$

The tensor $h_{ij}(P)$ projects normal to P^a and gives us the possibility to decompose any tensor field in irreducible parts in a direction parallel to P^a and another normal to P^a.

Proposition 1 *Any vector field R^i can be 1+3 decomposed wrt P^i as follows:*

$$R^i = \alpha P^i + \beta^j h^i_j, \tag{21}$$

where

$$\alpha = \tfrac{\varepsilon(P)}{P} R^i P_i,$$
$$\beta^i = R^i.$$

[4] The results apply to the $1 + (n-1)$ decomposition by making the necessary adjustments e,g replace 3 with $n - 1$ etc.

Proof. We have $R^i = R^j g^i_j$. Using (20) this gives:

$$\begin{aligned} R^i &= R^j g^i_j, \\ &= R^j \left(\frac{\varepsilon(P)}{P^2} P_j P^i + h^i_j \right), \\ &= \frac{\varepsilon(P)}{P^2} (R^j P_j) P^i + h^i_j R^j. \end{aligned}$$

■

Proposition 2 *A second rank tensor Y_{ij} is 1+3 decomposed wrt the vector P^i by means of the identity:*

$$Y_{ij} = \alpha P_i P_j + \varepsilon(P) \beta_k h^k_j P_i + \varepsilon(P) \gamma_k h^k_i P_j + \epsilon_{ij}, \tag{22}$$

where

$$\alpha = \frac{1}{P^4} Y_{ij} P^i P^j \ , \ \beta_i = \frac{1}{P^2} Y_{ij} P^j \ , \ \gamma_i = \frac{1}{P^2} Y_{ji} P^j \ , \ \epsilon_{ij} = Y_{kr} h^k_i h^r_j.$$

Proof. *We write $Y_{ij} = Y_{kr} g^k_i g^r_j$ and using (20) we have:*

$$\begin{aligned} Y_{ij} &= Y_{kr} g^k_i g^r_j, \\ &= Y_{kr} \left(\left(\frac{1}{P^2} \varepsilon(P) P^k P_i + h^k_i \right) \left(\frac{1}{P^2} \varepsilon(P) P^r P_j + h^r_j \right), \\ &= \left(\frac{1}{P^4} Y_{kr} P^k P^r \right) P_i P_j + \frac{1}{P^2} \varepsilon(P) \left(Y_{kr} P^k \right) P_i h^r_j + \frac{1}{P^2} \varepsilon(P) \left(Y_{kr} P^r \right) h^k_i P_j + Y_{kr} h^k_i h^r_j, \\ &= \alpha P_i P_j + \varepsilon(P) \beta_k h^k_j P_i + \varepsilon(P) \gamma_k h^k_i P_j + \epsilon_{ij}. \end{aligned}$$

■

We note that a symmetric 2-index tensor is specified (and specifies) in terms of five different quantities: one scalar, two vector and one projected second rank tensor.

5. 1+3 decomposition wrt a timelike unit vector field

In the following we consider P^i to be a normalized timelike vector field ($\varepsilon(P) = -1$) (e.g. a 4-velocity) and we denote it u^i. Then equations (20), (21), (22) give the 1+3 decomposition wrt u^i ($u^i u_i = -1$) :

$$h_{ij}(u) = g_{ij} + u_i u_j, \tag{23}$$

$$R^i = -\left(R^j u_j \right) u^i + R^j h^i_j(u), \tag{24}$$

$$Y_{ij} = \left(Y_{kr} u^k u^r \right) u_i u_j - \left(Y_{kr} u^k \right) u_i h^r_j - \left(Y_{kr} u^r \right) h^k_i u_j + Y_{kr} h^k_i h^r_j. \tag{25}$$

For a **symmetric** tensor of type (0,2) decomposition (25) is written as follows:

$$\begin{aligned} Y_{ab} &= (Y^{rs} u_r u_s) u_a u_b - Y^{rs} u_r h_{sb} u_a - Y^{rs} u_s h_{ra} u_b, \\ &+ \frac{1}{3} (Y^{rs} h_{rs}) h_{ab} + (h^r_a h^s_b - \frac{1}{3} h_{ab} h^{rs}) Y_{rs}, \end{aligned} \tag{26}$$

that is we brake further the symmetric part in a trace and a traceless part.

We consider now various applications of the 1+3 decomposition wrt the vector field u^a.

5.1. The kinematic variables of the four-velocity

We 1+3 decompose the tensor $u_{i;j}$. From (23) we have the following identity/decomposition:

$$u_{i;j} = \left(u_{k;r}u^k u^r\right) u_i u_j - \left(u_{k;r}u^k\right) u_i h^r_j - \left(u_{k;r}u^r\right) h^k_i u_j + u_{k;r}h^k_i h^r_j. \tag{27}$$

But $u_{i;j}u^i = \frac{1}{2}\left(u_i u^i\right)_{;j} = 0$ and $u_{i;j}u^i = \dot{u}_j = h^r_j \dot{u}_r$. Therefore equation (27) becomes:

$$u_{i;j} = -\dot{u}_i u_j + u_{k;r}h^k_i h^r_j. \tag{28}$$

We continue by decomposing the space-like part $u_{k;r}h^k_i h^r_j$ in an antisymmetric and a symmetric part as follows:

$$u_{k;r}h^k_i h^r_j = \omega_{ij} + \theta_{ij}, \tag{29}$$
$$\omega_{ij} = u_{k;r}h^k_{[i}h^r_{j]}, \tag{30}$$
$$\theta ij = u_{k;r}h^k_{(i}h^r_{j)}. \tag{31}$$

The symmetric part can be decomposed covariantly further to a trace and a traceless part, that is we write:

$$\theta_{ij} = \sigma_{ij} + \frac{1}{3}\theta h_{ij}, \tag{32}$$

where

$$\theta = \theta^i_i = h^{ij}u_{i;j},$$
$$\sigma_{ij} = \theta_{ij} - \frac{1}{3}\theta h_{ij} = \left[h^r_{(i}h^k_{r)} - \frac{1}{3}h^{rk}h_{ij}\right] u_{r;k}.$$

The term ω_{ij} is called the vorticity tensor of u^i, σ_{ij} the shear tensor of u^i,, θ the expansion of u^i and $\dot{u}^i$ the four-acceleration of the timelike vector field u^i. These are the kinematic variables of u^i, considering u^i to be the 4-velocity of a relativistic fluid.

We infer by their definition that the kinematic variables satisfy the properties:

$$\omega_{ij} = -\omega_{ji}\,,\ \omega_{ij}u^j = 0,$$
$$\sigma_{ij} = \sigma_{ji}\,,\ \sigma^i_i = \sigma_{ij}u^j = 0,$$
$$\omega_{ij} = h^k_i h^r_j \omega_{kr},$$
$$\sigma_{ij} = h^k_i h^r_j \sigma_{kr}.$$

The geometric meaning of each kinematic term is obtained from the study of the integral curves of u^a. We shall not comment further on that at this point. The physical meaning of each term comes from the physical interpretation of u^a as a four velocity. Therefore we interpret each quantity in terms of relative motion. In this respect we say that θ is expansion (isotropic strain), σ_{ab} is shear (anisotropic strain), ω_{ab} relative rotation and $\dot{u}_a$ 4-acceleration. These quantities (i.e. $\sigma_{ab}, \omega_{ab}, \dot{u}_a, \theta$) are the fundamental physical quantities of relativistic (and Newtonian) kinematics.

We introduce the two scalars:

$$\sigma^2 = \frac{1}{2}\sigma_{ab}\sigma^{ab}, \tag{33}$$
$$\omega^2 = \frac{1}{2}\omega_{ab}\omega^{ab}. \tag{34}$$

Let us see an application.

Proposition 3 *Necessary and sufficient conditions for the vector field u^i to be a Killing vector is:*

$$\sigma_{ij} = 0\,,\ \theta = 0\,, \dot{u}^i = 0,$$

Proof. We have:

$$L_u g_{ij} = 2u_{(i;j)} = 2\sigma_{ij} + \frac{2}{3}\theta h_{ij} - 2\dot{u}_{(i}u_{j)}, \tag{35}$$

In order u^i to be a killing vector it should satisfy the condition $L_u g_{ij} = 0$. Combining these two relations we find:

$$\sigma_{ij} = 0\,,\ \theta = 0\,, \dot{u}^i = 0.$$

■

5.2. The propagation and the constraint equations

As we remarked above the vector field of the four-velocity decomposes both the tensor fields defining the geometric objects of the theory as well as covariant equations between them. As we have seen in spacetime we have two main geometric identities: Ricci identity and Bianchi identities. It is possible that we 1+3 decompose these identities and obtain simpler relations which involve the kinematic quantities for the four-velocity. The resulting equations are constraint equations which *must* be satisfied by all the kinematic quantities involved.

The 1+3 decomposition of the Ricci identity leads to two sets of equations each set containing nine equations. The first set contains the derivatives of the kinematic quantities along u^a which we call the propagation equations. The second set of equations is called the constraint equations. The two sets of equations have as follows:

Propagation equations[5]:

$$h^b_a\dot{\omega}^a = (\sigma^a_d - \frac{2}{3}\theta h^a_d)\omega^d + \frac{1}{2}\eta^{abcd}u_b\dot{u}_{c;d}, \quad \text{(three equations)} \tag{36}$$

$$\dot{\theta} + \frac{1}{3}\theta^2 + 2(\sigma^2 - \omega^2) = -R_{ab}u^a u^b + \dot{u}^a{}_{;a}, \quad \text{(one equation)} \tag{37}$$

$$\begin{aligned} -E_{st} + \frac{1}{2}\left(h^a_s h^b_t - \frac{1}{3}h_{st}h^{ab}\right)R_{ab} &= h^a_s h^b_t\left[\dot{\sigma}_{ab} - \dot{u}_{(a;b)}\right] + \sigma_{sc}\sigma^c{}_t + \text{(five equations)} \\ &+ \frac{2}{3}\sigma_{st}\theta + \omega_s\omega_t - \dot{u}_s\dot{u}_t - \frac{1}{3}\left(2\sigma^2 + \omega^2 - \dot{u}^b_{;b}\right)h_{st}. \end{aligned} \tag{38}$$

Constraint equations:

$$h^a_b\omega^b_{;a} = \dot{u}^a\omega_a, \quad \text{(one equation)} \tag{39}$$

$$h^c_s\left[\frac{2}{3}\theta_{,c} - h^{ab}\sigma_{ca;b} - \eta_{camn}u^a\left(\omega^{m;n} + 2\omega^m\dot{u}^n\right)\right] = -h^c_s R_{cd}u^d \ ,(three equations) \tag{40}$$

$$-h^a_{(s}h^b_{l)}\left[\sigma_b{}^{c;d} + \omega_b{}^{c;d}\right]\eta_{arcd}u^r + 2\dot{u}_{(s}\omega_{l)} = H_{sl}. \quad \text{(five equations)} \tag{41}$$

[5] Note that the propagation equation of ω_{ab} (36) contains only kinematic terms and it is independent of R_{ab} hence of the dynamical variables to be introduced later on.

5.3. The 1+3 decomposition of the Bianchi identity (4)

We write the 1+3 decomposition of this geometric identity in terms of the Weyl tensor. It is well known that the Riemann tensor can be decomposed in the following irreducible parts:

$$R_{abcd} = C_{abcd} + \frac{1}{2}\left(g_{ac}R_{bd} + g_{bd}R_{ac} - g_{ad}R_{bc} - g_{bc}R_{ad}\right) - \frac{1}{6}Rg_{abcd}, \tag{42}$$

where:

$$g_{abcd} = g_{ac}g_{bd} - g_{ad}g_{bc}, \tag{43}$$

and C_{abcd} is the Weyl tensor. Equation (42) is a mathematical identity. The Weyl tensor is decomposed further in terms of the electric part E_{ab} and the magnetic part H_{ab} wrt the vector field u^a as follows:

$$C_{abcd} = (g_{abrs}g_{cdmt} - \eta_{abrs}\eta_{cdmt})u^r u^m E^{st} - (\eta_{abrs}g_{cdmt} + g_{abrs}\eta_{cdmt})u^r u^m H^{st}. \tag{44}$$

We note the relations:

$$E_{ac} = C_{abcd}u^b u^d, \quad H_{ac} = \frac{1}{2}\eta_{am.}{}^{kl}C_{klbt}u^m u^t. \tag{45}$$

The tensors E_{ac}, H_{ac} are symmetric, traceless and satisfy the property:

$$E_{ac}u^c = H_{ac}u^c = 0. \tag{46}$$

If in (4) we substitute the Weyl tensor C^{abcd} in terms of the electric and the magnetic parts E^{ab}, H^{ab} defined in (45), we obtain four identities which have a form similar to Maxwell equations for the electric and the magnetic field:

∇E :

$$\begin{aligned} h^t_a h^d_s E^{as}{}_{;d} - \eta^{tbpq}u_b\sigma_p{}^d H_{qd} + 3H^t{}_s\omega^s &= \frac{1}{3}h^{tb}\mu_{;b} - \frac{1}{2}h^t_c\pi^{cb}{}_{;b} - \frac{3}{2}\omega^t_{.b}q^b \\ &+ \frac{1}{2}\sigma^t_{.b}q^b + \frac{1}{2}\pi^t_{.b}\dot{u}^b - \frac{1}{3}\theta q^t, \end{aligned} \tag{47}$$

∇H :

$$\begin{aligned} h^t_a h^d_s H^{as}{}_{;d} + \eta^{tbpq}u_b\sigma^d_{.p}E_{qd} - 3E^t_{.s}\omega^s &= (\mu + p)\,\omega^t + \frac{1}{2}\eta^{tbsf}u_b q_{s;f} \\ &+ \frac{1}{2}\eta^{tbsf}u_b\pi_{sc}\left(\omega^{;c}_f + \sigma^{;c}_f\right), \end{aligned} \tag{48}$$

$\dot{E}_{ab}$:

$$\begin{aligned} \left[\begin{array}{c} h^m_a h^t_c \dot{E}^{ac} + h^{(m}_a\eta^{t)rsd}u_r H^a_{.s;d} - 2H^{\cdot(t}_q\eta^{m)bpq}u_b\dot{u}_p + \\ +h^{mt}\sigma^{ab}E_{ab} + \theta E^{mt} - 3E^{(m}_s\sigma^{t)s} - E^{(m}_s\omega^{t)s} \end{array}\right] &= -\frac{1}{2}(\mu + p)\,\sigma^{tm} - \dot{u}^{(t}q^{m)} \\ &- \frac{1}{2}h^t_a h^m_c q^{(a;c)} - \frac{1}{2}h^t_a h^m_c \dot{\pi}^{ac} \\ &- \frac{1}{2}\pi^{b(m}\left(\omega^{.t)}_b + \sigma^{.t)}_b\right) - \frac{1}{6}\pi^{tm}\theta \\ &+ \frac{1}{6}h^{mt}\left(q^a_{.;a} + \dot{u}_a q^a + \pi^{ab}\sigma_{ab}\right), \end{aligned} \tag{49}$$

$\dot{H}_{ab}$:

$$\begin{aligned} \left[\begin{array}{c} h^{ma}h^{tc}\dot{H}_{ac} - h^{(m}_a\eta^{t)rsd}E^a_{.s;d} + 2E^{(t}_q\eta^{m)bpq}u_b\dot{u}_p + \\ +h^{mt}\sigma^{ab}H_{ab} + \theta H^{mt} - 2H^{;(m}_s\sigma^{t)s} - H^{;(m}_s\omega^{t)s} \end{array}\right] &= \tfrac{1}{2}\sigma^{(t}_c\eta^{m)bcf}u_b q_f - \tfrac{1}{2}h^{(t}_c\eta^{m)bef}u_b\pi^c_{.e;f} \\ &+ \tfrac{1}{2}\left(h^{mt}\omega_c q^c - 3\omega^{(m}q^{t)}\right). \end{aligned} \tag{50}$$

The contracted Bianchi identity $G^{ab}_{..;b} = 0$ is contained in these identities.

6. Propagation of the kinematic quantities along the collineation vector

The propagation equations give the derivative of the kinematic quantities along the four-velocity u^a of the observers. However when a collineation is present then one wants to know how the kinematic quantities propagate along the collineation vector. Because the collineation is defined in terms of the Lie derivative (Lie transport) one is interested in the quantities $L_\xi\{u^a, \omega_{ab}, \sigma_{ab}, \theta, \dot{u}^a\}$. To do that we give the following general result.

Proposition 4 *Let X^a be a non-null vector field with index $\varepsilon(X) = \pm 1$ i.e. $X^a X_a = \varepsilon(X) X^2$ $(X > 0)$ and let ξ^a be an arbitrary vector field with collineation parameters(see [6]) ψ, H_{ab} that is:*

$$\xi_{(a;b)} = \psi g_{ab} + H_{ab} = \frac{1}{2} L_\xi g_{ab}, \qquad L_\xi g^{ab} = -2\psi g^{ab} - 2H^{ab}. \tag{51}$$

Then $L_\xi X^a$ is 1+3 decomposed wrt the four-velocity u^a as follows:

$$L_\xi X^a = \left[(\ln X)_{,b} \xi^b - \psi - \frac{\varepsilon(X)}{X^2} H_{cb} X^c X^b \right] X^a + V^a(X). \tag{52}$$

where $V^a(X) = h^a_b(X)(L_\xi X^b)$ a vector field normal to X^a. For the covariant quantity $L_\xi X_a$ holds:

$$L_\xi X_a = \left[(\ln X)_{,b} \xi^b + \psi + \frac{\varepsilon(X)}{X^2} H_{cb} X^c X^b \right] X_a + \hat{V}_a(X), \tag{53}$$

where:

$$\hat{V}_a(X) = 2h^b_a(X) H_{bc} X^c + V_a(X). \tag{54}$$

Note that $\hat{V}_a(X) \neq V_a(X)$!!

In case $X^2 = 1$ formulae (52) and (53) reduce to:

$$L_\xi X^a = -\left[\psi + \varepsilon(X) H_{cb} X^c X^b \right] X^a + V^a(X), \tag{55}$$

$$L_\xi X_a = \left[\psi + \varepsilon(X) H_{cb} X^c X^b \right] X_a + \hat{V}_a(X). \tag{56}$$

In the special case $X^a = u^a$ we find:

$$L_\xi u^a = -(\psi - H_{cd} X^c X^d) u^a + V^a(u), \tag{57}$$

$$L_\xi u_a = (\psi - H_{cd} X^c X^d) u_a + \hat{V}_a(u). \tag{58}$$

We have the following (easy to prove) result.

Proposition 5 *(i) $\hat{V}_a(X) = V^a(X)$ iff $h^{ab}(X) H_{bc} X^c = 0$ iff $H_{bc} X^c = a X_b$ i.e. X^a is an eigenvector of H_{bc}.*

(ii) $\hat{V}^a(X) = V^a(X)$ for all non-null X^a iff ξ^a is at most a CKV.

(iii) For $X^a = u^a$ we have $\hat{V}^a(u) = V^a(u)$ iff either u^a is an eigenvector of H_{bc} or ξ^a is at most a CKV. Furthermore in that later case holds:

$$L_\xi u^a = -\psi u^a + V^a(u), \tag{59}$$

$$L_\xi u_a = \psi u_a + V_a(u). \tag{60}$$

6.1. The case $X^2 = 1$

In the following we assume that X^a is **UNIT** ($\varepsilon(X) = \pm 1$) and compute the quantity $V_a(u)$ in terms of the "kinematic" quantities of X^a. We find:

$$L_\xi X_a = \left[2\omega(X)_{ab} + \varepsilon(X)(\overset{*}{X}_a X_b - \overset{*}{X}_b X_a)\right]\xi^b + (X_b\xi^b)_{;a} \tag{61}$$

where a star "*" over a symbol indicates derivation along X^a i.e. $\overset{*}{X}_a = X_{a;b}X^b$. We also write $2\omega(X)_{ab} = h^c_a(X)h^d_b(X)(X_{c;d} - X_{d;c})$. We have according to the general 1+3 decomposition formula (note that $X^2 = 1$):

$$X_{a;b} = h(X)^c_a h(X)^d_b X_{a;b} + \varepsilon(X)\overset{*}{X}_a X_b. \tag{62}$$

We 1+3 decompose $L_\xi X_a$ along X^a. For the parallel component we find[6]:

$$X^a L_\xi X_a = -(\overset{*}{X}_b \xi^b) + (X_b\xi^b)^* = X_b \overset{*b}{\xi}. \tag{63}$$

This implies the formula[7]:

$$L_\xi X_a = \varepsilon(X)\left[-(\overset{*}{X}_b\xi^b) + (X_b\xi^b)^*\right] + \hat{V}_a(X). \tag{64}$$

Comparing (61) and (64) we find:

$$\hat{V}_a(X) = 2\omega(X)_{ab}\xi^b + \varepsilon(X)\overset{*}{X}_a(X_b\xi^b) + (X_b\xi^b)_{;a} - \varepsilon(X)(X_b\xi^b)^* X_a. \tag{65}$$

Noting that:

$$(X_b\xi^b)_{;a} - \varepsilon(X)(X_b\xi^b)^* X_a = h^b_a(X_c\xi^c)_{;b} \tag{66}$$

we obtain the final result:

$$\hat{V}_a(X) = 2\omega(X)_{ab}\xi^b + \varepsilon(X)\overset{*}{X}_a(X_b\xi^b) + h^b_a(X_c\xi^c)_{;b}. \tag{67}$$

It is possible to express the parallel component of the $L_\xi X_a$ in terms of the collineation components. Indeed, by Lie differentiating:

$$g^{ab}X_a X_b = \varepsilon(X), \tag{68}$$

we find:

$$\begin{aligned}(L_\xi g^{ab})X_a X_b + 2g^{ab}X_a L_\xi X_b = 0 \Rightarrow \\ (-2\psi g^{ab} - 2H^{ab})X_a X_b + 2g^{ab}X_a L_\xi X_b = 0 \Rightarrow \\ X^b L_\xi X_b = \varepsilon(X)\psi + H^{ab}X_a X_b.\end{aligned} \tag{69}$$

From (63) and (69) follows:

$$\varepsilon(X)\psi + H^{ab}X_a X_b = -(\overset{*}{X}_b\xi^b) + (X_b\xi^b)^* = X_b\overset{*b}{\xi}. \tag{70}$$

[6] Note that $\overset{*b}{\xi}$ does not in general coincide with $\overset{*}{\xi}_b$

[7] Compare with (56). Also note that $L_\xi X_a = \varepsilon(X)(X^b L_\xi X_b)X_a + h^b_a L_\xi X_b$.

This relation implies the identity/decomposition:

$$L_\xi X_a = \left[\psi + \varepsilon(X) H^{cd} X_c X_d\right] X_a + 2\omega(X)_{ab}\xi^b + \varepsilon(X)\overset{*}{X}_a(X_b\xi^b) + h_a^b(X_c\xi^c)_{;b}. \tag{71}$$

Next, we compute the Lie derivative $L_\xi X^a$. We find:

$$L_\xi X^a = \left[-\psi + \varepsilon(X) H^{cd} X_c X_d\right] X^a - 2H^{ab}X_b + 2\omega(X)^a_{\ .b}\xi^b + \varepsilon(X)\overset{*}{X}{}^a(X_b\xi^b) + h^{ab}(X_c\xi^c)_{;b}. \tag{72}$$

Now ($X^2 = 1$):

$$H^{ab}X_b = \varepsilon(X)(H^{cd}X_cX_d)X^a + h_c^a H^{cd}X_d, \tag{73}$$

therefore:

$$L_\xi X^a = \left[-\psi - \varepsilon(X) H^{cd} X_c X_d\right] X^a - 2h_c^a H^{cd}X_d + 2\omega(X)^a_{\ .b}\xi^b + \varepsilon(X)\overset{*}{X}{}^a(X_b\xi^b) + h^{ab}(X_c\xi^c)_{;b}. \tag{74}$$

Comparing with (56) we find:

$$V_a(X) = -2h_c^a H^{cd}X_d + 2\omega(X)^a_{\ .b}\xi^b + \varepsilon(X)\overset{*}{X}{}^a(X_b\xi^b) + h^{ab}(X_c\xi^c)_{;b}. \tag{75}$$

Exercise 6 *Using the results (71),(74) prove that* $L_\xi(X^aX_a) = 0$.

6.2. The special case $X^a = u^a$

An important special case is $X^a = u^a$, that is , X^a is unit and timelike (the four-velocity). In this case the general formulae (52) and (55) give:

$$L_\xi u^a = (-\psi(\xi) + H_{cd}(\xi)u^cu^d)u^a + V^a(u), \tag{76}$$

$$L_\xi u_a = (\psi(\xi) - H_{cd}(\xi)u^cu^d)u_a + \hat{V}_a(u). \tag{77}$$

We write $L_\xi u_a$ and $L_\xi u^a$ in terms of the kinematic quantities of u^a. From (71) follows:

$$L_\xi u_a = \left[\psi - H^{cd}u_cu_d\right] u_a + 2\omega_{ab}\xi^b - \dot{u}_a\,(u_b\xi^b) + h_a^b(u_c\xi^c)_{;b}, \tag{78}$$

and from (72):

$$L_\xi u^a = -\left[\psi - H^{cd}u_cu_d\right] u^a - 2h_b^a H^{bc}u_c + 2\omega^a_{\ .b}\xi^b - \dot{u}^a(u_b\xi^b) + h^{ab}(u_c\xi^c)_{;b}. \tag{79}$$

These imply that:

$$\hat{V}_a(u) = 2\omega_{ab}\xi^b - \dot{u}_a\,(u_b\xi^b) + h_a^b(u_c\xi^c)_{;b}, \tag{80}$$

$$V^a(u) = -2h_b^a H^{bc}u_c + 2\omega^a_{\ .b}\xi^b - \dot{u}^a(u_b\xi^b) + h^{ab}(u_c\xi^c)_{;b}. \tag{81}$$

From (78) we draw the following conclusions:

1. If $u_a\xi^a$ is an acceleration potential (that is the acceleration is of the form $a = -(u_b\xi^b)^{-1}$) then for any collineation ξ^a:

$$L_\xi u_a = \left[\psi - H^{cd}u_cu_d\right] u_a + 2\omega_{ab}\xi^b. \tag{82}$$

One special case of this is when $u_a\xi^a = 0$ i.e. the symmetry vector is normal to the flow vector u^a (see [4]).

2. If ξ^a is a KV we have:

$$L_\xi u_a = 2\omega_{ab}\xi^b - \dot{u}_a(u_b\xi^b) + h_a^b(u_c\xi^c)_{;b}, \tag{83}$$

from which we conclude:

$$L_\xi u_a = 0 \Longleftrightarrow 2\omega_{ab}\xi^b - \dot{u}_a(u_b\xi^b) + h_a^b(u_c\xi^c)_{;b} = 0. \tag{84}$$

A CKV is inherited[8] by the four-velocity field u^a if $u_b\xi^b = 0$ and the fluid is either irrotational (i.e. $\omega_{ab} = 0$) or the vorticity vector $\omega^a \parallel \xi^a$ (see [4]).

3. For a CKV the inheritance of the symmetry by u^a (i.e.$L_\xi u_a = \psi u_a$ or $L_\xi u^a = -\psi u^a$) is equivalent to the identity $h_a^b L_\xi dx^a = 0$ (see [22]). Obviously for a KV/HKV/SCKV the result remains true.

We note that the requirement of surface forming of ξ^a with a unit timelike Killing vector (a static spacetime) results in the kinematic constraint (84) which is by no means trivial. For example when $\xi^a u_a = 0$ then $\omega_a\xi^a = 0$ i.e. $\omega_a // \xi^a$.

7. The 1+3 decomposition of the Lie derivative $L_\xi X_{a;b}$ wrt X^a

This decomposition is useful because allows us to compute the Lie derivative of the kinematic quantities along the collineation vector ξ^a. We have the following identity for any vector fields[9] X^a, ξ^a :

$$L_\xi X_{a;b} = (L_\xi X_a)_{;b} - (L_\xi \Gamma^c_{ab})X_c, \tag{85}$$

and also the identity:

$$L_\xi \Gamma^c_{ab} = \frac{1}{2}g^{cd}\left[(L_\xi g_{da})_{;b} + (L_\xi g_{db})_{;a} - (L_\xi g_{ab})_{;d}\right]. \tag{86}$$

Assuming ξ^a to be a collineation the second identity gives:

$$L_\xi \Gamma^c_{ab} = g^{cd}\left[\psi_{,b}\, g_{da} + \psi_{,a}\, g_{db} - \psi_{,d}\, g_{ab} + \quad H_{da;b} + H_{db;a} - H_{ab;d}\right].$$

Therefore:

$$L_\xi X_{a;b} = (L_\xi X_a)_{;b} - \left[\psi_{,b}\, g_{da} + \psi_{,a}\, g_{db} - \psi_{,d}\, g_{ab} + H_{da;b} + H_{db;a} - H_{ab;d}\right] X^d. \tag{87}$$

A different expression is found as follows:

$$\begin{aligned} L_\xi \Gamma^c_{ab} &= \frac{1}{2}g^{cd}\left[(L_\xi g_{da})_{;b} + (L_\xi g_{db})_{;a} - (L_\xi g_{ab})_{;d}\right], \\ &= g^{cd}\left[(\xi_{(d;a)})_{;b} + (\xi_{(d;b)})_a - (\xi_{(a;b)})_{;d}\right], \\ &= \frac{1}{2}g^{cd}\left[\xi_{d;ab} + \xi_{a;db} + \xi_{d;ba} + \xi_{b;da} - \xi_{a;bd} - \xi_{b;ad}\right], \\ &= g^{cd}\left[\xi_{d;(ab)} + \xi_{a;[db]} + \xi_{b;[da]}\right], \\ &= g^{cd}\left[\xi_{d;(ab)} + \frac{1}{2}R_{tadb}\xi^t + \frac{1}{2}R_{tbda}\xi^t\right], \\ &= g^{cd}\left[\xi_{d;(ab)} + R_{t(a|d|b)}\xi^t\right]. \end{aligned} \tag{88}$$

This gives:

$$L_\xi X_{a;b} = (L_\xi X_a)_{;b} - g^{cd}\left[\xi_{d;(ab)} + R_{t(a|d|b)}\xi^t\right] X_c, \tag{89}$$

where we have defined the Riemann tensor with the Ricci identity:

$$2\xi_{a;[bc]} = R_{tabc}\xi^t. \tag{90}$$

[8] The symmetry is inherited if $L_\xi u^a = \lambda u^a$.

[9] This identity gives the commutation of the Lie and the covariant derivative in a Riemannian space. It can be found in [5].

7.1. The case $X^2 = 1$

We consider the case $X^2 = 1$ and find from (55):

$$(L_\xi X_a)_{;b} = \left[\psi + \varepsilon(X) H_{cd} X^c X^d] X_a + \hat{V}_a(X)\right]_{;b}, \\ = \left[\psi + \varepsilon(X) H_{cd} X^c X^d\right]_{;b} X_a + \left[\psi + \varepsilon(X) H_{cd} X^c X^d\right] X_{a;b} + \hat{V}_a(X)_{;b}. \quad (91)$$

where:

$$\hat{V}_a(X) = 2\omega(X)_{ab}\xi^b + \varepsilon(X)\overset{*}{X}_a(X_b\xi^b) + h_a^b(X_c\xi^c)_{;b}. \quad (92)$$

To save writing we set:

$$K(X) = \psi + \varepsilon(X) H_{cd} X^c X^d, \quad (93)$$

and have:

$$(L_\xi X_a)_{;b} = K(X)_{;b} X_a + K(X) X_{a;b} + \hat{V}_a(X)_{;b}.$$

Then, using (91) we get:

$$L_\xi X_{a;b} = \left[K(X)_{;b} X_a + K(X) X_{a;b} + \hat{V}_a(X)_{;b}\right] + \\ - \left[\psi_{,b}\, g_{da} + \psi_{,a}\, g_{db} - \psi_{,d}\, g_{ab} + H_{da;b} + H_{db;a} - H_{ab;d}\right] X^d, \quad (94)$$

and also:

$$L_\xi X_{a;b} = \left[K(X)_{;b} X_a + K(X) X_{a;b} + \hat{V}_a(X)_{;b}\right] - \left[\xi_{d;(ab)} + R_{t(a|d|b)}\xi^t\right] X^d. \quad (95)$$

7.1.1. <u>The case $X^a = u^a$</u> For this special case we find from (94):

$$L_\xi u_{a;b} = K(u)_{;b} u_a + K(u) u_{a;b} + \hat{V}_a(u)_{;b} + \quad (96)$$
$$- \left[\psi_{,b}\, g_{da} + \psi_{,a}\, g_{db} - \psi_{,d}\, g_{ab} + H_{da;b} + H_{db;a} - H_{ab;d}\right] u^d, \quad (97)$$

where:

$$K = \psi - H_{cd} u^c u^d, \ \hat{V}_a = \hat{V}_a(u). \quad (98)$$

We consider the decomposition $u_{a;b} = \sigma_{ab} + \omega_{ab} + \frac{1}{2}\theta h_{ab} - \dot{u}_a u_b$ and the lhs becomes:

$$L_\xi \sigma_{ab} + L_\xi \omega_{ab} + \frac{1}{2} L_\xi \theta h_{ab} + \frac{1}{2}\theta L_\xi h_{ab} - (L_\xi \dot{u}_a) u_b - \dot{u}_a L_\xi u_b,$$

therefore we can compute the quantities $L_\xi \sigma_{ab}, L_\xi \omega_{ab}, L_\xi \theta, L_\xi \dot{u}_a$ in terms of the collineation parameters ψ, H_{ab} by taking the irreducible parts of equation (96). Before we do that we need to calculate the Lie derivatives of the projection tensor h_{ab}.

7.2. Calculation of $L_\xi h_{ab}$

For the Lie derivative of the projection tensor we find the results:

$$L_\xi h_{ab}(X) = L_\xi \left(g_{ab} - \frac{\varepsilon(X)}{X^2} X_a X_b\right) = 2\psi h_{ab}(X) + 2H_{ab} - 2\frac{1}{X^4}\left(H_{cd} X^c X^d\right) X_{(a} X_{b)} + \quad (99) \\ - 2\frac{\varepsilon(X)}{X^2} X_{(a} \hat{V}_{b)}(X) - 2\frac{\varepsilon(X)}{X^2}\left[(\ln X)_{,c}\xi^c\right] X_{(a} X_{b)},$$

$$\begin{aligned} L_\xi h^a_b(X) &= L_\xi\left(g^a_b - \frac{\varepsilon(X)}{X^2}X^aX_b\right), \\ &= -\frac{\varepsilon(X)}{X^2}\left[2(\ln X)_{,c}\xi^cX^aX_b + V^a(X)X_b + X^a\hat{V}_b(X)\right]. \end{aligned} \tag{100}$$

In the case where the vector field is unit $(X^2 = 1)$ equations (99) and (100) reduce to:

$$L_\xi h_{ab}(X) = 2\psi h_{ab}(X) + 2H_{ab} - 2\left(H_{cd}X^cX^d\right)X_{(a}X_{b)} - 2\varepsilon(X)X_{(a)}\hat{V}_{b)}(X), \tag{101}$$

$$L_\xi h^a_b(X) = -\varepsilon(X)\left[V^a(X)X_b + X^a\hat{V}_b(X)\right]. \tag{102}$$

In the special case X^a is the 4-velocity u^a the above relations become:

$$L_\xi h_{ab} = 2\psi h_{ab} + 2H_{ab} - 2\left(H_{cd}u^cu^d\right)u_{(a}u_{b)} + 2u_{(a}\hat{V}_{b)}, \tag{103}$$

$$L_\xi h^a_b = u^a\hat{V}_b + V^au_b. \tag{104}$$

Replacing $\hat{V}_b, V^a$ from (80), (81) we find:

$$L_\xi h_{ab} = 2\psi h_{ab} + 2H_{ab} - 2\left(H_{cd}u^cu^d\right)u_{(a}u_{b)} + 2u_{(a}[2\omega_{b)c}\xi^c - \dot{u}_{b)}(u_c\xi^c) + h^d_{b)}(u_c\xi^c)_{;d}], \tag{105}$$

$$L_\xi h^a_b = u^a[2\omega_{bc}\xi^c - \dot{u}_b(u_c\xi^c) + h^d_b(u_c\xi^c)_{;d}] + [-2h^a_dH^{dc}u_c + 2\omega^a_{\cdot c}\xi^c - \dot{u}^a(u_c\xi^c) + h^{ad}(u_c\xi^c)_{;d}]u_b. \tag{106}$$

Concerning h^{ab} we find:

$$\begin{aligned} L_\xi h^{ab} &= -g^{ac}g^{bd}L_\xi h_{cd}, \\ &= -2\psi h^{ab} - 2H^{ab} \\ &\quad - g^{ac}g^{bd}\left[-2\left(H_{mn}u^mu^n\right)u_cu_d + 2u_{(c}[2\omega_{d)m}\xi^m - \dot{u}_{d)}(u_n\xi^n) + h^m_{d)}(u_n\xi^n)_{;m}]\right]. \end{aligned} \tag{107}$$

Exercise 7 *Prove that for a spacelike CKV ξ^a the following are true:*

$$L_\xi h_{ab} = 2\psi h_{ab} + 4u_{(a}\omega_{b)c}\xi^c, \tag{108}$$
$$L_\xi h^a_b = 2(u^a\omega_{bc} + \omega^a_{\cdot c}u_b)\xi^c, \tag{109}$$
$$L_\xi h^{ab} = -2\psi h^{ab} - 2H^{ab} - 4g^{ac}g^{bd}u_{(c}\omega_{d)m}\xi^m. \tag{110}$$

7.3. Calculation of $L_\xi\omega_{ab}, L_\xi\sigma_{ab}, L_\xi\theta, L_\xi\dot{u}_a$

We compute now the quantities $L_\xi\sigma_{ab}, L_\xi\omega_{ab}, L_\xi\theta, L_\xi\dot{u}_a$ in terms of the collineation parameters ψ, H_{ab}. We have[10] $(\hat{V}=\hat{V}(u), V=V(u))$:

$$\begin{aligned}
L_\xi\left(h_a^c h_b^d u_{c;d}\right) &= h_a^c h_b^d\left(L_\xi u_{c;d}\right)+h_a^c(L_\xi h_b^d)u_{c;d}+h_b^d(L_\xi h_a^c)u_{c;d},\\
&= h_a^c h_b^d\left[\left[\psi-H_{mn}u^m u^n\right]_{;d}u_c+\left[\psi-H_{mn}u^m u^n\right]u_{d;c}+\hat{V}_{c;d}\right]\\
&-h_a^c h_b^d\left[\psi_{,c}\,g_{ed}+\psi_{,d}\,g_{ec}-\psi_{,e}\,g_{cd}+H_{ec;d}+H_{ed;c}-H_{cd;e}\right]u^e\\
&+h_a^c\left[u^d\hat{V}_b+V^d u_b\right]u_{c;d}+h_b^d\left[u^c\hat{V}_a+V^c u_a\right]u_{c;d},\\
&= h_a^c h_b^d\left(\psi-H_{mn}u^m u^n\right)u_{c;d}+h_a^c h_b^d\hat{V}_{c;d}\\
&+h_a^c u_b u_{c;d}V^d+h_b^c u_a u_{d;c}V^d+\dot{u}_a\hat{V}_b\\
&+\dot{\psi}h_{ab}-h_a^c h_b^d\left[H_{ec;d}+H_{ed;c}-H_{cd;e}\right]u^e.
\end{aligned} \tag{115}$$

We consider the symmetric and the antisymmetric parts of $u_{a;b}$. For the antisymmetric part we find:

$$\begin{aligned}
L_\xi\omega_{ab} &= h_a^c h_b^d L_\xi u_{[c;d]} \Rightarrow\\
L_\xi\omega_{ab} &= \left(\psi-H_{mn}u^m u^n\right)\omega_{ab}+h_a^c h_b^d\hat{V}_{[c;d]}+\dot{u}_{[a}\hat{V}_{b]}-2u_{[a}\omega_{b]c}\hat{V}^c.
\end{aligned} \tag{116}$$

For the symmetric part $\theta_{ab}=u_{(a;b)}$ we find:

$$\begin{aligned}
L_\xi\theta_{ab} &= L_\xi\left(h_a^c h_b^d u_{(c;d)}\right) \Rightarrow\\
L_\xi\theta_{ab} &= \left(\psi-H_{mn}u^m u^n\right)\theta_{ab}+h_a^c h_b^d\hat{V}_{(c;d)}+2u_{(a}\theta_{b)c}V^c+\dot{u}_{(a}\hat{V}_{b)}\\
&+\dot{\psi}h_{ab}-h_{(a}^c h_{b)}^d\left[H_{ec;d}+H_{ed;c}-H_{cd;e}\right]u^e.
\end{aligned} \tag{117}$$

We decompose θ_{ab} in trace and traceless part as follows $\theta_{ab}=\sigma_{ab}+\frac{1}{3}h_{ab}\theta$. For the trace we find:

$$\begin{aligned}
L_\xi\theta &= -\left(\psi+H_{mn}u^m u^n\right)\theta-2H^{ab}\theta_{ab}+h^{cd}\hat{V}_{(c;d)}+2\omega_{ab}\dot{u}^a\xi^b-\left(\dot{u}_a\dot{u}^a\right)\left(u_b\xi^b\right)\\
&+\dot{u}^a(u_c\xi^c)_{;a}+3\dot{\psi}-h^{cd}[H_{ec;d}+H_{ed;c}-H_{cd;e}]u^e.
\end{aligned} \tag{118}$$

We can simplify this expression further. The term:

$$2H^{ab}\theta_{ab}=2H^{ab}\sigma_{ab}+\frac{2}{3}H^{ab}h_{ab}\theta=2H^{ab}\sigma_{ab}+\frac{2}{3}H^{mn}u_m u_n\theta,$$

[10] Recall that

$$L_\xi u_a=\left[\psi-H^{cd}u_c u_d\right]u_a+2\omega_{ab}\xi^b-\dot{u}_a\left(u_b\xi^b\right)+h_a^b(u_c\xi^c)_{;b}, \tag{111}$$

and from (72):

$$L_\xi u^a=-\left[\psi+H^{cd}u_c u_d\right]u^a-2h_b^a H^{bc}u_c+2\omega^a_{\ .b}\xi^b-\dot{u}^a(u_b\xi^b)+h^{ab}(u_c\xi^c)_{;b}, \tag{112}$$

where

$$\hat{V}_a(u)=2\omega_{ab}\xi^b-\dot{u}_a\left(u_b\xi^b\right)+h_a^b(u_c\xi^c)_{;b}, \tag{113}$$

$$V^a(u)=-2h_b^a H^{bc}u_c+2\omega^a_{\ .b}\xi^b-\dot{u}^a(u_b\xi^b)+h^{ab}(u_c\xi^c)_{;b}. \tag{114}$$

therefore:

$$L_\xi \theta = -\left(\psi + \frac{5}{3} H_{mn} u^m u^n\right) \theta - 2H^{ab}\sigma_{ab} + h^{cd}\hat{V}_{(c;d)} + 2\omega_{ab}\dot{u}^a \xi^b$$
$$- (\dot{u}_a \dot{u}^a)(u_b \xi^b) + \dot{u}^a (u_c \xi^c)_{;a} + 3\dot{\psi} - h^{cd}[H_{ec;d} + H_{ed;c} - H_{cd;e}]u^e. \quad (119)$$

Concerning the 4-acceleration we have:

$$L_\xi \dot{u}_a = L_\xi \left(u_{a;b} u^b\right) = (L_\xi u_{a;b})u^b + (L_\xi u^b)u_{a;b} \Rightarrow$$

$$L_\xi \dot{u}_a = \dot{K}(u)\ u_a + \hat{V}_a(u)_{;b} u^b + V^b(u) u_{a;b} + \psi_{,a} - [H_{da;b} + H_{db;a} - H_{ab;d}]\, u^d u^b,$$

where $K(u) = \psi - H_{mn} u^m u^n$.

There remains the shear $\sigma_{ab.}$ We find after a standard calculation:

$$L_\xi \sigma_{ab} = K(u)\sigma_{ab} - \frac{1}{3}\left(\frac{2}{3} H_{mn} u^m u^n\right) \theta h_{ab} + h_a^c h_b^d \hat{V}_{(c;d)} + 2u_{(a}\theta_{b)c} V^c + \dot{u}_{(a}\hat{V}_{b)}$$
$$- h_{(a}^c h_{b)}^d \left[H_{ec;d} + H_{ed;c} - H_{cd;e}\right] u^e - \frac{1}{3}\left[-2H_{ab} - 2\left(H_{cd} u^c u^d\right) u_{(a} u_{b)} + 2u_{(a}\hat{V}_{b)}\right] \theta$$
$$- \frac{1}{3}\left[\begin{array}{c} -2H^{mn}\sigma_{mn} + h^{mn}\hat{V}_{(m;n)} + 2\omega_{mn}\dot{u}^m \xi^n - (\dot{u}_m \dot{u}^m)(u_n \xi^n) \\ +\dot{u}^m (u_n \xi^n)_{;m} - h^{mn}[H_{em;n} + H_{en;m} - H_{mn;e}]u^e \end{array}\right] h_{ab}. \quad (120)$$

8. The matter - Dynamics

General Relativity approaches the gravitational field using the model of a fluid. That is one considers the stress tensor T_{ab} on the fluid to be determined by the gravitational field created by the material content of spacetime. This stress tensor determines the metric of spacetime via Einstein field equations:

$$G_{ab} = T_{ab}, \quad (121)$$

where G_{ab} is the Einstein tensor. Subsequently the metric determines the timelike geodesics which when affinely parameterized become the four-velocity of a special class of observers called the **intrinsic observers** of the metric. The world lines of these observers are identified with the flow lines of the gravitational fluid. In that sense the particles of the fluid are the intrinsic observers.

This is the scenario of General Relativity [7]. We note that Einstein field equations (121) are independent of the observer and they relate the geometry of the background space with the matter content of the universe. Furthermore they cannot be solved because one neither knows the metric, therefore cannot compute the G_{ab} and one does not know the mater content in terms of some physical parameters. In other words equation (121) is used only to frame the general set up of the theory and cannot give any information until a class of observers is considered. The ideal choice would be the intrinsic observers but only 'God' knows who are they. What it is left to us is to chose an arbitrary class of observers by means of a timelike unit vector field and adding some more simplifying assumptions write down (121) in terms of some set of differential equations whose solutions will give us an indication of spacetime as seen by these observers. The experiment and the observations will show us how close to the intrinsic observers we are and how close to 'reality' our simplifying assumptions are. In conclusion we cannot solve the field equations but we can use them as the vehicle to make up scientific pictures of the gravitational field and get some answers which we hope they will be as close as possible to our measurements (=reality!). The ultimate truth is hidden within the intrinsic observers with whom we have no touch or communication.

Journal of Physics: Conference Series **189** (2009) 012041 doi:10.1088/1742-6596/189/1/012041

The major question is:

To what degree the matter content and the geometry of spacetime are interconnected as a result oft the field equations (121)?

The answer to this question has two parts. The first comes form the identities of Level G1 and the second form those of level G2. Indeed form level G1 we have that the symmetric tensor G_{ab} satisfies the contracted Bianchi identity (5). Form Einstein field equations (121) we infer that T_{ab} is symmetric[11] and must satisfy the constraint / identity:

$$T^{ab}{}_{;b} = 0. \tag{122}$$

This identity gives the conservation equations, which are independent of the observers we choose!

Concerning the identities from level G2 we note that $L_\xi G_{ab}$ can be expressed in terms of the generic collineation $L_\xi g_{ab}$ and subsequently in terns of the collineation parameters ψ, H_{ab}. The exact expression in terms of trace and traceless parts is:

$$\begin{aligned} L_\xi G_{ab} = & -(n-2)\,A_{ab} + 2K_{ab} - \Box H_{ab} - RH_{ab} + \\ & + \frac{1}{n} g_{ab}\left[(n-1)(n-2)\Box\psi - (n-2)\,H^{cd}_{.;cd} + nR_{cd}H^{cd}\right], \end{aligned} \tag{123}$$

where:

$$A_{ab} = \psi_{;ab} - \frac{1}{n} g_{ab}\Box\psi \text{ and } K_{ab} = H^{d}_{.(a;b)d} - \frac{1}{n} g_{ab} H^{cd}_{.;cd}. \tag{124}$$

Then, Einstein field equations (121) relate $L_\xi T_{ab}$ in terms of the collineation parameters ψ, H_{ab} which implies that the 'symmetries' of T_{ab} as a two index tensor are related to the symmetries of the metric g_{ab}. In this manner when we make a symmetry assumption in spacetime we simultaneously impose a constraint as a symmetry of the energy momentum tensor, therefore we restrict the possible forms of matter this spacetime can support. In a sense by an assumption at the level G2 we constraint both the geometry and the matter content of spacetime.

To bring all the above at a level which can be used in practice and make Physics we must introduce a class of observers u^a. These observers are our choice and there is no guarantee whatsoever that they are or they will be the intrinsic observers of spacetime. However this is not to discourage us because it is the most we can do. The introduction of observers immediately introduces a kinematics in spacetime and all the considerations of the previous sections apply.

However the introduction of u^a allows the 1+3 decomposition of the tensor objects and the covariant equations in the same way it did for the kinematics. The new tensor is the energy momentum tensor T_{ab} which describes the matter content of spacetime and the new equation is the conservation equations (122).

8.1. The 1+3 decomposition of the energy momentum tensor: The dynamic variables

We apply the general formula (22) in the case $Y_{ab} = T_{ab}$ where $T_{ab} = T_{ba}$ is the energy momentum tensor. We define the irreducible parts (tensors):

$$\mu = T_{ab}u^a u^b, \tag{125}$$

$$p = \frac{1}{3} h^{ab} T_{ab}, \tag{126}$$

$$q^a = h^{ab} T_{bc} u^c, \tag{127}$$

$$\pi_{ab} = (h^r_a h^s_b - \frac{1}{3} h_{ab} h^{rs}) T_{rs}, \tag{128}$$

[11] Or at least only the symmetric part of T_{ab} enters the field equations and determines the gravitational field.

and we have:

$$T_{ab} = \mu u_a u_b + p h_{ab} + 2q_{(a}u_{b)} + \pi_{ab}. \tag{129}$$

We note that in this decomposition T_{ab} is described by two scalar fields (μ, p) one spacelike vector $(q^a, q_a u^a = 0)$ and a traceless symmetric 2-tensor $(\pi_{ab}, g_{ab}\pi^{ab} = 0)$.

These quantities we call the physical variables and assume that they represent the mass density, the isotropic pressure, the heat flux and the traceless stress tensor respectively as measured by the observer u^a. Following this decomposition and physical interpretation we consider the types of 'gravitating fluids' given in the following Table:

TABLE: Types of Energy - Momentum Tensors

μ	p	q^a	π^{ab}	T_{ab}	Type of fluid
0	0	0	0	0	Empty space
$\neq 0$	0	0	0	$T_{ab} = \mu u_a u_b$	Dust
$\neq 0$	$\neq 0$	0	0	$T_{ab} = \mu u_a u_b + p h_{ab}$	Perfect fluid
$\neq 0$	$\neq 0$	$\neq 0$	0	$T_{ab} = \mu u_a u_b + p h_{ab} + 2q_{(a}u_{b)}$	Isotropic non-perfect fluid
$\neq 0$	$\neq 0$	0	$\neq 0$	$T_{ab} = \mu u_a u_b + p h_{ab} + \pi_{ab}$.	Anisotropic fluid without heat flux
$\neq 0$	$\neq 0$	$\neq 0$	$\neq 0$	$T_{ab} = \mu u_a u_b + p h_{ab} + 2q_{(a}u_{b)} + \pi_{ab}$.	General anisotropic fluid

8.2. The 1+3 decomposition of the conservation equations

The conservation equations follow from the contracted Bianchi identity and Einstein field equations. By 1+3 decomposing $T^{ab}_{;b} = 0$ wrt u^a we find the conservation equations as a set of two equations, one set resulting form the projection along the direction u^a and one normal to u^a. We replace T^{ab} form (129) in terms of the physical variables and find the two conservation equations[12]:

$$\dot{\mu} + (\mu + p)\theta + q^b_{;b} + q^a \dot{u}_a + \pi^{ab}\sigma_{ab} = 0, \tag{130}$$

$$(\mu + p)\dot{u}_a + h^c_a(p_{,c} + \dot{q}_c + \pi^d_{\ c;d}) + q^c\left(\omega_{ac} + \sigma_{ac} + \frac{4}{3}\theta h_{ac}\right) = 0. \tag{131}$$

9. The physical role of the propagation and the constraint equations

The propagation and the constraint equations have been written in terms of the kinematic variables and the Ricci tensor. However with the introduction of the physical variables by the 1+3 decomposition of the energy momentum tensor we can use Einstein equations and replace R_{ab} in terms of the physical variables. Then we have a complete physical role of the propagation and the constraint equations.

Einstein field equations give:

$$R_{ab} = T_{ab} - \frac{1}{2}g_{ab}T + \Lambda g_{ab}, \tag{132}$$

where $T_{ab} = \mu u_a u_b + p h_{ab} + 2q_{(a}u_{b)} + \pi_{ab}$ by (129) and the trace is

$$T = T^a{}_a = -\mu + 3p. \tag{133}$$

Replacing in (132) we find R_{ab} in terms of the physical variables:

$$R_{ab} = \mu u_a u_b + p h_{ab} + 2q_{(a}u_{b)} + \pi_{ab} + \frac{1}{2}g_{ab}(\mu - 3p + 2\Lambda), \tag{134}$$

[12] We write as usual the derivative along u^a by a dot, e.g. $\dot{A}^a = A^a_{;b}u^b$.

or:

$$R_{ab} = (\mu + p)u_a u_b + \frac{1}{2}(\mu - p + 2\Lambda)g_{ab} + 2q_{(a}u_{b)} + \pi_{ab}. \quad (135)$$

From (134) follows:

$$R_{ab}u^a u^b = \mu - \frac{1}{2}(\mu - 3p) - \Lambda = \frac{1}{2}\mu + \frac{3}{2}p - \Lambda, \quad (136)$$

$$R_{ab}u^a = -\mu u_b - \frac{1}{2}q_b + \frac{1}{2}u_b(\mu - 3p + 2\Lambda), \quad (137)$$

$$R = R^a{}_a = \mu - 3p + 4\Lambda, \quad (138)$$

where $R = g^{ab}R_{ab}$ is the curvature scalar.

We turn now to the constraint and propagation equations and we divide them in two sets. One set contains the ones which do not contain the Curvature tensor and the Ricci tensor, therefore they are independent of the field equations (and the matter content of the universe) and they are kinematic equations. The second set contains the equations which depend on the Ricci tensor and the field equations and are dynamical equations.

9.1. *The dynamical propagation equations*

The dynamical propagation equations[13] are equation (37) and equation (38). Replacing the Ricci tensor from (134) we find for the first:

$$\dot{\theta} + \frac{1}{3}\theta^2 + 2(\sigma^2 - \omega^2) - \dot{u}^a{}_{;a} = -\frac{1}{2}(\mu + 3p) + \Lambda. \quad (139)$$

This is known as the **Raychaudhuri equation**.

Concerning the propagation equation (38) we have:

$$h^a_s h^b_t\left[\dot{\sigma}_{ab} - \dot{u}_{(a;b)}\right] + \sigma_{(s|c|}\sigma^c{}_{t)} + \tfrac{2}{3}\sigma_{st}\theta + \omega_s\omega_t - \dot{u}_s\dot{u}_t - \tfrac{1}{3}\left(2\sigma^2 + \omega^2 - \dot{u}^b_{;b}\right)h_{st}$$
$$= -E_{st} + \tfrac{1}{2}\left(h^a_s h^b_t - \tfrac{1}{3}h_{st}h^{ab}\right)R_{ab},$$
$$= -E_{st} + \tfrac{1}{2}\left(h^a_s h^b_t - \tfrac{1}{3}h_{st}h^{ab}\right)\left(\mu u_a u_b + p h_{ab} + 2q_{(a}u_{b)} + \pi_{ab} + \tfrac{1}{2}g_{ab}(\mu - 3p + 2\Lambda)\right),$$
$$= -E_{st} + \tfrac{1}{2}\left(h^a_s h^b_t - \tfrac{1}{3}h_{st}h^{ab}\right)\pi_{ab} = -E_{st} + \tfrac{1}{2}\pi_{st}.$$

Therefore in terms of the physical variables the propagation equation (38) reads:

$$h^a_s h^b_t\left[\dot{\sigma}_{ab} - \dot{u}_{(a;b)}\right] + \sigma_{(s|c|}\sigma^c{}_{t)} + \frac{2}{3}\sigma_{st}\theta + \omega_s\omega_t - \dot{u}_s\dot{u}_t - \frac{1}{3}\left(2\sigma^2 + \omega^2 - \dot{u}^b_{;b}\right)h_{st} = -E_{st} + \frac{1}{2}\pi_{st}. \quad (140)$$

This equation gives the propagation of shear.

9.1.1. *The dynamical constraint equations*

The only dynamic constraint equation is equation (40). We find:

$$h^c_s\left[\frac{2}{3}\theta_{,c} - h^{ab}\sigma_{ac;b} - \eta_{camn}u^a\left(\omega^{m;n} + 2\omega^m\dot{u}^n\right)\right] = q_b. \quad (141)$$

[13] The remaining propagation equation (36) giving the propagation of the vorticity, does not involve the Ricci tensor hence it is a kinematical equation.

Journal of Physics: Conference Series **189** (2009) 012041 doi:10.1088/1742-6596/189/1/012041

10. The dynamical role of collineations

As we remarked in Section 8 the collineation parameters restrict the possible forms of T_{ab}. Indeed, Einstein field equations and (123) give:

$$\begin{aligned} L_\xi T_{ab} = & -(n-2)\,A_{ab} + 2K_{ab} - \Box H_{ab} - RH_{ab} + \\ & + \frac{1}{n} g_{ab}\left[(n-1)(n-2)\Box\psi - (n-2)\,H^{cd}_{.;cd} + nR_{cd}H^{cd}\right], \end{aligned} \tag{142}$$

where the tensors A_{ab}, K_{ab} are given in (124). From this equation it is possible to give the effect of collineation directly to the physical parameters of the observers u^a. Indeed if we replace in (142) T_{ab} in $L_\xi T_{ab}$ using (129) and take the irreducible parts, we express the Lie derivative along the collineation vector of the dynamical variables in terms of the collineation parameters. This form of the field equations will relate directly the physical and the geometric variables thus enabling one to draw direct conclusions of the symmetry assumptions. Furthermore the approach is completely general and the results hold for any type of matter, all observers and all collineations.

From (135) we compute the $L_\xi R_{ab}$ in irreducible parts of the 1+3 decomposition wrt u^a :

$$\begin{aligned} L_\xi R_{ab} = & \left[\frac{1}{2}(\mu+3p)^\circ - 2\mathring{q}_c u^c + \mathring{\pi}_{cd}u^c u^d - (\mu+3p-2\Lambda)\left(\dot{\xi}_c u^c\right) - 2q_c\dot{\xi}^c\right] u_a u_b + \\ & - 2\begin{bmatrix} -(\mu+p)\,\mathring{u}_d h^d_c - \mathring{q}_d h^d_c + \mathring{\pi}_{de}u^d h^e_c - \frac{1}{2}(\mu+3p-2\Lambda)\,u_c\xi^c_{;d}h^d_e + \\ -q_c\xi^c_{;d}h^d_e + \frac{1}{2}(\mu-p+2\Lambda)\,\dot{\xi}_d h^d_e + \left(q_d h^d_e\right)\left(\dot{\xi}^c u_c\right) - h^d_e\pi_{cd}\dot{\xi}^c \end{bmatrix} u_{(a}h^c_{e)} + \\ & + 2\left[\frac{3}{2}(\mu-p)^\circ + 2q_c\mathring{u}^c + \mathring{\pi}_{cd}h^{cd} + (\mu-p+2\Lambda)\,(4\psi+\dot{\xi}^c u_c) + 2u_c\xi^c_{;d}q^d + 2\pi_{cd}H^{cd}\right] h_{ab} \\ & + 2\begin{bmatrix} \frac{1}{2}(\mu-p)^\circ h_{ef} + 2q_{(c}\mathring{u}_{d)}h^c_e h^d_f + \mathring{\pi}_{cd}h^c_e h^d_f + \\ +(\mu-p+2\Lambda)\,\xi_{d;r}h^d_e h^r_f + 2q_r u_c\xi^c_{;d}h^d_e h^r_f + 2\pi_{cr}\xi^c_{;d}h^d_e h^r_f \end{bmatrix}\left[h^e_{(a}h^f_{b)} - \frac{1}{3}h_{ab}h^{ef}\right], \end{aligned} \tag{143}$$

where $\circ$ means covariant derivation along ξ^a e.g. $\mathring{\mu} = \mu_{;a}\xi^a$. Also from (17) we have (for $n=4$!):

$$L_\xi R_{ab} = -2\psi_{;ab} - g_{ab}\Box\psi + 2H^d_{.(a;b)d} - \Box H_{ab}. \tag{144}$$

We 1+3 decompose this expression and equate with (143) from which we find the field equations in the form of Lie derivative of physical variables in terms of kinematic variables and the collineation parameters.

To do that we 1+3 decompose first $\psi_{;ab}$. We write:

$$\psi_{;ab} = \lambda_\psi u_a u_b + p_\psi h_{ab} + 2q_{\psi(a}u_{b)} + \pi_{\psi ab}, \tag{145}$$

where:

$$\mu_\psi = \psi_{;ab}u^a u^b,\ p_\psi = \frac{1}{3}\psi_{;ab}h^{ab},\ q_{\psi a} = -\psi_{;bc}h^b_a u^c,\ \pi_{\psi ab} = (h^r_a h^s_b - \frac{1}{3}h_{ab}h^{rs})\psi_{;rs}. \tag{146}$$

We also have:

$$\Box\psi = \psi_{;ab}g^{ab} = -\mu_\psi + 3p_\psi. \tag{147}$$

It is also possible to 1+3 decompose the collineation tensor H_{ab} as follows:

$$H_{ab} = \mu_H u_a u_b + p_H h_{ab} - 2q_{H(a}u_{b)} + \pi_{Hab}, \tag{148}$$

where:

$$\mu_H = H_{ab}u^a u^b,\ p_H = \frac{1}{3}H_{ab}h^{ab},\ q_{Ha} = -H_{bc}h_a^b u^c,\ \pi_{Hab} = (h_a^r h_b^s - \frac{1}{3}h_{ab}h^{rs})H_{rs}. \tag{149}$$

We compute:

$$\begin{aligned} -2\psi_{;ab} - g_{ab}\Box\psi &= -2\left(\mu_\psi u_a u_b + p_\psi h_{ab} + 2q_{\psi(a}u_{b)} + \pi_{\psi ab}\right) + (\mu_\psi - 3p_\psi)\left(-u_a u_b + h_{ab}\right), \\ &= -3(\mu_\psi - p_\psi)u_a u_b + (\mu_\psi - 5p_\psi)h_{ab} - 4q_{\psi(a}u_{b)} - 2\pi_{\psi ab}. \end{aligned}$$

Concerning the second term in the rhs of (144) we define:

$$K_{ab} = 2H^d_{.(a;b)d} - \Box H_{ab}, \tag{150}$$

and we write:

$$K_{ab} = \mu_K u_a u_b + p_K h_{ab} - 2q_{K(a}u_{b)} + \pi_{Kab}, \tag{151}$$

where:

$$\mu_K = K_{ab}u^a u^b,\ p_K = \frac{1}{3}K_{ab}h^{ab},\ q_{Ka} = -K_{bc}h_a^b u^c,\ \pi_{Kab} = (h_a^r h_b^s - \frac{1}{3}h_{ab}h^{rs})K_{rs}. \tag{152}$$

Then we have:

$$L_\xi R_{ab} = (-3\mu_\psi + 3p_\psi + \mu_K)u_a u_b + (\mu_\psi - 5p_\psi + p_K)h_{ab} + 2(-2q_{\psi(a} + q_{K(a})u_{b)} - 2\pi_{\psi ab} + \pi_{Kab}. \tag{153}$$

Comparing with (143) we write the field equations in the following form:

$$\frac{1}{2}(\mu + 3p)^\circ - 2\mathring{q}_c u^c + \mathring{\pi}_{cd}u^c u^d - (\mu + 3p - 2\Lambda)\left(\dot{\xi}_c u^c\right) - 2q_c\dot{\xi}^c = -3\mu_\psi + 3p_\psi + \mu_K, \tag{154}$$

$$\left[\begin{array}{c} -(\mu + p)\,\mathring{u}_d h_e^d - \mathring{q}_d h_e^d + \mathring{\pi}_{dc}u^d h_e^c - \frac{1}{2}(\mu + 3p - 2\Lambda)\,u_c\xi^c_{;d}h_e^d + \\ -q_c\xi^c_{;d}h_e^d + \frac{1}{2}(\mu - p + 2\Lambda)\,\dot{\xi}_d h_e^d + \left(q_d h_e^d\right)\left(\dot{\xi}^c u_c\right) - h_e^d\pi_{cd}\dot{\xi}^c \end{array}\right] = -(-2q_{\psi e} + q_{Ke}), \tag{155}$$

$$\begin{aligned} &\frac{3}{2}(\mu - p)^\circ + 2q_c\mathring{u}^c + \mathring{\pi}_{cd}h^{cd} + (\mu - p + 2\Lambda)\left(4\psi + \dot{\xi}^c u_c\right) + 2u_c\xi^c_{;d}q^d + 2\pi_{cd}H^{cd} \\ &= \frac{3}{2}(\mu_\psi - 5p_\psi + p_K), \end{aligned} \tag{156}$$

$$2\left[\begin{array}{c}\frac{1}{2}(\mu-p)^{\circ}h_{ef}+2q_{(c}\mathring{u}_{d)}h^c_eh^d_f+\mathring{\pi}_{cd}h^c_eh^d_f+\\ +(\mu-p+2\Lambda)\,\xi_{d;r}h^d_eh^r_f+2q_ru_c\xi^c_{;d}h^d_eh^r_f+2\pi_{cr}\xi^c_{;d}h^d_eh^r_f\end{array}\right]\left[h^e_{(a}h^f_{b)}-\frac{1}{3}h_{ab}h^{ef}\right]$$
$$=-2\pi_{\psi ab}+\pi_{Kab}. \quad (157)$$

We simplify these equations. We note that:

$$\begin{aligned}\dot{\xi}^c u_c &= \xi_{c;d}u^cu^d=-\psi+H_{cd}u^cu^d,\\ q^d\left(u_c\xi^c_{;d}+\dot{\xi}_d\right) &= q^d(u^c\xi_{c;d}+\xi_{d;c}u^c)=2q^du^cH_{cd}.\end{aligned}$$

Then, (154) is written as follows:

$$\begin{aligned}LHS &= \frac{1}{2}\left\{(\mu+3p)^{\circ}-4\mathring{q}_cu^c+2\mathring{\pi}_{cd}u^cu^d-2\,(\mu+3p-2\Lambda)\left(\dot{\xi}_cu^c\right)-4q_c\dot{\xi}^c\right\},\\ &= \frac{1}{2}\left\{(\mu+3p)^{\circ}+4q_c\mathring{u}^c+2\mathring{\pi}_{cd}u^cu^d-2\,(\mu+3p-2\Lambda)\left(-\psi+H_{cd}u^cu^d\right)-4q_c\dot{\xi}^c\right\},\\ &= \frac{1}{2}\left\{(\mu+3p)^{\circ}+4q_c(\mathring{u}^c-\dot{\xi}^c)+2\mathring{\pi}_{cd}u^cu^d-2\,(\mu+3p-2\Lambda)\left(-\psi+H_{cd}u^cu^d\right)\right\},\\ &= \frac{1}{2}\left\{(\mu+3p)^{\circ}+4q_cL_\xi u^c+2\mathring{\pi}_{cd}u^cu^d-2\,(\mu+3p-2\Lambda)\left(-\psi+H_{cd}u^cu^d\right)\right\},\end{aligned}$$

hence (154) becomes:

$$(\mu+3p)^{\circ}+4q_cL_\xi u^c+2\mathring{\pi}_{cd}u^cu^d-2\,(\mu+3p-2\Lambda)\left(-\psi+H_{cd}u^cu^d\right)=2(-3\mu_\psi+3p_\psi+\mu_K). \quad (158)$$

Similarly, for (156) we have:

$$\begin{aligned}&LHS\\ &= 2\left[\frac{3}{2}(\mu-p)^{\circ}+2q_c\mathring{u}^c+\mathring{\pi}_{cd}h^{cd}+(\mu-p+2\Lambda)\,(4\psi+\dot{\xi}^cu_c)+2u_c\xi^c_{;d}q^d+2\pi_{cd}H^{cd}\right],\\ &= 2\left[\frac{3}{2}(\mu-p)^{\circ}+2q_c\mathring{u}^c+\mathring{\pi}_{cd}u^cu^d+(\mu-p+2\Lambda)\,(3\psi+H_{cd}u^cu^d)+2u^cq^d(\psi g_{cd}+H_{cd}-\xi_{d;c})\right.\\ &\quad\left.+2\pi_{cd}H^{cd}\right],\\ &= 2\left[\frac{3}{2}(\mu-p)^{\circ}+2q_c(\mathring{u}^c-\dot{\xi}^c)+\mathring{\pi}_{cd}u^cu^d+(\mu-p+2\Lambda)\,(3\psi+H_{cd}u^cu^d)+2u^cq^dH_{cd}+2\pi_{cd}H^{cd}\right],\\ &= 3\,(\mu-p)^{\circ}+4q_cL_\xi u^c+2\mathring{\pi}_{cd}u^cu^d+2\,(\mu-p+2\Lambda)\,(3\psi+H_{cd}u^cu^d)+42u^cq^dH_{cd}+4\pi_{cd}H^{cd},\end{aligned}$$

hence (156) becomes:

$$\begin{aligned}&3\,(\mu-p)^{\circ}+4q_cL_\xi u^c+2\mathring{\pi}_{cd}u^cu^d+2\,(\mu-p+2\Lambda)\,(3\psi+H_{cd}u^cu^d)+4H_{cd}u^cq^d+4\pi_{cd}H^{cd}\\ &=3(\mu_\psi-5p_\psi+p_K).\end{aligned} \quad (159)$$

Adding the two new equations we get:

$$\begin{aligned}LHS &= (\mu+3p)^{\circ}+3\,(\mu-p)^{\circ}+8q_cL_\xi u^c+4\mathring{\pi}_{cd}u^cu^d\\ &\quad+2\,(3\mu-3p+6\Lambda+\mu+3p-2\Lambda)\,\psi\\ &\quad+2\,(\mu-p+2\Lambda-\mu-3p+2\Lambda)\,H_{cd}u^cu^d\\ &\quad+4H_{cd}u^cq^d+4\pi_{cd}H^{cd},\\ &= 4\mathring{\mu}+8q_cL_\xi u^c+4\mathring{\pi}_{cd}u^cu^d+8(\mu+\Lambda)\psi\\ &\quad+8(-p+\Lambda)H_{cd}u^cu^d+4H_{cd}u^cq^d+4\pi_{cd}H^{cd},\end{aligned}$$

$$RHS = 3(\mu_\psi - 5p_\psi + p_K) + 2(-3\mu_\psi + 3p_\psi + \mu_K) = -3\mu_\psi - 9p_\psi + 3p_K + 2\mu_K.$$

Hence:

$$\begin{aligned} &4\mathring{\mu} + 8q_c L_\xi u^c + 4\mathring{\pi}_{cd}u^c u^d + 8(\mu+\Lambda)\psi \\ &+8(-p+\Lambda)H_{cd}u^c u^d + 4H_{cd}u^c q^d + 4\pi_{cd}H^{cd} \\ &= -3\mu_\psi - 9p_\psi + 3p_K + 2\mu_K, \end{aligned}$$

$$\begin{aligned} \mathring{\mu} = &\frac{1}{4}\left(-3\mu_\psi - 9p_\psi + 3p_K + 2\mu_K\right) - 2q_c L_\xi u^c - \mathring{\pi}_{cd}u^c u^d - 2(\mu+\Lambda)\psi \\ &+ 2(p-\Lambda)H_{cd}u^c u^d - H_{cd}u^c q^d - \pi_{cd}H^{cd}. \end{aligned} \tag{160}$$

This equation expresses the derivative of μ along the collineation vector ξ^a. To find $\mathring{p}$ we use (158) and replace $\mathring{\mu}$. We have:

$$\begin{aligned} &3\mathring{p} + \tfrac{1}{4}\left(-3\mu_\psi - 9p_\psi + 3p_K + 2\mu_K\right) - 2(\mu+\Lambda)\psi \\ &+2(p-\Lambda)H_{cd}u^c u^d - H_{cd}u^c q^d - \pi_{cd}H^{cd} \\ &+2q_c L_\xi u^c + \mathring{\pi}_{cd}u^c u^d - 2\left(\mu + 3p - 2\Lambda\right)\left(-\psi + H_{cd}u^c u^d\right) \\ &= 2(-3\mu_\psi + 3p_\psi + \mu_K), \end{aligned}$$

$$\begin{aligned} &3\mathring{p} - H_{cd}u^c q^d - \pi_{cd}H^{cd} \\ &+2q_c L_\xi u^c + \mathring{\pi}_{cd}u^c u^d + \tfrac{1}{4}\left(-3\mu_\psi - 9p_\psi + 3p_K + 2\mu_K\right) \\ &+2\left(\mu + 3p - 2\Lambda - \mu - \Lambda\right)\psi + 2(p - \Lambda - \mu - 3p + 2\Lambda)H_{cd}u^c u^d \\ &= 2(-3\mu_\psi + 3p_\psi + \mu_K), \end{aligned}$$

$$\begin{aligned} &3\mathring{p} - H_{cd}u^c q^d - \pi_{cd}H^{cd} \\ &+2q_c L_\xi u^c + \mathring{\pi}_{cd}u^c u^d \\ &+6(p-\Lambda)\psi - 2(2p + \mu - \Lambda)H_{cd}u^c u^d \\ &= 2(-3\mu_\psi + 3p_\psi + \mu_K) - \tfrac{1}{4}\left(-3\mu_\psi - 9p_\psi + 3p_K + 2\mu_K\right), \end{aligned}$$

$$\begin{aligned} &3\mathring{p} - H_{cd}u^c q^d - \pi_{cd}H^{cd} + 2q_c L_\xi u^c + \mathring{\pi}_{cd}u^c u^d \\ &+6(p-\Lambda)\psi - 2(2p + \mu - \Lambda)H_{cd}u^c u^d \\ &= \tfrac{3}{4}\left(7\mu_\psi + 11p_\psi + 2\mu_K - p_K\right), \end{aligned}$$

$$\begin{aligned} 3\mathring{p} = &H_{cd}u^c q^d + \pi_{cd}H^{cd} - 2q_c L_\xi u^c - \mathring{\pi}_{cd}u^c u^d \\ &- 6(p-\Lambda)\psi + 2(2p + \mu - \Lambda)H_{cd}u^c u^d \\ &+ \frac{1}{4}\left(-3\mu_\psi - 9p_\psi + 3p_K + 2\mu_K\right). \end{aligned} \tag{161}$$

We concentrate now on (155):

$$\begin{aligned} LHS \;&= -\left(\mu + p\right)\mathring{u}_d h^d_e - \mathring{q}_d h^d_e + \mathring{\pi}_{dc}u^d h^c_e - \tfrac{1}{2}\left(\mu + 3p - 2\Lambda\right)u_c\xi^c_{;d}h^d_e \\ &-q_c\xi^c_{;d}h^d_e + \tfrac{1}{2}\left(\mu - p + 2\Lambda\right)\dot{\xi}_d h^d_e + \left(q_d h^d_e\right)\left(\dot{\xi}^c u_c\right) - h^d_e\pi_{cd}\dot{\xi}^c. \end{aligned}$$

Journal of Physics: Conference Series **189** (2009) 012041 doi:10.1088/1742-6596/189/1/012041

The terms:

$$-\tfrac{1}{2}(\mu+3p-2\Lambda)\,u_c\xi^c_{;d}h^d_e+\tfrac{1}{2}(\mu-p+2\Lambda)\,\dot{\xi}_d h^d_e$$
$$=-\tfrac{1}{2}(\mu+3p-2\Lambda)\,H_{cd}u^c h^d_e+\tfrac{1}{2}(\mu+3p-2\Lambda)\,\dot{\xi}_d h^d_e+\tfrac{1}{2}(\mu-p+2\Lambda)\,\dot{\xi}_d h^d_e,$$
$$=-\tfrac{1}{2}(\mu+3p-2\Lambda)\,H_{cd}u^c h^d_e+(\mu+p)\dot{\xi}_d h^d_e,$$

hence:

$$\begin{aligned}LHS &= -(\mu+p)h_{ed}L_\xi u^d-\left[\mathring{q}_d h^d_e+q_c\xi^c_{;d}h^d_e\right]\\ &\quad +\mathring{\pi}_{dc}u^d h^c_e-h^d_e\pi_{cd}\dot{\xi}^c.\end{aligned}$$

The terms:

$$\begin{aligned}\mathring{q}_d h^d_e+q_c\xi^c_{;d}h^d_e &= \mathring{q}_d h^d_e+q_c(\psi\delta^c_d+H^c_d-\xi_d^{\cdot\cdot;c})h^d_e,\\ &= (\mathring{q}_d-\xi_d^{\cdot\cdot;c}q_c)h^d_e+\psi q_e+H_{cd}q^c h^d_e,\\ &= h^d_e L_\xi q_d+\psi q_e+H_{cd}q^c h^d_e.\end{aligned}$$

Also the Lie derivative:

$$L_\xi\pi_{cd}=\pi_{cd;f}\xi^f+\pi_{fd}\xi^f_{;c}+\pi_{cf}\xi^f_{;d}=\mathring{\pi}_{cd}+\pi_{fd}\xi^f_{;c}+\pi_{cf}\xi^f_{;d},$$

hence:

$$u^d h^c_e L_\xi\pi_{cd}=\mathring{\pi}_{cd}u^d h^c_e+\pi_{cf}\dot{\xi}^f h^c_e$$

Therefore:

$$LHS=-(\mu+p)h_{ed}L_\xi u^d-h^d_e L_\xi q_d+u^d h^c_e L_\xi\pi_{cd}-\psi q_e-H_{cd}q^c h^d_e$$

and (155) becomes:

$$-(\mu+p)h_{ed}L_\xi u^d-h_{de}L_\xi q^d+u^d h^c_e L_\xi\pi_{cd}-\psi q_e-H_{cd}q^c h^d_e=-(-2q_{\psi e}+q_{Ke}),$$

or:

$$-h_{de}L_\xi q^d+u^d h^c_e L_\xi\pi_{cd}=(\mu+p)h_{ed}L_\xi u^d+\psi q_e+H_{cd}q^c h^d_e-(-2q_{\psi e}+q_{Ke}). \tag{162}$$

We continue with (157). We have:

$$\begin{aligned}LHS &= 2\left[h^e_{(a}h^f_{b)}-\frac{1}{3}h_{ab}h^{ef}\right]\\ &\quad\left[\frac{1}{2}(\mu-p)^\circ h_{ef}+2q_{(c}\mathring{u}_{d)}h^c_e h^d_f+\mathring{\pi}_{cd}h^c_e h^d_f+(\mu-p+2\Lambda)(\psi g_{dr}+H_{dr})h^d_e h^r_f\right.\\ &\quad\left.+2q_r u_c\xi^c_{;d}h^d_e h^r_f+2\pi_{cr}\xi^c_{;d}h^d_e h^r_f\right]\end{aligned}$$

The terms:

$$2\left[h^e_{(a}h^f_{b)}-\frac{1}{3}h_{ab}h^{ef}\right]\frac{1}{2}(\mu-p)^\circ h_{ef}=0,$$

$$\left[h^e_{(a}h^f_{b)}-\tfrac{1}{3}h_{ab}h^{ef}\right](\mu-p+2\Lambda)\,\xi_{d;r}h^d_e h^r_f$$
$$=\left[h^e_{(a}h^f_{b)}-\tfrac{1}{3}h_{ab}h^{ef}\right](\mu-p+2\Lambda)\,(\psi g_{dr}+H_{dr})h^d_e h^r_f,$$
$$=(\mu-p+2\Lambda)\left[h^e_{(a}h^f_{b)}-\tfrac{1}{3}h_{ab}h^{ef}\right]H_{dr}h^d_e h^r_f,$$
$$=(\mu-p+2\Lambda)\left[h^e_{(a}h^f_{b)}-\tfrac{1}{3}h_{ab}h^{ef}\right]H_{ef},$$

i.e. the traceless part of the projection of the traceless tensor H_{ab}. The term:

$$\begin{aligned}
&\left[h^e_{(a}h^f_{b)} - \tfrac{1}{3}h_{ab}h^{ef}\right] 2q_r u^c \xi_{c;d} h^d_e h^r_f \\
&= \left[h^e_{(a}h^f_{b)} - \tfrac{1}{3}h_{ab}h^{ef}\right] 2q_r u^c (\psi g_{cd} + H_{cd}) h^d_e h^r_f, \\
&= \left[h^e_{(a}h^f_{b)} - \tfrac{1}{3}h_{ab}h^{ef}\right] 2q_r u^c H_{cd} h^d_e h^r_f, \\
&= \left[h^e_{(a}h^f_{b)} - \tfrac{1}{3}h_{ab}h^{ef}\right] 2q_f u^c H_{cd} h^d_e, \\
&= 2q_{(a}h^d_{b)} H_{cd} u^c - \tfrac{2}{3}\left(H_{cd}u^c q^d\right) h_{ab}.
\end{aligned}$$

The term:

$$\begin{aligned}
&\left[h^e_{(a}h^f_{b)} - \tfrac{1}{3}h_{ab}h^{ef}\right] 2\pi_{cr} \xi^c_{;d} h^d_e h^r_f \\
&= 2\left[h^e_{(a}h^f_{b)} - \tfrac{1}{3}h_{ab}h^{ef}\right] \pi^c_{.r} (\psi g_{cd} + H_{cd}) h^d_e h^r_f, \\
&= 2\left[h^e_{(a}h^f_{b)} - \tfrac{1}{3}h_{ab}h^{ef}\right] (\psi \pi_{ef} + \pi^c_{.f} H_{cd} h^d_e), \\
&= 2\left[h^e_{(a}h^f_{b)} - \tfrac{1}{3}h_{ab}h^{ef}\right] \psi \pi_{ef} + 2\left[h^e_{(a}h^f_{b)} - \tfrac{1}{3}h_{ab}h^{ef}\right] \pi^c_{.f} H_{cd} h^d_e, \\
&= 2\left[h^e_{(a}h^f_{b)} - \tfrac{1}{3}h_{ab}h^{ef}\right] \pi^c_{.f} H_{cd} h^d_e, \\
&= 2h^e_{(a}h^f_{b)} \pi^c_{.f} H_{cd} h^d_e - \tfrac{2}{3}h_{ab}(\pi^{cd}H_{cd}).
\end{aligned}$$

Therefore we have:

$$\begin{aligned}
&2q_{(c}\mathring{u}_{d)} h^c_e h^d_f + \mathring{\pi}_{cd} h^c_e h^d_f + (\mu - p + 2\Lambda)\left[h^e_{(a}h^f_{b)} - \frac{1}{3}h_{ab}h^{ef}\right] H_{ef} \\
&+ 2q_{(a}h^d_{b)} H_{cd} u^c - \frac{2}{3}\left(H_{cd}u^c q^d\right) h_{ab} \\
&+ 2h^e_{(a}h^f_{b)} \pi^c_{.f} H_{cd} h^d_e - \frac{2}{3}h_{ab}(\pi^{cd}H_{cd}) \\
&= \frac{1}{2}\left[\pi_{\psi ab} + \pi_{Kab}\right]
\end{aligned} \tag{163}$$

The gravitational field equations are Euqations (158),(162),(159) and (163). With these field equations we have completed the scenario for the generic General Relativistic model.

11. Application Example: The Bianchi I model

We consider a model spacetime on which we make the following assumptions[3]:

Geometric assumption:

A diagonal Bianchi I space-time is a spatially homogeneous space-time which admits an Abelian group of isometries G_3, acting on spacelike hypersurfaces, generated by the spacelike KVs $\xi_1 = \partial_x, \xi_2 = \partial_y, \xi_3 = \partial_z$ and a timelike gradient KV the $u^a = \frac{\partial}{\partial t}$ which is normal to the homogeneous 3-d hypersurfaces.

In synchronous coordinates $\{t, x, y, z\}$ the above assumptions imply that the metric of this spacetime is:

$$ds^2 = -dt^2 + A^2_\mu(t)(dx^\mu)^2, \tag{164}$$

where the metric functions $A_1(t), A_2(t), A_3(t)$ are functions of the time coordinate only. When two of the functions $A_\mu(t)$ are equal (e.g. $A_2 = A_3$) the Bianchi I space-times reduce to the important class of plane symmetric space-times (a special class of the Locally Rotational Symmetric space-times [8],[9] which admit a G_4 group of isometries acting multiply transitively on the spacelike hypersurfaces of homogeneity generated by the vectors ξ_1, ξ_2, ξ_3 and $\xi_4 = x^2\partial_3 - x^3\partial_2$.

For economy of writing in the following we write A_μ instead of $A^2_\mu(t)$. Furthermore we shall be interested only in *proper diagonal* Bianchi I space-times (which in the following will be referred for convenience simply as Bianchi I space-times), hence all metric functions are assumed to be different and the dimension of the group of isometries acting on the spacelike hypersurfaces is three.

The implications of the geometric assumption are:

1. The computation of the Ricci tensor

$$R_{tt} = -\frac{\ddot{A}_1 A_2 A_3 + \ddot{A}_2 A_1 A_3 + \ddot{A}_3 A_1 A_2}{A_1 A_2 A_3}, \tag{165}$$

$$R_{xx} = \frac{A_1\left(\ddot{A}_1 A_2 A_3 + \dot{A}_1 \dot{A}_2 A_3 + \dot{A}_1 \dot{A}_3 A_2\right)}{A_2 A_3}, \tag{166}$$

$$R_{yy} = \frac{A_2\left(\ddot{A}_2 A_1 A_3 + \dot{A}_1 \dot{A}_2 A_3 + \dot{A}_2 \dot{A}_3 A_1\right)}{A_1 A_3}, \tag{167}$$

$$R_{zz} = \frac{A_3\left(\ddot{A}_3 A_1 A_2 + \dot{A}_1 \dot{A}_3 A_2 + \dot{A}_2 \dot{A}_3 A_1\right)}{A_1 A_2}. \tag{168}$$

From the Ricci tensor we compute the Einstein tensor and then use Einstein field equations to write the energy momentum tensor in terms of the metric functions $A_1(t), A_2(t), A_3(t)$ and their derivatives. However that does not mean that we are able to discuss anything about the physical variables (energy density, isotropic pressure etc.) because in order to do that we need to have observers. The expression of T_{ab} we find is the matter content of this spacetime, the same for all observers.

2. The computation of the Weyl tensor

The Weyl tensor is important because it is involved in the second Bianchi identity. More specifically form this tensor one computes the electric and the magnetic parts E^{ab}, H^{ab} defined[14] in (45) which enter into the 1+3 decomposition of the second Bianchi identity given by equations (47) - (50). We compute:

$$E_{xx} = \frac{A_1\left(-2\,\ddot{A}_1 A_2 A_3 + \dot{A}_1 \dot{A}_2 A_3 + \dot{A}_1 \dot{A}_3 A_2 + \ddot{A}_2 A_1 A_3 + \ddot{A}_3 A_1 A_2 - 2\,\dot{A}_2 \dot{A}_3 A_1\right)}{6 A_2 A_3}, \tag{169}$$

$$E_{yy} = \frac{A_2\left(-2\,\ddot{A}_2 A_1 A_3 + \dot{A}_1 \dot{A}_2 A_3 + \dot{A}_2 \dot{A}_3 A_1 + \ddot{A}_1 A_2 A_3 + \ddot{A}_3 A_1 A_2 - 2\,\dot{A}_1 \dot{A}_3 A_2\right)}{6 A_1 A_3}, \tag{170}$$

$$E_{zz} = -\frac{A_3\left(2\,\ddot{A}_3 A_1 A_2 - \dot{A}_1 \dot{A}_3 A_2 - \dot{A}_2 \dot{A}_3 A_1 - \ddot{A}_1 A_2 A_3 - \ddot{A}_2 A_1 A_3 + 2\,\dot{A}_1 \dot{A}_2 A_3\right)}{6 A_1 A_2}, \tag{171}$$

[14] $E_{ac} = C_{abcd} u^b u^d, \;\; H_{ac} = \frac{1}{2}\eta_{am.}{}^{kl} C_{klbt} u^m u^t$

and the magnetic part $H_{ab} = 0$.

Observers.

The choice of observers is open and independent of the choice of the symmetries (i.e. the model) spacetime. However as it has been noted the kinematic quantities they define must satisfy the propagation, the constraint equations and the Bianchi second identity.

We choose the observers (this is one choice, any other would do provided it satisfies the aforementioned identities) to be the ones defined by the time coordinate t i.e. we take $u^a = \delta_0^a$ in the synchronous coordinate system.

The implications of this choice of observers are:

1. Kinematics

We 1+3 decompose $u_{a;b}$ and find the kinematic variables:

$$\theta = [\ln(A_1 A_2 A_3)]^{\cdot}, \tag{172}$$

$$\omega_{ab} = 0, \dot{u}^a = 0, \tag{173}$$

$$\sigma_{ab} = \frac{1}{3} diag\left(0, A_1^2\left[\ln\left(\frac{A_1^2}{A_2 A_3}\right)\right]^{\cdot}, A_2^2\left[\ln\left(\frac{A_2^2}{A_3 A_1}\right)\right]^{\cdot}, A_3^2\left[\ln\left(\frac{A_3^2}{A_1 A_2}\right)\right]^{\cdot}\right). \tag{174}$$

a. The propagation equations give[15]:

$$\dot{\theta} + \frac{1}{3}\theta^2 + 2\sigma^2 = \frac{\ddot{A}_1 A_2 A_3 + \ddot{A}_2 A_1 A_3 + \ddot{A}_3 A_1 A_2}{A_1 A_2 A_3}, \tag{175}$$

$$-E_{st} = h_s^a h_t^b \dot{\sigma}_{ab} + \sigma_{sc}\sigma^c{}_t + \frac{2}{3}\sigma_{st}\theta - \frac{2}{3}2\sigma^2 h_{st}. \tag{176}$$

b. The constraint equations give:

$$\frac{2}{3}h_s^c \theta_{,c} = h_s^c h^{ab}\sigma_{ac;b}, \text{(three equations)} \tag{177}$$

$$0 = -h_{(s}^a h_{t)}^b \sigma_b{}^{c;d}\eta_{arcd}u^r. \text{(five equations)} \tag{178}$$

Equations (175) - (178) must be satisfied identically by the kinematic quantities. It is easy to show that this is true for equations (175) and (176). Equation (177) is trivially satisfied because θ, σ_{ab} are functions of t only. Equation (178) is also trivially satisfied because it contains derivatives of the components of σ_{ab} along the space coordinates only (due to the term $\eta_{arcd}u^r$). We conclude that the propagation and the constraint equations do not give any conditions on the metric functions $A_1(t), A_2(t), A_3(t)$.

c. The Bianchi identities give the derivatives of E_{st}, H_{st} therefore they do not add new constraints on the metric functions. They are only compatibility conditions.

d. The propagation of the kinematic quantities along the symmetry vectors.

From (76) and (77) and the fact the ξ_μ^a ($\mu = 1, 2, 3$) are Killing vectors (hence $\psi = 0, H_{ab} = 0$) we find:

$$V^a(u) = \hat{V}_a(u) = 0. \tag{179}$$

[15] Note that $\sigma^2 = \frac{1}{2}\sigma_{ab}\sigma^{ab}$. For the case we are considering we calculate:

$$\sigma^2 = \frac{1}{3}\left[\sum_{I=1}^{3}\left(\frac{\dot{A}_I}{A_I}\right)^2 - \sum_{I \neq J=1}^{3}\frac{\dot{A}_I}{A_I}\frac{\dot{A}_J}{A_J}\right]$$

Journal of Physics: Conference Series **189** (2009) 012041 doi:10.1088/1742-6596/189/1/012041

From (119) and (120) we have taking into account the above results:

$$L_\xi \theta = 0 \Rightarrow \theta = \theta(t),$$
$$L_\xi \sigma_{ab} = 0 \Rightarrow \sigma_{ab} = \sigma_{ab}(t).$$

These equations give nothing new because we have already computed θ, σ_{ab} and have found that they are functions of t only.

Dynamics

a. We compute the physical parameters for the chosen observers. We find:

$$\mu = \frac{\dot{A}_1 \dot{A}_2 A_3 + \dot{A}_1 \dot{A}_3 A_2 + \dot{A}_2 \dot{A}_3 A_1}{A_1 A_2 A_3} = \frac{\dot{A}_1 \dot{A}_2}{A_1 A_2} + \frac{\dot{A}_1 \dot{A}_3}{A_1 A_3} + \frac{\dot{A}_2 \dot{A}_3}{A_2 A_3}, \tag{180}$$

$$p = -\frac{2\ddot{A}_2 A_1 A_3 + 2\ddot{A}_3 A_1 A_2 + \dot{A}_2 \dot{A}_3 A_1 + 2\ddot{A}_1 A_2 A_3 + \dot{A}_1 \dot{A}_3 A_2 + \dot{A}_1 \dot{A}_2 A_3}{3 A_1 A_2 A_3}, \tag{181}$$

$$\pi_{xx} = -\frac{A_1 \left(-2\ddot{A}_1 A_2 A_3 + \ddot{A}_2 A_1 A_3 + \ddot{A}_3 A_1 A_2 + 2\dot{A}_2 \dot{A}_3 A_1 - \dot{A}_3 \dot{A}_1 A_2 - \dot{A}_1 \dot{A}_2 A_3 \right)}{3 A_2 A_3}, \tag{182}$$

$$\pi_{yy} = -\frac{A_2 \left(-2\ddot{A}_2 A_1 A_3 + \ddot{A}_3 A_1 A_2 + \ddot{A}_1 A_2 A_3 + 2\dot{A}_1 \dot{A}_3 A_2 - \dot{A}_2 \dot{A}_3 A_1 - \dot{A}_1 \dot{A}_2 A_3 \right)}{3 A_1 A_3}, \tag{183}$$

$$\pi_{zz} = -\frac{A_3 \left(-2\ddot{A}_3 A_1 A_2 + \ddot{A}_2 A_1 A_3 + \ddot{A}_1 A_2 A_3 + 2\dot{A}_1 \dot{A}_2 A_3 - \dot{A}_2 \dot{A}_3 A_1 - \dot{A}_1 \dot{A}_3 A_2 \right)}{3 A_1 A_2}. \tag{184}$$

The momentum transfer vector is $q^a = 0$[16]. We note that the second equation can be written:

$$\frac{\ddot{A}_1}{A_1} + \frac{\ddot{A}_2}{A_2} + \frac{\ddot{A}_3}{A_3} = -\frac{1}{2}(\mu + 3p). \tag{185}$$

We introduce the notation:

$$\begin{aligned} I_2 &= \tfrac{\ddot{A}_1}{A_1} + \tfrac{\ddot{A}_2}{A_2} + \tfrac{\ddot{A}_3}{A_3} = -\tfrac{1}{2}(\mu + 3p), \\ I_1 &= \tfrac{\dot{A}_1 \dot{A}_2}{A_1 A_2} + \tfrac{\dot{A}_1 \dot{A}_3}{A_1 A_3} + \tfrac{\dot{A}_2 \dot{A}_3}{A_2 A_3} = \mu, \end{aligned}$$

and we have:

$$\mu = I_1, \tag{186}$$

$$p = -\frac{1}{3}\left(2I_2 + I_1\right), \tag{187}$$

$$\pi_{xx} = \frac{A_1^2}{3}\left(3\frac{\ddot{A}_1}{A_1} - 3\frac{\dot{A}_2 \dot{A}_3}{A_2 A_3} - I_2 + I_1 \right), \tag{188}$$

$$\pi_{yy} = \frac{A_2^2}{3}\left(3\frac{\ddot{A}_2}{A_2} - 3\frac{\dot{A}_1 \dot{A}_3}{A_1 A_3} - I_2 + I_1 \right), \tag{189}$$

$$\pi_{zz} = \frac{A_3^2}{3}\left(3\frac{\ddot{A}_3}{A_3} - 3\frac{\dot{A}_1 \dot{A}_2}{A_1 A_2} - I_2 + I_1 \right). \tag{190}$$

[16] This is expected from the symmetries of the metric and the non-degeneracy of the T_{ab}. Note that both R_{ab} and T_{ab} are of the same form as the metric. This is to be expected because they can be considered as metrics (a metric is a symmetric tensor of type (0,2) and nothing more or less) and they admit the same KVs with the metric.

b. Conservation equations.

From (130) and (131) we have for our case:

$$\dot{\mu} + (\mu + p)\theta + \pi^{ab}\sigma_{ab} = 0, \tag{191}$$

$$h^c_a(p_{,c} + \pi^d_{\ c\,;d}) = 0. \tag{192}$$

Equation (192) is trivially satisfied because all quantities are functions of t. Equation (191) is also trivially satisfied if we replace the expressions of μ, p, θ, π_{ab} from the corresponding expressions.

We see that there are no field equations to solve! Indeed we have solved them in terms of three arbitrary functions which are the metric functions $A_1(t), A_2(t), A_3(t)$! Therefore for the observers $u^a = \delta^a_0$ we have solved the problem completely[17].

We are free to select special solutions from the three parameter family of solutions we have found by imposing various requirements. In the following section we make *one such* requirement and consider those Bianchi I spacetime which *for the observers we have chosen* give rise to a special type of mater which we call string fluid. Needless to say that one could consider any other requirement and select other types of matter for the *same spacetime* and the *same observers.* Every specification / condition on the metric functions $A_1(t), A_2(t), A_3(t)$ will produce a model (physical or not).

12. The string fluid

The connection between strings and vortices is well known [10], [11], [12], [13]. In particular a geometric or Nambu string is a two-dimensional timelike surface in spacetime. Letelier [14] has considered a fluid represented by a combination of geometric strings with particles attached to them so that both have the same four velocity. He called such a fluid a string fluid and he studied the gravitational field it produces in given spacetime backgrounds. In a series of papers various authors [15], [16], [19], [17], [18], [20] [21] have considered various types of collineations for a string fluid and derived the conditions which must be satisfied in order the string fluid to admit a given collineation.

In this work we consider the string fluid as an imperfect fluid and show that the energy momentum tensor of a string fluid corresponds to an anisotropic fluid with vanishing heat conduction. We use the 1+3 to show that the string fluid is in fact the simplest anisotropic fluid possible. Applying previous studies concerning the effects of collineations on anisotropic fluids we study many types of symmetries of a string fluid and give the results in a unified and geometric/systematic way. In this manner we recover, correct and systematize the results of the many previous works and most important we set the methodology for a similar study of other anisotropic fluids.

The energy momentum tensor for a string fluid with particles attached to the strings is given by the expression[18] [12],[13]:

$$T_{ab} = \rho(u_a u_b - n_a n_b) + q p_{ab}, \tag{193}$$

where:

- $\rho = \rho_p + \rho_s$ is the sum of the mass density of the strings (ρ_s) and the mass density of the particles (ρ_p),
- u^a is the common four velocity ($u^a u_a = -1$) of the string and the attached particles
- n^a is a unit spacelike vector ($n^a n_a = 1$) normal to u^a ($u^a n_a = 0$), which specifies the direction of the string (and the direction of anisotropy of the string fluid)
- q is a parameter contributing to the dynamic and kinematic properties of the string

[17] For another class of observers we could have more constraint equations which would have to be solved. Obviously in this case all dynamic and kinematic variables will (in general) be different.

[18] This expression is found form equation (2.28a) of [12] if we set $\rho = -\sigma$and $\pi \to -\pi$.

- $p_{ab} = h_{ab} - n_a n_b$ is the screen projection operator defined by the vectors u^a, n^a.

By rewriting the energy momentum tensor as:

$$T_{ab} = \rho u_a u_b + \frac{1}{3}(2q - \rho)h_{ab} + (q + \rho)(\frac{1}{3}h_{ab} - n_a n_b), \tag{194}$$

(or otherwise) we compute its 1+3 decomposition. It follows that for a string fluid:

$$\mu = \rho, \;\; p = \frac{1}{3}(2q - \rho), \;\; q^a = 0, \;\; \pi_{ab} = (q + \rho)(\frac{1}{3}h_{ab} - n_a n_b). \tag{195}$$

We conclude that a string fluid is an anisotropic fluid with vanishing heat flux. Furthermore we note that n^a is an eigenvector of the anisotropic stress tensor π_{ab} with eigenvalue $-\frac{2}{3}(q+\rho)$. We assume $q + \rho \neq 0$ otherwise the string fluid reduces to a perfect fluid with energy momentum tensor $T_{ab} = qg_{ab}$. This fluid has the unphysical equation of state $\mu + p = 0$.

From the above we note that the structure of the energy momentum tensor is compatible with the general expression (180) - (184) therefore the model of a string fluid is included in the model of Bianchi I spacetime with observers $u^a = \delta_0^a$ we considered in the previous section. To find the specific metric functions $A_1(t), A_2(t), A_3(t)$ which correspond (or select) the string fluid model we equate the dynamical parameters of (180) - (184) with those of (195). We assume the string direction to be along the $x - axis$ i.e. we take ($A_1 \neq 0$):

$$n^a = \frac{1}{\sqrt{A_1}}\delta_1^a, \tag{196}$$

From the mass density and the pressure we find:

$$\mu + 3p = 2q \Rightarrow I_2 = -q, \tag{197}$$

$$\pi_{ab} = (q + \rho)(\frac{1}{3}h_{ab} - A_1^2\delta_a^1\delta_b^1) = (\rho - I_2)(\frac{1}{3}h_{ab} - A_1^2\delta_a^1\delta_b^1).$$

In the coordinates we use:

$$h_{ab} = diag(0, A_1^2, A_2^2, A_3^2). \tag{198}$$

Replacing in the expression of π_{ab} we find:

$$\pi_{ab} = \frac{1}{3}(\rho - I_2)diag(0, -2A_1^2, A_2^2, A_3^2). \tag{199}$$

Equating the two expressions of π_{ab}, (188)–(190) and (199) we find the field equations:

$$\frac{\ddot{A}_1}{A_1} - \frac{\dot{A}_2\dot{A}_3}{A_2A_3} = -\rho + I_2 = -(\rho + q), \tag{200}$$

$$\frac{\ddot{A}_2}{A_2} - \frac{\dot{A}_1\dot{A}_3}{A_1A_3} = 0, \tag{201}$$

$$\frac{\ddot{A}_3}{A_3} - \frac{\dot{A}_1\dot{A}_2}{A_1A_2} = 0. \tag{202}$$

The last three equations are dependent (one follows form the other two). Eventually we have the following system of four simultaneous equations for the five unknowns $A_1(t), A_2(t), A_3(t), q, \rho$:

$$\frac{\dot{A}_1\dot{A}_2}{A_1A_2} + \frac{\dot{A}_1\dot{A}_3}{A_1A_3} + \frac{\dot{A}_2\dot{A}_3}{A_2A_3} = \rho, \tag{203}$$

$$\frac{\ddot{A}_1}{A_1} + \frac{\ddot{A}_2}{A_2} + \frac{\ddot{A}_3}{A_3} = -q, \tag{204}$$

$$\frac{\ddot{A}_2}{A_2} - \frac{\dot{A}_1\dot{A}_3}{A_1A_3} = 0, \tag{205}$$

$$\frac{\ddot{A}_3}{A_3} - \frac{\dot{A}_1\dot{A}_2}{A_1A_2} = 0. \tag{206}$$

This means that we still have the freedom to specify one more condition. This condition could be an equation of state (in the board sense).

Note that the kinematic variables are not effected. Therefore when we take the equation of state and we determine the functions $A_1(t), A_2(t), A_3(t)$ then we can compute the kinematic variables θ, σ_{ab} and draw conclusions on the kinematics of the string fluid.

12.1. An alternative energy momentum tensor

Instead of the energy momentum tensor (193) Letelier [14] considers the energy momentum tensor:

$$T_{ab} = \rho u_a u_b - \lambda n_a n_b. \tag{207}$$

This tensor reduces to the one we have considered if we set $q = 0$ and $\rho = \lambda$. In general it is different to the one we have discussed above. However our analysis applies the same. Considering the same observers and the same direction n^a we have the following 1+3 decomposition of T_{ab} :

$$T_{ab} = \rho u_a u_b - \frac{1}{3}\lambda h_{ab} + \lambda(\frac{1}{3}h_{ab} - n_a n_b), \tag{208}$$

from which follows:

$$\mu = \rho, p = -\frac{1}{3}\lambda, q_a = 0, \tag{209}$$

$$\pi_{ab} = \lambda(\frac{1}{3}h_{ab} - n_a n_b) = \frac{\lambda}{3}(0, -2A_1^2, A_2^2, A_3^2). \tag{210}$$

Then:

$$I_2 = -\frac{1}{2}(\mu + 3p) = -\frac{1}{2}(\rho - \lambda).$$

Equating the two expressions of π_{ab} (188)- (190) and (210) we find the field equations:

$$\begin{aligned}
\frac{\ddot{A}_1}{A_1} - \frac{\dot{A}_2\dot{A}_3}{A_2A_3} &= -\lambda + I_2 = -\tfrac{1}{6}(5\lambda + 3\rho), \\
\frac{\ddot{A}_2}{A_2} - \frac{\dot{A}_1\dot{A}_3}{A_1A_3} &= I_2 = \tfrac{1}{2}(\tfrac{\lambda}{3} - \rho), \\
\frac{\ddot{A}_3}{A_3} - \frac{\dot{A}_1\dot{A}_2}{A_1A_2} &= \tfrac{\ddot{A}_2}{A_2} - \tfrac{\dot{A}_1\dot{A}_3}{A_1A_3}.
\end{aligned}$$

The last three equations are dependent (one follows form the other two). Eventually we have again the following system of four simultaneous equations for the five unknowns $A_1(t), A_2(t), A_3(t), \lambda(t), \rho(t)$:

$$\frac{\dot{A}_1\dot{A}_2}{A_1A_2} + \frac{\dot{A}_1\dot{A}_3}{A_1A_3} + \frac{\dot{A}_2\dot{A}_3}{A_2A_3} = \rho, \tag{211}$$

$$\frac{\ddot{A}_1}{A_1} + \frac{\ddot{A}_2}{A_2} + \frac{\ddot{A}_3}{A_3} = \frac{1}{2}(\lambda - \rho), \tag{212}$$

$$\frac{\ddot{A}_2}{A_2} - \frac{\dot{A}_1\dot{A}_3}{A_1A_3} = \frac{1}{2}(\lambda - \rho), \tag{213}$$

$$\frac{\ddot{A}_3}{A_3} - \frac{\dot{A}_1\dot{A}_2}{A_1A_2} = \frac{1}{2}(\lambda - \rho). \tag{214}$$

We note that in this approach one has still the freedom to consider an extra condition / equation of state. The solution of the field equations will be the starting point of making Physics in this model.

References

[1] Anderson J L 1973, "Principles of Relativity Physics " (Academic Press N.Y.)

[2] Tsamparlis M (1992) "Geometrization of a general Collineation" Jour. Math. Phys. **33**, 1472 - 1479.

[3] Greek indices take the space values $1, 2, 3$ and Latin indices the space-time values $0, 1, 2, 3$. We define the sign of the curvature tensor form the identity $A^a_{;bc} - A^a_{;cb} = R^a_{bcd}A^d$ or $A_{a;bc} - A_{a;cb} = R_{dabc}A^d$.In terms of the connection coefficients $R^a_{bcd} = \Gamma^a_{db},_c - \Gamma^a_{cb},_d + \Gamma^a_{cf}\Gamma^f_{db} - \Gamma^a_{df}\Gamma^f_{cb}$.

[4] Herrera L., Jemenez J, Leal L, Ponce de Leon J, Esculpi M, Galina V (1984) J. Math. Phys. **25**, 3274

[5] Yano K 1956 The theory of Lie derivatives and its Applications (Amsterdam: North Holland)

[6] Tsamparlis M 1992, "Geometrization of a General Collineation" J. Math. Phys. **33**, 1472 - 1479

[7] Saridakis E and Tsamparlis M 1991, "Symmetry Inheritance of Conformal Killing Vectors" J. Math. Phys. **32**, 1541 - 1551

[8] Ellis G F R 1967 *J. Math. Phys.* **8** 1171

[9] Stewart J M and Ellis G F R 1968 *J. Math. Phys.* **9** 1072

[10] Ray D (1978) "Some solutions for relativistic vortices interacting through a scalar field" Phys Rev D, **18**, 3879 - 3880

[11] Lund F and Regge T (1976), "Unified approach to strings and vortices with soliton solutions" Phys Rev D, **14**, 1524 - 1535

[12] Letelier P (1980), "Anisotropic fluids with two perfect fluid components" Phys Rev D **22**, 807 - 813

[13] Letelier P (1981), "" Nuovo Cimento **B63**, 519

[14] Letelier P (1983), "String Cosmologies" Phys Rev D **28**, 2414 - 2419

[15] I Yavuz and I Yilmaz "Inheriting Conformal and Special Conformal Killing Vectors in String Cosmology" (1997) Gen Rel Grav **9**, 1295 - 1307

[16] Yilmaz I (2001), "Timelike and Spacelike Ricci Collineation Vectors in String Cosmology" Inter Jour Modern Physics **10**, 681 - 690

[17] Baysal H Yilmaz I (2002), "Spacelike Ricci Inheritance vectors in a model of string cloud and string fluid stress tensor" Class Quantum Grav **19**, 6435 - 6443

[18] Camci U (2002), "Conformal Collineations and Ricci Inheritance symmetry in String Cloud and String Fluids" Inter Jour Modern Physics **11**, 353 - 366

[19] Baysal H, Camci U, Tarhan I and Yilmaz I (2002), "Conformal Collineations in String Cosmology" Inter Jour Modern Physics **11**, 463 - 469

[20] Sharif M Sheikh U (2005), "Timelike and Spacelike Matter Inheritance Vectors in Specific Forms of Energy-Momentum Tensor" Inter Jour Modern Physics (to appear), gr-qc/0504101

[21] "General Symmetries of a String Fluid space-time" , Gen. Relativ. Grav. **38**(2006) 311 - 324

[22] Maartens R, Mason D P and Tsamparlis M 1986, "Kinematic and Dynamic Properties of Conformal Killing Vectors in Anisotropic Fluids" J. Math. Phys. **27**, 2987 - 2994

Journal of Physics: Conference Series **189** (2009) 012042 doi:10.1088/1742-6596/189/1/012042

Lie symmetries of the geodesic equations and projective collineations

Michael Tsamparlis and Andronikos Paliathanasis
Department of Physics, Section Astrophysics Astronomy Mechanics, University of Athens, University of Athens, Zografos 15783, Athens, Greece
E-mail: mtsampa@phys.uoa.gr, paliathanasis@gmail.com

Abstract. We study the Lie symmetries of the geodesic equations in a Riemannian space and show that they coincide with the projective symmetries of the Riemannian metric. We apply the result to the spaces of constant curvature.

1. Motivation

Consider on the Euclidean plane a family of straight lines parallel to the $x-$axis. These curves can be considered either as the integral curves of the ODE $\ddot{y} = 0$ or as the geodesics of the Euclidean metric $dx^2 + dy^2$. Consider a symmetry operation defined by a reshuffling of these lines. According to the first interpretation this symmetry operation is a Lie symmetry of the ODE $\ddot{y} = 0$ and according to the second interpretation it is a projective symmetry of the Euclidean plane. This motivates one to examine to what extend the Lie symmetries of the equation of autoparallels in an affine space endowed with a symmetric connection Γ^i_{jk} are related to the projective collineations of that space. In the special case of a Riemannian connection this question reduces to the identification of the Lie symmetries of the geodesic equations with the projective symmetries of the metric of the space.

1.1. Preliminary results

Consider a C^∞ manifold M of dimension n, endowed with a symmetric[1] connection.

Let X^a be a vector field on the manifold. The connection defines a family of curves on the manifold, called **autoparallels**, by the requirement that the covariant derivative of the tangent to these curves is parallel to itself, that is:

$$\nabla_{\dot{x}(t)}\dot{x}(t) = \phi(t)\dot{x}(t), \tag{1}$$

where t is a parameter along the curves. When the parametrization is such that ϕ vanishes we say that the autoparallel is affinely parameterized and in this case t is called an **affine parameter**.

In a local coordinate system $\{x^i|i = 1, \ldots, n\}$ equation (1) is written as the system of ODEs:

$$\ddot{x}^i(t) + \Gamma^i_{jk}(x(t))\dot{x}^j(t)\dot{x}^k(t) = \phi(t)\dot{x}^i(t)\ , \quad i = 1, \ldots, n \tag{2}$$

[1] The coefficients Γ^i_{jk} in general are not symmetric. In the autoparallel equation (1) the antisymmetric part of Γ^i_{jk} (the torsion) does play a role.

where $\Gamma^i_{jk}\partial_i = \nabla_j\partial_k$. The Lie derivative (of a not-necessarily symmetric connection) Γ^i_{jk} with respect to a vector field $X = X^a\partial_a$ is defined as follows (see Yano [1] eqn. (2.16)):

$$\mathcal{L}_X\Gamma^i_{jk} = X^i_{,jk} + \Gamma^i_{jk,l}X^l + X^l_{,k}\Gamma^i_{lj} + X^l_{,j}\Gamma^i_{lk} - X^i_{,l}\Gamma^l_{jk}. \quad (3)$$

We say that the vector field X is an **affine collineation** or **affine motion** iff

$$\mathcal{L}_X\Gamma^i_{jk} = 0. \quad (4)$$

In flat space equation (4) implies the condition:

$$X_{a,bc} = 0, \quad (5)$$

whose solution is:

$$X_a = B_{ab}x^b + C_a, \quad (6)$$

where B_{ab}, C_a are constants.

The geometric property/definition of affine collineations is that they preserve the set of autoparallels of the connection together with their affine parametrization (that is, by an affine symmetry an affinely parameterized autoparallel goes over to an affinely parameterized autoparallel of the same connection). From (6) we infer that in an n-dimensional space there are at most $n + n^2 = n(n+1)$ Affine Collineation Vectors and when this is the case, it can be shown that the space is a space of constant curvature.

Note that the equation of the autoparallels for affine parametrization takes the form:

$$\ddot{x}^i(t) + \Gamma^i_{jk}(x(t))\dot{x}^j(t)\dot{x}^k(t) = 0, \quad i = 1, \ldots, n. \quad (7)$$

We say that a vector field X is a projective collineation of the connection if there exists a function $f(x^i)$ so that the following condition holds:

$$\mathcal{L}_X\Gamma^i_{jk} = f_{,j}\delta^i_k + f_{,k}\delta^i_j. \quad (8)$$

In flat space condition (8) implies that:

$$X_{a,b} = B_{ab} + (A_c x^c)g_{ab} + C_b x_a, \quad (9)$$

which has the solution:

$$X_a = B_{ab}x^b + (A_b x^b)x_a + C_a, \quad (10)$$

where again the various coefficients are constants. In an n- dimensional space there are at most $n^2+n+n = n(n+2)$ projective collineations of the connection and when this is the case, it can be shown that the space is a space of constant curvature. This holds in any space irrespective of the signature of the metric and the (finite) dimension of space. The dimension of the space fixes the range of values of the indices $a, b, c, \ldots$.

The geometric property/definition of the (proper) projective collineations is that they preserve the set of autoparallels, but they do not preserve the affine parametrization. Obviously the affine collineations are not proper projective collineations.

The affine collineations form a group which is an invariant subgroup of the projective collineations. Concerning the algebras we have that the algebra of affine collineations contains the proper subalgebras of the Killing vectors, the homothetic Killing vector and does not contain the subalgebra of conformal motions.

The issue we wish to address in this work is:

Consider the autoparallel equation (2) as a system of second order ODEs and study its Lie symmetries. Then consider the quantities Γ^i_{jk} to be the components of a connection on the manifold and establish a relation between the Lie symmetries of the geodesic equations and the projective collineations of this connection. If such a relation exists then one has established a dual perspective for the symmetries of an ODE, that is, either as an non-geometric property (Lie symmetry) and as a geometric collineation (geometric approach). This double approach can be used to establish an interrelation between the study of Lie symmetries through Differential Geometry (where many aspects of the collineations have been addressed) and conversely the study of geometric problems, using the rather formal approach of the Lie method[2].

2. Lie point symmetries of the geodesic equations

We consider the system of ODEs (2) and determine its Lie point symmetries generated by the vector field X. Because the ODEs are of second order, we must prolong X up to the second prolongation. Let $G^{[1]i}$ be the coefficients of the first prolongation. From the prolongation formula we have

$$G^{[1]i} = \frac{d}{dt}\eta^i - \dot{x}^i\frac{d}{dt}\xi = \eta^i_{,t} + \eta^i_{,j}\dot{x}^j - \xi_{,t}\dot{x}^i - \xi_{,j}\dot{x}^i\dot{x}^j \ .$$

For the second prolongation coefficient $G^{[2]i}$ of X we have:

$$\begin{aligned} G^{[2]i} &= \frac{d}{dt}G^{[1]i} - \ddot{x}^i\frac{d}{dt}\xi \\ &= \eta^i_{,tt} + 2\eta^i_{,tj}\dot{x}^j + \eta^i_{,kj}\dot{x}^k\dot{x}^j - \xi_{,tt}\dot{x}^i - 2\xi_{,tj}\dot{x}^j\dot{x}^i - \xi_{jk}\dot{x}^j\dot{x}^k\dot{x}^i \\ &+ \eta^i_{,j}\ddot{x}^j - 2\xi_{,t}\ddot{x}^i - \xi_{,j}\ddot{x}^j\dot{x}^i - 2\xi_{,j}\ddot{x}^i\dot{x}^j \ . \end{aligned} \tag{11}$$

The computation of Lie symmetries can be done in two different ways : Either by using the linear operator $\mathbf{A}$ associated with the system of ODEs and demanding the symmetry condition $[X^{[1]}, \mathbf{A}] = \lambda\mathbf{A}$, or by computing directly the action of the prolonged generator on the equation and demanding that $X^{[1]}(ODE) = 0 \bmod (ODE = 0)$ where ODE is given by (2).

3. Calculation of Lie symmetries using the associated linear operator

We write the system of ODEs in the form $\ddot{x}^i = \omega^i(x, \dot{x}, t)$ where:

$$\omega^i(x, \dot{x}, t) = -\Gamma^i_{jk}(x)\dot{x}^j\dot{x}^k - \phi(x)\dot{x}^i. \tag{12}$$

The associated linear operator defined by this system of ODEs is:

$$\mathbf{A} = \frac{\partial}{\partial t} + \dot{x}^i\frac{\partial}{\partial x^i} + \omega^i(t, x^j, \dot{x}^j)\frac{\partial}{\partial \dot{x}^i}. \tag{13}$$

The condition for a Lie symmetry for the system of equations is [3]:

$$[X^{[1]}, \mathbf{A}] = \lambda(x^a)\mathbf{A}, \tag{14}$$

where $X^{[1]}$ is the first prolongation of the symmetry vector $X = \xi(t,x)\partial_t + \eta^i(t,x)\partial_{x^i}$ defined as follows:

$$X^{[1]} = \xi(t, x, \dot{x})\partial_t + \eta^i(t, x, \dot{x})\partial_{x^i} + G^{[1]i}\partial_{\dot{x}^i}. \tag{15}$$

[2] A question, which could be addressed further, is if one can define a specific set of connection coefficients for a given ODE so that the solution curves of the ODE coincide with the autoparallels of that connection. If this is done then we have geometrized the ODE.

It is a standard result [3] that (14) leads to the following three conditions:

$$-\mathbf{A}\xi = \lambda, \tag{16}$$

$$G^{[1]i} = \mathbf{A}\eta^i - \dot{x}^i\mathbf{A}\xi, \tag{17}$$

$$X^{[1]}(\omega^i) - \mathbf{A}(G^{[1]i}) = -\omega^i\mathbf{A}\xi. \tag{18}$$

For any function $f(t,x^i)$ the $\mathbf{A}f = \frac{df}{dt}$ where $\frac{df}{dt} = f_{,t} + f_{,i}\dot{x}^i$ is the total derivative of f. Using this result we write the symmetry conditions as follows:

$$\lambda = -\frac{d\xi}{dt}, \tag{19}$$

$$G^{[1]i} = \frac{d\eta^i}{dt} - \dot{x}^i\frac{d\xi}{dt}, \tag{20}$$

$$X^{[1]}(\omega^i) - A(G^{[1]i}) = -\omega^i\frac{d\xi}{dt}. \tag{21}$$

We note that the second condition (20) defines the first extension $G^{[1]i}$. The first equation (19) gives the factor λ, therefore the essential symmetry condition is equation (21).

To compute the symmetry condition first we have to compute the quantities $X^{[1]}(\omega^i)$ and $A(G^{[1]i}) - \omega^i\frac{d\xi}{dt}$ taking into consideration (20) and (12). The result of this formal calculation is:

$$\begin{aligned} X^{[1]}(\omega^i) &= (\xi\partial_t + \eta^i\partial_{x^i} + G^{[1]i}\partial_{\dot{x}^i})(-\Gamma^i_{jk}(x)\dot{x}^j\dot{x}^k - \phi(x)\dot{x}^i) \\ &= -\eta^i_{,t}\phi \\ &+ (-\xi\phi_{,t}\delta^i_j - \phi_{,k}\eta^k\delta^i_j - \eta^k_{,t}\Gamma^i_{kj} - \eta^k_{,t}\Gamma^i_{jk} - \phi\eta^i_{,j} + \phi\xi_{,t}\delta^i_j)\dot{x}^j \\ &+ (-\xi\Gamma^i_{(kj),t} - \eta^l\Gamma^i_{(kj),l} - \eta^l_{,k}\Gamma^i_{(lj)} - \eta^l_{,k}\Gamma^i_{(jl)} + \phi\xi_{,k}\delta^i_j + 2\xi_{,t}\Gamma^i_{(kj)})\dot{x}^j\dot{x}^k \\ &+ \xi_{,(k}\Gamma^i_{jl)}\dot{x}^j\dot{x}^k\dot{x}^l. \end{aligned} \tag{22}$$

$$\begin{aligned} A(G^{[1]i}) - \omega^i\frac{d\xi}{dt} = \eta^i_{,tt} & \\ &(2\eta^i_{,tj} + \phi\eta^i_{,j} - 2\phi\xi_{,t}\delta^i_j - \xi_{,tt}\delta^i_j)\dot{x}^j \\ &+ (\eta^i_{,(jk)} - 2\xi_{,t(j}\delta^i_{k)} - \eta^i_{,l}\Gamma^l_{(jk)} + 2\xi_{,t}\Gamma^i_{(jk)} - 2\phi\xi_{(,j}\delta^i_{k)})\dot{x}^j\dot{x}^k \\ &+ (\xi_{,(j}\Gamma^i_{kl)} + \xi_{,m}\Gamma^m_{(kl}\delta^i_{j)} - \xi_{,(jk}\delta^i_{l)})\dot{x}^j\dot{x}^k\dot{x}^l. \end{aligned}$$

Substituting in the symmetry condition (21) and collecting terms of the same order in $\dot{x}^j$ we find the following equations: $(i = 1, \ldots, n)$:
$(\dot{x})^0$ terms:

$$\eta^i_{,tt} + \eta^i_{,t}\phi = 0, \tag{23}$$

$(\dot{x})^1$ terms:

$$\xi_{,tt}\delta^i_j - \xi\phi_{,t}\delta^i_j - 2[\eta^i_{,tj} + \eta^k_{,t}\Gamma^i_{(kj)}] - [\phi\xi_{,t} + \phi_{,k}\eta^k]\delta^i_j = 0, \tag{24}$$

$(\dot{x})^2$ terms:

$$(-\eta^i_{,(jk)} - \eta^l\Gamma^i_{(jk),l} - \eta^l_{,k}\Gamma^i_{lj} - \eta^l_{,k}\Gamma^i_{jl} + \eta^i_{,l}\Gamma^l_{jk}) + 2\xi_{,t(j}\delta^i_{k)} - 2\phi\xi_{(,j}\delta^i_{k)} - \xi\Gamma^i_{(kj),t} = 0 \quad \Rightarrow$$

$$\mathcal{L}_\eta\Gamma^i_{(jk)} = -2\phi\xi_{(,j}\delta^i_{k)} + \xi\Gamma^i_{(kj),t} + 2\xi_{,t(j}\delta^i_{k)}, \tag{25}$$

$(\dot{x})^3$ terms:

$$(\xi_{,(jk} - \xi_{,|e|}\Gamma^e_{(jk})\delta^i_{l)} = 0\ . \tag{26}$$

Define the quantity:

$$\Phi = \xi_{,t} - \phi\xi. \tag{27}$$

Then condition (25) is written:

$$\mathcal{L}_\eta \Gamma^i_{(jk)} = 2\Phi_{(,j}\delta^i_{k)} - \xi\Gamma^i_{(kj),t}. \tag{28}$$

If we consider the vector $\xi = \xi\partial_t$ (which does not have components along ∂_i) we compute that:

$$\mathcal{L}_\xi \Gamma^i_{(jk)} = \xi\Gamma^i_{(kj),t}, \tag{29}$$

hence (28) is written:

$$\mathcal{L}_{\mathbf{X}} \Gamma^i_{(jk)} = 2\Phi_{(,j}\delta^i_{k)}, \tag{30}$$

where $X = \xi + \eta = \xi\partial_t + \eta^i(t,x)\partial_{x^i}$. We observe that this condition is precisely condition (8) for a projective collineation of the connection $\Gamma^i_{(jk)}$ along the symmetry vector $\mathbf{X}$ and with projecting function Φ. Concerning the other conditions we note that (24) can be written in covariant form (relevant to the indices $a = 1, 2, ..., n$) as follows:

$$\Phi_{,t}\delta^i_j - 2\eta^i_{,t|j} = 0, \tag{31}$$

where $\eta^i_{,t|j} = \eta^i_{,tj} + \eta^k_{,t}\Gamma^i_{(kj)}$ is the covariant derivative with respect to $\Gamma^i_{(kj)}$ of the vector $\eta^i_{,t}$. Similarly condition (26) can be written:

$$\xi_{|(jk}\delta^i_{l)} = 0.$$

Contracting the indices i, j we find the final form:

$$\xi_{|(jk)} = 0. \tag{32}$$

This implies $\xi_{,i} = f(t)$ (where $f(t)$ is an arbitrary function of its argument) is a gradient KV of the metric of the space x^i.

Condition (23) is obviously in covariant form with respect to the indices a.

We arrive at the following conclusion:

(i) The Lie symmetries of the autoparallel equations (12) (not necessarily affinely parameterized) for a general connection defined on a C^∞ manifold are the following:

$$\eta^i_{,tt} + \eta^i_{,t}\phi = 0, \tag{33}$$

$$\xi_{|(jk)} = 0, \tag{34}$$

$$\Phi_{,t}\delta^i_j - 2\eta^i_{,t|j} = 0, \tag{35}$$

$$\mathcal{L}_{\mathbf{X}}\Gamma^i_{(jk)} = 2\Phi_{(,j}\delta^i_{k)}. \tag{36}$$

These are covariant equations because if we consider the connection in the augmented $1+n$ space $\{x^i, t\}$, all components of Γ which contain an index along the direction of t vanish, therefore the partial derivatives with respect to t can be replaced with covariant derivative with respect to t.

(ii) Equation (33) gives the functional dependence of η^i on t and the parametrization

(iii) Equation (34) gives that the vector $\xi_{,i}$ is a gradient Killing vector of the $n-$dimensional space $\{x^i\}$.

(iv) Equation (35) relates the functional dependence of η^i and ξ in terms of t.

(v) Equation (36) is the most important equation to our purpose, because it states that *the symmetry vector X is an affine collineation in the jet space* $\{t, x^i\}$ because it preserves both the geodesic an its parametrization. In the space $\{x^i\}$ the vectors $\eta^i(t,x)\partial_{x^i}$ *are projective collineations* because they preserve the geodesics and not necessarily their parametrization.

In the following we restrict our considerations to the case of Riemannian connections that is the Γ^i_{jk} are symmetric and the covariant derivative of the metric vanishes.

4. Calculation of the symmetry vector for a Riemannian connection

We compute the Lie symmetry vectors for the case of affine parametrization ($\Leftrightarrow \phi = 0$) and the assumption $\Gamma^i_{jk,t} = 0$ i.e. the Γ^i_{jk} are independent of the parameter t. The later is a logical assumption because the Γ^i_{jk} are computed in terms of the metric which does not depend on the parameter t. Under these assumptions the symmetry conditions (33) - (36) read:

$$\eta^i_{,tt} = 0, \tag{37}$$

$$\xi_{|(jk)} = 0, \tag{38}$$

$$\xi_{,tt}\delta^i_j - 2\eta^i_{,t|j} = 0, \tag{39}$$

$$\mathcal{L}_\eta \Gamma^i_{jk} = 2\xi_{,t(j}\,\delta^i_{k)}. \tag{40}$$

We proceed with the solution of this system of equations. Equation (37) implies:

$$\eta^i(t,x) = A^i(x)t + B^i(x), \tag{41}$$

where $A^i(x), B^i(x)$ are arbitrary differentiable vector fields.

The solution of equation (38) is:

$$\xi(t,x) = C_J(t)S^J(x) + D(t), \tag{42}$$

where $C_J(t), D(t)$ are arbitrary functions of the affine parameter t and $S^J(x)$ is a function whose gradient is a gradient KV i.e.:

$$S(x)_{|(i,j)} = 0. \tag{43}$$

The index J runs through the number of gradient KVs of the metric.

Condition (39) gives:

$$2A(x)^i_{|j} = \left[C_J(t)_{,tt}\, S^J(x) + D(t)_{,tt}\right]\delta^i_j. \tag{44}$$

Because the left hand side is a function of x only we must have:

$$D(t)_{,tt} = M \Rightarrow D(t) = \frac{1}{2}Mt^2 + Kt + L, \quad \text{where } M, K, L \text{ constants} \tag{45}$$

$$C_J(t)_{,tt} = G_J = \text{constant} \Rightarrow C_J(t) = \frac{1}{2}G_J t^2 + E_J t + F_J, \quad \text{where } G_J, E_J, F_J \text{ constants.} \tag{46}$$

Replacing in (44) we find:

$$2A(x)^i_{;j} = \left(G_J S^J(x) + M\right)\delta^i_j \Rightarrow A(x)_{i;j} = \frac{1}{2}\left(G_J S^J(x) + M\right) g_{ij}, \tag{47}$$

where we have lowered the index because the connection is metric (i.e. $g_{ij|k}=0$). The last equation implies that the vector $A(x)^i$ is a conformal Killing vector with conformal factor $\psi=\frac{1}{2}(G_J S^J(x)+M)$. Because $A(x)_{[i;j]}=0$ this vector is a gradient CKV.

We continue with condition (40) and replace $\eta^i(t,x)$ from (41):

$$\mathcal{L}_{\mathbf{A}}\Gamma^i_{jk}t+\mathcal{L}_{\mathbf{B}}\Gamma^i_{jk}=2\xi_{,t\,(,j}\delta^i_{k)}=2\left[(G_Jt+E_J)\,S^J(x)+Mt+K\right]_{|(j}\delta^i_{k)}=2\,(G_Jt+E_J)\,S^J(x)_{|(j}\delta^i_{k)}\Rightarrow$$

$$\mathcal{L}_{\mathbf{A}}\Gamma^i_{jk}=2G_JS^J(x)_{,(j}\delta^i_{k)} \tag{48}$$

$$\mathcal{L}_{\mathbf{B}}\Gamma^i_{jk}=2E_JS^J(x)_{,(j}\delta^i_{k)}. \tag{49}$$

The last two equations imply that the vectors $A^i(x)$, $B^i(x)$ are *projective collineations* of the metric - or one of their specializations - with projective functions $G_JS^J(x)$ and $E_JS^J(x)$ respectively. Note that relations (48), (49) remain true if we add a KV to the vectors $A^i(x)$, $B^i(x)$, therefore these vectors are determined up to a KV.

Now it is well known that in a Riemannian space a CKV K^i with conformal factor $\psi(x^i)$ satisfies the following identity:

$$\mathcal{L}_{\mathbf{K}}\Gamma^i_{(jk)}=g^{is}\left[\psi_{,j}g_{ks}+\psi_{,k}g_{js}-\psi_{,s}g_{jk}\right]. \tag{50}$$

Applying this identity to the CKV A^i we find:

$$G_JS^J(x)_{,k}=0\Rightarrow G_JS^J(x)=2\rho=\text{constant}. \tag{51}$$

This implies that A^i *is a gradient HKV with homothetic factor* $\rho+\frac{1}{2}M$. Furthermore (47) implies:

$$2A^i=(2\rho+M)x^i+2L^i\Rightarrow$$

$$A^i=(\rho+\frac{1}{2}M)x^i+L^i, \tag{52}$$

where L^i is a KV.

We continue with the projective collineation vector B^i. For this vector we use the property that for a symmetric connection the following identity holds:

$$\mathcal{L}_{\mathbf{B}}\Gamma^i_{(jk)}=B^i_{;jk}-R^i_{jkl}B^l.$$

Replacing the left hand side from (49) we find:

$$B^i_{;jk}-R^i_{jkl}B^l=2E_JS^J(x)_{,(j}\delta^i_{k)}. \tag{53}$$

Contracting the indices i,j we find:

$$\left(B^i_{;i}-(n+1)E_JS^J(x)\right)_{;k}=0, \tag{54}$$

which implies that:

$$B^i_{;i}=(n+1)E_JS^J(x)+2b, \tag{55}$$

where b =constant.

Using the above results we find for $\xi(t,x)$:

$$\begin{aligned}\xi(t,x)&=C_J(t)S^J+D(t),\\&=\left(\frac{1}{2}G_Jt^2+E_Jt+F_J\right)S^J+\frac{1}{2}Mt^2+Kt+L,\\&=\frac{1}{2}(G_JS^J+M)t^2+(E_JS^J+K)t+F_JS^J+L.\end{aligned}$$

We summarize the above results in the following Theorem

Theorem 1 *1. The Lie symmetry vector $X = \xi + \eta = \xi(t,x)\partial_t + \eta^i(t,x)\partial_{x^i}$ of the equation of geodesics (7) in a Riemannian space involves all symmetry vectors, that is, (gradient) KVs, (gradient) HKVs, PCs and their degeneracies HKVs and ACs as follows:*
a. The function:

$$\xi(t,x) = \left(\rho + \frac{1}{2}M\right)t^2 + \left[E_J S^J + K\right]t + F_J S^J + L,$$

where ρ, M, b, K, F_J, L are constants, the index J running along the number of gradient KVs.
b. The vector:

$$\eta^i(t,x) = A^i(x)t + B^i(x) + D^i(x),$$

where the vector $A^i(x)$ is a gradient HKV with conformal factor $\psi = \rho + \frac{1}{2}M$, $D^i(x)$ is a KV of the metric and $B^i(x)$ is a projective collineation with projection function $E_J S^J(x)$ whose divergence $B^i_{;i}$ must satisfy condition (55).

5. The Lie symmetries of geodesic equations of spaces of constant curvature

The metric of a space of constant curvature $K = 0, \pm 1$ and dimension n in Cartesian (stereographic) coordinates is:

$$ds^2 = U(x^i)\eta_{ij}dx^i dx^j, \tag{56}$$

where $U(x^i) = \frac{1}{1+\frac{Kx\cdot x}{4}}$, $\eta_{ij} = diag(\pm 1, \pm 1, ..., \pm 1)$. It is easy to show that the connection coefficients of this metric are:

$$\Gamma^i_{jk} = -\frac{KU}{2}(x_k\delta^i_j + x_j\delta^i_k - x^i\eta_{jk}), \tag{57}$$

hence the geodesics equations for affine parametrization are:

$$\ddot{x}^i - \frac{KU}{2}(2x_k\dot{x}^k\dot{x}^i - (\eta_{jk}\dot{x}^j\dot{x}^k)\dot{x}^i) = 0, \tag{58}$$

where a dot indicates derivation with respect to the affine parameter s (say). We shall apply Theorem 1 to determine the Lie symmetries of the geodesic equations (58). Feroze et all [9] have computed the Lie symmetries of these equations for some well known spaces of constant curvature and essentially they considered the Killing vectors of the corresponding metrics. In the following we shall extent and complete their results to all spaces of constant curvature. Furthermore we clarify the general and important results of Aminova [6], [7].

In order to apply Theorem 1 we need to know the conformal and the projective collineations of spaces of constant curvature. From the literature we have the following results.

A. Flat spaces

a. Conformal algebra

$\frac{n(n-1)}{2}$ Non-gradient Killing vectors:

$$X_{IJ} = \delta^j_{[I}\delta^i_{J]}x_j\partial_i,$$

n Gradient Killing vectors:

$$\delta^i_I\partial_i,$$

1 Gradient Homothetic Killing vector:

$$x^i\partial_i \text{ with homothety factor } \psi = 1,$$

n non-gradient special CKVs:

$$\left[2x_I x^i - \delta^i_I (x^k x_k)\right] \partial_i \text{ with conformal factor } \psi = 2x^i,$$

b. $n(n+1)$ Proper Affine Collineations [1]:

$$\left[b^i_j x^j + c^i\right] \partial_i,$$

c. $n(n+2)$ Proper Projective Collineations [2]:

$$\left[(c_j x^j)x^i + b^i_j x^j + a^i\right] \partial_i.$$

B. Spaces of constant non-vanishing curvature (the indices I, J count vectors):
a. Conformal algebra [10]:
n Non-gradient Killing vectors:

$$\frac{1}{U}\left[(2U-1)\,\delta^i_I + \frac{KU}{2} x_I x^i\right] \partial_i,$$

$\frac{n(n-1)}{2}$ Non-gradient Killing vectors:

$$X_{IJ} = \delta^j_{[I}\delta^i_{J]}\partial_i,$$

1 Gradient Homothetic Killing vector:

$$x^i \partial_i,$$

n Gradient Conformal Killing vectors:

$$X_I = \frac{1}{U}\left[\delta^i_I - \frac{1}{2} K U x_I x^i\right] \partial_i.$$

b. Affine Colliniations [11]:
Not admitted.
c. $n(n+2)$ Proper Projective Collineations [11]:

$$\phi_{,i} + D_i \; (n \geq 2),$$

where $\phi = c + \frac{(1-Kr^2/4)a_i x^i + b_{ij}x^i x^j}{(1-Kr^2/4)^2} a_i, b_{ij} = b_{ji}, c$ and D_i is a Killing vector. The constant c is irrelevant because it does not give a proper projective collineation.

5.1. The case of a flat space

In this case the equation of geodesics when affinely parameterized is:

$$\ddot{x}^i = 0 \; i = 1, 2.,, n.$$

We know that these equations admit the maximal number of Lie symmetries. We determine these symmetries using Theorem 1. The gradient KV $\delta^i_I \partial_i$ implies:

$$S^J(x) = \delta^J_i x^i,$$

(plus an irrelevant constant). The gradient homothetic vector $x^i\partial_i$ implies:

$$A^i = x^i, L^i = 0.$$

and the homothetic factor $\psi = 1$:

$$\rho + \frac{1}{2}M = 1.$$

Equation (47) gives:

$$2\delta^i_j = \left(G_J S^J(x) + M\right)\delta^i_j \Rightarrow G_J = 0, M = 2.$$

The condition (55) on the proper projective collineations gives:

$$b^i_{;i} + (n+1)(c_i x^i) = (n+1)E_i x^i + 2b \Rightarrow E_i = c_i, b = \frac{1}{2}b^i_i.$$

We have now for the symmetry function $\xi(t,x)$:

$$\xi(t,x) = t^2 + (c_i x^i + K)t + F_i x^i + L$$

where K, F_i, L are constants. Concerning the symmetry vector n^i we have:

$$n^i = x^i t + \left[(c_j x^j)x^i + b^i_j x^j + a^i\right] + \delta^j_{[I}\delta^i_{J]}x_j + \delta^i_{I'}.$$

These symmetry quantities must be a solution of the symmetry conditions (37) - (40). It is an easy exercise to show that (37) - (39) are satisfied. Concerning (40) we find:

$$\mathcal{L}_\eta \Gamma^i_{jk} = 2c_{(j}\delta^i_{k)} = n^i_{;jk},$$

which is correct. Therefore the solution we have found satisfies he Lie symmetry conditions.

5.2. The case of a space of constant non-vanishing curvature

The system of equations we want to find the Lie symmetries are (58). This time we do not have gradient KVs therefore $S^J(x) = 0$ and the index J is suppressed. The space does not admit a gradient homothetic vector therefore we set $\psi(A) = 0$. This implies $\rho + \frac{1}{2}M = 0$. Condition (55) cannot be satisfied therefore the proper projective collineations $\phi_{,i}$ are not admitted. The answer is then:

$$\xi(t,x) = Kt + L,$$
$$n^i = \delta^j_{[I}\delta^i_{J]}x_j + \frac{1}{U}\left[(2U-1)\,\delta^i_{I'} + \frac{KU}{2}x_{I'}x^i\right].$$

It is easy to show that these quantities satisfy all the conditions (37) - (40).

References

[1] Yano K 1956 *The theory of Lie derivatives and its Applications*, (Amsterdam: North Holland)
[2] Scouten KJ A 1954 *Ricci Calculus*, (Berlin: Springer)
[3] Stephani, H. 1989 , *Differential Equations: Their Solutions using Symmetry*, Cambridge University Press, NY
[4] Aminova, A. V. 1978, "Concircular vector fields and group symmetries in spaces of constant curvature" (in Russian), *Gravit.i. Teoriya Otmotisel' n* **14**, 4 - 16
[5] Aminova, A. V. 1994, " Automorphisms of geometric structures as symmetries of diferential equations" (in Russian), *Izv. - Vyssh. - Uchebn. - Zaved. - Mat.* **2**, 3 - 10
[6] Aminova, A. V. 1995, "Projective transformations and symmetries of differential equations", Sbornik Mathematics **186** (12), 1711 - 1726

[7] Aminova, A. V. 2000, "Projective geometry of systems of differential equations: General conceptions", *Tensor, N.S.*, 65 - 86
[8] We define the sign of the curvature tensor form the identity $A^a_{;bc} - A^a_{;cb} = R^a_{bcd}A^d$ or $A_{a;bc} - A_{a;cb} = R_{dabc}A^d$.In terms of the connection coefficients $R^a_{bcd} = \Gamma^a_{db,c} - \Gamma^a_{cb,d} + \Gamma^a_{cf}\Gamma^f_{db} - \Gamma^a_{df}\Gamma^f_{cb}$.
[9] -Feroze, T., Mahomed, F.M., Qadir, A. 2006, "The connection betweeb isometries and symetries of geodesic equations of the underlying spaces", *Nonlinear Dynamics* 65 - 74
[10] Tsamparlis, M., Nikolopoulos, D., Apostolopoulos, P. S. 1998, *Class. Quantum Grav.*, **15**, 2909
[11] Barnes, A. 1993, "Projective collineations in Einstein spaces" *Class. Quantum Grav.*, **10**, 1139 - 1145

Codimension-2 black hole solutions on a thin 3-brane and their extension into the bulk

Minas Tsoukalas

Department of Physics, National Technical University of Athens,
Zografou Campus GR 157 73, Athens, Greece.

E-mail: minasts@central.ntua.gr

Abstract. In this talk we discuss black hole solutions in six-dimensional gravity with a Gauss-Bonnet term in the bulk and an induced gravity term on a thin 3-brane of codimension-2. It is shown that these black holes can be localized on the 3-brane, and they can further be extended into the bulk by a warp function. These solutions have regular horizons and no other curvature singularities appear apart from the string-like ones. The projection of the Gauss-Bonnet term on the brane imposes a constraint relation which requires the presence of matter in the extra dimensions, in order to sustain our solutions.

1. Introduction

Taking a look at the last decade, a lot of attention has been drawn to brane-world models. These are mainly string inspired models [1], where our universe is trapped on a brane, embedded in a higher dimensional space-time, the bulk. All matter fields are trapped on the brane and only gravity can propagate to the bulk. If we have n dimensions perpendicular to the brane, then we have a codimension-n brane-world [2], [3].

Of special interest, in the context of brane-worlds, is the study of black holes residing on the brane, mainly in the set up of codimension-1 models. The issue was first addressed by Chamblin et.al. [4]. A four-dimensional black hole on a 3-brane can be stretched to the extra dimension, forming something like a black cigar due to Gregory-Laflame instability [5], while at the same time solving the five-dimensional Einstein equations. Still instabilities can occur [6]. Furthermore one can look into the effective four-dimensional Einstein equations on the brane [7], and try to solve them [8]. Also recently there has been an investigation of brane-world black holes in the context of heterotic brane-world scenario [9]. Investigating the case of a 2-brane in a four-dimensional universe there is a way out. Based on the form for an accelerating four-dimensional black hole (C-metric) [10] , Emparan et.al. [11] found a BTZ black hole [12] residing on a 2-brane, which can be extended into the bulk, forming a BTZ black string. Their thermodynamic analysis showed that there is an instability similar to the Gregory-Laflamme instability [5].

Expanding the number of extra dimensions, one can investigate higher codimension models. The first attempt of course is to examine codimension-2 brane-worlds. Just like in three-dimensional gravity, where a point particle can induce a deficit angle, [13], [14], the same can happen if we imbed a 3-brane into a six-dimensional space time. The vacuum energy of the brane, instead of curving the brane-world volume, induces a deficit angle in the bulk solution around the brane [15]. This property was the driving force, in order to solve the cosmological problem as was done by many authors [16]. Still due to near brane singularities, the brane's energy-momentum has to be proportional to its induced metric, if one wants to find nonsingular solutions [17]. In order to avoid this obstacle one has to introduce some thickness to the brane [18, 19, 20] , or modify the gravitational action in the thin brane limit, introducing either a Gauss-Bonnet term [21] or an induced gravity term on the brane [22], recovering in this way four-dimensional gravity on the brane.

Unlike the situation in codimension-1 brane-world models, there is no clear understanding of time-dependent cosmological solutions in codimension-2 set ups. In the thin brane limit the brane equation of state and energy density are tuned, due to a relation between the energy-momentum tensors on the brane and in the bulk, giving this way a non-standard cosmology on the brane [23, 24]. In order to resolve this problem one has to regularize the codimension-2 branes by introducing some thickness and then consider matter on them [25, 26, 27, 28]. To have a cosmological evolution on the regularized branes the brane world-volume should be expanding and in general the bulk space should also evolve in time. Alternatively [29, 30] one has to consider a codimension-1 brane moving in the regularized static background. The resulting cosmology, however, is unrealistic having a negative Newton's constant (for a review on the cosmology in six dimensions see [31]).

Deficit angles can also appear in the context of cosmic strings [32]. For example, if a black hole is pierced by a cosmic string, a deficit angle appears in the space outside the string [33, 34]. Taking advantage of that property, and adjusting it, a six-dimensional black hole localized on a 3-brane of codimension-2 [35] was constructed, with a conical structure in the bulk and deformations accommodating the deficit angle. Still due to near brane instabilities we mentioned earlier, it is not easy to realize these solutions in the thin brane limit, where high curvature terms are needed to accommodate matter on the brane. Finally application of rotation [36] and perturbative analysis was performed [37, 38, 39, 40].

In [41] black holes on an infinitely thin conical 2-brane and their extension into a five-dimensional bulk with a Gauss-Bonnet term were studied. Two classes of solutions were found. The first class consists of the familiar BTZ black hole which solves the junction conditions on a conical 2-brane in vacuum. These solutions in the bulk are BTZ string-like objects with regular horizons and no pathologies. The warping to five-dimensions depends on the length $\sqrt{\alpha}$ where α is the Gauss-Bonnet coupling, and this length scale defines the shape of the horizon. Consistency of the bulk solutions requires a fine-tuned relation between the Gauss-Bonnet coupling and the five-dimensional cosmological constant. The second class of solutions consists of BTZ black holes with short distance corrections. These solutions correspond to a BTZ black hole conformally dressed with a scalar field [42, 43]. Localization of these black holes on the 2-brane leads to the interesting result that the energy-momentum tensor required to support such solutions on the brane corresponds to the energy-momentum tensor of a scalar field in the limit $r/L_3 << 1$, where L_3 is the length scale of the three-dimensional AdS space and r the radial distance on the brane. Also these solutions have black string-like extensions into the bulk.

In these notes we will present an extension of the work done in [41]. Now instead of having a 2-brane in a five-dimensional space, we will have a 3-brane into a six-dimensional space-time [44]. It is shown that four-dimensional Schwarzschild-AdS black holes on the brane can have an extension into the bulk with a warp factor. These look like black string-like objects with regular horizons. The warping to the extra dimensions, again depends on the Gauss-Bonnet coupling which is again fine-tuned to the six-dimensional cosmological constant. Keeping the deficit angle constant, the presence of a four dimensional black holes on the brane, requires matter in the two extra dimensions.

In the following, first we will present briefly the BTZ string-like solutions of the five-dimensional case. In section 3 we discuss the black holes on a 3-brane in a six-dimensional space-time and in section 4 we discuss the special role played by the Gauss-Bonnet term. Finally, in section 5 we conclude.

2. BTZ string-like solutions in five-dimensional braneworlds of codimension-2

We consider the following gravitational action in five dimensions with a Gauss-Bonnet term in the bulk and an induced three-dimensional curvature term on the brane

$$\begin{aligned} S_{\rm grav} &= \frac{M_5^3}{2}\left\{\int d^5x\sqrt{-g^{(5)}}\left[R^{(5)}+\alpha\left(R^{(5)2}-4R^{(5)}_{MN}R^{(5)MN}+R^{(5)}_{MNKL}R^{(5)MNKL}\right)\right]\right. \\ &+ \left. r_c^2\int d^3x\sqrt{-g^{(3)}}\,R^{(3)}\right\}+\int d^5x\mathcal{L}_{bulk}+\int d^3x\mathcal{L}_{brane}\,, \end{aligned} \tag{2.1}$$

where $\alpha\,(\geq 0)$ is the GB coupling constant and $r_c = M_3/M_5^3$ is the induced gravity "cross-over" scale (marking the transition from 3D to 5D gravity). We also consider the following bulk metric

$$ds_5^2 = g_{\mu\nu}(x,\rho)dx^\mu dx^\nu + a^2(x,\rho)d\rho^2 + L^2(x,\rho)d\theta^2\,, \tag{2.2}$$

where $g_{\mu\nu}(x,0)$ is the brane-world metric and x^μ denote three dimensions, $\mu = 0,1,2$ whereas ρ,θ denote the radial and angular coordinates of the two extra dimensions. Capital M, N indices will take values in the five-dimensional space.

The Einstein equations resulting from the variation of the action (2.1) are

$$G^{(5)N}_M + r_c^2 G^{(3)\nu}_\mu g^\mu_M g^N_\nu \frac{\delta(\rho)}{2\pi L} - \alpha H^N_M = \frac{1}{M_5^3}\left[T^{(B)N}_M + T^{(br)\nu}_\mu g^\mu_M g^N_\nu \frac{\delta(\rho)}{2\pi L}\right]\,, \tag{2.3}$$

where

$$\begin{aligned} H^N_M &= \left[\frac{1}{2}g^N_M(R^{(5)\,2}-4R^{(5)\,2}_{KL}+R^{(5)\,2}_{ABKL})-2R^{(5)}R^{(5)N}_M\right. \\ &\left.+4R^{(5)}_{MP}R^{NP}_{(5)}\quad+4R^{(5)\quad N}_{KMP}R^{KP}_{(5)}-2R^{(5)}_{MKLP}R^{NKLP}_{(5)}\right]\,. \end{aligned} \tag{2.4}$$

To obtain the brane-world equations we expand the metric around the brane as $L(x,\rho) = \beta(x)\rho + O(\rho^2)$. Function L behaves as $L'(x,0) = \beta(x)$, where a prime denotes derivative with respect to ρ. We also demand that the space in the vicinity of the conical singularity is regular which imposes the supplementary conditions that $\partial_\mu\beta = 0$ and $\partial_\rho g_{\mu\nu}(x,0) = 0$ [21].

The extrinsic curvature is given by $K_{\mu\nu} = g'_{\mu\nu}$. Furthermore the second derivatives of the metric functions contain δ-function singularities at the position of the brane [21].

$$\frac{L''}{L} = -(1-L')\frac{\delta(\rho)}{L} + \mathrm{non-singular\ terms}\,, \tag{2.5}$$

$$\frac{K'_{\mu\nu}}{L} = K_{\mu\nu}\frac{\delta(\rho)}{L} + \mathrm{non-singular\ terms}\,. \tag{2.6}$$

From the above singularity expressions and using the Gauss-Codazzi equations, we can match the singular parts of the Einstein equations (2.3) and get the following "boundary" Einstein equations

$$G^{(3)}_{\mu\nu} = \frac{1}{M^3_{(5)}(r_c^2+8\pi(1-\beta)\alpha)}T^{(br)}_{\mu\nu} + \frac{2\pi(1-\beta)}{r_c^2+8\pi(1-\beta)\alpha}g_{\mu\nu}\,. \tag{2.7}$$

Table 1. BTZ string-like solutions in five-dimensional braneworlds of codimension-2

$n(r)$	$f(\rho)$	$L(\rho)$	$-\Lambda_5$	Constraints
BTZ	$\cosh\left(\frac{\rho}{2\sqrt{\alpha}}\right)$	$\forall L(\rho)$	$\frac{3}{4\alpha}$	$L_3^2 = 4\alpha$
BTZ	$\cosh\left(\frac{\rho}{2\sqrt{\alpha}}\right)$	$2\beta\sqrt{\alpha}\sinh\left(\frac{\rho}{2\sqrt{\alpha}}\right)$	$\frac{3}{4\alpha}$	-
BTZ	$\cosh\left(\frac{\rho}{2\sqrt{\alpha}}\right)$	$2\beta\sqrt{\alpha}\sinh\left(\frac{\rho}{2\sqrt{\alpha}}\right)$	$\frac{3}{4\alpha}$	$L_3^2 = 4\alpha$
BTZ	±1	$\frac{1}{\gamma}\sinh(\gamma\rho)$	$\frac{3}{l^2}$	$\gamma = \sqrt{-\frac{2\Lambda_5}{3+4\alpha\Lambda_5}}$
$\forall n(r)$	$\cosh\left(\frac{\rho}{2\sqrt{\alpha}}\right)$	$2\beta\sqrt{\alpha}\sinh\left(\frac{\rho}{2\sqrt{\alpha}}\right)$	$\frac{3}{4\alpha}$	-
$\sqrt{-M+\frac{r^2}{L_3^2}-\frac{\varsigma}{r}}$	$\cosh\left(\frac{\rho}{2\sqrt{\alpha}}\right)$	$2\beta\sqrt{\alpha}\sinh\left(\frac{\rho}{2\sqrt{\alpha}}\right)$	$\frac{3}{4\alpha}$	$L_3^2 = 4\alpha$
$\sqrt{-M+\frac{r^2}{L_3^2}-\frac{\varsigma}{r}}$	±1	$2\beta\sqrt{\alpha}\sinh\left(\frac{\rho}{2\sqrt{\alpha}}\right)$	$\frac{1}{4\alpha}$	$\Lambda_5 = -\frac{1}{4\alpha} = -\frac{3}{L_3^2}$

We assume that there is a localized (2+1) black hole on the brane. The brane metric is

$$ds_3^2 = \left(-n(r)^2dt^2 + n(r)^{-2}dr^2 + r^2d\phi^2\right) . \tag{2.8}$$

We will look for black string solutions of the Einstein equations (2.3) using the five-dimensional metric (2.2) in the form

$$ds_5^2 = f^2(\rho)\left(-n(r)^2dt^2 + n(r)^{-2}dr^2 + r^2d\phi^2\right) + a^2(r,\rho)d\rho^2 + L^2(r,\rho)d\theta^2 . \tag{2.9}$$

The space outside the conical singularity is regular, therefore, we demand that the warp function $f(\rho)$ is also regular everywhere. We assume that there is only a cosmological constant Λ_5 in the bulk and we take $a(r,\rho) = 1$. Then, from the bulk Einstein equations

$$G^{(5)}_{MN} - \alpha H_{MN} = -\frac{\Lambda_5}{M_5^3} g_{MN} , \tag{2.10}$$

combining the $(rr, \phi\phi)$ equations we get

$$\left(\dot{n}^2 + n\ddot{n} - \frac{n\dot{n}}{r}\right)\left(1 - 4\alpha\frac{L''}{L}\right) = 0 , \tag{2.11}$$

while a combination of the $(\rho\rho, \theta\theta)$ equations gives

$$\left(f'' - \frac{f'L'}{L}\right)\left[3 - 4\frac{\alpha}{f^2}\left(\dot{n}^2 + n\ddot{n} + 2\frac{n\dot{n}}{r} + 3f'^2\right)\right] = 0 , \tag{2.12}$$

where a dot denotes derivatives with respect to r. The solutions of the equations (2.11) and (2.12) are summarized in the following table [41]

In tab. 1, L_3 is the length scale of AdS_3 space. Note that in all solutions there is a fine-tuned relation between the Gauss-Bonnet coupling α and the five-dimensional cosmological constant Λ_5, except for the solution in the fourth row [41].

To introduce a brane we must solve the corresponding junction conditions given by the Einstein equations on the brane (2.7) using the induced metric on the brane given by (2.8). In the case of a BTZ black hole ($n^2(r) = -M + \frac{r^2}{L_3^2}$), and a brane cosmological constant given by $\Lambda_3 = -1/L_3^2$, we found that the energy-momentum tensor is null.

When $n(r)$ is of the form given by $n(r) = \sqrt{-M + \frac{r^2}{L_3^2} - \frac{\zeta}{r}}$, which is the BTZ black hole solution with a short distance correction term, we find that the matter source necessary to sustain such a solution on the brane is given by $T_\alpha^\beta = \text{diag}\left(\frac{\zeta}{2r^3}, \frac{\zeta}{2r^3}, -\frac{\zeta}{r^3}\right)$, which is conserved on the brane [45]. Interesting enough, for a scalar field conformally coupled to BTZ [42, 43], the energy-momentum tensor needed to support such a solution at a certain limit reduces to (2) which is necessary to localize this black hole on the conical 2-brane.

These solutions extend the brane BTZ black hole into the bulk. Calculating the square of the Riemann tensor we find that at the AdS horizon ($\rho \to \infty$) all solutions give finite result and hence the only singularity is the BTZ-corrected black hole singularity extended into the bulk. The warp function $f^2(\rho)$ gives the shape of a 'throat' to the horizon of the BTZ string-like solution. The size of the horizon is defined by the scale $\sqrt{\alpha}$ and this scale is fine-tuned to the length scale of the five-dimensional AdS space.

3. Black string-like solutions in six-dimensional braneworlds of codimension-2

The gravitational action similar to (2.1), but in six dimensions reads as

$$\begin{aligned} S_{grav} &= \frac{M_6^4}{2}\left\{\int d^6x\sqrt{-g^{(6)}}\left[R^{(6)} + \alpha\left(R^{(6)2} - 4R^{(6)}_{MN}R^{(6)MN} + R^{(6)}_{MNKL}R^{(6)MNKL}\right)\right]\right. \\ &+ \left. r_c^2\int d^4x\sqrt{-g^{(4)}}R^{(4)}\right\} + \int d^6x\mathcal{L}_{bulk} + \int d^4x\mathcal{L}_{brane}\,. \end{aligned} \tag{3.1}$$

The metric as in the five-dimensional case is

$$ds_6^2 = g_{\mu\nu}(r,\chi)dx^\mu dx^\nu + a^2(r,\chi)d\chi^2 + L^2(r,\chi)d\xi^2\ , \tag{3.2}$$

now with $\mu = 0, 1, 2, 3$ whereas χ, ξ denote the radial and angular coordinates of the two extra dimensions (the χ direction may or may not be compact and the ξ coordinate ranges form 0 to 2π).

The corresponding Einstein equations are

$$G_M^{(6)N} + r_c^2 G_\mu^{(4)\nu} g_M^\mu g_\nu^N \frac{\delta(\chi)}{2\pi L} - \alpha H_M^N = \frac{1}{M_6^4}\left[-\Lambda_6 + T_M^{(B)N} + T_\mu^{(br)\nu} g_M^\mu g_\nu^N \frac{\delta(\chi)}{2\pi L}\right]\ , \tag{3.3}$$

where H_M^N is the corresponding six-dimensional term of (2.4) To obtain the braneworld equations we expand the metric around the 3-brane as $L(r,\chi) = \beta(r)\chi + O(\chi^2)$, and as in the five-dimensional case the function L behaves as $L'(r,0) = \beta(r)$, where a prime now denotes derivative with respect to χ. The "boundary" Einstein equations are

$$\begin{aligned} G_{\mu\nu}^{(4)}\left(r_c^2 + 8\pi(1-\beta)\alpha\right)|_0 &= \frac{1}{M_6^4}T_{\mu\nu}^{(br)}|_0 + 2\pi(1-\beta)g_{\mu\nu}|_0 \\ &+ \pi L(r,\chi)\,E_{\mu\nu}|_0 - 2\pi\beta\alpha\,W_{\mu\nu}|_0\ , \end{aligned} \tag{3.4}$$

where the term

$$E_{\mu\nu}|_0 = (K_{\mu\nu} - g_{\mu\nu}\,K)\,|_0\ , \tag{3.5}$$

appears because of the presence of the induced gravity term in the gravitational action (we remind that $K_{\mu\nu} = g'_{\mu\nu}$), while the term

$$\begin{aligned} W_{\mu\nu}|_0 &= g^{\lambda\sigma}\partial_\chi g_{\mu\lambda}\partial_\chi g_{\nu\sigma}|_0 - g^{\lambda\sigma}\partial_\chi g_{\lambda\sigma}\partial_\chi g_{\mu\nu}|_0 \\ &+ \frac{1}{2}g_{\mu\nu}\left[\left(g^{\lambda\sigma}\partial_\chi g_{\lambda\sigma}\right)^2 - g^{\lambda\sigma}g^{\delta\rho}\partial_\chi g_{\lambda\delta}\partial_\chi g_{\sigma\rho}\right]\Big|_0\ , \end{aligned} \tag{3.6}$$

is the Weyl term due to the presence of the Gauss-Bonnet term in the bulk [21].

If we demand that the space in the vicinity of the conical singularity is regular ($\partial_\mu\beta = 0$) then (3.4) simply becomes [21, 22]

$$G^{(4)}_{\mu\nu}\left(r_c^2 + 8\pi(1-\beta)\alpha\right)|_0 = \frac{1}{M_6^4}T^{(br)}_{\mu\nu}|_0 + 2\pi(1-\beta)g_{\mu\nu}|_0\ . \tag{3.7}$$

We will look for black string solutions of the Einstein equations (3.3) using the following six-dimensional metric

$$ds_6^2 = F^2(\chi)\left(-A(r)^2dt^2 + A(r)^{-2}dr^2 + r^2d\theta^2 + r^2\,\sin^2\theta\,d\phi^2\right) + a^2(r,\chi)d\chi^2 + L^2(r,\chi)d\xi^2\ . \tag{3.8}$$

Under the assumption that $T^\chi_\chi = T^\xi_\xi$, combining the $rr - \theta\theta$ and $\chi\chi - \xi\xi$ components of the six-dimensional bulk equations we get similar expressions as in (2.11), and (2.12). The solutions are summarized in tab. 2.

Where

$$A^2(r) = 1 + \frac{r^2}{L_4^2} \pm \sqrt{1 + \frac{C_3}{L_4^2} + \frac{C_4}{L_4^4}r}\ , \tag{3.9}$$

and

$$A(r)^2 = 1 + \frac{r^2}{4\alpha} - \frac{\sqrt{3}}{12\alpha}\sqrt{2\,r^4 - 3\,C_4\,r + 48\,\alpha\,(\alpha - C_3)}. \tag{3.10}$$

4. The role of the Gauss-Bonnet term

In codimension-2 brane-worlds there is a relation connecting the Gauss-Bonnet term projected on the brane with the components of the bulk energy-momentum tensor corresponding to the extra dimensions [22]. In six dimensions it reads [1].

$$-\frac{1}{2}\,R^{(4)}|_0 - \frac{1}{2}\alpha\left(R^{(4)\,2} - 4R^{(4)\,2}_{\mu\nu} + R^{(4)\,2}_{\mu\nu\kappa\lambda}\right)\Big|_0 = \frac{1}{M_6^4}\,T^{(B)\,\chi}_{\ \chi}|_0 - \frac{\Lambda_6}{M_6^4}|_0\,. \tag{4.1}$$

All bulk solutions have to satisfy this relation which acts as a consistency relation. For the Schwarzschild-AdS solution the square of the Riemann tensor reads

$$R^2_{\mu\nu\kappa\lambda} = \frac{192\zeta^2 e^{\frac{4\chi}{L_4}}}{(1+e^{\frac{2\chi}{L_4}})^4 r^6} + \frac{60}{L_4^4}\ , \tag{4.2}$$

[1] A similar relation involving the Gauss-Bonnet term was presented in [46] in a different context.

Table 2. Black string-like solutions in six-dimensional braneworlds of codimension-2

$A^2(r)$	$F(\chi)$	$L(\chi)$	$-\Lambda_6$	Constraints & $T^{(B)}$
$1+\frac{r^2}{L_4^2}-\frac{\zeta}{r}$	$\cosh\left(\frac{\chi}{2\sqrt{3\alpha}}\right)$	$\forall L(\chi)$	$\frac{5}{12\alpha}$	$\alpha=\frac{L_4^2}{12}$, $T_\chi^\chi=T_\xi^\xi=-\frac{6\alpha\zeta^2}{r^6F(\chi)^4}$
$1+\frac{r^2}{L_4^2}-\frac{\zeta}{r}$	$\cosh\left(\frac{\chi}{2\sqrt{3\alpha}}\right)$	$2\sqrt{3\alpha}\beta\sinh\left(\frac{\chi}{2\sqrt{3\alpha}}\right)$	$\frac{5}{12\alpha}$	$\alpha=\frac{L_4^2}{12}$, $T_\chi^\chi=T_\xi^\xi=-\frac{6\alpha\zeta^2}{r^6F(\chi)^4}$
$1+\frac{r^2}{L_4^2}-\frac{\zeta}{r}$	±1	$\frac{\beta}{\gamma}\sinh(\gamma\chi)$	$\frac{6}{L_4^2}\left(1-\frac{2\alpha}{L_4^2}\right)$	$\gamma=\frac{1}{L_4}\sqrt{\frac{1-\frac{L_4^2}{4\alpha}}{1-\frac{L_4^2}{12\alpha}}}$, $T_\chi^\chi=T_\xi^\xi=-\frac{6\alpha\zeta^2}{r^6}$
$1+\frac{r^2}{L_4^2}-\frac{\zeta}{r}$	±1	$\frac{\beta}{\gamma}\chi\sinh\gamma$	$\frac{6}{L_4^2}\left(1-\frac{2\alpha}{L_4^2}\right)$	$\gamma=\frac{1}{L_4}\sqrt{\frac{1-\frac{L_4^2}{4\alpha}}{1-\frac{L_4^2}{12\alpha}}}$, $T_\chi^\chi=T_\xi^\xi=-\frac{6\alpha\zeta^2}{r^6}$, $T_t^t=T_r^r=T_\theta^\theta=T_\phi^\phi=\frac{3(4\alpha-L_4^2)}{L_4^2}$
(3.9)	$\cosh\left(\frac{\chi}{2\sqrt{3\alpha}}\right)$	$2\sqrt{3\alpha}\,\beta\sinh\left(\frac{\chi}{2\sqrt{3\alpha}}\right)$	$\frac{5}{12\alpha}$	$\alpha=\frac{L_4^2}{12}$
(3.10)	±1	$2\sqrt{\alpha}\,\beta\sinh\left(\frac{\chi}{2\sqrt{\alpha}}\right)$	$\frac{1}{4\alpha}$	$\alpha=\frac{L_4^2}{4}$

while the Ricci scalar and Ricci tensor are constants. Therefore, for the relation (4.1) to be satisfied the bulk energy-momentum tensor $T_\chi^{(B)\,\chi}|_0$ has to scale as $1/r^6$ with the right coefficients. This is actually what happens considering the result shown in the table. Thus, the presence of the Gauss-Bonnet term in the bulk, which acts as a source term because of its divergenceless nature, dictates the form of matter that must be introduced in the bulk in order to sustain a black hole on the brane[2]. Investigating the nature of this matter we can see that at large χ goes to zero. Furthermore, on the brane, and at large distances we recover conventional four-dimensional gravity, while at small scales strong modifications appear.

5. Conclusions

The issue of localization of black holes in the context of brane-worlds is very difficult and still remains open. Here we discussed black holes residing on a thin brane of codimension-2 and their extension into the bulk, in the presence of a Gauss-Bonnet term in the bulk and an induced gravity term on the brane. Unlike the case of a 2-brane in a five-dimensional space-time filled with a cosmological constant, where a BTZ black hole can be smoothly extended to the bulk, the case of a 3-brane in a six-dimensional space-time is more tricky. Four-dimensional Schwarzschild-AdS black hole solutions can be extended into the bulk with a warp function, but now additional matter is needed to the transverse space in order to sustain the solution.

The Gauss-Bonnet term alongside with the induced gravity term on the brane, which are needed in order to avoid pure tensional branes and to reproduce gravity on the brane, dictate

[2] Black hole solutions in codimension-2 braneworlds were also recently discussed in [47].

the form of the matter which is needed. Still the nature of this matter remains undetermined. Furthermore there is the issue of stability. The Gauss-Bonnet term and the presence of matter make the task difficult to handle even in the case of the five-dimensional set up.

Acknowledgments

The talk is based on work done in collaboration with B.Cuadros-Melgar, E. Papantonopoulos and V. Zamarias. We would like to thank the organizers of the 13th NEB Conference, held in Thessaloniki, for their kind hospitality and and perfect organization of the conference. This work was supported by the NTUA research program PEVE07.

References

[1] Horava P and Witten E 1996 *Nucl. Phys.* B **460** 506 (*Preprint* hep-th/9510209)
[2] Arkani-Hamed N, Dimopoulos S and Dvali G R 1998 *Phys. Lett. B* **429** 263 (*Preprint* hep-ph/9803315)
[3] Randall L and Sundrum R 1999 *Phys. Rev. Lett.* **83** 3370 (*Preprint* hep-ph/9905221)
Randall L and Sundrum R 1999 *Phys. Rev. Lett.* **83** 4690 (*Preprint* hep-th/9906064)
[4] Chamblin A, Hawking S W and Reall H S 2000 *Phys. Rev. D* **61** 065007 (*Preprint* hep-th/9909205)
[5] Gregory R and Laflamme R 1993 *Phys. Rev. Lett.* **70** 2837 (*Preprint* hep-th/9301052)
[6] Gregory R 2000 *Class. Quant. Grav.* **17** L125 (*Preprint* hep-th/0004101)
Gibbons G and Hartnoll S A 2002 *Phys. Rev. D* **66** 064024 (*Preprint* hep-th/0206202)
Gregory R 2008 (*Preprint* 0804.2595 [hep-th])
Harmark T, Niarchos V and Obers N A 2007 *Class. Quant. Grav.* **24** R1 (*Preprint* hep-th/0701022)
[7] Shiromizu T, Maeda K I and Sasaki M 2000 *Phys. Rev. D* **62** 024012 (*Preprint* gr-qc/9910076)
[8] Dadhich N, Maartens R, Papadopoulos P and Rezania V 2000 *Phys. Lett. B* **487:1-6** (*Preprint* hep-th/0003061)
Germani C and Maartens R 2001 *Phys. Rev. D* **64** 124010 (*Preprint* hep-th/0107011)
Bruni M, Germani C and Maartens R 2001 *Phys. Rev. Lett.* **87** 231302 (*Preprint* gr-qc/0108013)
Kofinas G, Papantonopoulos E and Pappa I 2002 *Phys. Rev. D* **66** 104014 (*Preprint* hep-th/0112019)
Kofinas G, Papantonopoulos E and Zamarias V 2002 *Phys. Rev. D* **66** 104028 (*Preprint* hep-th/0208207)
Kanti P and Tamvakis K 2002 *Phys. Rev. D* **65** 084010 (*Preprint* hep-th/0110298)
[9] Gregory R, Leksono A and Mistry B 2009 (*Preprint* 0901.1067 [hep-th])
[10] Kinnersley W and Walker M 1970 *Phys. Rev. D* **2** 1359
[11] Emparan R, Horowitz G T and Myers R C 2000 *JHEP* **0001** 007 (*Preprint* hep-th/9911043)
Emparan R, Horowitz G T and Myers R C 2000 *JHEP* **0001** 021 (*Preprint* hep-th/9912135
[12] Banados M, Teitelboim C and Zanelli J 1992 *Phys. Rev. Lett.* **69** 1849 (*Preprint* hep-th/9204099)
[13] Deser S, Jackiw R and Hooft G 't 1984 *Ann. Phys.* 152 220.
[14] Deser S and Jackiw R 1988 *Commun. Math. Phys.* 118 495
[15] Chen J W, Luty M A and Ponton E 2000 *JHEP* **0009** 012 (*Preprint* hep-th/0003067)
[16] Carroll S M and Guica M M 2003 (*Preprint* hep-th/0302067)
Navarro I 2003 *JCAP* **0309** 004 (*Preprint* hep-th/0302129)
Aghababaie Y, Burgess C P,Parameswaran S L and Quevedo F 2004 *Nucl. Phys. B* **680** 389 (*Preprint* hep-th/0304256)
Navarro I 2003 Class. Quant. Grav. **20** 3603
Aghababaie Y *et al.* 2003 *JHEP* **0309** 037 (*Preprint* hep-th/0308064)
Nilles H P, Papazoglou A and Tasinato G 2004 *Nucl. Phys. B* **677** 405 (*Preprint* hep-th/0309042)
Lee H M 2004 *Phys. Lett. B* **587** 117 (*Preprint* hep-th/0309050)
Graesser M L, Kile J E and Wang P 2004 *Phys. Rev. D* **70** 024008 (*Preprint* hep-th/0403074)
Kehagias A 2004 *Phys. Lett. B* **600** 133 (*Preprint* hep-th/0406025)
Garriga K and Porrati M 2004 *JHEP* **0408** 028 (*Preprint* hep-th/0406158)
Randjbar-Daemi S and Rubakov V 2004 *JHEP* **0410** 054 (*Preprint* hep-th/0407176)
Lee H M and Papazoglou A 2005 *Nucl. Phys. B* **705** 152 (*Preprint* hep-th/0407208)
Nair V P and Randjbar-Daemi S 2005 *JHEP* **0503** 049 (*Preprint* hep-th/0408063)
Redi M 2005 *Phys. Rev. D* **71** 044006 (*Preprint* hep-th/0412189)
[17] Cline J M, Descheneau J, Giovannini M and Vinet J 2003 *JHEP* **0306** 048 (*Preprint* hep-th/0304147)
[18] Kanno S and Soda J 2004 *JCAP* **0407** 002 (*Preprint* hep-th/0404207)

[19] Vinet J and Cline J M 2004 *Phys. Rev. D* **70** 083514 (*Preprint* hep-th/0406141)
Vinet J and Cline J M 2005 *Phys. Rev. D* **71**, 064011 (*Preprint* hep-th/0501098)
[20] Navarro I and Santiago J 2005 *JHEP* **0502** 007 (*Preprint* hep-th/0411250)
[21] Bostock P, Gregory R, Navarro I and Santiago J 2004 *Phys. Rev. Lett.* **92** 221601 (*Preprint* hep-th/0311074)
[22] Papantonopoulos E and Papazoglou A 2005 *JCAP* **0507** 004 (*Preprint* hep-th/0501112)
[23] Kofinas G 2006 *Phys. Lett. B* **633** 141 (*Preprint* hep-th/0506035)
[24] Papantonopoulos E and Papazoglou A 2005 *JHEP* **0509** 012 (*Preprint* hep-th/0507278)
[25] Carter B, Battye R A and Uzan J P 2003 *Commun. Math. Phys.* **235** 289 (*Preprint* hep-th/0204042)
Kolanovic M, Porrati M and Rombouts J W 2003 *Phys. Rev. D* **68** 064018 (*Preprint* hep-th/0304148)
Peter P, Ringeval C and Uzan J P 2005 *Phys. Rev. D* **71** 104018 (*Preprint* hep-th/0301172)
Gott J R I 1985 *Astrophys. J.* **288** 422.
[26] Peloso M, Sorbo L and Tasinato G 2006 *Phys. Rev. D* **73** 104025 (*Preprint* hep-th/0603026)
[27] Papantonopoulos E, Papazoglou A and Zamarias V 2007 *JHEP* **0703** 002 (*Preprint* hep-th/0611311)
[28] Tolley A J, Burgess C P, Rham C de and Hoover D 2006 *New J. Phys.* **8** 324 (*Preprint* hep-th/0608083)
Himmetoglu B and Peloso M 2007 *Nucl. Phys. B* **773** 84 (*Preprint* hep-th/0612140)
Kobayashi T and Minamitsuji M 2007 *Phys. Rev. D* **75** 104013 (*Preprint* hep-th/0703029)
Burgess C P, Hoover D and Tasinato G 2007 *JHEP* **0709** 124 (*Preprint* 0705.3212 [hep-th])
Kobayashi T and Minamitsuji M 2007 *JCAP* **0707** 016 (*Preprint* 0705.3500 [hep-th])
Kobayashi T and Takamizu Y I 2007 *Class. Quant. Grav.* **25** 015007 (*Preprint* 0707.0894 [hep-th])
Copeland E J and Seto O 2007 *JHEP* **0708** 001 (*Preprint* 0705.4169 [hep-th])
Fujii S, Kobayashi T and Shiromizu T 2007 *Phys. Rev. D* **76** 104052 (*Preprint* 0708.2534 [hep-th])
Arroja F, Kobayashi T, Koyama K and Shiromizu T 2007 *JCAP* **0712** 006 (*Preprint* 0710.2539 [hep-th])
Lee H M and Papazoglou A 2008 *JHEP* **0801** 008 (*Preprint* 0710.4319 [hep-th])
Chen F, Cline J M and Kanno S 2008 *Phys. Rev. D* **77** 063531 (*Preprint* 0801.0226 [hep-th])
Corradini O, Koyama K and Tasinato G 2008 *Phys. Rev. D* **78** 124002 (*Preprint* 0803.1850 [hep-th])
Minamitsuji M 2008 *Class. Quant. Grav.* **25** 075019 (*Preprint* 0801.3080 [hep-th])
[29] Papantonopoulos E, Papazoglou A and Zamarias V 2008 *Nucl. Phys. B* **797** 520 (*Preprint* 0707.1396 [hep-th])
[30] Minamitsuji M and Langlois D 2007 *Phys. Rev. D* **76** 084031 (*Preprint* 0707.1426 [hep-th])
[31] Papantonopoulos E 2006 (*Preprint* gr-qc/0601011)
[32] Vilenkin A and Shellard E P S 1994 *Cosmic Strings and Other Topological Defects* (Cambridge: Cambridge University Press)
[33] Aryal M, Ford L H and Vilenkin A 1986 *Phys. Rev. D* **34** 2263.
[34] Achucarro A, Gregory R and Kuijken K 1995 *Phys. Rev. D* **52** 5729. (*Preprint* gr-qc/9505039)
[35] Kaloper N and Kiley D 2006 *JHEP* **0603** 077 (*Preprint* hep-th/0601110)
[36] Kiley D 2007 *Phys. Rev. D* **76** 126002 (*Preprint* 0708.1016 [hep-th])
[37] Chen S, Wang B and Su R K 2007 *Phys. Lett. B* **647** 282 (*Preprint* hep-th/0701209)
[38] Al-Binni UA and Siopsis G 2007 *Phys. Rev. D* **76** 104031 (*Preprint* 0708.3363 [hep-th])
[39] Chen S, Wang B, Su R K and Hwang W Y 2008 *JHEP* **0803** 019 (*Preprint* 0711.3599 [hep-th])
[40] Rogatko M and lowska A S 2008 (*Preprint* 0812.1644 [hep-th])
[41] Cuadros-Melgar B, Papantonopoulos E, Tsoukalas M and Zamarias V 2008 *Phys. Rev. Lett.* **100** 221601 (*Preprint* 0712.3232 [hep-th])
[42] Martinez C and Zanelli J 1996 *Phys. Rev. D* **54** 3830 (*Preprint* gr-qc/9604021)
[43] Henneaux M, Martinez C, Troncoso R and Zanelli J 2002 *Phys. Rev. D* **65** 104007 (*Preprint* hep-th/0201170)
[44] Cuadros-Melgar B, Papantonopoulos E, Tsoukalas M and Zamarias V 2009 *Nucl. Phys. B* **810** 246 (*Preprint* 0804.4459 [hep-th])
[45] Kofinas G 2005 *Class. Quant. Grav.* **22** L47 (*Preprint* hep-th/0412299)
[46] A. Molina and N. Dadhich 2008 arXiv:0804.1194 [gr-qc].
[47] C. Charmousis and A. Papazoglou 2008 *JHEP* **0807** 062 (*Preprint* 0804.2121 [hep-th])

Journal of Physics: Conference Series **189** (2009) 012044 doi:10.1088/1742-6596/189/1/012044

Non-standard symmetries and Killing tensors

Mihai Visinescu

Department of Theoretical Physics, National Institute for Physics and Nuclear Engineering, P.O.Box MG-6, Magurele, Bucharest, Romania

E-mail: mvisin@theory.nipne.ro

Abstract. Higher order symmetries corresponding to Killing tensors are investigated. The intimate relation between Killing-Yano tensors and non-standard supersymmetries is pointed out. The gravitational anomalies are absent if the hidden symmetry is associated with a Killing-Yano tensor. In the Dirac theory on curved spaces, Killing-Yano tensors generate Dirac type operators involved in interesting algebraic structures as dynamical algebras or even infinite dimensional algebras or superalgebras. The general results are applied to space-times which appear in modern studies. The 4-dimensional Euclidean Taub-NUT space and its generalizations introduced by Iwai and Katayama are analyzed from the point of view of hidden symmetries. One presents the infinite dimensional superalgebra of Dirac type operators on Taub-NUT space that can be seen as a twisted loop algebra. The axial anomaly, interpreted as the index of the Dirac operator, is computed for the generalized Taub-NUT metrics. The existence of the conformal Killing-Yano tensors is investigated for some spaces with mixed Sasakian structures.

1. Introduction

One of the key concepts in physics is that of symmetries, Noether's theorem giving a correspondence between symmetries and conserved quantities. For the geodesic motions on a space-time the usual conserved quantities are related to the isometries which correspond to Killing vectors. Sometimes a space-time could admit higher order symmetries described by Killing tensors. These are known as *hidden symmetries* and the typical example is the Runge-Lenz vector in the Kepler/Coulomb problem. The corresponding conserved quantities are quadratic, or, more general, polynomial in momenta. Their existence guarantees the integrability of the geodesic motions and is intimately related to separability of Hamilton-Jacobi (see, e. g. [1]) and the Klein-Gordon equation [2] at the quantum level.

In the case of the gravitational interactions, a consistent perturbative quantization is not available, even if there are no fermions. The implementation of fermions on curved space-times represent an additional difficulty. Fermions are essentially quantum objects while the space-time backgrounds are classic. Sometimes it can appear anomalies representing discrepancies between the conservation laws at the classical level and the corresponding ones at the quantum level.

The above two problems are correlated. First of all the symmetric Killing tensors, called Stäckel-Killing (S-K), are used to construct the conserved quantities, polynomials in momenta. The next most simple objects that can be studied in connection with the symmetries of a manifold after the S-K tensors are the Killing-Yano tensors (K-Y) [3]. Their physical interpretation remained obscure until Floyd [4] and Penrose [5] showed that the Killing tensor $K^{\mu\nu}$ of the 4−dimensional Kerr-Newman space-time admits a certain square-root which defines a K-Y tensor. Subsequently it was observed [6] that a K-Y tensor generates additional *supercharges* in the dynamics of pseudo-classical spinning particles being the natural geometrical objects to be coupled with the fermionic degrees of freedom [6, 7]. In this way it was realized the significant connection between K-Y tensors and *non-standard supersymmetries.* Passing to quantum Dirac equation it was discovered [8] that K-Y tensors generate conserved *non-standard* Dirac operators which commute with the *standard* one.

The conformal extension of the Killing tensor equation determines the conformal Killing tensors [9] which define first integrals of the null geodesic equation. Investigation of the properties of higher-dimensional space-times has pointed out the role of the conformal K-Y (CKY) tensors to generate background metrics with black-hole solutions (see, e. g. [10]).

The aim of this paper is to investigate a few examples of curved spaces endowed with special structures admitting K-Y tensors which could be relevant in the theories of modern physics [7, 6, 11, 12].

The first example is represented by the 4-dimensional Euclidean Taub-Newman-Unti-Tamburino (Taub-NUT) space. The motivation to carry out this example is twofold. First of all, in the Taub-NUT geometry there are known to exist four K-Y tensors [13]. From this point of view the Taub-NUT manifold is an exceedingly interesting space to exemplify the effective construction of the conserved quantities in terms of geometric ones. On the other hand, the Taub-NUT geometry is involved in many modern studies in physics. For example the Kaluza-Klein monopole of Gross and Perry [14] and of Sorkin [15] was obtained by embedding the Taub-NUT gravitational instanton [16] into five-dimensional Kaluza-Klein theory. Remarkably the same object has re-emerged in the study of monopole scattering. In the long distance limit, neglecting radiation, the relative motion of slow Bogomolny-Prasad-Sommerfield monopoles is described by the geodesics of this space [17, 18]. The dynamics of well-separated monopoles is completely soluble and has a Kepler type symmetry [13].

In the second example we investigate the existence of CKY tensors in higher dimensional space-times [19]. Investigations of the properties of space-times of higher dimensions ($D > 4$) have recently attracted considerable attention as a result of their appearance in theories of unification such as string and M theories.

Versions of $M-$theory could be formulated in space-times with various number of time dimensions giving rise to exotic space-time signatures. The $M-$theory in $10+1$ dimensions is linked via dualities to a M^* theory in $9+2$ dimensions and a M' theory in $6+5$ dimensions. Various limits of these will give rise to $IIA-$ and $IIB-$like string theories in many variants of dimensions and signatures [20].

The paraquaternionic structures arise in a natural way in theoretical physics, both in string theory and integrable systems ([21], [22], [23], [24]). The counterpart in odd dimension of a paraquaternionic structure was introduced in [25]. It is called mixed 3-structure, which appears in a natural way on lightlike hypersurfaces in paraquaternionic manifolds. A compatible metric with a mixed 3-structures is necessarily semi-Riemann and mixed 3-Sasakian manifolds are Einstein [26, 27], hence the possible importance of these structures in theoretical physics.

The plan of the paper is as follows: in Section 2 we present the Killing tensors which generalize the Killing vectors, while Section 3 explains how the hidden symmetries associated with Killing tensors do not generate gravitational anomalies. In Section 4 we describe the Dirac-type operators generated by K-Y tensors. In Section 5 the general results are applied to the 4-dimensional Euclidean Taub-NUT space [28, 29] and its generalizations [30, 31, 32]. The CKY tensors on manifolds with mixed 3−structures are presented in the last Section. Some details concerning the geometrical properties of these metrics are summarized in the Appendices.

2. Killing vector fields and their generalizations

Let (M, g) be a semi-Riemannian manifolds. A vector field X on M is said to be a Killing vector field if the Lie derivative with respect to X of the metric g vanishes.

It is clear that a Killing vector field X on a semi-Riemannian manifold (M, g) preserves the metric. Killing vector fields can be generalized to conformal Killing vector fields [3], i.e. vector fields with a flow preserving a given conformal class of metrics.

A natural generalization of conformal Killing vector fields is given by the CKY tensors [33].

Definition 1 *A CKY tensor of rank p on a semi-Riemannian manifold (M, g) is a p-form f which satisfies:*

$$\nabla_X f = \frac{1}{p+1} X \lrcorner f - \frac{1}{n-p+1} X^* \wedge d^* f, \tag{1}$$

for any vector field X on M.

Here ∇ is the Levi-Civita connection of g, n is the dimension of M, X^* is the 1-form dual to the vector field X with respect to the metric g, $\lrcorner$ is the operator dual to the wedge product and d^* is the adjoint of the exterior derivative d. If f is co-closed in (1), then we obtain the definition of a K-Y tensor (introduced by Yano [3]). We can easily see that for $p = 1$, they are dual to Killing vector fields.

A K-Y tensor can be characterized in several ways. In a equivalent manner a differential $p-$form f is called a K-Y tensor if its covariant derivative $\nabla_\lambda f_{\mu_1 \dots \mu_p}$ is totally antisymmetric. As a consequence of the antisymmetry a K-Y tensor satisfy the equation

$$\nabla_{(\lambda} f_{\mu_1) \dots \mu_p} = 0\,. \tag{2}$$

Let us remark that for covariantly constant K-Y tensors each term of the l. h. s. of (2) vanishes. The covariantly constant K-Y tensors represent a particular class of K-Y tensors and they play a special role in the theory of Dirac operators as it will be seen in Section 4.

We mention that K-Y tensors are also called Yano tensors or Killing forms, and CKY tensors are sometimes referred as conformal Yano tensors, conformal Killing forms or twistor forms [34, 35, 36].

For generalizations of the Killing vectors one might also consider higher order symmetric tensors.

Definition 2 *A symmetric tensor of $K_{\mu_1 \dots \mu_r}$ of rank $r > 1$ satisfying a generalized Killing equation*

$$\nabla_{(\lambda} K_{\mu_1 \dots \mu_r)} = 0\,. \tag{3}$$

is called a S-K tensor.

The relevance in physics of the S-K tensors is given by the following proposition which could be easily proved:

Proposition 1 *A symmetric tensor K on M is a S-K tensor if and only if the quantity*

$$K = K_{\mu_1 \dots \mu_r} \dot{s}^{\mu_1} \cdots \dot{s}^{\mu_r}, \tag{4}$$

is constant along every geodesic s in M.

Here the over-dot denotes the ordinary proper time derivative and the proposition ensures that K is a first integral of the geodesic equation.

These two generalizations (3) and (2) of the Killing vector equation could be related. Let $f_{\mu_1 \dots \mu_p}$ be a K-Y tensor, then the tensor field

$$K_{\mu\nu} = f_{\mu \mu_2 \dots \mu_p} f^{\mu_2 \dots \mu_p}{}_{\nu}, \tag{5}$$

is a S-K tensor and it sometimes refers to this S-K tensor as the associated tensor with f. However, the converse statement is not true in general: not all S-K tensors of rank 2 are associated with a K-Y tensor. That is the case of the generalized Taub-NUT spaces (Appendix A) which admit S-K tensors but no K-Y tensors.

3. Gravitational anomalies

Let us consider a non-trivial S-K tensor of rank 2 with a quadratic constant along the geodesic flow constructed as in (4). The generalized Killing equation (3) represents the necessary and sufficient condition for the existence of a quadratic constant of motion as follows from the Poisson bracket of K with the Hamiltonian. Passing from the classical motion to the hidden symmetries of a quantized system, the corresponding quantum operator analog of the quadratic function (4) is [8, 2, 7]:

$$\mathcal{K} = \Delta_\mu K^{\mu\nu} \Delta_\nu, \tag{6}$$

where Δ_μ is the covariant differential operator on the manifold M. Working out the commutator of (6) with the scalar Laplacian

$$\mathcal{H} = \Delta_\mu \Delta^\mu, \tag{7}$$

we get after some calculations

$$[\mathcal{H}, \mathcal{K}] = -\frac{4}{3}\{K_\lambda{}^{[\mu} R^{\nu]\lambda}\}_{;\nu} \Delta_\mu, \tag{8}$$

which means that in general the quantum operator $\mathcal{K}$ does not define a genuine quantum mechanical symmetry [7]. On a generic curved space-time there appears a *gravitational quantum anomaly* proportional to a contraction of the S-K tensor $K_{\mu\nu}$ with the Ricci tensor $R_{\mu\nu}$.

It is obvious that for a Ricci-flat manifold this quantum anomaly is absent. However, a more interesting situation is represented by the manifolds in which the S-K tensor $K_{\mu\nu}$ can be written as a product of K-Y tensors [8] as in (5). On the other hand the integrability condition for any solution of (2) implies the vanishing of the commutator (8) for S-K tensors which admit a decomposition in terms of K-Y tensors. That is the case of the standard Euclidean Taub-NUT space, but not for its generalizations [30, 31, 32]. In the case of generalized Taub-NUT metrics, there are S-K tensors, but no K-Y tensors [37] and consequently for these spaces there are quantum gravitational anomalies [38].

4. Dirac-type operators

For a quantum relativistic description of a spin-1/2 particle on a curved space we use the *standard* Dirac operator

$$D_s = i\gamma^\mu \nabla_\mu, \tag{9}$$

where ∇_μ are the spin covariant derivatives including spin-connection, while γ^μ are the standard Dirac matrices carrying natural indices.

We note that for any isometry with a Killing vector R^μ there is an appropriate operator [8]

$$X_k = -i(R^\mu \nabla_\mu - \frac{1}{4}\gamma^\mu \gamma^\nu R_{\mu;\nu}), \tag{10}$$

which commutes with D_s.

Moreover each K-Y tensor $f_{\mu\nu}$ produces a *non-standard* Dirac operator of the form [8]

$$D_f = i\gamma^\mu (f_\mu{}^\nu \nabla_\nu - \frac{1}{6}\gamma^\nu \gamma^\rho f_{\mu\nu;\rho}), \tag{11}$$

which anticommutes with the standard Dirac operator D_s and can be involved in new types of genuine or hidden (super)symmetries.

4.1. Covariantly constant K-Y tensors

Remarkable superalgebras of Dirac-type operators can be produced by special second-order K-Y tensors that represent square roots of the metric tensor [39, 40, 41].

Definition 3 *The non-singular real or complex-valued K-Y tensor f of rank 2 defined on M which satisfies*

$$f^{\mu}{}_{\alpha} f_{\mu\beta} = g_{\alpha\beta}\,, \tag{12}$$

is called an unit root of the metric tensor of M, or simply an unit root of M.

Let us observe that (12) is a particular case of (3) with the metric tensor as an ordinary S-K tensor. It was shown that any K-Y tensor that satisfy (12) is covariantly constant [39], i. e. $f_{\mu\nu;\sigma} = 0$.

In what follows we look for manifolds admitting families of unit roots $\mathbf{f} = \{f^i | i = 1, 2, ...N_f\}$ having supplementary properties which should guarantee that: (I) the linear space $L_{\mathbf{f}} = \{\rho | \rho = \rho_i f^i,\ \rho_i \in \mathbf{R}\}$ is isomorphic with a real Lie algebra, and (II) each element of $L_{\mathbf{f}} - 0$ is a root (of arbitrary norm). In these circumstances we have the following theorem [41]:

Theorem 1 *The unique type of family of unit roots with $N_f > 1$ having the properties (I) and (II) are the triplets $\mathbf{f} = \{f^1, f^2, f^3\}$ which satisfy*

$$\langle f^i \rangle \langle f^j \rangle = -\delta_{ij} 1_n + \varepsilon_{ijk} \langle f^k \rangle\,, \quad i, j, k ... = 1, 2, 3\,. \tag{13}$$

Proof: We have denoted by $\langle f \rangle$ the matrix form of the K-Y tensor f. Taking into account that ε_{ijk} is the antisymmetric tensor with $\varepsilon_{123} = 1$ we recognize that equations (13) are the well-known multiplication rules of the quaternion units or similar algebraic structures (e.g. the Pauli matrices). Consequently, the matrices $\langle f^i \rangle$ and 1_n generate a matrix representation of the quaternionic algebra $\mathbf{H}$. Other choices are forbidden by the Frobenius theorem. ■

If the unit roots f^i have only real-valued components we recover the hypercomplex structures that obey (13) and these are connected with the hyper-Kähler structure of the space. An example of hyper-Kähler manifold is the Euclidean Taub-NUT space (see Appendix A) which is equipped with only one family of real unit roots.

The main geometric feature of all manifolds admitting a triplet of unit roots is given by

Theorem 2 *If a manifold M allows a triplet of unit roots then this must be Ricci flat (i.e. $R_{\mu\nu} = 0$).*

Proof: As in the case of the hyper-Kähler manifolds, we start with the identity

$$0 = f_{\mu\nu;\alpha;\beta} - f_{\mu\nu;\beta;\alpha} = f_{\mu\sigma} R^{\sigma}{}_{\nu\alpha\beta} + f_{\sigma\nu} R^{\sigma}{}_{\mu\alpha\beta},$$

giving

$$R_{\mu\nu\alpha\beta} f^{\mu}{}_{\sigma} f^{\nu}{}_{\tau} = R_{\sigma\tau\alpha\beta}\,,$$

and calculate

$$\begin{aligned} R_{\mu\nu\alpha\beta} f^{1\,\alpha\beta} &= R_{\mu\nu\sigma\beta} f^{3\,\sigma}{}_{\alpha} (\langle f^3 \rangle \langle f^1 \rangle)^{\alpha\beta} \\ &= R_{\mu\nu\sigma\beta} f^{3\,\sigma}{}_{\alpha} f^{2\,\alpha\beta} = -R_{\mu\nu\alpha\beta} f^{1\,\alpha\beta} = 0\,. \end{aligned}$$

Then, permutating the first three indices of R we find the identity

$$2R_{\mu\alpha\nu\beta} f^{1\,\alpha\beta} = R_{\mu\nu\alpha\beta} f^{1\,\alpha\beta} = 0\,.$$

Finally, writing

$$R_{\mu\nu} = R_{\mu\alpha\nu\beta} f^{1\,\alpha}_{\ \ \tau} f^{1\,\beta\tau} = -R_{\mu\alpha\sigma\beta} f^{1\,\sigma}_{\ \ \nu} f^{1\,\alpha\beta} = 0\,,$$

we draw the conclusion that the manifold is Ricci flat. The same procedure holds for $f^{(2)}$ or $f^{(3)}$ leading to similar identities. ■

Note that the manifolds possessing only single unit roots (as the Kähler ones) are not forced to be Ricci flat.

It is worthy to be noted that the covariantly constant K-Y tensors give rise to Dirac-type operators of the form (11) connected with the standard Dirac operators as follows:

Theorem 3 *The Dirac-type operator D_f produced by the K-Y tensor f satisfies the condition*

$$(D_f)^2 = D_s^2, \tag{14}$$

if and only if f is an unit root.

Proof: The arguments of Ref. [39] show that the condition (14) is equivalent with (12) f being a covariantly constant K-Y tensor. ■

5. Dirac operators on Taub-NUT space

To make things more specific let us consider the Taub-NUT space which is hyper-Kähler and possesses many non-standard symmetries expressed in terms of four K-Y tensors and three S-K tensors.

From the covariantly constant K-Y tensors f^i (A.5), using prescription (11), we can construct three Dirac-type operators $D^{(i)}$ which anticommute with standard Dirac operator D_s (9). It is convenient to define [42]

$$Q_i = iH^{-1}D^{(i)}, \tag{15}$$

where $H = -\gamma^0 D_s$ is the massless Hamiltonian operator. These operators form a representation of the quaternionic units:

$$Q_i\, Q_j = \delta_{ij} I + i\varepsilon_{ijk} Q_k\,. \tag{16}$$

On the other hand Dirac-type operator constructed from the K-Y tensor f^Y (A.6) is D^Y and again it is convenient to define a new operator $Q^Y = HD^Y$.

The conserved Runge-Lenz operator of the Dirac theory is

$$\mathcal{K}_i = \frac{\mu}{4}\{Q^Y,\, Q_i\} + \frac{1}{2}(\mathcal{B} - P_4)Q_i - \mathcal{J}_i P_4\,, \tag{17}$$

where $\mathcal{B}^2 = {P_4}^2 - H^2$, $J_i\,, (i = 1, 2, 3)$ are the components of the total angular momentum, while $P_4 = -i\partial_4$ corresponding to the fourth Cartesian coordinate $x^4 = -4m(\chi + \varphi)$.

The operators J_i and $\mathcal{K}_i$ are involved in the following system of commutation relations:

$$\begin{aligned}
[\mathcal{J}_i,\, \mathcal{J}_j] &= i\varepsilon_{ijk}\mathcal{J}_k\,,\\
[\mathcal{J}_i,\, \mathcal{K}_j] &= i\varepsilon_{ijk}\mathcal{K}_k\,,\\
[\mathcal{K}_i,\, \mathcal{K}_j] &= i\varepsilon_{ijk}\mathcal{J}_k\mathcal{B}^2\,,
\end{aligned} \tag{18}$$

and commute with the operators Q_i (15)

$$[\mathcal{J}_i,\, Q_j] = i\varepsilon_{ijk}Q_k\,, \qquad [\mathcal{K}_i,\, Q_j] = i\varepsilon_{ijk}Q_k\mathcal{B}\,. \tag{19}$$

The algebra (18) does not close as a finite Lie algebra because of the factor $\mathcal{B}^2$. In the standard treatment one concentrates on individual subspaces of the whole Hilbert space which belong to definite eigenvalues of $\mathcal{B}^2$. This is similar to the dynamical algebra of the hydrogen atom which can be identified in a natural way with an infinite dimensional twisted loop algebra [43].

The dynamical algebras of the Dirac theory have to be obtained by replacing this operator $\mathcal{B}^2$ with its eigenvalue $q^2 - E^2$ and rescaling the operators $\mathcal{K}_i$. The same kind of problems appears for the anticommutators involving the fermionic operators Q_i and Q^Y. In what follows, in order to keep the presentation as simple as possible, we shall only give the briefest account of the algebra of operators connected with hidden symmetries in the bosonic sector. For the algebra of operators from the fermionic sector the reader should consult [44].

In the bosonic sector of conserved operators let us define the new operators "absorbing" the operator $\mathcal{B}$ [45]:

$$J^i_n = \mathcal{J}_i\mathcal{B}^n\,, \quad K^i_n = \mathcal{K}_i\mathcal{B}^n\,, \tag{20}$$

for any $n = 0, 1, 2....$

Non-trivial commutators of the bosonic sector (18) become

$$\begin{aligned} \left[J^i_n, J^j_m\right] &= i\varepsilon_{ijk}J^k_{n+m}\,, \\ \left[J^i_n, K^j_m\right] &= i\varepsilon_{ijk}K^k_{n+m}\,, \\ \left[K^i_n, K^j_m\right] &= i\varepsilon_{ijk}J^k_{n+m+2}\,. \end{aligned} \tag{21}$$

We should like to show that this algebra can be seen as an infinite dimensional twisted loop algebra. The simplest way to achieve a Lie algebra of the Kac-Moody type is to assign grades to each operators

$$A^i_{2n} := \mathcal{J}_i\mathcal{B}^n\,, \quad B^i_{2n+2} := \mathcal{K}_i\mathcal{B}^n\,. \tag{22}$$

In this way the commutation relations of the bosonic sector are

$$\begin{aligned} \left[A^i_{2n}, A^j_{2m}\right] &= i\varepsilon_{ijk}A^k_{2(n+m)}\,, \\ \left[A^i_{2n}, B^j_{2m+2}\right] &= i\varepsilon_{ijk}B^k_{2(n+m+1)}\,, \\ \left[B^i_{2n+2}, B^j_{2m+2}\right] &= i\varepsilon_{ijk}A^k_{2(n+m+2)}\,, \end{aligned} \tag{23}$$

and in this Kac-Moody type algebra the sum of the grades is conserved under commutation as it is expected.

The same kind of construction can be done in the fermionic sector for the anticommutation relations of the operators Q_i, Q^Y.

5.1. *Index formulas and axial anomalies*

Atiyah, Patodi and Singer [46] (APS) discovered an index formula for first-order differential operators on manifolds with boundary with a non-local boundary condition. Their index formula contains two terms, none of which is necessarily an integer, namely a bulk term (the integral of a density in the interior of the manifold) and a boundary term defined in terms of the spectrum of the boundary Dirac operator. Endless trouble is caused in this theory by the condition that the metric and the operator be of "product type" near the boundary.

In [38] we computed the index of the Dirac operator on annular domains and on disk, with the non-local APS boundary condition. For the generalized Taub-NUT metrics [30, 31, 32], we found that the index is a number-theoretic quantity which depends on the metrics. In particular, our formula shows that the index vanishes on balls of sufficient large radius, but can be non-zero for some values of the parameters c, d (see Appendix A).

Theorem 4 *If $c > -\frac{\sqrt{15d}}{2}$ then the extended Taub-NUT metric does not contribute to the axial anomaly on any annular domain (i.e., the index of the Dirac operator with APS boundary condition vanishes).*

Proof: The proof is given in [38]. ■

The result is natural since the index of an operator is unchanged under continuous deformations of that operator. In our case this would amount to a continuous change in the metric. The absence of axial anomalies is due to the fact there exists an underlying structure that does not depend on the metric. However for larger deformations of the metric there could appear discontinuities in the boundary conditions and therefore the index could present jumps. Our formula for the index involves a computable number-theoretic quantity depending on the parameters of the metric.

We also examined the Dirac operator on the complete Euclidean space with respect to this metric, acting in the Hilbert space of square-integrable spinors. We found that this operator is not Fredholm, hence even the existence of a finite index is not granted.

We mentioned in [38] some open problems in connection with unbounded domains. The paper [47] brings new results in this direction. First we showed that the Dirac operator on $\mathbf{R}^4$ with respect to the standard Taub-NUT metric does not have L^2 harmonic spinors. This follows rather easily from the Lichnerowicz formula, since the standard Taub-NUT metric has vanishing scalar curvature. In particular, the index vanishes.

Theorem 5 *For the standard Taub-NUT metric on $\mathbf{R}^4$ the Dirac operator does not have L^2 solutions.*

Proof: Recall that the standard Taub-NUT metric is hyper-Kähler, hence its scalar curvature κ vanishes.

By the Lichnerowicz formula,

$$D^2 = \nabla^*\nabla + \frac{\kappa}{4} = \nabla^*\nabla.$$

Let $\phi \in L^2$ be a solution of D in the sense of distributions. Then, again in distributions, $\nabla^*\nabla\phi = 0$. The operator $\nabla^*\nabla$ is essentially self-adjoint with domain $\mathcal{C}_c^\infty(\mathbf{R}^4, \Sigma_4)$, which implies that its kernel equals the kernel of ∇. Hence $\nabla\phi = 0$. Now a parallel spinor has constant pointwise norm, hence it cannot be in L^2 unless it is 0, because the volume of the metric ${ds_K}^2$ is infinite. Therefore $\phi = 0$. ■

6. CKY tensors on manifolds with mixed 3-structures

An almost para-hypercomplex structure on a smooth manifold M is a triple $H = (J_\alpha)_{\alpha=\overline{1,3}}$, where J_1 is an almost complex structure on M and J_2, J_3 are almost product structures on M, satisfying: $J_1J_2J_3 = -Id$. In this case (M, H) is said to be an almost para-hypercomplex manifold.

A semi-Riemannian metric g on (M, H) is said to be para-hyperhermitian if it satisfies:

$$g(J_\alpha X, J_\alpha Y) = \epsilon_\alpha g(X, Y), \ \alpha \in \{1, 2, 3\}, \tag{24}$$

for all $X, Y \in \Gamma(TM)$, where $\epsilon_1 = 1, \epsilon_2 = \epsilon_3 = -1$. In this case, (M, g, H) is called an almost para-hyperhermitian manifold. Moreover, if each J_α is parallel with respect to the Levi-Civita connection of g, then (M, g, H) is said to be a para-hyper-Kähler manifold.

Theorem 6 *Let (M, g) be a semi-Riemannian manifold. Then the following five assertions are mutually equivalent:*

(1) (M, g) admits a mixed 3-Sasakian structure.

(2) The cone $(C(M), \overline{g}) = (M \times \mathbf{R}_+, dr^2 + r^2 g)$ admits a para-hyper-Kähler structure.

(3) There exists three orthogonal Killing vector fields $\{\xi_1, \xi_2, \xi_3\}$ on M, with ξ_1 unit space-like vector field and ξ_2, ξ_3 unit time-like vector fields satisfying

$$[\xi_\alpha, \xi_\beta] = -2\epsilon_\gamma \xi_\gamma, \tag{25}$$

where (α, β, γ) is an even permutation of (1,2,3) and $\epsilon_1 = 1, \epsilon_2 = \epsilon_3 = -1$, such that the tensor fields ϕ_α of type (1,1), defined by: $\phi_\alpha X = \nabla_X \xi_\alpha$, $\alpha \in \{1, 2, 3\}$, satisfies the conditions (B.7), (B.8) and (B.9).

(4) There exists three orthogonal Killing vector fields $\{\xi_1, \xi_2, \xi_3\}$ on M, with ξ_1 unit space-like vector field and ξ_2, ξ_3 unit time-like vector fields satisfying (25), such that:

$$R(X, \xi_\alpha)Y = g(\xi_\alpha, Y)X - g(X, Y)\xi_\alpha, \ \alpha \in \{1, 2, 3\}, \tag{26}$$

where R is the Riemannian curvature tensor of the Levi-Civita connection ∇ of g.

(5) There exists three orthogonal Killing vector fields $\{\xi_1, \xi_2, \xi_3\}$ on M, with ξ_1 unit space-like vector field and ξ_2, ξ_3 unit time-like vector fields satisfying (25), such that the sectional curvature of every section containing ξ_1, ξ_2 or ξ_3 equals 1.

Proof: (1) $\Rightarrow$ (2) If M^{4n+3} is a manifold endowed with a mixed 3-Sasakian structure (see Appendix B) $((\phi_\alpha, \xi_\alpha, \eta_\alpha)_{\alpha=\overline{1,3}}, g)$, then we can define a para-hyper-Kähler structure $\{J_\alpha\}_{\alpha=\overline{1,3}}$ on the cone $(C(M), \overline{g}) = (M \times \mathbf{R}_+, dr^2 + r^2 g)$, by:

$$J_\alpha X = \phi_\alpha X - \eta_\alpha(X)\Phi, \ J_\alpha \Phi = \xi_\alpha, \tag{27}$$

for any $X \in \Gamma(TM)$ and $\alpha \in \{1, 2, 3\}$, where $\Phi = r\partial_r$ is the Euler field on $C(M)$.

(2) $\Rightarrow$ (1) If the cone $(C(M), \overline{g}) = (M \times \mathbf{R}_+, dr^2 + r^2 g)$ admits a para-hyper-Kähler structure $\{J_\alpha\}_{\alpha=\overline{1,3}}$, then we can identify M with $M \times \{1\}$ and we have a mixed 3-Sasakian structure $((\phi_\alpha, \xi_\alpha, \eta_\alpha)_{\alpha=\overline{1,3}}, g)$ on M given by:

$$\xi_\alpha = J_\alpha(\partial_r), \ \phi_\alpha X = \nabla_X \xi_\alpha, \ \eta_\alpha(X) = g(\xi_\alpha, X), \tag{28}$$

for any $X \in \Gamma(TM)$ and $\alpha \in \{1, 2, 3\}$.

(2) $\Leftrightarrow$ (3) This equivalence is clear (see also [48]).

(3) $\Leftrightarrow$ (4) This equivalence follows from direct computations.

(4) $\Leftrightarrow$ (5) This equivalence follows using the formula of the sectional curvature. ∎

From the above Theorem we can easily obtain the next properties (see also [36]).

Corollary 1 *Let M^{4n+3} be a manifold endowed with a mixed 3-Sasakian structure $((\phi_\alpha, \xi_\alpha, \eta_\alpha)_{\alpha=\overline{1,3}}, g)$. Then:*

(1) ξ_1 is unit space-like Killing vector field and ξ_2, ξ_3 are unit time-like Killing vector fields on M;

(2) η_1, η_2, η_3 are CKY tensors of rank 1 on M;

(3) $d\eta_1, d\eta_2, d\eta_3$ are CKY tensors of rank 2 on M.

(4) M admits K-Y tensors of rank $(2k + 1)$, for $k \in \{0, 1, ..., 2n + 1\}$.

Corollary 2 *Let M^{4n+3} be a manifold endowed with a mixed 3-Sasakian structure $((\phi_\alpha, \xi_\alpha, \eta_\alpha)_{\alpha=\overline{1,3}}, g)$. Then the distribution spanned by $\{\xi_1, \xi_2, \xi_3\}$ is integrable and defines a 3-dimensional Riemannian foliation on M, having totally geodesic leaves of constant curvature 1.*

Acknowledgments
The author would like to thank the organizers for the excellent NEB-XIII Conference *Recent Developments in Gravity* held in Thessaloniki, Greece in June 2008. The author is grateful to I. I. Cotăescu, S. Ianuş, S. Moroianu, G. E. Vîlcu for sharing their thoughts on the subject of this paper. This work is supported in part by CNCSIS Programs, Romania.

References

[1] Benenti S 1977 *Rep. Math. Phys.* **12** 311
[2] Carter B 1977 *Phys. Rev.* D **16** 3395
[3] Yano K 1952 *Ann. Math.* **55** 328
[4] Floyd R 1973 *The dynamics of Kerr fields* (PhD Thesis: London University)
[5] Penrose R 1973 *Ann. NY Acad. Sci.* **224** 125
[6] Gibbons G W, Rietdijk R and van Holten J W 1993 *Nucl. Phys.* B **404** 42
[7] Cariglia M 2004 *Class. Quantum Grav.* **21** 1051
[8] Carter B and McLenaghan R G 1979 *Phys. Rev.* D **19** 1093
[9] Jost J 2002 *Riemannian Geometry and Geometric Analysis* (Berlin: Springer)
[10] Frolov V P and Kubiznak D 2008 *Class. Quantum Grav.* **25** 154005
[11] Tanimoto M 1995 *Nucl. Phys.* B **442** 549
[12] Vaman D and Visinescu M 1998 *Phys. Rev.* D **57** 3790
[13] Gibbons G W and Ruback P J 1988 *Comm. Math. Phys.* **115** 267
[14] Gross D and Perry M J 1983 *Nucl. Phys.* B **226** 29
[15] Sorkin R D 1983 *Phys. Rev. Lett.* **51** 87
[16] Hawking S W 1977 *Phys. Lett.* A **60** 81
[17] Manton N S 1985 *Phys.Lett.* B **110** 54
[18] Atiyah M F and Hitchin N 1985 *Phys. Lett.* A **107** 21
[19] Ianuş S, Visinescu M and Vîlcu G E 2008 Conformal Killing-Yano tensors on manifolds with mixed 3-structures (to be published)
[20] Hull C M 1998 *J. High Energy Phys.* **9811** 017
[21] Cortés V, Mayer C, Mohaupt T and Saueressig F 2005 *J. High Energy Phys.* **0506** 025
[22] Dunajski M and West S 2006 Anti-self-dual conformal structures in neutral signature (*Preprint* 0610280 [math.DG])
[23] Hull C M 1998 *Nucl. Phys.* B **509** 252
[24] Ooguri H and Vafa C 1991 *Nucl. Phys.* B **361** 469
[25] Ianuş S, Mazzocco R and Vîlcu G E 2006 *Mediterr. J. Math.* **3** 581
[26] Caldarella A and Pastore A M 2003 Mixed metric 3-contact manifolds and paraquaternionic Kahler manifolds (*Preprint* 0803.2974 [math.DG])
[27] Ianuş S and Vîlcu G E 2008 *Int. J. Geom. Methods Mod. Phys.* **5** 893
[28] Taub A H 1951 *Ann. Math.* **53** 472
[29] Newman E, Tamburino L and Unti T 1963 *J. Math. Phys.* **4** 915
[30] Iwai T and Katayama N 1993 *J. Geom. Phys.* **12** 55
[31] Iwai T and Katayama N 1994 *J. Math. Phys.* **35** 2914
[32] Iwai T and Katayama N 1994 *J. Phys. A: Math. Gen.* **27** 3179
[33] Kashiwada T 1968 *Natur. Sci. Rep. Ochanomizu Univ.* **19** 67
[34] Belgun F, Moroianu A and Semmelmann U 2006: *Differ. Geom. Appl.* **24** 215
[35] Moroianu A and Semmelmann U 2003 *Ann. Sc. Norm. Super. Pisa, Cl. Sci.* **2** 823
[36] Semmelmann U 2003 *Math. Z.* **245** 503
[37] Visinescu M 2000 *J. Phys. A: Math. Gen.* **33** 4383
[38] Cotăescu I I, Moroianu S and Visinescu M 2005 *J. Phys. A: Math. Gen.* **38** 7005
[39] Klishevich V V 2000 *Class. Quantum Grav.* **17** 305
[40] Cotăescu I I and Visinescu M 2007 Symmetries and supersymmetries of the Dirac operators in curved spacetimes *Frontiers in General Relativity and Quantum Cosmology Research* (Horizons in World Physic vol 235) ed V H Marselle (New York: Nova Science) pp 109-166
[41] Cotăescu I I and Visinescu M 2006 *Fortschr. Phys.* **54** 1142
[42] Cotăescu I I and Visinescu M 2008 *Fortschr. Phys.* **56** 400
[43] Daboul J, Slodowy P and Daboul C 1993 *Phys. Lett.* B **317** 321
[44] Cotăescu I I and Visinescu M 2007 *J. Phys. A: Math. Theor.* **40** 11987
[45] Visinescu M 2008 *J. Phys. A: Math. Theor.* **41** 164072
[46] Atiyah M F, Patodi V K and Singer I M 1975 *Math. Proc. Cambridge Phil. Soc.* **77** 43

[47] Moroianu S and Visinescu M 2006 *J. Phys. A: Math. Gen.* **39** 6575
[48] Boyer C and K. Galicki K 1999 *Suppl. J. Differ. Geom.* **6** 123
[49] Gibbons G W and Manton N S 1986 *Nucl. Phys.* B **274** 183
[50] van Holten J W 1995 *Phys. Lett.* B **342** 47
[51] Feher L Gy and Horvathy P A 1987 *Phys. Lett.* B **183** 182
[52] Cordani B, Feher L Gy and Horvathy P A 1988 *Phys. Lett.* B **201** 481
[53] Hitchin N 1974 *Adv. Math.* **14** 1
[54] D. E. Blair D E 1976 *Contact manifolds in Riemannian Geometry* (Lectures Notes in Math. vol 509) (Berlin: Springer)
[55] Matsumoto K 1989 *Bull. Yamagata Univ. Nat. Sci.* **12** 151

Appendix A. Euclidean Taub-NUT space

Let us consider the Taub-NUT space [28, 29] and the chart with Cartesian coordinates $x^\mu(\mu,\nu = 1,2,3,4)$ having the line element

$$ds^2 = g_{\mu\nu}dx^\mu dx^\nu = f(r)(d\vec{x})^2 + \frac{g(r)}{16m^2}(dx^4 + A_i dx^i)^2, \tag{A.1}$$

where $\vec{x}$ denotes the three-vector

$$\vec{x} = (r,\theta,\varphi)\,, (d\vec{x})^2 = (dx^1)^2 + (dx^2)^2 + (dx^3)^2,$$

and $\vec{A}$ is the gauge field of a monopole

$$\text{div}\vec{A} = 0\,, \quad \vec{B} = \text{rot}\,\vec{A} = 4m\frac{\vec{x}}{r^3}.$$

The real number m is a parameter of the theory which enter in the form of the functions

$$f(r) = g^{-1}(r) = V^{-1}(r) = \frac{4m+r}{r}, \tag{A.2}$$

and the so called NUT singularity is absent if x^4 is periodic with period $16\pi m$. Sometimes it is convenient to make the coordinate transformation $x^4 = -4m(\chi+\varphi)$, with $0 \le \chi < 4\pi$.

In the Taub-NUT geometry there are four Killing vectors [13, 49]

$$D_A = R^\mu_A\,\partial_\mu, \qquad A = 1,2,3,4\,. \tag{A.3}$$

Three Killing vectors correspond to the invariance of the metric (A.1) under spatial rotations ($A = 1,2,3$), obeying an $SU(2)$ algebra with $[D_1, D_2] = -D_3, etc.$ D_4 generates the $U(1)$ of χ translations, commuting with the other Killing vectors.

In the scalar case these invariances would correspond to the conservation of angular momentum and the so called "relative electric charge":

$$\vec{J} = \vec{r}\times\vec{p} + q\frac{\vec{r}}{r}\quad , \quad q = g(r)(\dot{\theta} + \cos\theta\dot{\varphi}), \tag{A.4}$$

where $\vec{p} = fr\dot{\vec{r}}$ is the mechanical momentum.

On the other hand in the Taub-NUT geometry there are known to exist four K-Y tensors of valence 2. The first three are covariantly constant

$$\begin{aligned} f_i &= 8m(d\chi + \cos\theta d\varphi)\wedge dx_i - \epsilon_{ijk}(1+\frac{4m}{r})dx_j\wedge dx_k, \\ \nabla_\mu f^\nu_{i\lambda} &= 0\,, \quad i,j,k = 1,2,3. \end{aligned} \tag{A.5}$$

The f^i define three anticommuting complex structures of the Taub-NUT manifold, their components realizing the quaternion algebra

$$f^i f^j + f^j f^i = -2\delta_{ij} \quad , \quad f^i f^j - f^j f^i = -2\varepsilon_{ijk} f^k .$$

The existence of these K-Y tensors is linked to the hyper-Kähler geometry of the manifold and shows directly the relation between the geometry and the $N = 4$ supersymmetric extension of the theory [6, 50].

The fourth K-Y tensor is

$$f_Y = 8m(d\chi + \cos\theta d\varphi) \wedge dr + 4r(r + 2m)(1 + \frac{r}{4m}) \sin\theta d\theta \wedge d\varphi, \tag{A.6}$$

having a non-vanishing covariant derivative

$$f_{Y\, r\theta;\varphi} = 2(1 + \frac{r}{4m}) r \sin\theta .$$

In Taub-NUT space there is a conserved vector analogous to the Runge-Lenz vector of the Kepler-type problem [13, 51, 52]

$$\vec{K} = \frac{1}{2}\vec{K}_{\mu\nu}\dot{x}^\mu \dot{x}^\nu = \vec{p} \times \vec{j} + \left(\frac{q^2}{4m} - 4mE\right)\frac{\vec{r}}{r}, \tag{A.7}$$

where the E is the conserved energy

$$E = \frac{1}{2} g_{\mu\nu}\dot{x}^\mu \dot{x}^\nu .$$

The components $K_{i\mu\nu}$ involved with the Runge-Lenz vector (A.7) are S-K tensors satisfying (3) and they can be expressed as symmetrized products of the K-Y tensors f_i, f_Y and Killing vectors R_A [12]

$$K_{i\mu\nu} - \frac{1}{8m}(R_{4\mu}R_{i\nu} + R_{4\nu}R_{i\mu}) = m\left(f_{Y\mu\lambda}{f_i}^\lambda{}_\nu + f_{Y\nu\lambda}{f_i}^\lambda{}_\mu\right). \tag{A.8}$$

In what follows we restrict ourselves to the *generalized* Taub-NUT manifolds whose metrics are defined on $\mathbf{R}^4 \setminus \{0\}$ by the line element [30, 31, 32]:

$$\begin{aligned} {ds_K}^2 &= g_{\mu\nu}(x) dx^\mu dx^\nu \\ &= f(r)(dr^2 + r^2 d\theta^2 + r^2 \sin^2\theta\, d\varphi^2) + g(r)(d\chi + \cos\theta\, d\varphi)^2, \end{aligned} \tag{A.9}$$

where the angle variables (θ, φ, χ) parametrize the sphere S^3 as mentioned above, while the functions

$$f(r) = \frac{a + br}{r} \quad , \quad g(r) = \frac{ar + br^2}{1 + cr + dr^2} . \tag{A.10}$$

depend on the arbitrary real constants a, b, c and d.

Here it is worth pointing out that the above metrics are related to the Berger family of metrics on 3-spheres [53]. These are introduced starting with the Hopf fibration $\pi_H : S^3 \to S^2$ that defines the vertical subbundle $V \subset TS^3$ and its orthogonal complement $H \subset TS^3$ with respect to the standard metric g_{S^3} on S^3. Denoting with g_H and g_V the restriction of g_{S^3} to the horizontal, respectively the vertical bundle, one finds that the corresponding line elements are ${ds_H}^2 = \frac{1}{4}{ds_2}^2$ and ${ds_V}^2 = \frac{1}{4}({ds_3}^2 - {ds_2}^2)$. For each constant $\lambda > 0$ the Berger metric on S^3 is defined by the formula

$$g_\lambda = g_H + \lambda^2 g_V .$$

This line element can be written in terms of the Berger metrics as

$$ds_K{}^2 = (ar + br^2)\left(\frac{dr^2}{r^2} + 4ds_{\lambda(r)}{}^2\right),$$

where $ds_{\lambda(r)}{}^2 = (g_{\lambda(r)})_{\mu\nu}dx^\mu dx^\nu$ and

$$\lambda(r) = \frac{1}{\sqrt{1 + cr + dr^2}}.$$

If one takes the constants

$$c = \frac{2b}{a}, \quad d = \frac{b^2}{a^2}, \tag{A.11}$$

the generalized Taub-NUT metric becomes the original Euclidean Taub-NUT metric (A.1), (A.2) up to a constant factor.

By construction, the spaces with the metric (A.9) have four Killing vectors. The corresponding constants of motion in generalized Taub-NUT backgrounds consist of a conserved quantity for the cyclic variable χ

$$q = g(r)(\dot{\chi} + \cos\theta\dot{\varphi}),$$

and the angular momentum vector

$$\vec{J} = \vec{x} \times \vec{p} + q\frac{\vec{x}}{r}, \quad \vec{p} = f(r)\dot{\vec{x}}.$$

The generalized Taub-NUT space admits the conserved quantities (A.4) involving the functions f and g given by (A.10).

The remarkable result of Iwai and Katayama [30, 31, 32] is that the generalized Taub-NUT space (A.9) admits a hidden symmetry represented by a conserved vector, quadratic in 4-velocities, analogous to the Runge-Lenz vector of the following form

$$\vec{K} = \vec{p} \times \vec{J} + \kappa\frac{\vec{x}}{r}. \tag{A.12}$$

The constant κ involved in the Runge-Lenz vector (17) is $\kappa = -a\,E + \frac{1}{2}c\,q^2$ where the conserved energy E is

$$E = \frac{\vec{p}^{\,2}}{2f(r)} + \frac{q^2}{2g(r)}.$$

As it is expected the components $K_i = k_i^{\mu\nu}p_\mu p_\nu$ of the vector $\vec{K}$ (17) involve three S-K tensors $k_i^{\mu\nu}$, $i = 1, 2, 3$ satisfying (3).

However, the generalized Taub-NUT space does not admit K-Y tensors [37]. Therefore a decomposition of the S-K tensors $k_i^{\mu\nu}$ in terms of K-Y tensors is not possible, in contrast to (A.8) which is applicable only to the standard Taub-NUT space.

Appendix B. Manifolds with mixed 3-structures

Let M be a differentiable manifold equipped with a triple (ϕ, ξ, η), where ϕ is a a field of endomorphisms of the tangent spaces, ξ is a vector field and η is a 1-form on M such that:

$$\phi^2 = -\epsilon I + \eta \otimes \xi, \ \eta(\xi) = \epsilon. \tag{B.1}$$

If $\epsilon = 1$ then (ϕ, ξ, η) is said to be an almost contact structure on M (see [54]), and if $\epsilon = -1$ then (ϕ, ξ, η) is said to be an almost paracontact structure on M (see [55]).

Definition 4 *[25] Let M be a differentiable manifold which admits an almost contact structure (ϕ_1, ξ_1, η_1) and two almost paracontact structures (ϕ_2, ξ_2, η_2) and (ϕ_3, ξ_3, η_3), satisfying the following conditions:*

$$\eta_\alpha(\xi_\beta) = 0, \forall \alpha \neq \beta, \tag{B.2}$$

$$\phi_\alpha(\xi_\beta) = -\phi_\beta(\xi_\alpha) = \epsilon_\gamma \xi_\gamma, \tag{B.3}$$

$$\eta_\alpha \circ \phi_\beta = -\eta_\beta \circ \phi_\alpha = \epsilon_\gamma \eta_\gamma, \tag{B.4}$$

$$\phi_\alpha \phi_\beta - \eta_\beta \otimes \xi_\alpha = -\phi_\beta \phi_\alpha + \eta_\alpha \otimes \xi_\beta = \epsilon_\gamma \phi_\gamma, \tag{B.5}$$

where (α, β, γ) is an even permutation of (1,2,3) and $\epsilon_1 = 1, \epsilon_2 = \epsilon_3 = -1$.

Then the manifold M is said to have a mixed 3-structure $(\phi_\alpha, \xi_\alpha, \eta_\alpha)_{\alpha=\overline{1,3}}$.

Definition 5 *If a manifold M with a mixed 3-structure $(\phi_\alpha, \xi_\alpha, \eta_\alpha)_{\alpha=\overline{1,3}}$ admits a semi-Riemannian metric g such that:*

$$g(\phi_\alpha X, \phi_\alpha Y) = \epsilon_\alpha g(X, Y) - \eta_\alpha(X)\eta_\alpha(Y), \; g(X, \xi_\alpha) = \eta_\alpha(X), \tag{B.6}$$

for all $X, Y \in \Gamma(TM)$ and $\alpha = 1, 2, 3$, then we say that M has a metric mixed 3-structure and g is called a compatible metric. Moreover, if $(\phi_1, \xi_1, \eta_1, g)$ is a Sasakian structure, i.e. (see [54]):

$$(\nabla_X \phi_1)Y = g(X, Y)\xi_1 - \eta_1(Y)X, \tag{B.7}$$

and $(\phi_2, \xi_2, \eta_2, g)$, $(\phi_3, \xi_3, \eta_3, g)$ are LP-Sasakian structures, i.e. (see [55]):

$$(\nabla_X \phi_2)Y = g(\phi_2 X, \phi_2 Y)\xi_2 + \eta_2(Y)\phi_2^2 X, \tag{B.8}$$

$$(\nabla_X \phi_3)Y = g(\phi_3 X, \phi_3 Y)\xi_3 + \eta_3(Y)\phi_3^2 X, \tag{B.9}$$

then $((\phi_\alpha, \xi_\alpha, \eta_\alpha)_{\alpha=\overline{1,3}}, g)$ is said to be a mixed Sasakian 3-structure on M.

It is easy to see that any manifold M with a mixed 3-structure admits a compatible semi-Riemannian metric g. Moreover, the signature of g is $(2n+1, 2n+2)$ and the dimension of the manifold M is $4n+3$. The main property of a manifold endowed with a mixed 3-Sasakian structure is the following (see [26, 27]):

Theorem 7 *Any $(4n+3)-$dimensional manifold endowed with a mixed 3-Sasakian structure is an Einstein space with Einstein constant $\lambda = 4n+2$.*

Emergent continuum spacetime from a random, discrete, partial order[1]

David Rideout
Perimeter Institute for Theoretical Physics
31 Caroline Street North, Waterloo, Ontario N2L 2Y5, Canada
E-mail: drideout@perimeterinstitute.ca

Petros Wallden
Raman Research Institute, Theoretical Physics Group
Sadashivanagar, Bangalore - 560 080, India
E-mail: petros.wallden@gmail.com

Abstract.
There are several indications (from different approaches) that Spacetime at the Plank Scale could be discrete. One approach to Quantum Gravity that takes this most seriously is the Causal Sets Approach. In this approach spacetime is fundamentally a discrete, random, partially ordered set (where the partial order is the causal relation). In this contribution, we examine how timelike and spacelike distances arise from a causal set (in the case that the causal set is approximated by Minkowski spacetime), and how one can use this to obtain geometrical information (such as lengths of curves) for the general case, where the causal set could be approximated by some curved spacetime.

1. Motivation

One of the main problems that any discrete/combinatoric approach to Quantum Gravity[2] faces is the so called *inverse problem* [1], which has two sides: (a) How do smooth continuum-like structures emerge from the underlying discrete dynamics and (b) how would we recognize that our discrete entity happens to be something that is smooth and continuum-like. Here we will focus on the second part of the inverse problem, for the specific approach to quantum gravity known as *causal sets* (see Ref. [2] or Refs. [3, 4] for reviews).

[1] Based on talk given by P. Wallden at the NEB XIII conference
[2] Loop quantum gravity, spin foams, causal dynamical triangulations, causal sets, etc.

2. This paper

This contribution is largely based on Ref. [5], where detailed numerics and other related work can be found. In Section 3 we introduce the causal sets approach to Quantum Gravity, in Section 4 we review previous attempts to recover continuum properties from a causal set, in Section 5 we introduce the new suggestion to recover spatial distance in a "flat" causal set and in Section 6 we mention how this can be used for a general (curved spacetime) causal set. We summarize and conclude in Section 7.

3. Causal sets

Causal sets is a fundamentally discrete approach to quantum gravity. There are several indications that spacetime at the Plank scale could be discrete. These involve the finite black hole entropy, the infinities in general relativity and quantum field theory, as well as indications from other approaches (the discrete volume spectrum in loop quantum gravity, effective minimum length due to dualities in string theory, and others). Causal sets take these indications literary and start with a spacetime that is not a differentiable manifold, but rather a locally finite partially ordered set. More specifically, a causal set is:

(i) Partially Ordered Set: a set $\mathcal{P}$ with relation $\prec$, such that $\forall \quad x, y, z \in \mathcal{P}$:
   (a) $x \prec y$ and $y \prec z \Rightarrow x \prec z$: Transitivity
   (b) $x \not\prec x$: Irreflexive (with (a) this forbids closed loops)

(ii) Locally Finite: $\{z|x \prec z \prec y\}$ is a finite set for all $x, y \in \mathcal{P}$.

(iii) The elements of the set correspond to spacetime points, the order relation ($\prec$) is the causal relation between spacetime points, and the number of elements corresponds to the spacetime volume.

The second condition is what makes the causal set approach a discrete approach to quantum gravity, since it guarantees that in a finite volume there will be only be a finite number of elements.

A theorem due to Malament (see Ref. [6]) states that one can recover the conformal metric of a spacetime (Lorentzian manifold) using solely the causal (partial) order. This means that all degrees of freedom are encoded in the causal structure, apart from a scale factor. However if we assume spacetime to be discrete, we can fix the scale factor, by counting the number of elements (corresponding to spacetime volume). This leads us to the central conjecture of causal sets "Hauptvermutung":

> Two distinct, non-isometric spacetimes cannot arise from a single causal set.

In order to be precise, we need to define when we say that a causal set is approximated accurately by a spacetime manifold. For this we define the concept of *faithful embedding*:

A faithful embedding is a map ϕ from a causal set $\mathcal{P}$ to a spacetime $\mathcal{M}$ that:

(i) preserves the causal relation ($i.e., x \prec y \iff \phi(x) \prec \phi(y)$) and

(ii) is "volume preserving", meaning that the number of elements mapped to every spacetime region is Poisson distributed, with mean the volume of the spacetime region in fundamental units, and

(iii) $\mathcal{M}$ does not possess curvature at scales smaller than that defined by the "intermolecular spacing" of the embedding (discreteness scale).

Thus the central conjecture reads as "a causal set cannot be faithfully embedded into two non-isometric spacetimes". One would like to use these causal sets as fundamental entities in order to construct a quantum theory of gravity. Work on the possible dynamics and phenomenology has been carried out recently (see Refs. [7, 8, 9] and Refs. [10, 11] respectively). However, this contribution deals with some kinematical questions, and in particular with the question "how can one derive effective continuum properties such as spatial distance starting from a causal set, when this causal set can be approximated by a spacetime". This would arguably, among other things, help us prove the central conjecture.

Before moving to the main topic of this contribution, let us make a remark about Lorentz invariance versus discreteness (see Ref. [12]). It is well known that these two concepts are not (easily) compatible. To have Lorentz invariance, we would like to have (approximately) equal number of elements for every volume of equal size (number-volume correspondence) and this should hold independently of frames. In particular, let us consider a regular lattice approximating a 2-dimensional Minkowski spacetime (see left side of Figure 1). The number-volume correspondence seems to hold, however if we consider the same in a boosted frame (right side of Figure 1), we can clearly see that there are some big "voids" and the lattice cannot be thought of as being approximated by Minkowski spacetime.

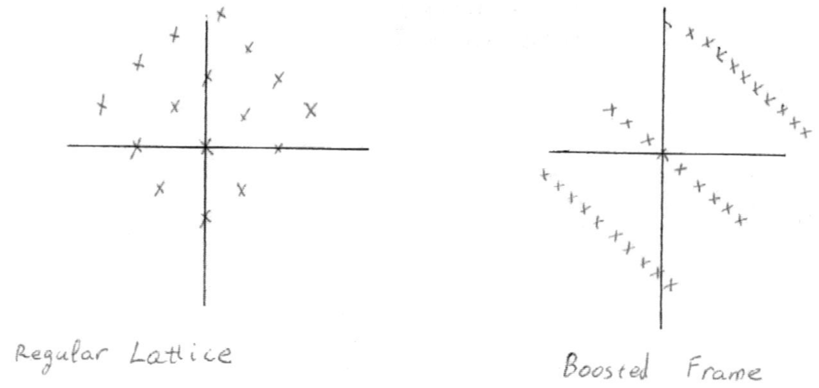

Figure 1. Two dimensional Minkowski spacetime (a) Left: Regular lattice in rest frame (b) Right: Same lattice in a boosted frame

It can be shown that the unique way[3] to achieve discreteness and Lorentz invariance, is by considering a *Poisson sprinkling*, i.e. sprinkle randomly elements in the spacetime in question, with probability determined by the poisson distribution with average number of elements proportional to the size of the spacetime volume of each region.

$$P(n) = \frac{(\rho V)^n e^{-\rho V}}{n!} , \tag{1}$$

where ρ is the density, usually fixed at one per Plank volume. Therefore, we can say that a causal set is a *random*, discrete, partially ordered set.

[3] Subject to some mathematical subtlety, which allows for some more essentially same solutions

4. Previous attempts

Much has been written about causal set kinematics (see Refs. [13, 14, 15, 16, 17, 18]); here we will focus on a way to achieve timelike and spacelike distances for a causal set that arises from a sprinkling in a Minkowski spacetime. Moreover we will consider how (some of) these can be generalized for curved spacetimes, such as determining the length of curves.

4.1. Timelike distance

It can be shown that the concept of "timelike" distance (or else the proper time between two timelike separated elements), can be easily defined for a "Minkowski" causal set and can also be generalized for a "curved" causal set. First let us define a *link*. It is the most basic relation between elements, a relation that cannot be deduced by transitivity:

$$x \prec y \text{ are linked if } \nexists \quad z| \quad x \prec z \prec y. \tag{2}$$

A chain is a collection of elements C such that for all $x, y \in C \quad x \prec y$ or $y \prec x$. Proper time $d(x, y)$, between two related elements $x \prec y$, we define to be (following Ref. [13]) the number of links L in the *longest* chain between (and including) x and y[4]:

$$d(x, y) := L\,. \tag{3}$$

In other words, one considers all chains starting from x and ending at y, and counts the number of links in a largest one[5]. This definition is intrinsic to the causal set, and does not depend on whether it can be faithfully embedded into a manifold, nor on the expected dimension of such a manifold. In [13] (and references therein) it is shown that, in the case of a casual set $\mathcal{P}$ which arises by a sprinkling of density ρ into d-dimensional Minkowski space (so one gets a faithful embedding $\phi : \mathcal{P} \to M$), the distance $d(x, y)$ is proportional to the proper time between the endpoints $\phi(x)$ and $\phi(y)$. In particular the authors state that

$$L(\rho V)^{-1/d} \to m_d \text{ as } \rho V \to \infty\,, \tag{4}$$

for some constant m_d which depends upon the dimension. Here V is the spacetime volume of the causal interval $J^+(x) \cap J^-(y)$ ($J^{\pm}(x)$ represents the causal future/past of x respectively). The exact value of m_2 is known to be 2, while for other dimensions some bounds exist: $1.77 \le m_d \le 2.62$. Note that this expression holds at the limit where ρV goes to infinity, however numerical analysis for the behavior before the asymptotic limit can be found in Ref. [5].

4.2. Naive spatial distance

In the continuum, for Minkowski spacetime, one can show the following: For two spacelike points x, y, their distance is identically the same as the minimum timelike-distance between a point $z \in J^+(x) \cap J^+(y)$ and a point $w \in J^-(x) \cap J^-(y)$, in other words the shortest time taken to go from the common past to the common future. Inspired by this one can define naive spatial distance on a causal set[6], to be:

$$d_{ns}(x, y) := \min d(w, z) \text{ where } w \in J^-(x) \cap J^-(y) \text{ and } z \in J^+(x) \cap J^+(y)\,. \tag{5}$$

[4] Note that the Lorentzian character of the partial order is manifested in this definition. Graphs (of 'finite valence'), on the other hand, naturally embed into Euclidean spaces, and hence one generally defines distance in terms of *shortest* paths on a graph.

[5] Typically there exist more than one biggest chain.

[6] This proposal had been made in the past (e.g. Ref. [13]), but was rejected by the original authors themselves, for essentially the same reasons that we present below.

It is clear that for *any* two unrelated elements x, y, the minumum possible distance is 2, when the relevant interval (w, z) is empty, and we shall call the the *trivial distance.* While the definition of Eq. (5) works for causal sets faithfully embedded into $\mathbf{M}^2$, it fails for higher dimensions (as first noted in Ref. [13] and numerically confirmed in Ref. [5]). We will explore the reasons for the failure so that we can then proceed to a new proposal that does not suffer from those problems.

4.3. Failure of naive spatial distance

We first consider the continuum. In 1+1 dimensions, there exists a unique pair of elements (w, z) from the common past to the common future of the spacelike separated points in question (x, y), such that the distance is minimum. The pair of points has w at the point of intersection of the past lightcones of x and y, and similarly z the point of intersection of the future lightcones.

In 2+1 dimensions, the locus of points of intersection of the two past lightcones forms a hyperbola (a one dimensional object). For *each* of the points of the intersection of the past lightcones (w), there exist a unique point in the intersection of the future lightcones (z), such that the timelike distance between the pair (w, z) is minimum (and thus equal to the spacelike distance between our target points x, y). We name such pairs *minimizing pairs* and (in the continuum) there are clearly infinite of those. Moreover, we can select a subset of those pairs (of infinite cardinality), such that the intervals between any two of those pairs has arbitrarily small overlap (we fix some small threshold ϵ). We name this collection of minimizing pairs *independent minimizing pairs* (IMP).

Now we return to the causal set that can be faithfully embedded in 2+1 dimensional Minkowski spacetime. It is not difficult to see that there will be elements of the causal set close[7] to each of the IMP. Each of these pairs, gives (on average) the correct distance, but since the causal set arises as a poisson distribution, there is a finite (but very small) chance that each pair will give a trivial distance. In other words there is a finite probability for each (w, z) IMP, that there will be no elements between w and z. Since we have infinite of those, we are guaranteed that we will find *at least* one IMP, that has no elements between w and z[8].

This is a very strong theoretical reason, spotted from the authors of Ref. [13]. However, to check this on a computer is much more difficult, because on a computer we cannot simulate a causal set faithfully emebedded into infinite Minkowski spacetime, only a finite region. The *independent* minimizing pairs are widely separated, so testing this effect is difficult, however in Ref. [5] it was clearly observed.

5. Spatial distance

As we can see, the trouble with the above attempt was that we minimized over all those pairs. Each of those had the correct *average* value, which means that taking an average over suitably chosen pairs would solve this problem. We therefore need:

(a) A mechanism to select causal set elements which lie close to a minimizing pair. Such a mechanism involves:
   (i) Finding elements which are close to the intersection of the future (or past) light cones of our unrelated pair of elements.
   (ii) For each such an element z in the future (say), select an element w in the common past which locates a pair (w, z) which is close to some continuum minimizing pair.

[7] The precise meaning of "close" is not important here, but to be accurate would require more care.

[8] The independent condition is required because we need the $d(w, z)$ variables to be independent. Otherwise, we would not be guaranteed to find one pair with zero elements in between.

(b) To take an **average** over minimizing pairs, and not minimize

For the first issue, we require the definition of a *2-link*: Given unrelated elements x, y of a causal set we define w to be a 2-link (of x, y) if the element w is linked to both x and y (see above for the definition of a link). Given an element x, the elements that are linked to x are elements that lie very close to the lightcone of x. Therefore, it is easy to see that the 2-links lie close to the intersection of the lightcones of x and y. It is easy to see that for dimensions greater than 2, the number of 2-links for any pair of unrelated elements x, y is infinite (corresponding to the (non-countably) infinite points in the continuum). However in 1+1 this is not the case, and most frequently there exist no 2-links at all. In what follows, we could define things slightly differently to account for this problem in 1+1 dimensions[9]. However, (a) we live in 3+1 dimensions and (b) in 1+1 dimensions the naive spatial distance works fine, so we are not going to go into more details on this issue here, and assume the existence of 2-links. We are now in position to define the following procedure that gives us a spatial distance that does not suffer from problems discussed earlier:

Step 1: Given spacelike elements x, y we find a future 2-link f_i.

Step 2: Find the element p_i in the common past of x and y that makes the timelike distance $d(p_i, f_i)$ minimum. This will select a minimizing pair.

Step 3: Store the timelike distance $d^i(x, y)$ for the future 2-link f_i.

Step 4: Repeat for all other future 2-links.

Step 5: Take the average over all future 2-links $\langle d^i(x, y)\rangle$ to be the spacelike distance between elements x and y. We call this average the *2-link distance* between x and y.

In Ref. [5] we can see how this distance performs numerically and compare it with naive spatial distance for both the cases that naive spatial distance is valid and where it fails. The results of the simulations agree with the intuitive theoretical expectations analyzed above.

6. Towards curved spacetime

In the previous section we considered only how to define spacelike distance between elements of a causal set approximated by a Minkowski spacetime. Using this, we can define a concept of closest spatial neighbor. We define a (symmetric) relation *s-link* between to unrelated elements x, y. This relation exists if the 2-link distance between x and y is less than some fixed threshold λ. This threshold is somewhat arbitrary, but it should be between 2 and 3 (note that with the definition we have above, the smallest conceivable spatial distance would be 2, however taking an average over all the IMP would give a value greater than 2 value no matter how close in the embedding the elements x, y are). Given a target element at the origin of 2+1 dimensional Minkowski spacetime, we can see in Figure 2, that the s-links indeed lie on a hyperboloid as one would expect.

Having defined the concept of an s-link, which is the most elementary relation between unrelated (spacelike) elements, it is straight forward to define what a continuous curve is: It is a collection of elements that can be ordered in such a way that any two consecutive elements are related either as links or as s-links. The length of such continuous curve can simply be derived by counting the number of links and s-links (or else the number of elements).

An important observation is that for spacetimes for which the curvature does not vary rapidly at scales just above the Plank scale, the prescription we define to identify closest

[9] A concept of closest 2-neighbor that reduces to a 2-link if such thing exists is one obvious attempt.

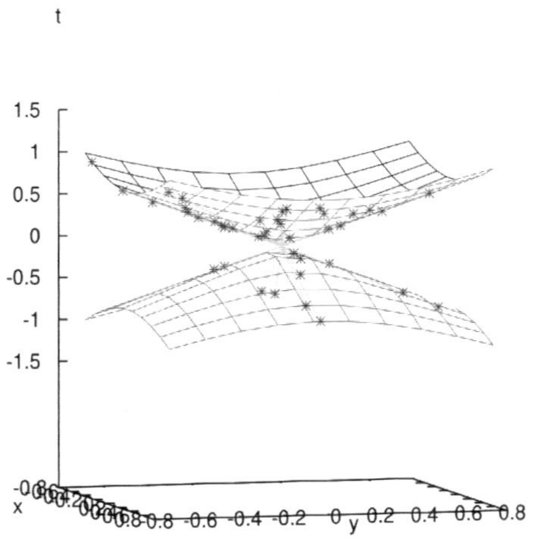

Figure 2. Spatial nearest neighbors of an element in a sprinkling into (a fixed cube in) 2+1 dimensional owski. $\langle N\rangle = 65\,536$. The future and past light cones of the "origin element" x are shown. The spacelike cyan lines are drawn between x and each neighbor, for emphasis.

neighbors would carry over[10]. Identifying closest neighbors in *any* (flat or curved) spacetime means that we can compute lengths of curves in those spacetimes. The very concept of spatial distance is not well defined for curved spacetimes, however the length of curves *is* and the prescription we gave can achieve this.

Detailed numerical analysis, and possible extensions of these ideas in order to recover the full metric, are left for further work. Recovering the full metric may also help in giving direction for deriving quantum dynamics of the causal set (e.g. by rewritting the Einstein-Hilbert action in terms of the relations of the causal set). However, already recovering the lengths of curves is an important achievement in recovering the effective spacetime that approximates our causal set and thus a major step forward in proving the central conjecture.

7. Summary and conclusions

We have reviewed how to recover timelike distance from a causal set and a failed attempt to do something similar for the spacelike distance on a Minkowski causal set. The reason for this failure was that there exist infinite independent "minimizing pairs" and the definition of naive spatial distance involved minimizing over all of them. Since they turn out to be random variables, minimizing over infinite of those would give a trivial result. To avoid this, we suggested taking an average over those minimizing pairs. However, to do so we need to define what is precisely meant by a minimizing pair in a causal set, which involved the concept of a 2-link. By considering, the closest spatial neighbors, we were able to define length of

[10] It essentially says that for spacetimes that are "locally flat", in the sense defined above, we can identify which are the closest neighbors using the prescription we mentioned above for flat spacetimes.

curves. Moreover, this relies only on closest neighbors, so for causal sets that do not have curvature at (or close to) the Plank scale, we can hope that the definition of closest neighbors carries over intact. Therefore we can define length of curves in *any* causal set (flat or curved) that can be approximated by a continuum spacetime. Further analysis of the curved spacetime case, and the possibility of fully recovering the metric, is left for future work.

Acknowledgments

We are especially grateful to Graham Brightwell for a number of discussions on this work. We are also grateful to Rafael Sorkin, David Meyer, and the attenders of 'relativity lunch' at Imperial College London, for numerous helpful discussions.

This research was supported by a number of grants/organizations, including the Marie Curie Research and Training Network ENRAGE (MRTN-CT-2004-005616), the Royal Society International Joint Project 2006-R2, and the Perimeter Institute for Theoretical Physics. Research at Perimeter Institute is supported by the Government of Canada through Industry Canada and by the Province of Ontario through the Ministry of Research & Innovation.

The numerical results were made possible by the facilities of the Shared Hierarchical Academic Research Computing Network (SHARCNET:www.sharcnet.ca).

PW thanks the Perimeter Institute for Theoretical Physics for hospitality during a visit in which a large part of this work was carried out, and the organizers of NEB XIII for giving the opportunity to present this talk.

References

[1] Lee Smolin, in Rickles, D. (ed.) et al.: *The structural foundation of quantum gravity* 196–239. (2005).
[2] L. Bombelli, J.H. Lee, D. Meyer, and R. Sorkin, *Phys. Rev. Lett.* **59** (1987) 521.
[3] R. D. Sorkin, in *Proceedings of the ninth Italian Conference on General Relativity and Gravitational Physics, Capri, Italy, September 1990*, R. Cianci, R. de Ritis, M. Francaviglia, G. Marmo, C. Rubano, and P. Scudellaro, eds., pp. 68–90. World Scientific, Singapore, 1991.
[4] F. Dowker, AIP Conf. Proc. **861**, 79 (2006).
[5] D. Rideout and P. Wallden, (2008) preprint [arXiv: 0810.1768v2], (gr-qc).
[6] David. Malament, J. Math. Phys. **18**, 1399 (1977).
[7] G. Brightwell, J. Henson and S. Surya, (2007), preprint [arXiv: 0706.0375], (gr-qc).
[8] Steven Johnston, (2008), preprint [arXiv: 0806.3083], (hep-th).
[9] Roman Sverdlov and Luca Bombelli, preprint [arXiv:0801.0240], (gr-qc).
[10] R. D. Sorkin, AIP Conf. Proc. **957**, 142 (2007).
[11] R. D. Sorkin, to appear in *Towards Quantum Gravity*, D. Oriti (ed.), Cambridge University Press.
[12] L. Bombelli, J. Henson and R. D. Sorkin, preprint [arXiv:gr-qc/0605006].
[13] G. Brightwell and R. Gregory, Phys. Rev. Lett. 66: 260-263 (1991).
[14] Jan Myrheim, "Statistical Geometry", CERN preprint Ref.TH.2538-CERN (1978).
[15] G. 't Hooft, in *Recent Developments in Gravitation (Proceedings of the 1978 Cargese Summer Institute)*, M. Levy and S. Deser, eds. Plenum, 1979.
[16] D.A. Meyer, *Order* **10** 227–237 (1993).
[17] R. D. Sorkin, in *Lectures on Quantum Gravity, Proceedings of the Valdivia Summer School, Valdivia, Chile, January 2002*, A. Gomberoff and D. Marolf, eds. Plenum, 2005. preprint [arXiv: gr-qc/0309009].
[18] S. Major, D. Rideout, and S. Surya, *J. Math. Phys.* **48**, 032501, (2007).
Sumati Surya, *Theoretical Computer Science* **405**, 1–2, pp. 188–197 (2008).

Journal of Physics: Conference Series **189** (2009) 012046 doi:10.1088/1742-6596/189/1/012046

Signature-change events in emergent spacetimes with anisotropic scaling

Silke Weinfurtner[1], Angela White[2], and Matt Visser[3]

[1] Department of Physics and Astronomy, University of British Columbia, Vancouver, BC, V6T 1Z1, Canada

[2] School of Mathematics and Statistics, Newcastle University Herschel Building, Newcastle upon Tyne, NE1 7RU, United Kingdom

[3] School of Mathematics, Statistics, and Operations Research, Victoria University of Wellington, PO Box 600, Wellington, New Zealand

E-mail: silke@physics.ubc.ca, angela.white@ncl.ac.uk, matt.visser@msor.vuw.ac.nz

Abstract. We investigate the behaviour of quantum fields coupled to a spacetime geometry exhibiting finite regions of Euclidean (Riemannian) signature. Although from a gravity perspective this situation might seem somewhat far fetched, we will demonstrate its direct physical relevance for an explicitly realizable condensed matter system whose linearized perturbations experience an effective emergent spacetime geometry with externally controllable signature. This effective geometry is intrinsically quantum in origin, and its signature is determined by the details of the microscopic structure. At the level of the effective field theory arising from our condensed matter system we encounter explicit anisotropic scaling in time and space. Here Lorentz symmetry is an emergent symmetry in the infrared. This anisotropic scaling of time and space cures some of the technical problems that arise when working within a canonical quantisation scheme obeying strict Lorentz invariance at all scales, and so is helpful in permitting signature change events to take place.

1. Motivation

Recently much attention has been paid to the possibility of ad-hoc construction of quantum field theories with a certain degree z of anisotropic scaling between the space and time dimensions — this anisotropic scaling often significantly improving the ultraviolet behaviour. See for example [1, 2, 3, 4] and [5, 6, 7, 8, 9, 10]. Wide-scale interest in these ideas was largely ignited by Hořava's specific model for quantum gravity in $3+1$ dimensions, based on anisotropic scaling at the $z=3$ Lifshitz point [3, 4], but the basic ideas have much wider applicability — for instance it is possible to extend these ideas to general spatial dimensions [1, 2, 3, 4, 5, 6, 7], such that the engineering dimensions of space and time are of degree $z=d$, such that

$$[\mathrm{d}x] = [\kappa]^{-1} \quad \text{and} \quad [\mathrm{d}t] = [\kappa]^{-z}. \tag{1}$$

Here d is the number of spatial dimensions, and κ is a placeholder with dimensions of momentum. Based on the anisotropy between space and time it is possible to modify the Einstein–Hilbert action of general relativity, adding one kinetic and several higher order spatial curvature terms, containing up to $2d$ spatial derivatives. See for example [3, 4, 6, 7]. It is worth emphasising that these models do not need (or even permit) more than 2 time derivatives. There are strong indications that such a quantum gravity model based on anisotropic scaling exhibits vastly improved behaviour at small scales, which alters the theory to be renormalisable (possibly even finite), at the cost of giving up Lorentz symmetry as an exact symmetry [3, 4].

Shortly after Hořava's papers on this quantum gravity candidate [3, 4], anisotropic scaling was adapted for use as a regulator for scalar polynomial quantum field theories, thereby throwing light on their renormalizability properties [5]. The free Lagrangian (for a scalar field with mass m_s) in this proposal is given by

$$S_{\text{free}} = \int \left\{ \dot{\phi}^2 - \phi \left[m_s^2 - c^2 \triangle + ... + (-\triangle)^z \right] \phi \right\} \mathrm{d}t\, \mathrm{d}^d x. \tag{2}$$

It has been shown that in $d+1$ spacetime dimensions subject to $z = d$ anisotropic scaling arbitrary polynomial interactions of the form $P(\phi) = \sum_{n=1}^{N} g_n \phi^n$ are perturbatively ultraviolet finite when normal ordering is applied. Even without normal ordering this quantum field theory is perturbatively renormalizable.

As already mentioned above, in both proposals the ultraviolet modifications are more or less chosen ad-hoc, with hindsight showing that one gets a better ultraviolet control over the quantum behaviour of the gravitational / scalar field. In this spirit we think it may be interesting to note that there is a physically realistic class of condensed matter systems that naturally exhibit some of the features predicted in equation (2). A specific example of such a system is an ultra-cold gas of bosons in its superfluid phase, referred to as a Bose–Einstein condensate (BEC), or more generally any condensed matter system exhibiting a Bogoliubov dispersion relation. It has been well established, see [11, 12, 13, 14, 15], that small perturbations around the condensate experience an effective / acoustic / emergent spacetime geometry. In this context Lorentz symmetry is not an exact symmetry, it is emergent [16, 17, 18, 19, 20, 21]. Previously, see [17, 22], we have shown that the signature of the emergent spacetime geometry is related to the nature of the interaction between the fundamental bosons, and that it is possible to drive (sudden) finite transitions between periods of different spacetime signature. With certain technical assumptions (including a translation invariant background condensate) we can show that the linearised BEC resembles a scalar field with anisotropic scaling, as suggested in [5],

$$S_{\text{BEC}}^{(1)} \approx \int \left\{ \dot{\phi}^2 - \phi \left[-c_0^2 \triangle + \gamma_{qp}^2 \triangle^2 \right] \phi \right\} \mathrm{d}t\, \mathrm{d}^d x. \tag{3}$$

Here c_0 is the speed of sound (which takes the role of the speed of light c within the context of emergent spacetimes), γ_{qp} is associated with the microscopic substructure, the elementary bosons, while the low-momentum speed of sound c_0 depends on the condensate. However in a more general emergent spacetime background (especially once the background is not translation invariant) the effective free field action is considerably more complicated and has to be treated with care [17, 23]. Since the degree of anisotropy is in this case (only) $z = 2$, the modifications are not sufficient to cure all problems associated with finite regions of Euclidean (Riemannian) signature. Without anisotropic scaling the particle number-density produced from finite regions of Euclidean (Riemannian) signature is given by

$$N_k = \sinh^2 \left[\sqrt{m_s^2 + k^2}\, \Delta\tau \right]. \tag{4}$$

This corresponds to a particle spectrum that is exponentially increasing both in wave-number k, and in the length ("duration") of the Euclidean region $\Delta\tau$. For a detailed discussion on the subject consult [17, 22]. This result shows that regardless of the number of spatial dimensions and the mass m_s of the particle, the number-density and energy density of the excitations is infinite, and thus the quantisation scheme at hand breaks down. We will see that adding anisotropic scaling in the ultraviolet greatly improves the situation.

2. Emergent spacetime with anisotropic scaling

In the following discussion we will introduce a "real life" example where the notion of geometry is emergent. The emergent spacetime we are going to study closely involves: (i) a microscopic system consisting of an ultra-cold weakly interacting gas of bosons; (ii) which exhibits (after a first-order phase transition) a mean-field regime where the microscopic degrees of freedom give way to collective variables; (iii) small perturbations above the ground state which are dominated by a 'geometrical object', a symmetric second rank tensor; (iv) Lorentz symmetry as an emergent symmetry in the infrared, while at higher energies the effective field theory shows anisotropic scaling in time and space; (v) the signature of the emergent spacetime is of microscopic origin and can be 'switched' by the application of external magnetic fields.

(i) Microscopic system

The microscopic degrees of freedom at hand are represented by individual quantum field operators that act on quantum states (Hilbert space of states; *e.g.*, Fock space), creating, $\hat{\psi}^\dagger(t,\mathbf{x})$, or destroying, $\hat{\psi}(t,\mathbf{x})$, an individual boson at a particular point in space and time. These operators satisfy the usual boson commutators [24, 25]:

$$\left[\hat{\psi}(t,\mathbf{x}),\hat{\psi}(t,\mathbf{x}')\right] = \left[\hat{\psi}^\dagger(t,\mathbf{x}),\hat{\psi}^\dagger(t,\mathbf{x}')\right] = 0\,; \quad \left[\hat{\psi}(t,\mathbf{x}),\hat{\psi}^\dagger(t,\mathbf{x}')\right] = \delta(\mathbf{x}-\mathbf{x}')\,. \tag{5}$$

The Hamiltonian of the microscopic system is chosen to be non-relativistic,

$$\hat{H} = \int d\mathrm{x} \left(-\hat{\psi}^\dagger \frac{\hbar^2}{2m} \nabla^2 \hat{\psi} + \hat{\psi}^\dagger V_{\mathrm{ext}} \hat{\psi} + \frac{U}{2}\, \hat{\psi}^\dagger \hat{\psi}^\dagger \hat{\psi} \hat{\psi} \right), \tag{6}$$

and obeys $SO(2)$ symmetry transformations of the form $\hat{\psi} \to \hat{\psi}\, \exp(i\alpha)$. The Hamiltonian is a sum of the kinetic energy of the boson field, and two potential energy contributions: the external trap potential V_{ext}, and the quasi-particle interactions. The extreme dilution of the gas (typically $10^{13} - 10^{15}$ atoms/cm^3) suppresses interactions involving more than two particles, and in the weakly interacting regime the actual inter-atom potential has been approximated by a pseudo-contact potential

$$U = \frac{4\pi\hbar^2 a}{m}\,. \tag{7}$$

Here m is the single-boson mass, and a the s-wave scattering length.

(ii) Mean-field regime

Under an appropriate cooling process (one that protects the gas from solidification through a first-order phase-transition) a new state of matter will occur, the Bose–Einstein condensate. The condensate is a complex-valued macroscopic mean-field that is a result of a spontaneous breaking of the $SO(2)$ symmetry, such that the field operators for topologically trivial regions (without zeros or singularities) acquire non-zero expectation values,

$$\langle \hat{\psi}(t,\mathbf{x}) \rangle = \psi(t,\mathbf{x}) = \sqrt{n_0(t,\mathbf{x})}\, \exp\left(i\, \phi_0(t,\mathbf{x})\right) \neq 0. \tag{8}$$

The individual microscopic degrees of freedom give way to collective variables, such as the condensate density $n_0 \equiv n_0(t,\mathbf{x})$, and the phase $\phi_0 \equiv \phi_0(t,\mathbf{x})$ [24, 25].

(iii) Emergent metric tensor

The emergence of a geometrical rank two tensor enters (in a by now quite standard manner) at the linear level after an expansion around the mean-field variables [26]. In terms of the

Hermitian density $\hat{n}$ and phase $\hat{\phi}$ fluctuations we have [16, 23, 17]:

$$\psi(t,\mathbf{x}) + \delta\hat{\psi}(t,\mathbf{x}) \;\rightarrow\; \psi(t,\mathbf{x})\left(1 + \frac{1}{2}\frac{\hat{n}}{n_0} + i\hat{\phi}\right), \tag{9}$$

$$\psi^*(t,\mathbf{x}) + \delta\hat{\psi}^\dagger(t,\mathbf{x}) \;\rightarrow\; \psi^*(t,\mathbf{x})\left(1 + \frac{1}{2}\frac{\hat{n}}{n_0} - i\hat{\phi}\right). \tag{10}$$

Form the usual boson commutators of equation (5), it is straightforward to show that the new operators are a pair of canonical variables:

$$\left[\hat{n}(t,\mathbf{x}),\hat{n}(t,\mathbf{x}')\right] = \left[\hat{\phi}(t,\mathbf{x}),\hat{\phi}(t,\mathbf{x}')\right] = 0\,; \quad \text{and} \quad \left[\hat{n}(t,\mathbf{x}),\hat{\phi}(t,\mathbf{x}')\right] = i\delta(\mathbf{x}-\mathbf{x}'). \tag{11}$$

By assuming the perturbation to be small, allowing us to neglect quadratic and higher-order products of the perturbation field $\delta\hat{\psi}$, and working within the hydrodynamic limit ('long-wavelength' limit), for low-momentum modes we are able to express the excitation spectrum in a very compact and insightful way:

$$\frac{1}{\sqrt{|\det(g_{ab})|}}\,\partial_a\left(\sqrt{|\det(g_{ab})|}\; g^{ab}\;\partial_b\hat{\phi}\right) = 0\,. \tag{12}$$

Here we have introduced the "acoustic metric"

$$g_{ab} = \left(\frac{c_0}{U/\hbar}\right)^{\frac{2}{d-1}} \begin{bmatrix} -\left(c_0^2 - \mathbf{v}^2\right) & -v_x & -v_y & -v_z \\ -v_x & 1 & 0 & 0 \\ -v_y & 0 & 1 & 0 \\ -v_z & 0 & 0 & 1 \end{bmatrix}, \tag{13}$$

a covariant rank two metric tensor, whose entries are purely collective variables. For a detailed derivation see [16, 17, 23]. The conformal factor depends on the spatial dimensionality, d, of the condensate cloud. Here we have introduced the background velocity of the condensate as

$$\mathbf{v} = \frac{\hbar}{m}\nabla\phi_0, \tag{14}$$

and the propagation speed for excitations within this hydrodynamic approximation simplifies to a common sound speed given by

$$c_0^2 = \frac{n_0 U}{m}\,. \tag{15}$$

In the same manner that we eliminated all (explicit) occurrences of density perturbations in the excitation spectrum, it is possible to show that the commutation relation (11) can be re-written solely in terms of the field $\hat{\phi}$ and its conjugate momentum $\hat{\Pi}_{\hat{\phi}}$ on the effective curved spacetime geometry:

$$\left[\hat{n}(t,\mathbf{x}),\hat{\phi}(t,\mathbf{x}')\right] = i\delta(\mathbf{x}-\mathbf{x}') \rightarrow \left[\hat{\phi}(t,\mathbf{x}),\hat{\Pi}_{\hat{\phi}}(t,\mathbf{x}')\right] = i\delta(\mathbf{x}-\mathbf{x}')\,. \tag{16}$$

(iv) Emergent Lorentz symmetry and anisotropic scaling

Within the context of emergent spacetimes Lorentz symmetry is not an exact symmetry. Low-energy / infrared excitations around the macroscopic field, which describes the vacuum state of the system, experience an emergent Lorentz symmetry. This symmetry will be broken in the high-energy / ultraviolet regime, that is at scales dominated by the underlying microscopic theory. These corrections are of a non-perturbative nature involving only higher spatial derivatives [16, 17, 18, 19, 20, 21, 23].

It is possible to incorporate these non-perturbative corrections into the derivation of the effective spacetime geometry by formally replacing the interaction strength U with an appropriate differential operator:

$$U \to \widetilde{U} = U - \frac{\hbar^2}{2m}\,\widetilde{D}_2\,; \quad \text{where} \quad \widetilde{D}_2 = \frac{1}{2}\left\{\frac{(\nabla n_0)^2 - (\nabla^2 n_0) n_0}{n_0^3} - \frac{\nabla n_0}{n_0^2}\nabla + \frac{1}{n_0}\nabla^2\right\}. \tag{17}$$

This replacement now leads to an integro-differential equation,

$$\partial_a \left(f^{ab}\,\partial_b \hat{\phi} \right) = 0\,, \tag{18}$$

where we have introduced a matrix f^{ab} with inverse-differential-operator-valued entries:

$$f^{ab} = \hbar \left[\begin{array}{c|c} -\,\widetilde{U}^{-1} & -\widetilde{U}^{-1} v^j \\ \hline -\,v^i \widetilde{U}^{-1} & \frac{n_0}{m}\delta^{ij} - v^i \widetilde{U}^{-1} v^j \end{array} \right]. \tag{19}$$

In general $\widetilde{U}^{-1}$ is an integral operator which can be formally expanded as [1]

$$\begin{aligned} \widetilde{U}^{-1} = U^{-1} \; &+ \; \left[\frac{\hbar^2}{2m}\right] U^{-1}\widetilde{D}_2\, U^{-1} \\ &+ \; \left[\frac{\hbar^2}{2m}\right]^2 U^{-1}\widetilde{D}_2\, U^{-1}\widetilde{D}_2\, U^{-1} \\ &+ \; \left[\frac{\hbar^2}{2m}\right]^3 U^{-1}\widetilde{D}_2\, U^{-1}\widetilde{D}_2\, U^{-1}\widetilde{D}_2\, U^{-1} + \ldots \end{aligned} \tag{20}$$

Note that we additionally require that there exists an (inverse) metric tensor g^{ab} such that

$$f^{ab} \equiv \sqrt{-g}\, g^{ab}\,, \tag{21}$$

where g is the determinant of g_{ab}. Then the connection is formally made to the field equation for a minimally coupled massless scalar field in a curved spacetime; see equation (13). Let us rewrite this (for more details see [23]) as

$$\widetilde{U} \to U \left(1 - \frac{\gamma_{qp}^2}{c_0^2}\,\widetilde{D} \right)\,; \quad \text{where} \quad \widetilde{D} = \left\{ \left(\frac{\nabla n_0}{n_0} \right)^2 - \frac{\nabla^2 n_0}{n_0} - \frac{\nabla n_0}{n_0}\nabla + \nabla^2 \right\}, \tag{22}$$

and further take $n_0(t, \mathbf{x}) \equiv$ const and $U(t, \mathbf{x}) \equiv U(t)$, so that the background is translation invariant. We then get $c_0(t, \mathbf{x}) \equiv c_0(t)$, and see that

$$\widetilde{U} \to U \left(1 - \frac{\gamma_{qp}^2}{c_0^2}\,\nabla^2 \right), \tag{23}$$

and

$$\widetilde{U}^{-1} \to U^{-1} \left(1 - \sum_{n=1}^{\infty} \left[\frac{\gamma_{qp}}{c_0} \right]^{2n} \Delta^n \right). \tag{24}$$

[1] The *formal* series converges only on the subspace of functions spanned by the eigenfunctions whose eigenvalues satisfy

$$\lambda\left(\frac{\hbar^2}{2m} U^{-1}\widetilde{D}_2\right) < 1\,.$$

Here γ_{qp} is constant and only depends on the mass of the fundamental bosons:

$$\gamma_{qp} = \frac{\hbar}{2m}. \tag{25}$$

Therefore it can be seen that the whole notion of an emergent spacetime geometry is, even under this very simplified situation, difficult to set-up if we are outside the validity of the Lorentz invariant regime, and one practically has to give up the notion of geometry. However, in cases of an isotropic spacetime, such as FRLW spacetime it is very simple to construct the effective field theory, since a mapping onto momentum space is trivial, $\triangle \to -k^2$, and

$$\widetilde{U}^{-1} \to U^{-1}\left(1 - \sum_{n=1}^{\infty}\left[\frac{\gamma_{qp}}{c_0}\right]^{2n}(-k^2)^n\right) \to \frac{1}{U\left(1+\frac{\gamma_{qp}^2}{c_0^2}k^2\right)}. \tag{26}$$

Altogether for $v^i \equiv 0$ on a translation invariant background, we have

$$f^{ab} = \hbar\,\mathrm{diag}\left(-\widetilde{U}^{-1}, \frac{n_0}{m}, \frac{n_0}{m}, \frac{n_0}{m}\right). \tag{27}$$

For the equation of motion we see

$$\partial_t\partial_t\hat{\phi} + (f^{tt})^{-1} f^{ij}\partial_i\partial_j\hat{\phi} = 0, \tag{28}$$

which we re-write as

$$\partial_t\partial_t\hat{\phi} - \left(c_0^2\,\triangle - \gamma_{qp}^2\,\triangle^2\right)\hat{\phi} = 0. \tag{29}$$

Subject to these approximations, and the simplifications imposed on the condensate background, we are able to associate the effective field theory of our BEC system with the specific case of anisotropic scaling at $z = 2$:

$$S^{(1)}_{\mathrm{BEC}} \approx \int \left\{\dot{\phi}^2 - \phi\left[-c_0^2\triangle + \gamma_{qp}^2\,\triangle^2\right]\phi\right\}\,\mathrm{d}t\,\mathrm{d}^d x. \tag{30}$$

A change to momentum space leads to the well-known Bogoliubov dispersion relation

$$\omega^2 = c_0^2\,k^2 + \gamma_{qp}^2\,k^4. \tag{31}$$

Therefore we are able to connect the anisotropic scaling at $z = 2$ with an emergent Lorentz symmetry for momenta below the effective Lorentz breaking scale,

$$k^2 \ll \frac{c_0^2}{\gamma_{qp}^2}. \tag{32}$$

For further information about the role of Lorentz symmetry breaking within the emergent spacetime / analogue models for gravity / acoustic metric programme see [16, 18, 19, 20, 21, 27, 28].

(v) Signature and signature change events
The signature of the emergent spacetime geometry can easily be read off by setting $v^i \to 0$. This can always be done locally, and we can read off the pattern of metric eigenvalues from equation (13):

$$\eta_{ab} \sim \begin{bmatrix} -c_0^2 & 0 & 0 & 0 \\ 0 & 1 & 0 & 0 \\ 0 & 0 & 1 & 0 \\ 0 & 0 & 0 & 1 \end{bmatrix}. \tag{33}$$

Since $c_0^2 \sim U$ we are able to relate the signature of the emergent spacetime geometry with the qualitative behaviour of the atom-atom interactions, here approximated with a purely contact potential only accounting for s-wave scattering processes [16, 17, 23]. (Certainly this particular approximation is a significant weakness of the whole ansatz, and it would be interesting to extend the analysis towards a more realistic interaction potential.) Experimentally it is possible to drive transitions between negative and positive values of U, where

$$\begin{aligned} U > 0 &\qquad \text{repulsive}\,; \\ U < 0 &\qquad \text{attractive}\,. \end{aligned} \tag{34}$$

Such changes in the *sign* of the interaction are accessible by tuning external magnetic fields that interact with the inter-atomic potential; this process is called Feshbach resonance [29]. This is not just an "in principle" possibility, since in 2001 a BEC experiment [30, 31] was carried out that can be viewed as the first emergent spacetime experiment involving particle production resulting from a signature change event, see [17, 22]. For a detailed discussion regarding the behaviour of the effective field theory exposed to a finite region of Riemannian signature consult [17, 22]. There it has been shown that at very high momenta, due to the anisotropic scaling of time and space, the number of quasi-particles being produced in sudden signature variations of the form Lorentzian–Euclidean–Lorentzian, $(-,+++) \to (+,+++) \to (-,+++)$, scales asymptotically as $N_k \sim k^{-4}$. Therefore

$$N_{\leqslant k} \sim 2^{d-1}\pi \int dk\, k^{d-1}\, k^{-4} \sim k^{d-4}. \tag{35}$$

That is, the total number of quasi-particles produced is finite, as expected on physical grounds. However, it is easy to see that the total energy emitted,

$$E_{\leqslant k} = 2^{d-1}\pi \int_k dk\, k^{d-1}\, N_k\, \omega_k \sim 2^{d-1}\pi \int_k dk\, k^{d-1}\, k^{-2} \sim k^{d-2}, \tag{36}$$

is still infinite. This interesting effect is due to the specific behaviour of the coefficients characterizing the anisotropic scaling — only the k^2-term in the effective Lagrangian is affected by the Feschbach resonance, while the k^4-term remains — at the linear level — unaffected:

$$\left\{\dot\phi^2 - \phi\left[-c_0^2\Delta + \gamma_{qp}^2\,\Delta^2\right]\phi\right\} \to \left\{\dot\phi^2 - \phi\left[+c_0^2\Delta + \gamma_{qp}^2\,\Delta^2\right]\phi\right\} \to \left\{\dot\phi^2 - \phi\left[-c_0^2\Delta + \gamma_{qp}^2\,\Delta^2\right]\phi\right\}. \tag{37}$$

Therefore at energy scales $k^2 \gg c_0^2/\gamma_{qp}^2$ the higher spatial derivatives (which in fact define the $z = 2$ Lifschitz point) are unaffected:

$$\left\{\dot\phi^2 - \phi\left[\gamma_{qp}^2\,\Delta^2\right]\phi\right\} \to \left\{\dot\phi^2 - \phi\left[\gamma_{qp}^2\,\Delta^2\right]\phi\right\} \to \left\{\dot\phi^2 - \phi\left[\gamma_{qp}^2\,\Delta^2\right]\phi\right\}. \tag{38}$$

However we would like to emphasise that this linear regime is only applicable to some limited extent, since ultimately three-body losses (and consequently backreaction effects) will play a rapidly growing role with the increasing duration of the attractive interaction / Riemannian signature regime.

3. Summary and conclusions

We have discussed the concept of a spacetime emergent from a Bose–Einstein condensate, and the possibility of signature change events within this context [17, 22, 32, 33, 34]. The recent interest in quantum gravity models sacrificing Lorentz symmetry as an exact symmetry in order to obtain better control of the ultraviolet behaviour of quantum gravitational / field excitations [1, 2, 3, 4, 5, 6, 7, 8, 9, 10], has motivated us to emphasise some of the features that 'anisotropic scaling' and the presence of Lorentz-violating Lifschitz points may contribute to the subject of condensed-matter emergent spacetimes. We have, in brief, demonstrated how spacetime geometry — as seen by quasiparticle excitations — emerges from a mean-field structure that is truly quantum in origin [12]. Although this model has no direct connection or application to the search of a quantum theory of gravity, we have shown that it nevertheless exhibits some (but not all) of the features that are essential to Hořava–Lifschitz gravity. The key features are that Lorentz symmetry is an emergent symmetry in the infrared, and that the non-perturbative terms added at the level of the effective field theory are anisotropic in space and time, involving only higher spatial derivatives. In fact, these two features seem to be common to all known emergent spacetimes, independent of the specific media from which they are emergent — *e.g.*, water, superfluids, electromagnetic waveguides, *etc.*

Coming back to the specific model at hand, the Bose–Einstein condensate, we would like to point out that the free-field description for small perturbations is only an approximation. We are, strictly speaking, dealing with an interacting field theory, but due to the anisotropic scaling naturally built into the effective field theory the ultraviolet regime has some advantages compared to conventional field theories. A specific way to explore this is by exposing the effective field theory to one of the most drastic gravitational scenarios possible (besides curvature singularities): signature change events. While for a $z = 1$ (Lorentz invariant) scaling of time and space the canonical quantisation scheme applied to quantum field theories on Riemannian manifolds breaks down, predicting infinite quasi-particle production per unit volume, we see that in contrast $z = 2$ anisotropic scaling is already sufficient to render quasiparticle production finite. It is straightforward to show that at $z = 3$ the energy density per volume will also be finite [16, 17, 18, 23, 35].

The main lesson to be taken from the discussion is that nature quite naturally supplies us with little role models equipped with both of the desired features, both emergent Lorentz symmetry and anisotropic scaling.

References

[1] Petr Hořava. Quantum Criticality and Yang-Mills Gauge Theory. arXiv:0811.2217, 2008.

[2] Petr Hořava. Membranes at Quantum Criticality. *JHEP*, 03:020, 2009.

[3] Petr Hořava. Spectral dimension of the universe in quantum gravity at a Lifshitz point. *Physical Review Letters*, 102(16):161301, 2009.

[4] Petr Hořava. Quantum Gravity at a Lifshitz Point. *Phys. Rev.*, D79:084008, 2009.

[5] Matt Visser. Lorentz symmetry breaking as a quantum field theory regulator. arXiv:0902.0590, 2009.

[6] Thomas P Sotiriou, Matt Visser, and Silke Weinfurtner. Quantum gravity without Lorentz invariance. arXiv:0905.2798, 2009.

[7] Thomas Sotiriou, Matt Visser, and Silke Weinfurtner. Phenomenologically viable Lorentz-violating quantum gravity. arXiv:0904.4464, 2009.

[8] G. E. Volovik. $z = 3$ Lifshitz-Hořava model and Fermi-point scenario of emergent gravity. arXiv:0904.4113, 2009.

[9] Christos Charmousis, Gustavo Niz, Antonio Padilla, and Paul M. Saffin. Strong coupling in Hořava gravity. arXiv:0905.2579, 2009.

[10] H. Lu, Jianwei Mei, and C. N. Pope. Solutions to Hořava Gravity. arXiv:0904.1595, 2009.

[11] W. G. Unruh. Experimental black hole evaporation. *Phys. Rev. Lett.*, 46:1351–1353, 1981.

[12] Carlos Barceló, Stefano Liberati, and Matt Visser. Analogue gravity. *Living Rev. Rel.*, 8:12, 2005.

[13] Grigory E. Volovik. *The universe in a Helium droplet.* International Series of Monographs on Physics. Oxford University Press, Jun 1 2003.

[14] M. Novello, M. Visser, and G. Volovik. *Artificial Black Holes.* World Scientific, Singapore; River Edge, U.S.A., 2002.

[15] Ralf Schützhold and W. G. Unruh. Quantum analogues: From phase transitions to black holes and cosmology. *Lecture Notes in Physics*, 718:300 p., 2007.

[16] Silke Weinfurtner. Emergent spacetimes. arXiv:0711.4416, 2007.

[17] Silke Weinfurtner, Angela White, and Matt Visser. Trans-Planckian physics and signature change events in Bose gas hydrodynamics. *Physical Review D*, 76:124008, 2007.

[18] Silke Weinfurtner, Stefano Liberati, and Matt Visser. Modelling Planck-scale Lorentz violation via analogue models. *J. Phys. Conf. Ser.*, 33:373–385, 2006.
[19] Stefano Liberati, Matt Visser, and Silke Weinfurtner. Analogue quantum gravity phenomenology from a two-component Bose–Einstein condensate. *Class. Quant. Grav.*, 23:3129–3154, 2006.
[20] Silke Weinfurtner, Stefano Liberati, and Matt Visser. Analogue model for quantum gravity phenomenology. *J. Phys.*, A39:6807–6814, 2006.
[21] Stefano Liberati, Matt Visser, and Silke Weinfurtner. Naturalness in emergent spacetime. *Phys. Rev. Lett.*, 96:151301, 2006.
[22] Angela White, Silke Weinfurtner, and Matt Visser. Signature change events: A challenge for quantum gravity? arXiv:0812.3744, 2008.
[23] Silke Weinfurtner, Piyush Jain, Matt Visser, and C. W. Gardiner. Cosmological particle production in rainbow spacetimes. arXiv:0801.2673v1, 2007.
[24] C. J. Pethick and H. Smith. *Bose–Einstein Condensation in Dilute Gases.* Cambridge University Press, 2001.
[25] Crispin Gardiner and Peter Zoller. *Quantum Noise.* Springer Series in Synergetics. Springer, Berlin, 2nd edition, 2000.
[26] Carlos Barceló, Stefano Liberati, and Matt Visser. Refringence, field theory, and normal modes. *Class. Quant. Grav.*, 19:2961–2982, 2002.
[27] Matt Visser. Acoustic propagation in fluids: An unexpected example of Lorentzian geometry. *gr-qc/9311028*, 1993.
[28] Matt Visser. Acoustic black holes: Horizons, ergospheres, and Hawking radiation. *Class. Quant. Grav.*, 15:1767–1791, 1998.
[29] S. Inouye, M. R. Andrews, J. Stenger, H. J. Miesner, D. M. Stamper-Kurn, and W. Ketterle. Observation of Feshbach resonances in a Bose–Einstein condensate. *Nature*, 392:152–154, 1998.
[30] Elizabeth A. Donley, Neil R. Claussen, Simon L. Cornish, Jacob L. Roberts, Eric A. Cornell, and Carl E. Wieman. Dynamics of collapsing and exploding Bose–Einstein condensates. *Nature*, 412(6844):295–299, 2001.
[31] J. L. Roberts, N. R. Claussen, S. L. Cornish, E. A. Donley, E. A. Cornell, and C. E. Wieman. Controlled collapse of a Bose–Einstein condensate. *Phys. Rev. Lett.*, 86(19):4211–4214, May 2001.
[32] E. A. Calzetta and B. L. Hu. Early universe quantum processes in BEC collapse experiments. *Int. J. Theor. Phys.*, 44:1691–1704, 2005.
[33] E. A. Calzetta and B. L. Hu. Bose–Einstein condensate collapse and dynamical squeezing of vacuum fluctuations. *Phys. Rev. A*, 68(043625), 2003.
[34] B. L. Hu and E. A. Calzetta. BEC collapse, particle production and squeezing of the vacuum. arXiv:cond-mat/0208569, 2002.
[35] Piyush Jain, Silke Weinfurtner, Matt Visser, and Crispin Gardiner. Analogue model of an expanding FRW universe in Bose–Einstein condensates: Application of the classical field method. *Physical Review A*, 76:033616, 2007.

Journal of Physics: Conference Series **189** (2009) 012047 doi:10.1088/1742-6596/189/1/012047

Conical singularities regularized in warped six-dimensional flux compactification and their induced brane cosmology

Vassilios Zamarias

Department of Physics, National Technical University of Athens, Zografou Campus GR 157 73, Athens, Greece
and
Department of Physical Sciences, Hellenic Military Academy, Vari, 16673, Attica, Greece

E-mail: zamarias@central.ntua.gr

Abstract.
We discuss the regularization of codimension-2 singularities in warped six-dimensional Einstein-Maxwell axisymmetric models by replacing them by codimension-1 branes of a ring form, situated around the axis of symmetry. Further we consider the case of one capped regularized conical brane of codimension one and study the cosmological evolution which is induced on it as it moves in between the known *static* bulk and cap solutions. We present the resulting brane Friedmann equation which gives a dominant five-dimensional ρ^2 energy density term at high energies and a term linear to the energy density at low energies with, however, negative coefficient in the small four-brane radius limit (*i.e.*, with negative effective Newton's constant).

1. Introduction

Cosmology in theories with branes embedded in extra dimensions has been the subject of intense investigation during the last years. The most detailed analysis has been done for braneworld models in five-dimensional space [1]. The effect of the extra dimension can modify the cosmological evolution, depending on the model, both at early and late times. The cosmology of this and other related models with one transverse to the brane extra dimension (codimension-1 brane models) is well understood [2, 3, 4, 5, 6, 7, 8] (for a review see [9]).

Braneworld models can also be extended in higher than five-dimensions. We can consider a $(n+1)$ brane embedded in $(n+2)$ spacetime (codimension-1 models) or a $(3+1)$ brane embedded in $(n+1)$ spacetime (codimension-$n-3$ models). Six or higher-dimensional braneworld models of codimension-1 are considered as generalizations of the Randall-Sundrum model [10].

Six-dimensional brane world models are rather interesting to be studied for a number of reasons. Firstly the large (sub-millimeter) extra dimensions proposal for the resolution of the electroweak hierarchy problem was provided [11] in the framework of brane theories with two extra transverse dimensions. Therefore the study of extra dimensional theories, and string theory in particular, become relevant to low energy phenomenology (colliders as well as astrophysical and cosmological observations) and is testable in the very near future. The different possibilities of realizing a brane world model in six dimensions must therefore be studied to allow for a comparison with experiment.

Another attractive motivation has been the proposal to ameliorate the cosmological constant problem, using codimension-2 branes (for a recent review on the subject see [12]). These branes have the interesting property that their vacuum energy instead of curving their world-volume, just introduces a deficit angle in the local geometry [13]. Models with this property which exhibit no fine-tuning between the brane and bulk quantities have been known as self-tuning (for early attempts to find similar models in five dimensions see [14]). Such self-tuning models with flux compactification [15, 16] have been extensively looked, but the flux quantization condition always introduces a fine tuning [17], unless one allows for singularities more severe than conical [18]. Alternative sigma-model compactifications have been shown to satisfy the self-tuning requirements [19]. However, the successful resolution of the cosmological constant problem would also require that there are no fine-tuning between bulk parameters themselves. No such self-tuning model has been found yet with all these properties.

A further motivation in studying such models with codimension-2 branes is that gravity on them is purely understood. The introduction of matter (*i.e.,* anything different from vacuum energy) on them, immediately introduces malicious non-conical singularities [20]. A way out of this problem is to complicate the gravity dynamics by adding a Gauss-Bonnet term in the bulk or an induced curvature term on the brane, in which case the singularity structure of the theory is altered and non-trivial matter is allowed [21]. However, the components of the energy-momentum tensor of the brane and the bulk are tuned artificially and the brane matter is rather restricted [22]. Alternatively, one can regularize the codimension-2 branes by introducing some thickness and then consider matter on them [23]. For example, one can mimic the brane by a six-dimensional vortex (as *e.g.,* in [24]), a procedure which becomes a rather difficult task if matter is added on it.

Even thought we do not fully understand black hole solutions in codimension-2 braneworlds they have been extensively discussed in [25, 26, 27, 33]. Lately a six-dimensional black hole localized on a 3-brane of codimension-2 [26, 27] was proposed. However, it is not clear how to realize these solutions in the thin brane limit where high curvature terms are needed to accommodate matter on the brane. The localization of a black hole on the brane and its extension to the bulk is a difficult task. In codimension-1 braneworlds the first attempt was to consider the Schwarzschild metric and study its black string extension into the bulk [28]. which is unfortunately unstable to classical linear perturbations [29]. Since then, several authors have attempted to find the full metric using numerical techniques [30]. Analytically, the brane metric equations of motion were considered with the only bulk input coming from the projection of the Weyl tensor [31] onto the brane. Since this system is not closed because it contains an unknown bulk dependent term, assumptions have to be made either in the form of the metric or on the Weyl term [32]. In codimension-2 braneworlds, recently black holes on a thin conical brane and their extension into a five and six-dimensional bulk with a Gauss-Bonnet term were found [33].

Moreover we are still lacking an understanding of time dependent cosmological solutions in codimension-2 braneworlds. To have a cosmological evolution we need regularized branes, the brane world-volume should be expanding and in general the bulk space should also evolve in time. This is a formidable task, so we follow an alternative approach [34, 35]. [1]

First for our setup we consider another way of regularization which was recently proposed and consists merely of the reduction of codimensionality of the brane. In this approach, the bulk around the codimension-2 brane is cut close to the conical tip and it is replaced by a codimension-1 brane which is capped by appropriate bulk sections [37] (see [38] for a similar regularization of cosmic strings in flat spacetime). This regularization has been applied to flux compactification systems in six dimensions for unwarped "rugby-ball"-like solutions in [37]. The case of warped solutions with conical branes (with or without supersymmetry) [39] and even more general warped solutions allowing non-conical branes [40] have been studied. More precisely we consider the case of one capped regularized conical brane of codimension one and give the *static* bulk and cap solutions.

Then we study the motion of the regularized codimension-1 brane in the space between the bulk and the brane-cap which remains static (see *e.g.,* [41]). In this way, a cosmological evolution will be induced on the brane in a similar way as in the mirage cosmology [42], but with the inclusion of the back-reaction of the brane energy density (*i.e.,* the brane is not considered merely a probe one). Since in the mirage cosmology, the four-dimensional scale factor descends from the warp factor in the four-dimensional part of the bulk metric, we will discuss the regularized brane in the case of warped bulk [39], rather than unwarped bulk [37]. It is worth noting that the above procedure provided in five dimensions the most general isotropic brane cosmological solutions [3].

We find the Friedmann equation on the brane by solving the Israel junction conditions, which play the rôle of the equations of motion of the codimension-1 brane. We find that at early times cosmology is dominated by an energy density term proportional to ρ^2, like in the Rundall-Sundrum model in five dimensions. However, at late times where the brane moves close to its equilibrium point, which in turn is close to the would-be conical singularity, the coefficient of the linear to the energy density term is negative (*i.e.,* we obtain negative effective Newton's constant). Thus, we cannot recover the standard four dimensional cosmology at late times. This seems to be the consequence of considering the bulk sections static. It is possible, that this behaviour is due to a ghost mode appearing among the perturbations of the system, after imposing the staticity of the bulk sections. Furthermore, the above result points out that there is a difference between the six-dimensional brane cosmology in comparison to the five-dimensional one. The study of brane cosmology in Einstein gravity in five dimensions, can be made either in a gauge where the bulk is time-dependent and the brane lies at a fixed position, or in a gauge where the bulk is static and the brane movement into the bulk induces a cosmological evolution on it [43]. This, however, does not seem to hold in six-dimensions anymore.

The talk is organized as follows. In Sec. 2 we present the static regularized brane solution in a bulk of general warping. In Sec. 3 we derive the equations of motion of the codimension-1 brane and in Sec. 4 we study the induced cosmological evolution on the moving brane. Finally in Sec. 5 we draw our conclusions.

2. Regularized static brane solutions

In this section we will briefly give the necessary results of the static solution which we will need in the following for the brane motion. The bulk theory that we use is a six-dimensional

[1] Recently, some works have studied the cosmological evolution of codimension-1 or 2 branes in six dimensional models [36]

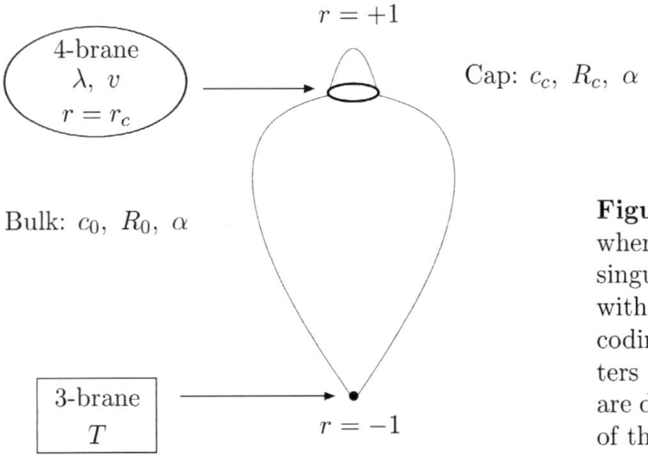

Figure 1. The internal space where the upper codimension-2 singularity has been regularized with the introduction of a ring-like codimension-1 brane. The parameters of the action and the solution are denoted in the appropriate part of the internal space.

Einstein-Maxwell system which in the presence of a positive cosmological constant and a gauge flux, spontaneously compactifies the internal space [44]. The known axisymmetric solutions have in general two codimension-2 singularities at the poles of a deformed sphere [45]. We study the case where only one (*e.g.*, the upper) codimension-2 brane is regularized by the introduction of a ring-like brane at $r = r_c$ with an appropriate cap. The dynamics of the system is given by the following action

$$S = \int d^6x\sqrt{-g}\left(\frac{M^4}{2}R - \Lambda_i - \frac{1}{4}\mathcal{F}^2\right) - \int d^5x\sqrt{-\gamma_+}\left(\lambda + \frac{v^2}{2}(\tilde{D}_{\hat{\mu}}\sigma)^2\right) - \int d^4x\sqrt{-\gamma_-}\,T\ , \quad (1)$$

where M is the six-dimensional fundamental Planck mass, Λ_i are the bulk ($i = 0$) and cap ($i = c$) cosmological constants, $\mathcal{F}_{MN}$ the gauge field strength, T the tension of the lower codimension-2 brane, λ the 4-brane tension, σ the 4-brane Goldstone scalar field necessary for the regularization and v the vev of the Higgs field from which the Goldstone field originates. For the coupling between the Goldstone field and the bulk gauge field we use the notation $\tilde{D}_{\hat{\mu}}\sigma = \partial_{\hat{\mu}}\sigma - e\, a_{\hat{\mu}}$, with $a_{\hat{\mu}} = \mathcal{A}_M\partial_{\hat{\mu}}X^M$ the pullback of the gauge field on the ring-like brane and e its coupling to the scalar field. In the above action we omitted the Gibbons-Hawking term. The configuration is shown in more detail in Fig.1.

The solution for the bulk and cap regions depends on a parameter α which is a measure of the warping of the space (for $\alpha = 1$ we obtain the unwarped case) and is given by [39]

$$ds_6^2 = z^2\eta_{\mu\nu}dx^\mu dx^\nu + R_i^2\left[\frac{dr^2}{f} + c_i^2 f\ d\varphi^2\right]\ , \quad (2)$$

$$\mathcal{F}_{r\varphi} = -c_i R_i M^2 S\cdot\frac{1}{z^4}\ , \quad (3)$$

with $R_i^2 = M^4/(2\Lambda_i)$ and the following bulk functions

$$z(r) = \frac{1}{2}[(1-\alpha)r + (1+\alpha)] \quad (4)$$

$$f(r) = \frac{1}{5(1-\alpha)^2}\left[-z^2 + \frac{1-\alpha^8}{1-\alpha^3}\cdot\frac{1}{z^3} - \alpha^3\frac{1-\alpha^5}{1-\alpha^3}\cdot\frac{1}{z^6}\right]\ , \quad (5)$$

with $S(\alpha) = \sqrt{\frac{3}{5}\alpha^3\frac{1-\alpha^5}{1-\alpha^3}}$. The range of the internal space coordinates is $-1 \leq r \leq 1$ and $0 \leq \varphi < 2\pi$. Taking into account that in the limit $r \to \pm 1$, it is $f \to 2(1 \mp r)X_\pm$ with the constants $X_\pm$ given by

$$X_+ = \frac{5+3\alpha^8-8\alpha^3}{20(1-\alpha)(1-\alpha^3)} \quad , \quad X_- = \frac{3+5\alpha^8-8\alpha^5}{20\alpha^4(1-\alpha)(1-\alpha^3)} \, , \tag{6}$$

the cap is smooth at $r = +1$ as long as is $c_c = 1/X_+$. Furthermore, the metric is continuous if $c_0 R_0 = c_c R_c$, which gives $R_c = \beta_+ R_0$ with $\beta_+ = X_+ c_0$ [2]. The conical singularity at $r = -1$ is supported by a codimension-2 brane with tension

$$T = 2\pi M^4(1 - c_0 X_-), \tag{7}$$

while the parameters of the 4-brane λ, v are fixed by the radii R_0, R_c and the brane position r_c [39]. Finally, the gauge field is quantized as

$$2c_0 R_0 M^2 e \, Y = N \quad , \quad N \in \mathbb{Z} \, , \tag{8}$$

with $Y = \frac{(1-\alpha^3)}{3\alpha^3(1-\alpha)} S$ and the brane scalar field has solution $\sigma = n\varphi$ with $n \in \mathbb{Z}$. The two quantum numbers n, N are related through the junction conditions as

$$n = \frac{N}{2}\frac{2}{(1-\alpha^3)}\left[\frac{5(1-\alpha^8)}{8(1-\alpha^5)} - \alpha^3\right] \, . \tag{9}$$

Since the quantities n, N are integers, the above relation imposes a restriction to the values of the admissible warpings α, which implies that static solutions are consistent only for discrete values of the warping α.

3. The cosmological dynamics of the 4-brane

To study the cosmological evolution on the brane we include matter on it which in general makes both the brane and the bulk to evolve in time. Unfortunately we do not know any time dependent solution for such setup. Therefore we will approach the problem in a simplest way, by making the approximation that the bulk remains *static* and that the brane matter merely makes the brane to move between the static bulk and the static cup, away from its equilibrium point $r = r_c$, thus obtaining cosmology on the brane, as in the mirage cosmology approach [42]. This is to be regarded as a first step towards understanding the generic brane cosmological evolution. In our case the brane is not merely a probe brane and we will use the junction conditions to derive the induced cosmology, thus taking into account the back-reaction of the brane energy density.

Firstly, to embed the brane in the static bulk, let us take the brane coordinates be $\sigma^{\hat{\mu}} = (\sigma, x^i, \varphi)$. [The brane-time coordinate σ is not to be confused with the Goldstone field σ which will not appear in our subsequent analysis.] Then the brane embedding X^M in the bulk is taken for both sections to be

$$X^i = x^i \quad , \quad X^r = \mathcal{R}(\sigma) \quad \text{and} \quad X^\varphi = \varphi \, , \tag{10}$$

[2] In this brief presentation of the background, we have taken $\xi = 1$ in comparison with [39]. The physical quantities, however, are ξ-independent and depend only on β_+.

while for the time coordinate embedding we choose for the outer bulk section

$$X^0_{(out)} = \sigma \ , \tag{11}$$

and for the inner cap section

$$X^0_{(in)} = T(\sigma) \ . \tag{12}$$

The continuity of the induced metric $\gamma_{\hat{\mu}\hat{\nu}} = g_{MN}\partial_{\hat{\mu}}X^M\partial_{\hat{\nu}}X^N$, apart from the relation $c_0R_0 = c_cR_c$ as in the static case, gives a relation of the time coordinate T in the upper cap region with the brane time coordinate σ (dots are with respect to σ)

$$\dot{T}^2\left(1-\beta_+^2\frac{\dot{\mathcal{R}}^2}{\dot{T}^2}\frac{R_0^2}{fz^2}\right)=\left(1-\dot{\mathcal{R}}^2\frac{R_0^2}{fz^2}\right) \ . \tag{13}$$

Then the induced metric $\gamma_{\hat{\mu}\hat{\nu}}$ on the brane reads

$$ds^2_{(5)} = -z^2\left(1-\dot{\mathcal{R}}^2\frac{R_0^2}{fz^2}\right)d\sigma^2 + z^2d\vec{x}^2 + c_0^2R_0^2fd\varphi^2 \ . \tag{14}$$

The continuity of the gauge field, on the other hand, is guaranteed by the fact that its only non-vanishing component is A_φ and X^φ is σ-independent.

Secondly, we introduce an energy momentum tensor of a perfect fluid on the brane $t_{\hat{\mu}}^{\hat{\nu}\,(br)} = -(2/\sqrt{-\gamma_+})\delta S_{br}/\delta\gamma_+{}^{\mu}{}_{\nu} = \mathrm{diag}(-\rho, P, P, P, \hat{P})$ (where S_{br} is the brane action). We also consider a possible coupling of the brane matter to the bulk gauge field (consistent with the cosmological symmetries) by $\delta S_{br}/\delta a^{\hat{\kappa}} = (l, L, L, L, \hat{L})$. Splitting the above quantities to one part responsible for the static solution and another expressing the presence of matter on the brane, we have

$$\begin{aligned}
\rho &= \lambda + \frac{v^2(n-e\mathcal{A}_\varphi^+)^2}{2c_0^2R_0^2f(r_c)} + \rho_m \equiv \rho_0 + \rho_m, && (15)\\
P &= -\lambda - \frac{v^2(n-e\mathcal{A}_\varphi^+)^2}{2c_0^2R_0^2f(r_c)} + P_m \equiv -\rho_0 + P_m, && (16)\\
\hat{P} &= -\lambda + \frac{v^2(n-e\mathcal{A}_\varphi^+)^2}{2c_0^2R_0^2f(r_c)} + \hat{P}_m, && (17)\\
l &= l_m, && (18)\\
L &= L_m, && (19)\\
\hat{L} &= ev^2(n-e\mathcal{A}_\varphi^+) + \hat{L}_m, && (20)
\end{aligned}$$

with all the other quantities vanishing and ρ_m, P_m, $\hat{P}_m$, l_m, L_m, $\hat{L}_m$ the matter contributions.

In the spirit of mirage cosmology approach [42] the brane matter merely makes the brane to move in the static bulk. As we have already stated this movement induces cosmology on the brane and warping of the bulk is necessary. Therefore it is convenient to rewrite the metric in the form

$$ds^2_{(5)} = -d\tau^2 + a^2(\tau)d\vec{x}^2 + b^2(\tau)d\varphi^2 \ , \tag{21}$$

with $a = z(\mathcal{R}(\tau))$ and $b = c_0 R_0 \sqrt{f(\mathcal{R}(\tau))}$. The brane proper time is given by

$$\dot{\tau}^2 = z^2 \left(1 - \dot{\mathcal{R}}^2 \frac{R_0^2}{f z^2}\right) . \tag{22}$$

From now on we will assume without loss of generality that $\dot{\tau} > 0$. It is evident from the above, that cosmological evolution from mere motion of the brane in the static bulk is possible only when there is warping in the bulk. In the unwarped version of this model [37], mirage cosmology is impossible and some bulk time-dependence is compulsory.

The Hubble parameters for the two scale factors are given by

$$H_a \equiv \frac{1}{a}\frac{da}{d\tau} = \frac{z'}{z^2} \frac{\dot{\mathcal{R}}}{\sqrt{1 - \dot{\mathcal{R}}^2 \frac{R_0^2}{f z^2}}} , \tag{23}$$

and

$$H_b \equiv \frac{1}{b}\frac{db}{d\tau} = \frac{f'}{2fz} \frac{\dot{\mathcal{R}}}{\sqrt{1 - \dot{\mathcal{R}}^2 \frac{R_0^2}{f z^2}}} . \tag{24}$$

Then the ratio of H_a and H_b gives a precise relation between them for our particular model. It is given by

$$H_b = \frac{z f'}{2 f z'} H_a , \tag{25}$$

and we notice that since in the model we study it is always $f' < 0$, in the neighborhood of $r = 1$, the two Hubble parameters have opposite sign. This means that if the four dimensional space expands, the internal space shrinks.

Apart from the continuity conditions we have to take into account the junction conditions for the derivatives of the metric and the gauge field, which read

$$\{\hat{K}_{\hat{\mu}\hat{\nu}}\} = -\frac{1}{M^4} t^{(br)}_{\hat{\mu}\hat{\nu}} , \tag{26}$$

$$\{n_M \mathcal{F}^M_{\ N} \partial_{\hat{\kappa}} X^N\} = -\frac{\delta S_{br}}{\delta a^{\hat{\kappa}}} . \tag{27}$$

We denote $\{H\} = H^{in} + H^{out}$ the sum of the quantity H from each side of each brane. The extrinsic curvatures are constructed using the normal to the brane n_M which points *inwards to the corresponding part of the bulk* each time (we use the conventions of [46]). [3].

So far we have 6 parameters: $\rho, P, \hat{P}, l, L, \hat{L}$ and with the use of the continuity relations of the induced metric, $c_0 R_0 = c_c R_c$ and (13), the (t) and (i) components of the junction conditions for the gauge field give for the coupling: $l = L = 0$, the (φ) component of the gauge field junction and the $(\varphi\varphi)$ component of the metric junction determine $\hat{P}$ and $\hat{L}$. Finally the two remaining metric junction conditions ($(\sigma\sigma)$ and (ij) components) give the Friedmann equation and the acceleration equation respectively.

The obtained Friedmann equation is

$$H_a^2 = \frac{1}{4M^8 \mathcal{C}^2 \mathcal{A}} \rho^2 + \frac{M^8 \mathcal{C}^2 (1-\beta_+^2)^2}{4R_0^4 \beta_+^4 \mathcal{A}} \cdot \frac{1}{\rho^2} - \frac{1+\beta_+^2}{2\beta_+^2 R_0^2 \mathcal{A}} , \tag{28}$$

[3] The left hand sides of the above equations are computed in detail in the Appendix A of [34]

where $\mathcal{A} = \frac{z^2}{fz'^2}$ and $\mathcal{C} = \sqrt{f}\left(3\frac{z'}{z} + \frac{f'}{2f}\right)$, both evaluated on the brane. It has unconventional dependence on the energy density, in particular the inverse square dependence is known to occur for motions in backgrounds of asymmetrical warping [47].

The equilibrium point r_c of the system is found if we set $H_a = 0$, which gives the brane energy density without matter

$$\rho_0 = -\frac{M^4 \mathcal{C}_c}{R_0 \beta_+}(1 - \beta_+) , \tag{29}$$

where $\mathcal{C}_c$ is the value of $\mathcal{C}$ at r_c and is the same appearing in (15). The behaviour of the function $\mathcal{C}$ is that, as $r \to -1$ it limits to $\mathcal{C} \to \infty$ and monotonically decreases and limits to $\mathcal{C} \to -\infty$ as $r \to 1$. On the other hand $\mathcal{A}$, is always positive with $\mathcal{A} \to \infty$ as $r \to \pm 1$, and $\mathcal{O}(1)$ in the intermediate region. In the unwarped limit $\alpha \to 1$ the Friedmann equation becomes as expected trivial, *i.e.*, $H_a = 0$.

Before studying various limits of the above equation, let us define the effective four dimensional matter energy density $\rho_m^{(4)}$ by averaging over the azimuthal direction (we assume that ρ_m is independent of φ)

$$\rho_m^{(4)} = \int d\varphi \sqrt{g_{\varphi\varphi}} \rho_m = \frac{2\pi\beta_+}{X_+} R_0 \sqrt{f} \rho_m , \tag{30}$$

with similar definitions for the other 4-brane quantities.

Let us suppose now that initially $-1 \ll \mathcal{R}(\sigma) < r_c < 1$, with $1 - r_c \ll 1$. The goal is to find how $\mathcal{R}(\sigma)$ behaves. To recover a four-dimensional Friedmann equation at late times we can assume that the brane energy density is small in comparison with the static case energy density, *i.e.*, $\rho_m^{(4)} \ll \rho_0$, so we can expand (28) in powers of $\rho_m^{(4)}$ and obtain the following four dimensional form of the Friedmann equation

$$H_a^2 = \frac{8\pi}{3} G_{eff} \rho_m^{(4)} + \Delta(a) + \mathcal{O}(\rho_m^{(4)\,2}) , \tag{31}$$

where the effective Newton's constant is

$$G_{eff} = \frac{3X_+ \mathcal{C}_c (1 - \beta_+)}{32\pi^2 R_0^2 \beta_+^2 M^4 \mathcal{A}\, \mathcal{C}^2 \sqrt{f}} \left[\frac{\mathcal{C}^4}{\mathcal{C}_c^4} \left(\frac{1 + \beta_+}{1 - \beta_+} \right)^2 - 1 \right] . \tag{32}$$

The quantity $\Delta(a)$ depends on the parameters of the bulk and plays the rôle of the the mirage matter induced on the brane from the bulk and it is given by

$$\Delta(a) = \frac{(1 - \beta_+)^2}{4R_0^2 \beta_+^2 \mathcal{A}} \left[\frac{\mathcal{C}_c^2}{\mathcal{C}^2} + \frac{\mathcal{C}^2}{\mathcal{C}_c^2} \left(\frac{1 + \beta_+}{1 - \beta_+} \right)^2 - 2\frac{1 + \beta_+^2}{(1 - \beta_+)^2} \right] . \tag{33}$$

The behaviour of G_{eff} as a function of $\mathcal{R}(\sigma)$ is given in Fig. 2 and has the following important features: At the points where the geometry becomes conical ($r \to \pm 1$) the effective Newton's constant is negative and diverging. In between, there is a point r_d, with

$$z_d = \left(\frac{3(1 - \alpha^8)}{8(1 - \alpha^3)} \right)^{1/5} , \tag{34}$$

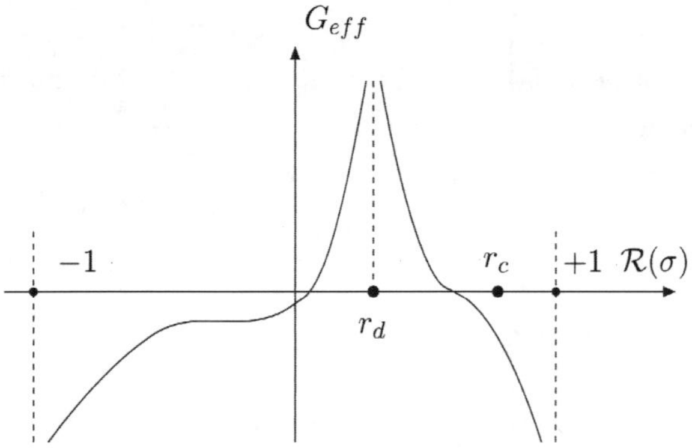

Figure 2. The generic form of the effective Newton's constant G_{eff} as a function of the brane position $\mathcal{R}(\sigma)$. As the brane approaches the equilibrium point r_c, we always have $G_{eff} < 0$. Additionally, G_{eff} diverges as $r \to \pm 1$ and at one point r_d in between.

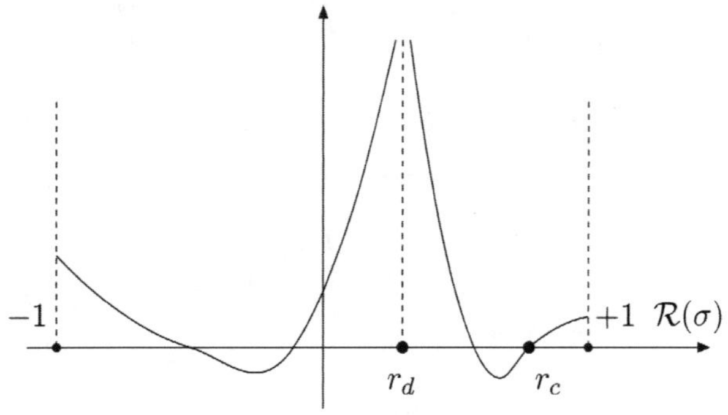

Figure 3. The generic form of the mirage matter contribution Δ as a function of the brane position $\mathcal{R}(\sigma)$. At the static equilibrium point r_c it vanishes and at r_d it diverges.

which is a root of $\mathcal{C}$ and G_{eff} diverges to $+\infty$. At this point the matter energy density is bound to vanish. It is important to note that even in the region where G_{eff} is positive, there is always a strong time variation of G_{eff}, which for

$$\frac{1}{G_{eff}} \frac{dG_{eff}}{d\tau} = H_a \delta \ , \tag{35}$$

has $\delta > \mathcal{O}(10)$, in contradiction with observations [48] which dictate that $\delta < 0.1$. But even more important is the fact that close to the static equilibrium point, where the cosmology is supposed to mimic best the one of a codimension-2 brane, we get *negative* Newton's constant

$$G_{eff}(\mathcal{R} = r_c) = \frac{3X_+}{8\pi^2 R_0^2 \beta_+ (1 - \beta_+) M^4 \mathcal{A}\ \mathcal{C}_c \sqrt{f}} < 0 \ , \tag{36}$$

since we have that for any value of the parameters and for r_c in the neighborhood of $r = +1$, it is $\mathcal{C}_c < 0$.

Let us also look at the mirage matter contribution Δ, which is depicted in Fig. 3. As expected, it vanishes for the static equilibrium point r_c, it is finite at the boundaries $r = \pm 1$ and diverges at the root r_d of $\mathcal{C}$. Again there is no region in the brane position interval where the contribution of Δ is constant enough to resemble a cosmological constant contribution to the four dimensional Friedmann equation.

This result is not altered by supersymmetrizing the model [16]. The bulk and cap solutions are different from the non-supersymmetric case due to the presence of the dilaton. However, the only difference in the Friedmann is a redefinition of the quantities $\mathcal{A}$ and $\mathcal{C}$. In the supersymmetric case, these quantities read

$$\mathcal{A}_{susy} = 4\frac{z^2}{fz'^2} \quad \text{and} \quad \mathcal{C}_{susy} = \sqrt{f}\left(\frac{3z'}{2z} + \frac{f'}{2f}\right) . \tag{37}$$

It is easy to see again that the effective Newton's constant is negative for the motion near the would-be conical tip and that even away from that point, it is very strongly varying.

On the opposite limit, that the matter energy density is much larger than the static case energy density, *i.e.*, $\rho_m^{(4)} \gg \rho_0$, we get the expected asymptotics

$$H_a^2 = \frac{1}{4M^8\mathcal{C}^2\mathcal{A}}\,\rho_m^2 , \tag{38}$$

which is a five-dimensional Friedmann law (with time-varying five-dimensional Newton's constant) at early times.

Taking under consideration the difficulties of the model to give a Fiedmann equation with the correct sign of G_{eff}, we will not proceed with the presentation of the analysis of the acceleration equation. With this equation, one finds even more difficulties towards obtaining a realistic four-dimensional evolution. For example, in the low energy limit, one gets a coefficient of the linear energy density term, which is not related to the G_{eff}, that we obtained from the Friedmann equation, in the way it does in standard four-dimensional General Relativity.

This pathological features of the low energy density limit, where the expansion does not appear to have an effective four-dimensional limit, can find a potential explanation when looking at the energy continuity equation on the brane. Taking the covariant divergence of the Israel junction condition (26), together with the Codazzi equation $\nabla^{(4)}_{\hat\mu}\hat K^{\hat\mu}_{\hat\nu} = G_{K\Lambda}n^\Lambda h^K_N\partial_{\hat\nu}X^N$ and the bulk Einstein equation, we arrive at the following simple equation [46]

$$\nabla^{(4)}_{\hat\mu}t^{\hat\mu\ (br)}_{\hat\nu} = -\{T^{(B)}_{K\Lambda}h^K_N n^\Lambda\partial_{\hat\nu}X^N\} . \tag{39}$$

Because of the jump of the bulk energy momentum tensor across the 4-brane, the energy-momentum tensor on the 4-brane is not conserved. This is a usual feature of moving brane cosmologies in asymmetrically warped backgrounds [49]. In more detail, the form of the above equation for the particular model is given by

$$\frac{d\rho}{d\tau} + 3(\rho+P)H_a + (\rho+\hat P)\frac{zf'}{2fz'}H_a = -\frac{S^2\rho H_a}{z'z^7\mathcal{C}\sqrt{f}} , \tag{40}$$

where in the right hand side we have used the Friedmann equation (28). In a straightforward but lengthy calculation, one can see that the latter equation can be derived from the $(\sigma\sigma)$, (ij) and $(\varphi\varphi)$ components of the junction conditions of (26). The problems in the four-dimensional limit, that we faced previously, can be traced to large energy dissipation off the brane as well as a large work done during the contraction of the ring-brane.

4. Conclusions

In the present presentation we made a first step towards the study of the cosmology of a codimension-2 brane. We regularized the codimension-2 singularities by the method of lowering its codimensionality. We cut the space close to the conical tip and replaced it by a ring brane with an appropriate cap. Unfortunately we do not know any time dependent solution for such setup. Therefore we will approach the problem in a simplest way, we assumed that the bulk and the cap remain static as the brane moves between them. The motion of the brane then induces a cosmological evolution for the matter on the brane. The junction conditions provide the Friedmann and acceleration equations on the brane.

From the Friedmann equation we can see that we cannot recover standard cosmological evolution of the brane at low energies. The effective Newton's constant is negative in the interesting limit that the brane approaches its equilibrium point, close to the would-be conical singularity. In other words, we obtain antigravity in this limit. Even away from this point, *i.e.,* when the brane moves away from its equilibrium point, the Newton's constant varies significantly in contradiction with standard cosmology. At one position of the internal space, the Newton's constant even diverges and forces the matter energy density to vanish. Taking all the above into account, we did not present the further analysis of the system by considering the acceleration equation.

The reason for this unconventional cosmological evolution, is the use of the specific restricted ansatz for the solution of the system's equation of motion. The staticity of the bulk was proved to be an oversimplification. We imagine that this restriction on the system may result to the appearance of some scalar mode in the perturbative analysis of [50], which for a certain region of the brane motion (close to the pole of the internal manifold) is ghost-like. This mode may then be responsible for the negative effective Newton's constant. The unrestricted perturbative analysis in [50] resulted in a linearized four-dimensional Einstein equation for distances larger than the compactification scale. The same happened in [37] in the unwarped model perturbation analysis[4] (see [51] for a related analysis with a brane induced gravity term), even though it would not have mirage evolution, as we studied it here, because of the absence of a warp factor. The gradient expansion technique was recently used in similar six dimensional models [52] to derive a low energy effective theory on the regularized brane and show that standard four-dimensional Einstein gravity is recovered at low energies. Consequently, this approach could be used in our case to derive the low energy effective theory [53]. The appearance of the standard four-dimensional linearized dynamics, shows that they are realized when the bulk is necessarily time-dependent.

Clearly, the next step should be the study of the system in a setup where the bulk is also time-dependent. In that respect, brane cosmology in six dimensions seems to be different for the one in five dimensions. In Einstein gravity in five dimensions, one can always work in a gauge where the bulk is static and the brane acquires a cosmological evolution by moving into the bulk. However, in six dimensions there are more degrees of freedom which make this gauge choice not general. In the present paper, we have frozen these degrees of freedom hoping to find consistent cosmological solutions, but it turned out that this does not give a viable cosmology. There are several known time-dependent backgrounds which can be used to look for realistic brane cosmologies [54]. We plan to address this issue in the near future.

Acknowledgments

This talk is based on the works done in collaboration with E. Papantonopoulos and A. Papazoglou. We thank the organizers of the 13th NEB Conference, held in Thessaloniki,

[4] The question of stability of the regularized models was not discussed in [37, 50] and it could be that these compactifications have modes with negative mass squared.

for their kind hospitality and perfect organization. This work was supported by the Hellenic Army Academy and the NTUA research program PEVE07.

References

[1] Randall L and Sundrum R 1999 *Phys. Rev. Lett.* **83** 3370, 1999 *Phys. Rev. Lett.* **83** 4690, Binetruy P, Deffayet C and Langlois D 2000 *Nucl. Phys.* B **565** 269, Binetruy P, Deffayet C, Ellwanger U and Langlois D 2000 *Phys. Lett.* B **477** 285, Csaki C, Graesser M, Randall L and Terning J 2000 *Phys. Rev.* D **62** 045015, Kanti P, Kogan I I, Olive K A and Pospelov M 2000 *Phys. Rev.* D **61** 106004, Maartens R 2000 *Phys. Rev.* D **62** 084023, Bowcock P, Charmousis C and Gregory R 2000 *Class. Quant. Grav.* **17** 4745

[2] Kaloper N 1999 *Phys. Rev.* D **60** 123506, Csaki C, Graesser M, Kolda C F and Terning J 1999 *Phys. Lett.* B **462** 34, Cline J M, Grojean C and Servant G 1999 *Phys. Rev. Lett.* **83** 4245, Kanti P, Kogan I I, Olive K A and Pospelov M 1999 *Phys. Lett.* B **468** 31, Flanagan E, Tye S H H and Wasserman I 2000 *Phys. Rev.* D **62** 044039, Charmousis C 2002 *Class. Quant. Grav.* **19** 83

[3] Kraus P 1999 *JHEP* **9912** 011 Ida D 2000 *JHEP* **0009** 014

[4] Collins H and Holdom B 2000 *Phys. Rev.* D **62** 105009, Dvali G R, Gabadadze G and Porrati M 2000 *Phys. Lett.* B **485** 208, Dvali G R and Gabadadze G 2001 *Phys. Rev.* D **63** 065007

[5] Shtanov Y V 2000 *Preprint* hep-th/0005193, Kofinas G 2001 *JHEP* **0108** 034 Kim N J, Lee H W and Myung Y S 2001 *Phys. Lett.* B **504** 323, Deffayet C, Dvali G R and Gabadadze G 2002 *Phys. Rev.* D **65** 044023,

[6] Zwiebach B 1985 *Phys. Lett.* B **156** 315, Zumino B 1986 *Phys. Rept.* **137** 109, Gross D J and Sloan J H 1987 *Nucl. Phys.* B **291** 41

[7] Kim J E, Kyae B and Lee H M 2000 *Nucl. Phys.* B **582** 296, 2000 *Erratum-ibid.* B **591** 587, Charmousis C and Dufaux J F 2002*Class. Quant. Grav.* **19** 4671, Mavromatos N E and Rizos J 2000 *Phys. Rev.* D **62** 124004, 2003 *Int. J. Mod. Phys.* A **18** 57, 2003 *Phys. Rev.* D **67** 024030, Gravanis E and Willison S 2003 *Phys. Lett.* B **562** 118, Mavromatos N E and Papantonopoulos E 2006 *Phys. Rev.* D **73** 026001

[8] Kofinas G, Maartens R and Papantonopoulos E 2003 *JHEP* **0310** 066, Brown R A, Maartens R, Papantonopoulos E and Zamarias V 2005 *JCAP* **0511** 008

[9] Papantonopoulos E 2004 *Preprint gr-qc/0410032*, Gunther U and Zhuk A 2004 *Preprint gr-qc/0410130*, Langlois D 2004 *Preprint gr-qc/0410129.* Maartens R 2004 *Living Rev. Rel.* **7** 7

[10] Kanti P, Madden R and Olive K A 2001 *Phys. Rev.* D **64** 044021, Arkani-Hamed N, Dimopoulos S, Kaloper N and March-Russell J 2000 *Nucl. Phys.* B **567** 189

[11] Arkani-Hamed N, Dimopoulos S and Dvali G R 1998 *Phys. Lett.* B **429** 263 Antoniadis I, Arkani-Hamed N, Dimopoulos S and Dvali G R 1998 *Phys. Lett.* B **436** 257 Arkani-Hamed N, Dimopoulos S and Dvali G R 1999 *Phys. Rev.* D **59** 086004

[12] Koyama K 2008 *Gen. Rel. Grav.* **40** 421

[13] Chen J W, Luty M A and Ponton E 2000 *JHEP* **0009** 012

[14] Arkani-Hamed N, Dimopoulos S, Kaloper N and Sundrum R 2000 *Phys. Lett.* B **480** 193, Kachru S, Schulz M B and Silverstein E 2000 *Phys. Rev.* D **62** 045021, Forste S, Lalak Z, Lavignac S and Nilles H P 2000 *Phys. Lett.* B **481** 360, Forste S, Lalak Z, Lavignac S and Nilles H P 2000 *JHEP* **0009** 034, Csaki C, Erlich J and Grojean C 2001 *Nucl. Phys.* B **604** 312, Kim J E, Kyae B and Lee H M 2001 *Nucl. Phys.* B **613** 306,

[15] Carroll S M and Guica M M 2003 *Preprint* hep-th/0302067, Navarro I 2003 *JCAP* **0309** 004, Aghababaie Y, Burgess C P, Parameswaran S L and Quevedo F 2004 *Nucl. Phys.* B **680** 389

[16] Gibbons G W, Guven R and Pope C N 2004 *Phys. Lett.* B **595** 498, Burgess C P, Quevedo F, Tasinato G and Zavala I 2004 *JHEP* **0411** 069

[17] Navarro I 2003 *Class. Quant. Grav.* **20** 3603, Nilles H P, Papazoglou A and Tasinato G 2004 *Nucl. Phys.* B **677** 405, Lee H M 2004 *Phys. Lett.* B **587** 117

[18] Burgess C P 2005 *AIP Conf. Proc.* **743** 417, Kaloper N and Kiley D 2006 *JHEP* **0603** 077

[19] Kehagias A 2004 *Phys. Lett.* B **600** 133, Randjbar-Daemi S and Rubakov V A 2004 *JHEP* **0410** 054, Lee H M and Papazoglou A 2005 *Nucl. Phys.* B **705** 152, Kehagias A and Mattheopoulou C 2008 *Nucl. Phys.* B **797** 117, Bogdanos C, Kehagias A and Tamvakis K 2007 *Phys. Lett.* B **656** 112

[20] Cline J M, Descheneau J, Giovannini M and Vinet J 2003 *JHEP* **0306** 048

[21] Bostock P, Gregory R, Navarro I and Santiago J 2004 *Phys. Rev. Lett.* **92** 221601, Charmousis C and Zegers R 2005 *Phys. Rev.* D **72** 064005
[22] Papantonopoulos E and Papazoglou A 2005 *JCAP* **0507** 004, Kofinas G 2006 *Phys. Lett.* B **633** 141, Papantonopoulos E and Papazoglou A 2005 *JHEP* **0509** 012
[23] Carter B, Battye R A and Uzan J P 2003 *Commun. Math. Phys.* **235** 289, Kolanovic M, Porrati M and Rombouts J W 2003 *Phys. Rev.* D **68** 064018, Kanno S and Soda J 2004 *JCAP* **0407** 002, Vinet J and Cline J M 2004 *Phys. Rev.* D **70** 083514, Navarro I and Santiago J 2005 *JHEP* **0502** 007, Vinet J and Cline J M 2005 *Phys. Rev.* D **71** 064011, Lee H M and Papazoglou A 2008 *JHEP* **0801** 008
[24] Peter P, Ringeval C and Uzan J P 2005 *Phys. Rev.* D **71** 104018
[25] Charmousis C and Papazoglou A 2008 *JHEP* **0807** 062
[26] Kaloper N and Kiley D 2006 *JHEP* **0603** 077
[27] Kiley D 2007 *Phys. Rev.* D **76** 126002
[28] Chamblin A, Hawking S W and Reall H S 2000 *Phys. Rev.* D **61** 065007
[29] Gregory R 2000 *Class. Quant. Grav.* **17** L125, Gregory R and Laflamme R 1993 *Phys. Rev. Lett.* **70** 2837
[30] Shiromizu T and Shibata M 2000 *Phys. Rev.* D **62** 127502, Wiseman T 2002 *Phys. Rev.* D **65** 124007
[31] Shiromizu T, Maeda K I and Sasaki M 2000 *Phys. Rev.* D **62** 024012, Maartens R 2001 *Preprint* gr-qc/0101059
[32] Dadhich N, Maartens R, Papadopoulos P and Rezania V 2000 *Phys. Lett.* B **487** 1, Bruni M, Germani C and Maartens R 2001 *Phys. Rev. Lett.* **87** 231302, Casadio R, Fabbri A and Mazzacurati L 2002 *Phys. Rev.* D **65** 084040, Kofinas G, Papantonopoulos E and Pappa I 2002 *Phys. Rev.* D **66** 104014, Kofinas G, Papantonopoulos E and Zamarias V 2002 *Phys. Rev.* D **66** 104028, Kanti P and Tamvakis K 2002 *Phys. Rev.* D **65** 084010, Ovalle J 2007 *Preprint* gr-qc/0703095
[33] Cuadros-Melgar B, Papantonopoulos E, Tsoukalas M and Zamarias V 2008 *Phys. Rev. Lett.* **100** 221601 Cuadros-Melgar B, Papantonopoulos E, Tsoukalas M and Zamarias V 2009 *Nucl. Phys.* B **810** 246
[34] Papantonopoulos E, Papazoglou A and Zamarias V 2008 *Nucl. Phys.* B **797** 520
[35] Minamitsuji M and Langlois D 2007 *Phys. Rev.* D **76** 084031
[36] Minamitsuji M 2008 *Phys. Lett.* B **666** 404, Corradini O, Koyama K and Tasinato G 2008 *Phys. Rev.* D **78** 124002, Chen F, Cline J M and Kanno S 2008 *Phys. Rev.* D **77** 063531
[37] Peloso M, Sorbo L and Tasinato G 2006 *Phys. Rev.* D **73** 104025
[38] Gott J R I 1985 *Astrophys. J.* **288** 422
[39] Papantonopoulos E, Papazoglou A and Zamarias V 2007 *JHEP* **0703** 002.
[40] Tolley A J, Burgess C P, de Rham C and Hoover D 2006 *New J. Phys.* **8** 324, Kobayashi T and Minamitsuji M 2007 *Phys. Rev.* D **75** 104013, 2007 *JCAP* **0707** 016, Kobayashi T and Takamizu Y I 2007 *Class. Quant. Grav.* **25** 015007 Fujii S, Kobayashi T and Shiromizu T 2007 *Phys. Rev.* D **76** 104052 Lee H M and Papazoglou A 2008 *JHEP* **0801** 008
[41] Cuadros-Melgar B and Papantonopoulos E 2005 *Phys. Rev.* D **72** 064008 Papantonopoulos E 2006 *Preprint* gr-qc/0601011
[42] Kehagias A and Kiritsis E 1999 *JHEP* **9911** 022
[43] Mukohyama S, Shiromizu T and Maeda K I 2000 *Phys. Rev.* D **62** 024028, 2001 *Erratum-ibid.* D **63** 029901
[44] Freund P G O and Rubin M A 1980 *Phys. Lett.* B **97** 233, Randjbar-Daemi S, Salam A and Strathdee J A 1983 *Nucl. Phys.* B **214** 491
[45] Mukohyama S, Sendouda Y, Yoshiguchi H and Kinoshita S 2005 *JCAP* **0507** 013, Yoshiguchi H, Mukohyama S, Sendouda Y and Kinoshita S 2006 *JCAP* **0603** 018
[46] Chamblin H A and Reall H S 1999 *Nucl. Phys.* B **562** 133
[47] Ida D 2000 *JHEP* **0009** 014
[48] Benvenuto O G, Garcia-Berro E and Isern J 2004 *Phys. Rev.* D **69** 082002
[49] Battye R A, Carter B, Mennim A and Uzan J P 2001 *Phys. Rev.* D **64** 124007, Yamauchi D and Sasaki M 2007 *Prog. Theor. Phys.* **118** 245
[50] Kobayashi T and Minamitsuji M 2007 *Phys. Rev.* D **75** 104013
[51] Kaloper N and Kiley D 2007 *JHEP* **0705** 045
[52] Kanno S and Soda J 2002 *Phys. Rev.* D **66** 043526, Shiromizu T and Koyama K 2003 *Phys. Rev.* D **67** 084022 Fujii S, Kobayashi T and Shiromizu T 2007 *Phys. Rev.* D **76** 104052 Kobayashi T, Shiromizu T and de Rham C 2008 *Phys. Rev.* D **77** 124012
[53] Papantonopoulos E, Papazoglou A and Zamarias V *Work in progress*

[54] Maeda K I and Nishino H 1985 *Phys. Lett.* B **154** 358, Maeda K I and Nishino H 1985 *Phys. Lett.* B **158** 381, Tolley A J, Burgess C P, de Rham C and Hoover D 2006 *New J. Phys.* **8** 324, Himmetoglu B and Peloso M 2007 *Nucl. Phys.* B **773** (2007) 84 Kobayashi T and Minamitsuji M 2007 *JCAP* **0707** 016 Copeland E J and Seto O 2007 *JHEP* **0708**, 001

AUTHOR INDEX

Akiyama, T. 189, 274
Andréasson, H. 1
Apostolatos, T. A. 4
Ashtekar, A. 9
Basilakos, S. 20, 341
Battisti, M. V. 29
Baugh, C. M. 122
Blagojevic, M. 85
Bresolin, F. 341
Carvalho, C. S. 34
Charmousis, C. 47
Christodoulakis, T. 53
Cole, S. 122
Contopoulos, G. 4
Creighton, J. D. E. 247
Cuadros-Melgar, B. 79
Cvetkovic, B. 85
Deligiannis, J. 4
Doulis, G. 53
Ellis, G. F. R. 91
Esteban, E. P. 106
Fanidakis, N. 122
Farakos, K. 296
Florides, P. S. 135
Frenk, C. S. 122
Friedman, J. L. 247
Frolov, V. P. 140
Gaertig, E. 153
Gambini, R. 371
Georgantopoulos, I. 341
Giambò, R. 163
Giannios, D. 175
Giannoni, F. 163
Grammenos, T. 53
Hata, M. 189, 274
Ikeda, Y. 189, 274
Kanti, P. 199
Kawamura, H. 189, 274
Kleidis, K. 216
Kokkotas, K. D. 153, 417
Kunze, K. E. 234
Lackey, B. D. 247
Lasky, P. 240
Lukes-Gerakopoulos, G. 4
Magli, G. 163
Markakis, C. 247
Mavromatos, N. E. 296
Melas, E. 53
Melnick, J. 341
Miritzis, J. 264
Murata, J. 189, 274
Narita, K. 189, 274
Ninomiya, K. 189, 274
Ogawa, N. 189, 274
Paliathanasis, A. 473
Papadopoulos, G. O. 53
Papakostas, T. 281
Papantonopoulos, E. 79
Papazoglou, A. 47
Pappas, G. 286
Pasipoularides, P. 296
Passamonti, A. 304
Perivolaropoulos, L. 310
Plionis, M. 20, 341
Psaltis, D. 360
Pullin, J. 371
Read, J. S. 247
Rideout, D. 507
Sakellariadou, M. 376
Sato, T. 189, 274
Seitaibashi, E. 189, 274
Sekiguchi, Y. 189, 274
Shibata, M. 247
Siopsis, G. 393
Skákala, J. 409
Sotani, H. 417
Sotiriou, T. P. 423
Spanou, A. 53
Spyrou, N. K. 434
Stergioulas, N. 417
Terlevich, E. 341
Terlevich, R. 341
Terzis, P. A. 53
Tsamparlis, M. 439, 473
Tsoukalas, M. 79, 484
Tsutsui, R. 189, 274
Uryu, K. 247
Visinescu, M. 493
Visser, M. 409, 515
Wallden, P. 507
Weinfurtner, S. 515
White, A. 515
Yazawa, K. 189, 274
Zamarias, V. 79, 524

Institute of Physics
Dirac House, Temple Back
Bristol BS1 6BE UK

CURRAN ASSOCIATES INC
proceedings.com